D1064707

The Enzymes of Biological Membranes

SECOND EDITION

Volume 2

Biosynthesis and Metabolism

THE ENZYMES OF BIOLOGICAL MEMBRANES
Second Edition

The Enzymes of Biological Membranes

SECOND EDITION

Volume 2

Biosynthesis and Metabolism

Edited by

Anthony N. Martonosi

State University of New York
Syracuse, New York

Plenum Press • New York and London

Library of Congress Cataloging in Publication Data

Main entry under title:

The Enzymes of biological membranes.

Includes bibliographical references and indexes.
Contents: v. 1. Membrane structure and dynamics—v. 2. Biosynthesis and metabolism.
1. Membranes (Biology)—Collected works. 2. Enzymes-Collected works. I. Martonosi, Anthony, 1928–
QH601.E58 1984 574.87′5 84-8423
ISBN 0-306-41452-X (v. 2)

A Division of Plenum Publishing Corporation
233 Spring Street, New York, N.Y. 10013

Printed in the United States of America

Contributors

Rob Akeroyd, Laboratory of Biochemistry, State University of Utrecht, University Centre "De Uithof," NL-3584 CH Utrecht, The Netherlands

James Baddiley, Department of Biochemistry, University of Cambridge, Cambridge CB1 1QW, England

Linda Brady, Departments of Pharmacology and Medicine, Case Western Reserve University School of Medicine, Cleveland, Ohio 44106

Inka Brockhausen, Department of Biochemistry, University of Toronto, and Division of Biochemistry Research, Hospital for Sick Children, Toronto, Ontario, Canada M5G 1X8

Marc G. Caron, Howard Hughes Medical Institute Research Laboratories, Departments of Medicine (Cardiology), Biochemistry, and Physiology, Duke University Medical Center, Durham, North Carolina 27710

James L. Gaylor, Central Research and Development Department, E. I. du Pont de Nemours and Company, Glenolden, Pennsylvania 19036

John Ghrayeb, Department of Biochemistry, State University of New York at Stony Brook, Stony Brook, New York 11794

Paul Gleeson, Department of Biochemistry, University of Toronto, and Division of Biochemistry Research, Hospital for Sick Children, Toronto, Ontario, Canada M5G 1X8

Trevor W. Goodwin, C.B.E., F.R.S., Department of Biochemistry, The University of Liverpool, Liverpool L69 3BX, England

Ian C. Hancock, Department of Microbiology, University of Newcastle upon Tyne, Newcastle upon Tyne NE1 7RU, England

Charles L. Hoppel, Medical Research Service, VA Medical Center, Cleveland, Ohio 44106

Masayori Inouye, Department of Biochemistry, State University of New York at Stony Brook, Stony Brook, New York 11794

P. G. Kostyuk, A. A. Bogomoletz Institute of Physiology, Academy of Sciences of the Ukrainian SSR, Kiev, USSR

Ten-ching Lee, Medical and Health Sciences Division, Oak Ridge Associated Universities, Oak Ridge, Tennessee 37831

Robert J. Lefkowitz, Howard Hughes Medical Institute Research Laboratories, Departments of Medicine (Cardiology), Biochemistry, and Physiology, Duke University Medical Center, Durham, North Carolina 27710

Saroja Narasimhan, Department of Biochemistry, University of Toronto, and Division of Biochemistry Research, Hospital for Sick Children, Toronto, Ontario, Canada M5G 1X8

Robert C. Nordlie, Department of Biochemistry, University of North Dakota School of Medicine, Grand Forks, North Dakota 58202

Harry Schachter, Department of Biochemistry, University of Toronto, and Division of Biochemistry Research, Hospital for Sick Children, Toronto, Ontario, Canada M5G 1X8

Giorgio Semenza, Laboratorium für Biochemie der ETH, ETH-Zentrum, CH-8092 Zürich, Switzerland

Robert G. L. Shorr, Howard Hughes Medical Institute Research Laboratories, Departments of Medicine (Cardiology), Biochemistry, and Physiology, Duke University Medical Center, Durham, North Carolina 27710

Fred Snyder, Medical and Health Sciences Division, Oak Ridge Associated Universities, Oak Ridge, Tennessee 37831

James K. Stoops, Verna and Marrs McLean Department of Biochemistry, Baylor College of Medicine, Houston, Texas 77030

Katherine A. Sukalski, Department of Biochemistry, University of North Dakota School of Medicine, Grand Forks, North Dakota 58202

Tom Teerlink, Laboratory of Biochemistry, State University of Utrecht, University Centre "De Uithof," NL-3584 CH Utrecht, The Netherlands

James M. Trzaskos, Central Research and Development Department, E. I. du Pont de Nemours and Company, Glenolden, Pennsylvania 19036

George Vella, Department of Biochemistry, University of Toronto, and Division of Biochemistry Research, Hospital for Sick Children, Toronto, Ontario, Canada M5G 1X8

George P. Vlasuk, Department of Biochemistry, State University of New York at Stony Brook, Stony Brook, New York 11794

Salih J. Wakil, Verna and Marrs McLean Department of Biochemistry, Baylor College of Medicine, Houston, Texas 77030

Karel W. A. Wirtz, Laboratory of Biochemistry, State University of Utrecht, University Centre "De Uithof," NL-3584 CH Utrecht, The Netherlands

Robert L. Wykle, Department of Biochemistry, Bowman Gray School of Medicine, Winston-Salem, North Carolina 27103

Preface to the Second Edition

In the first edition of *The Enzymes of Biological Membranes*, published in four volumes in 1976, we collected the mass of widely scattered information on membrane-linked enzymes and metabolic processes up to about 1975. This was a period of transition from the romantic phase of membrane biochemistry, preoccupied with conceptual developments and the general properties of membranes, to an era of mounting interest in the specific properties of membrane-linked enzymes analyzed from the viewpoints of modern enzymology. The level of sophistication in various areas of membrane research varied widely; the structures of cytochrome c and cytochrome b_5 were known to atomic detail, while the majority of membrane-linked enzymes had not even been isolated.

In the intervening eight years our knowledge of membrane-linked enzymes expanded beyond the wildest expectations. The purpose of the second edition of *The Enzymes of Biological Membranes* is to record these developments. The first volume describes the physical and chemical techniques used in the analysis of the structure and dynamics of biological membranes. In the second volume the enzymes and metabolic systems that participate in the biosynthesis of cell and membrane components are discussed. The third and fourth volumes review recent developments in active transport, oxidative phosphorylation and photosynthesis.

The topics of each volume represent a coherent group in an effort to satisfy specialized interests, but this subdivision is to some extent arbitrary. Several subjects of the first edition were omitted either because they were extensively reviewed recently or because there was not sufficient new information to warrant review at this time. New chapters cover areas where major advances have taken place in recent years. As a result, the second edition is a fundamentally new treatise that faithfully and critically reflects the major transformation and progress of membrane biochemistry in the last eight years. For a deeper insight into membrane function, the coverage includes not only well-defined enzymes, but several membrane proteins with noncatalytic functions.

We hope that *The Enzymes of Biological Membranes* will catalyze the search for general principles that may lead to better understanding of the structure and function of membrane proteins. We ask for your comments and criticisms that may help us to achieve this aim.

My warmest thanks to all who contributed to this work.

Anthony N. Martonosi

Syracuse, New York

Contents of Volume 2

19. Membrane-Bound Enzymes in Plant Sterol Biosynthesis

Trevor W. Goodwin, C.B.E., F.R.S.

24. Multifunctional Glucose-6-Phosphatase: A Critical Review

Robert C. Nordlie and Katherine A. Sukalski

25. The β-Adrenergic Receptor: Elucidation of Its Molecular Structure

Robert G. L. Shorr, Robert J. Lefkowitz, and Marc G. Caron

26. *Ionic Channels and Their Metabolic Control*

P. G. Kostyuk

Contents of Volume 1

Contents of Volume 3

Contents of Volume 4

14

Ether-Linked Glycerolipids and Their Bioactive Species: Enzymes and Metabolic Regulation

Fred Snyder, Ten-ching Lee, and Robert L. Wykle

I. INTRODUCTION

The pioneering studies of those scientists who first discovered (1915–1924) and ultimately identified the precise nature of the chemical structure of the alkyl and alk-1-enyl (plasmalogens) linkages in glycerolipids are described in exquisite detail in a review of the history (the years 1915–1960) of ether lipids by Debuch (1972). It is noteworthy that the chemical nature of the alkyl linkage in glycerolipids was firmly established 29 years before the precise chemical structure of the alk-1-enyl linkage in plasmalogens became known.

There was a considerable lag period between the completion of the structural studies and the discovery of the enzymes that catalyze the biosynthesis of the ether linkages. In the 1968–1969 era, the alkyl group in glycerolipids was found to originate from long-chain fatty alcohols (Snyder *et al.*, 1969a,b). This was subsequently shown to occur via a unique reaction in which the alcohol chain (RO-) formed the *O*-alkyl bond by replacing the acyl moiety of acyl-dihydroxyacetone phosphate (DHAP) (Hajra, 1970; Wykle *et al.*, 1972b). It soon became apparent (Wykle *et al.*, 1970; Blank *et*

Fred Snyder and Ten-ching Lee • Medical and Health Sciences Division, Oak Ridge Associated Universities, Oak Ridge, Tennessee 37831. *Robert L. Wykle* • Department of Biochemistry, Bowman Gray School of Medicine, Winston-Salem, North Carolina 27103.

al., 1970, 1971; Snyder *et al.*, 1971a; Paltauf, 1972b) that *O*-alkyl glycerolipids serve as the precursor of plasmalogens (the *O*-alk-1-enyl type). Wykle *et al.* (1972a) obtained the first enzymatic evidence that showed 1-alkyl-2-acyl-*sn*-glycero-3-phosphoethanolamine was converted to plasmalogens by a cytochrome b_5-dependent mixed-function oxidase (alkyl desaturase) in microsomal preparations from Fischer sarcoma R-3259. This was independently confirmed by Paltauf and Holasek (1973) with a microsomal alkyl desaturase preparation from the hamster small intestine. Paltauf *et al.* (1974b) finally used an antibody to definitively prove that cytochrome b_5 was involved, as was shown earlier for the acyl-CoA desaturase system (see Holloway, 1975 for a review of historical aspects).

Developments in the chemistry and biosynthesis of the alkyl lipids emerged earlier than for the alk-1-enyl type and this also occurred with regard to their possible cellular functions. Whereas the biological role of plasmalogens still remains obscure, a major breakthrough in the ether lipid field occurred in 1979 when an acetylated form of an alkyl phospholipid was found to possess potent biological activities that involved platelet and neutrophil aggregation, antihypertensive effects, hypersensitivity, and degranulation reactions, including the cellular release of other bioactive components, e.g., serotonin (see reviews by Pinckard *et al.*, 1980; Vargaftig *et al.*, 1981; Chignard *et al.*, 1980; Snyder, 1982). This newly identified group of phospholipid mediators has led to the discovery of a number of unique enzyme activities involved in their metabolism (see Section IV).

Both the alkyl and alk-1-enyl phospholipids appear to serve as important structurally stable lipid components of membranes, since the ether linkage can impede the hydrolysis of the acyl moiety at the *sn*-2 position by phospholipase A_2. This protective structural feature is also inferred from *in vivo* experiments that have demonstrated that ether lipids preferentially retain arachidonate during severe dietary deficiency of essential fatty acids (Blank *et al.*, 1973; Wykle *et al.*, 1973). However, the precise relationship of the ether lipids and the metabolism of polyunsaturated fatty acids is still lacking.

In this update of our earlier chapter on the enzymatic pathways of ether-linked lipids (Wykle and Snyder, 1976), we have included a section on the newly identified group of enzyme activities responsible for the metabolism of the bioactive molecules that contain the *O*-alkyl linkage. In addition, major recent developments on the mechanism of *O*-alkyl bond synthesis by alkyl-DHAP synthase has provided important new information about this unique enzyme.

For those enzymes of the ether lipid pathway where little or no new information is available, only the references concerned with significant and direct enzymatic studies are cited. The reader is referred to earlier reviews (Snyder, 1969; Horrocks, 1972; Snyder, 1972a,b; Thompson 1972a,b; Hajra, 1973; Wood, 1973; Wykle and Snyder, 1976; Hajra *et al.*, 1978) for more extensive treatment of the historical background and a review of the occurrence of ether lipids in cancer cells (Snyder and Snyder, 1975). Our intent in this chapter is to present a concise integrated view of the metabolic pathways and a detailed description of the properties of the various enzymes, mostly of microsomal origin, that are involved in the biosynthesis and catabolism of ether-linked glycerolipids.

II. ETHER LIPID PRECURSORS

A. Acyl-CoA Reductase

In earlier studies of the cleavage of 1-*O*-alkylglycerol (Tietz *et al.*, 1964; Pfleger *et al.*, 1967) and the degradation of sphingosine or dihydrosphingosine (Stoffel *et al.*, 1970), fatty alcohols, fatty aldehydes, and fatty acids were observed in the products under certain conditions. The initial lipid product formed in these systems is the fatty aldehyde, which is either reduced to the alcohol or oxidized to the acid depending on the presence of reduced or oxidized nucleotides, respectively. However, these degradative pathways are not considered major sources of fatty alcohols.

In mammalian tissues, the long-chain alcohols are synthesized from the corresponding fatty acids (Snyder, 1972c; Wykle and Snyder, 1976) as they are in birds (Buckner and Kolattukudy, 1976), marine organisms (Sargent *et al.*, 1976), plants (Kolattukudy *et al.*, 1976), and bacteria (Day *et al.*, 1970). The fatty alcohols produced serve as precursors of waxes in most of these organisms as well as in certain mammalian tissues (Snyder *et al.*, 1969a; Downing, 1976; Grigor and Harris, 1977) and, in addition, they are precursors of both the alkyl- and alk-1-enyl-linked glycerolipids (Wykle and Snyder, 1976).

Direct enzymatic evidence for the conversion of fatty acids to fatty alcohols was obtained by Kolattukudy (1970) in cell-free preparations of *Euglena gracilis*. The synthetic pathway for fatty alcohols occurs mainly in the microsomal fraction where it appears to be tightly coupled to a microsomal fatty acid synthetase (Khan and Kolattukudy, 1973a,b, 1975); the reductase acts only on the activated fatty acid, acyl-CoA, and requires NADH. An aldehyde intermediate is formed that can be trapped by addition of phenylhydrazine to the incubation mixture.

Snyder and Malone (1970) demonstrated that fatty acyl-CoAs are reduced to fatty alcohols by microsomal preparations from mouse preputial gland tumors. They showed that NADPH is required for the reduction. A small amount of fatty aldehyde synthesis was observed in the tumor system (Wykle *et al.*, 1979). The mouse preputial gland was subsequently shown to contain a similar microsomal reductase (Moore and Snyder, 1982a,b) that acts only on acyl-CoAs, has a specific requirement for NADPH, and forms an aldehyde intermediate that can be trapped by the addition of semicarbazide to the incubation mixture. The reductase activity is sensitive to sulfhydryl and serine reagent modifications and is stimulated by bovine serum albumin. The apparent K_m for the acyl-CoA reductase is 14 μM; however, the maximum velocity varies with the concentration of albumin used. Topographical experiments demonstrated that the active center of acyl-CoA reductase was located at the cytoplasmic surface of the preputial gland microsomal vesicles since it was readily accessible to trypsin. A low-molecular-weight (10,000–12,000) protein was isolated from the cytosol of the preputial gland and shown to stimulate the activity of the reductase. The level of the stimulatory cytosolic protein and the activity of acyl-CoA reductase were shown to increase during development of puberty in male mice. The reductase activity first appeared at 4 weeks of age, peaked at 6 weeks of age, and thereafter decreased to the level found in mature male mice. These studies suggest that the concentrations of acyl-CoA, cytoplasmic acyl-CoA binding protein, and the acyl-CoA reductase regulate the level of fatty alcohol synthesis *in vivo* and that the reductase activity and binding protein have similar patterns of development during puberty (Moore and Snyder, 1982b).

A microsomal acyl-CoA reductase has been described in brain that is very similar to the reductase of the preputial gland (Natarajan and Sastry, 1976; Bourre and Daudu, 1978; Bishop and Hajra, 1981). Thus, formation of acyl-CoA is required before the fatty acid can be reduced and reduction requires NADPH. Natarajan and Sastry (1976) reported that fatty aldehydes were recovered from the system in the absence of trapping agents. The acyl-CoA reductase activity of rat brain was shown to increase severalfold during development and to reach a maximum at 25 days when myelination was at its peak. Bishop and Hajra (1981) found that albumin greatly stimulated the reductase and they concluded that albumin stimulates the reductase by protecting the acyl-CoA from an acyl hydrolase present in the system. The enzyme specifically utilizes only B side hydrogen of [4-^{3}H]-NADPH. Bishop and Hajra (1981) observed little free aldehyde but they were able to trap the intermediate; these investigators suggested that the aldehyde intermediate may be the free aldehyde itself or a hemithioacetal derivative of either the active site sulfhydryl group or of CoASH. In the brain microsomal system, polyunsaturated fatty acids were not converted to fatty alcohols (Natarajan and Schmid, 1977a, 1978; Bishop and Hajra, 1978); the reductase was fairly specific for palmitic, stearic, and oleic acids. Since alkyl-DHAP synthase incorporates both polyunsaturated and shorter-chain alcohols, the acyl-CoA reductase appears to determine the distribution of ether chains found in tissues. Although a wider distribution of ether chains is found in mouse preputial gland tumors, the rates of conversion of various fatty acids (acyl-CoAs) to alcohols closely parallel the distribution of alkyl chains found in the tumor lipids (Wykle *et al.*, 1979). Alcohol formation was highest with the following substrates listed in decreasing order of their rates of conversion: 16:0 > 17:0 > 18:0 > 14:0 > 18:1 *(trans)* > 18:2 > 18:3 > 18:1 *(cis)*. The fatty alcohol content (Grigor, 1976) and synthesis of alcohols (Grigor and Harris, 1977) vary with age in the mouse preputial gland tumor and increase markedly between the twentieth and thirtieth days of tumor growth.

Rock *et al.* (1978) described the synthesis of fatty alcohols in the pink portion of the rabbit harderian gland. The system consisted of soluble fatty acid synthetase from the gland which was employed to generate fatty acids, and microsomes which contained the NADPH-requiring reductase. The system converted only acyl groups formed by the fatty acid synthetase and did not utilize acyl-CoA or free fatty acid. Thus, the synthesis of fatty alcohols in the gland appears to be closely coupled to the fatty acid synthetase; the substrate of the reductase may be the fatty acid bound to the acyl carrier protein portion of the synthetase. Evidence was recently reported that fatty aldehyde synthesis in the luminescent bacteria *Photobacterium phosphoreum* requires the activity of an acyl protein synthetase activity as well as an NADPH-dependent reductase (Riendeau *et al.*, 1982).

A soluble acyl-CoA reductase has been described in bovine cardiac muscle (Johnson and Gilbertson, 1972; Kawalek and Gilbertson, 1973, 1976). The first enzyme reduces fatty acyl-CoA to the corresponding aldehyde and requires NADH as the reductant. A second soluble enzyme catalyzes the reduction of long-chain aldehydes to alcohols and is specific for NADPH. Ferrell and Yao (1976) reported that homogenates of human heart convert palmitaldehyde to the alcohol but the enzymes were not characterized. Other soluble aldehyde reductases have been described in the brain (Tabakoff and Erwin, 1970; Erwin *et al.*, 1972). These reductases in brain appear to be nonspecific and are probably not involved in the synthesis of fatty alcohols (Bishop and Hajra, 1981). However, Stoffel *et al.* (1970) have shown that the degradation product of sphingosine, fatty aldehydes, can be converted by an alcohol dehydrogenase to a fatty alcohol, which could then enter the lipid pathway.

Bishop and Hajra (1981) recently presented evidence that the acyl-CoA reductase of brain may be localized in the catalase-containing particles (microperoxisomes) rather than in the microsomal fraction. Based on susceptibility of the reductase to trypsin, it was concluded that the acyl-CoA reductase is localized on the external (cytosolic) side of the membrane. In the same experiments, it was concluded that both DHAP acyltransferase and alkyl-DHAP synthase are localized on the internal side of the microperoxisomal membrane.

B. *Acyl-CoA:Dihydroxyacetone-P Acyltransferase (EC 2.3.1.42)*

The initial step in the biosynthesis of ether-linked lipids from fatty alcohols and DHAP is formation of acyl-DHAP catalyzed by acyl-CoA:DHAP acyltransferase (Figure 1). The finding that acyl-DHAP is the immediate precursor of the ether-linked lipids (Hajra, 1970; Wykle *et al.*, 1972b) explains the requirement of ATP, CoA, and Mg^{2+} in systems where DHAP is used as the precursor. These factors are required to activate endogenous fatty acids in such a system and are not required when acyl-DHAP is used as a substrate. The synthesis of acyl-DHAP was discovered in crude mitochondrial fractions from guinea pig liver (Hajra, 1968a; Hajra and Agranoff, 1968a; Hajra *et al.*, 1968) but was also demonstrated in similar preparations from kidney, brain, and heart (Hajra, 1968a). In addition to serving as a precursor of the ether-containing lipids, acyl-DHAP can be converted to 1-acyl-2-lyso-*sn*-glycero-3-P through reduction of the keto group by NADPH:acyl-DHAP oxidoreductase (Hajra, 1968b; Hajra and Agranoff, 1968b). Acylation of the 1-acyl-2-lyso-*sn*-glycero-3-P yields phosphatidic acid (Figure 1), which can be channeled into triglycerides and diacyl phospholipids. This sequence of reactions thus provides an alternate pathway to the well-known glycerolipid synthesis route in which two acyl groups are added to *sn*-glycero-3-P to form phosphatidic acid (Kennedy, 1953; Kornberg and Pricer, 1953).

The acyl-DHAP pathway has now been demonstrated in a wide variety of tissues and organisms (LaBelle and Hajra, 1972; Puleo *et al.*, 1970; Fisher *et al.*, 1976; Mason, 1978; Rock *et al.*, 1977b; Agranoff and Hajra, 1971; Hajra and Burke, 1978; Schlossman and Bell, 1976, 1977; Coleman and Bell, 1980; Coleman and Haynes, 1983). Since acyl-DHAP appears to be an obligatory precursor of alkyl glycerolipids in animals, the acyl-DHAP pathway is presumably present in all tissues that synthesize alkyl glycerolipids. However, the relative importance of the pathways initiated by acylation of *sn*-glycero-3-P or DHAP for the synthesis of non-ether-linked glycerolipid has not been clearly established (Hajra, 1977).

In recent years, Hajra and co-workers (Jones and Hajra, 1977; Hajra *et al.*, 1979;

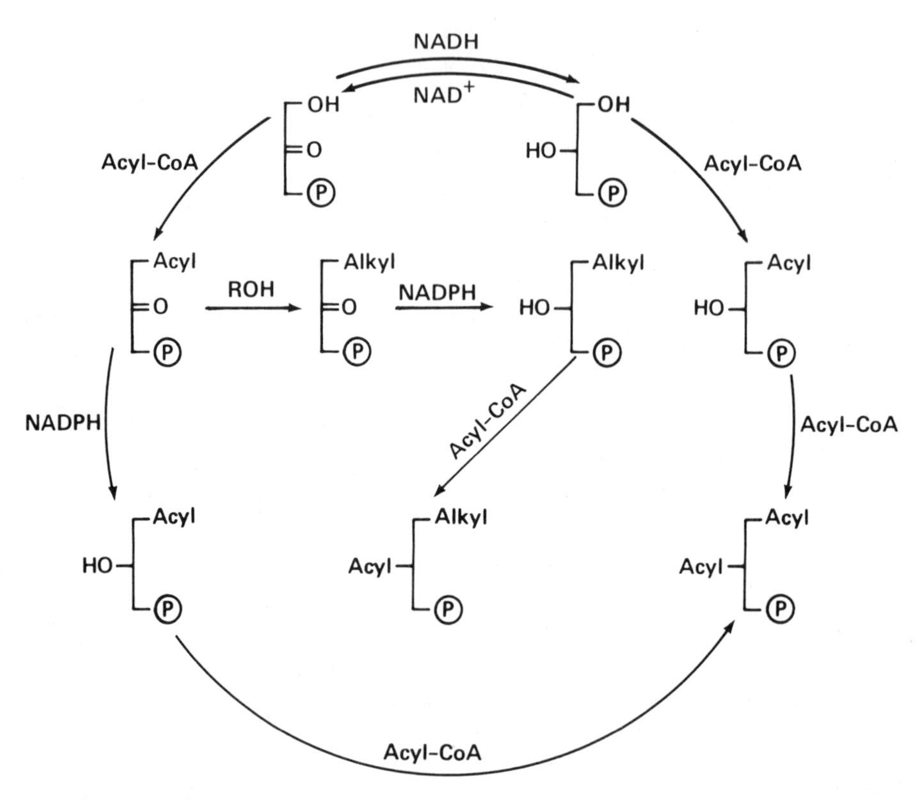

Figure 1. Biosynthesis of phosphatidic acid and its alkyl ether analog of phosphatidic acid from *sn*-glycerol-3-P and dihydroxyacetone-P.

Jones and Hajra, 1980; Hajra and Bishop, 1982) have obtained convincing evidence that acyl-CoA:DHAP acyltransferase of guinea pig liver and kidney is actually localized in peroxisomes rather than in mitochondria or lysosomes. Bats and Saggerson (1979) reported similar findings for the rat liver enzyme. More recently, Hajra and Bishop (1982) presented evidence that the brain enzyme is localized in the microperoxisomes; these are analogous to the peroxisomes of the liver and kidney but are normally recovered in the microsomal fraction rather than in the mitochondrial fraction. Localization of the enzyme in peroxisomes and microperoxisomes explains the apparently anomalous results of earlier studies indicating that the enzyme of certain tissues is mitochondrial but in others, it is microsomal. Unless special procedures are used, the large peroxisomes of liver and kidney sediment with mitochondria, whereas the microperoxisomes of brain and other tissues are recovered in the microsomal fraction. The conclusion that the enzyme is localized in the peroxisomes or microperoxisomes is based on the cofractionation with catalase and uricase (peroxisome markers) rather than with microsomal or lysosomal marker enzymes during careful subcellular fractionation. It appears that other enzymes involved in the synthesis of alkyl-linked lipids might also be localized in the peroxisomes and microperoxisomes.

The topographical distribution of acyl-CoA:DHAP acyltransferase has been investigated in microsomes of the pink portion of the rabbit harderian gland (Rock *et al.*, 1977b). Although a microsomal fraction was used in these studies, it is possible the activity described was actually present in microperoxisomes. The studies indicated that acyl-DHAP is present in the luminal compartment of the vesicles rather than on the cytoplasmic surface. Several lines of evidence support this view. The enzyme activity is latent and is stimulated approximately 20-fold by addition of detergents at concentrations below the critical micellar concentrations. It thus appeared that detergent increased substrate accessibility to the acyltransferase but did not solubilize the enzymes. In addition, the acyltransferase was found to be sensitive to proteolytic digestion only in the presence of detergent concentrations that render the vesicles permeable to the protease. Bishop *et al.* (1982) recently reported similar findings in a fraction enriched in microperoxisomes from 12-day-old rat brain. On the basis of susceptibility of the acyltransferase to trypsin in the presence and absence of detergent, it was concluded that the enzyme of the brain is also on the internal side of the membrane. See Bell *et al.* (1981) for review of these and related studies.

It has been apparent that at least two DHAP acyltransferase activities exist in mammalian cells. The enzyme found in peroxisomes and microperoxisomes is specific for DHAP and does not acylate *sn*-glycero-3-P. This enzyme is not significantly inhibited by *N*-ethylmaleimide nor competitively inhibited by *sn*-glycero-3-P (Jones and Hajra, 1980; Bates and Saggerson, 1979). DHAP acyltransferase of the harderian gland has properties similar to those of the peroxisomal enzyme (Rock *et al.*, 1977b; Rock and Snyder, 1978). In contrast, the microsomal acyl-CoA:DHAP acyltransferase of adipocytes, liver, and several other tissues appears to be the same enzyme as that responsible for the acylation of *sn*-glycero-3-P (Schlossman and Bell, 1976, 1977). Both the *sn*-glycero-3-P and DHAP activities are inhibited by *N*-ethylmaleimide. Acylation of DHAP is competitively inhibited by *sn*-glycero-3-P, and the activities have similar thermolabilities and pH dependence. Further evidence that the microsomal activities are catalyzed by the same enzyme was obtained in studies of the pattern of enzyme activities in differentiating 3T3-L1 preadipocytes (Coleman and Bell, 1980) and fetal and postnatal rat liver (Coleman and Haynes, 1983).

III. BIOSYNTHESIS OF ALKYL AND ALK-1-ENYL (PLASMALOGEN) GLYCEROLIPIDS

A. Alkyldihydroxyacetone-P Synthase (EC 2.5.1.26)

In 1969, a cell-free system from preputial gland tumors was discovered that could incorporate a long-chain fatty alcohol into the *O*-alkyl linkage of glycerolipids (Snyder *et al.*, 1969a). [^{14}C]Hexadecanol, ATP, Mg^{2+}, CoA, and glyceraldehyde-3-P were necessary for the formation of the [^{14}C]alkyl glycerolipids, and neither fatty aldehydes nor fatty acids were effective in replacing the hexadecanol in this system (Snyder *et al.*, 1969b). Alkyl-DHAP synthase formed alkyl-DHAP, the first detectable etherlinked product in the ether lipid biosynthetic pathway (Figure 1) (Snyder *et al.*, 1970b; Hajra,

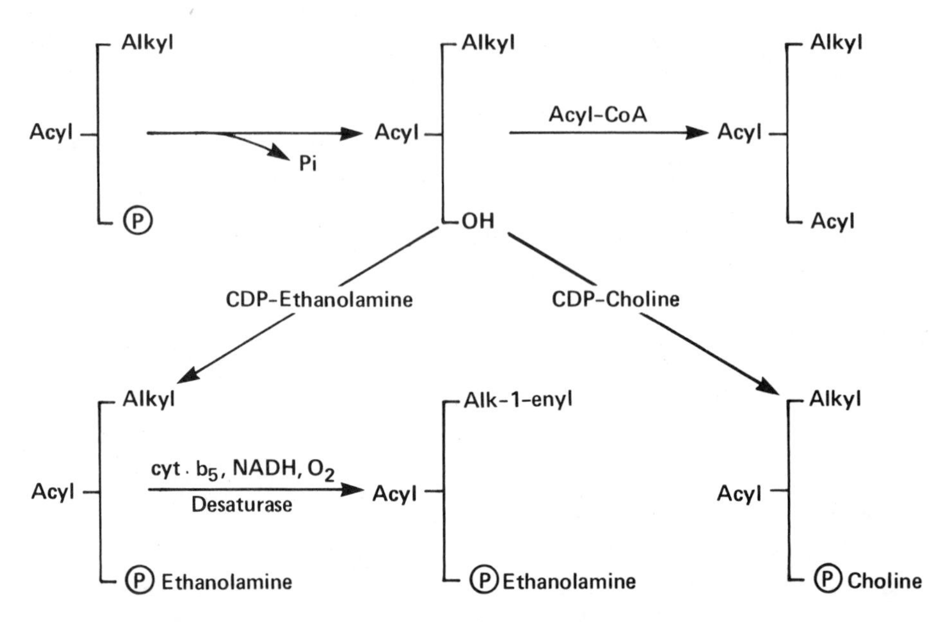

Figure 2. Biosynthesis of alkyl and alk-1-enyl (plasmalogen) types of ether-linked phospholipids.

1970) that leads to the production of ether-linked glycerolipids found in various tissues (Figure 2).

In the early investigations of *O*-alkyl lipid biosynthesis with microsomal preparations, a significant activity of triose-P isomerase was present and, therefore, it was difficult to distinguish whether DHAP or glyceraldehyde-3-P was the triose-P source for the alkyl glycerolipids (Snyder *et al.*, 1969b,c; Hajra, 1969; Wykle and Snyder, 1969; Snyder *et al.*, 1970a,e). However, this problem was soon solved by kinetic and inhibitor studies. Hajra (1969) demonstrated that DHAP was the source of the glycerol backbone for ether lipids based on data from time course kinetics with ^{32}P-labeled triose-P and from the effects of glyceraldehyde-P dehydrogenase and glycerol-P dehydrogenase on the system. At the same time, Wykle and Snyder (1969) reported that when an inhibitor of triose-P isomerase (1-hydroxy-3-chloro-2-propane-P) was added to the *O*-alkyl synthesizing systems, only DHAP, *not* glyceraldehyde-3-P, could be utilized in the biosynthesis of *O*-alkyl lipids.

The requirement of ATP, CoA, and Mg^{2+} in the initial reports was explained when Hajra (1970) found that acyl-DHAP could react with fatty alcohols to produce alkyl-DHAP in enzyme preparations from guinea pig liver mitochondria or mouse brain microsomes. These findings were later confirmed in the tumor system (Wykle *et al.*, 1972b; Snyder *et al.*, 1973a; Hajra, 1973) and in *Tetrahymena* (Friedberg and Heifetz, 1975). The formation of acyl-DHAP in the crude enzyme preparations containing ATP, CoA, and Mg^{2+} was due to the presence of acyl-CoA synthetase and acyl-CoA:DHAP acyltransferase.

Exchange of the ester-linked fatty acid attached to DHAP and a fatty alcohol by alkyl-DHAP synthase is unique in higher biological systems. The unusual nature of this reaction was first demonstrated in experiments with ^{18}O-labeled hexadecanol that showed that oxygen of the ether linkage in *O*-alkyl lipids was donated by the alcohol (Snyder *et al.*, 1970c; Bell *et al.*, 1971).

Important features of the forward reaction (which forms alkyl-DHAP) catalyzed by alkyl-DHAP synthase are (1) the source of the ether-linked oxygen is the oxygen of the fatty alcohol substrate (Snyder *et al.*, 1970c; Bell *et al.*, 1971), (2) the cleavage of the acyl moiety of acyl-DHAP occurs with both oxygens of the original acyl group being found in the fatty acid produced (Friedberg *et al.*, 1983), (3) the *pro-R* hydrogen attached to C-1 of the DHAP moiety of acyl-DHAP is exchanged with water during the reaction (Friedberg *et al.*, 1971, 1972; Friedberg and Heifetz, 1973) without any modification of the configuration at the C-1 carbon atom (Davis and Hajra, 1977b, 1979), (4) when acyl-DHAP is incubated in the presence of tritiated water and hexadecanol, one atom of hydrogen is incorporated from water into alkyl-DHAP for each mole of hexadecanol (Friedberg and Heifetz, 1975), and (5) no Schiff base intermediate is formed between the ketone portion of acyl-DHAP and a functional grouping of the enzyme (Brown and Snyder, 1980; Friedberg *et al.*, 1980; Davis and Hajra, 1981). Specific details of these observations are discussed below with regard to the mechanism of the reaction.

In addition to the forward reaction that produces alkyl-DHAP, two exchange reactions are also catalyzed by alkyl-DHAP synthase. An acyl exchange was shown to occur by the incorporation of [1-^{14}C]palmitic acid into palmitoyl-DHAP by microsomes (Davis and Hajra, 1977a) and by a purified enzyme fraction (Brown and Snyder, 1982) obtained from Ehrlich ascites cells. In an analogous manner, an alkyl exchange was documented by the incorporation of [1-^{14}C]hexadecanol into hexadecyl-DHAP (Brown and Snyder, 1981, 1982) under the same experimental conditions used to study the acyl exchange.

The reaction catalyzed by alkyl-DHAP synthase is reversible. However, the back reaction, which catalyzes the formation of acyl-DHAP when alkyl-DHAP and [^{14}C]-fatty acid are the substrates, occurs at only 1–2% of the rate observed for the forward reaction (Brown and Snyder, 1980).

An important characteristic of the alkyl-DHAP synthase reaction is the exchange of the *pro-R* hydrogen at the C-1 position of the DHAP portion of acyl-DHAP with water. This was first demonstrated in studies of the formation of the ether bond with microsomal preparations from *Tetrahymena pyriformis* (Friedberg *et al.*, 1971, 1972; Friedberg and Heifetz, 1973), in which the acyl[^{3}H]-DHAP substrate was generated from [^{3}H]-DHAP *in situ* via endogenous acyl-CoA acyltransferase activity. The *pro-S* hydrogen is not exchanged under these experimental conditions. It should be noted that the exchangeable hydrogen *(pro-R)* is the same hydrogen exchanged from DHAP during the reaction catalyzed by triose-P isomerase (Friedberg *et al.*, 1972). The configuration of the carbon where the exchange of the *pro-R* hydrogen occurs is retained during the reaction (Friedberg and Alkek, 1977; Davis and Hajra, 1979). Experiments by Friedberg and Heifetz (1973) also demonstrated that the acylation of [^{3}H]-DHAP did not involve an exchange of ^{3}H.

Friedberg and co-workers (Friedberg *et al.*, 1980; Friedberg and Gomillion, 1981) reported that alkyl-DHAP synthase could cleave the fatty acid ester of acyl-DHAP in the absence of the fatty alcohol substrate. They found that when [1-R-^{3}H, 1,3-^{14}C]-DHAP (acylated endogenously by acyltransferases) was incubated with Ehrlich ascites cell microsomes, the ^{3}H of DHAP exchanged with the aqueous medium. The rate of exchange was the same in the presence or absence of fatty alcohols and the DHAP isolated after the incubation had a lower ^{3}H:^{14}C ratio than did the starting substrate. The amount of ^{3}H lost from DHAP during the reaction equaled that recovered as [^{3}H]-H_2O. Under these experimental conditions, the rate of hydrolysis of acyl-DHAP, based on the ^{3}H exchanged with the water, was about 15 times faster than alkyl-DHAP synthesis, even with saturating levels of the fatty alcohol. Friedberg's group concluded from these results with intact microsomes that alkyl-DHAP synthase reacted with acyl-DHAP by behaving as a hydrolase, and that either fatty alcohols or fatty acids could react in place of water to produce alkyl-DHAP and acyl-DHAP, respectively, albeit at slower rates than with water.

Based on the above observations with microsomal preparations as the enzyme source, Friedberg *et al.* (1980) proposed a detailed sequential molecular mechanism for alkyl-DHAP synthese. It accounted for the stereospecific exchange of the *pro-R* hydrogen at carbon-1 with retention of configuration, the reaction with either alkoxide or hydroxide ($R' = H$) ions to form alkyl-DHAP or DHAP, respectively, the incorporation of the entire alkoxide moiety including the oxygen into alkyl-DHAP, and the lack of Schiff's base or intermediate during the reaction.

However, Davis and Hajra (1977a) had proposed a ping-pong mechanism for alkyl-DHAP synthase in which the enzyme first binds acyl-DHAP and then releases the fatty acid to form an enzyme-DHAP complex. The latter reacts with a fatty alcohol to produce alkyl-DHAP. The formation of binary-enzyme intermediates instead of the ternary complex for alkyl-DHAP synthase as suggested by Friedberg and co-workers (Friedberg *et al.,* 1980; Friedberg and Gomillion, 1981), was based on the exchange of the acyl group of acyl-DHAP with [1-^{14}C]palmitic acid and its inhibition by hexadecanol with Ehrlich ascites cell microsomes as the enzyme source. However, the validity of the proposal by Davis and Hajra (1977a) was also open to question because the complex microsomal system contained competing enzymes and endogenous substrates to complicate the interpretations. Since acyl-CoA:DHAP acyltransferase is present in microsomes, it was necessary to rule out the involvement of this enzyme and acyl-CoA synthetase in the observed formation of [^{14}C]acyl-DHAP. Davis and Hajra (1977a) accomplished this by showing that ATP and CoA did not stimulate the reaction and that 1 mM 5,5′dithiobis(2-nitrobenzoic acid) (DTNB) did not affect the acyl exchange. However, Brown and Snyder (1980) found that 1 mM DTNB inhibited (>90%) both the acyl exchange and the forward reaction catalyzed by alkyl-DHAP synthase.

The question of the molecular mechanism of alkyl-DHAP synthase was recently resolved when the enzyme was successfully solubilized from Ehrlich ascites cell microsomes and then subsequently purified for mechanistic studies (Brown and Snyder, 1982, 1983). Solubilization of the enzyme (>95% of the microsomal activity) was accomplished by using 1% Triton-X 100 and acetone (ten volume excess) precipitation. The latter step removed 80–85% of the total phospholipids and approximately 95% of the neutral lipids. Under these conditions, the molecular weight of the solubilized alkyl-DHAP synthase was 300,000 daltons. After Triton-X 100 (1%) treatment, the enzyme aggregates as a large form (>4 × 10^6 daltons) whereas the lower-molecular-weight form can be attained by subsequent acetone (Brown and Snyder, 1982) or phospholipase C treatment (Brown and Snyder, 1979). The initial use of phospholipase C treatment to remove microsomal lipids had to be abandoned because it contained contaminating proteases that ultimately interfered in the subsequent chromatographic purification steps.

Solubilized alkyl-DHAP synthase is labile unless ethylene glycol or glycerol (20%) is added as a stabilizer. Such a preparation has been successfully purified by a multi-step chromatographic procedure developed by Brown and Snyder (1982). This involved successive column elutions from DEAE-cellulose, QAE-Sephadex, Matrex Red, and hydroxylapatite. A new rapid assay method with DEAE-cellulose disks facilitates monitoring the enzyme activity during purification. Acyl- or alkyl-DHAP oxidoreductase, DHAP acyltransferase, alkyl-DHAP phosphohydrolase, and a DNFB-insensitive acyl-DHAP acylhydrolase activities were removed during purification, whereas the DNFB-sensitive acylhydrolase was still associated with this 97-fold purified preparation of alkyl-DHAP synthase (Brown and Snyder, 1982). However, in a subsequent 1000-fold purification of alkyl-DHAP synthase from a second batch of microsomes with a larger column bed volume and better resolution than described earlier, the DNFB-sensitive acylhydrolase was resolved from the alkyl-DHAP synthase activity (Brown and Snyder, 1983).

Two activities that copurify with alkyl-DHAP synthase during the purification as described above are those responsible for (1) the acyl exchange reaction between [1-^{14}C]palmitic acid and palmitoyl-DHAP, and (2) the alkyl exchange reaction between [1-^{14}C]hexadecanol and hexadecyl-DHAP, where palmitic acid is a competitive inhibitor. Such exchange reactions could only be associated with a ping-pong type of reaction mechanism and not with those involving the sequential addition of substrates to the enzyme.

The molecular ping-pong mechanism for alkyl-DHAP synthase shown in Figure 3 was proposed by Brown and Snyder (1982, 1983) on the basis of data obtained with purified alkyl-DHAP synthase preparations from Ehrlich ascites cells. It takes into account the following characteristics of the alkyl-DHAP synthase reaction: (1) the *pro-R* hydrogen at C-1 of the DHAP moiety in acyl-DHAP exchanges with water, with

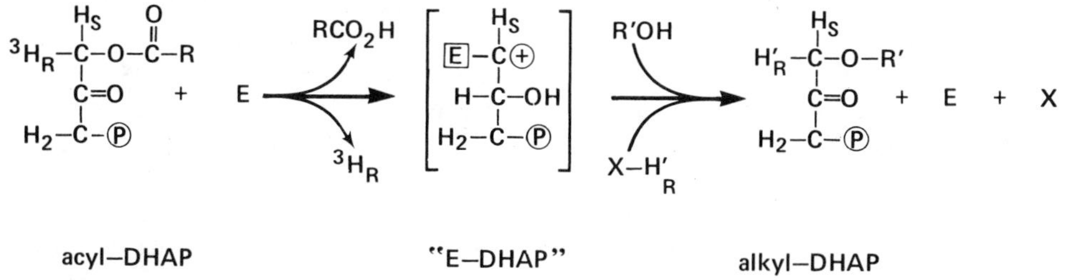

acyl–DHAP + E → (− RCO_2H, − ^3H_R) → "E–DHAP" → (+ R′OH, + X–H′$_R$) → alkyl–DHAP + E + X

acyl–DHAP "E–DHAP" alkyl–DHAP

Figure 3. Ping-pong mechanism for reaction catalyzed by purified alkyldihydroxyacetone-P synthase.

no configurational change at the C-1 atom, (2) the acyl group of acyl-DHAP is cleaved before the fatty alcohol addition, (3) fatty acids or fatty alcohols can interact via exchange reactions with the activated enzyme-DHAP intermediate to produce acyl-DHAP or alkyl-DHAP, respectively, (4) no Schiff's base intermediate is formed, and (5) the oxygen of the ether linkage is donated by the alcohol. A nucleophilic residue, perhaps an amino acid functional group at the active site of the enzyme, covalently binds with the DHAP moiety of the substrate.

The hydrogen exchange that occurs during the reaction catalyzed by alkyl-DHAP synthase was recently reinvestigated with palmitoyl-[1-R-^{3}H]-DHAP and a 1000-fold purified preparation of the enzyme (Brown and Snyder, 1983). A small but significant exchange occurred in the absence of fatty alcohols. However, in the presence of [^{14}C]hexadecanol, the increase in the exchange of the *pro-R* ^{3}H equaled the [^{14}C]hexadecyl-DHAP produced and with [^{14}C]palmitic acid as the cosubstrate, the exchange of the *pro-R* ^{3}H equaled the amount of [^{14}C]palmitoyl-DHAP formed (via the acyl exchange activity) by alkyl-DHAP synthase. These results further support the proposed ping-pong mechanism (Figure 3).

Hixson and Wolfenden (1981) have described the preparation of analogs of Friedberg's proposed intermediate alkyl-DHAP analogs as inhibitors where the ether oxygen was replaced with a methylene group. Although these structures (monopalmitoyl-1,2,3-trihydroxyeicosane-1-P) inhibited the formation of alkyl-DHAP in microsomes from Ehrlich ascites cells, the high levels of inhibitor and the complexity of the system used raises questions about the specificity of such analogs in blocking ether lipid synthesis. For example, it is known that phospholipids, in general, are inhibitory toward a alkyl-DHAP synthase (Brown and Snyder, 1982). Moreover, since Hixson and Wolfenden (1981) started with DHAP, ATP, CoA, and Mg^{2+} to generate the acyl-DHAP precursor of ether lipids, it is not possible to rule out whether the inhibition of *O*-alkyl glycerolipid biosynthesis was due to effects on acyl-CoA synthetase or acyl-CoA:DHAP acyltransferase instead of a direct effect on alkyl-DHAP synthase.

Alkyl-DHAP synthase has a very broad substrate specificity for fatty alcohols. Of the *n*-alkanols, those with 12-, 14-, and 16-carbon chains reacted similarly, whereas the 10- and 18-carbon alcohols were poorer substrates with microsomal preparations from mouse preputial gland tumors (Snyder *et al.*, 1970a). Hajra *et al.* (1978) have also investigated the relative activities (compared to hexadecanol with an activity of 100%) of a variety of fatty alcohols as substrates for alkyl-DHAP synthase. These workers reported that octanol (<1%), decanol (23%), eicosanol (15%), docosanol (14%), phytol (21%), hexadecylglycerol (8%), octadecylglycerol (4%), and 1-octadec-9-enyl-glycerol (23%) were poor substrates for alkyl-DHAP synthase when compared to normal long-chain fatty alcohols possessing 12–18 carbon atoms. Alkyl-DHAP synthase also uses unsaturated species of fatty alcohols as substrates in the formation of ether bonds (Snyder *et al.*, 1973a; Hajra *et al.*, 1978; Bandi *et al.*, 1971). Hexadecane-1,16-diol and hexadecane-1,2-diol were not incorporated into ether lipids by mouse preputial gland tumor microsomes (Snyder *et al.*, 1973a). In contrast, microsomes from the pink portion of the rabbit harderian gland, which contains approximately 80% hydroxyalkylglycerol-type lipids, readily incorporated [1-^{3}H]octadecane-1,12-diol into the hydroxyalkylglycerol fraction via the alkyl-DHAP synthase reaction

(Kasama *et al.*, 1973). Two branched chain alcohols, 14-methylpentadecanol and 12-methylyltetraetradecanol were better substrates than hexadecanol in the formation of the alkyl glycerolipids. Hexadecanol, octadecanol, and octadecenol (49%, 21%, and 14%, respectively), which represent the most common *O*-alkyl chains of ether lipids, are rapidly incorporated into alkyldiacylglycerols of Ehrlich ascites cells (Wood and Snyder, 1967).

Natarajan and Schmid (1977a) reported that octadecenol, octadecadienol, octadecatrienol, and eicosatetraenol were readily incorporated into alkyl-DHAP by microsomal preparations obtained from 19-day-old rat brains, however, they found that fatty acids containing more than one *cis*-double bond were poor substrates for acyl-CoA reductase, the enzyme that produces the fatty alcohol precursor. Since the assessment of alcohol formation from the fatty acids involved an acyl-CoA-generating system (Natarajan and Schmid, 1977a), it is possible that acyl-CoA synthetase could also account in part for the observed results accorded to the substrate specificity of the reductase. In other experiments with the rat brain microsomes, using fatty alcohols or fatty acids with 12- to –22 carbon chains, Natarajan and Schmid (1978) observed that chain length selectivity in the reduction of the acid to alcohol is less pronounced than the specificity of the alkyl-DHAP synthase for the alcohol substrate; again this conclusion differs from others who used acyl-DHAP and acyl-CoA to directly measure the specificity of alkyl-DHAP synthase and acyl-CoA reductase reactions, respectively.

Davis and Hajra (1981) reported that Ehrlich ascites cell microsomes incorporated fatty alcohols ranging in chain lengths from $C_{10:0}$ to $C_{22:0}$ into alkyl-DHAP, however, the highest rates were with alcohols containing chains of 12–18 carbon atoms. Somewhat higher activities were seen when 18:1, 18:2, or 18:3 fatty alcohols were the substrates for alkyl-DHAP synthase (Davis and Hajra, 1981). With suspensions of L-1210 and S-180 murine ascites cells, Richter and Weber (1981) found that there was no specificity for double bond location in the carbon chain of a series of isomeric *cis*-octadecen-1-ols that were incorporated into ether-linked glycerolipids. Other *in vivo* results (Weber and Richter, 1982) based on the labeling of lipids in rat brains and tumors indicate that there was no selectivity in the incorporation of *cis*- versus *trans*-octadecenols into the ether chains of alkylacylglycerols.

In vivo studies with *cis*-9-[1-^{14}C]octadecenol and [1-^{14}C]docosanol confirmed the enzymatic experiments that showed significant amounts of the 18:1 alcohol were incorporated into the alkyl and alk-1-enyl moieties of brains in 19-day-old rats, whereas essentially none of the 22:0 alcohol was incorporated (Natarajan and Schmid, 1977b). It is also of interest that Schmid and co-workers (Muramatsu and Schmid, 1971, 1972, 1973; Chang and Schmid, 1975) found that 1,2-alkanediol and 1-hydroxy-2-alkanone could be incorporated into the ether linkage of phospholipids in myelinating rat brain and these data support the concept that the specificity of alkyl-DHAP synthase is rather broad.

Specificity of alkyl-DHAP synthase for acyl-DHAPs with various acyl groups is not as well defined due to the unavailability of the substrates. Davis and Hajra (1981) reported that tetradecanoyl-, hexadecanoyl-, and octadecanoyl-DHAP (at 100 μM concentrations) gave activities of 0.42, 1.20, and 0.95 nmol/min/mg proteins, respectively, with Ehrlich ascites cell microsomes. They found, as have others (Wykle *et al.*, 1972b), that the enzyme activity increased with hexadecanoyl-DHAP concentrations up to 100 μM, but then decreased rapidly at higher concentrations. Davis and Hajra (1981) did not show that 100 μM is the optimal concentration for tetradecanoyl- and octadecanoyl-DHAP, and therefore the activities obtained relative to the other substrates could reflect different optimal concentrations.

The ketone of acyl-DHAP is an essential structural feature of this cosubstrate. When NADPH is added to microsomal systems at the beginning of assays of alkyl-DHAP synthase, ether lipid synthesis is blocked (Snyder *et al.*, 1969b) since the microsomal acyl-DHAP oxidoreductase (see Section IIB and IIIB) that is also present, reduces acyl-DHAP to 1-acyl-2-lyso-glycero-P, which is not a substrate for alkyl-DHAP synthase. 1-Acylpropandiol-3-P, an acyl-DHAP analog with no ketone group and a dimethylketal analog of acyl-DHAP, also did not serve as a substrate for alkyl-DHAP synthase (Wykle and Snyder, 1976). It is of interest to note that Murooka *et al.* (1970) have reported that certain bacteria can exchange short chain alcohols (ethanol, methanol, *n*-propanol, or *n*-butanol) with the ester group of *O*-acetylhomoserine in the formation of an ether linkage. In this isolated example of a similar type of ether-bond-forming reaction, no ketone group is involved.

The highest activities of alkyl-DHAP synthase occur in tumor tissues; in normal tissues such as brain, the enzyme activity appears to vary with the stage of development. Comparisons of various tissues or enzyme preparations are summarized in Table 1. The cell-free system for the biosynthesis of alkyl glycerolipids by *Tetrahymena* was first described by Kapoulas and Thompson (1969); Malins and Sargent also (1971) described a similar system prepared from the liver of dogfish (*Squalus acanthias*), but specific activities of the enzyme activities were not specified.

Most investigations of alkyl-DHAP synthase have been reported with microsomal preparations as the enzyme source. However, subcellular studies by Hajra *et al.* (1978) have indicated that alkyl-DHAP synthase and acyl-CoA:DHAP acyltransferase (Jones and Hajra, 1977, 1980; Hajra *et al.*, 1979; Hajra and Bishop, 1982; Bishop *et al.*, 1982) are highest in peroxisomes of guinea pig liver. Whether the primary site of alkyl-DHAP synthase activity is located in peroxisomes of other tissues remains to be seen since data for other tissues, including tumors, are lacking.

Membrane topographic studies by Rock *et al.* (1977a) and Rock and Snyder (1978) demonstrated that alkyl-DHAP synthase was located asymmetrically in microsomes from Ehrlich ascites cells and the pink portion of the rabbit harderian gland. The conclusion from these experiments was that the catalytic important segment of alkyl-DHAP synthase in these cells is located on the luminal side of the microsomal vesicles. DHAP acyltransferase was also found to have a luminal site in microsomes from the harderian glands (Rock *et al.*, 1977b). However, with liver microsomes, Ballas and Bell (1981) reported DHAP acyltransferase to be located at the membrane surface, based on its accessibility to proteases in the absence of detergent.

It is noteworthy that Ferrell and Desmyter (1974) have reported that a thioether bond can also be formed in glycerolipids by microsomal preparations from livers of mice. The reaction involved appeared similar to the one catalyzed by alkyl-DHAP synthase, except that [^{14}C]hexadecyl mercaptan substituted for the [^{14}C]hexadecanol in the formation of the *S*-alkyl ethers. However, in experiments with a purified alkyl-

Table 1. Relative Specific Activities of Alkyl-DHAP Synthase in Various Tissues with Hexadecanol as a Cosubstrate

Source	Specific activity (nmoles/mg protein per hr)	Reference
Mouse preputial gland microsomes	21	Snyder *et al.* (1970a)
Ehrlich ascites cells microsomes	20	Wykle and Snyder (1970)
	72	Davis and Hajra (1981)
	7.6	Friedberg *et al.* (1980) (calculated from Figure 1)
Starfish microsomes		Snyder *et al.* (1969c)
Digestive gland	11	
Gonads	7.7	
Mouse fibroblasts (L-M cell) microsomes	10	Snyder *et al.* (1970e)
Hamster small intestinal mucosa microsomes	5.7	Paltauf (1972a)
Tetrahymena microsomes	5.6	Friedberg *et al.* (1971)
Mouse brain microsomes	0.52	Hajra (1969)
Guinea pig mitochondria	3.4	Hajra (1969)
Rat brain microsomes		Snyder *et al.* (1971b)
1 Day before and 5 days after birth	1.3	
45 Days after birth	0.2	
Rat liver microsomes		Snyder *et al.* (1971b)
1 Day before and 5 days after birth	1.5	
45 Days after birth	0.2	
Purified (97-fold) alkyl-DHAP synthase from Ehrlich ascites cells	1248	Brown and Snyder (1982)
Purified (1000-fold) alkyl-DHAP synthase from Ehrlich ascites cells	4020	Brown and Snyder (1983)
Guinea pig liver		Hajra *et al.* (1978)
Peroxisomes	108	
Microsomes	48	
Morris hepatoma 5123C homogenate	0.32	Lee *et al.* (1980)
Fischer R-3259 sarcoma homogenate	0.57	Lee *et al.* (1980)

DHAP synthase from Ehrlich ascites cells, hexadecyl mercaptan exhibited no competition with hexadecanol as the substrate in the reaction (Brown and Snyder, unpublished observations).

B. NADPH:Alkyldihydroxyacetone-P Oxidoreductase (EC 1.1.1.100)

The ketone group of alkyl-DHAP (the first ether-linked intermediate formed by alkyl-DHAP synthase) can be reduced by an NADPH-linked oxidoreductase to form 1-alkyl-2-lyso-*sn*-glycero-3-P (Snyder *et al.*, 1969b; Hajra, 1970; Wykle *et al.*, 1972b), the alkyl analog of lysophosphatidic acid (Figures 1 and 4). Only the B hydrogen of the nicotinamide ring of NADPH is utilized for the reduction of alkyl-DHAP (LaBelle and Hajra, 1972). Although NADPH is required in this reduction step as mentioned earlier, NADPH can also inhibit the alkyl-DHAP synthase reaction through reduction

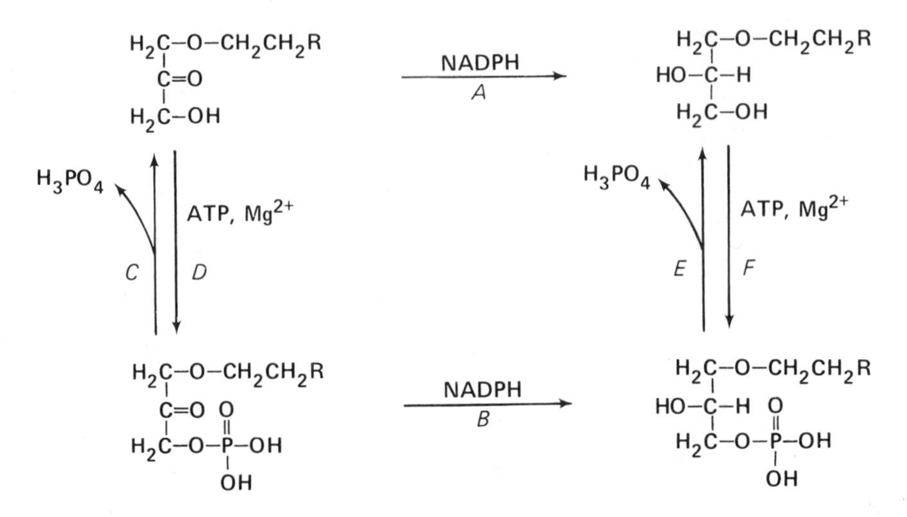

Figure 4. Biosynthesis and interconversions of 1-alkyl-*sn*-glycerol-3-P, 1-alkyl-*sn*-glycerol, alkyldihydroxyacetone, and alkyldihydroxyacetone-P. R designates alkyl chains, e.g., $CH_3(CH_2)_{13}$. (See Section III B, C, D, and E, and Section VD)

of acyl-DHAP to form 1-acyl-*sn*-glycero-3-P, which is not a substrate for alkyl-DHAP synthase.

Evidence by LaBelle and Hajra (1974) indicates that the NADPH-oxidoreductase involved in the reduction of alkyl-DHAP and acyl-DHAP is the same enzyme. In their studies of NADPH:acyl-DHAP oxidoreductase, they showed alkyl-DHAP and acyl-DHAP were competitive inhibitors. Also, thermolability curves for the enzyme were identical with both substrates, and the tissue distribution of the enzyme activities were similar when either alkyl-DHAP or acyl-DHAP were the substrates. Moreover, the apparent K_m for hexadecyl-DHAP (3.5 μM) and palmitoyl-DHAP (4.9 μM) were similar. The oxidoreductase is specific for NADPH at concentrations in the 40–160-μM range (LaBelle and Hajra, 1972; Chae *et al.*, 1973b). At high concentrations (>1 mM), NADH substitutes for NADPH (LaBelle and Hajra 1972; Wykle *et al.*, 1972b).

With microsomal preparations obtained from mouse preputial gland tumors, the apparent K_m values for alkyl-DHAP oxidoreductase were 0.12 mM for NADPH and 3.2 mM for NADH (Chae *et al.*, 1973b). With a five-fold purified preparation of the oxidoreductase from Ehrlich ascites cell microsomes, the K_m values were for NADPH 7.8 μM and 11.3 μM for alkyl-DHAP and acyl-DHAP, respectively (LaBelle and Hajra, 1974).

$NADP^+$ inhibits the oxidoreductase catalyzed reaction and alkyl-DHA and acyl-DHA behave as competitive substrates (LaBelle and Hajra, 1974); 1-acyl- or 1-alkyl-*sn*-glycero-3-P were not inhibitors. The kinetic results and product inhibition data from these experiments suggested that enzyme ternary complexes are formed.

NADPH:alkyl-DHAP oxidoreductase has been found in microsomes from mouse preputial gland tumors (Snyder *et al.*, 1970a; Chae *et al.*, 1973b), Ehrlich ascites tumors (Wykle and Snyder, 1970; LaBelle and Hajra, 1972; Wykle *et al.*, 1972b; LaBelle and Hajra, 1974), suspension cultures of L-M cell fibroblasts (Snyder *et al.*, 1970e), and brain (LaBelle and Hajra, 1972; El-Bassiouni *et al.*, 1975). LaBelle and Hajra (1972) have also described oxidoreductase activities for alkyl-DHAP in both microsomes and mitochondria from liver, brain, kidney, heart, and adipose tissue of rats and in mitochondria from guinea pig liver, rat spleen, rat lung, and rat testes.

C. NADPH:Alkyldihydroxyacetone Oxidoreductase

Alkyl-DHAP can be dephosphorylated to alkyl-DHA by a microsomal phosphohydrolase (Figure 4; see Section V-D). Although the metabolic significance of alkyl-dihydroxyacetone (DHA) is unknown, it can serve as a substrate for an NADPH:alkyl-DHA oxidoreductase that forms 1-alkyl-*sn*-glycerol (Figure 4). The fact that NADH cannot replace NADPH as the reduced cosubstrate even at high concentrations (Chae *et al.*, 1973b), suggested that this enzyme might differ from the oxidoreductase that utilizes alkyl-DHAP or acyl-DHAP. However, since LaBelle and Hajra (1974) found that both alkyl-DHA and acyl-DHA competitively inhibit the NADPH-linked reduction of alkyl-DHAP by the oxidoreductase, they suggested that the reduction of alkyl-DHAP and alkyl-DHA might be catalyzed by the same enzyme.

D. ATP:1-Alkyl-sn-glycerol Phosphotransferase (EC 2.7.1.93)

Conversion of alkylglycerol to 1-alkyl-*sn*-glycerol-3-P (Figure 4) by a phosphotransferase has been demonstrated in mouse preputial gland tumors (Chae *et al.*, 1973a) and in the pink portion of the rabbit harderian gland (Rock and Snyder, 1974). Transfer of the phosphate from ATP to the *sn*-3 position of 1-alkyl-*sn*-glycerol is stereospecific. The phosphotransferase is of microsomal origin and has optimal activity at pH 7.1. ATP and Mg^{2+} are absolute requirements (K_m = 1.6 mM for Mg^{2+}-ATP) in this reaction (Rock and Snyder, 1974). When Mn^{2+} was substituted for Mg^{2+}, the reaction rate was one-half that obtained in the presence of Mg^{2+}. High levels of monoacylglycerols, diacylglycerols, 2-alkyl-*sn*-glycerol, alkylethyleneglycol, or 1-*S*-alkylglycerols had no effect on the alkylglycerol phosphotransferase in the microsomal system from the harderian gland.

E. ATP:Alkyldihydroxyacetone Phosphotransferase (EC 2.7.1.84)

Experiments with microsomal preparations (in the presence of 40 mM NaF) obtained from preputial gland tumors revealed that alkyl-DHA can be phosphorylated (Figure 4) directly by an ATP-Mg^{2+}-dependent phosphotransferase (Chae *et al.*, 1973b). The K_m for ATP was 3.6 mM when the enzyme was assayed at pH 7.0. This phosphotransferase would permit the reentry of alkyl-DHA into the ether lipid biosynthetic pathway (Figure 4).

F. Acyl-CoA:1-Alkyl-2-lyso-sn-glycero-3-P Acyltransferase (EC 2.3.1.63)

Once the ketone group of alkyl-DHAP is reduced by NADPH:alkyl-DHAP oxidoreductase (Figures 1 and 4), the product, alkylglycerophosphate, can then be utilized as a substrate by an acyltransferase (EC 2.3.1.63) to form the *O*-alkyl analog of phosphatidic acid (Figure 1). The 1-acyl-2-lyso-*sn*-glycero-P acyltransferase was originally described by Hill and Lands (1968) and its analogy in the ether lipid pathway was demonstrated by others (Snyder *et al.*, 1969b, 1970b; Hajra, 1969; Wykle and Snyder, 1970; Fleming and Hajra, 1977). The acyltransferase activity, which is highest in microsomal fractions, has been described in a variety of normal tissues, e.g., brain, heart, spleen, liver, lung (Fleming and Hajra, 1977), as well as tumor cells (Wykle and Snyder, 1970). The acylation step is stereospecific, and, based on kinetics, tissue distribution, and competition studies, the acyltransferase that utilizes the ether-linked substrate appears to be a different enzyme from the one that utilizes the acyl analog (Fleming and Hajra, 1977).

G. Acyl-CoA:1-Alkyl-2-acyl-sn-glycerol Acyltransferase

1-Alkyl-2,3-diacyl-*sn*-glycerols (the ether analog of triacylglycerols) are produced (Snyder *et al.*, 1970b) by a pathway analogous to that involved in triacylglycerol biosynthesis (Figure 2). Blank *et al.* (1974) first demonstrated that a microsomal acyl-CoA acyltransferase could catalyze the acylation of 1-[1-^{14}C]alkyl-2-acyl-*sn*-glycerols to form 1-alkyl-2,3-diacyl-*sn*-glycerols. This acyltransferase activity was detected in rat liver, 7777 Morris hepatomas, and mouse preputial gland tumors. Since liver contains no alkyldiacylglycerols, it would appear that the acyltransferase activity responsible for the formation of the ether analog of triacylglycerols is probably the same enzyme that utilizes diacylglycerols as substrates.

In the pink portion of the harderian gland, the *sn*-3 position of the alkyldiacylglycerols contains only isovaleric acid (Blank *et al.*, 1972b). Rock and Snyder (1975) showed that the unique fatty acid composition of the alkyldiacylglycerols in this gland was due to a specific acyltransferase that only utilizes isovaleryl-CoA. Neither acetyl-CoA nor long-chain acyl-CoAs served as substrates in this system.

H. Acyl-CoA:Alkylglycerol Acyltransferase

Snyder *et al.* (1970f) have described acyltransferases in homogenates of several tumors or rat liver that utilize 1-alkyl-*sn*-glycerols and acyl-CoAs as substrates. The major product found with the *sn*-1 isomer was 1-alkyl-3-acyl-*sn*-glycerol; only traces of 1-alkyl-2,3-diacyl-*sn*-glycerol was produced. With the 2-isomer (2-[1-^{14}C]hexadecylglycerol), both 1- and 3-acyl-2-hexadecyl-*sn*-glycerol and 1,3-diacyl-2-alkyl-*sn*-glycerol were formed. Microsomes isolated from the hamster intestinal mucosa (Paltauf and Johnston, 1971) produced alkyldiacylglycerols with [1-^{14}C]palmitic acid and either 1-alkyl-*sn*-glycerol or 2-alkyl-*sn*-glycerol as substrates but not with 3-alkyl-*sn*-glycerol as an acyl acceptor. 1-Alkyl-2-acyl-*sn*-glycerol was also a substrate in this reaction. Acyltransferases in homogenates of rat liver or preputial gland tumors appear to utilize alkylethyleneglycols as substrates in the same manner as alkylglycerols (Snyder *et al.*, 1974).

I. 1-Alkyl-2-acyl-sn-glycerol:CDP-Choline (or CDP-ethanolamine) Choline(ethanolamine)phosphotransferase (EC 2.7.8.2 and EC 2.7.8.1)

The requirement for cytidine diphosphate derivatives of choline or ethanolamine and magnesium, and the involvement of choline (EC 2.7.8.2)- or ethanolamine (EC 2.7.8.1)-phosphotransferases in the synthesis of 1-alkyl-2-acyl-*sn*-glycero-3-phosphocholine-(GPC) or 1-alkyl-2-acyl-*sn*-glycero-3-phosphoethanolamine-(GPE) (Figure 2), respectively, was originally documented with microsomal preparations from Ehrlich ascites cells or preputial gland tumors (Snyder *et al.*, 1970d). The pathway (Figure 2) is analogous to the one established much earlier for the diacyl phospholipid counterparts (Kennedy and Weiss, 1956) and, in fact, the enzymes might indeed be identical. However, the reactions with ether lipids as substrates appear to occur at slower rates than with diacylglycerols. Utilization of alkylacylglycerols by ethanolamine phosphotransferase to form GPE in microsomal fractions of rat brain (Radominska-Pyrek and Horrocks, 1972; Radominska-Pyrek *et al.*, 1977) and in neuronal and glial cell preparations from rabbit brain cortex (Roberti *et al.*, 1975) have also been reported. Ether-linked diradylglycerols as substrates for cholinephosphotransferase and ethanolaminephosphotransferase in microsomal preparation from brains and livers of rats have been described (Radominska-Pyrek *et al.*, 1977).

Indirect evidence was obtained by Goracci *et al.* (1977, 1978) to indicate ethanolaminephosphotransferase activity is reversible. Addition of CMP to rat brain microsomes inhibited the incorporation of [^{3}H]alkylacylglycerol (plus CDP-ethanolamine) into alkylacyl-GPE by approximately 51%. However, CMP had essentially no effect on the cholinephosphotransferase in an identical system with CDP-choline substituted for CDP-ethanolamine (Goracci *et al.*, 1977). The effect of CMP on the reverse reactions with either GPC or alkylacyl-GPE per se as the labeled substrates has not been reported. However, it is well established in other systems that the reactions catalyzed by choline- and ethanolamine phosphotransferases with diacylglycerols as substrates are reversible (Weiss *et al.*, 1958).

Studies of the microsomal diacylglycerol cholinephosphotransferase from chicken liver have revealed that oleic acid markedly stimulates this enzyme activity (Sribney and Lyman, 1973). However, when alkylacylglycerols were the substrate, oleic acid, stearic acid, or linoleic acid were inhibitory toward the synthesis of both alkylacyl-GPC and alkylacyl-GPE by microsomes from brains and livers of mature rats (Radominska-Pyrek *et al.*, 1976). Mixed results were obtained when the free fatty acids were added to systems containing only endogenous diradylglycerols. The authors concluded from these experiments that free fatty acids might change the conformation of the phosphotransferases so that the utilization of diradylglycerols would differ according to the nature of the radyl group at the *sn*-1 position of the substrate. The possibility of two isoenzymes for the ethanolaminephosphotransferase and cholinephosphotransferase was proposed (Radominska-Pyrek *et al.*, 1976) with the suggestion that only one form was affected by the free fatty acids. The role of free fatty acids could be important in regulating the amounts of diacyl, alkylacyl, and alk-1-enyl types of phospholipids in tissues via the phosphotransferase reactions, but this remains to be validated *in vivo*.

Both the choline- and ethanolaminephosphotransferases have been shown to be responsible for the synthesis of the corresponding ether-linked glycerophospholipids in the white and pink portions of the rabbit harderian gland (Radominska-Pyrek *et al.*, 1979). The ethanolaminephosphotransferase was more active with diacylglycerols than with alkylacylglycerols, whereas the cholinephosphotransferase exhibited equal activities with both types of diradylglycerols.

For years, it appeared that the cholinephosphotransferase and ethanolaminephosphotransferase activities for diacylglycerols were catalyzed by a single enzyme in microsomes. However, a substantial amount of recent evidence based on the tissue distribution of the activities, detergent inhibitions, phospholipase or protease treatments, responses to albumin, palmitoyl-CoA or $MgCl_2$, and kinetic analyses has supported the notion that the cholinephosphotransferase and ethanolaminephosphotransferase are separate enzymes (Bell and Coleman, 1980).

On the other hand, it is not totally clear whether identical enzymes are responsible for the transfer of diacyl-, alkylacyl-, and alk-1-enylacyl-glycerols into phospholipids. Several studies have suggested that the same enzymes are involved (Radominska-Pyrek and Horrocks, 1972; Porcellati *et al.*, 1970; Lee *et al.*, 1982a). This conclusion is based on the similar K_m values for the substrates, inhibition results by alkylacylglycerols, pH optima, thermolabilities, and effects of free fatty acids, ATP, Mn^{2+}, or dithiothreitol (Porcellati *et al.*, 1970; Radominska-Pyrek *et al.*, 1976; Lee *et al.*, 1982a). Apparent K_m values of 0.28 mM for CDP-ethanolamine and 1.9 mM for 1-alkyl-2-acyl-*sn*-glycerols were reported for the enzyme from the microsomal fraction of rat brains (Radominska-Pyrek and Horrocks, 1972). In rat liver microsomes, the apparent K_m values were 48.4 μM for CDP-choline and 14.3 μM for 1-alkyl-2-acyl-*sn*-glycerols (Lee *et al.*, 1982a). Only minor differences in K_m values were found between alkylacylglycerols, alk-1-enyl-acylglycerols, and diacylglycerols in all three investigations.

In vivo data obtained from experiments with rat brain, leukemia cells 1210, and sarcoma cells 180 (Weber and Richter, 1982) indicate that cholinephosphotransferase

Radyl
HO— + Acyl-CoA —Acyltransferase→ Acyl— + CoASH
P-Choline (or P-Ethanolamine)

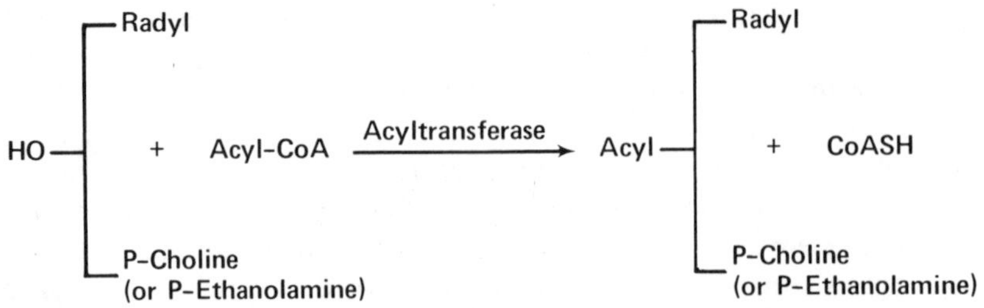

Figure 5. Molecular remodeling reaction of the *sn*-2 acyl position in phospholipid species.

prefers alkylacylglycerols with *cis*-8, *cis*-9, and *cis*-10-octadecenylglycerols and has no preference for double-bond positions in *trans*-octadecenylacylglycerols. In contrast, the ethanolaminephosphotransferase appeared to be nonselective with regard to position of the double bond in the *cis*- or *trans*-octadecenyl moieties of alkylglycerols (Richter and Weber, 1981; Weber and Richter, 1982). These experiments also suggested that the enzymes involved in the synthesis of the alkyl analogs of phosphatidylcholine and phosphatidylethanolamine exhibit a selectivity for the chain length of the fatty alcohols incorporated into the ether chains.

J. 1-Alk-1′-enyl-2-acyl-sn-glycerol:CDP-Choline (or CDP-Ethanolamine) Choline(ethanolamine)phosphotransferase

The conversion of 1-alk-1′enyl-2-acyl-*sn*-glycerol (plasmalogenic diacylglycerols) to 1-alk-1′-enyl-2-acyl-GPC (or GPE) by reaction with either CDP-[^{14}C]choline or CDP-[^{14}C]ethanolamine (Figure 2), respectively, as cosubstrates was first demonstrated by Kiyasu and Kennedy (1960) in a particulate fraction (30,000g × 40 min pellet) from rat liver. They compared the synthesis of 1-alk-1′-enyl-2-acyl-GPC and 1,2-diacyl-GPC from the corresponding diacylglycerols and found that reactions with the two substrates were similar in many respects, including inhibition of the activity by Ca^{2+}. These results suggested that the same cholinephosphotransferase is responsible for the synthesis of both 1-alk-1′-enyl-2-acyl-GPC and 1,2-diacyl-GPC. It was pointed out that the 1-alk-1′-enyl-2-acyl-*sn*-glycerol might only be acting as an analog of 1,2-diacyl-*sn*-glycerol and may not be physiologically significant. McMurray (1964) demonstrated that similar reactions of 1-alk-1′-enyl-2-acyl-*sn*-glycerol with CDP-[^{14}C]choline and CDP-[^{14}C]ethanolamine yield the corresponding plasmalogens in brain homogenates.

Ansell and Spanner (1967) injected [^{14}C]ethanolamine in rat brain and found that after 16–20 hr, 40% of the injected ethanolamine was incorporated into ethanolamine containing phosphoglycerides; no evidence of methylation to choline-linked species was observed. The incorporation of ethanolamine was attributed to the cytidine pathway, although the possible role of base-exchange reactions (Gaiti *et al.*, 1972; Brunetti *et al.*, 1979) was not discussed. Subsequently, Ansell and Metcalfe (1968, 1971) found the highest ethanolamine phosphotransferase activity in the microsomal fraction of brain homogenates. The transferase activities toward 1-alk-1′-enyl-2-acyl-*sn*-glycerol and 1,2-diacyl-*sn*-glycerol were similar in all respects. The K_m for CDP-ethanolamine was approximately 2.5×10^{-4} M with ether lipid substrates, and the K_ms for both lipid classes were 1.6×10^{-3} M. Either Mg^{2+} or Mn^{2+} were required for enzyme activity, and optimal activity required 10 mM Mn^{2+} or 40 mM Mg^{2+}. Phospholipid synthesis was highest in 16-day-old rats. The rates of synthesis were 70 and 270 nmol/mg microsomal protein/hour for 1-alk-1′-enyl-2-acyl-GPC and 1,2-diacyl-GPE, respectively. The relative rates of synthesis of these two products were the same in several tissues. These studies indicated the same phosphotransferase is responsible for the synthesis of both glycerolipids. Evidence obtained by Porcellati *et al.* (1970) indicated that the two activities are catalyzed by the same enzyme in microsomal preparations from chicken brain. On the other hand, differences in response of the activities to cAMP and biogenic amines have led some workers to speculate that there may be isozymes specific for each type of diacylglycerol class (Freysz *et al.*, 1978; Strosznajder *et al.*, 1979).

The cholinephosphotransferase reaction has been demonstrated in a particulate fraction from ox heart (Poulos *et al.*, 1968), a tissue rich in choline plasmalogens (Horrocks, 1972). The amount of choline-linked plasmalogens synthesized from CDP-[^{14}C]choline was increased by 50-fold after the addition of 1-alk-1′-enyl-2-acyl-*sn*-glycerol (11 mM) to the incubation mixture. This reaction may be involved in the synthesis of choline- and ethanolamine-linked plasmalogens or the interconversion of the two species, but presently, the metabolic significance of this reaction for the synthesis of plasmalogens is unclear. The desaturation step forming the alk-1′-enyl linkage takes place on 1-alkyl-2-acyl-GPE, whereas desaturation of 1-alkyl-2-acyl-GPC is nil (see Section III-K). 1-Alk-1′-enyl-2-acyl-*sn*-glycerol for the synthesis of choline plasmalogens could possibly be derived from ethanolamine plasmalogens by a reversal of the phosphotransferase reaction (Kanoh and Ohno, 1973, 1975).

A high rate of conversion of ^{32}P-labeled 1-alkyl-2-acyl-GPE to plasmalogens by a microsomal fraction of rat brain was reported by Horrocks and Radominska-Pyrek (1972). The conversion required Mg^{2+} and was inhibited by NADPH. This incorporation of label into plasmalogens now appears to be attributable to a reversal of the CDP-ethanolamine:1-alk-1′-enyl-2-acyl-*sn*-glycerol ethanolamine phosphotransferase reaction (Goracci *et al.*, 1977, 1978). Reversal of the choline and ethanolamine phosphotransferase activities can be stimulated by the addition of CMP (Goracci *et al.*, 1977). In studies employing microsomes from the rabbit harderian gland, Radominska-Pyrek *et al.* (1979) found that the incorporation of radiolabel from CDP-[^{14}C]choline or CDP-[^{14}C]ethanolamine into corresponding plasmalogens could be increased several-fold by the addition of alkylacylglycerols but not by diacylglycerols. It has been suggested that the reverse reactions of these phosphotransferases may play an important role in the conversion of ethanolamine- to choline-linked plasmalogens (Goracci *et al.*, 1981).

K. *1-Alkyl-2-acyl-sn-glycero-3-phosphoethanolamine Desaturase (EC 1.14.99.19)*

Evidence that plasmalogens are derived from 1-*O*-alkyl glycerolipids in animals was first obtained from *in vivo* studies. However, many of these studies yielded conflicting results; the earlier studies have been thoroughly reviewed elsewhere (Snyder, 1969, 1972a,b; Thompson 1972a,b; Paltauf, 1973; Hajra, 1973; Wood, 1973; Wykle and Snyder, 1976). Some of the strongest *in vivo* evidence for the conversion of 1-*O*-alkyl lipids to plasmalogens was obtained in studies utilizing 1-[1-^{14}C]hexadecyl-[1-^{3}H]glycerol (Thompson, 1968; Blank *et al.*, 1970), hexadecanol labeled with ^{14}C and ^{3}H (Wood and Healy, 1970; Stoffel and LeKim, 1971), or [^{18}O]hexadecanol (Snyder *et al.*, 1970c; Bell *et al.*, 1971) as precursors. Both Debuch *et al.* (1971) and Paltauf (1972c) found that 1-*O*-alkyl-2-lyso-GPE labeled in the *O*-alkyl chain can be desaturated to yield the ethanolamine plasmalogens. Debuch *et al.* (1971) concluded that the desaturation step occurs on 1-alkyl-2-lyso-GPE, whereas Paltauf (1971) concluded that it occurs on the acylated derivative, 1-alkyl-2-acyl-GPE. As discussed below, enzymatic studies indicate that only the acylated form is desaturated (Figure 2). Stoffel and LeKim (1971), using specifically-labeled [^{14}C,^{3}H]hexadecanols, demonstrated that the erythro-1(*S*),2(*S*)tritiums are removed during the desaturation of the 1-alkyl chain.

After development of a cell-free system that synthesized plasmalogens (Wykle *et al.*, 1970; Snyder *et al.*, 1971a) from [1-^{14}C]hexadecanol and other components required for alkyl lipid synthesis, it became apparent that of the many alkyl-linked intermediates of the biosynthetic pathway formed in the system, none contained detectable alk-1-enyl chains except the final product, 1-radyl-2-acyl-GPE. The synthesis of plasmalogens in the system was stimulated by CDP-ethanolamine. In further studies of the system (Blank *et al.*, 1971), it was shown that 1-[9,10-^{3}H]hexadecyl-2-acyl-[U-^{14}C]glycero-3-P is incorporated into ethanolamine plasmalogens with no change in the $^{3}H/^{14}C$ ratio. As with the labeled alcohol, only 1-radyl-2-acyl-GPE contained significant levels of radioactivity in the alk-1-enyl chains. These studies were extended to cell fractions from rat brain (Blank *et al.*, 1972a), Fischer R-3259 rat sarcomas (Wykle *et al.*, 1972a), and hamster small intestine (Paltauf, 1972b; Paltauf and Holasek, 1973). In these experiments, 1-alkyl-2-acyl-GPE and 1-alkyl-2-lyso-GPE were used as substrates. As in the *in vivo* studies (Debuch *et al.*, 1971), 1-alkyl-2-lyso-GPE was a better precursor of plasmalogens than 1-alkyl-2-acyl-GPE. However, the 1-alkyl-2-lyso-GPE was rapidly acylated in the system, and it was concluded from the kinetic data that desaturation occurs only after the lysophospholipid is acylated and that only 1-alkyl-2-acyl-GPE is desaturated. It was suggested in subsequent studies that the acylation step may position the substrate so that it becomes more accessible to the alkyl desaturase complex (Wykle and Schremmer, 1979). The desaturase required O_2 and either NADH or NADPH (Wykle *et al.*, 1972a; Paltauf and Holasek, 1973) and was inhibited by KCN but not by CO. These properties are similar to the stearoyl-CoA desaturase system (Oshino *et al.*, 1966; Holloway and Katz, 1972; Oshino and Omuura, 1973) and suggest that the cytochrome b_5-dependent microsomal electron transport system is responsible for plasmalogen synthesis. Paltauf *et al.* (1974b) concluded from studies with an antibody to cytochrome b_5 that plasmalogen synthesis by hog spleen microsomes requires cytochrome b_5. Brain microsomes were shown to synthesize ethanolamine plasmalogens by the same system as the tumor and spleen (Wykle and Lockmiller, 1975). Thus, brain appears to desaturate 1-alkyl-2-acyl-GPE by the cytochrome b_5-requiring electron transport system. Woelk *et al.* (1976) reported that antibodies to cytochrome b_5 and to NADPH or NADH cytochrome b_5 reductase inhibit the alkyl desaturase in preparations from brain neuronal and glial cells. Based on the substrate specificity and requirement for CoA with 1-alkyl-2-lyso-GPE, they concluded that only 1-alkyl-2-acyl-GPE is desaturated by the brain cell preparations.

Lee *et al.* (1973) found that the desaturase activities of rat Fischer R-3259 tumors responsible for the desaturation of stearoyl-CoA and 1-alkyl-2-acyl-GPE respond differently to fat-free diets. The stearoyl-CoA desaturase activity was increased by the fat-free diet, while no increase in the desaturation of 1-alkyl-2-acyl-GPE was observed. The studies suggested that the two desaturase systems are not identical and probably depend on different cyanide-sensitive factors. This conclusion was based primarily on evidence that dietary control of stearoyl-CoA desaturase in rat liver is mediated by the level of cyanide-sensitive factor(s) (Oshino and Sato, 1972).

The 1-alkyl-2-acyl-GPE desaturase is a microsomal enzyme, but its activity is stimulated by soluble fraction (Snyder *et al.*, 1971a; Wykle *et al.*, 1972a; Paltauf and Holasek, 1973). The soluble factor stimulating the desaturase of the tumor microsomes was identified as catalase (Baker *et al.*, 1976). Catalase appeared to act primarily by protecting cytochrome b_5 against H_2O_2. In contrast, the activity of pig spleen microsomes was not significantly stimulated by catalase but was stimulated three- to tenfold by two other proteins that were isolated from pig kidney (Paltauf, 1978a,b). The two partially purified proteins have identical molecular weights (27,000) but have different isoelectric points (pH 5.1 and 4.9). It was concluded that the soluble proteins have no enzymatic activity themselves and do not act simply as detergents. Rather, it was suggested that proteins function as specific mediators between the membrane-bound enzyme system and the 1-alkyl-2-acyl-GPE, perhaps forming a ternary complex to provide the correct steric arrangement for the desaturase system and the substrate. The brain microsomal desaturase was not significantly stimulated by the soluble fraction (Wykle and Lockmiller, 1975).

The influence of specific 2-acyl residues on the desaturation of 1-alkyl-2-acyl-GPE was investigated by Wykle and Schremmer (1979). The following 2-acyl residues were tested: 2:0, 4:0, 7:0, 12:0, 18:1, 18:2, and 20:4. Variation of the 2-acyl chains had little influence on the desaturation of the various homologs. However, the acyl groups added *in situ* to 1-alkyl-2-lyso-GPE were highly unsaturated. Tjiong *et al.* (1976) compared 1-alkyl-2-acyl-GPEs containing different fatty acids as precursors of plasmalogens by injecting the precursors into brain, but low conversions were obtained with the acylated compounds and it was suggested that acyl groups may have been hydrolyzed before being converted to plasmalogens.

Enzymatic studies with brain preparations employing substrates labeled in the alkyl chain indicate conversion of 1-alkyl-2-acyl-GPE (as in tumors, rat small intestine, and hog spleen) is dependent on the cytochrome b_5-containing electron transport chain (Blank *et al.*, 1972a; Wykle and Lockmiller, 1975; Woelk *et al.*, 1976). The studies reported on the conversion of ^{32}P-labeled 1-alkyl-2-acyl-GPE to plasmalogens by brain preparations (Horrocks and Radominska-Pyrek, 1972) probably did not reflect significant desaturation of the added substrate but rather measured transfer of the [^{32}P]ethanolamine into endogenous alk-1-enyl species by a reversal of the CDP-ethanolamine phosphotransferase reaction (Goracci *et al.*, 1977, 1978, 1981). Debuch and co-workers (Debuch *et al.*, Tjiong *et al.*, 1976; Gunawan and Debuch, 1977) have consistently observed higher conversion of 1-alkyl-2-lyso-GPE than 1-alkyl-2-acyl-GPE to plasmalogens when the precursors are injected into brain or incubated with cultured glial cells (Debuch *et al.*, 1982). These findings are in agreement with the results obtained with microsomal systems where higher rates are also obtained using 1-alkyl-2-lyso-GPE as a precursor (Wykle *et al.*, 1972a; Wykle and Lockmiller, 1975; Wykle and Schremmer, 1979; Paltauf and Holasek, 1973). Since all the enzymatic studies indicate that only 1-alkyl-2-acyl-GPE is desaturated, it appears that the higher rate of synthesis obtained with the lyso precursor is related to differences in uptake by the cells or positioning of the substrates for accessibility to the enzyme in the cell membranes (Wykle and Schremmer, 1979).

The desaturation of 1-alkyl-2-lyso-GPC or 1-alkal-2-acyl-GPC has not been reported. Although the compounds have been tested in enzyme systems that converted up to 50% of the corresponding ethanolamine precursors to plasmalogens, no synthesis of choline plasmalogens was observed. Several reactions have been reported that might bring about the conversion of ethanolamine plasmalogens to choline plasmalogens: a Ca^{2+}-stimulated base exchange reaction (Gaiti *et al.*, 1972), reversal of the CDP-choline:1-alk-1′enyl-2-acyl-*sn*-glycerol cholinephosphotransferase (Goracci *et al.*, 1977, 1978, 1981), and *N*-methylation of 1-alk-1′-enyl-2-acyl-*sn*-GPE (Mogelson and Sobel, 1981; Mozzi *et al.*, 1981).

L. *Acyl-CoA:1-Radyl-2-lyso-sn-glycero-3-phosphocholine (or Phosphoethanolamine) Acyltransferase (EC 2.3.1.25)*

Acylation of the lyso form of ether-linked phospholipids is catalyzed by a membrane-bound acyltransferase that utilizes acyl-CoAs as the cosubstrate (Figure 5). Lands and Hart (1965) originally described an acyl-CoA:alk-1-enyl-2-lyso-GPC acyltransferase activity that was extremely low in rat liver. In contrast, relatively high activities were detected in human erythrocytes and in muscle and testes from rabbits (Waku and Lands, 1968). 1-Alk-1-enyl-2-lyso-GPE is also a substrate for the acyltransferase from rabbit muscle. Acylation rates were lower when 1-alkyl-2-lyso-GPC was the substrate, and in all instances, the rates obtained with the lyso acyl phospholipid analogs were highest (Waku and Lands, 1968).

Acyl-CoA:1-alkyl-2-lyso-GPC acyltransferase activity (assayed with polyunsaturated acyl-CoAs) in rabbit sarcoplasmic reticulum has been described by Waku and Nakazawa (1970), but the activity was considerably lower than that found with the 1-acyl-2-lyso- and 1-alk-1′-enyl-2-lyso-GPC analogs. Under the same conditions, 1-alkyl-2-lyso-GPE was a poor acyl acceptor. Microsomal acyltransferase activities have been compared in Ehrlich ascites cells using 1-acyl-2-lyso-, 1-alkyl-2-lyso-, or 1-alk-1′-enyl-2-lyso-GPC as phospholipid acceptors and [1-^{14}C]linoleoyl-CoA as the acyl donor (Waku and Nakazawa, 1972). Acylation rates for both ether lipid substrates were only 12–15% of the rate obtained with 1-acyl-2-lyso-GPC.

Acylation of 1-alkyl-2-lyso-GPC and 1-alk-1′-enyl-2-lyso-GPC with [1-^{14}C]oleoyl-CoA as a substrate was reported to occur in a particulate fraction from intestinal mucosal cells (Subbaiah *et al.*, 1970). Again, the activity was much lower with the ether-linked substrates than with 1-acyl-2-lyso-GPC. Matsumoto and Suzuki (1973) observed fairly active acylation of both 1-alk-1′-enyl-2-lyso-GPC and 1-alk-1′-enyl-2-lyso-GPE in homogenates of cultured human amnion cells (FL 60), but the former was a better substrate than the 1-alk-1′-enyl-2-lyso-GPE. In similar studies, Natarajan and Sastry (1973) described the acylation of ^{32}P-labeled 1-alk-1′-enyl-2-lyso-GPE by microsomes and mitochondria obtained from rat brain. Acylation of 1-alkyl-2-lyso-GPE also occurred in studies of plasmalogen biosynthesis in microsomes from Fischer R-3259 sarcoma (Wykle *et al.*, 1972a), brain (Blank *et al.*, 1972a; Wykle and Lockmiller, 1975), and in mucosal preparations from the hamster small intestine (Paltauf and Holasek, 1973).

M. Ca^{2+}-Dependent Base Exchange Reactions

Non-energy-requiring Ca^{2+}-base exchange reactions have been reported between L-[^{14}C]serine or [^{14}C]ethanolamine and endogenous ethanolamine plasmalogens in rat brain microsomes (Gaiti *et al.*, 1972). These experiments also demonstrated that 1-alkyl-2-acyl-GPE and 1-alkyl-2-acyl-GPS can be formed by base exchange. The properties of the Ca^{2+}-stimulated "phospholipase D type" exchange reaction for ether lipids appear to be same as those described for their diacyl counterparts (van den Bosch, 1974).

IV. BIOLOGICALLY ACTIVE ALKYL PHOSHOLIPIDS (PLATELET-ACTIVATING FACTOR)

A. Acetyl-CoA:1-Alkyl-2-lyso-sn-glycero-3-phosphocholine Acetyltransferase (EC 2.3.1.67)

Unstimulated cells contain little if any platelet activating factor 1-alkyl-2-acetyl-GPC or PAF, but a number of cells synthesize and secrete this bioactive phospholipid when appropriately stimulated. The synthesis of 1-alkyl-2-acetyl-GPC, at least in some cells, requires extracellular Ca^{2+} (Lynch *et al.*, 1979; Betz and Henson, 1980; Tence *et al.*, 1980) and is stimulated by the Ca^{2+} ionophore, A23187 (for review see O'-Flaherty and Wykle, 1983). These and other studies (Lee *et al.*, 1982b) suggest that Ca^{2+} plays an important role in the control of 1-alkyl-2-acetyl-GPC production.

The enzymatic synthesis of 1-alkyl-2-acetyl-GPC by acetyl-CoA:1-alkyl-2-lyso-GPC acetyltransferase (Figure 6) was first demonstrated in microsomal preparations of rat spleens (Wykle *et al.*, 1980b). Activity was also found in several other rat tissues. The acetyltransferase was found to be much more sensitive to detergents than palmitoyl-CoA:1-alkyl-2-lyso-GPC acyltransferase. Furthermore, high levels of acetyl-CoA did not competitively inhibit the long-chain acyltransferase activity. The evidence thus indicates that the acetyltransferase is selective for a short-chain acyl-CoA and differs from the long-chain acyltransferases. Ninio *et al.* (1982) found that the acetyltransferase of murine macrophages is not competitively inhibited by malonyl-CoA and is only weakly inhibited by nonanoyl- and butyryl-CoA. However, the transferase did utilize propionyl-CoA (Ninio *et al.*, 1982). Both the spleen and macrophage enzymes are strongly inhibited by long-chain acyl-CoAs, presumably by detergent effects. Higher concentrations of 1-alkyl-2-lyso-GPC ($>$30–40 μM) also lead to decreased activity (Wykle *et al.*, 1980b; Ninio *et al.*, 1982). The acetyltransferase does not appear to be highly selective for 1-alkyl-2-lyso-GPC since, although less effective, 1-acyl-2-lyso-GPC also serves as a substrate (Wykle *et al.*, 1980b). Ninio *et al.* (1982) reported that the addition of 1-acyl-2-lyso-GPC to macrophage preparations also stimulates the production of 1-alkyl-2-acetyl-GPC. The acetyltransferases of macrophages and spleen have very similar characteristics except the K_m for acetyl-CoA differs (182 μM and 67 μM respectively) and, although EDTA inhibits the activity in both systems, the inhibition can be overcome by Ca^{2+} in the macrophage system but not in the

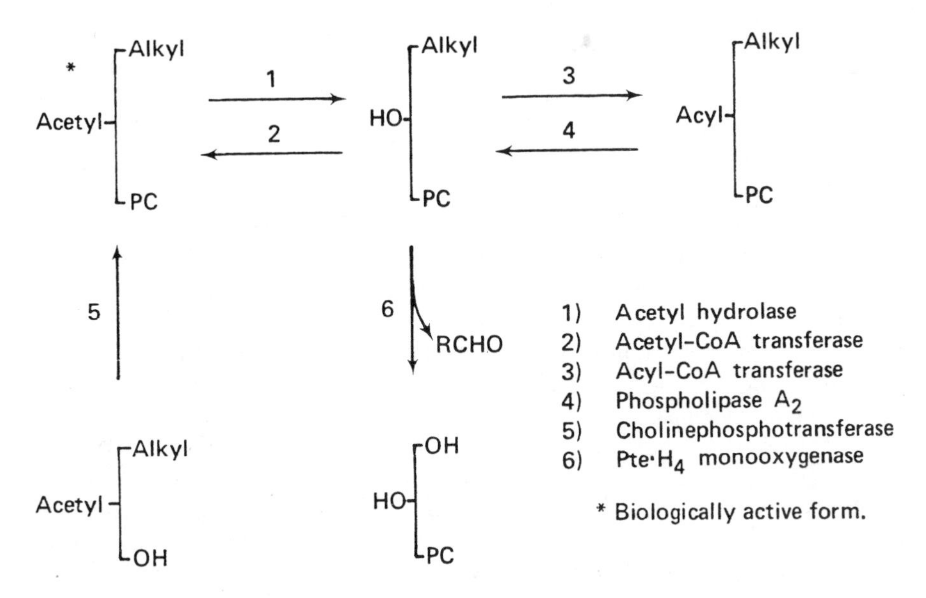

Figure 6. Pathway and enzymes involved in the biosynthesis and catabolism of the bioactive phospholipid 1-alkyl-2-acetyl-*sn*-glycero-3-phosphocholine (platelet-activating factor).

spleen system. The pH optimum was found to be 6.9 for the acetyltransferase of the spleen and 7.0 for the macrophage enzyme.

In human neutrophils, Alonso *et al.* (1982) found that the acetyl-CoA:1-alkyl-2-lyso-GPC acetyltransferase activity increases up to ten-fold during phagocytosis. There was a close correlation between the generation of 1-alkyl-2-acetyl-GPC in response to various doses of opsonized zymosan and the level of acetyltransferase activity. In addition, the temporal relationships between the production of 1-alkyl-2-acetyl-GPC and acetyltransferase activation were consistent with acetyltransferase being the enzyme responsible for the synthesis. The cholinephosphotransferase activity was not increased during phagocytosis. Lee *et al.* (1982b) found a similar increase in acetyltransferase in human neutrophils stimulated by the Ca^{2+} ionophore A23187. Eosinophils from patients with eosinophilia also had elevated acetyltransferase levels compared to the levels in normal eosinophils. A lower release of 1-alkyl-2-acetyl-GPC is observed from thioglycollate-elicited mouse macrophages than from normal macrophages. This decreased release was recently attributed to lower levels of acetyltransferase in the thioglycollate-elicited cells (Roubin *et al.*, 1982). Albert and Snyder (1983) recently found that the acetyltransferase activity of rat alveolar macrophages is increased during phagocytosis or upon ionophore stimulation.

In addition to the above findings, several other lines of evidence are consistent with acetyl-CoA:1-alkyl-2-lyso-GPC acetyltransferase being a key enzyme responsible for the synthesis of 1-alkyl-2-acetyl-GPC. The 1-alkyl-2-lyso-GPC required for the reaction is likely derived from 1-alkyl-2-acyl-GPC containing a long-chain acyl residue in the 2-position. Several cells known to synthesize and release 1-alkyl-2-acetyl-GPC have now been shown to contain a high percentage of the choline-containing phosphoglycerides as the 1-alkyl-linked species: rabbit peritoneal neutrophils, 46% (Mueller *et al.*, 1982), human neutrophils, 45% (Mueller *et al.*, unpublished), rabbit alveolar macrophages, 33% (Sugiura *et al.*, 1983), rat alveolar macrophages, 35% (Albert and Snyder, 1983), guinea pig neutrophils, 16%, and peritoneal macrophages, 14% (Sugiura *et al.*, 1982). These cells thus contain a store of the precursor that upon stimulation can be acted on by phospholipase A_2 to release 1-alkyl-2-lyso-GPC, and indeed such a release has been observed in hog leukocytes (Polonsky *et al.*, 1980), rabbit platelets (Benveniste *et al.*, 1982), and rat alveolar macrophages (Albert and Snyder, 1983). Furthermore, phospholipase A_2 inhibitors block the synthesis of 1-alkyl-2-acetyl-GPC (Mencia-Huerta *et al.*, 1981; Benveniste *et al.*, 1982; Albert and Snyder, 1983). Both labeled acetate and 1-alkyl-2-lyso-GPC have been shown to be incorporated into 1-alkyl-2-acetyl-GPC by stimulated intact cells (Chap *et al.*, 1981; Mencia-Huerta *et al.*, 1981, 1982; Albert and Snyder, 1983; Mueller *et al.*, (1983). The *de novo* synthesis of choline-containing alkyl-linked lipids was inhibited by ionophore A23187 stimulation in rabbit neutrophils (Mueller *et al.*, 1983).

Overall, the above studies provide strong evidence for the synthesis of 1-alkyl-2-acetyl-GPC by a deacylation–reacetylation pathway. The key enzymes required for the pathway are a phospholipase A_2 to release 1-alkyl-2-lyso-GPC, and the acetyltransferase to activate the lyso derivative (Figures 6 and 9). Presently, evidence favors this pathway over the cholinephosphotransferase pathway (Renooij and Snyder, 1981) for the synthesis of 1-alkyl-2-acetyl-GPC.

B. 1-Alkyl-2-acetyl-sn-glycerol:CDP-Choline Cholinephosphotransferase (EC 2.7.8.16)

Platelet-activating factor can also be biosynthesized from 1-alkyl-2-acetyl-*sn*-glycerol by a highly specific microsomal cholinephosphotransferase (Figure 6) found in liver, spleen, heart, kidney, and lung of rats (Renooij and Snyder, 1981) and in neutrophils from humans (Alonso *et al.*, 1982). Alkylacetylglycerols of several alkyl chains (16:0, 18:0, and 18:1) served as substrates for the enzyme, and it appears that the cholinephosphotransferase catalyzing this reaction differs from the one that transfers the phosphocholine moiety of CDP-choline to diacylglycerols. The latter activity that utilizes long-chain diradylglycerols as a substrate is inhibited by dithiothreitol (5 mM), whereas when 1-alkyl-2-acetylglycerol was the substrate, the cholinephosphotransferase activity was enhanced slightly and stabilized in the presence of the dithiothreitol (Renooij and Snyder, 1981). The sharp optimum pH for the alkylacetylglycerol cholinephosphotransferase of 8.0 also differed from the pH optimum of 8.5 found for the long-chain diacylglycerol cholinephosphotransferase. Both enzyme activities required Mg^{2+} for activity and were inhibited by Ca^{2+} or centrophenoxine.

The metabolic significance of 1-alkyl-2-acetylglycerol:CDP-choline cholinephosphotransferase is still questionable since no enzyme activity has yet been reported for the synthesis of 1-alkyl-2-acetylglycerol, the substrate required for the highly specific cholinephosphotransferase that forms PAF. However, the cholinephosphotransferase that forms PAF could be important in reversing a phospholipase C-type pathway involved in the catabolism of PAF. Thus, utilization of an alkylacetylglycerol species would provide an alternate precursor to 1-alkyl-2-acetyl-GPC in the formation of the bioactive phospholipid, PAF.

C. *1-Alkyl-2-acetyl-sn-glycero-3-phosphocholine Acetylhydrolase (EC 3.1.1.48)*

1-Alkyl-2-acetyl-GPC acetylhydrolase (Figure 6) catalyzes the stoichiometric conversion of 1-alkyl-2-acetyl-GPC to 1-alkyl-2-lyso-GPC and acetate (Blank *et al.*, 1981). This enzyme is located in the cytosol of rat liver and has a broad pH optimum between 7.5 and 8.5. It is present in a variety of rat tissues (Blank *et al.*, 1981) and human blood cells (Lee *et al.*, 1982b). The apparent K_m and V_{max} of acetylhydrolase in the soluble fraction of rat kidney cortex were 3.1 μM and 17.8 nmol/min/mg protein, respectively. The activity of acetylhydrolase was (1) unaffected by the addition of Ca^{2+} or Mg^{2+}, (2) slightly stimulated by EDTA or dithiothreitol, and (3) inhibited by deoxycholate or diisopropylfluorophosphate. Other phosphatides having an acetyl group at the *sn*-2 position significantly inhibited the hydrolysis of 1-alkyl-2-acetyl-GPC, whereas long chain diacylphospholipids, i.e., egg lecithin, at comparable concentrations had no effect on the acetylhydrolase activity. These results show that the properties of acetylhydrolase distinctly differ from those normally associated with the phospholipase A_2 that utilizes phospholipids having long-chain acyl groups (Waite and van Deenen, 1967; Brokerhoff and Jensen, 1974). Acetylhydrolase in rat kidney cortex, partially purified by DEAE-cellulose, and Sephadex G-200 column chromatography, had a molecular weight of 100,000 (Lee, unpublished data).

Normal serum from rabbits and humans contains an acid-labile factor that rapidly and irreversibly destroys the functional activity of PAF (Pinckard *et al.*, 1979; Farr *et al.*, 1980). This factor was recently shown to be an acylhydrolase with properties similar to the one described above (Farr *et al.*, 1982). The PAF acetylhydrolase activity in rat plasma is higher in hypertensive rats than in normotensive controls (Blank *et al.*, 1983). The intracellular and serum forms of acetylhydrolase appear to be important in the metabolism of 1-alkyl-2-acetyl-GPC, since these activities control the circulating and tissue levels of the biologically active form.

D. *Pte · H_4-Dependent Alkyl Monooxygenase (EC 1.14.16.5)*

1-Alkyl-2-acetyl-GPC cannot serve as a substrate for the microsomal Pte·H_4-dependent alkyl monooxygenase (Figure 6) in rat liver (Lee *et al.*, 1981). However, the alkyl moiety of 1-alkyl-2-lyso-GPC, the product of the acetylhydrolase reaction, is cleaved to form an aldehyde and glycero-3-phosphocholine. Results on the Pte·H_4 requirement, tissue distribution, responses to thermal inactivation, catalase, and inhibitors, and substrate specificity indicate that the enzyme responsible for the hydrolysis of 1-alkyl-2-lyso-GPC is identical to the monooxygenase that cleaves the *O*-alkyl moiety of alkylglycerols (see Section V-B).

V. CATABOLISM OF ALKYL AND ALK-1-ENYL (PLASMALOGEN) GLYCEROLIPIDS

A. Fatty Alcohol:NAD^+ Oxidoreductase

The enzymic oxidation of a fatty alcohol to a fatty acid was first demonstrated in microsomes from mouse preputial gland tumors (Snyder and Malone, 1970). NAD^+ was required for the conversion, whereas $NADP^+$ had little effect. Ferrell and Kessler (1971) reported that hexadecanol was oxidized to the corresponding aldehyde by both soluble and particulate fractions from mouse liver when NAD^+ and $NADP^+$ were the respective cofactors; no evidence was obtained to indicate that fatty acids were formed. Detailed characterization of this enzyme in rat liver (Lee, 1979) and the caecum of gourami (Thyagarajan *et al.*, 1979) confirmed that hexadecanol was oxidized to hexadecanoic acid with NAD^+ as an absolute requirement in rat liver and with $NADP^+$ having 60% of the activity obtained with NAD^+ in gourami caeca. Fatty alcohol:NAD^+ oxidoreductase, located in the endoplasmic reticulum of rat liver, has a pH optimum of 8.4 (pyrophosphate buffer) or 8.8–9.0 (glycine or barbital buffer), whereas in the microsomal fractions of gourami caeca the pH optimum is 9.5.

Possibilities that the reaction in rat liver was catalyzed by contaminating enzymes such as an alcohol dehydrogenase, a combination of the oxidase and the peroxidatic activity of catalase, or a flavoprotein were ruled out by the fact that the activity was not affected by pyrazole (an inhibitor of alcohol dehydrogenase), sodium azide (an inhibitor of catalase), or AMP, ATP, and quinacrine (analogs or inhibitors of flavin coenzymes), respectively. Fatty alcohols of various chain lengths can serve as substrates, but ethanol had no inhibiting effect on the reaction (Lee, 1979). Isolation and identification of fatty aldehydes as intermediates in the reaction was accomplished by trapping aldehydes as the semicarbazide derivative (Lee, 1979).

NAD^+:Fatty alcohol oxidoreductase activities are found in most tissues, but the highest activities occur in rat liver and gourami caeca. The apparent K_m for hexadecanol is 0.67 μM and the V_{max} is 5.98 nmoles/min/mg protein for the enzyme from rat liver, whereas the apparent K_m for hexadecanol is 33 μM and the V_{max} is 42.9 nmol/min/mg protein with gourami caecum microsomes.

B. Pte·H_4-Dependent Alkyl Monooxygenase (EC 1.14.16.5)

Pte·H_4-dependent alkyl monooxygenase catalyzes the oxidative cleavage of the *O*-alkyl ether linkage in glycerolipids. Initial studies (Tietz *et al.*, 1964) indicated that the reaction occurs in microsomes of rat livers according to the following scheme:

$$\text{Hexadecylglycerol} + O_2 + \text{Pte}\cdot H_4 \rightarrow \text{hexadecanal} + \text{glycerol} + \text{Pte}\cdot H_2 + H_2O$$

The Pte·H_4 is regenerated by an NADPH-linked pteridine reductase in the soluble fraction. Tietz *et al.* (1964) proposed that the mechanism of the oxidative attack on the ether bond is similar to that described by Kaufman (1959) for the hydroxylation of phenylalanine. The first step in the sequence is the hydroxylation of the 1-carbon atom of the *O*-alkyl chain. The unstable hemiacetal intermediate that is formed spontaneously degrades to yield glycerol and a fatty aldehyde. The aldehyde is then oxidized to the carboxylic acid or reduced to the alcohol by an enzyme(s) found both in the microsomal and supernatant fractions of rat livers (Tietz *et al.*, 1964; Pfleger *et al.*, 1967). Addition of NAD^+ to the incubation system favors the oxidation step and NADH increases the production of fatty alcohols (Soodsma *et al.*, 1972).

Optimum pH for the reaction is 9.0 and the enzyme requires ammonium ions, sulfhydryl groups, and a heat labile, nondialyzable component from the soluble fraction for maximal activity (Soodsma *et al.*, 1972). The stimulating activity of this soluble factor can be totally replaced by catalase. Catalase functions to protect the enzyme from inactivation by H_2O_2 in addition to its ability to retard the nonenzymatic oxidation of the pterin cofactor (Rock *et al.*, 1976).

The substrate specificities of the alkyl monooxygenase in rat liver microsomes have been studied extensively. It was found that this enzyme cleaves alkylglycerols containing alkyl residues of chain lengths C_{18}, C_{16}, and C_{14} at the same rate, whereas the C_{12} alkyl chain was cleaved at only about one-third that rate and the C_8 alkylglycerol was not oxidized. The only aromatic compound tested, DL-α-benzyl glyceryl ether, was not attacked (Tietz *et al.*, 1964). The ether linkage in racemic and 1-,2-,3-isomeric forms of alkylglycerols is readily cleaved by the enzyme (Pfleger *et al.*, 1967; Snyder *et al.*, 1973b). Structural requirements for alkyl glycerolipids that serve as substrates for the cleavage enzyme are a free OH group at the *sn*-2 position and a free OH or phospho-base group at the *sn*-3 position of the glycerol moiety (Snyder *et al.*, 1973b; Lee *et al.*, 1981). When the hydroxyl group is substituted by a ketone group or an ester (acetate or long-chain acyl group) at the *sn*-2 position, or when phosphate is located at the *sn*-3 position, the *O*-alkyl linkage is not cleaved. On the other hand, alkyl moieties of alkyllysophosphatidylethanolamine and alkyllysophosphatidylcholine are cleaved under similar experimental conditions (see Section IV-D for details). Furthermore, the ether-linked diol lipids, alkylethyleneglycols, are metabolized in a manner similar to alkylglycerols (Snyder *et al.*, 1974).

Livers and intestines have the highest alkyl cleavage activities of various rat tissues investigated, and high enzyme activities were also found in livers from mice, rabbits, dogs, guinea pigs, gerbils, and hamsters (Pfleger *et al.*, 1967). Kapoulas *et al.* (1969) reported a similar cleavage enzyme in *Tetrahymena pyriformis,* but it differs from the liver system in that Pte·H_4 was not required. In a broad survey of various transplantable tumors, only the slow-growing Morris hepatoma (7794A) exhibited cleavage activity similar to that of normal livers (Soodsma *et al.*, 1970). Furthermore, the highest cleavage activities are found in tissues containing negligible quantities of ether lipids, and lower activities are associated with tissues containing relatively high quantities of ether lipids (Pfleger *et al.*, 1967; Soodsma *et al.*, 1970). It appears that the Pte·H_4-dependent alkyl monooxygenase could be an important regulatory enzyme in the metabolism of ether-linked glycerolipids.

C. Alkenylglycerophosphocholine(ethanolamine)hydrolase (EC 3.3.2.2) (Plasmalogenase)

Warner and Lands (1961) first demonstrated the hydrolysis of 1-alk-1′-enyl-2-lyso-GPC to free aldehyde and GPC using microsomal preparations from rat liver. The liver plasmalogenase did not hydrolyze the corresponding ethanolamine-linked analog, 1-alk-1′-enyl-2-lyso-GPE, or the acylated derivatives 1-alk-1′-enyl-2-acyl-GPC or -GPE. The plasmalogenase had a pH optimum of 7.1 and required no cofactors. The activity was abolished by heat, acid, alkali, or chymotrypsin treatment and was inhibited by imidazole (Ellingson and Lands, 1968). Activity of the enzyme was dependent on a lipid environment. Treatment of the microsomes with phospholipase A or C resulted in loss of activity, which could be partially restored by addition of sphingomyelin, diacyl-GPC, or, though less effective, diacyl-GPE (Ellingson and Lands, 1968). Freezing and thawing the preparation also resulted in loss of activity. Gunawan and Debuch (1981) recently reportd that the liver plasmalogenase also hydrolyzes 1-alk-1′-enyl-2-lyso-GPE. The K_m observed for the substrate was 1.66×10^{-4} M. Otherwise, the properties of the plasmalogenase appeared to be the same as observed by Warner and Lands (1961). It did not act on 1-alk-1′-enyl-2-acyl-GPE, required no Mg^{2+} or other cofactors, and was strongly inhibited by sulfhydryl reagents but not by diisopropylfluorophosphate. An SH group therefore appears important for activity of the plasmalogenase.

Ansell and Spanner (1965) found a plasmalogenase activity in rat brain that differs significantly from that of the liver. The enzyme was recovered in extracts of acetone-dried brain powder, and it required Mg^{2+} for activity and had a pH optimum of 7.4. The brain plasmalogenase hydrolyzed 1-alk-1′-enyl-2-acyl-GPE and to a lesser extent, 1-alk-1′-enyl-2-lyso-GPE.

In further studies of the plasmalogenase present in acetone-dried powder from bovine brain, the K_ms for ethanolamine- and choline-linked plasmalogens were 180 μM and 208 μM, respectively (D'Amato *et al.*, 1975). It was concluded from the kinetic properties and competitive inhibition studies that the same enzyme may cleave both choline and ethanolamine plasmalogens. Dorman *et al.* (1977) measured the plasmalogenase activity of intact neuronal perikarya, astroglia, and oligodendroglia cells isolated from bovine brain and found the highest activity in oligodendroglia cells. Using 1-alk-1′-enyl-2-acyl-GPE dispersed in myelin as the substrate, the rates of hydrolysis by the respective cells were 0.6, 1.1, and 6.7 μmoles/hr/mg protein. Ansell and Spanner (1968) observed higher plasmalogenase activity in white matter than in gray matter.

A factor in the soluble fraction of rat brain preparations that catalyzes the oxygen-dependent hydrolysis of alk-1′-enyl chains of choline- and ethanolamine-linked species (Yavin and Gatt, 1972a) was subsequently identified as a ferrous-ascorbate complex (Yavin and Gatt, 1972b).

Plasmalogenase activity is elevated in demyelinating CNS tissue (Horrocks *et al.*, 1978). Tissues studied include brains of Jimpy and Quaking mice (Dorman *et al.*, 1978) and brains of dogs with lesions associated with canine distemper virus-associated demyelinating encephalomyelitis (Fu *et al.*, 1980). These studies and others (Horrocks *et al.*, 1978; Horrocks and Fu, 1978) have led Horrocks and co-workers to suggest that plasmalogenase may play an important role in inflammation in nervous tissue and perhaps other tissues by releasing arachidonate from the 2-position of plasmalogens through the combined actions of the plasmalogenase and a lysophospholipase. See Horrocks and Fu (1978) for further review of plasmalogenase studies.

D. Phosphohydrolases

Both acyl- and alkyl-DHAP phosphohydrolases (Figure 4) that remove the phosphate group have been described in Ehrlich ascites cell microsomes and guinea pig liver mitochondria (Wykle *et al.*, 1972b; LaBelle and Hajra, 1974). Identification of alkyl-DHA was firmly established in the early studies of alkyl-DHAP synthase catalyzed reaction (Snyder *et al.*, 1970b). Alkyl-DHAP and alkylglycerol-P have been shown to be substrates for the phosphohydrolase activities in the microsomes of mouse preputial gland tumors (Chae *et al.*, 1973b). These phosphohydrolase activities were maximally inhibited at NaF concentrations greater than 40 mM. Phosphohydrolase activities associated with rat brain microsomes exhibited two optimal pH maxima at 5.2–5.6 and at 7.5–7.8 for alkyl-DHAP, alkylglycerol-P, and alkylacylglycerol-P. The alkylglycerol-P and alkyl-DHAP phosphohydrolases differ from alkylacylglycerol-P phosphohydrolase in that the latter is modulated by magnesium ions. Alkylglycerol-P was the most active substrate under all experimental conditions (El-Bassiouni *et al.*, 1975). Properties of these phosphohydrolases have not been characterized in detail, but it appears some specificities might be involved since it is known that a bacterial alkaline phosphatase hydrolyzes the phosphate of alkyl lipids only if hydroxyl or ketone groups are at the *sn*-2 position, whereas an acid phosphatase exhibited no specificity (Blank and Snyder, 1970).

There are two phosphatases in the lysosomes of rat liver that hydrolyze alkylacylglycerol-P, an acidic phosphatase and a phosphatase with a pH optimum of 6–7 (Stoffel and Heimann, 1973). In addition, the same authors (Stoffel and Heimann, 1973) demonstrated the presence of three phosphatases that can utilize alkylglycerol-P as a substrate with distinct pH optima at 4.6, 6.2, and 8.2. The major phosphohydrolase activity in lysosomal membranes was around pH 7.5–8.5. The soluble fraction exhibited two phosphatase activities, one at pH 4.6 and the other with a broad pH optimum around 6.0. The phosphatase with the pH optimum at 4.6 is completely inhibited by tartrate, whereas the membrane-bound phosphatase is inhibited by Triton WR-1339. The soluble phosphatase activity at pH 6.2 is not affected by either tartrate or Triton WR-1339. Presumably, the phosphatases that remove the phosphate moiety from alkylglycerol-P are the same enzymes that degrade alkylacylglycerol-P, although the two enzymes may have different metabolic functions. Alkylacylglycerol-P phosphohydrolase generates an important intermediate, alkylacylglycerol, which is a precursor for the synthesis of complex ether-linked lipids. On the other hand, alkylglycerol-P phosphohydrolase produces alkylglycerol, which is the most active substrate for the Pte·H_4-dependent alkyl monooxygenase in the catabolic pathway.

E. Lipases

The occurrence of acyl-DHAP acylhydrolase was first suggested by the findings that acyl-DHAP was cleaved to form DHAP during the assay of alkyl-DHAP synthase in Ehrlich ascites tumor cell microsomes (Friedberg *et al.*, 1980; Friedberg and Gomillion, 1981). Brown and Snyder (1982) provided evidence for the presence of two acyl-DHAP lipases in the same tissue; one is sensitive and the other is insensitive to inhibition by dinitrofluorobenzene (DNFB). However, neither of these activities appear to be associated with alkyl-DHAP synthase that has been purified >1000-fold (Brown and Snyder, 1983). The DNFB-insensitive lipase is not stable under the storage conditions used for the purification of alkyl-DHAP synthase (Brown and Snyder, 1982).

The possible presence of a lysosomal lipase that hydrolyzes 1-alkyl-2-acylglycerol to yield alkylglycerol was mentioned by Stoffel and Heimann (1973), but this has not yet been clearly established. Paltauf and co-workers (Paltauf *et al.*, 1974a; Paltauf and Wagner, 1976) demonstrated that lipoprotein lipase in postheparin plasma from humans and rats, rat adipose tissue, and bovine skim milk preferentially hydrolyzes only the ester at the *sn*-1 position when both 1-alkyl-2,3-diacyl-*sn*-glycerol and 1,2-diacyl-3-alkyl-*sn*-glycerol are presented as substrates. In contrast, pancreatic lipase cleaves the ester bond at the *sn*-1 and *sn*-3 positions at equal rates and has been shown to hydrolyze the *sn*-3 acyl moieties of both alkyldiacylglycerols and alk-1-enyldiacylglycerols (Snyder and Piantadosi, 1968). With either lipoprotein lipase or pancreatic lipase, the 1-alkyl-2-acyl-*sn*-glycerol isomer is almost completely resistant to hydrolysis, whereas the 2-acyl-3-alkyl isomer is readily hydrolyzed (Paltauf *et al.*, 1974a). However, the authors suggested that the presence of another esterase in the crude pancreatic lipase preparation could be responsible for the hydrolysis of the 2-acyl-3-aklyl-*sn*-glycerol. These results indicate that the lysosomal lipase described by Stoffel and Heimann (1973) is the only known lipase to hydrolyze the 1-alkyl-2-acyl-*sn*-glycerols.

F. Phospholipases

Few studies have been carried out on the susceptibility of alkyl ether-containing phospholipids to phospholipases in mammalian cells. Stoffel and Heimann (1973) found that there are two phospholipases, one in the soluble fraction of lysosomes with a pH optimum at 4.5 and the other in the lysosomal membrane fraction with a pH optimum at 8.0, that can hydrolyze 1-alkyl-2-acyl-GPE to 1-alkyl-2-lyso-GPE. The lyso alkyl derivatives formed cannot be further degraded by other lysosomal enzymes. Both phospholipases were completely inhibited by Ca^{2+} or Triton WR-1339. Since the phospholipase A_2 that hydrolyzes the fatty acid at the *sn*-2 position of diacyl-GPE or alkylacyl-glycero-P also has an optimal pH of 4.5 and is strongly inhibited by Ca^{2+} (Stoffel and Heimann, 1973; Waite *et al.*, 1969), Stoffel and Heimann (1973), it is suggested that the same enzyme catalyzes the hydrolysis of 1-alkyl-2-acyl-GPE, diacyl-GPE, and alkylacyl-glycero-P. Mitochondrial fractions from rat brain and an acetone powder preparation from human cerebral cortex were also found to contain a phospholipase A_2 activity that could remove the fatty acid at the *sn*-2 position of 1,2-

diacyl-GPC and 1-alkyl-2-acyl-GPC (Woelk *et al.*, 1974a). However, with both enzyme sources, 1-alkyl-2-acyl-GPC was hydrolyzed at only 40–50% the rate observed for diacyl-GPC. In the mitochondrial fractions of brains from rats afflicted with experimental allergic encephalomyelitis, both the 1,2-diacyl-GPC and the 1-alkyl-2-acyl-GPC phospholipase A_2 activities were increased approximately 25% over the corresponding activities from normal rat brains (Woelk *et al.*, 1947b).

Lysophospholipase-D (EC 3.1.4.39), an enzyme that appears specific for ether lipid substrates, catalyzes the removal of the base moiety, i.e., choline or ethanolamine, from 1-hexadecyl-2-lyso-GPC, 1-hexadecyl-2-lyso-GPE, or 1-hexadec-1-enyl-2-lyso-GPE (Figure 7; Wykle and Schremmer, 1974). This enzyme is present in the microsomal fractions of rat brain, kidney, intestine, lung, testes, and liver from rats; liver and testes have the highest activities (Wykle *et al.*, 1977). It is most active at pH 7.2 and requires Mg^{2+} for activity. Ca^{2+} at low concentrations (1.3 mM) failed to stimulate the reaction, whereas higher concentrations (25 mM) inhibited the reaction. The enzyme is tightly bound to the microsomes, and a sulfhydryl group appears to be required for activity (Wykley and Schremmer, 1974; Wykle *et al.*, 1977). Lysophospholipase-D is not active toward 1-acyl-2-lyso-GPC and 1-acyl-2-lyso-GPE (Wykle *et al.*, 1980a). When the *sn*-2 position of the 1-alkyl-2-lyso-GPC and 1-alkyl-2-lyso-GPE were acyl-

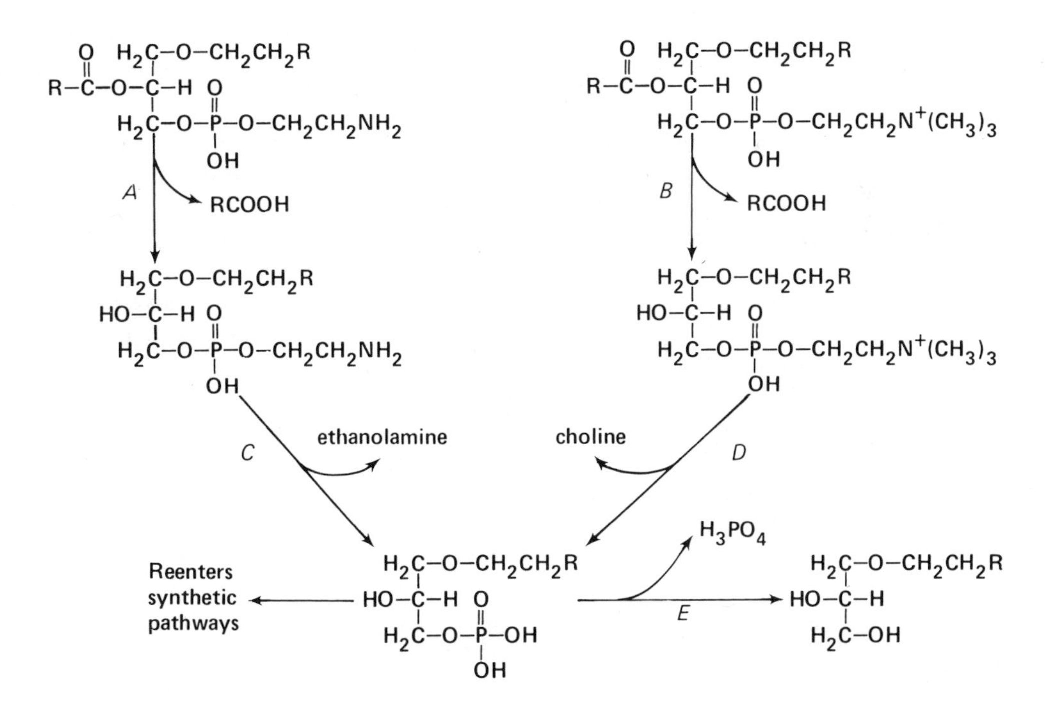

Figure 7. Action of phospholipase A_2 and lysophospholipase D on 1-alkyl-2-acyl-*sn*-glycero-3-phosphoethanolamine and 1-alkyl-2-acyl-*sn*-glycero-3-phosphocholine. Reactions A and B are catalyzed by phospholipase A_2, reactions C and D by lysophospholipase D. The removal of the phosphate moiety in reaction E is catalyzed by 1-alkyl-*sn*-glycerol-3-P phosphohydrolase.

ated, only minimal hydrolysis of the base group occurred (Wykle and Schremmer, 1974). The lysophospholipase-D pathway may be metabolically important for the removal of alkyl lysophospholipids, since these lipids are potent cell lysing agents and cannot be degraded by lysophospholipase or phospholipase A_1.

The presence of lysophospholipase-D activity in rat brain microsomes has been confirmed by Vierbuchen *et al.* (1979), who also showed that there was a phospholipase C-type activity in the same fractions that can hydrolyze 1-alkyl-2-lyso-GPE to alkylglycerol and phosphoethanolamine. The 1-alkyl-2-lyso-GPE was hydrolyzed at a faster rate than 1-alk-1-enyl-2-lyso-GPE (Gunawan *et al.,* 1979). 1-Alkyl-2-lyso-GPE exhibited noncompetitive inhibition kinetics in the hydrolysis of the 1-alk-1-enyl substrate. In contrast, 1-alk-1-enyl-2-lyso-GPE had no effect on the hydrolysis of 1-alkyl-2-lyso-GPE (Gunawan *et al.,* 1979).

Saito and Kanfer (1975) observed that the terminal phosphate diester bond of diacyl phosphatidylcholine is cleaved by a phospholipase-D activity associated with a solubilized preparation from a rat brain particulate fraction to yield phosphatidic acid. This D activity seems to be different from that of lysophospholipase-D in that the phospholipase-D activity is stimulated by Ca^{2+} or Mg^{2+}, but neither was an absolute requirement.

VI. ASSESSMENT OF THE PROMINENCE OF THE ACYL-DHAP AND ETHER LIPID PATHWAYS IN COMPLEX SYSTEMS

Several techniques have been employed to evaluate the contribution of the acyl-DHAP and ether lipid pathways in various tissues. All are based on specific enzymes in the initial steps of these pathways.

Agranoff and Hajra (1971) and Hajra (1973) used the nucleotide specificities of NADPH : acyl-DHAP oxidoreductase and *sn*-glycero-3-P dehydrogenase (*sn*-glycero-3-P : NAD^+ oxidoreductase, EC 1.1.1.8) to quantitatively measure the relative importance of the acyl-DHAP vs. *sn*-glycero-3-P pathways in homogenates of Ehrlich ascites tumor cells. They reasoned that tritium from [B-^{3}H]-NADH could only be incorporated into lipids by the *sn*-glycero-3-P pathway after reduction of DHAP to *sn*-glycero-3-P, while that from [B-^{3}H]-NADPH would be incorporated only through the reduction of acyl-DHAP. Their results indicated that the acyl-DHAP route is the major pathway for glycerolipid synthesis in Ehrlich ascites cells. However, the accuracy of the method is questionable because of possible isotope effects, lack of absolute nucleotide specificity, and the influence of endogenous levels of substrates and nucleotides. Another drawback of the method is that it cannot be used *in vivo*.

Manning and Brindley (1972) studied the relative rates of incorporation of a mixture of [2-^{3}H]- or [1,3-^{3}H]glycerol and [1-^{14}C]glycerol by rat liver slices, taking into account isotope effects. They concluded that the acyl-DHAP pathway is the major pathway in rat liver. Rognstad *et al.* (1974) challenged the interpretation of Manning and Brindley's data and concluded from similar experiments with liver cells that the acyl-DHAP pathway is only a minor pathway under their experimental conditions. Pollock *et al.* (1975) compared the synthesis of phosphatidic acid from DHAP and *sn*-glycero-3-P in homogenates from 13 different tissues, in which most were deficient in *sn*-glycero-3-P dehydrogenase. DHAP was incorporated into phosphatidic acid via acyl-DHAP, more rapidly than through its conversion to *sn*-glycero-3-P. All of the homogenates possessed an apparently greater capacity to synthesize phosphatidate from *sn*-glycero-3-P than from acyl-DHAP.

Lumb and Snyder (1971) showed that the use of a 1-[^{3}H]hexadecanol as an ether lipid precursor was effective in evaluating *O*-alkylglycerolipid biosynthesis in complex systems (*in vivo* and cell cultures). The tritium on the 1-carbon of the alcohol can only be incorporated into the ether-linked alcohol moieties of glycerolipids or the alcohol portion of wax esters, since otherwise the alcohol is oxidized to the corresponding fatty acid, whereby the tritium is removed from the lipid pool. Thus, the appearance of tritium in any glycerolipid class indicates that an ether linkage is present.

VII. REGULATION OF ETHER LIPID METABOLIC PATHWAYS

A. Turnover of Ether-Linked Glycerolipid Species

Several mammalian systems have been used to assess the metabolic stability and interrelationships among different subclasses of diradyl glycerolipids *in vivo*. The distribution of radioactivity from intravenously administered *cis*-9-[1-^{14}C]octadecenol into various tissues of the rat was studied as a function of time (Mukherjee *et al.*, 1980). Oxidation of the fatty alcohol and esterification of the resulting fatty acid into a variety of lipids were by far the most predominant reactions. Acylation of the long chain alcohol was observed in liver, which appears to be the major site for the biosynthesis of wax esters. Incorporation of the fatty alcohol into ether-linked lipids occurred in most tissues, however, the alkyl and alk-1-enyl moieties of ethanolamine glycerolipids in the heart contained the highest amounts of radioactivity. The incorporation of labeled fatty alcohol into ether-linked phospholipids continued to increase between 1–24 hr in all tissues (except in the intestine), which suggested a high metabolic stability for these type lipids. A similar conclusion was reached by Scott *et al.* (1979a) when the metabolic turnover of ether-linked lipids was determined with palmitic acid as a radioactive tracer in cultured primary rat tracheal epithelial cells and B 2-1 squamous carcinoma cells.

In a series of experiments, Waku and co-investigators (Waku *et al.*, 1976; Waku and Nakazawa, 1978a,b; and Waku and Nakazawa, 1979) measured the turnover rates of diacyl-, alkylacyl-, and alk-1-enylacyl-GPC and -GPE in Ehrlich ascites cells by injecting various radioactive precursors of phospholipids, i.e., ^{32}Pi, [1-^{14}C]acetate, or [1-^{14}C]glycerol, into the peritoneal cavity of the tumor-bearing mice. The alkylacyl-GPE was found to have the most rapid turnover rate. The turnover rate of alk-1-enylacylphospholipids was slower than the rates for diacyl and alkylacyl phospholipids. When the turnover rates of different molecular species of diacyl and alkylacyl species of ethanolamine- and choline-phosphoglycerides were compared, the results demonstrated that hexaenoic molecular species from alkylacyl-GPE and disaturated molecular species from both diradyl choline-phosphoglycerolipids and ethanolamine-phosphoglycerolipids had fast turnover rates (Waku and Nakazawa, 1978a,b). For alk-1-enyl-acyl-GPE, the hexaenoic molecular species turned over most rapidly and disaturated and tetraenoic species turned over at a high rate, while monoenoic and dienoic molecular species turned over very slowly (Waku and Nakazawa, 1979). The significance of the fast turnover rate for hexaenoic molecular species of alkylacyl-GPE is not clear at present. However, Waku and Nakazawa (1978a,b) proposed that the rapid synthesis of the hexaenoic species of the ethanolamine-phosphoglycerolipids may be the consequence of a higher affinity for the hexaenoic diradylglycerols in the diradylglycerol : CDP-ethanolamine ethanolaminephosphotransferase system.

B. Ether Lipid Precursors and Alkyl Glycerolipids

A biochemical characteristic of a variety of animal and human tumors is the presence of higher quantities of ether-linked lipids (especially alkyldiacylglycerols) than found in most normal tissues (Snyder and Wood, 1968, 1969; Howard *et al.*, 1972; Lin *et al.*, 1978). The overall enzymatic pathway depicted in Figure 8 emphasizes important regulatory steps in ether lipid metabolism, however, knowledge of factors responsible for their regulation is scarce and sometimes even contradictory. Important control points to consider in the regulation of ether lipids are the formation of their precursors (fatty alcohols and DHAP), the rate of synthesis of the first ether-linked intermediate (alkyl-DHAP formed by alkyl-DHAP synthase), and the rate of ether lipid degradation.

Howard *et al.* (1972) observed a correlation between the elevated level of ether lipids, the high rate of glycolysis, and the low glycerol-P dehydrogenase activities in a number of transplantable hepatomas. It is quite plausible from a consideration of the metabolic pathways (Figure 8) that a decrease in glycerol-P dehydrogenase could result in an increase in ether lipid biosynthesis. A decrease in glycerol-P dehydrogenase activity would lead to increased levels of DHAP and, therefore, a greater proportion of glycerolipids would be synthesized from DHAP instead of glycerol-P.. Since acyl-DHAP is the precursor of alkyl-DHAP, increased formation of acyl-DHAP could result in an increased proportion of ether lipids.

However, several lines of evidence suggest that the change in ether lipid levels is probably not an essential characteristic of the neoplastic process itself, even though the amount of ether-linked lipids is higher in both virally- and carcinogen-transformed cells grown in culture than in their normal counterparts (WI-38 vs. WI-38VA13H, C3H/10T1/2 clone 8 vs. 16; Howard *et al.*, 1972; Cabot and Welsh, 1981). Scott *et al.* (1979a) found that alkyldiacylglycerols are undetectable in tracheal epithelium *in vivo,* and yet, high amounts of alkyldiacylglycerols accumulated when normal primary tracheal cells were grown in culture. In addition, the WI-38VA13H cells (derived from WI-38 by transformation with SV-40 virus) contain ether-lipid levels about three times higher than the WI-38 cells (human fibroblasts with a limited lifespan of 40–50 cell generations in culture), but their glycerol-P dehydrogenase levels are similar (Howard *et al.*, 1972).

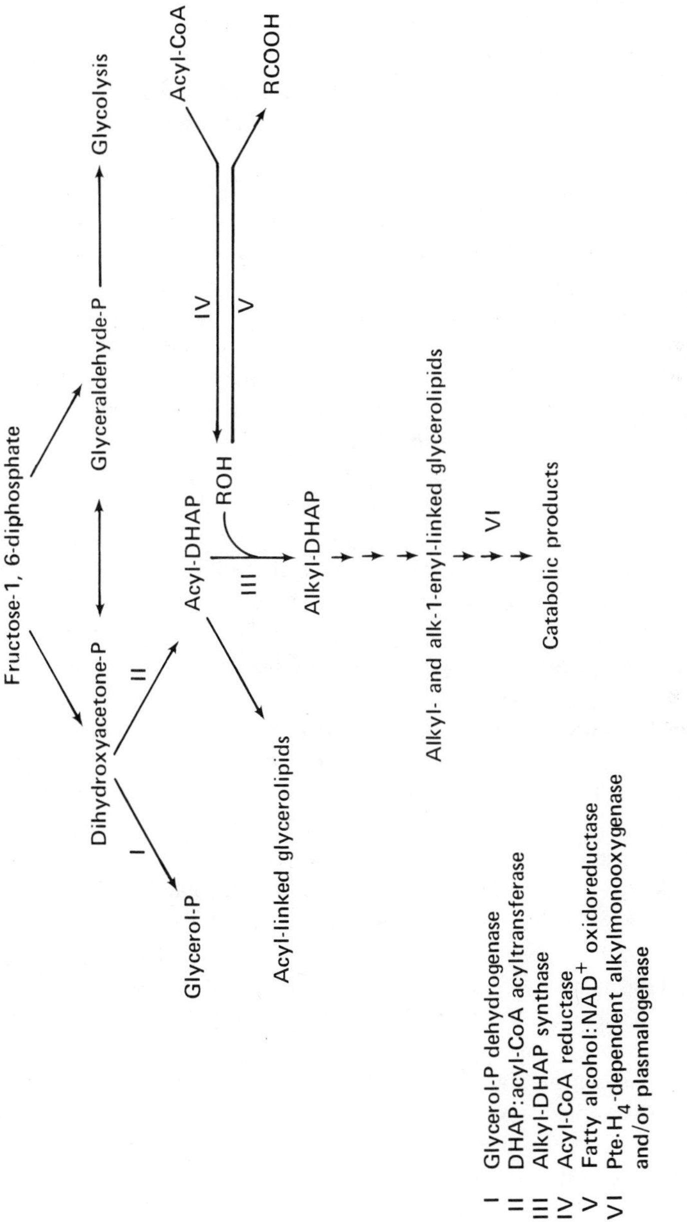

Figure 8. Key regulatory control points in the metabolic pathways for ether-linked glycerolipids.

Furthermore, Scott *et al.* (1979b) were able to manipulate the level of alkyldiacylglycerols in a neoplastic cell line (B 2-1 rat squamous carcinoma) by growing cells either in the presence of 5.5 mM galactose or 5.5 mM glucose. The cells grown in 5.5 mM galactose in the absence of glucose produced very low levels of alkyldiacylglycerols. Increasing concentrations of glucose caused a progressive increase in alkyldiacylglycerols (up to a level ten-fold higher than that found in the absence of glucose) after a 4-hr lag. This lag period could be shortened when hexadecanol was added into the culture media along with glucose. The data suggest that under these conditions the level of fatty alcohols was rate-limiting in the formation of alkyldiacylglycerols. In contrast, the addition of hexadecanol alone to the galactose-grown cells had little effect on their content of alkyldiacylglycerols. The extent of uptake and oxidation of hexadecanol was similar in both the glucose- and galactose-grown cells, a finding also reflected by the similar activities of NAD^+ : fatty alcohol oxidoreductase in homogenates of both the glucose- and galactose-grown cells. The level of free fatty alcohols and the activity of acyl-CoA reductase (the enzyme responsible for the synthesis of fatty alcohols) were not determined. The other important precursor for the synthesis of ether-linked lipids, DHAP, was barely detectable in galactose-grown B 2-1 cells but increased rapidly after glucose supplementation and reached a maximum at 2 hr, followed by a gradual decline (Scott *et al.*, 1979b). The increase in DHAP content was not correlated with any change in activity of *sn*-glycero-3-P dehydrogenase, which was similar in both the glucose- and galactose-grown cells. Glucose supplementation of galactose-grown cells had little influence on the alkyl and alk-1-enyl content of the phospholipids. In view of these results, it seems that levels of alkyldiacylglycerols in neoplastic cells are regulated by the extent their precursors are formed from glucose. The synthesis of hexadecanol by itself was not sufficient to increase the level of ether lipids.

Lee and co-workers (Lee *et al.*, 1979; Lee and Stephens, 1982) observed that supplementation of culture medium with *trans* fatty acids, i.e., elaidate, *trans* vaccenate, and linolelaidate, in L-M cells induced the accumulation of fatty alcohols (six-fold over unsupplemented cells) and alkyldiacylglycerols. Yet, the concentration of total alcohol-containing lipids, i.e., total alkyl and alk-1-enyl lipids, was slightly decreased due to a major decrease in ether-linked phospholipids, even though there was a net increase in the production of free fatty alcohols.

In addition to *sn*-glycero-3-P dehydrogenase, several other key enzymes involved in the metabolism of ether-linked lipids have been determined in normal rat liver and a variety of transplantable tumors with different levels of ether lipids. The activity of $Pte \cdot H_4$-dependent alkyl monooxygenase (cleaves the ether bond) is low in tumors and other tissues that contain high levels of ether-linked lipids (Pfleger *et al.*, 1967; Soodsma *et al.*, 1970). In other experiments, Lee *et al.* (1980) found that the activity of fatty alcohol:NAD^+ oxidoreductase was high and the activities of acyl-CoA reductase and alkyl-DHAP synthase were low in rat liver, which has low amounts of ether glycerolipids. On the other hand, in a Fischer R-3259 sarcoma with high concentrations of ether-linked lipids, the reverse pattern for these enzyme activities was observed. In Morris hepatomas 5123C, with an intermediate level of ether lipids, the activities of all three enzymes ranged between that found in liver and Fischer sarcomas (Lee *et al.*, 1980). A negative correlation between fatty alcohol:NAD^+ oxidoreductase activity and the content of ether-linked lipids was also found in normal tissues (Lee, 1979). Furthermore, the activities of alkyl-DHAP synthase and acyl-CoA reductase seem to be high in several normal and neoplastic tissues, such as developing rat brain (Snyder *et al.*, 1971b; Bishop and Hajra, 1981), preputial glands (Moore and Snyder, 1982a), harderian glands, and Ehrlich ascites carcinoma (Rock *et al.*, 1977a) that are rich in ether-linked lipids. Therefore, it would appear that the level of ether-linked lipids in tissues is regulated in a coordinated manner by the increased biosynthetic and decreased catabolic activities of the enzymes involved in their metabolism.

It is interesting to note that a similar or identical CDP-choline:diradylglycerol cholinephosphotransferase with similar V_{max} values seems to be responsible for the formation of 1,2-diacyl- and 1-alkyl-2-acyl-GPC in rat liver (Radominska-Pyrek *et al.*, 1977; Lee *et al.*, 1982a). However, since there is essentially no 1-alkyl-2-acyl-GPC in liver, it appears that the level of ether-linked phospholipids is controlled by the availability of diradylglycerols and the turnover rate of this phospholipid intermediate (Lee *et al.*, 1982a).

The alkyl chains at the *sn*-1 position of ether-linked lipids consist mainly of $C_{16:0}$, $C_{18:0}$, and $C_{18:1}$ moieties (Horrocks, 1972). Alkyl-DHAP synthase that forms the ether bond by catalyzing the reaction between acyl-DHAP and a fatty alcohol is relatively nonspecific. A wide variety of long-chain fatty alcohols can act as substrates for this enzyme, both *in vivo* and *in vitro* (see Section III-A). It appears that the substrate specificity of acyl-CoA reductase can explain in part the narrow composition of alkyl chains in ether glycerolipids. Other kinetic parameters of acyl-CoA reductase and alkyl-DHAP synthase could also exert an influence on the composition of alkyl chains. The long-chain saturated fatty alcohols (20:0 and 22:0) are formed at low rates by acyl-CoA reductase, and they have been found as the free alcohols in rat brain (Natarajan and Schmid, 1977c; Bishop and Hajra, 1981). These 20:0 and 22:0 fatty alcohols are poor substrates for alkyl-DHAP synthase and, therefore, the combined action of synthase and acyl-CoA reductase in the utilization of substrates with carbon chains 20 or longer is probably responsible for the complete exclusion of the 20:0 and 22:0 moieties from the brain ether lipids. Appreciable acyl-CoA reductase activity was observed when acyl (18 : 2)-CoA was used as a substrate. Yet, the high K_m for linoleoyl-CoA and the virtual absence of linoleic acid in brain make it unlikely that this reductase is of physiological significance (Bishop and Hajra, 1981). When an equimolar mixture of homologous saturated (13:0, 15:0, 17:0, 19:0) and vinylogous (19:0, 19:1, 19:2) fatty alcohols were fed to two groups of rats, significant quantitative differences were observed in the distribution of the various odd-numbered alkyl and alk-1-enyl moieties of the diradyl species of phosphatidylcholines and diradylphosphatidylethanolamines in the rat small intestine (Bandi and Mangold, 1978). These results suggest that there is some compositional selectivity in the biosynthesis of the ether-linked chains of glycerolipids based on the long-chain fatty alcohols available *in vivo*.

C. Alk-1-enyl Glycerolipids (Plasmalogens)

Few enzymatic studies of the regulation of plasmalogen synthesis (Figure 2) have been reported. The alkyl desaturase activity was not increased in tumors of rats maintained on fat-free diets, although stearoyl-CoA desaturase was significantly increased (Lee *et al.*, 1973) as discussed in Section III-K. Thus, it appears the alkyl and stearoyl-CoA desaturases are regulated independently. The rates of synthesis of plasmalogens in various membrane preparations are difficult to compare with certainty since there is endogenous 1-alkyl-2-acyl-GPE in the preparations that may lower the specific radioactivity of added substrates.

Plasmalogenase must play an important role in the turnover of plasmalogens. In brain, the plasmalogenase activity increases during myelination (Horrocks *et al.*, 1978; Horrocks and Fu, 1978), and plasmalogenase activity is elevated in demyelinating lesions and increases in injured white matter (Horrocks *et al.*, 1978). Properties of plasmalogenase are discussed in Section V-C.

Korey and Orchen (1959) found that the most rapid increase in plasmalogens of rat brain occurs during the 10–22-day postnatal period, which corresponds with the time of most active myelin deposition. On the basis of concentration (μmol/g), plasmalogens mainly increase in the human brain between the thirty-second week of gestational age and about the sixth postnatal month (Martinez and Ballabriga, 1978).

Roots and Johnston (1968) found that the plasmalogen content of lipids from goldfish brain varies when the fish are acclimated to temperatures ranging from 5–30°C. The percent of plasmalogens in phospholipids varied from 35% at 5°C to 46% at 30°C (Roots and Johnston, 1968). When individual phospholipid classes were examined, the highest levels of plasmalogens were found in the ethanolamine-containing fraction, which varied from approximately 43% plasmalogen at 5°C to 59% at 30°C (Driedzic *et al.*, 1976). The changes in the aliphatic chains of the plasmalogens and diacyl phospholipids of myelin and other brain fractions in response to environmental temperature were examined in more detail in a subsequent study (Selivonchik *et al.*, 1977). It was concluded from these experiments that changes in plasmalogen content play an important role in maintaining the proper fluidity of membranes in fish brains acclimated to different temperatures. The molecular basis for such a function was discussed (Selivonchick *et al.*, 1977) in terms of the model of membrane lipid interaction proposed by Brockerhoff (1974). In hibernating Syrian hamster, changes in the plasmalogen composition of brain lipids were also observed (Goldman, 1975). The most striking change observed was an increase in the 18:1 alk-1′-enyl content in the hibernating animals, which increased from 14.4–27.2% during hibernation. It is apparent that the regulation of plasmalogen composition is likely of key importance in membrane function, however, little is known of the molecular mechanisms that control the cellular levels of plasmalogens.

D. *Bioactive Alkyl Phospholipids*

At the present time, little is known about the regulation of enzymes involved in the metabolism of alkylacetyl-GPC. Yet, it is clear that cellular activation states increase PAF biosynthesis by stimulation of the acetyl-CoA : alkyllyso-GPC acetyltransferase reaction (Figures 6 and 9). Chap *et al.* (1981) reported that when rabbit platelets are treated with A23187, [^{3}H]acetate was rapidly incorporated into alkylacetyl-GPC. In other experiments involving direct measurements of enzyme activities, ace-

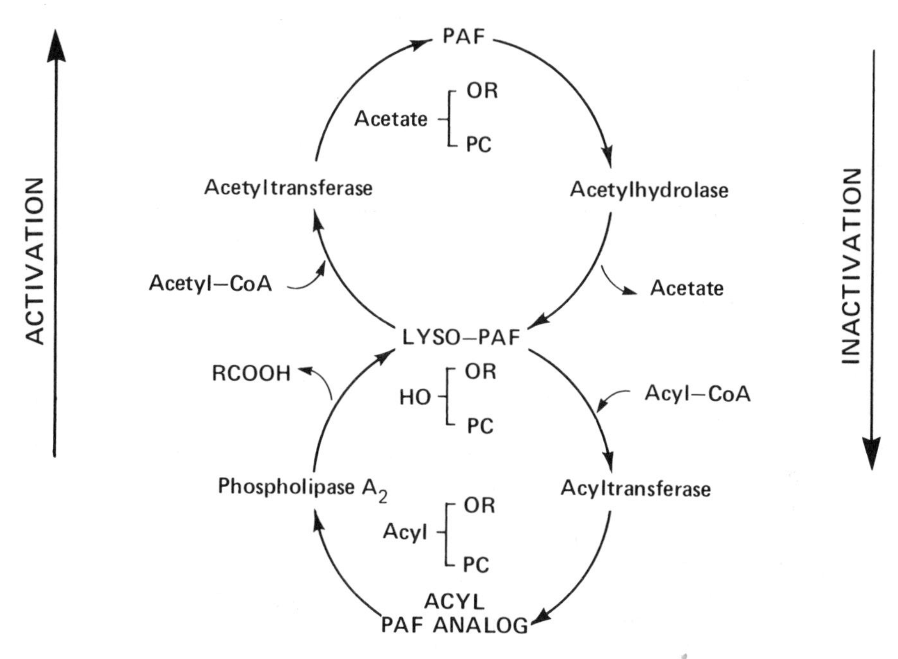

Figure 9. The activation–inactivation cycle for bioactive 1-alkyl-2-acetyl-*sn*-glycero-3-phosphocholine (platelet-activating factor).

tyltransferase was stimulated several-fold in human neutrophils by ionophore A23187 (Lee *et al.*, 1982b) or zymosan (Alonso *et al.*, 1982), in rat alveolar macrophages by ionophore A23187 or zymosan (Albert and Snyder, 1983), and in human eosinophils from patients with eosinophilia (Lee *et al.*, 1982b). Acetylhydrolase activities were also stimulated in human neutrophils by ionophore treatment, but to a much smaller extent than the acetyltransferase under the same conditions (Lee *et al.*, 1982b).

Mencia-Huerta *et al.* (1982) observed similar results for the acetylation step after zymosan stimulation of peritoneal macrophages from rats. Their data based on PAF release indicated that the addition of acetyl-CoA was better than acetate itself, and that the addition of malonyl-CoA, butyryl-CoA, or palmitoyl-CoA were not effective in stimulating the release of PAF after zymosan treatment. In a separate investigation with peritoneal macrophages, Roubin *et al.* (1982) showed that the decreased release of PAF by peritoneal macrophages (after zymosan exposure) obtained from thioglycollate-injected mice compared to macrophages from untreated mice was due to a defect in the acetylation step in the thioglycollate-exposed macrophages.

It has also been documented in rat alveolar macrophages that phospholipase A_2 activities (Figures 6 and 9) are stimulated by both zymosan or ionophore A23187, and that the stimulation could be blocked by either mepacrine or bromophenacylbromide (Albert and Snyder, 1983). Although these inhibitors can interact with cells in various ways, the effect on the *sn*-2 deacylation of phospholipids observed in this macrophage study is consistent with the ability of mepacrine and bromophenacylbromide to prevent fatty acid release via phospholipase A_2 (Hofmann *et al.*, 1982; Lapetina *et al.*, 1981;

Vargaftig, 1977). Indirect evidence has also implicated the involvement of phospholipase A_2 in the formation of PAF as seen by the inhibition of the release of alkylacetyl-GPC from rabbit platelets (after stimulation with A23187 or thrombin) by known phospholipase A_2 inhibitors, including EDTA, EGTA, dibutyryl-cAMP, *p*-bromophenacylbromide, or 874CB (Benveniste *et al.*, 1982).

Alonso *et al.* (1982) showed that the specific cholinephosphotransferase that forms PAF from alkylacetylglycerols (Figure 6) was unaffected by zymosan stimulation of human neutrophils. It also should be noted that alkyl-DHAP synthase activities were not altered after treatment of human neutrophils with ionophore A23187 (Lee *et al.*, 1982b).

Although it is clear from the studies described above that acetyltransferase, phospholipase A_2, and acetylhydrolase are important regulatory control points in the metabolism of alkylacetyl-GPC, i.e., the inactivation-activation cycle (Figure 9), knowledge of the physiological regulation of these important enzymes is still lacking. Also, understanding the regulatory factors involved in PAF metabolism is complicated by the close interaction of PAF and arachidonate metabolites generated by the lipoxygenase pathway (Chilton *et al.*, 1982; Voelkel *et al.*, 1982; Smith and Bowman, 1982; Lapetina, 1982; O'Flaherty *et al.*, 1983; Lee *et al.*, 1983) or cyclooxygenase pathway, i.e., stimulation of thromboxane B_2 formation in rabbit platelets (Shaw *et al.*, 1981). Other aspects of potential regulatory controls for PAF pathways can be gleaned from a recent review of developments in the PAF field (Snyder, 1982).

ACKNOWLEDGMENTS

This work was supported by the Office of Energy Research, U.S. Department of Energy (Contract No. DE-AC05-76OR00033), the American Cancer Society (Grant BC-70N), the National Cancer Institute (Grant CA-11949-13), and the National Heart, Lung, and Blood Institute (Grant HL-27109-03) to F.S.; and the National Heart, Lung, and Blood Institute (Grant HL-26818-03) and the National Allergy and Infectious Diseases Institute (Grant AI-17287-02) to R.L.W.

REFERENCES

Agranoff, B. W., and Hajra, A. K., 1971, The acyl dihydroxyacetone phosphate pathway for glycerolipid biosynthesis in mouse liver and Ehrlich ascites tumor cells, *Proc. Natl. Acad. Sci. USA* **68:**411–415.

Albert, D. H., and Snyder, F., 1983, Biosynthesis of 1-alkyl-2-acetyl-*sn*-glycero-3-phosphocholine (platelet activating factor) from 1-alkyl-2-acyl-*sn*-glycero-3-phosphocholine by rat alveoler macrophages: Ionophore stimulation, *J. Biol. Chem.* **258:**97–102.

Alonso, F., Gil, M. G., Sanchez-Crespo, M., and Mato, J. M., 1982, Activation of 1-alkyl-2-lysoglycero-3-phosphocholine. Acetyl-CoA transferase during phagocytosis in human polymorphonuclear leukocytes, *J. Biol. Chem.* **257:**3376–3378.

Ansell, G. B., and Metcalfe, R. F., 1968, The labelling of brain phosphatidylethanolamine and ethanolamine plasmalogen from cytidine diphosphate ethanolamine *in vitro, Biochem. J.* **109:**29P.

Ansell, G. B., and Metcalfe, R. F., 1971, Studies on the CDP-ethanolamine-1,2-diglyceride ethanolamine-phosphotransferase of rat brain, *J. Neurochem.* **18:**647–665.

Ansell, G. B., and Spanner, S., 1965, The magnesium-ion-dependent cleavage of the vinyl ether linkage of brain ethanolamine plasmalogen, *Biochem. J.* **94:**252–258.

Ansell, G. B., and Spanner, S., 1967, The metabolism of labeled ethanolamine in the brain of the rat *in vivo*, *J. Neurochem.* **14**:873–885.

Ansell, G. B., and Spanner, S., 1968, Plasmalogenase activity in normal and demyelinating tissue of the central nervous system, *Biochem. J.* **108**:207–209.

Baker, R. C., Wykle, R. L., Lockmiller, J. S., and Snyder, F., 1976, Identification of a soluble protein stimulator of plasmalogen biosynthesis and stearoly-CoA desaturase, *Arch. Biochem. Biophys.* **177**:299–306.

Ballas, L. M., and Bell, R. M., 1981, Topography of glycerolipid synthetic enzymes: Synthesis of phosphatidylserine, phosphatidylinositol and glycerolipid intermediates occurs on the cytoplasmic surface of rat liver microsomal vesicles, *Biochim. Biophys. Acta* **665**:586–595.

Bandi, Z. L., and Mangold, H. K., 1978, Substrate specificity of enzymes catalyzing the biosynthesis of ionic alkoxylipids from alcohol, *Nutr. Metabol.* **22**:190–199.

Bandi, Z. L., Aaes-Jørgensen, E., and Mangold, H. K., 1971, Metabolism of unusual lipids in the rat. I. Formation of unsaturated alkyl and alk-1-enyl chains from orally administered alcohols, *Biochim. Biophys. Acta* **239**:357–367.

Bates, E. J., and Saggerson, E. D., 1979, A study of the glycerol phosphate acyltransferase and dihydroxyacetone phosphate acyltransferase activities in rat liver mitochondrial and microsomal fractions. Relative distribution in parenchymal and non-parenchymal cells, effects of *N*-ethylmaleimide, palmitoyl-coenzyme A concentration, starvation, adrenalectomy and anti-insulin serum treatment, *Biochem. J.* **182**:751–762.

Bell, R. M., and Coleman, R. A., 1980, Enzymes of glycerolipid synthesis in eukaryocytes, *Annu. Rev. Biochem.* **49**:459–487.

Bell, O. E., Jr., Blank, M. L., and Snyder, F., 1971, The incorporation of ^{18}O and ^{14}C from long chain alcohols into the alkyl and alk-1-enyl ethers of phospholipids of developing rat brain, *Biochim. Biophys. Acta* **231**:579–583.

Bell, R. M., Ballas, L. M., and Coleman, R. A., 1981, Lipid topogenesis, *J. Lipid Res.* **22**:391–403.

Benveniste, J., Chignard, M., LeCouedic, J. P., and Vargaftig, B. B., 1982, Biosynthesis of platelet-activating factor (PAF-acether). II. Involvement of phospholipase A_2 in the formation of PAF-acether and lyso-PAF-acether from rabbit platelets, *Thromb. Res.* **25**:375–385.

Betz, S. J., and Henson, P. M., 1980, Production and release of platelet-activating factor (PAF); dissociation from degranulation and superoxide the production in human neutrophil, *J. Immunol.* **125**:2756–2763.

Bishop, J. E., and Hajra, A. K., 1978, Specificity of reduction of fatty acids to long chain alcohols by rat brain microsomes, *J. Neurochem.* **30**:643–647.

Bishop, J. E., and Hajra, A. K., 1981, Mechanism and specificity of formation of long chain alcohols by developing rat brain, *J. Biol. Chem.* **256**:9542–9550.

Bishop, J. E., Salem, M., and Hajra, A. K., 1982, Topographical distribution of lipid biosynthetic enzymes on peroxisomes (microbodies), *Ann. N.Y. Acad. Sci.* **386**:411–413.

Blank, M. L., and Snyder, F., 1970, Specificities of alkaline and acid phosphatases in the dephosphorylation of phospholipids, *Biochemistry* **9**:5034–5036.

Blank, M. L. Wykle, R. L., Piantadosi, C., and Snyder, F., 1970, The biosynthesis of plasmalogens from labeled *O*-alkylglycerols in Ehrlich ascites cells, *Biochim. Biophys. Acta* **210**:442–447.

Blank, M. L., Wykle, R. L., and Snyder, F., 1971, Enzymic synthesis of ethanolamine plasmalogens from an *O*-alkyl glycerolipid, *FEBS Lett.* **18**:92–94.

Blank, M. L., Wykle, R. L., and Snyder, F., 1972a, The biosynthesis of ethanolamine plasmalogens by a postmitochondrial fraction from rat brain, *Biochem. Biophys. Res. Commun.* **47**:1203–1208.

Blank, M. L., Kasama, K., and Snyder, F., 1972b, Isolation and identification of an alkyldiacylglycerol containing isovaleric acid, *J. Lipid Res.* **13**:390–395.

Blank, M. L., Wykle, R. L., and Snyder, F., 1973, The retention of arachidonic acid in ethanolamine plasmalogens of rat testes during esential fatty acid deficiency, *Biochim. Biophys. Acta* **316**:28–34.

Blank, M. L., Wykle, R. L., Alper, S., and Snyder, F., 1974, Microsomal synthesis of the ether analogs of triacylglycerols. Acyl CoA : alkylaoylglycerol and aoyl CoA : alk-1-enylacylglycerol acyltransferases in tumors and liver, *Biochim. Biophys. Acta* **348**:397–403.

Blank, M. L., Lee, T-c., Fitzgerald, V., and Snyder, F., 1981, A specific acetylhydrolase for 1-alkyl-2-acetyl-*sn*-glycero-3-phosphocholine a (hypotensive and platelet-activating lipid), *J. Biol. Chem.* **256**:175–178.

Blank, M. L., Hall, M. N., Cress, E. A., and Snyder, F., 1983, Inactivation of 1-alkyl-2-acetyl-*sn*-glycero-3-phosphocholine by a plasma acetylhydrolase: higher activities in hypertensive rats. *Biochem. Biophys. Res. Commun.* **113:**666–671.

Bourre, J. M., and Daudu, O., 1978, Stearyl-alcohol biosynthesis from stearyl-CoA in mouse brain microsomes in normal and dysmyelinating mutants (Quaking and Jimpy), *Neurosci. Lett.* **7:**225–230.

Brockerhoff, H., 1974, Model of interaction of polar lipids, cholesterol, and proteins in biological membranes, *Lipids* **9:**645–650.

Brockerhoff, H., and Jensen, R. G., 1974, Phospholipases: Carboxyl esterases, in : *Lipolytic Enzymes,* Academic Press, New York, pp. 194–265.

Brown, A. J., and Snyder, F., 1979, Solubilization of alkyldihydroxyacetone-P synthase from Ehrlich ascites cell microsomal membranes, *Biochem. Biophys. Res. Commun.* **90:**278–284.

Brown, A. J., and Snyder, F., 1980, Properties of membrane-bound and solubilized alkyldihydroxyacetone-P synthase, *Fed. Proc.* **39:**1993 (Abstract 2032).

Brown, A. J., and Snyder, F., 1981, Partial purification of alkyl synthase, *Fed. Proc.* **40:**1806 (Abstract 1535).

Brown, A. J., and Snyder, F., 1982, Alkyldihydroxyacetone-P synthase: Solubilization, partial purification, new assay method, and evidence for a ping-pong mechanism, *J. Biol. Chem.* **257:**8835–8839.

Brown, A. J., and Snyder, F., 1983, The mechanism of alkyldihydroxyacetone-P synthase. Formation of [^{3}H]H_2O from acyl-[1-R-^{3}H]dihydroxyacetone-P by purified alkyldihydroxyacetone-P synthase in the absence of acylhydrolase activity, *J. Biol. Chem.,* **258:**4184–4189.

Brunetti, M., Gaiti, A., and Porcellati, G., 1979, Synthesis of phosphatidylcholine and phosphatidylethanolamine at different ages in the rat brain *in vitro, Lipids* **14:**925–931.

Buckner, J. S., and Kolattukudy, P. E., 1976, Biochemistry of bird waxes, in: *Chemistry and Biochemistry of Natural Waxes* (P. E. Kolattukudy, ed.), Elsevier, New York, pp. 147–200.

Cabot, M. C., and Welsh, C. J., 1981, Ether lipid studies in mouse C3H/10T1/2 cells and a 3-methylcholanthrene-transformed clone, *Arch. Biochem. Biophys.* **211:**240–244.

Chae, K., Piantadosi, C., and Snyder, F., 1973a, An alternate enzymic route for the synthesis of the alkyl analog of phosphatidic acid involving alkylglycerol, *Biochem. Biophys. Res. Commun.* **51:**119–124.

Chae, K., Piantadosi, C., and Snyder, F., 1973b, Reductase, phosphatase, and kinase activities in the metabolism of alkyldihydroxyacetone phosphate and alkyldihydroxyacetone, *J. Biol. Chem.* **248:**6718–6723.

Chang, N-C., and Schmid, H. H. O., 1975, Structural specificity in ether lipid biosynthesis. Formation of hydroxyalkyl and oxoalkyl glycerophosphatides, *J. Biol. Chem.* **250:**4877–4882.

Chap, H., Mauco, G., Simon, M. F., Benveniste, J., and Douste-Blazy, L., 1981, Biosynthetic labelling of platelet activating factor from radioactive acetate by stimulated platelets, *Nature* **289:**312–314.

Chignard, M., Vargaftig, B. B., Benveniste, J., and LeCouedic, J. P., 1980, L'agregation plaquettaire et le platelet-activating factor, *J. Pharmacol.* **11:**371–377.

Chilton, F. H., O'Flaherty, J. T., Walsh, C. E., Thomas, M. J., Wykle, R. L., DeChatelet, L. R., and Waite, B. M., 1982, Platelet Activating factor. Stimulation of the lipoxygenase pathway in polymorphonuclear leukocytes, by 1-*O*-alkyl-2-*O*-acetyl-*sn*-glycero-3-phosphocholine, *J. Biol. Chem.* **257:**5402–5407.

Coleman, R. A., and Bell, R. M., 1980, Selective changes in enzymes of the *sn*-glycerol-3-phosphate and dihydroxyacetonephosphate pathways of triacylglycerol biosynthesis during differentiation of 3T3-L1 preadipocytes, *J. Biol. Chem.* **255:**7681–7687.

Coleman, R. A., and Haynes, E. B., 1983, Selective changes in microsomal enzymes of triacylglycerol and phosphatidylcholine synthesis in fetal and postnatal rat liver. Induction of microsomal *sn*-glycerol-3-phosphate and dihydroxyacetonephosphate acyltransferase activities, *J. Biol. Chem.* **258:**450–456.

D'Amato, R. A., Horrocks, L. A., and Richardson, K. E., 1975, Kinetic properties of plasmalogenase from bovine brain, *J. Neurochem.* **24:**1251–1255.

Davis, P. A., and Hajra, A. K., 1977a, The enzymatic exchange of the acyl group of acyl dihydroxyacetone phosphate with free fatty acid, *Biochem. Biophys. Res. Commun.* **74:**100–105.

Davis, P. A., and Hajra, A. K., 1977b, Stereospecific exchange of hydrogen from 1-acyl dihydroxyacetone-3-P with water during the formation of ether lipid, *Fed. Proc.* **36:**850 (Abstract 3050).

Davis, P. A., and Hajra, A. K., 1979, Stereochemical specificity of the biosynthesis of the alkyl ether bond in alkyl ether lipids, *J. Biol. Chem.* **254:**4760–4763.

Davis, P. A., and Hajra, A. K., 1981, Asay and properties of the enzyme catalyzing the biosynthesis of 1-*O*-alkyl dihydroxyacetone 3-phosphate, *Arch. Biochem. Biophys.* **211:**20–29.

Day, J. I. E., Goldfine, H., and Hagen, P-O., 1970, Enzymic reduction of long-chain acyl-CoA to fatty aldehyde and alcohol by extracts of *Clostridium butyricum, Biochim. Biophys. Acta* **218:**179–182.

Debuch, H., 1972, The history of ether-linked lipids through 1960, in: *Ether Lipids: Chemistry and Biology* (F. Snyder, ed.), Academic Press, New York, pp. 1–24.

Debuch, H., Miller, J., and Furniss, H., 1971, The synthesis of plasmalogens at the time of myelination in the rat, IV: The incorporation of ^{14}C-labelled *O*-(1-alkyl-*sn*-glycero-3-phosphoryl)-ethanolamine—A direct precursor of plasmalogens, *Hoppe-Seyler's Z. Physiol Chem.* **352:**984–990.

Debuch, H., Witter, B., Illig, H. K., and Gunawan, J., 1982, On the metabolism of etherphospholipid in glial cell cultures, in: *Phospholipids in the Nervous System, Vol. 1: Metabolism* (L. A. Horrocks, G. B. Ansell, and G. Porcellati, eds.), Raven Press, New York, pp. 199–210.

Dorman, R. V., Toews, A. D., and Horrocks, L. A., 1977, Plasmalogenase activities in neuronal perikarya, astroglia, and oligodendroglia isolated from bovine brain, *J. Lipid Res.* **18:**115–117.

Dorman, R. V., Freysz, L., Mandel, P., and Horrocks, L. A., 1978, Plasmalogenase activities in the brains of Jimpy and Quaking mice, *J. Neurochem.* **30:**157–159.

Downing, D. T., 1976, Mammalian waxes. in: *Chemistry and Biochemistry of Natural Waxes* (P. E. Kolattukudy, ed.), Elsevier, New York, pp. 17–48.

Driedzic, W., Selivonchick, D. P., and Roots, B. I., 1976, Alk-1-enyl ether-containing lipids of goldfish (*Carassius auratus* L.). Brain and temperature acclimation, *Comp. Biochem. Physiol.* **53B:** 311–314.

El-Bassiouni, E. A., Piantadosi, C., and Snyder, F., 1975, Metabolism of alkyldihydroxyacetone phosphate in rat brain, *Biochim. Biophys. Acta* **388:**5–11.

Ellingson, J. S., and Lands, W. E. M., 1968, Phospholipid reactivation of plasmalogen metabolism, *Lipids* **3:**111–120.

Erwin, V. G., Heston, W. D. W., and Tabakoff, B., 1972, Purification and characterization of an NADH-linked aldehyde reductase from bovine brain, *J. Neurochem. 19:*2269–2278.

Farr, R. S., Cox, C. P., Wardlow, M. L., and Jorgensen, R., 1980, Preliminary studies of an acid-labile factor (ALF) in human sera that inactivates platelet-activating factor (PAF), *Clin. Immunol. Immunopathol.* **15:**318–330.

Farr, R. S., Wardlow, M. L., Cox, C. P., Meng, K. E., and Greene, D. E., 1982, Human serum acid-labile factor (ALF) is an acylhydrolase that inactivates platelet activating factor (PAF), *Fed. Proc.* **41:**733 (Abstract 2704).

Ferrell, W. J., and Desmyter, E., 1974, Studies on the biosynthesis of *S*-alkyl bonds of glycerolipids, *Physiol. Chem. Phys.* **6:**497–503.

Ferrell, W. J., and Kessler, R. J., 1971, Enzymic relationship of free fatty acids, aldehydes and alcohols in mouse liver, *Physiol. Chem. Phys.* **3:**549–558.

Ferrell, W. J., and Yao, K-C., 1976, Metabolism of palmitaldehyde in human cardiac muscle, *J. Mol. Cell. Cardiol.* **8:**1–13.

Fisher, A. B., Huber, G. A., Furia, L., Bassett, D., and Rabinowitz, J. L., 1976, Evidence for lipid synthesis by the dihydroxyacetone phosphate pathway in rabbit lung subcellular fractions, *J. Lab. Clin. Med.* **87:**1033–1040.

Fleming, P. J., and Hajra, A. K., 1977, 1-Alkyl-*sn*-glycero-3-phosphate : acyl-CoA acyltransferase in rat brain microsomes, *J. Biol. Chem.* **252:**1663–1672.

Freysz, L., Horrocks, L. A., and Mandel, P., 1978, Ethanolamine and choline phosphotransferases of chicken brain, *Adv. Exp. Med. Biol.* **101:**253–268.

Friedberg, S. J., and Alkek, R. D., 1977, Absolute configuration of tritiated *O*-alkylglycerol synthesized enzymatically from [1,3-$^3H_2$,1,3-$^{14}C_2$]dihydroxyacetone phosphate, *Biochemistry* **16:**5291–5294.

Friedberg, S. J., and Gomillion, M., 1981, Hydrogen exchange in the formation of dihydroxyacetone phosphate from acyldihydroxyacetone phosphate in *O*-alkyl lipid synthesis in Ehrlich ascites tumor cell microsomes, *J. Biol. Chem.* **256:**291–295.

Friedberg, S. J., and Heifetz, A., 1973, Hydrogen exchange in the synthesis of glyceryl ether and in the formation of dihydroxyacetone in *Tetrahymena pyriformis*, *Biochemistry* **12:**1100–1106.

Friedberg, S. J., and Heifetz, A., 1975, The formation of tritiated *O*-alkyl lipid from acyldihydroxyacetone phosphate in the presence of tritiated water, *Biochemistry* **14:**570–574.

Friedberg, S. J., Heifetz, A., and Greene, R. C., 1971, Loss of hydrogen from dihydroxyacetone phosphate during glyceryl ether synthesis, *J. Biol. Chem.* **246:**5822–5827.

Friedberg, S. J., Heifetz, A., and Greene, R. C., 1972, Studies on the mechanism of *O*-alkyl lipid synthesis, *Biochemistry* **11:**297–301.

Friedberg, S. J., Gomillion, D. M., and Strotter, P. L., 1980, The mechanism of ether bond formation in *O*-alkyl lipid synthesis, *J. Biol. Chem.* **255:**1074–1079.

Friedberg, S. J., Weintraub, S. T., Singer, M. R., and Greene, R. C., 1983, The mechanism of ether bond formation in *O*-alkyl lipid synthesis in Ehrlich ascites tumor. Unusual cleavage of fatty acid moiety of acyl dihydroxyacetone phosphate, *J. Biol. Chem.* **258:**136–142.

Fu, S. C., Mozzi, R., Krakowka, S., Higgin, R. J., and Horrocks, L. A., 1980, Plasmalogenase and phospholipase A_1, A_2, and L_1 activities in white matter in canine distemper virus-associated demyelinating encephalomyelitis, *Acta Neuropathol. (Berlin)* **49:**13–18.

Gaiti, A., Goracci, G., de Medio, G. E., and Porcellati, G., 1972, Enzymic synthesis of plasmalogen and *O*-alkyl glycerolipid by base-exchange reaction in the rat brain, *FEBS Lett.* **27:**116–120.

Goldman, S. S., 1975, Cold resistance of the brain during hibernation. III. Evidence of a lipid adaptation, *Am. J. Physiol.* **228:**834–838.

Goracci, G., Horrocks, L. A., and Porcellati, G., 1977, Reversibility of ethanolamine and choline phosphotransferases (EC 2.7.8.1 and EC 2.7.8.2) in rat brain microsomes with labelled alkylacylglycerols, *FEBS Lett.* **80:**41–44.

Goracci, G., Horrocks, L. A., and Porcellati, G., 1978, Studies of rat brain choline ethanolamine phosphotransferases using labeled alkylacylglycerol as substrate with evidence for reversibility of the reactions, *Adv. Exp. Med. Biol.* **101:**269–278.

Goracci, G., Francescangeli, E., Horrocks, L. A., and Porcellati, G., 1981, The reverse reaction of cholinephosphotransferase in rat brain microsomes. A new pathway for degradation of phosphatidylcholine, *Biochim. Biophys. Acta* **664:**373–379.

Grigor, M. R., 1976, The age-related occurrence of wax esters in the mouse preputial gland tumor, *Biochim. Biophys. Acta* **431:**157–164.

Grigor, M. R., and Harris, E. L., 1977, Wax ester synthesis in the mouse preputial gland tumor, *Biochim. Biophys. Acta* **488:**121–127.

Gunawan, J., and Debuch, H., 1977, On the biosynthesis of plasmalogens during myelination in the rat. IX. Incorporation of radioactivity from 1-[^{3}H]alkyl-2-[1-^{14}C]octadecenoyl-*sn*-glycero-3-phosphoethanolamine, *Hoppe-Seyler's Z. Physiol. Chem.* **358:**537–543.

Gunawan, J., and Debuch, H., 1981, Liberation of free aldehyde from 1-(1-alkenyl)-*sn*-glycero-3-phosphoethanolamine (lysoplasmalogen) by rat liver microsomes, *Hoppe-Seyler's Z. Physiol. Chem.* **362:**445–452.

Gunawan, J., Vierbuchem, M., and Debuch, H., 1979, Studies on the hydrolysis of 1-alk-1′-enyl-*sn*-glycero-3-phosphoethanolamine by microsomes from myelinating rat brain, *Hoppe-Seyler's Z. Physiol. Chem.* **360:**971–978.

Hajra, A. K., 1968a, Biosynthesis of acyl dihydroxyacetone phosphate in guinea pig liver mitochondria, *J. Biol. Chem.* **243:**3458–3465.

Hajra, A. K., 1968b, Biosynthesis of phosphatidic acid from dihydroxyacetone phosphate, *Biochem. Biophys. Res. Commun.* **33:**929–935.

Hajra, A. K., 1969, Biosynthesis of alkyl-ether containing lipid from dihydroxyacetone phosphate, *Biochem. Biophys. Res. Commun.* **37:**486–492.

Hajra, A. K., 1970, Acyl dihydroxyacetone phosphate: Precursor of alkyl ethers, *Biochem. Biophys. Res. Commun.* **39:**1037–1044.

Hajra, A. K., 1973, The role of acyl dihydroxyacetone phosphate in tumor lipid metabolism, in: *Tumor Lipids: Biochemistry and Metabolism* (R. Wood, ed.), The American Oil Chemists Society, Champaign, Illinois, pp. 183–199.

Hajra, A. K., 1977, Biosynthesis of glycerolipids via acyldihydroxyacetone phosphate, *Biochem. Soc. Trans.* **5:**34–36.

Hajra, A. K., and Agranoff, B. W., 1968a, Acyl dihydroxyacetone phosphate. Characterization of a ^{32}P-labeled lipid from guinea pig liver mitochondria, *J. Biol. Chem.* **243:**1617–1622.

Hajra, A. K., and Agranoff, B. W., 1968b, Reduction of palmitoyl dihydroxyacetone phosphate by mitochondria, *J. Biol. Chem.* **243:**3542–3543.

Hajra, A. K., and Bishop, J. E., 1982, Glycerolipid biosynthesis in peroxisomes via the acyl dihydroxyacetone phosphate pathway, *Ann. N.Y. Acad. Sci.* **386:**170–182.

Hajra, A. K., and Burke, C., 1978, Biosynthesis of phosphatidic acid in rat brain via acyl dihydroxyacetone phosphate, *J. Neurochem.* **31:**125–134.

Hajra, A. K., Seguin, E. B., and Agranoff, B. W., 1968, Rapid labeling of mitochondrial lipids by labeled orthophosphate and adenosine triphosphate, *J. Biol. Chem.* **243:**1609–1616.

Hajra, A. K., Jones, C. L., and Davis, P. A., 1978, Studies on the biosynthesis of the *O*-alkyl bond in glycerol ether lipids, *Adv. Exp. Med. Biol.* **101:**369–378.

Hajra, A. K., Burke, C. L., and Jones, C. L., 1979, Subcellular localization of acyl coenzyme A: Dihydroxyacetone phosphate acyltransferase in rat liver peroxisomes (microbodies), *J. Biol. Chem.* **254:**10896–10900.

Hill, E. E., and Lands, W. E. M., 1968, Incorporation of long-chain and polyunsaturated acids into phosphatidate and phosphatidylcholine, *Biochim. Biophys. Acta* **152:**645–648.

Hixson, S., and Wolfenden, R., 1981, Inhibitors of ether lipid biosynthesis, *Biochem. Biophys. Res. Commun.* **101:**1064–1070.

Hofmann, S. L., Prescott, S. M., and Majerus, P. W., 1982, The effects of mepacrine and *p*-bromophenacyl bromide on arachidonic acid release in human platelets, *Arch. Biochem. Biophys.* **215:**237–244.

Holloway, P. W., 1975, Desaturation of long-chain fatty acids by animal liver, *Meth. Enzymol.* **35:**253–262.

Holloway, P. W., and Katz, J. T., 1972, A requirement for cytochrome b_5 in microsomal stearyl coenzyme A desaturation, *Biochemistry* **11:**3689–3695.

Horrocks, L. A., 1972, Content, composition, and metabolism of mammalian and avian lipids that contain ether groups, in: *Ether Lipids Chemistry and Biology* (F. Snyder, ed.), Academic Press, New York, pp. 177–272.

Horrocks, L. A., and Fu, S. C., 1978, Pathway for hydrolysis of plasmalogens in brain, *Adv. Exp. Med. Biol.* **101:**397–406.

Horrocks, L. A., and Radominska-Pyrek, A., 1972, Enzymic synthesis of ethanolamine plasmalogens from 1-alkyl-2-acyl-*sn*-glycero-3-(^{32}P)phosphorylethanolamines by microsomes from rat brain, *FEBS Lett.* **22:**190–192.

Horrocks, L. A., Spanner, S., Mozzi, R., Fu, S. C., D'Amato, R. A., and Krakowka, S., 1978, Plasmalogenase is elevated in early demyelinating lesions, *Adv. Exp. Med. Biol.* **100:**423–438.

Howard, B. V., Morris, H. P., and Bailey, J. M., 1972, Ether-lipids, α-glycerol phosphate dehydrogenase, and growth rate in tumors and cultured cells, *Cancer Res.* **32:**1533–1538.

Johnson, R. C., and Gilbertson, J. R., 1972, Isolation, characterization, and partial purification of a fatty acyl coenzyme A reductase from bovine cardiac muscle, *J. Biol. Chem.* **247:**6991–6998.

Jones, C. L., and Hajra, A. K., 1977, The subcellular distribution of acyl CoA : dihydroxyacetone phosphate acyl transferase in guinea pig liver, *Biochem. Biophys. Res. Commun.* **76:**1138–1143.

Jones, C. L., and Hajra, A. K., 1980, Properties of guinea pig liver perioxsomal dihydroxyacetone phosphateacyltransferase, *J. Biol. Chem.* **255:**8289–8295.

Kanoh, H., and Ohno, K., 1973, Studies of 1,2-diglycerides formed from endogenous lecithins by the back-reaction of rat liver microsomal CDP-choline : 1,2-diacylglycerol cholinephosphotransferase, *Biochim. Biophys. Acta* **326:**17–25.

Kanoh, H., and Ohno, K., 1975, substrate-selectivity of rat liver microsomal 1,2-diacylglycerol : CDP-choline(ethanolamine) choline(ethanolamine)-phosphotransferase in utilizing endogenous substrates, *Biochem. Biophys. Acta* **380:**199–207.

Kapoulas, V. M., and Thompson, G. A., Jr., 1969, The formation of glyceryl ethers by cell-free Tetrahymena extracts, *Biochem. Biophys. Acta* **187:**594–597.

Kapoulas, V. M., Thompson, G. A., Jr., and Hanahan, D. J., 1969, Metabolism of α-glyceryl ethers by *Tetrahymena pyriformis,* II. Properties of a cleavage system *in vitro. Biochim. Biophys. Acta* **176:**250–264.

Kasama, K., Rainey, W. T., Jr., and Snyder, F., 1973, Chemical identification and enzymatic synthesis of a newly discovered lipid class—hydroxyalkylglycerols, *Arch. Biochem. Biophys.* **154:**648–658.

Kaufman, S., 1959, Studies on the mechanism of enzymatic conversion of phenylalanine to tyrosine, *J. Biol. Chem.* **234:**2677–2682.

Kawalek, J. C., and Gilbertson, J. R., 1973, Enzymic reduction of free fatty aldehydes in bovine cardiac muscle, *Biochem. Biophys. Res. Commun.* **51:**1027–1033.

Kawalek, J. C., and Gilbertson, J. R., 1976, Partial purification of the NADPH-dependent aldehyde reductase from bovine cardiac muscle, *Arch. Biochem. Biophys.* **173:**649–657.

Kennedy, E. P., 1953, Synthesis of phosphatides in isolated mitochondria, *J. Biol. Chem.* **201:**399–412.

Kennedy, E. P., and Weiss, S. B., 1956, The function of cytidine coenzymes in the biosynthesis of phospholipides, *J. Biol. Chem.* **222:**193–214.

Kahn, A. A., and Kolattukudy, P. E., 1973a, A microsomal fatty acid synthetase coupled to acyl-CoA reductase in *Euglena gracilis, Arch. Biochem. Biophys.* **158:**411–420.

Kahn, A. A., and Kolattukudy, P. E., 1973b, Control of synthesis and distribution of acyl moieties in etiolated *Euglena gracilis, Biochemistry* **12:**1939–1948.

Kahn, A. A., and Kolattukudy, P. E., 1975, Solubilization of fatty acid synthetase, acyl-CoA reductase, and fatty acyl-CoA alcohol transacylase from the microsomes of *Euglena gracilis, Arch. Biochem. Biophys.* **170:**400–408.

Kiyasu, J. Y., and Kennedy, E. P., 1960, The enzymatic synthesis of plasmalogens, *J. Biol. Chem.* **235:**2590–2594.

Kolattukudy, P. E., 1970, Reduction of fatty acids to alcohols by cell-free preparations of *Euglena gracilis, Biochemistry* **9:**1095–1102.

Kolattukudy, P. E., Croteau, R., and Buckner, J. S., 1976, Biochemistry of plant waxes, in: *Chemistry and Biochemistry of Natural Waxes* (P. E. Kolattukudy, ed.), Elsevier, New York, pp. 289–347.

Korey, S. R., and Orchen, M., 1959, Plasmalogens of the nervous system. I. Deposition in developing rat brain and incorporation of ^{14}C isotope from acetate and palmitate into the α,β-unsaturated ether chain, *Arch. Biochem. Biophys.* **83:**381–389.

Kornberg, A., and Procer, W. E., Jr., 1953, Enzymatic esterification of α-glycerophosphate by long chain fatty acids, *J. Biol. Chem.* **204:**345–357.

LaBelle, E. F., Jr., and Hajra, A. K., 1972, Enzymatic reduction of alkyl and acyl derivatives of dihydroxyacetone phosphate by reduced pyridine nucleotides, *J. Biol. Chem.* **247:**5825–5834.

LaBelle, E. F., Jr., and Hajra, A. K., 1974, Purification and kinetic properties of acyl and alkyl dihydroxyacetone phosphate oxidoreductase, *J. Biol. Chem.* **249:**6936–6944.

Lands, W. E. M., and Hart, P., 1965, Metabolism of plasmalogen III. Relative reactivities of acyl and alkenyl derivatives of glycerol-3-phosphorylcholine, *Biochim. Biophys. Acta* **98:**532–538.

Lapetina, E. G., 1982, Platelet-activating factor stimulates the phosphatidylinositol cycle. Appearance of phosphatidic acid is associated with the release of serotonin in horse platelets, *J. Biol. Chem.* **257:**7314–7317.

Lapetina, E. G., Billah, M. M., and Cuatrecasas, P., 1981, The initial action of thrombin on platelets. Conversion of phosphatidylinositol to phosphatidic acid preceding the production of arachidonic acid, *J. Biol. Chem.* **256:**5037–5040.

Lee, T-c., 1979, Characterization of fatty alcohol : NAD^+ oxidoreductase from rat liver, *J. Biol. Chem.* **254:**2892–2896.

Lee, T-c., and Stephens, N., 1982, The modification of lipid composition in L-M cultured cells supplemented with elaidate. Increased formation of fatty alcohols, *Biochim. Biophys. Acta* **712:**299–304.

Lee, T-c., Wykle, R. L., Blank, M. L., and Snyder, F., 1973, Dietary control of stearyl CoA and alkylacylglycerophosphorylethanolamine desaturase in tumor, *Biochem. Biophys. Res. Commun.* **55:**574–579.

Lee, T-c., Houghland, A. E., and Stephens, N., 1979, Perturbation of lipid metabolism in L-M cultured cells by elaidic acid supplementation: Formation of fatty alcohols, *Biochem. Biophys. Res. Commun.* **91:**1497–1503.

Lee, T-c., Fitzgerald, V., Stephens, N., and Snyder, F., 1980, Activities of enzymes involved in the metabolism of ether-linked lipids in normal and neoplastic tissues of rat, *Biochim. Biophys. Acta* **619:**420–423.

Lee, T-c., Blank, M. L., Fitzgerald, V., and Snyder, F., 1981, Substrate specificity in the biocleavage of the *O*-alkyl bond: 1-alkyl-2-acetyl-*sn*-glycero-3-phosphocholine (a hypotensive and platelet-activating lipid) and its metabolites, *Arch. Biochem. Biophys.* **208:**353–357.

Lee, T-c., Blank, M. L., Fitzgerald, V., and Snyder, F., 1982a, Formation of alkylacyl- and diacylglycerophosphocholines via diracylglycerol cholinephosphotransferase in rat liver, *Biochim. Biophys. Acta* **713**:479–483.

Lee, T-c., Malone, B., Wasserman, S. I., Fitzgerald, V., and Snyder, F., 1982b, Activities of enzymes that metabolize platelet-activating factor (1-alkyl-2-acetyl-*sn*-glycero-3-phosphocholine) in neutrophils and eosinophils from humans and the effect of a calcium ionophore, *Biochem. Biophys. Res. Commun.* **105**:1303–1308.

Lee, T-c., Malone, B., and Snyder, F., 1983, Stimulation of calcium uptake by 1-alkyl-2-acetyl-*sn*-glycero-3-phosphocholine (platelet-activating factor) in rabbit platelets: Possible involvement of the lipoxygenase pathway, *Arch. Biochem. Biophys.*, **223**:33–39.

Lin, H. J., Ho, F. C. S., and Lee, C. L. H., 1978, Abnormal distribution of *O*-alkyl groups in the neutral glycerolipids from human hepatocellular carcinomas, *Cancer Res.* **38**:946–949.

Lumb, R. H., and Snyder, F., 1971, A rapid isotopic method for assessing the biosynthesis of ether linkages in glycerolipids of complex systems, *Biochim. Biophys. Acta* **244**:217–221.

Lynch, J. M., Lotner, G. Z., Betz, S. J., and Henson, P. M., 1979, The release of a platelet-activating factor by stimulated rabbit neutrophils, *J. Immunol.* **123**:1219–1226.

Malins, D. C., and Sargent, J. R., 1971, Biosynthesis of alkyldiacylglycerols and triacylglycerols in a cell-free system from the liver of dogfish (Squalus acanthias), *Biochemistry* **10**:1107–1110.

Manning, R., and Brindley, D. N., 1972, Tritium isotope effects in the measurement of the glycerol phosphate and dihydroxyacetone phosphate pathways of glycerolipid biosynthesis in rat liver, *Biochem. J.* **130**:1003–1012.

Martinez, M., and Ballabriga, A., 1978, A chemical study on the development of human forebrain and cerebellum during the brain "growth spurt" period. I. Gangliosides and plasmalogens, *Brain Res.* **159**:351–362.

Mason, R. J., 1978, Importance of the acyl dihydroxyacetone phosphate pathway in the synthesis of phosphatidylglycerol and phosphatidylcholine in alveolar type II cells, *J. Biol. Chem.* **253**:3367–3370.

Matsumoto, M., and Suzuki, Y., 1973, Acylation of lysophospholipids including lysoplasmalogen by cultured human amnion cells (FL cells), *J. Biochem. (Tokyo)* **73**:793–802.

McMurray, W. C., 1964, Metabolism of phosphatides in developing rat brain. II. Labeling of plasmalogens and other alkali-stable lipids from radioactive cytosine nucleotides, *J. Neurochem.* **11**:315–326.

Mencia-Huerta, J. M., Ninio, E., Roubin, R., and Benveniste, J., 1981, Is platelet-activating factor (PAF-acether) synthesis by murine peritoneal cells (PC) a two-step process? *Agents Actions* **11**:556–558.

Mencia-Huerta, J. M., Roubin, R., Morgat, J. L., and Benveniste, J., 1982, Biosynthesis of platelet-activating factor (PAF-acether). III. Formation of PAF-acether from synthetic substrates by stimulated murine macrophages, *J. Immunol.* **129**:804–808.

Mogelson, S., and Sobel, B. E., 1981, Ethanolamine plasmalogen methylation by rabbit myocardial membranes, *Biochem. Biophys. Acta* **666**:205–211.

Moore, C., and Snyder, F., 1982a, Properties of microsomal acyl coenzyme A reductase in mouse preputial glands, *Arch. Biochem. Biophys.* **214**:489–499.

Moore, C., and Snyder, F., 1982b, Regulation of acyl coenzyme A reductase by a heat-stable cytosolic protein during preputial gland development, *Arch. Biochem. Biophys.* **214**:500–504.

Mozzi, R., Siepi, D., Andreoli, V., and Porcellati, G., 1981, The synthesis of choline plasmalogen by the methylation pathway in rat brain, *FEBS Lett.* **131**:115–118.

Mueller, H. W., O'Flaherty, J. T., and Wykle, R. L., 1982, Ether lipid content and fatty acid distribution in rabbit polymorphonuclear neutrophil phospholipids, *Lipids* **17**:72–77.

Mueller, H. W., O'Flaherty, J. T., and Wylke, R. L., 1983, Biosynthesis of platelet activating factor in rabbit polymorphonuclear neutrophils, *J. Biol. Chem.*, in press.

Mukherjee, K. D., Weber, N., Mangold, H. K., Volm, M., and Richter, I., 1980, Competing pathways in the formation of alkyl, alk-1-enyl and acyl moieties in the lipids of mammalian tissues, *Eur. J. Biochem.* **107**:289–294.

Muramatsu, T., and Schmid, H. H. O., 1971, Formation of 1-*O*-2′-hydroxyalkyl glycerophosphatides from 1,2-heptadecanediol in myelating brain, *J. Lipid Res.* **12**:740–746.

Muramatsu, T., and Schmid, H. H. O., 1972, On the formation of 1-*O*-2′-hydroxyalk-1′-enyl or 1-*O*-2′-ketoalkyl glycerophosphatides from 1,2-alkanediol in myelinating brain, *Biochim. Biophys. Acta* **260**:365–368.

Muramatsu, T., and Schmid, H. H. O., 1973, Metabolism of 1-hydroxy-2-ketoheptadecane in myelinating brain, *Biochem. Biophys Acta* **296:** 265–270.

Murooka, Y., Seto, K., and Harada, T., 1970, *O*-Alkylhomoserine synthesis from *O*-acetylhomoserine and alcohol, *Biochem. Biophys. Res. Commun.* **41:**407–417.

Natarajan, V., and Sastry, P. S., 1973, *In vitro* studies on the acylation of 1-*O*-alkenyl glycero-3-phosphorylethanolamine by rat brain preparations, *FEBS Lett.* **32:**9–12.

Natarajan, V., and Sastry, P. S., 1976, Conversion of [1-^{14}C]palmitic acid to [1-^{14}C]hexadecanol by developing rat brain cell-free preparations, *J. Neurochem.* **26:**107–113.

Natarajan, V., and Schmid, H. H. O., 1977a, Substrate specificities in ether lipid biosynthesis. Metabolism of polyunsaturated fatty acids and alcohols by rat brain microsomes, *Biochem. Biophys. Res. Commun.* **79:**411–416.

Natarajan, V., and Schmid, H. H. O., 1977b, Chain length specificity in the utilization of long chain alcohols for ether lipid biosynthesis in rat brain, *Lipids* **12:**872–875.

Natarajan, V., and Schmid, H. H. O., 1977c, 1-Docosanol and other long chain alcohols in developing rat brain, *Lipids* **12:**128–130.

Natarajan, V., and Schmid, H. H. O., 1978, Biosynthesis and utilization of long-chain alcohols in rat brain: Aspects of chain length specificity, *Arch. Biochem. Biophys.* **187:**215–222.

Ninio, E., Mencia-Huerta, J. M., Heymans, F., and Benveniste, J., 1982, Biosynthesis of platelet-activity factor. I. Evidence for an acetyl-transferase activity in murine macrophages, *Biochim. Biophys. Acta* **710:**23–31.

O'Flaherty, J. T., and Wykle, R. L., 1983, Biology and biochemistry of platelet activating factor, *Clin. Rev. Allergy,* in press.

O'Flaherty, J. T., Thomas, M. J., Hammett, M. J., Carroll, C., McCall, C. E., and Wykle, R. L., 1983, 5-L-Hydroxy-6,8,11,14-eicosatetraenoate potentiates the human neutrophil degranulating action of platelet-activating factor, *Biochem. Biophys. Res. Commun.* **111:**1–7.

Oshino, N., and Omura, T., 1973, Immunochemical evidence for the participation of cytochrome b_5 in microsomal stearyl-CoA desaturation reaction, *Arch. Biochem. Biophys.* **157:**395–404.

Oshino, N., and Sato, R., 1972, The dietary control of the microsomal stearyl-CoA desaturation enzyme system in rat liver, *Arch. Biochem. Biophys.* **149:**369–377.

Oshino, N., Imai, Y., and Sato, R., 1966, Electron-transfer mechanism associated with fatty acid desaturation catalyzed by liver microsomes, *Biochim. Biophys. Acta* **128:**13–28.

Paltauf, F., 1971, Biosynthesis of plasmalogens from alkyl- and alkyl-acylglycerophosphoryl ethanolamine in the rat brain, *FEBS Lett.* **17:**118–120.

Paltauf, F., 1972a, Biosynthesis of alkyl-glycerol lipids in the intestinal mucosa *in vivo* and in a cell-free system, *Biochim. Biophys. Acta* **260:**345–351.

Paltauf, F., 1972b, Plasmalogen biosynthesis in a cell-free system. Enzymic desaturation of 1-*O*-alkyl (2-acyl) glycerophosphoryl ethanolamine, *FEBS Lett.* **20:**79–82.

Paltauf, F., 1972c, Intestinal uptake and metabolism of alkyl acyl glycerophospholipids and of alkyl glycerophospholipids in the rat. Biosynthesis of plasmalogens from [^{3}H]alkyl glycerophosphoryl [^{14}C]ethanolamine, *Biochim. Biophys. Acta* **260:**352–364.

Paltauf, F., 1973, Synthesis of alkoxylipids, *Chem. Phys. Lipids* **11:**270–294.

Paltauf, F., 1978a, Studies on soluble proteins stimulating plasmalogen biosynthesis, *Adv. Exp. Med. Biol.* **101:**378–395.

Paltauf, F., 1978b, Isolation of soluble proteins capable of stimulating aerobic plasmalogen biosynthesis, *Eur. J. Biochem.* **85:**263–270.

Paltauf, F., and Holasek, A., 1973, Enzymatic synthesis of plasmalogens. Characterization of the 1-*O*-alkyl-2-acyl-*sn*-glycero-3-phosphorylethanolamine desaturase from mucosa of hamster small intestine, *J. Biol. Chem.* **248:**1609–1615.

Paltauf, F., and Johnston, J. M., 1971, The metabolism *in vitro* of enantiomeric 1-*O*-alkyl glycerols and 1,2- and 1,3-alkyl acyl glycerols in the intestinal mucosa, *Biochim. Biophys. Acta* **239:**47–56.

Paltauf, F., and Wagner, E., 1976, Stereospecificity of lipases. Enzymatic hydrolysis of enantiomeric alkyldiacyl- and dialkylacylglycerols by lipoprotein lipase, *Biochim. Biophys. Acta* **431:**359–362.

Paltauf, F., Esfandi, F., and Holasek, A., 1974a, Stereospecificity of lipases. Enzymic hydrolysis of enantiomeric alkyl diacylglycerols by lipoprotein lipase, lingual lipase and pancreatic lipase, *FEBS Lett.* **40:**119–123.

Paltauf, F., Prough, R. A., Masters, B. S. S., and Johnston, J. M., 1974b, Evidence for the participation of cytochrome b_5 in plasmalogen biosynthesis, *J. Biol. Chem.* **249:**2661–2662.

Pfleger, R. C., Piantadosi, C., and Snyder, F., 1967, The biocleavage of isomeric glyceryl ethers by soluble liver enzymes in a variety of species, *Biochim. Biophys. Acta* **144:**633–648.

Pinckard, R. N., Farr, R. S., and Hanahan, D. J., 1979, Physicochemical and functional identity of rabbit platelet-activating factor (PAF) released *in vivo* during IgE anaphylaxis with PAF released *in vitro* from IgE sensitized basophils, *J. Immunol.* **123:**1847–1857.

Pinckard, R. N., McManus, L. M., Demopoulos, C. A., Halonen, M., Clark, P. O., Shaw, J. O., Kniker, W. T., and Hanahan, D. J., 1980, Molecular pathobiology of acetyl glyceryl ether phosphorylcholine: Evidence for the structural and functional identity with platelet-activating factor, *J. Reticuloendothel. Soc.* **28**(Suppl.):95S–103S.

Pollock, R. J., Hajra, A. K., and Agranoff, B. W., 1975, The relative utilization of the acyl dihydroxyacetone phosphate and glycerol phosphate pathways for synthesis of glycerolipids in various tumors and normal tissues, *Biochim. Biophys. Acta* **380:**421–435.

Polonsky, J., Tence, M., Varenne, P., Das, B. C., Lunel, J., and Benveniste, J., 1980, Release of 1-*O*-alkylglyceryl-3-phosphorylcholine, *O*-deacetyl platelet-activating factor, from leukocytes: Chemical ionization mass spectrometry of phospholipids, *Proc. Natl. Acad. Sci. USA* **77:**7019–7023.

Porcellati, G., Biasion, M. G., and Pirotta, M., 1970, The labeling of brain ethanolamine phosphoglycerides from cytidine diphosphate ethanolamine *in vitro, Lipids* **5:**734–742.

Poulos, A., Hughes, B. P., and Cumings, J. N., 1968, The biosynthesis of choline plasmalogens by ox heart, *Biochim. Biophys. Acta* **152:**629–632.

Puleo, L. E., Rao, G. A., and Reiser, R., 1970, Triose phosphates as precursors of glyceride biosynthesis by rat liver microsomes, *Lipids* **5:**770–776.

Radominska-Pyrek, A., and Horrocks, L. A., 1972, Enzymic synthesis of 1-alkyl-2-acyl-*sn*-glycerol-3-phosphorylethanolamines by the CDP-ethanolamine : 1-radyl-2-acyl-*sn*-glycerol ethanolaminephosphotransferase from microsomal fraction of rat brain, *J. Lipid Res.* **13:**580–587.

Radominska-Pyrek, A., Strosznajder, J., Dabrowiecki, Z., Chojnacki, T., and Horrocks, L. A., 1976, Effects of free fatty acids on the enzymic synthesis of diacyl and ether types of choline and ethanolamine phosphoglycerides, *J. Lipid Res.* **17:**657–662.

Radominska-Pyrek, A., Strosznajder, J., Dabrowiecki, Z., Goracci, G., Chojnacki, T., and Horrocks, L. A., 1977, Enzymic synthesis of ether types of choline and ethanolamine phosphoglycerides by microsomal fractions from rat brain and liver, *J. Lipid Res.* **18:**53–58.

Radominska-Pyrek, A., Dabrowiecki, Z., and Horrocks, L. A., 1979, Synthesis and content of ether-linked glycerophospholipids in the harderian gland of rabbits, *Biochim. Biophys. Acta* **574:**248–257.

Renooij, W., and Snyder, F., 1981, Biosynthesis of 1-alkyl-2-acetyl-*sn*-glycero-3-phosphocholine (platelet activating factor and a hypotensive lipid) by choline phosphotransferase in various rat tissues, *Biochim. Biophys. Acta* **663:**545–556.

Richter, I., and Weber, N., 1981, Formation of ether lipids from isomeric cis-octadecen-1-ols in normal and neoplastic cells: Substrate specificity of enzymes with regard to position of double bonds, *Hoppe-Seyler's Z. Physiol. Chem.* **362:**1163–1166.

Riendeau, D., Rodriguez, A., and Meighen, E., 1982, Resolution of the fatty acid reductase from *Photobacterium phosphoreum* into acyl protein synthetase and acyl-CoA reductase activities, *J. Biol. Chem.* **257:**6908–6915.

Roberti, R., Binaglia, L., Francescangeli, E., Goracci, G., and Porcellati, G., 1975, Enzymatic synthesis of 1-alkyl-2-acyl-*sn*-glycero-3-phosphorylethanolamine through ethanolaminephosphotransferase activity in the neuronal and glial cells of rabbit *in vitro, Lipids* **10:**121–127.

Rock, C. O., and Snyder, F., 1974, Biosynthesis of 1-alkyl-*sn*-glycero-3-phosphate via adenosine triphosphate : 1-alkyl-*sn*-glycerol phosphotransferase, *J. Biol. Chem.* **249:**5382–5387.

Rock, C. O., and Snyder, F., 1975, A short-chain acyl-CoA : 1-alkyl-2-acyl-*sn*-glycerol acyltransferase from a microsomal fraction of the rabbit harderian gland, *Biochim. Biophys. Acta* **388:**226–230.

Rock, C. O., and Snyder, F., 1978, Asymmetric localization of alkyldihydroxyacetone-P synthase and acyldihydroxyacetone-P acyltransferase in microsomal vesicles, *Adv. Exp. Med. Biol.* **101:**379–385.

Rock, C. O., Baker, R. C., Fitzgerald, V., and Snyder, F., 1976, Stimulation of the microsomal alkylglycerol monooxygenase by catalase, *Biochim. Biophys. Acta* **450:**469–473.

Rock, C. O., Fitzgerald, V., and Snyder, F., 1977a, Activation of alkyldihydroxyacetone phosphate synthase by detergents, *Arch. Biochem. Biophys.* **181:**172–177.

Rock, C. O., Fitzgerald, V., and Snyder, F., 1977b, Properties of dihydroxyacetone phosphate acyltransferase in the harderian gland, *J. Biol. Chem.* **252:**6363–6366.

Rock, C. O., Fitzgerald, V., and Snyder, F., 1978, Coupling of the biosynthesis of fatty acids and fatty alcohols, *Arch. Biochem. Biophys.* **186:**77–83

Rognstad, R., Clark, D. G., and Katz, J., 1974, Pathways of glyceride glycerol synthesis, *Biochem. J.* **140:**249–251.

Roots, B. I., and Johnston, P. V., 1968, Plasmalogens of the nervous system and environmental temperature, *Comp. Biochem. Physiol.* **26:**553–560.

Roubin, R., Mencia-Huerta, J. M., Landes, A., and Benveniste, J., 1982, Biosynthesis of platelet-activating factor (PAF-acether). IV. Impairment of acetyl-transferase activity in thioglycollate-elicited mouse macrophages, *J. Immunol.* **129:**809–813.

Saito, M., and Kanfer, J., 1975, Phosphatidohydrolase activity in a solubilized preparation from rat brain particulate fraction, *Arch. Biochem. Biophys.* **169:**318–323.

Sargent, J. R., Lee, R. F., and Nevenzel, J. C., 1976, Marine waxes, in: *Chemistry and Biochemistry of Natural Waxes* (P. E. Kolattukudy, ed.), Elsevier, New York, pp. 49–91.

Schlossman, D. M., and Bell, R. M., 1976, Triacylglycerol synthesis in isolated fat cells. Evidence that the *sn*-glycerol-3-phosphate and dihydroxyacetone phosphate acyltransferase activities are dual catalytic functions of a single microsomal enzyme, *J. Biol. Chem.* **251:**5738–5744.

Schlossman, D. M., and Bell, R. M., 1977, Microsomal *sn*-glycerol-3-phosphate and dihydroxyacetone phosphate acyltransferase activities from liver and other tissues. Evidence for a single enzyme catalyzing both reactions, *Arch. Biochem. Biophys.* **182:**732–742.

Scott, C. C., Heckman, C. A., Nettesheim, P., and Snyder, F., 1979a, Metabolism of ether-linked glycerolipids in cultures of normal and neoplastic rat respiratory tract epithelium, *Cancer Res.* **39:**207–214.

Scott, C. C., Heckman, C. A., and Snyder, F., 1979b, Regulation of ether lipids and their precursors in relation to glycolysis in cultured neoplastic cells, *Biochim. Biophys. Acta* **575:**215–224.

Selivonchick, D. P., Johnston, P. V., and Roots, B. I., 1977, Acyl and alkenyl group composition of brain subcellular fractions of goldfish (*Carassius auratus* L.) acclimated to different environmental temperatures, *Neurochem. Res.* **2:**379–393.

Shaw, J. O., Klusick, S. J., and Hanahan, D. J., 1981, Activation of rabbit platelet phospholipase and thromboxane synthesis by 1-*O*-hexadecyl/octadecyl-2-acetyl-*sn*-glyceryl-3-phosphorylcholine (platelet activating factor), *Biochim. Biophys. Acta* **663:**222–229.

Smith, R. J., and Bowman, B. J., 1982, Stimulation of human neutrophil degranulation with 1-*O*-octadecyl-2-*O*-acetyl-*sn*-glyceryl-3-phosphorylcholine: Modulation by inhibitors of arachidonic acid metabolism, *Biochem. Biophys. Res. Commun.* **104:**1495–1501.

Snyder, F., 1969, The biochemistry of lipids containing ether bonds, *Prog. Chem. Fats Other Lipids,* **10:**287–335.

Snyder, F. (ed.), 1972a, *Ether Lipids: Chemistry and Biology,* Academic Press, New York.

Snyder, F., 1972b, The enzymic pathways of ether-linked lipids and their precursors, in: *Ether Lipids: Chemistry and Biology* (F. Snyder, ed.), Academic Press, New York, pp. 121–156.

Snyder, F., 1972c, Ether-linked lipids and fatty alcohol precursors in neoplasms, in: *Ether Lipids: Chemistry and Biology* (F. Snyder, ed.), Academic Press, New York, pp. 273–295.

Snyder, F., 1982, Platelet activating factor (PAF), a novel type of phospholipid with diverse biological properties, *Annu. Rep. Med. Chem.* **17:**243–252.

Snyder, F., and Malone, B., 1970, Enzymic interconversion of fatty alcohols and fatty acids, *Biochem. Biophys. Res. Commun.* **41:**1382–1387.

Snyder, F., and Piantadosi, C., 1968, Deacylation of isomeric diacyl[1-^{14}C]alkoxyglycerols by pancreatic lipase, *Biochim. Biophys. Acta* **152:**794–797.

Snyder, F., and Snyder, C., 1975, Glycerolipids and cancer, *Prog. Biochem. Pharmacol.* **10:**1–41.

Snyder, F., and Wood, R., 1968, The occurrence and metabolism of alkyl and alk-1-enyl ethers of glycerol in transplantable rat and mouse tumors, *Cancer Res.* **28:**972–978.

Snyder, F., and Wood, R., 1969, Alkyl and alk-1-enyl ethers of glycerol in lipids from normal and neoplastic human tissues, *Cancer Res.* **29:**251–257.

Snyder, F., Malone, B., and Wykle, R. L., 1969a, The biosynthesis of alkyl ether bonds in lipids by a cell-free system, *Biochim. Biophys. Res. Commun.* **34:**40–47.

Snyder, F., Wykle, R. L., and Malone, B., 1969b, A new metabolic pathway: Biosynthesis of alkyl ether bonds from glyceraldehyde-3-phosphate and fatty alcohols by microsomal enzymes, *Biochem. Biophys. Res. Commun.* **34:**315–321.

Snyder, F., Malone, B., and Blank, M. L., 1969c, The biosynthesis of alkyl glyceryl ethers by microsomal enzymes of digestive glands and gonads of the starfish, *Asterias forbesi, Biochim. Biophys. Acta* **187:**302–306.

Snyder, F., Malone, B., and Blank, M. L., 1970a, Enzymic synthesis of *O*-alkyl bonds in glycerolipids, *J. Biol. Chem.* **245:**1790–1799.

Snyder, F., Blank, M. L., Malone, B., and Wykle, R. L., 1970b, Identification of *O*-alkyldihydroxyacetone phosphate, *O*-alkyldihydroxyacetone, and diacyl glyceryl ethers after enzymic synthesis, *J. Biol. Chem.* **245:**1800–1805.

Snyder, F., Rainey, W. T., Jr., Blank, M. L., and Christie, W. H., 1970c, The source of oxygen in the ether bond of glycerolipids: ^{18}0 studies, *J. Biol. Chem.* **245:**5853–5856.

Snyder, F., Blank, M. L., and Malone, B., 1970d, Requirement of cytidine derivatives in the biosynthesis of *O*-alkyl phospholipids, *J. Biol. Chem.* **245:**4016–4018.

Snyder, F., Malone, B., and Cumming, R. B., 1970e, Synthesis of glyceryl ethers by microsomal enzymes derived from fibroblasts (L-M cells) grown in suspension cultures, *Can. J. Biochem.* **48:**212–215.

Snyder, F., Piantadosi, C., and Malone, B., 1970f, The participation of 1- and 2-isomers of *O*-alkylglycerols as acyl acceptors in cell-free systems, *Biochim. Biophys. Acta* **202:**244–249.

Snyder, F., Blank, M. L., and Wykle, R. L., 1971a, The enzymic synthesis of ethanolamine plasmalogens, *J. Biol. Chem.* **246:**3639–3645.

Snyder, F., Hibbs, M., and Malone, B., 1971b, Enzymic synthesis of *O*-alkyl glycerolipids in brain and liver of rats during fetal and postnatal development, *Biochim. Biophys. Acta* **231:**409–411.

Snyder, F., Clark, M., and Piantadosi, C., 1973a, Biosynthesis of alkyl lipids: Displacement of the acyl moiety of acyldihydroxyacetone phosphate with fatty alcohol analogs, *Biohem. Biophys. Res. Commun.* **53:**350–356.

Snyder, F., Malone, B., and Piantadosi, C., 1973b, Tetrahydropteridine-dependent cleavage enzyme for *O*-alkyl lipids: Substrate specificity, *Biochim. Biophys. Acta* **316:**259–265.

Snyder, F., Malone, B., and Piantadosi, C., 1974, Enzymic studies of glycol and glycerol lipids containing *O*-alkyl bonds in liver and tumor tissues, *Arch. Biochem. Biophys.* **161:**402–407.

Soodsma, J. F., Piantadosi, C., and Snyder, F., 1970, The biocleavage of alkyl glyceryl ethers in Morris hepatomas and other transplantable neoplasms, *Cancer Res.* **30:**309–311.

Soodsma, J. F., Piantadosi, C., and Snyder, F., 1972, Partial characterization of the alkylglycerol cleavage enzyme system of rat liver, *J. Biol. Chem.* **247:**3923–3929.

Sribney, M., and Lyman, E. M., 1973, Stimulation of phosphorylcholine glyceride transferase activity by unsaturated fatty acids, *Can. J. Biochem.* **51:**1479–1486.

Stoffel, W., and Heimann, G., 1973, Action of lysosomal lipolytic enzymes on alkyl ether-containing phospholipids, *Hoppe-Seyler's Z. Physiol. Chem.* **355:**651–659.

Stoffel, W., and LeKim, D., 1971, Studies on the biosynthesis of plasmalogens. Precursors in the biosynthesis of plasmalogens: on the stereospecificity of the biochemical dehydrogenation of the 1-*O*-alkyl glyceryl to the 1-*O*-alk-1′-enyl glyceryl ether bond, *Hoppe-Seyler's Z. Physiol. Chem.* **352:**501–511.

Stoffel, W., LeKim, D., and Heyn, G., 1970, Spinganine (dihydrophingosine), an effective donor of the alk-1′ enyl chain of plasmalogens, *Hoppe-Seyler's Z. Physiol. Chem.* **351:**875–883.

Strosznajder, J., Radominska-Pyrek, A., and Horrocks, L. A., 1979, Choline and ethanolamine glycerophospholipid synthesis in isolated synaptosomes of rat brain, *Biochim. Biophys. Acta* **574:**48–56.

Subbaiah, P. V., Sastry, P. S., and Ganguly, J., 1970, Acylation of lysolecithin in the intestinal mucosa of rats, *Biochem. J.* **118:**241–246.

Sugiura, T., Onuma, Y., Sekiguchi, N., and Waku, K., 1982, Ether phospholipids in guinea pig polymorphonuclear leukocytes and macrophages. Occurrence of high levels of 1-*O*-alkyl-2-acyl-*sn*-glycero-3-phosphocholine, *Biochim. Biophys. Acta* **712:**515–522.

Sugiura, T., Nakajima, M., Sekiguchi, N., Nakagawa, Y., and Waku, K., 1983, Different fatty chain compositions of alkenylacyl, alkylacyl and diacyl phospholipids in rabbit alveolar macrophages: High amounts of arachidonic acid in ether phospholipids, *Lipids* **18:**125–129.

Tabakoff, B., and Erwin, V. G., 1970, Purification and characterization of a reduced nicotinamide adenine dinucleotide phosphate-linked aldehyde reductase from brain, *J. Biol. Chem.* **245:**3263–3268.

Tence, M., Polonsky, J., Le Couedic, J. P., and Benveniste, J., 1980, Release, purification, and characterization of platelet-activating factor (PAF), *Biochimie* **62:**251–259.

Thompson, G. A., Jr., 1968, The biosynthesis of ether-containing phospholipids in the slug, *Arion ater.* III. Origin of the vinylic ether bond of plasmalogens, *Biochim. Biophys. Acta* **152:**409–411.

Thompson, G. A., Jr., 1972a, Ether-linked lipids in molluscs, in: *Ether Lipids: Chemistry and Biology* (F. Snyder, ed.), Academic Press, New York, pp. 313–320.

Thompson, G. A., Jr., 1972b, Ether-linked lipids in protozoa, in: *Ether Lipids: Chemistry and Biology* (F. Snyder, ed.), Academic Press, New York, pp. 321–327.

Thyagarajan, K., Sand, D. M., Brockman, H. L., and Schlenk, H., 1979, Oxidation of fatty alcohols to acids in the caecum of a gourami (*Trichogaster cosby*), *Biochim. Biophys. Acta* **575:**318–326.

Tietz, A., Lindberg, M., and Kennedy, E. P., 1964, A new pteridine-requiring enzyme system for the oxidation of glyceryl ethers, *J. Biol. Chem.* **239:**4081–4090.

Tjiong, H. B., Gunawan, J., and Debuch, H., 1976, On the biosynthesis of plasmalogens during myelination in the rat. VIII. Incorporation of 1-[1-^{14}C]alkyl-2-acyl-3-*sn*-glycerophosphoethanolamine with different fatty acids, *Hoppe-Seyler's Z. Physiol. Chem.* **357:**707–712.

van den Bosch, H., 1974, Phosphoglyceride metabolism, *Annu. Rev. Biochem.* **43:**243–277.

Vargaftig, B. B., 1977, Carrageenan and thrombin trigger prostaglandin synthetase-independent aggregation of rabbit platelets: Inhibition by phospholipase A_2 inhibitors, *J. Pharm. Pharmacol.* **29:**222–228.

Vargaftig, B. B., Chignard, M., Benveniste, J., Lefort, J., and Wal, F., 1981, Background and present status of research on platelet-activating factor (PAF-acether), *Ann. N.Y. Acad. Sci.* **370:**119–137.

Vierbuchen, M., Gunawan, J., and Debuch, H., 1979, Studies on the hydrolysis of 1-alkyl-*sn*-glycero-3-phosphoethanolamine in subcellular fractions of rat brain, *Hoppe-Seyler's Z. Physiol. Chem.* **360:**1091–1097.

Voelkel, N. F., Worthen, S., Reeves, J. T., Henson, P. M., and Murphy, R. C., 1982, Nonimmunological production of leukotrienes induced by platelet-activating factor, *Science* **218:**286–288.

Waite, M., and van Deenen, L. L. M., 1967, Hydrolysis of phospholipids and glycerides in rat-liver preparations, *Biochim. Biophys. Acta* **137:**498–517.

Waite, M., Scherphof, G. L., Boshouwers, F. M. G., and Van Deenen, L. L. M., 1969, Differentiation of phospholipases A in mitochondria and lysosomes of rat liver, *J. Lipid Res.* **10:**411–420.

Waku, K., and Lands, W. E. M., 1968, Acyl coenzyme A : 1-alkenyl-glycero-3-phosphorycholine acyltransferase action in plasmalogen biosynthesis, *J. Biol. Chem.* **243:**2654–2659.

Waku, K., and Nakazawa, Y., 1970, Acyltransferase activity of 1-*O*-alkyl-glycero-3-phosphorylcholine in sarcoplasmic reticulum, *J. Biochem.* **68:**459–466.

Waku, K., and Nakazawa, Y., 1972, Acyltransferase activity to 1-acyl-1-*O*-alkenyl-, and 1-*O*-alkyl-glycero-3-phosphorylcholine in Ehrlich ascites tumor cells, *J. Biochem.* **72:**495–497.

Waku, K., and Nakazawa, Y., 1978a, Incorporation rates of [1^{14}C]glycerol into the molecular species of alkyl ether phospholipids of Ehrlich ascites tumor cells *in vivo, Eur. J. Biochem.* **88:**489–494.

Waku, K., and Nakazawa, Y., 1978b, Turnover of molecular species of diacyl and alkyl ether phospholipids in Ehrlich ascites tumor cells, *Adv. Exp. Med. Biol.* **101:**407–413.

Waku, K., and Nakazawa, Y., 1979, Turnover rates of the molecular species of alkenyl ether phospholipids of Ehrlich ascites tumor cells, *Eur. J. Biochem.* **100:**317–320.

Waku, K., Nakazawa, Y., and Mori, W., 1976, Phospholipid metabolism in Ehrlich ascites tumor cells, II. Turnover rate of ether phospholipids, *J. Biochem.* **80:**711–716.

Warner, H. R., and Lands, W. E. M., 1961, The metabolism of plasmalogen: Enzymatic hydrolysis of the vinyl ether, *J. Biol. Chem.* **236:**2404–2409.

Weber, N., and Richter, I., 1982, Formation of ether lipids and wax esters in mammalian cells. Specificity of enzymes with regard to carbon chains of substrates, *Biochim. Biophys. Acta* **711:**197–207.

Weiss, S. B., Smith, S. W., and Kennedy, E. P., 1958, The enzymatic formation of lecithin from cytidine diphosphate choline and D-1,2-diglyceride, *J. Biol. Chem.* **231:**53–64.

Woelk, H., Goracci, G., and Porcellati, G., 1974a, The action of brain phospholipases A_2 on purified, specifically labelled 1,2-diacyl-, 2-acyl-1-alk-1′-enyl- and 2-acyl-1-alkyl-*sn*-glycero-3-phosphorylcholine, *Hoppe-Seyler's Z. Physiol. Chem.* **355:**75–81.

Woelk, H., Kanig, K., and Peiler-Ichikawa, K., 1974b, Phospholipid metabolism in experimental allergic encephalomyelitis: Activity of mitochondrial phospholipase A_2 of rat brain towards specifically labeled 1,2-diacyl-, 1-alk-1′-enyl-2-acyl- and 1-alkyl-2-acyl-*sn*-glycero-3-phosphorylcholine, *J. Neurochem.* **23:**745–750.

Woelk, H., Porcellati, G., and Goracci, G., 1976, Metabolic studies on 1-alkyl- and 1-alk-1′enyl-glycerophosphatides in the brain, *Adv. Exp. Med. Biol.* **72:**55–61.

Wood, R., 1973, Tumor lipids: Structural and metabolism studies of Ehrlich ascites cells, in: *Tumor Lipids: Biochemistry and Metabolism* (R. Wood, ed.), American Oil Chemists Society Press, Champaign, Illinois, pp. 139–182.

Wood, R., and Healy, K., 1970, Tumor lipids. Biosynthesis of plasmalogens, *J. Biol. Chem.* **245:**2640–2648.

Wood, R. and Snyder, F., 1967, Characterization and identification of glyceryl ether diesters present in tumor cells, *J. Lipid Res.* **8:**494–500.

Wykle, R. L., and Schremmer Lockmiller, J. M., 1975, The biosynthesis of plasmalogens by rat brain: Involvement of the microsomal electron transport system, *Biochim. Biophys. Acta* **380:**291–298.

Wykle, R. L., and Schremmer, J. M., 1974, A lysophospholiase D pathway in the metabolism of ether-linked lipids in brain microsomes, *J. Biol. Chem.* **249:**1742–1746.

Wykle, R. L., and Schremmer, J. M., 1979, Biosynthesis of plasmalogens by the microsomal fraction of Fischer R-3259 sarcoma. Influence of specific 2-acyl chains on the desaturation of 1-alk-2-acyl-*sn*-glycero-3-phosphoethanolamine, *Biochemistry* **18:**3512–3517.

Wykle, R. L., and Snyder, F., 1969, The glycerol source for the biosynthesis of alkyl glyceryl ethers, *Biochem. Biophys. Res. Commun.* **37:**658–662.

Wykle, R. L., and Snyder, F., 1970, Biosynthesis of an *O*-alkyl analogue of phosphatidic acid and *O*-alkylglycerols via *O*-alkyl ketone intermediates by microsomal enzymes of Ehrlich ascites tumor, *J. Biol. Chem.* **245:**3047–3058.

Wykle, R. L., and Snyder, F., 1976, Microsomal enzymes involved in the metabolism of ether-linked glycerolipids and their precursors in mammals, in: *The Enzymes of Biological Membranes,* Vol. 2, Chap. 2 (A. Martonosi, ed.), Plenum Press, New York, pp. 87–117.

Wykle, R. L., Blank, M. L., and Snyder, F., 1970, The biosynthesis of plasmalogens in a cell-free system, *FEBS Lett.* **12:**57–60.

Wykle, R. L., Blank, M. L., Malone, B., and Snyder, F., 1972a, Evidence for a mixed function oxidase in the biosynthesis of ethanolamine plasmalogens from 1-alkyl-2-acyl-*sn*-glycero-3-phosphorylethanolamine, *J. Biol. Chem.* **247:**5442–5447.

Wykle, R. L., Piantadosi, C., and Snyder, F., 1972b, The role of acyldihydroxyacetone phosphate, NADH, and NADPH in the biosynthesis of *O*-alkyl glycerolipids by microsomal enzymes of Ehrlich ascites tumor, *J. Biol. Chem.* **247:**2944–2948.

Wykle, R. L., Blank, M. L., and Snyder, F., 1973, The enzymic incorporation of arachidonic acid into ether-containing choline and ethanolamine phosphoglycerides by deacylation-acylation reactions, *Biochim. Biophys. Acta* **326:**26–33.

Wykle, R. L., Kraemer, W. F., and Schremmer, J. M., 1977, Studies of lysophospholipase D rat liver and other tissues, *Arch. Biochem. Biophys.* **184:**149–155.

Wykle, R. L., Malone, B., and Snyder, F., 1979, Acyl-CoA reductase specificity and synthesis of wax esters in mouse preputial gland tumors, *J. Lipid Res.* **20:**890–896.

Wykle, R. L., Kraemer, W. F., and Schremmer, J. M., 1980a, Specificity of lysophospholipase D, *Biochim. Biophys. Acta* **619:**58–67.

Wykle, R. L., Malone, B., and Snyder, F., 1980b, Enzymatic synthesis of 1-alkyl-2-acetyl-*sn*-glycero-3-phosphocholine, a hypotensive and platelet-aggregating lipid, *J. Biol. Chem.* **255:**10256–10260.

Yavin, E., and Gatt, S., 1972a, Oxygen-dependent cleavage of the vinyl-ether linkage of plasmalogens. 1. Cleavage by rat brain supernatant, *Eur. J. Biochem.* **25:**431–436.

Yavin, E., and Gatt, S., 1972b, Oxygen-dependent cleavage of the vinyl-ether linkage of plasmalogens. 2. Identification of the low-molecular-weight active component and the reaction mechanism, *Eur. J. Biochem.* **25:**437–446.

Yavin, E., and Gatt, S., 1972a, Oxygen-dependent cleavage of the vinyl-ether linkage of plasmalogens. 1. Cleavage by rat brain supernatant, *Eur. J. Biochem.* **25:**431–436.

Yavin, E., and Gatt, S., 1972b, Oxygen-dependent cleavage of the vinyl-ether linkage of plasmalogens. 2. Identification of the low-molecular-weight active component and the reaction mechanism, *Eur. J. Biochem.* **25:**437–446.

15

Fatty Acid Synthetases of Eukaryotic Cells

Salih J. Wakil and James K. Stoops

I. INTRODUCTION

The synthesis of long-chain fatty acids, such as palmitic, occurs in the living cell by head-to-tail condensation of C_2 units as acetyl-CoA. The condensation occurs in two steps. First, acetyl-CoA is converted to malonyl-CoA by acetyl-CoA carboxylase, a multienzyme system containing biotin [Eq. (1); for a recent review see Wakil *et al.*, 1983]. The second step (Eq. 2) is the conversion of acetyl-CoA and malonyl-CoA to palmitate, catalyzed by fatty acid synthetases (FAS) in the presence of NADPH.

$$CH_3COS\text{-}CoA + CO_2 + ATP \rightleftharpoons HOOCCH_2COS\text{-}CoA + ADP + P_i \qquad (1)$$

$$CH_3COS\text{-}CoA + 7\ HOOCCH_2COS\text{-}CoA + 14\ NADPH + 14H^+ \rightarrow \qquad (2)$$

$$CH_3CH_2(CH_2CH_2)_6CH_2COOH + 7\ CO_2 + 14\ NADP^+ + 8\ CoA\text{-}SH + 6\ H_2O$$

Studies of FAS in cell-free extracts of *Escherichia coli* (Wakil, 1971; Volpe and Vagelos, 1977) have shown that numerous sequential reactions and acyl intermediates are involved in the synthesis of long-chain fatty acids. The coenzyme which binds all acyl intermediates is a protein (M_r 10,000) with a 4′-phosphopantetheine prosthetic group [acyl carrier protein (ACP)]. Individual enzymes were isolated and then used in reconstitution of the FAS system. The enzymes and reactions involved in the synthesis of palmitate, the major product, are shown in the following equations:

Salih J. Wakil and James K. Stoops • Verna and Marrs McLean Department of Biochemistry, Baylor College of Medicine, Houston, Texas 77030.

Acetyl transacylase

$$CH_3COS\text{-}CoA + ACP\text{-}SH \rightleftharpoons CH_3COS\text{-}ACP + CoA\text{-}SH \quad (3)$$

Malonyl transacylase

$$\underset{\displaystyle CH_2COS\text{-}CoA}{\overset{COOH}{|}} + ACP\text{-}SH \rightleftharpoons \underset{\displaystyle CH_2COS\text{-}ACP}{\overset{COOH}{|}} + CoA\text{-}SH \quad (4)$$

β-Ketoacyl-ACP synthetase (condensing enzyme)

$$CH_3COS\text{-}ACP + Enz\text{-}SH \rightleftharpoons CH_3COS\text{-}Enz + ACP\text{-}SH \quad (5a)$$

$$CH_3COS\text{-}Enz + \underset{\displaystyle CH_2COS\text{-}ACP}{\overset{COOH}{|}} \rightarrow CH_3COCH_2COS\text{-}ACP + CO_2 + Enz\text{-}SH \quad (5b)$$

β-Ketoacyl-ACP reductase

$$CH_3COCH_2COS\text{-}ACP + NADPH + H^+ \rightleftharpoons D\text{-}CH_3CHOHCH_2COS\text{-}ACP + NADP^+ \quad (6)$$

β-Hydroxyacyl-ACP dehydratase

$$CH_3CHOHCH_2COS\text{-}ACP \rightleftharpoons \mathit{trans}\text{-}CH_3CH{=}CHCOS\text{-}ACP + H_2O \quad (7)$$

Enoyl-ACP reductase

$$CH_3CH{=}CHCOS\text{-}ACP + NADPH + H^+ \rightarrow CH_3CH_2CH_2COS\text{-}ACP + NADP^+ \quad (8)$$

Thioesterase

$$CH_3(CH_2)_{14}COS\text{-}ACP + H_2O \rightarrow CH_3(CH_2)_{14}COOH + ACP\text{-}SH \quad (9)$$

FAS may be separated into two groups according to structural and physicochemical properties (Table 1). Type I are multifunctional enzymes and may be further divided into subclasses IA and IB. Type IA synthetases are isolated from animal tissues and consist of two identical subunits (α_2 complex) with a total molecular weight near 500,000. The product is free palmitic acid. The IB synthetases are also multifunctional, but may have either identical subunits, as found in *Mycobacterium smegmatis,* or nonidentical subunits, as found in yeast FAS ($\alpha_6\beta_6$). Type IB enzymes require flavin mononucleotide (FMN) for activity, whereas type IA do not, and yield a CoA derivative, palmitoyl-CoA, rather than palmitic acid.

Type II FAS are found in prokaryotes, e.g., *E. coli,* and in eukaryotic plant cells. There is no evidence that the component enzymes are associated, as they are readily separated by conventional procedures. The synthetase system of plant tissues is similar to prokaryotic synthetases, and unlike other eukaryotic systems, is exclusively located in organelles such as the chloroplast in leaves and the proplastids in seeds (Goldberg and Bloch, 1972; Shimakata and Stumpf, 1982a,b). The *E. coli* enzymes, which have been reviewed extensively (Wakil, 1971; Volpe and Vagelos, 1977), and the similar

Table 1. Classification of Fatty Acid Synthetases

Type	Source	Molecular weight	Major product
IA: Multifunctional enzymes (dimeric, α)[a]	Liver (rat, dog, human, pig, chicken, pigeon)	$4–5 \times 10^5$	Palmitate
	Adipose (rat)	$4–5 \times 10^5$	Palmitate
	Uropygial gland (goose)	$4–5 \times 10^5$	Palmitate
	Mammary gland (bovine, rabbit, guinea pig, goat, rat)	$4–5 \times 10^5$	Palmitate
	Oviduct (chicken)	$4–5 \times 10^5$	Palmitate
IB: Aggregated multifunctional enzymes (oligomeric, α_6, or α, β)[b]	Fungi (yeast, *Neurospora*, penicillin)	2.3×10^6	Palmitoyl-CoA
	Higher bacteria (*Mycobacterium smegmatis*)	$>1 \times 10^6$	Acyl-CoA (long-chain fatty acids)
	Corynebacterium, streptomyces algae (etiolated *Euglena*)	$\sim 10^6$	Palmitate
II: Unassociated enzymes[c]	Lower bacteria (*E. coli, Clostridium* sp, *Bacillus, Pseudomonas*)	—	Palmitate, acyl-ACP
	Plants and algae (*Chlamydomonas*, avocado, safflower seeds, lettuce, photosynthesizing *Euglena*, spinach)	—	Acyl-ACP

[a] Liver: rat, Burton *et al.*, 1968 and Stoops *et al.*, 1979; dog, Roncari, 1974a; human, Roncari, 1974b; chicken, Stoops *et al.*, 1978a and Hsu and Yun, 1970; pigeon, Burton *et al.*, 1968. Adipose: rat, Stoops *et al.*, 1979. Uropygial gland: goose, Buckner and Kolattukudy, 1976. Mammary gland: bovine, Kinsella *et al.*, 1975 and Maitra and Kumar, 1974; rabbit, Grunnet and Knudsen, 1978; guinea pig, Strong and Dils, 1972; goat, Grunnet and Knudsen, 1978; rat, Smith and Abraham, 1970. Oviduct: chicken, Aprahamian *et al.*, 1979.

[b] Fungi, Lynen, 1961 and Stoops *et al.*, 1978b; *Neurospora*, Elovson, 1975; penicillin, Holtermuller *et al.*, 1970. Higher bacteria: *Mycobacterium smegmatis*, Wood *et al.*, 1978; *Corynebacterium*, Knoche and Koths, 1973; streptomyces algae, Rossi and Corcoron, 1973; etiolated *Euglena*, Delo *et al.*, 1971.

[c] Lower bacteria: *E. coli*, Lennarz *et al.*, 1962, Goldman *et al.*, 1963 and Pugh *et al.*, 1966; *Clostridium sp*, Butterworth and Bloch, 1970; *Bacillus*, Brindley *et al.*, 1969; *Pseudomonas*, Lennarz *et al.*, 1962. Plants and algae: *Chlamydomonas*, Sirevag and Levine, 1972; avocado, Overath and Stumpf, 1964; safflower seeds, Shimakata and Stumpf, 1982a; lettuce, Brooks and Stumpf, 1966; photosynthesizing *Euglena*, Dutler *et al.*, 1971 and Ernst-Fonberg and Bloch, 1971; spinach, Shimakata and Stumpf, 1982b.

plant synthetases will not be considered here. This chapter will review recent studies of FAS of eukaryotic cells found in animal tissues and yeast.

II. FATTY ACID SYNTHETASES OF EUKARYOTES

Until the early 1970s, it was believed that synthetases of eukaryotic cells were multienzyme complexes since all the component enzymes catalyzing Eqs. (3–9) were found to be associated oligomers of molecular weight 500,000 to 2.4×10^6 (Stoops *et al.*, 1977). Research efforts were concentrated on dissociation of these enzyme complexes and the isolation of their component enzymes. Evidence that these synthetases were multienzyme aggregates seemed beyond question at the time. The yeast enzyme was reported to yield seven different NH_2-terminal amino acids in equal amounts, six different proteins and a peptide of molecular weight 16,000, containing 4′-phosphopantetheine (Lynen, 1964; Willecke *et al.*, 1969). Similar results were reported for the pigeon liver enzyme. The finding of five NH_2-terminal amino acids and eight protein bands on phenolacetic acid–urea gels prompted the proposal that the multienzyme complex consisted of eight proteins (Yang *et al.*, 1967). The concept was supported by the isolation of a protein of molecular weight near 10,000 containing 4′-phosphopantetheine from FAS of liver of dog, pigeon, chicken, human, and rat (Roncari, 1974a,b; Lornitzo *et al.*, 1974; Bratcher and Hsu, 1975; Qureshi *et al.*, 1976). Separation of active components from chicken liver enzyme was also reported (Bratcher and Hsu, 1975).

Contrary to the concept just described, the independent investigations of yeast synthetases in Schweizer's laboratory (Schweizer *et al.*, 1973, 1975) and of both yeast and animal synthetases in our laboratory (Stoops *et al.*, 1975) showed that these enzymes are composed of multifunctional subunits. Genetic studies by Schweizer *et al.* (1973, 1975) suggested that yeast FAS (type IB) is encoded by two unlinked structural genes. They attributed the presence of multiple components in yeast enzyme preparations to unspecific proteolysis and proposed that the yeast enzyme is composed of two nonidentical subunits.

A. Animal Fatty Acid Synthetases

Stoops *et al.* (1975) demonstrated the multifunctional enzyme nature of the FAS (type IA) isolated from animal tissues; they found that the release of lower molecular weight peptides from the complex during sodium dodecyl sulfate (SDS) was more extensive at high protein concentration (10 mg/ml) than at lower (1 mg/ml). Such a result could not be explained on the basis of an incomplete dissociation in SDS. They demonstrated that proteolysis occurred during the preparation of the enzyme and was prominent in the presence of SDS as well as other denaturing agents (guanidinium chloride, urea). If, however, proteolysis was avoided, the synthetase subunits migrated in sodium dodecyl sulfate–polyacrylamide gel electrophoresis (SDS–PAGE), as a single polypeptide of molecular weight 250,000 as shown in Figure 1. The extensive evidence that type IA and type IB FAS are multifunctional enzymes has been reviewed elsewhere (Stoops *et al.*, 1977).

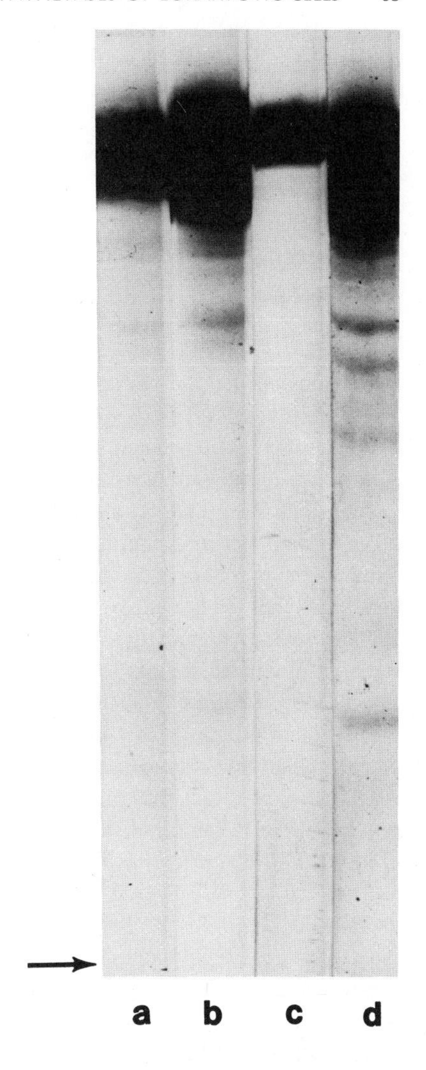

Figure 1. SDS–PAGE (5%) pattern of FAS from rat and chicken livers. The arrow indicates the approximate position of the dye front. (a, b) 23 and 46 μg of rat FAS, (c, d) 20 and 50 μg of chicken FAS.

The finding that these enzymes are multifunctional was important in understanding type I FAS since it was then possible to carry out meaningful physicochemical and mechanistic studies.

1. *Identity of Subunits*

The finding that the animal synthetases consist of a single polypeptide raised the question whether the two subunits were identical. In addition to having the same size, they migrated as a single band on nondenaturing gels (Stoops *et al.*, 1979), indicating that they also have the same charge and shape. No free NH_2-terminal residues were found, suggesting blocked NH_2-terminal residues (Stoops *et al.*, 1975).

The probability that the subunits are identical prompted reinvestigation of 4′-phosphopantetheine content of the enzyme. Chicken and rat synthetases analyzed by several methods yielded values of 1.4–1.8 per mole of enzyme, or about 1 mole per subunit (Stoops *et al.*, 1978a; Arslanian *et al.*, 1976). Similar values were obtained from goose uropygial gland and *Mycobacterium smegmatis* (Buckner and Kolattukudy, 1976; Wood *et al.*, 1978). These results contradicted previous reports of values of one prosthetic group per mole of synthetase of pigeon (Jacob *et al.*, 1968), rat (Roncari *et al.*, 1972), dog liver (Roncari, 1974a), and rat lactating mammary gland (Smith and Abraham, 1970). The finding of one prosthetic group per subunit was additional evidence that the subunits are identical, a conclusion upheld by further studies discussed in this section.

Electron microscopic studies of negatively stained rat liver FAS (Stoops *et al.*, 1979) showed that the enzyme subunit is a linear structure 200 Å long, containing at least four lobes 50 Å in diameter (Figure 2), having a mass of approximately 200,000 daltons, a weight consistent with estimation of the molecular weight of the subunit. A set of pseudotetrahedral images was also observed, their dimension and form suggesting that the pseudotetrahedral mass may be composed of four lobes, each about 50 Å in diameter and, therefore, of the same mass as the linear structure (Figure 2). Because the numbers of the linear and pseudotetrahedral structures are unequal, it was concluded that the linear form may fold into the pseudotetrahedron, and that the dimer is composed of chemically identical chains. The well-defined lobes seen in the electron micrograph are analogous to beads on a string and may be the catalytic domains of the complex. These conclusions are compatible with the chemical and enzymological properties of animal FAS.

Isolation and characterization of the synthetase mRNA and its translation product support the proposal that animal synthetase is a multifunctional enzyme (Zehner *et al.*, 1980; Mattick *et al.*, 1981; Flick *et al.*, 1978; Nepokroeff and Porter, 1978; Morris *et al.*, 1982). The FAS mRNA of goose (Zehner *et al.*, 1980) and rat (Mattick *et al.*, 1981) were isolated. Poly(A)$^+$ RNA was then isolated by affinity chromatography on an oligo(dT) cellulose column. Translation of poly(A)$^+$ RNA in rabbit reticulocyte lysate produced a polypeptide (M_r 250,000) which was recognized by FAS antibodies. Binding to the antibodies was blocked by excess native synthetase. Further purification of the mRNA by sucrose-density centrifugation of the poly(A)$^+$ RNA gave sedimentation values of 35 S and 27 S for goose and rat mRNA, respectively. Gel electrophoresis of FAS mRNA after denaturation with methylglyoxal or methyl mercury hydroxide showed molecular weight of 2.9×10^6 (goose) and 3.5×10^6 (rat). There are therefore 2000–3000 more nucleotides than are required to code for the entire FAS subunit.

Goodridge's group (Morris *et al.*, 1982) cloned gene sequences for partially purified mRNA of goose FAS. Seven clones containing sequences complementary to FAS mRNA were identified by colony hybridization and the identities confirmed by hybrid-selection translation. Two plasmids, pFAS1 and pFAS3, containing FAS sequences had inserts of 1400 and 1700 base pairs, respectively. A precise size determination followed by identification yielded a value of 16 kilobases, considerably larger than previous estimates (Zehner *et al.*, 1980; Mattick *et al.*, 1981). In each case, the mRNA was large enough to code for the synthetase subunit of 250,000 daltons, or about 2300 amino acids. The synthetase mRNA is possibly the largest vertebrate mRNA identified as yet.

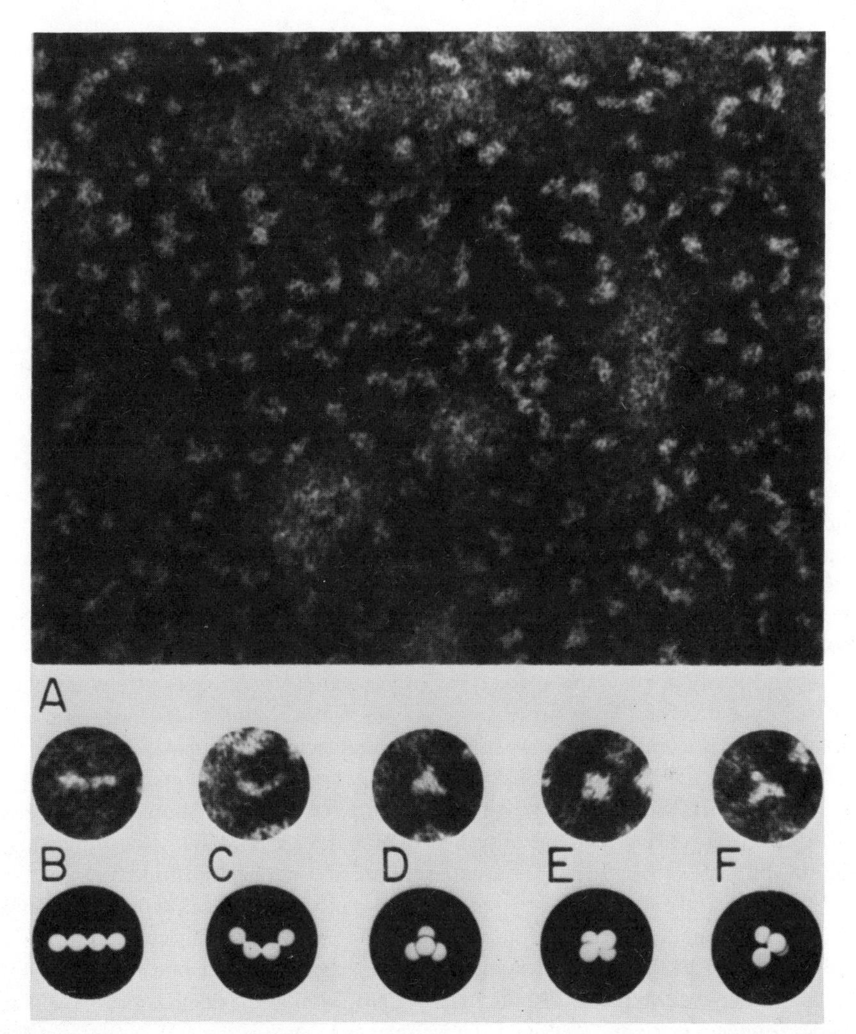

Figure 2. (A) Electron micrograph of rat liver FAS negatively stained with methylamine tungstate, ×300,000. (B, C) Selected images of the linear form of the subunit, ×400,000. (D, E, F) Selected images of the pseudotetrahedral form of the subunit, ×400,000. Accompanying each subunit image is a photograph of a model of the subunit that has been folded and oriented to present a correlative view with the image.

These results confirm the concept of the multifunctional nature of the synthetase and indicate that the subunit arises as a single polypeptide chain synthesized from one contiguous mRNA.

2. *Structural Organization of the Subunit*

Evidence presented in the previous section indicates that the synthetase subunits are identical and that each contains sites of partial reactions required for palmitate synthesis. The monomers retain six of the enzymatic activities: acetyl transacylase, malonyl transacylase, β-ketoacyl reductase, β-hydroxyacyl dehydratase, enoyl reductase, and thioesterase (Stoops *et al.*, 1979; Butterworth *et al.*, 1967; Muesing *et al.*, 1975; Yung and Hsu, 1972; Stoops and Wakil, 1981a). The one activity absent is β-ketoacyl synthetase (condensing enzyme) which has been shown to depend on the presence of two juxtapositioned thiols (Stoops and Wakil, 1981a), one from each subunit (see Section III-B).

There is increased evidence that multifunctional proteins are arranged as a series of globular domains (sites of catalytic or regulatory activity), connected by polypeptide bridges that are more sensitive to proteolytic attack than the domains themselves (Kirschner and Bisswanger, 1976; Wetlauter, 1973). Such domains are readily discernible in the electron micrograph of rat liver FAS (Figure 2). Thus, it has been possible to isolate active fragments from a number of multifunctional proteins, such as DNA polymerase I, immunoglobulins, and the "CAD" protein of pyrimidine biosynthesis, among others (Porter, 1959; Setlow *et al.*, 1972; Pabo *et al.*, 1979; Davidson *et al.*, 1981). Indications are that the FAS is organized in a similar fashion and, hence, is amenable to controlled analysis using proteolytic dissection. Treatment of the synthetase with trypsin, elastase, or subtilisin has been employed by several laboratories to separate the thioesterase activity from the remainder of the complex (Smith and Stern, 1979; Guy *et al.*, 1978; Crisp and Wakil, 1982; Agradi *et al.*, 1976; Crisp, 1976; Bedord *et al.*, 1978; Tsukamoto *et al.*, 1982).

The proteolysis of chicken liver FAS by a variety of proteases (chymotrypsin, elastase, trypsin, *Myxobacter* protease, subtilisin A and B, and kallikrein), used either individually or in combination, gave fragments that were analyzed with respect to both the kinetics and the size to establish the precursor–product relationships required for mapping (Mattick *et al.*, 1983b). Chymotrypsin, for instance, showed the most restricted cleavage of the synthetase by hydrolyzing its subunits into two fragments of 230,000 and 33,000 (Figure 3). The smaller fragment contained the thioesterase activity and could be readily separated from the 230,000-dalton fragment. Trypsin and elastase cleaved the synthetase subunits into the 230,000- and 33,000-dalton thioesterase domains, and the 230,000-dalton fragment was then centrally split into two major fragments of molecular weights 127,000 and 107,000. Similar cleavage was also obtained with *Myxobacter* protease. At higher trypsin concentrations, the 107,000-dalton fragment was further degraded to yield a 94,000-dalton polypeptide and a 15,000-dalton fragment. When the synthetase was labeled with [^{14}C]pantethenate, the ^{14}C-labeled prosthetic group was sequentially found in 230,000-, 107,000-, and 15,000-dalton polypeptides but not in the other tryptic fragments, suggesting that the acyl carrier protein (ACP) domain was associated with these fragments. The results of these cleavage patterns are summarized in Figure 4. When the [^{14}C]pantetheine-labeled synthetase was cleaved with kallikrein and subtilisin, a new set of peptide fragments was obtained that contained the [^{14}C]pantetheine as shown in Figure 4.

The results of individual and combined proteolytic digestions strongly uphold the contention that the subunits of chicken liver FAS are identical, and that the polypeptides consist of domains linked by polypeptides susceptible to proteolysis. Also, a reasonably detailed map of the synthetase subunit has been constructed (Figure 5). Analyses of all the fragment patterns and summation of their molecular weights consistently gave

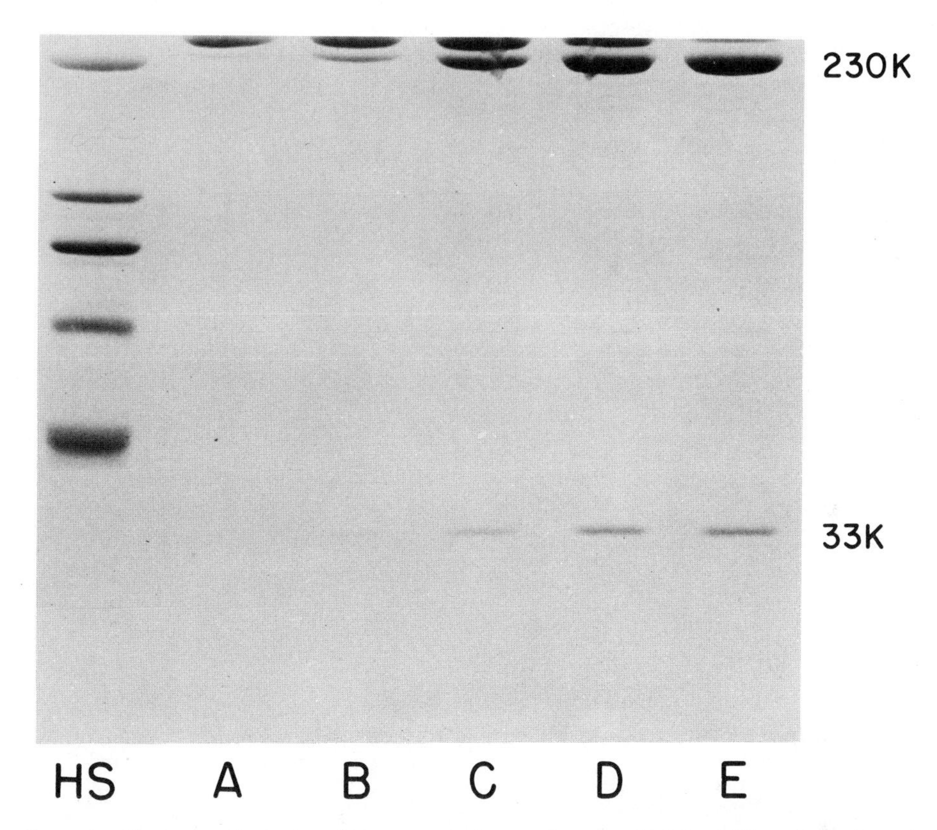

Figure 3. Electrophoretic display of chymotryptic digests of chicken FAS (1 : 1000, wt./wt.). Lane A contains undigested synthetase (2.5 μg), and lanes B–E contain hydrolysates after proteolysis for 7 min (4 μg), 15 min (7.5 μg), 30 min (10 μg), and 60 min (12.5 μg), respectively. The lane designated HS contains a mixture of high-molecular-weight standard proteins: myosin (200K), β-galactosidase (116K), phosphorylase B (94K), bovine serum albumin and ovalbumin (43K). The numbers suffixed by K refer to the molecular weights of corresponding peptides, × 1000.

a value of 267,000 for the molecular weight of the intact synthetase subunit, irrespective of the size or source of the individual peptides. α-Chymotrypsin specifically cleaves synthetase into a terminal 33,000-dalton thioesterase (domain III) and a large 230,000-dalton multifunctional complex (Figure 5). Other proteases attack the same site in addition to hydrolyzing the enzyme at other sites. Cleavage of the FAS by α-chymotrypsin into only two fragments made possible the determination of the N–C orientation of the subunit. Since intact synthetase has a blocked NH_2-terminus (Stoops *et al.*, 1975), NH_2-terminal sequence analysis of the chymotryptic fragments identified those with a free amino terminus, thereby specifying the orientation of the protein. These studies showed that the thioesterase (M_r 33,000) has a free NH_2-terminus, and that the other fragment (M_r 230,000) has a blocked NH_2-terminus, as does intact synthetase. The sequence of the thioesterase at the NH_2-terminus is H_2N-Lys-Thr-Gly-

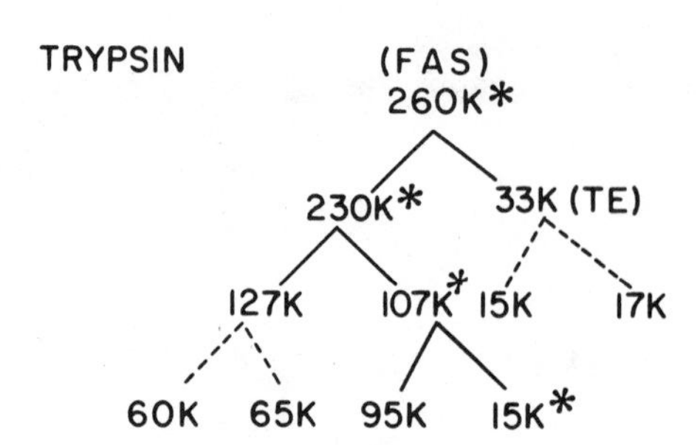

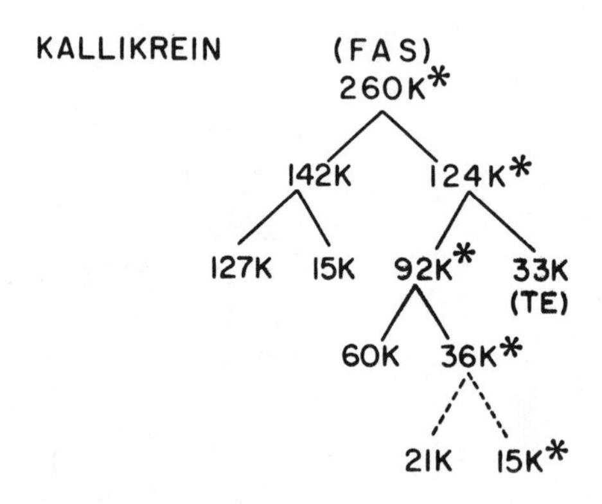

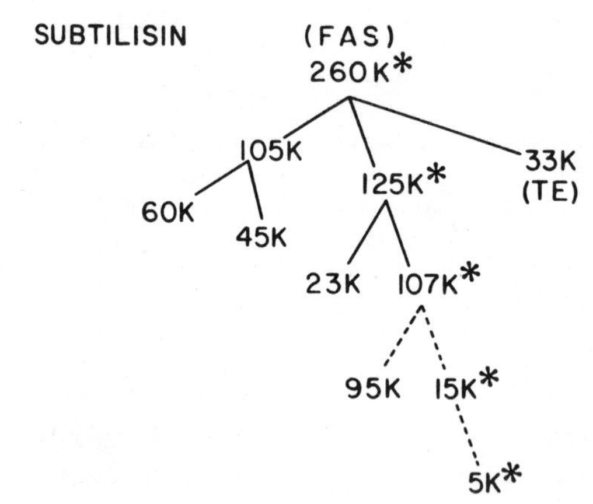

Figure 4. Summary of the cleavage patterns of [^{14}C]pantetheine-labeled chicken liver FAS by proteases. Asterisks in the figure indicate the radioactive fragments that contain [^{14}C]phosphopantetheine of the ACP. Numbers are molecular weights of polypeptides and K stands for thousand. A continuous line indicates the main pathway of proteolysis; a dotted line indicates the hydrolysis by longer incubation. TE is thioesterase.

Pro-Gly-Glu-Pro-Pro. Therefore, the thioesterase is placed at the COOH-terminus of the subunit.

The thioesterase fragment is released intact and catalytically active. However, upon prolonged incubation with any of the proteases, it is degraded into fragments of 18,000 and 15,000 mol. wt. (Mattick *et al.*, 1983a). The breakdown is accompanied by loss of catalytic activity and is apparent at later stages of the cleavage.

The core protein of 230,000 daltons can be separated into two domains, designated I (M_r, 127,000) and II (M_r, 107,000). Domain II is adjacent to the thioesterase (domain

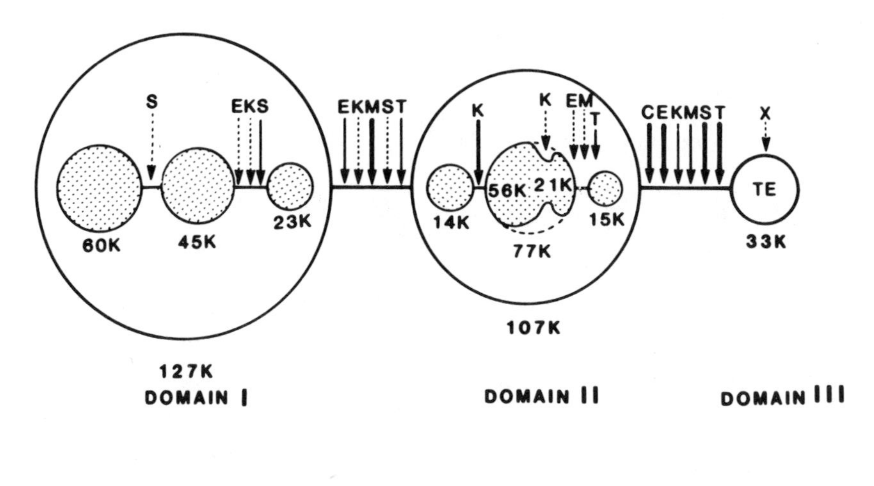

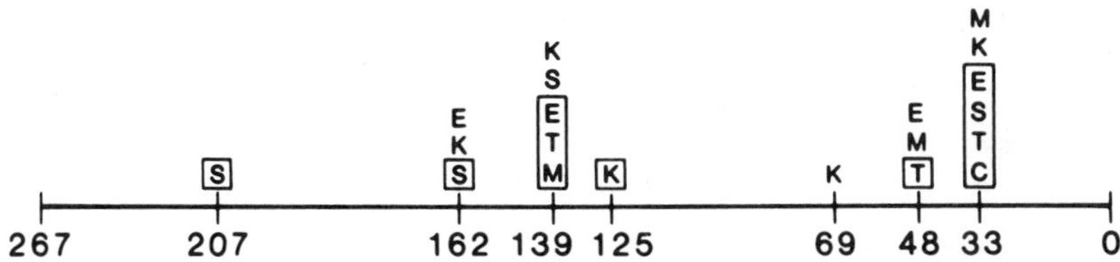

Figure 5. Proteolytic map of the chicken FAS. In the upper figure, thick arrows (→) indicate primary cleavage sites by individual proteases; thin arrows (→) indicate secondary cleavage sites; and dashed arrows (---▶), substantial cleavage sites at longer time courses. Abbreviations for proteases are as follows: T, trypsin; E, elastase; M, *Myxobacter* protease; S, subtilisin A or B; C, α-chymotrypsin; K, kallikrein; X, all of these proteases. The molecular weights of the fragments are as indicated with K = 1000. The lower figure shows the actual distances (in daltons × 10^{-3}) of the protease cleavage sites from the thioesterase terminus of the monomer. These distances were derived by summation of the molecular weights of individual peptides released by proteolysis. Proteases in boxes are capable of inflicting primary cleavage at the sites indicated.

III) and contains the primary kallikrein site, a 92,000-dalton peptide from the thioesterase junction. There is a secondary tryptic site at which ACP, a terminal 15,000-dalton segment, is released.

Placement of the thioesterase at the COOH-terminal of the synthetase polypeptide led to mapping of the functional centers of the synthetase subunit (Tsukamoto *et al.*, 1983). Known properties of the active sites of the component activities were employed when possible. The ACP site of the synthetase was labeled with radioactive pantetheine, which was then followed throughout the course of proteolysis (Figure 4). Similar methods were used with assays of catalytic activities, labeled substrates, and either monoclonal or polyclonal antibodies developed against components of the domains. Antithioesterase antibodies completely inhibit FAS activity, binding both isolated thioesterase and intact synthetase. By visualizing the antibody-binding sites with ^{125}I-labeled protein A, we found that all fragments of molecular weight 33,000 produced by the various proteases (chymotrypsin, trypsin, elastase, and subtilisin) are related, if not identical, and represent the thioesterase moiety (Figures 6A and B). The anti-

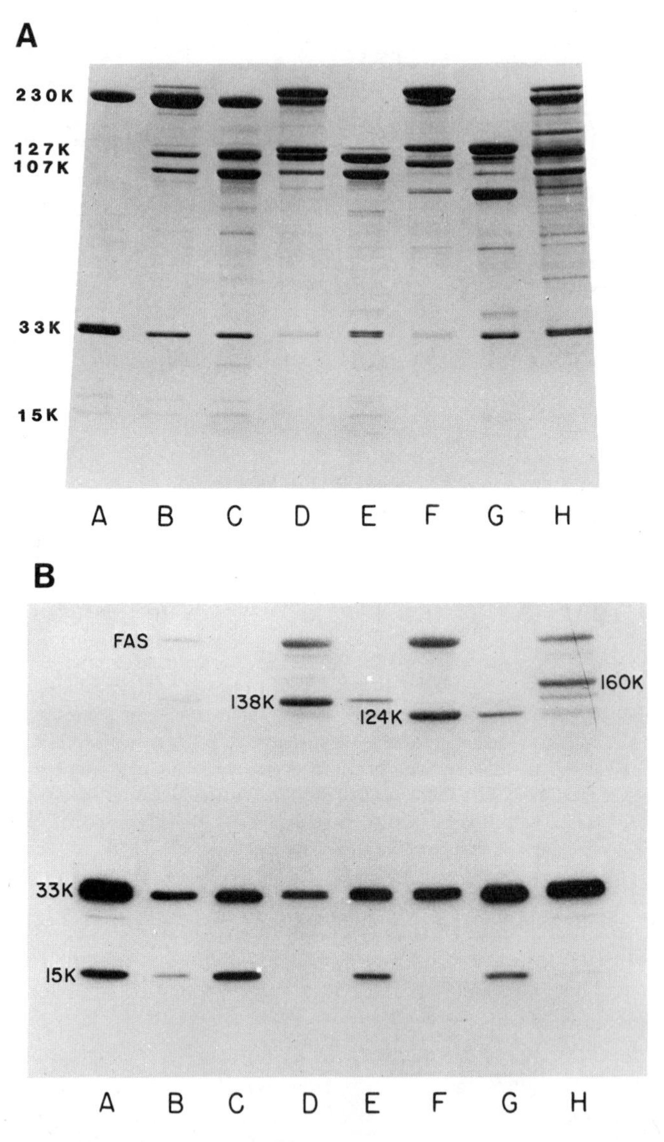

Figure 6. Antithioesterase antibodies binding to proteolyzed FAS fragments detected by "western" transfer analysis. (A) Electrophoretic display of proteolyzed FAS fragments stained for protein that was transferred to nitrocellulose paper electrophoretically. The tracts contain the following: (A) 2 hr, chymotrypsin digest; (B) 2 hr, tryptic digest; (C) 1 hr, elastase digest (D) 30 min, *Myxobacter* digest; (E) 4 hr, *Myxobacter* digest; (F) 1 hr, kallikrein digest (1 : 500); (G) 1 hr, kallikrein digest (1 : 100); (H) 30 min, subtilisin A digest. (B) The peptides shown in (A) were transferred electrophoretically to nitrocellulose paper treated with antithioesterase antibody followed by ^{125}I-labeled protein A.

thioesterase antibody reacts only with the 33,000-dalton peptides, their breakdown products, and the intermediate fragments of the synthetase which were predicted from proteolytic mapping to include the 33,000-dalton domain (Figure 5).

The specificity of the antibody preparation is most clearly shown by the chymotryptic digest (Figure 6, track A), which shows that the antibody recognizes only the thioesterase domain and its degradative product of molecular weight 15,000. This degradation product is, in fact, observed with all of the proteases, indicating a common susceptible internal cleavage site in the thioesterase domain. The reason for failure of the other breakdown product (M_r 18,000) to bind antibody is unclear. This fragment may not regain its original native conformation after SDS denaturation, or this region may not have elicited an antigenic response in the first place. There is, however, no binding whatsoever of the antiserum to the 230K core peptide, demonstrating the specificity of the antibodies for the thioesterase domain.

A similar pattern is observed in the cases of the well-characterized trypsin and elastase digestions, where the 230K peptide has been partially broken down into two (unlabeled) fragments of 127K and 107K (Figure 6, tracks B and C). Analysis of the *Myxobacter* digests (Figure 6, tracks D and E) reveals a progression of antibody binding from whole synthetase to the 138K intermediate (107K + 33K), and ultimately to the 33K thioesterase and its degradation product. Again, the 127K and 107K fragments are unlabeled. These results confirm the conclusion of the proteolytic mapping studies that the thioesterase domain lies directly adjacent to the 107K fragment of the 230K core peptide, and that this 107K peptide is the central region of the FAS subunit (see Figure 5). Also in agreement with the mapping assignments are the kallikrein digests (Figure 6, tracks F and G), which show that the thioesterase is contained in the 124K intermediate produced by this protease; as proteolysis proceeds, it is then released along with a 90K fragment, a component of the 107K domain.

Successive proteolytic digestion of the chicken FAS by trypsin and subtilisin yielded 6–8 polypeptides, with molecular weights of 15,000–94,000 (Wong *et al.*, 1983). Fractionation of the digest by ammonium sulfate and chromatography on a Procion Red HE3B affinity column permitted the isolation of a polypeptide (M_r 94,000) containing the β-ketoacyl reductase activity but no other partial activities normally associated with the synthetase. The specific activity of the β-ketoacyl reductase increased 2–3 times in this fraction, an increase that is within the expected range based on its relative molecular weight. Another fragment (M_r 36,000) containing the β-ketoacyl reductase activity was also isolated from the synthetase after double digestion of the synthetase with kallikrein and subtilisin. Mapping studies such as those summarized in Figures 4 and 5 have shown that this fragment lies adjacent to the C-terminal thioesterase domain and overlaps the tryptic 94,000-dalton peptide by approximately 21,000 daltons. This fragment, but not the 94,000-dalton fragment, was found to contain the phosphopantetheine prosthetic group, indicating that the ACP moiety is located in the 15,000-dalton segment that separates the β-ketoacyl reductase from the thioesterase domain.

The enoyl reductase activity of the FAS is also located in domain II. This reductase is quite sensitive to proteolytic digestion and its location in domain II was deduced by using a specific inhibitor of the enoyl reductase activity, pyridoxal phosphate, to label the synthetase in the presence or absence of the (protective) substrate NADPH. Proteolytic analysis of the labeled synthetase showed two sites of pyridoxal binding, one located in domain I, which is not affected by NADPH, and another located in domain II which can be protected by NADPH. The latter site is, therefore, associated with the enoyl reductase, and is present in both the 95,000-dalton tryptic fragment and the 36,000-dalton kallikrein fragment derived from domain II (see Figure 4). These are the same peptides shown to contain the β-keto reductase activity. This result was confirmed by labeling the synthetase with a ^{14}C-photoaffinity analogue of NADP$^+$ (in the presence or absence of pyridoxal phosphate). Only the tryptic fragments 107K → 94K and the kallikrein fragments 125K → 94K → 36K (Figure 4) possessed NADPH-binding sites, which is in agreement with the conclusion that both the β-keto and enoyl reductase activities occur in the same region of the synthetase molecule.

The location of the acetyl and malonyl transacylase activities was determined from the binding of labeled acetyl and malonyl groups to the enzyme. Apart from the ACP segment, the only fragment labeled by each of these acyl-CoAs is a 60K peptide derived from domain I (Figure 4). This binding site was found to be insensitive to thiol reagents, consistent with the earlier evidence that the transacylase reactions involve an active serine hydroxyl group (Stoops *et al.*, 1977). It may be concluded, therefore, that both acetyl and malonyl transacylases are located in the same 60K region. Whether these centers are indeed separate or share a common site is not known.

These studies have established the position of six of the eight functions known to be present in the FAS subunit, that is, all except the dehydratase and β-ketoacyl synthetase activities. In the former case, no specific reagent is available, and attempts to isolate and characterize active fragments are under way. Because this activity occurs between the two reductase steps in the reaction sequence, it seems likely that it also is located within the central region of domain II.

Domain I contains the β-ketoacyl synthetase (condensing enzyme) activity, which is well-characterized by a reactive cysteine-SH, a site of attachment for acetyl groups and one known to be attacked by a number of alkylating reagents. Dibromopropanone reacts readily to crosslink the cysteine-SH of the β-keto synthetase of one subunit and the pantetheine-SH of the other subunit (see Section III-B). Cleavage of the synthetase by trypsin followed immediately by treatment with dibromopropanone resulted in crosslinking of the fragments containing the cysteine-SH with that containing the pantetheine-SH (ACP). In doing so, one would predict that the 127K tryptic fragment would crosslink with the 107K, or 15K ACP fragment, yielding 234K or 142K fragments, respectively. Indeed, the results shown in Figures 7A and B support this prediction. After 7 min digestion of the synthetase by trypsin fragments, 230, 127, 107, 33, and 15K were produced as expected (Figures 4 and 7A, lane 5). Treatment of this digest with dibromopropanone yielded new fragments with approximate molecular weights of 460, 360, 240, and 140K with a corresponding decrease in the staining intensities of the 230, 127, 107, and 15K (Figure 7A, lane 6) suggesting that the newly formed bands are products of crosslinking of 230K + 230K, 230K + 127K, 127K + 107K, and 127K + 15K, respectively. The presence of acetyl-CoA or malonyl-CoA in the tryptic digest prior to the addition of dibromopropanone significantly reduced the intensities of the newly formed bands (Figure 7A, lanes 7 and 8) indicating that the crosslinking occurred between the cysteine-SH of the β-ketoacyl synthetase

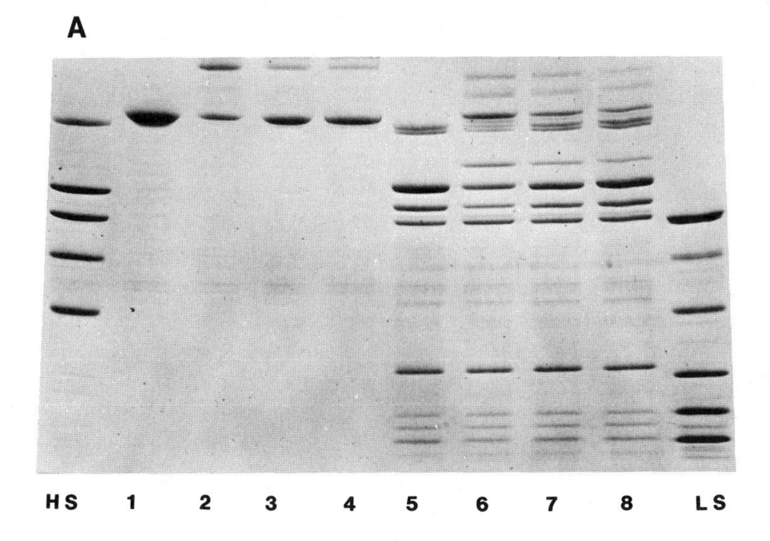

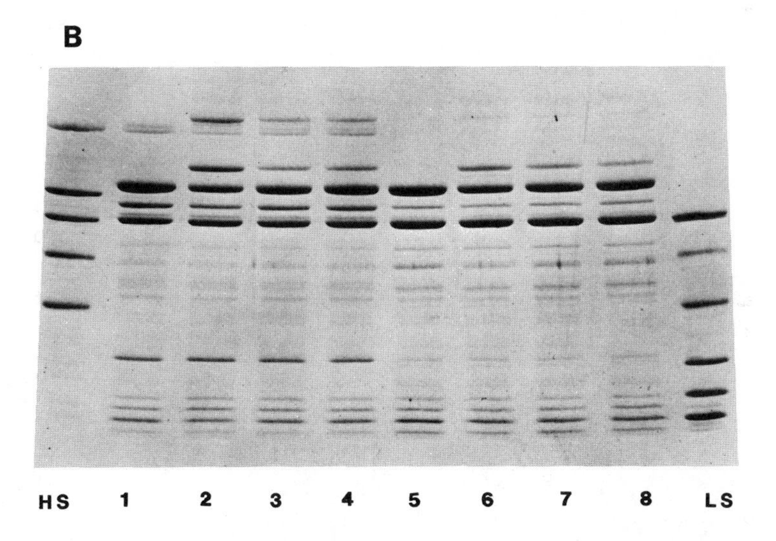

Figure 7. SDS–PAGE display of tryptic fragments of synthetase crosslinked with dibromopropanone. The synthetase was dialyzed overnight at room temperature against 0.05 M sodium phosphate, pH 7.5, containing 1 mM EDTA and 1 mM mercaptoethanol. The reducing agent was removed by gel filtration, and all subsequent operations were carried out in an O_2-free atmosphere. Samples of synthetase (1 mg/ml) were treated with dibromopropanone (6 μM) for 60 sec in the presence or absence of either acetyl-CoA (0.5 mM) or malonyl-CoA (0.5 mM). Proteolysis by trypsin (1 : 100, wt./wt.) was carried out at 25°C and stopped by addition of soybean trypsin inhibitor (20 μg/ml) after 7, 30, and 60 min of incubation. Each sample was then treated with dibromopropanone as before, boiled and analyzed by gradient SDS–PAGE (Tsukamoto *et al.*, 1983). (A) lane 1, synthetase; lane 2, synthetase treated with dibromopropanone; lanes 3 and 4, same as lane 2 except the synthetase was incubated with 0.5 mM of acetyl-CoA or malonyl-CoA, respectively. Synthetase digested with trypsin for 7 min and lane 5, applied to gel; lane 6, after treatment with dibromopropanone; lane 7, after treatment with acetyl-CoA and dibromopropanone; lane 8, and after treatment with malonyl-CoA and dibromopropanone. (B) Lanes 1–4 and lanes 5–8 are respectively the same as lanes 5–8 of (A) except synthetase was digested for 30 and 60 min, respectively, Lanes HS contain high molecular weight protein standards as in Figure 3; LS contain low molecular weight standards: phosphorylase (94K), bovine serum albumin (68K), ovalbumin (43K), carbonic anhydrase (30K), soybean trypsin inhibitor (21K), and lysozyme (14K).

site and the pantetheine-SH site of ACP. Longer exposure of the synthetase to trypsin eliminated the 230K fragment, and subsequently its crosslinked products, and intensified the 240K and 140K crosslinked bands (Figure 7B, lanes 1 and 2). Further proteolysis with trypsin resulted in cleavage of the 107K fragment to 95K + 15K ACP fragments (Figure 4). Accordingly, the complete loss of the 240K crosslinked product but not the 140K was observed as shown in Figure 7B (lanes 5 and 6). Again, the presence of acetyl-CoA and malonyl-CoA in the hydrolysates prior to crosslinking with dibromopropanone reduced significantly the intensity of the crosslinked bands (Figure 7B). Since the pantetheine prosthetic group were identified with tryptic fragments 230, 107, and 15K (Figure 4), it is reasonable to conclude that the cysteine-SH is located in fragments 230K and 127K, thus placing the β-ketoacyl synthetase site in domain I of the synthetase, i.e., distal to the 15K ACP region, which is located

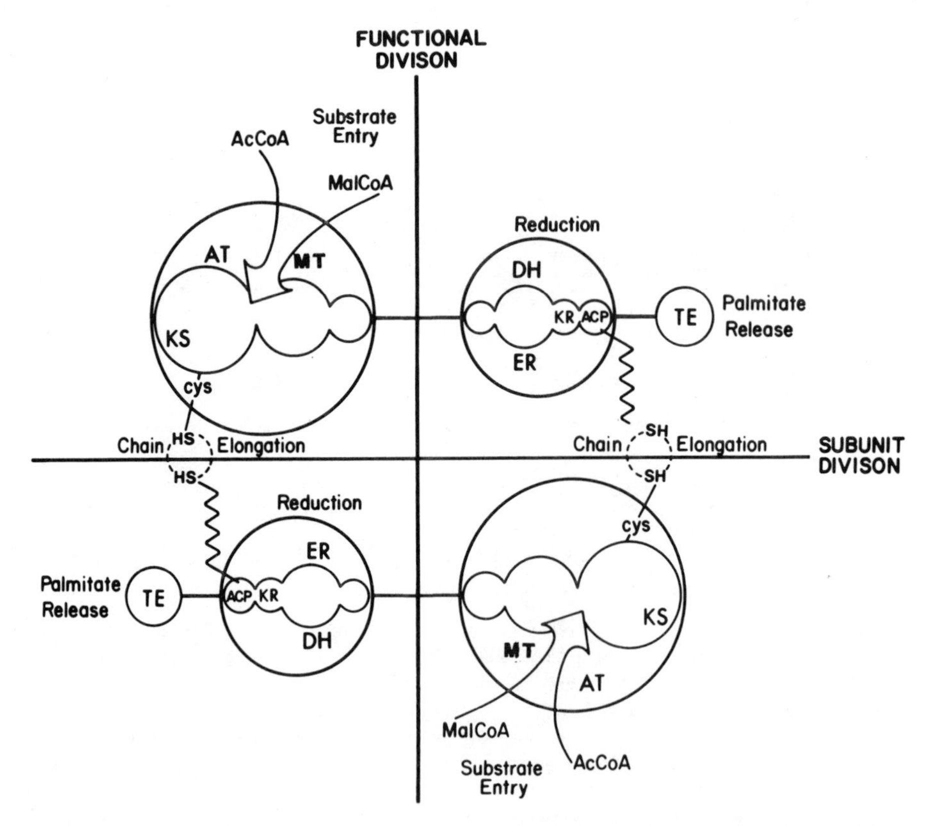

Figure 8. Proposed functional map of the chicken FAS. The model is based on the map shown in Figure 7 and the results obtained, where possible, from assays of catalytic activities and the binding of substrates, specific inhibitors, or antibodies. Two subunits are drawn in head-to-tail arrangement (subunit division) so that two sites of palmitate synthesis are constructed (functional division). The abbreviations for partial activities used are AT, acetyltransacylase; MT, malonyl transacylase; KS, β-ketoacyl synthetase; KR, β-ketoacyl reductase; DH, dehydratase, ER, enoyl reductase; TE, thioesterase; and ACP, acyl carrier protein. The wavy line represents the 4′-phosphopantetheine prosthetic group.

at the other end of the synthetase subunit, adjacent to the carboxy-terminal thioesterase domain.

On the basis of these studies, a two-dimensional diagram was proposed (Figure 8) which shows the relative sizes of the domains and the associated activities. Two subunits are arranged head-to-tail, thus forming two sites of palmitate synthesis (see Section IV-A). The β-ketoacyl synthetase (KS) and the ACP sites are in separate domains of the synthetase subunit, at a distance from each other. Domain I, containing the acetyl and malonyl transacylases and the condensing enzyme (KS), is, therefore, the domain of substrate entry and chain elongation. Domain II, containing β-ketoacyl reductase, dehydratase, and enoyl reductase partial activities, is the processing domain for NADPH reduction of the carbonyl carbon to its methylene analogue. ACP with its 4′-phosphopantetheine arm is located next to domain II and connects it to domain III (thioesterase), the site of palmitate release and chain termination (for details, see Section IV-A).

B. Yeast Fatty Acid Synthetase

Yeast FAS, like animal FAS, catalyzes generally the same chemical reactions, leading to the formation of long-chain fatty acids from acetyl-CoA, malonyl-CoA, and NADPH. However, there are differences in the structure of the enzyme and in the mechanism of some of the reactions. Enoyl reductase requires FMN as a cofactor and produces palmitoyl-CoA rather than free palmitate. Previous reports suggested that yeast synthetase is a multienzyme complex of eight enzymes held together by non-covalent interactions (Lynen, 1967a,b). According to Schweizer *et al.* (1973, 1975), those studies were misinterpreted because of partial proteolysis of the synthetase during isolation. When proteolysis was reduced, SDS–PAGE studies indicated that the enzyme consists of two subunits of molecular weights 185,000 and 180,000. We had reached similar conclusions with respect to variable degradation of the enzyme by proteolysis and to the need for protease inhibitors during isolation (Stoops *et al.*, 1978b). We also isolated the enzyme from yeast mutant with reduced protease levels. Enzyme prepared without inhibitors was found to have the same SDS–PAGE patterns as those of enzyme isolated in the presence of protease inhibitors. Tris-glycine SDS–PAGE of yeast synthetase produced two protein bands (designated α and β) in equal amounts (Figure 9), with estimated molecular weights of 213,000 and 203,000 respectively. Lower relative values were reported by Schweizer *et al.* (1973, 1975), perhaps because different protein standards were used. Our studies led to the conclusion that the native yeast synthetase is an $\alpha_6\beta_6$ complex.

Pantetheine content of yeast synthetase ranges from 3.8–5.0 per mole of synthetase (M_r 2.3 × 10^6) or approximately 1 mole of pantetheine for two subunits (Stoops *et al.*, 1978b; Schweizer *et al.*, 1970). Radioautography of the SDS–PAGE of the [^{14}C]pantetheine-labeled synthetase showed that the α-subunit bears the ACP site (Stoops *et al.*, 1978b; Schweizer *et al.*, 1973). The peptides containing 4′-phosphopantetheine have been isolated from yeast (Schreckenbach *et al.*, 1977) and rat liver (Roncari *et al.*, 1972) synthetases, and the sequence of amino acids around the serine-phosphopantetheine diester determined. As may be seen in Figure 10, there is a distinct

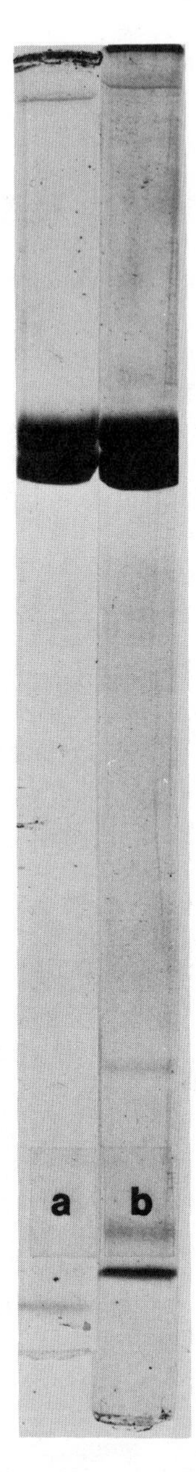

Figure 9. SDS–PAGE of yeast FAS. The Tris-glycine gels were loaded with 10 μg (a) and 15 μg (b) of enzyme isolated from Fleishmann and pep4-1 yeasts, respectively (Stoops *et al.*, 1978b).

correlation between the amino acid sequences of peptides of ACP of yeast, rat, and *E. coli* (Vanaman *et al.*, 1968). Of the 18 amino acids around the 4′-phosphopantetheine-carrying serine of ACP from yeast and *E. coli,* there are five identical amino acids (boxes) in analogous positions and eight pairs of amino acids (arrows) whose codons vary by a single point mutation (Lynen, 1980). Homology between animal

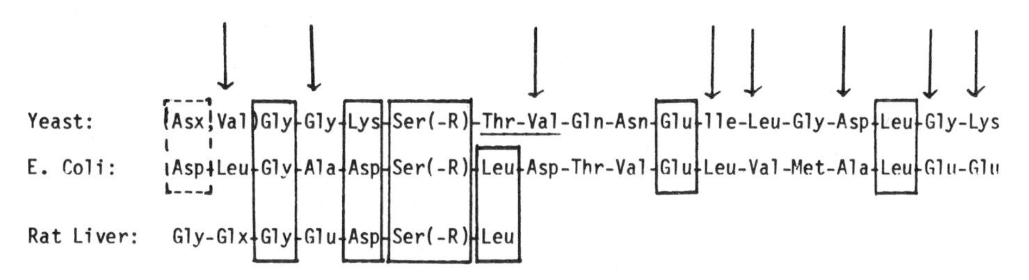

Figure 10. Sequence homologies of the ACP from yeast, *E. coli,* and rat liver FAS in the neighborhood of the prosthetic group, 4′-phosphopantetheine (R) (Lynen, 1980).

and *E. coli* sequences is also good, despite the restricted number of amino acids sequenced in the animal enzyme. These observations support the thesis that the ACP of all three organisms arises from the same ancestral gene, with some noncritical mutation.

Electron microscopy of negatively stained yeast synthetase revealed two distinct structures, arches and plates, suggesting a model of the enzyme (Stoops *et al.*, 1978b): an ovate structure with plate-like proteins on its short axis and six arch-like proteins equally distributed on either side. Measurements of the subunits indicated an approximate molecular weight of 200,000, in agreement with values obtained by physico-

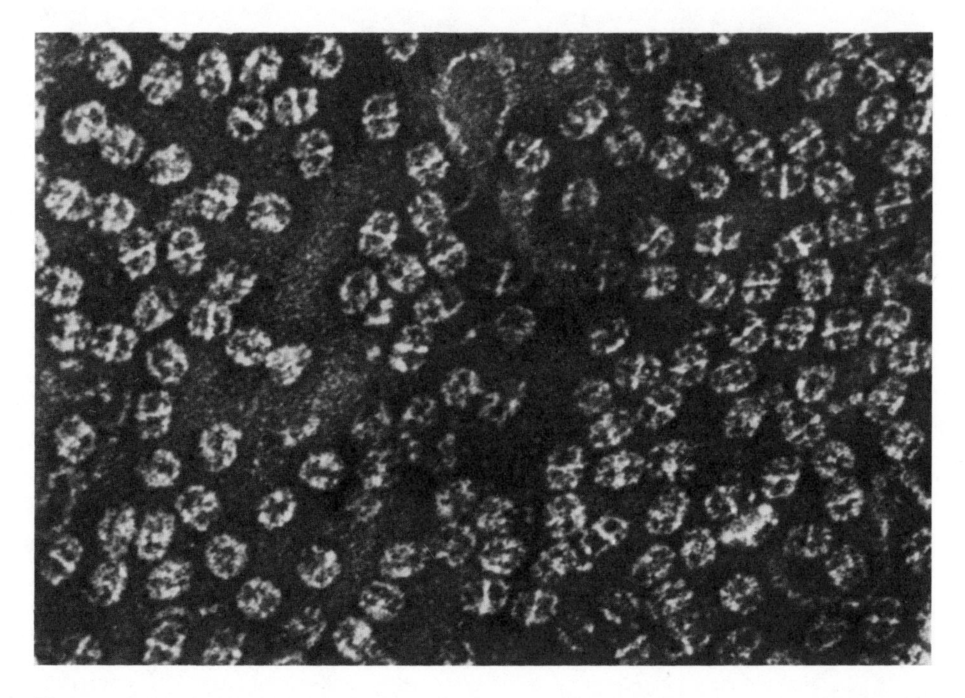

Figure 11. Micrograph of FAS from yeast negatively stained with methylamine tungstate: ×200,000. The protein was crosslinked with glutaryldehyde prior to exposure to the stain (Stoops *et al.*, 1978b).

chemical methods (Stoops *et al.*, 1978b). The structural organization of yeast synthetase became apparent when stereoscopic images of particles were studied (Figure 11). These particles appear to contain three arches on each side of the plate. An arch begins on one side of the plate and terminates on the opposite side of an adjacent plate subunit. This arrangement of arches suggested that the plates are alternately reversed (black and white shown in Figure 12) in their orientation with respect to their unique axis. This conclusion was later found to be compatible with (and supportive of) the results obtained from our studies of the mechanism of action of the β-ketoacyl synthetase component of the yeast FAS (see Section III-B).

A similar model was proposed by Wieland *et al.* (1979), with the difference that α and β form a V-shaped complex with the peptides protruding upward and downward from the plate proteins.

The α- and β-subunits of the yeast FAS can be separated by first modifying the protein with either 3,4,5,6-tetrahydrophthalic anhydride followed by ion-exchange chromatography on a DEAE Biogel A column (Stoops and Wakil, 1978), or with cyclic anhydrides (such as citraconic and dimethylmaleic anhydrides) followed by sucrose-gradient centrifugation (Wieland *et al.*, 1978). After mild acid treatment of the separated subunits, the acylating agent is removed and the subunits slowly regain some of the activities. Assays of such fractions showed that α contains, in addition to the phosphopantetheine site, the active thiol of the condensing site (β-ketoacyl

Figure 12. Model of the yeast FAS. The arch-like structures are the β-subunits; the plate-like structures in the center are the α-subunits. The black and white faces of the α-subunits represent the alternate arrangement of these structures in the complex $\alpha_6\beta_6$.

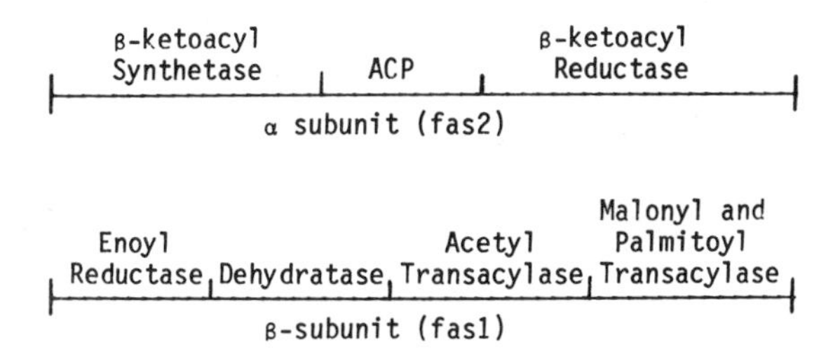

Figure 13. Distribution of partial activities on the α- and β-subunits (Schweizer *et al.*, 1975).

synthetase) and the β-ketoacyl reductase site, whereas, the β-subunit contains the transacylases, the dehydratase, and the enoyl reductase sites. These findings confirm the earlier conclusions by Schweizer *et al.* (1975) based on genetic studies (Figure 13).

Antibodies against α and β polypeptides were prepared and utilized to identify the α- and β-subunits in the model based on electron microscopic studies (Wieland *et al.*, 1978; Figure 14). The results show that α-subunits are the plates, and the β-subunits are the arches, a conclusion fully supported by crosslinking studies of Stoops and Wakil (see Section III-B).

Kuziora and co-workers (Kuziora *et al.*, 1983; Wakil and Kuziora, 1983) used the yeast transformation technique to isolate DNA clones of the genes which code for the α- and β-subunits of the synthetase through complementation of fatty acid auxotrophs of *Saccharomyces cerevisiae*. Plasmids YEpFAS1 and YEpFAS2 were selected from a bank of yeast DNA sequences in the vector YEp13 by their ability to complement mutations in the *fas1* or *fas2* locus, respectively.

Confirming evidence that plasmids YEpFAS1 and YEpFAS2 contain DNA sequences which code for the subunits of yeast synthetase was provided by showing homology to plasmids 33F1 and 102B5 isolated from a yeast genomic bank (Kuziora *et al.*, 1983). Using the antibody selection method, approximately 5000 clones were screened that contain randomly sheared yeast DNA inserted into the ColE1 vector. Two clones (33F1 and 102B5) that expressed FAS-related antigen were identified (Kuziora *et al.*, 1983; Chalmers and Hitzeman, 1980). The Southern blotting technique

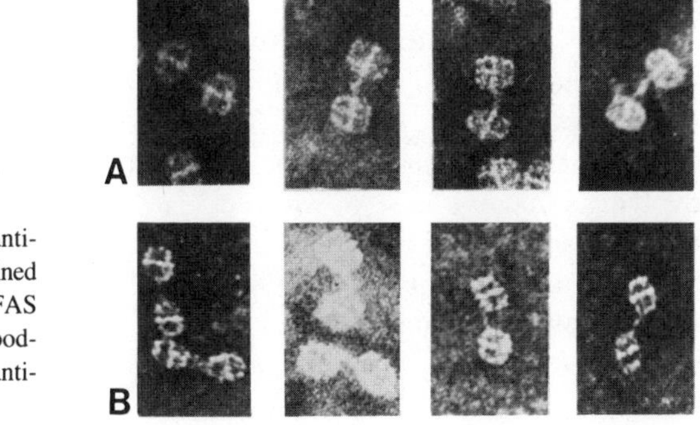

Figure 14. Electron micrographs of antibody-crosslinked yeast FAS negatively stained with phosphotungstate, ×240,000. (A) FAS crosslinked plate-to-plate by anti-α antibodies; (B) FAS crosslinked arch-to-arch by anti-β antibodies (Wieland *et al.*, 1978).

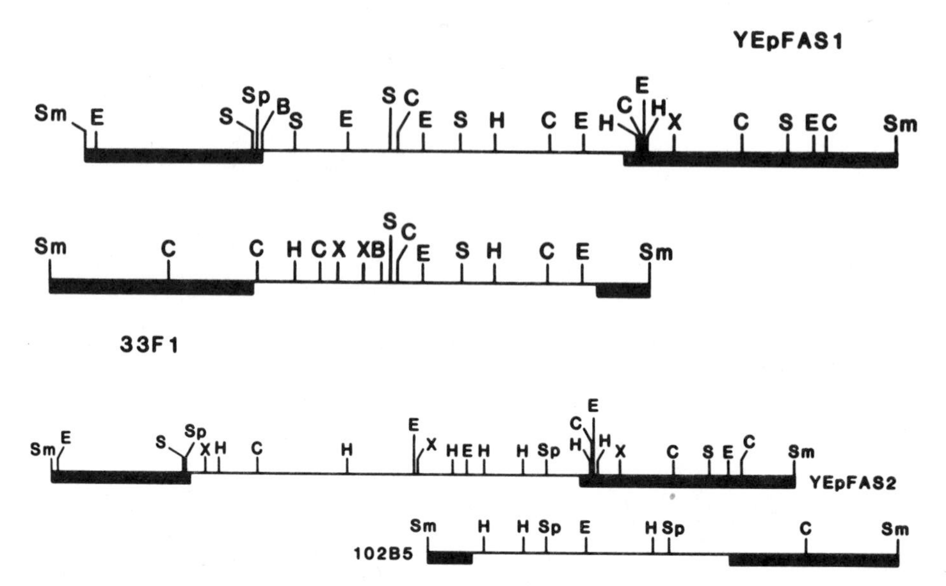

Figure 15. Comparison of the restriction endonuclease maps of YEpFAS1 and 33F1, and YEpFAS2 and 102B5, respectively. E, EcoRI; H, HindIII; X, XbaI; C, ClaI; B, BamHI; Sm, SmaI; S, SalI; Sp, SphI; Heavy lines indicate vector DNA (YEp13 in YEpFAS1 and YEpFAS2, and ColE$_1$ in 33F1 and 102B5); thin lines indicate yeast DNA inserts.

was used to show that 33F1 hybridized to a ^{32}P-labeled nick-translated probe from YEpFAS1, and that 102B5 hybridized to a ^{32}P-probe from YEpFAS2, showing that these clones contained DNA from the *fas1* and *fas2* loci, respectively. Restriction endonuclease mapping of the two clones further identified the regions of homology between the two sets of plasmids (Figure 15). A comparison of restriction endonuclease maps of 33F1 and YEpFAS1 suggests that a region of 5.3 kilobase pairs of cloned yeast DNA is identical. However, within a region of 3.1 kilobase pairs located to the left of the *Bam*HI site of 33F1 in Figure 15, the restriction endonuclease sites do not align. The restriction maps of 102B5 and YEpFAS2 are compared to Figure 15. Based on the alignment of endonuclease sites, a segment of 3.4 kilobase pairs of yeast DNA appears to be common to both plasmids.

III. COMPONENT ACTIVITIES

Synthesis of palmitate from acetyl-CoA and malonyl-CoA is a cyclical process involving seven enzymes, of which five are active eight times in the sequence. Therefore, at least 30 acyl intermediates are formed and covalently bound in turn to the enzyme. Properties and interrelationships of the component enzymes are summarized in the following section.

A. Acetyl and Malonyl Transacylases

Acetyl and malonyl transacylases prime FAS with the carbon atoms necessary for fatty acid synthesis [Eqs. (3 and 4)]. Malonyl transacylase is specific for the malonyl group. Acetyl transacylase normally uses acetyl-CoA but is species-dependent and manifests some variability toward acyl-CoA substrates. Avian liver FAS shows preference for acetyl-CoA as a primer, while bovine and rodent synthetases utilize butyryl-CoA as well as acetyl-CoA (Aprahamian *et al.*, 1982; Bressler and Wakil, 1962). The latter group of enzymes releases 25–40% of the fatty acid product as butyryl-CoA while the remaining product is free fatty acid.

It has been proposed that butyryl-*S* enzyme formed as an intermediate during the course of fatty acid synthesis may either undergo chain elongation by condensation with malonyl-CoA or that the butyryl group is transferred to CoA, presumably by acetyl transacylase of the FAS. This hypothesis is supported by studies showing that an increase in malonyl-CoA concentration favors chain elongation at the expense of butyryl-CoA formation, whereas increased CoA concentration increases the amount of butyryl-CoA and decreases the amount of long-chain fatty acid (Abdinejad *et al.*, 1981). The physiological importance of this variation is unknown.

Transacylases which use butyryl-CoA as primer also transacylate acetoacetyl or crotonyl groups from their CoA derivatives, thus becoming primers of FAS after reduction by synthetase to butyryl derivatives. These synthetases thereby use acetoacetyl-CoA or crotonyl-CoA, 20–50 times more efficient as primers in the synthesis of palmitate than do avian or yeast enzymes (Dodds *et al.*, 1980). This variation in transacylase specificity may be of particular importance in bovine tissues, which are rich in these acids.

There is an active serine residue in the transacylases of all synthetases where the acetyl and malonyl groups are bound as an *O*-ester linkage. Lynen and co-workers (Engeser *et al.*, 1979b; Ziegenhorn *et al.*, 1972) isolated the serine-containing peptide of acetyl and malonyl transacylases of yeast FAS. As may be seen in Figure 16, the neighboring amino acids of octapeptides containing active sites from acetyl and malonyl transacylases have three pairs of identical amino acids (boxed residues) and three pairs of amino acids whose codons differ by only one base (starred residues). Though no conclusions can be drawn from such short sequences, a common origin for the two transacylases has been suggested (Lynen, 1980). It appears that the palmitoyl-transferase of yeast FAS and the malonyl transacylase share the same active serine site.

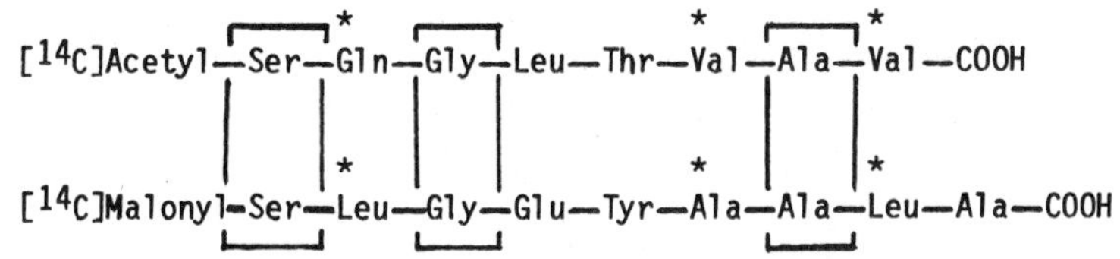

[^{14}C]Acetyl—Ser—Gln*—Gly—Leu—Thr—Val*—Ala—Val*—COOH

[^{14}C]Malonyl—Ser—Leu*—Gly—Glu—Tyr—Ala*—Ala—Leu*—Ala—COOH

Figure 16. Sequences of active-site peptides from malonyl or palmitoyl transacylases and acetyl transacylase (Engeser *et al.*, 1979b; Ziegenhorn *et al.*, 1972).

The same sequences of amino acids are adjacent to the active serine residue, and genetic analyses show that the two enzymes are coded by the same gene (Knobling *et al.*, 1975). Moreover, binding studies of radiolabeled malonyl and palmitoyl groups show that malonyl binding excludes that of palmitoyl, and vice versa (Engeser *et al.*, 1979b).

Acetyl transacylase was isolated from yeast FAS after partial proteolysis by elastase of the yeast complex (Lynen, 1980). Acetyl transacylase activity was almost unaffected by this procedure, although overall FAS activity and remaining partial activities were lost rapidly. After further separation by ultracentrifugation and gel filtration of the proteolysis products, the enzyme retained all of its activity. The molecular weight determined by SDS–PAGE and analytical centrifugation studies was 56,000. No other component activities were associated with the protein of this molecular weight. Acetyl pantetheine and CoA substrates used to assay the transacylase had K_m values ten times greater with isolated enzyme than with transacylase present in native FAS, indicating that release of the enzyme from the complex is accompanied by changes in protein conformation.

It was previously thought that the loading of an acetyl or malonyl group to animal FAS was an ordered process. When charged with an acetyl group, the enzyme underwent a change in conformation so that malonyl group was preferentially loaded next, favoring the condensation and chain elongation to palmitate (Kumar *et al.*, 1972). Apparently this is not the case. Removal of CoA by a scavenging system (phosphotransacylase, acetyl-CoA synthetase, or ATP citrate lysase) causes fatty acid synthesis to cease. Addition of either CoA or thioesterase III reactivates synthesis (Linn and Srere, 1980). It was proposed that CoA is required for formation of palmitoyl-CoA prior to its hydrolysis to free palmitate by synthetase-bound thioesterase. Studies by Smith and co-workers (Smith, 1982; Stern *et al.*, 1982) and by Poulose and Kolattukudy (1982) confirmed the CoA requirement for FAS but did not support the requirement for thioesterase. Stern *et al.* (1982) suggest that free CoA participates in a continuous exchange of acetyl and malonyl moieties between CoA thioesters and the enzyme itself. The acetyl and malonyl transacylases operate independently, their respective acyl substrates competing for the pantetheine component of the synthetase. If free CoA is not present, the synthetase is loaded with either acetyl or malonyl groups at the condensing site and is inactivated due to the presence of an incorrect condensing partner. This conclusion was based on the observation that the presence of a CoA scavenging system blocks unloading of acetyl and malonyl groups, thereby preventing reloading of the appropriate substrate to the synthetase. Inhibition is relieved by addition of CoA or pantetheine. The uptake of acetyl and malonyl groups seems then to be a random process rather than a sequential one. Kinetic studies are needed to distinguish between the two processes and to give more information on the action of free CoA in fatty acid synthesis.

B. β-Ketoacyl Synthetase (Condensing Enzyme)

β-Ketoacyl synthetase catalyzes the coupling of acyl and malonyl groups to form β-ketoacyl derivatives. Studies of the condensation of acetyl-ACP and malonyl-ACP

[Eqs. (5a and b)] by the *E. coli* enzyme (Toomey and Wakil, 1966; D'Agnolo *et al.*, 1975) have shown that β-ketoacyl synthetase contains an active cysteine-SH which forms an acyl-*S*-enzyme intermediate before coupling with malonyl-*S*-ACP to yield β-ketoacyl-ACP and CO_2. The reaction occurs in both animal (Stoops and Wakil, 1981a) and yeast FAS (Oesterhelt *et al.*, 1977; Stoops and Wakil, 1981b). In each, an active cysteine-SH was identified. Alkylation with iodoacetamide inhibited only the condensing activity. The enzyme was protected from inhibition by acetyl-CoA but not by malonyl-CoA, suggesting formation of acetyl-*S*-enzyme intermediate. The binding site of the acetyl group was identified by isolating a carboxymethyl peptide after treatment of yeast synthetase with iodo[^{14}C]acetamide and subsequent proteolysis. The sequence of amino acids in this peptide was found to be N_2N-Thr-Pro-Val-Gly-Ala-Cys-COOH. A similar acetate-containing peptide was also isolated and shown to have identical mobilities by high-voltage electrophoresis and paper chromatography (Kresze *et al.*, 1977).

Evidence indicates that in eukaryotic synthetases, the acetyl and other acyl groups form primarily the acyl derivative of the cysteine-SH of the condensing enzyme and malonyl group from the acyl derivative of the pantetheine-SH before their condensation to the β-ketoacyl derivative (for review, see Stoops *et al.*, 1977). Stoops and Wakil (1981a,b, 1982a,b) studied the structure and mechanism of the β-ketoacyl synthetase of animal and yeast FAS. There is solid experimental evidence for the identification of the residues to which the acetyl and malonyl groups are bound at the β-ketoacyl synthetase site of animal FAS. Furthermore, the studies support previous assignment of attachment sites of the acyl groups of yeast FAS, and reveal a novel and common arrangement of the β-ketoacyl synthetase site in both animal and yeast FAS.

The state of aggregation of animal and yeast synthetases during catalysis was determined by active enzyme centrifugation in the pH range of 6.5–7.5 (Stoops *et al.*, 1979). Values of the sedimentation coefficients, measured in both presence and absence of substrates, were S_{20w} 15.0–16.5 S, indicating that the dimer is the active form of animal synthetase. Sedimentation equilibrium experiments show these values to correspond with the dimer form; the monomer form (S_{20w} 9.1 S) was inactive.

Reasons for the requirement of the dimer form of the enzyme became evident from studies by Stoops and Wakil (1981a, 1982a) of the role of active thiols in fatty acid synthesis. FAS of chicken liver was completely inhibited by 0.5 mM iodoacetamide in pseudo-first-order process. When synthetase was treated with iodo[^{14}C]acetamide, more than 80% of the ^{14}C-label was recovered as ^{14}C-labeled *S*-carboxymethylcysteine but none as ^{14}C-labeled *S*-carboxymethylcysteamine after HCl hydrolysis. These results indicate that inhibition of the enzyme was due to alkylation of the active cysteine-SH, but not of pantetheine-SH (Stoops and Wakil, 1981a). Since only this partial activity was lost, the cysteine residue was identified as an essential component of the β-ketoacyl synthetase site. Preincubation of the synthetase with acetyl-CoA protects the enzyme from inhibition by iodoacetamide, suggesting that this thiol is the site of binding of the acetyl group to the β-ketoacyl synthetase site. However, preincubation of the enzyme with malonyl-CoA prior to treatment with iodoacetamide did not protect the FAS against iodoacetamide inhibition, suggesting that the site of binding of the malonyl group is not the cysteine-SH.

In contrast to the slow inhibition of the chicken FAS by iodoacetamide, the bifunctional reagent 1,3-dibromo-2-propanone inhibits the enzyme rapidly (within 30 sec) and completely (Stoops and Wakil, 1981a). The loss of synthetase activity is due to inhibition of only the β-ketoacyl synthetase activity. Preincubation of the synthetase with acetyl-CoA, though not with malonyl-CoA, protects the enzyme against inhibition by dibromopropanone. These results are similar to those found for iodoacetamide inhibition and clearly show that like iodoacetamide, the dibromopropanone competes with acetyl-CoA for the same thiol in the β-ketoacyl synthetase site.

When the dibromopropanone-inhibited synthetase was analyzed on SDS–PAGE, the synthetase subunit of molecular weight 250,000 was nearly absent, with a concomitant appearance of oligomers of higher molecular weight (450,000–550,000) (Figure 17). These observations suggested that the synthetase subunits were crosslinked by the bifunctional reagent dibromopropanone. Preincubation of the synthetase with acetyl-CoA or malonyl-CoA prevented the crosslinking. A similar result was obtained when the synthetase was treated with iodoacetamide prior to its reaction with the dibromopropanone (Figure 17). The stoichiometry of inhibition was determined by binding studies, which indicated that the binding of about 1.8 moles of dibromopropanone per mole of enzyme was required for complete inactivation of the FAS (Stoops and Wakil, 1981a). Altogether, these results indicated that the dibromopropanone is

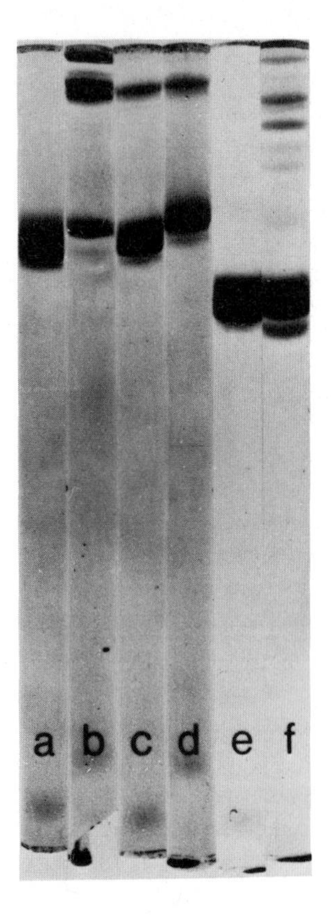

Figure 17. SDS–PAGE of chicken liver FAS, untreated enzyme (a), or treated enzyme with 2 moles of dibromopropanone per mole of enzyme (1 mg/ml) (b). (c and d), the enzyme was treated with dibromopropanone in the presence of 0.5 mM acetyl-CoA or malonyl-CoA, respectively. (e) The enzyme was reacted with iodoacetamide to obtain 90% inhibition. (f) The iodoacetamide-inhibited enzyme was reacted with dibromopropanone as in (b). Electrophoresis of gels (e) and (f) were performed for a longer time than those of gels (a)–(d) (Stoops and Wakil, 1981a).

reacting as a bifunctional reagent, crosslinking the two subunits that comprise the enzymically active FAS dimer.

Finally, when dibromo[^{14}C]propanone was used as the crosslinking reagent, the crosslinked oligomers separated by SDS–PAGE contained over 85% of the protein-bound radioactivity. Hydrolysis of the ^{14}C-labeled oligomers with HCl after oxidation with performic acid yielded ^{14}C-labeled sulfones as outlined in Figure 18. The ^{14}C-labeled sulfones of the hydrolysate cochromatographed with standard *S*-carboxymethylcysteine and *S*-carboxymethylcysteamine sulfones (Figure 19) and after correction for destruction loss during HCl hydrolysis, were present in equal amounts. These results indicated that the dibromopropanone crosslinks the two synthetase subunits by reacting with a cysteine-SH of one subunit and the cysteamine-SH of the adjacent subunit.

The vicinal sulfhydryl groups resulting from the cysteine residue of one subunit being juxtapositioned with the pantetheine residue of the adjacent subunit predicted that the Ellman's reagent 5,5′-dithiobis(2-nitrobenzoic acid) (DTNB) would oxidize the two SH groups, forming the mixed disulfide (Figure 20). The studies (described in the following paragraphs) with this highly specific sulfhydryl reagent bear out this

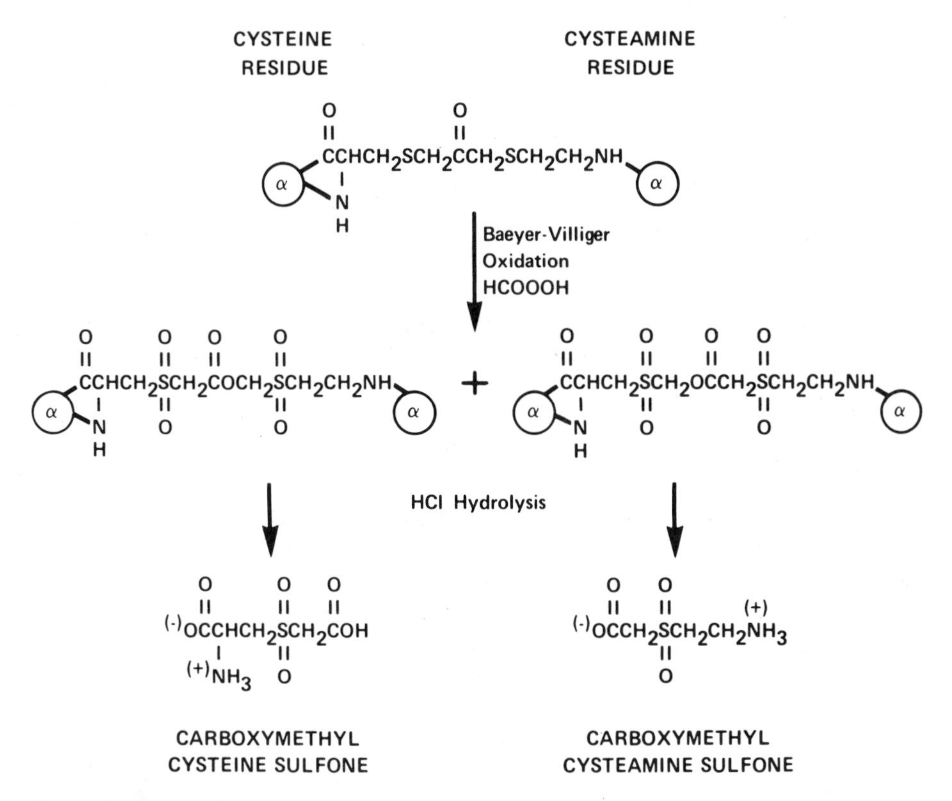

Figure 18. Outline of the Baeyer–Villiger oxidation and hydrolysis reactions which result in the formation of the carboxymethyl derivatives of cysteine and cysteamine sulfones.

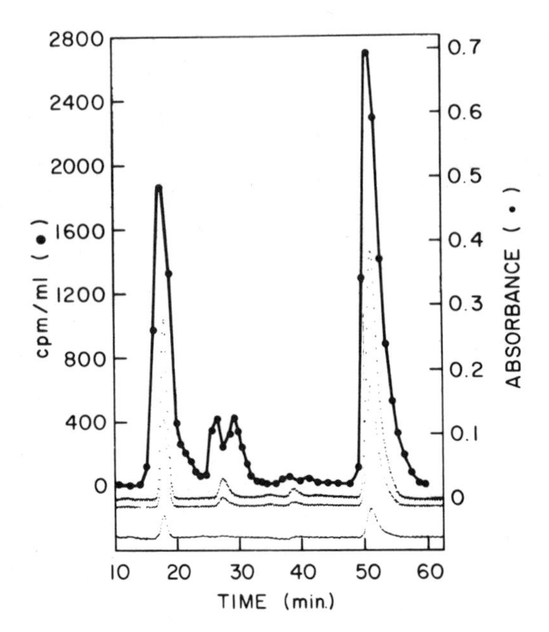

Figure 19. The identification of the residues derivatized by dibromo[^{14}C]propanone. The elution profile from an amino acid analyzer column of the standards (----) carboxymethylcysteine sulfone (18 min) and carboxymethylcysteamine sulfone (51 min) and the ^{14}C-labeled residues (●) derived from the treatment of the chicken liver FAS with 1,3-dibromo-2-[2-^{14}C]propanone followed by performic acid oxidation and HCl hydrolysis (Stoops and Wakil, 1981a,b).

prediction and further support the assignment of this novel arrangement of the β-ketoacyl synthetase site. DTNB (10^{-5} M) rapidly inhibits the synthetase; inhibition is prevented by acetyl-CoA (0.5 mM), but not malonyl-CoA (0.5 mM; Stoops and Wakil, 1982b). These results indicate that in the presence of malonyl-CoA, the free cysteine residue reacts with DTNB to form the mixed disulfide of 5-thio, 2-nitrobenozic acid, thus inhibiting the enzyme.

The reaction of the enzyme with DTNB resulted in the crosslinking of the two subunits as shown by SDS–PAGE. Our studies indicate that the crosslinking reaction involved the formation of a mixed disulfide from the cysteine and pantetheine thiols of the β-ketoacyl synthetase site. Acetyl- or malonyl-CoA prevented the crosslinking reaction, and when the enzyme was inhibited to 90% by iodoacetamide, the crosslinking reaction was again prevented. A survey of the component activities showed only the condensing activity was affected by DTNB. Binding analyses of the DTNB inhibition showed that inactivation of the synthetase required 2 moles of DTNB per mole of enzyme (Stoops and Wakil, 1982b). If the subunits are in fact crosslinked by a disulfide bridge, treatment of the crosslinked enzyme with reducing agent should produce on denaturing gels the same pattern as the active enzyme; such was found to be the case. These results are consistent with those of the dibromopropanone experiments, further demonstrating the proximity of SH groups of the two residues, and placing these SH groups within bonding distance, about 2 Å.

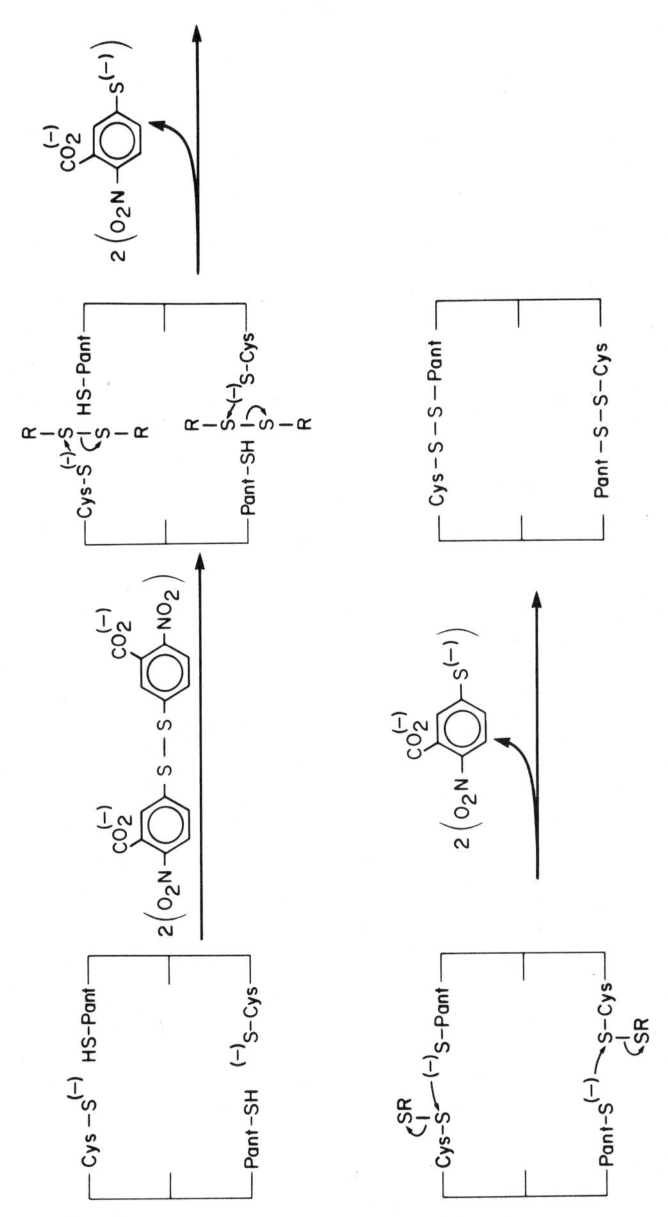

Figure 20. Proposed reaction of DTNB with the FAS homodimer. Cys-$S^{(-)}$ is cysteine-$S^{(-)}$ and Pant-SH is pantetheine-SH.

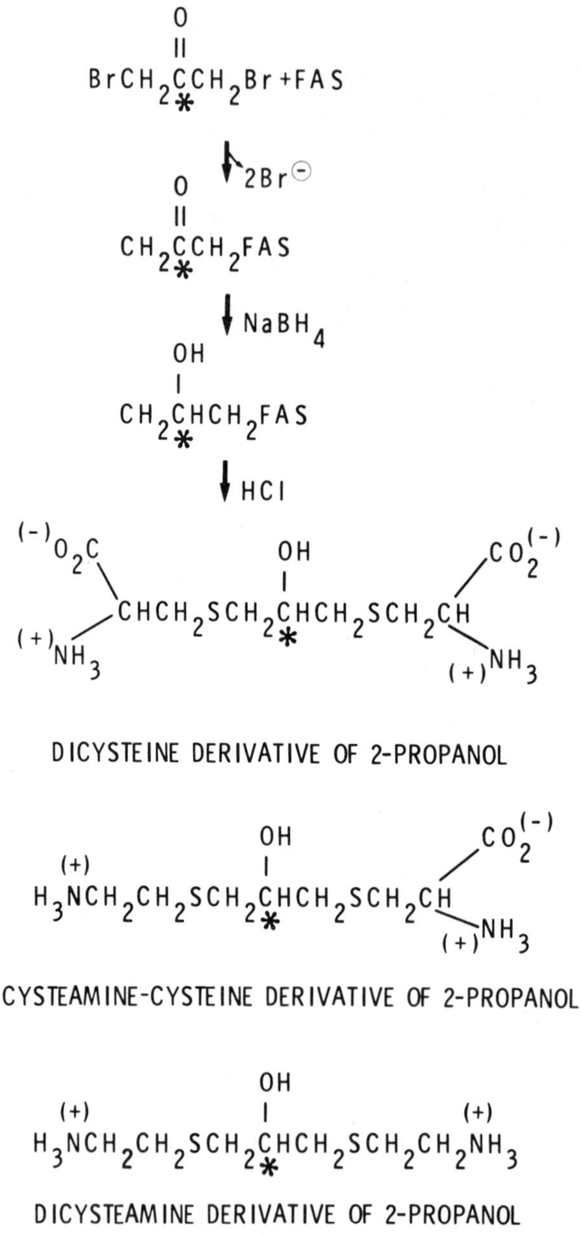

$$BrCH_2\overset{*}{C}(=O)CH_2Br + FAS \xrightarrow{} 2Br^{\ominus}$$

$$CH_2\overset{*}{C}(=O)CH_2FAS$$

$$\downarrow NaBH_4$$

$$CH_2\overset{*}{C}H(OH)CH_2FAS$$

$$\downarrow HCl$$

$$^{(-)}O_2C(^{(+)}NH_3)CHCH_2SCH_2\overset{*}{C}H(OH)CH_2SCH_2CH(CO_2^{(-)})(NH_3^{(+)})$$

DICYSTEINE DERIVATIVE OF 2-PROPANOL

$$^{(+)}H_3NCH_2CH_2SCH_2\overset{*}{C}H(OH)CH_2SCH_2CH(CO_2^{(-)})(NH_3^{(+)})$$

CYSTEAMINE-CYSTEINE DERIVATIVE OF 2-PROPANOL

$$^{(+)}H_3NCH_2CH_2SCH_2\overset{*}{C}H(OH)CH_2SCH_2CH_2NH_3^{(+)}$$

DICYSTEAMINE DERIVATIVE OF 2-PROPANOL

Figure 21. Reaction of dibromopropanone with FAS and the manner in which the crosslinked residues were released from the enzyme. The propanol derivatives of these residues are stable to HCl hydrolysis for 24 hr and are readily separated by the amino acid analyzer. Asterisks denote the position of the ^{14}C-label.

The requirement for proximity of pantetheine and cysteine residues on separate subunits for the β-ketoacyl synthetase reaction explains the loss of this partial activity and that of FAS upon partial dissociation of the homodimer. The requirement explained the cold inactivation of the enzyme. After incubation at 0°C for 12 hr, pigeon and chicken liver enzymes lost more than 90% of their activity (Muesing *et al.*, 1975; Stoops and Wakil, 1982a). Loss of activity was not due to dissociation of the dimer, as was shown by ultracentrifuge analyses and by the observation that protein concentration did not affect activity. Full activity was restored when the temperature was raised to 25°C over a 2-hr period. Full reversibility of the process was further demonstrated by the restoration of full activity after incubation of the enzyme for 5 min at 25°C in the presence of NADPH. The rate of reactivation was not increased by addition of acetyl-CoA, malonyl-CoA, or NADH. It appears, therefore, that proper positioning of the phosphate residue of NADPH is important in reactivation of the enzyme.

Dibromo[^{14}C]propanone was used to elucidate the molecular basis for cold inactivation of the enzyme (Stoops and Wakil, 1982a). At 0°C, it inhibited the synthetase but did not crosslink the subunits. When the ^{14}C-labeled enzyme was subjected to performic acid oxidation and HCl hydrolysis, 85% of the ^{14}C appeared in carboxymethylcysteine sulfone, less than 7% in carboxymethylcysteamine sulfone. When the active enzyme is so treated, radioactivity is equally distributed between the two products. The cold inactivation appears to be the result of a change in conformation which eliminates the proximal arrangment of the cysteine and pantetheine residues. In the inactive conformation, the bromoketo derivative of cysteine cannot react with the pantetheine residue. Upon reactivation, the vicinal conformation is restored, as is FAS activity and the crosslinking activity of the two residues by dibromopropanone.

The nature of the reaction of dibromopropanone with the enzyme at 0°C is not apparent from these studies. The reagent may crosslink two cysteine residues, or the bromoketo derivative of the cysteine residue may produce primarily the *S*-carboxymethylcysteine sulfone from performic acid oxidation. Either reaction results in a high yield of *S*-carboxymethylcysteine sulfone when the cold-inactivated enzyme reacts with dibromopropanone. However, if the propanone derivative of the enzyme is reduced with sodium borohydride, rather than being oxidized with performic acid, the crosslinked residues can be isolated in good yield (Figure 21; Stoops *et al.*, 1983). These procedures confirmed the proposed crosslinking of the residues by dibromopropanone and identified an additional cysteine residue in the β-ketoacyl synthetase site.

The cysteine-cysteamine derivative of the ketone was the predominant product of the crosslinking reaction when the enzyme was inactivated at room temperature (Table 2). However, about 17% of the dicysteine derivative was formed, and this derivative was the predominant product of the cold-inactivated enzyme (70%). The dicysteine derivative apparently was formed at the expense of formation of the cysteine-cysteamine derivative, indicating that there are competing reactions; the mixed derivative is favored at 25°C and the dicysteine at 0°C. The effect of temperature on these reactions is consistent with the isolation of the *S*-carboxymethylcysteine sulfone and *S*-carboxymethylcysteamine sulfone at the two temperatures described above. However, in this study, we were able to determine the predominant reaction of dibromo-

Table 2. Effect of Temperature on the Product Distribution of the Reaction of Dibromopropanone with Chicken Liver Fatty Acid Synthetase

Dibromopropanone derivative[a]	Radioactivity at 0°C		Radioactivity at 25°C	
	Recovered[b] (dpm)	% Yield	Recovered[b] (dpm)	% Yield
Dicysteine	10,500	70	2500	17
Cysteine-cysteamine	700	5	6000	40
Dicysteamine	0	0	0	0

[a] Isolated as the derivatives of 2-propanol. Dicysteine, cysteine-cysteamine and dicysteamine correspond to these constitutents crosslinked by the 2-propanone bridge.
[b] Represents radioactivity recovered of the 15,000 dpm loaded on the columns.

propanone at 0°C with the enzyme. It is therefore proposed that there are two cysteine residues on one subunit juxtapositioned with the pantetheine residue on the adjacent subunit. The arrangement of the catalytic group in the β-ketoacyl synthetase site that emerged from these studies are summarized in Figure 22.

The rate constant for the reaction of iodoacetamide with the cysteine residue of the β-ketoacyl synthetase site is 66 M^{-1} min^{-1} at pH 6.5 at 25°C (Stoops *et al.*, 1983). This rate constant is comparable to the reaction of iodoacetamide with ionized thiol groups (Stoops and Wakil, 1981b) and suggests that the p*K* of the cysteine residue of the β-ketoacyl synthetase is perturbed. Such a perturbation has been reported for the cysteine residue in the active site of papain, which resulted from the stabilization of cysteine S^- by the positive charge of the imidazolium ion of a nearby histidine residue (Hussain and Lowe, 1968). The vicinal arrangement of the histidine and cysteine residues was determined using dibromopropanone and was later confirmed by X-ray crystallographic studies of the enzyme (Dreuth *et al.*, 1968). However, our studies of the chicken liver FAS with dibromopropanone gave no evidence of such a vicinal arrangement. Thus, when the dibromo[^{14}C]propanone-treated enzyme was subjected to performic acid oxidation and HCl hydrolysis and 12,000 dpm of the acid hydrolysate was chromatographed on the amino acid analyzer column, no radioactivity eluted in the position of either N-1 or N-3 carboxymethylhistidine (Stoops *et al.*, 1983).

Since there was no evidence for the histidine–cysteine interaction, the bifunctional reagent *o*-phthalaldehyde was employed to determine whether a lysine residue provides

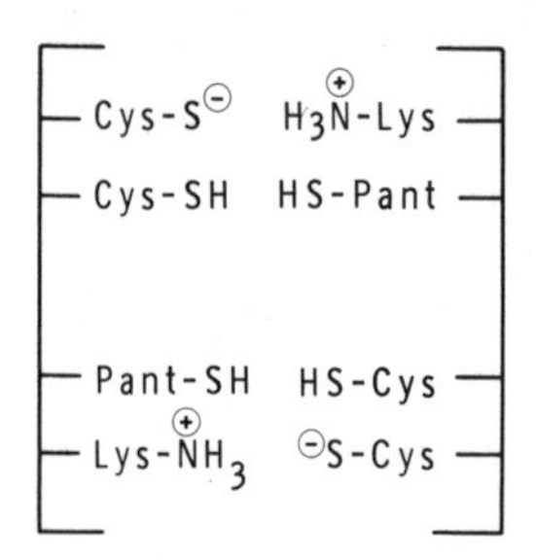

Figure 22. Arrangement of some of the residues in the β-ketoacyl synthetase site. The model depicts the homodimer form of the complex with two β-ketoacyl synthetase sites. Pantetheine-SH (Pant-SH), cysteine-SH (Cys-SH), and ε-amino group of a lysine residue (Lys-$\overset{\oplus}{N}H_3$).

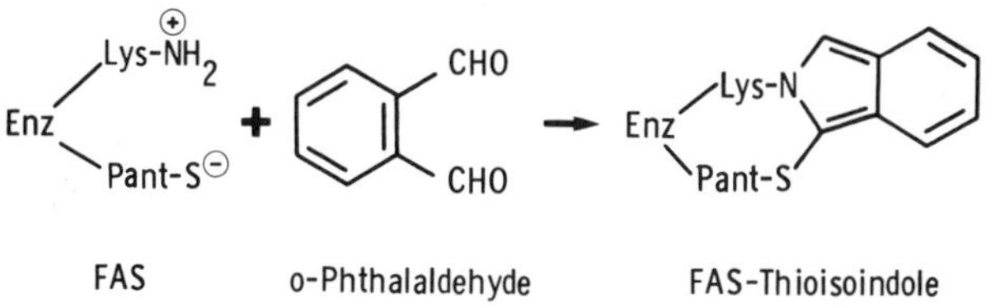

FAS o-Phthalaldehyde FAS-Thioisoindole

Figure 23. Crosslinking of the ε-amino group of lysine (Lys-NH_3^+) with the thiol group of pantetheine (Pant-S^-) of the β-ketoacyl synthetase site by *o*-phthalaldehyde to form the thioisoindole ring.

the positively charged group (Stoops *et al.*, 1983). *o*-Phthalaldehyde is known to form a highly fluorescent thioisoindole ring with primary amines and thiol groups (Simons and Johnson, 1978). Such a reaction with the enzyme would require the vicinal arrangement of an ε-amino group of a lysine residue and a thiol group. The possibility that the primary amino group resulted from a NH_2-terminal residue was remote, and in the case of the fatty acid synthetase, the NH_2-terminal residue is blocked (Stoops *et al.*, 1975). Four moles of *o*-phthalaldehyde per mole of enzyme inhibit the FAS primarily by inhibiting the β-ketoacyl synthetase activity. Malonyl-CoA decreased the rate of inhibition, indicating that the aldehyde reaction at the β-ketoacyl synthetase site involved thioisoindole formation, which in turn involved the pantetheine-SH and an ε-amino group of lysine, as depicted in Figure 23.

Excitation of the reaction product of the *o*-phthalaldehyde with the enzyme at 340 nm resulted in the emission maximum of 430 nm (Figure 24). These results

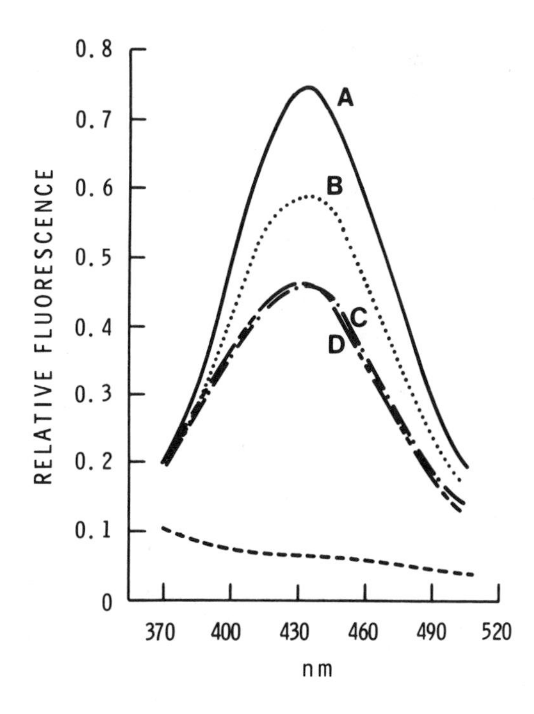

Figure 24. Evidence for the formation of the thioisoindole derivative of chicken liver FAS. Fluorescent spectra of chicken liver FAS (---) and after reaction with *o*-phthalaldehyde (A). Alternatively, the enzyme was inhibited by reacting the β-ketoacyl synthetase site with iodoacetamide (B), dibromopropanone (C), or DTNB (D) before treatment with *o*-phthalaldehyde.

indicate that *o*-phthalaldehyde reacts with the FAS to form a thioisoindole ring (Simons and Johnson, 1978). Fluorescence studies of *o*-phthalaldehyde-treated enzyme in which only the β-ketoacyl synthetase site was inhibited by iodoacetamide, dibromopropanone, and DTNB show a marked decrease in the fluorescence at 430 nm (Figure 24). As previously discussed, iodoacetamide reacts with a cysteine residue, whereas dibromopropanone and DTNB react with, and crosslink, a cysteine, and pantetheine residue (Stoops and Wakil, 1981a, 1982a,b). The enzyme preparations containing the crosslinked residues exhibit lower fluorescence when treated with *o*-phthalaldehyde than do preparations in which only the cysteine residue is blocked with the carboxyamidomethyl group. These results indicate that *o*-phthalaldehyde may have the choice of reacting with more than one thiol group in the β-ketoacyl synthetase site. The possibility that the crosslinked thiol and amino groups are on separate subunits was examined. When the enzyme was inhibited to 90% with *o*-phthalaldehyde and immediately subjected to SDS gel electrophoresis, there was no crosslinking of subunits, indicating that the crosslinked residues are on the same subunit, as depicted in Figure 22. These results support the partial activity studies and further support the theory that the aldehyde reacts primarily with thiol group(s) at the β-ketoacyl synthetase site, crosslinking them to an adjacent lysine residue.

A mechanistic role for the ε-amino group of lysine is depicted in Figure 25. In

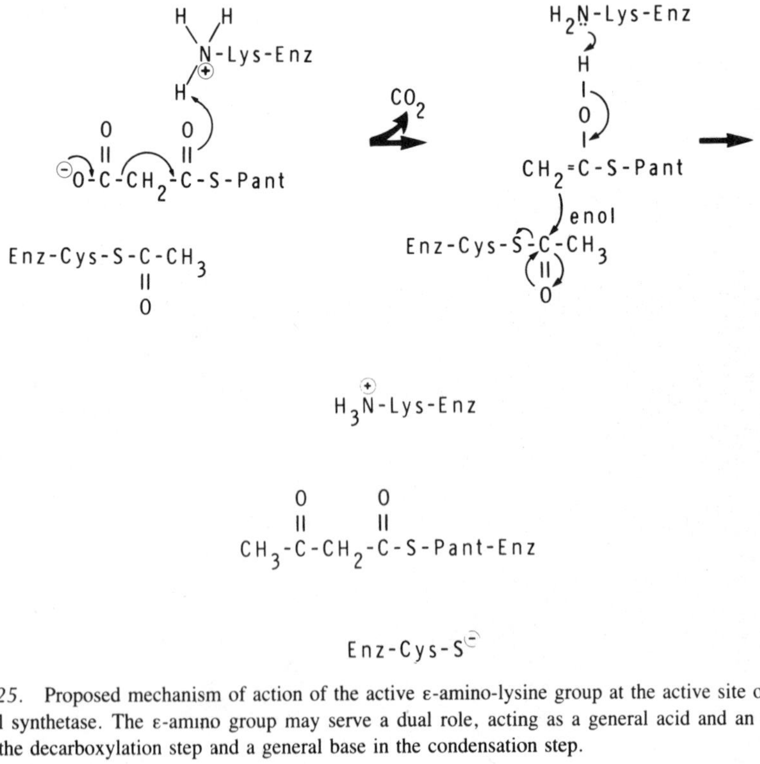

$H_2N^{\oplus}(H)$-Lys-Enz ; $^{\ominus}O-C(=O)-CH_2-C(=O)-S$-Pant ; Enz-Cys-S-C(=O)-$CH_3$

CO_2

$H_2\ddot{N}$-Lys-Enz ; H-O ; $CH_2=C-S$-Pant (enol) ; Enz-Cys-S-C(=O)-CH_3

$H_3N^{\oplus}$-Lys-Enz

$CH_3-C(=O)-CH_2-C(=O)-S$-Pant-Enz

Enz-Cys-$S^{\ominus}$

Figure 25. Proposed mechanism of action of the active ε-amino-lysine group at the active site of the β-ketoacyl synthetase. The ε-amino group may serve a dual role, acting as a general acid and an electron sink in the decarboxylation step and a general base in the condensation step.

this mechanism the lysine serves as a proton donor and an electron sink to stabilize the enolate anion formed during the decarboxylation step. This is the first evidence that a lysine residue can serve as an electron sink in an enzyme-catalyzed decarboxylation reaction that does not involve imine formation with the substrate (Warren *et al.*, 1966) and is analogous to metal-ion-promoted decarboxylation found in some acetoacetate decarboxylases (Steinberger and Westheimer, 1951). Furthermore, the free base may promote the condensation step (Figure 25) and thus play a dual role in this reaction. Another enzyme that may share this mechanistic feature with the FAS is 6-methylsalicyclic acid synthetase (Dimrother *et al.*, 1976). This enzyme catalyzes the condensation of acetyl- and malonyl-CoA, in the presence of NADPH, to form 6-methylsalicyclic acid. In addition to making use of the same substrates, neither enzyme utilizes divalent metal ions, thiamin pyrophosphate, or pyridoxal phosphate, which are involved in most enzyme-catalyzed decarboxylation reactions.

As previously mentioned in this section, yeast FAS is very sensitive to inhibition by iodoacetamide, which reacts specifically with the active cysteine-SH of the β-ketoacyl-synthetase-component activity. Loss of FAS activity relative to the amount of the carboxamidomethyl groups bound to the enzyme was studied employing iodo[^{14}C]acetamide (Stoops and Wakil, 1981b). Analyses of the data showed that 4–5 moles of carboxamidomethyl groups bound per mole of enzyme resulted in complete inhibition of the synthetase. This value is somewhat lower than the expected value for six β-ketoacyl synthetase sites present in the $\alpha_6\beta_6$ structure of yeast synthetase and may be explained on the basis that there are usually 4–5 prosthetic groups (4′-phosphopantetheine) per mole of synthetase (Stoops *et al.*, 1978b; Schweizer *et al.*, 1970). Thus, there are 1–2 sites of β-ketoacyl synthetase that are nonfunctional because they lack the prosthetic group and, therefore, are not manifest in the binding analyses, even though their cysteine-SH reacts with the iodoacetamide. The proposal that there are six condensing sites per complex is supported by the finding that six carboxamidomethyl residues bind to the enzyme after complete reaction with iodo[^{14}C]acetamide (Stoops and Wakil, 1981b). This result disagrees with the value of three reported earlier, and is inconsistent with the proposal of half-site reactivity (Oesterhelt *et al.*, 1977). Instead it supports the concept of full-site activity in the $\alpha_6\beta_6$ yeast structure and is consistent with the results obtained from studies with dibromopropanone discussed below.

1,3-Dibromo-2-propanone inhibits yeast synthetase by reacting rapidly ($t_{1/2}$ = 7 sec) with two juxtapositioned active sulfhydryl groups (Stoops and Wakil, 1980). SDS–PAGE of the dibromopropanone-inhibited synthetase shows the β-subunit to be intact and the α-subunit nearly absent, with a concomitant appearance of oligomers with an estimated molecular weight of 400,000 to 1.2×10^6 (Figure 26). These results indicate that the α-subunits are crosslinked by the bifunctional reagent. Since the active centers of the dibromopropanone are 5 Å apart, it was concluded that the α-subunits are so closely arranged that the reacting thiols of the adjacent α-subunits are within 5 Å of each other. Furthermore, since the plate-like structures in our model (Figure 12) are the only components arranged closely enough to satisfy this requirement, it was proposed that the α-subunits are the plates and the β-subunits the arches.

Yeast synthetase was treated with dibromopropanone and analyzed in the same manner as the chicken enzyme. From these studies, Stoops and Wakil (1981b) deduced

Figure 26. SDS–PAGE of yeast FAS treated with (a) iodoacetamide and then dibromopropanone; (b) with dibromopropanone; and (c) with acetyl-CoA or malonyl-CoA prior to treatment with dibromopropanone.

that the site of action of dibromopropanone is the active cysteine-SH of the β-ketoacyl synthetase of one α-subunit and the pantetheine-SH of the ACP moiety of an adjacent α-subunit. The active center of the β-ketoacyl synthetase would then consist of an acyl group-cysteine-SH complex of one α-subunit (plate) and a malonyl-pantetheine-SH complex of an adjacent subunit (Figure 27). These structural arrangements seem to be required for the coupling of the acyl and β-carbon of the malonyl group to yield CO_2 and the β-ketoacyl product. They may also explain the requirement for the $\alpha_6\beta_6$

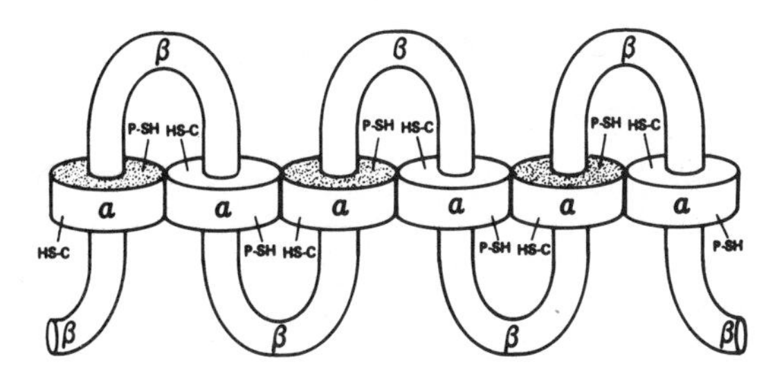

Figure 27. A linear drawing of the model shown in Figure 12 depicting the six sites of fatty acid synthesis and the complementary arrangement of the 4′-phosphopantetheine-SH (P-SH) and the active cysteine-SH (C-SH) at the β-ketoacyl synthetase centers in each of the sites.

form to be the only active form (Stoops *et al.*, 1978b) as determined by active-enzyme centrifugation. Such a structure has six active sites for β-ketoacyl synthetase all functioning simultaneously (Figure 27). This arrangement is a novel feature of our mechanism for the condensation reaction and for synthesis of fatty acids by yeast FAS.

C. *β-Ketoacyl and Enoyl Reductases*

β-Ketoacyl reductase catalyzes the transfer of a hydride ion from NADPH to the keto group of the β-keto ester derivative of the 4′-phosphopantetheine [Eq. (6)]. In the same way, enoyl reductase catalyzes the transfer of a hydride ion to the 2,3 double bond of the *trans*-unsaturated fatty acid ester derivative of pantetheine [Eq. (7)]. These reactions exhibit high specificity for NADPH over NADH in all fatty acid synthetases, regardless of source [Eq. (6)]. The intermediate alcohol is always the D-isomer.

Because of the requirement for NADPH in these reactions several arginine inhibitors have been used to probe the catalytic site of the two reductases in a way similar to their use with other NADPH-requiring enzymes. Poulose and Kolattukudy (1980b) found that phenylglyoxal and 2,3-butanedione inactivated goose FAS. These reagents also inactivated the β-ketoacyl reductase and the enoyl reductase activities. $NADP^+$ slowed the rate of inhibition of these component activities, as well as the FAS activity, but NADH did not. It was therefore concluded that the 2′-phosphate of NADPH is important in the binding of this substrate to the enzyme. Arginine was identified as the residue modified by phenylglyoxal since the arginine adduct was obtained.

Binding analyses showed that inactivation required 4 moles of phenylglyoxal bound per subunit. The first 2 moles bound per subunit did not affect the partial activities or those of FAS, showing that the reagent is not specific for arginine residues at the reductase sites. The requirement for 4 moles of phenylglyoxal per subunit led the authors to conclude that one residue binds at the β-keto reductase site and the other at the enoyl reductase site. The finding that there are two such sites per subunit further supports the proposal that the dimer consists of two subunits, each having a keto reductase and an enoyl reductase domain (Poulose and Kolattukudy, 1980b).

These studies indicate that an arginine residue participates in the binding of NADPH to β-keto and enoyl reductase sites. Studies of the inhibition of uropygial gland enzyme with pyridoxal 5′-phosphate indicate that a lysine residue in or near the enoyl reductase site may take part in the binding of NADPH. Only the enoyl reductase activity was inhibited by pyridoxal phosphate. NADPH prevented such inhibition. Labeled pyridoxal phosphate was found to form an N^6-pyridoxal lysine, thus identifying the reactive residue as lysine. It was shown also that pyridoxal forms a Schiff base with the ε-amino group of lysine by reduction of the derivatized enzyme with NaB_3H_4. Binding analyses showed modification of two lysine residues per subunit. One residue has no effect on activity and presumably is not active in catalysis. On the basis of these findings, Poulose and Kolattukudy proposed that each subunit contains an enoyl reductase domain and further proposed that the lysine residue reacts with the pyrophosphate bridge of the NADPH molecule and the arginine residue with the 2′-phosphate (Poulose and Kolattukudy, 1980a). This assignment seems tentative as there must be multiple sites of interaction.

Similar studies of yeast enzyme (Shoukry *et al.*, 1983) showed that pyridoxal phosphate inhibits β-keto reductase activity, though not that of enoyl reductase, the reverse of its action in animal synthetase. Structural differences in the two enzymes may account for the difference in reaction sites. Pyridoxal phosphate was found to react with a lysine residue and six β-keto reductase sites were found to be associated with the α-subunit of the $\alpha_6\beta_6$ complex. This assignment is consistent with previously presented physicochemical and genetic studies. It is proposed that the lysine residue is active in NADPH binding at the β-keto reductase site, since NADPH prevents inhibition of the enzyme.

It has been assumed that the acyl derivative of pantetheine is the substrate for component activities from the condensing enzyme to the thioesterase [Eqs. (5–9)]. However, Stram and Kumar (1979) proposed that the α,β-unsaturated acid is bound to the cysteine residue of the condensation site, rather than to the pantetheine residue, during reduction by NADPH. The basis for this proposal was the finding that sulfhydryl reagents *N*-ethylmaleimide, DTNB, and 4-chloro-7-nitrobenzo-2-oxa-1,3-diazole are potent inhibitors of the crotonyl-CoA reductase component of bovine mammary fatty acid synthetase. According to their proposal the crotonyl group is transferred to the enzyme before reduction. Competitive inhibition of the reductase reaction by acetyl- and butyryl-CoA supports this theory and further supports previous studies which indicated that butyryl, crotonyl, and acetoacetyl groups are readily transferred by certain fatty acid synthetases to the enzyme. Since malonyl-CoA was found to be a noncompetitive inhibitor of the crotonyl-CoA reductase activity, it was concluded that the malonyl and crotonyl groups are bound to different sites. Previous studies indicated that the malonyl group is bound to the pantetheine residue; thus, it was concluded that the crotonyl group is bound to the cysteine residue. Stram and Kumar (1979) interpret the inhibition of the reductase activity by the various sulfhydryl reagents, as well as the competitive inhibition observed with acetyl- and butyryl-CoA, as supportive of this argument. These results, however, do not give adequate support for the proposed binding site of the crotonyl group. The site of reaction of the sulfhydryl reagents was not identified, and we have shown that DTNB results in the reaction of not only the cysteine-SH but also the pantetheine-SH. The other sulfhydryl reagents used in this study may react with the pantetheine-SH as well. The fact that malonyl-CoA acts as a noncompetitive inhibitor of the reductase does not serve to identify the site of binding of the crotonyl group as the cysteine residue. The noncompetitive inhibition may, instead, result from interdomain interaction, as proposed by Poulose and Kolattukudy (1981).

As mentioned, $NADP^+$ protects both the β-ketoacyl and enoyl reductases from inhibition by the lysine- and arginine-directed inhibitors. It is not surprising then that $NADP^+$ is a competitive inhibitor of both activities as well as that of FAS activity. Surprisingly, the condensing activity was inhibited in a noncompetitive manner with a K_i value of 2.5 μM when malonyl-CoA and hexanoyl-CoA were used as substrates of the condensing reaction (Poulose and Kolattukudy, 1981). This value corresponds to the K_i value obtained for the reduction of crotonyl-CoA. If the enoyl reductase domain was inactivated by pyridoxal phosphate, $NADP^+$ no longer inhibits the condensing activity. From these results, it was proposed that the binding of $NADP^+$ to the enoyl reductase site causes a conformational change which inactivates the β-ketoacyl synthetase. It was further suggested that the functional interaction of the two domains may play a role in regulating fatty acid synthesis (Poulose and Kolattukudy, 1981b).

The β-keto reductase component of chicken liver FAS has been isolated after proteolysis of the complex (Wong *et al.*, 1983). This enzyme proved to be more difficult to separate from the complex than the thioesterase component (see Section III-E). Unlike the thioesterase, the β-keto reductase was found to have considerable physical interaction with the complex after proteolysis of the FAS with trypsin and subtilisin. As a result, the reductase could not be separated by ammonium sulfate fractionation or gel filtration. The activity could be separated by affinity chromatography, after which a 94,000-dalton protein was obtained with the β-keto reductase activity, but no other component activities. A kallikrein-subtilisin digest yielded an even smaller β-keto reductase of molecular weight 36,000 which had the 4′-phosphopantetheine group attached. Evidently, a 60,000-dalton portion of the 94,000-dalton protein was not required for enzyme activity. The 94,000-dalton enzyme had a specific activity 2–3 times higher than that of the component in the native FAS when *N*-acetyl-*S*-acetoacetylcysteamine was the substrate. The two- to threefold increase in specific activity is consistent with the separation of the enzyme from the complex without loss of activity, demonstrating that this component can function without the complex even though there is considerable physical interaction of this domain with the remainder of the complex. The fluorescent probe etheno-$NADP^+$ was found to bind to the β-keto reductase, and Scatchard analysis showed there was one binding site per 94,000-dalton protein. The binding of the photoaffinity analogue of $NADP^+$ to the β-keto reductase also yielded about 1 mole bound per mole of enzyme. Both of these binding studies indicate that the binding site for the enoyl reductase is not present on the 94,000-dalton enzyme and no such activity associated with it could be measured.

The kinetic parameters for the enzyme-catalyzed reduction of *N*-acetyl-*S*-acetoacetylcysteamine by NADPH were compared between the FAS and the isolated β-keto reductase (Table 3). The K_m and V_{max} values of both NADPH and *N*-acetyl-*S*-acetoacetylcysteamine were nearly the same. However, *S*-acetoacetyl-CoA was not a substrate for the isolated reductase even though it is reduced by the component activity in the complex.

The flavoenzyme enoyl reductase of yeast FAS has been characterized (Simons

Table 3. Comparison of Kinetic Constants for Synthetase and Purified β-Ketoacyl Reductase[a]

Substrates	Synthetase V_{max} (μmoles · min^{-1} mg^{-1})	Synthetase K_m (M)	Reductase V_{max} (μmoles · min^{-1} mg^{-1})	Reductase K_m (M)
NADPH	3.9	4.1×10^{-6}	3.5	4.8×10^{-6}
N-Acetyl-*S*-aceto-acetyl cysteamine	5.0	2.6×10^{-2}	5.8	6.6×10^{-2}
Acetoacetyl-SCoA	4.7	5×10^{-4}	Inactive	Inactive

[a] The above values were derived from Lineweaver–Burk type plots for fatty acid synthetase and purified β-ketoacyl reductase. The constants for NADPH (varied between concentrations of 2.3 and 300 μM) were derived with the substrate *N*-acetyl-*S*-acetoacetylcysteamine maintained at a concentration of 0.24 M. The constants for *N*-acetyl-*S*-acetoacetylcysteamine (varied between concentrations of 0.018 and 0.480 M) and acetoacetyl-*S*-CoA (varied between concentrations of 0.18 and 4.8 M) were derived with NADPH maintained at a concentration of 75 μM.

and Johnson, 1978). The FMN mediates the reduction of the double-bond of the fatty acid utilizing NADPH. Removal of FMN from the enzyme results in the loss of only the enoyl reductase activity, demonstrating that the keto reductase does not utilize this constituent (Warren *et al.*, 1966). Six moles of FMN is bound per mole of enzyme and the binding studies indicate that the binding sites are independent of each other. This result is consistent with the proposed $\alpha_6\beta_6$ structure of the complex. As previously mentioned, the enoyl reductase domain is associated with the β-subunit. Chemical reduction titrations with sodium dithionite, NADPH, and NADH showed that each flavin required one reductant molecule ($2e^-$) for reduction. Attempts to detect the flavin-free radical were not successful, and it appears that the reductase is characterized as a transhydrogenase, while the hydride ion from NADPH is shuttled through FMN to the double bond of the fatty acid. The redox potential for FMN in the yeast enzyme is little perturbed ($E\eta = -0.193$V), indicating that there is little apoenzyme–flavin interaction with the isoalloxazine ring.

D. β-Hydroxyacyl Dehydratase

No studies have been reported to date of the mechanism for removal of H_2O from the β-hydroxy fatty acid to form the α,β-*trans* double bond in the fatty acids. The activity of this component has evidently been the least studied of the seven.

E. Palmitoyl Thioesterase

The thioesterase activity which is the terminal reaction of fatty acid synthesis was the first to be separated during limited proteolysis of the multifunctional enzyme. Later, an acetyl transacylase component of yeast enzyme and β-keto reductase components of chicken liver enzyme were separated. From the isolation of thioesterase activity of FAS from different sources, it is known that the remaining core protein cannot synthesize fatty acid. This finding was surprising since all the component activities are present, as well as the clipped thioesterase. The covalent attachment of the thioesterase component to the multifunctional enzyme is essential for the release of fatty acids from the complex. It is significant also that some, if not all, of the component enzymes exist in discrete and separate domains in the complex.

The thioesterase component was isolated from rat mammary gland after trypsin treatment (Smith *et al.*, 1976). The other component activities were not affected by the trypsin treatment and subsequent separation of the thioesterase; the same was true for isolation of thioesterase of goose uropygial gland (Bedord *et al.*, 1978) and from chicken liver (Crisp and Wakil, 1982). Trypsin cleavage released thioesterase from the mammary and uropygial gland; subtilisin was used to release the enzyme from the enzyme of chicken liver. Molecular weights of these enzymes were nearly the same (33,000–35,000). The fact that FAS from different sources yield thioesterases of similar size after cleavage by proteases of different specificities indicates that the region of the polypeptide chain which separates the thioesterase from the core complex is highly susceptible to proteolysis. Thioesterases of both goose (Bedord *et al.*, 1978) and chicken (Mattick *et al.*, 1983a) are inhibited by phenylmethanesulfonyl fluoride and diisopropylphosphofluoridate. Complete inhibition requires 2 moles of diisopropyl-phosphate to be bound to the native complex, indicating that both subunits contain the thioesterase domain. It was proposed that the inhibiting agents react with an active-site serine residue, and such was shown to be the case. The active-site peptide was isolated from uropygial gland enzyme containing the diisopropyl derivative of serine and the sequence Ser-Phe-Gly-Ala-Cys-Val-Phe found. The sequence is similar to that of active-site regions of plasmin, trypsin, and carboxylesterase (Poulose *et al.*, 1981).

Enzymes of uropygial gland and chicken liver are inhibited by the sulfhydryl reagents *p*-chloromercuribenzoate and *N*-ethylmaleimide (Crisp and Wakil, 1982; Bedord *et al.*, 1978). The substrate decanoyl-CoA does not protect chicken liver enzyme from *N*-ethylmaleimide inhibition; therefore, SH group(s) may not be necessary for the catalytic process (Crisp and Wakil, 1982). Both enzymes show marked specificity for palmitoyl- and stearoyl-CoA, with lower reactivity for myristoyl-CoA and no activity for shorter-chain fatty acid esters (Crisp and Wakil, 1982; Bedord *et al.*, 1978). This marked discrimination of the thioesterases suggest that it may be the component activity most important in determining the chain length of the fatty acid produced.

Yeast synthetase has no thioesterase activity, but contains an acyl transferase activity. This component enzyme requires CoA-SH and produces palmitoyl-CoA. There is evidence that this activity shares the same active site as malonyl transacylase (see Section III-A).

IV. MECHANISM OF ACTION OF FATTY ACID SYNTHETASES

The observations of structure and function of the synthetase subunits call for reevaluation of the mechanism of action of animal and yeast FAS. The proposed models for the animal and yeast synthetases have the same basic pattern and are consistent with earlier data, including data from electron microscopy.

A. Animal

Since the subunits of the animal FAS are identical (with each subunit containing the same catalytic domains, including the active cysteine-SH of the β-ketoacyl synthetase and 4′-phosphopantetheine-SH of the ACP domain), it is proposed from the crosslinking studies with dibromopropanone or the Ellman's reagent that in the dimer state the two subunits are arranged in head-to-tail fashion (Stoops and Wakil, 1981b) as shown in Figure 8. The head-to-tail arrangement of the two subunits of the FAS of animal tissues predicts the presence of two centers of β-ketoacyl synthetase (Stoops and Wakil, 1981b) and, therefore, two centers of palmitate synthesis; this is consistent with the stoichiometry of binding of dibromopropanone and DTNB. Studies using the core complex of the 230,000-dalton peptide allowed the estimation of the stoichiometry of NADPH oxidation and fatty acids synthesized relative to the pantetheine content of the 230,000-dalton core dimer. The results show that in the absence of thioesterase the core enzyme continues the chain elongation and reduction processes until fatty acids of C_{20} and C_{22} limit the synthesis (Singh *et al.*, 1984). As shown in Table 4, little or no palmitate or stearate, usually synthesized, is formed under these conditions, indicating that these fatty acyl groups are still attached to the pantetheine-SH and are further elongated to the C_{20} and C_{22} acids (Figure 28). The chain-terminating process in the native synthetase, therefore, is dictated by the thioesterase, which has the highest activity for palmitoyl and stearoyl thioesters. In any case, quantitation of the fatty acids bound to the enzyme and the stoichiometry of the NADPH oxidation show that 1 mole of long-chain fatty acids synthesized per mole of pantetheine associated with the core dimer; therefore, the two centers of palmitate synthesis are active simultaneously.

Libertini and Smith (1979) studied fatty acid synthesis with rat mammary gland synthetase modified by removal of thioesterase or by inhibition of thioesterase with phenylmethanesulfonyl fluoride. They found that the modified enzyme synthesized

Table 4. Stoichiometry of Fatty Acids Synthesized de Novo by Chymotrypsin Proteolyzed Fatty Acid Synthetase

Fatty acid	Mol/mol enzyme dimer[a]	Mol/mol 4′-phosphopantetheine[b]
$C_{14:0}$	0.01	0.01
$C_{16:0}$	0.02	0.02
$C_{18:0}$	0.06	0.05
$C_{20:0}$	0.55	0.42
$C_{22:0}$	0.51	0.39
$C_{24:0}$	0.10	0.08
Total	1.25	0.97

[a] Calculated by normalizing amount of each fatty acid recovered relative to the added internal standard fatty acids $C_{17:0}$ and $C_{21:0}$.
[b] Calculated using the separately determined value of 1.3 moles taurine/mole enzyme dimer.

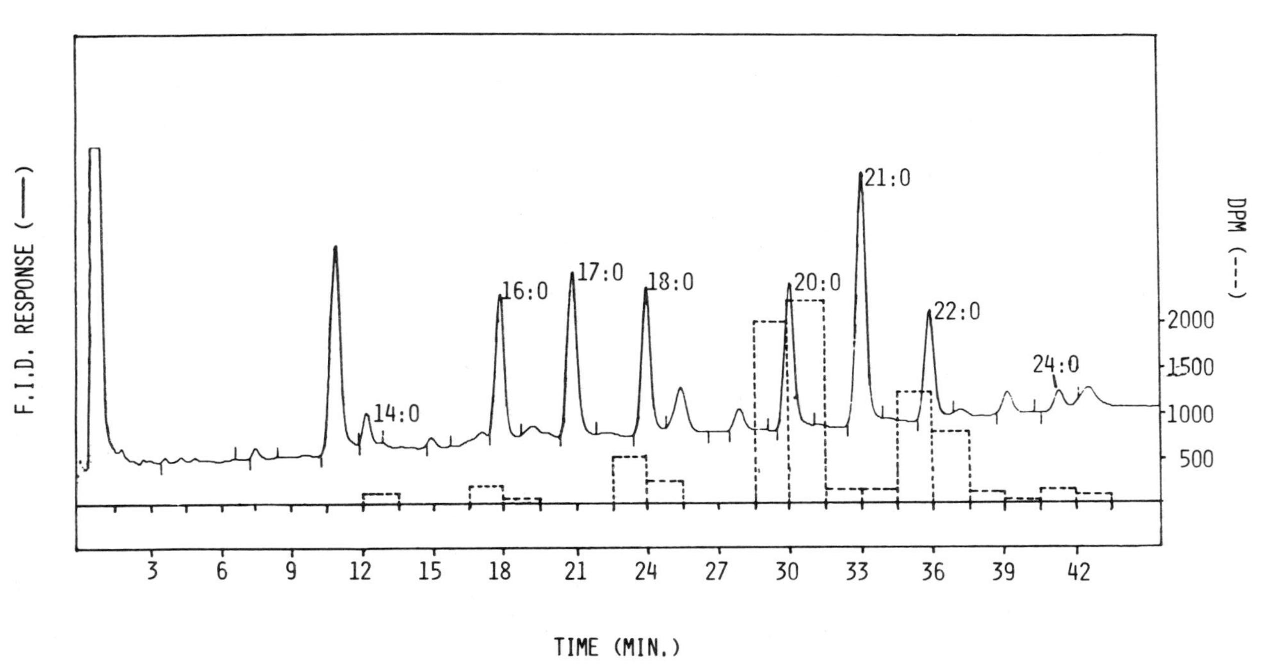

Figure 28. Distribution of radioactivity from [1-^{14}C]acetyl-CoA in fatty acids synthesized by thioesterase deficient fatty acid synthetase. Chymotrypsinized fatty acid synthetase, 4.5 ml of 0.99 mg/ml, was mixed with an equal volume of a solution containing NADPH (300 μM), malonyl-CoA (400 μM), and [1-^{14}C]acetyl-CoA (100 μM, 2.95 Ci/mol), by driving through the stopped-flow apparatus. After adding $C_{17:0}$ and $C_{21:0}$ internal standard fatty acids, the total fatty acids were extracted and subjected to gas liquid chromatography. The plotted radioactivity represents total DPM in each indicated fraction.

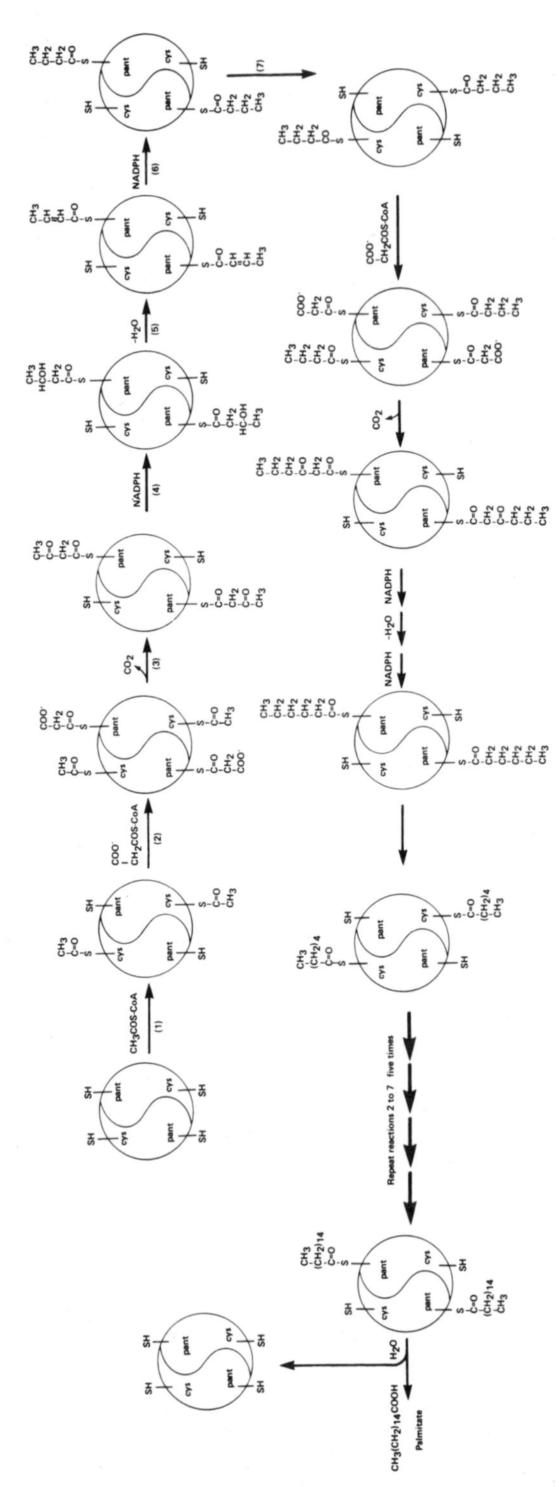

Figure 29. Proposed mechanism for palmitate synthesis. The circles represent the multifunctional subunits of the FAS in its homodimer form. The cys-SH represents the active cysteine thiol of the β-ketoacyl synthetase site and pant-SH represents the pantetheine thiol of the ACP site. Other catalytic domains are not shown and are presumed to be present in both subunits [see Eqs. (3)–(8) and Figure 8].

C_{16}–C_{22} fatty acid covalently bound to the protein and that a single enzyme-bound long-chain acyl thioester was formed by each molecule of modified synthetase dimer. Since they did not determine the pantetheine content of their preparation, it is not possible to answer the question of half-site or full-site reactivity from their studies.

Our studies suggest that two centers may function independently of each other and engage catalytic domains on the two subunits. In this arrangement (Figure 8), each center has the entire complement of enzymes. Therefore, the following mechanism for palmitate synthesis (Figure 29) was proposed (Singh *et al.*, 1984; Wakil *et al.*, 1981, 1983). The active FAS in the dimer form interacts with the substrates acetyl-CoA and malonyl-CoA. The active cysteine-SH of the condensing site is charged with the acetyl group and the cysteamine-SH of the acyl carrier site is charged with the malonyl group via their respective transacylases (steps 1 and 2, Figure 29). The acetyl group of one subunit is coupled to the β-carbon of the malonyl group of the second subunit with a simultaneous release of CO_2 and the formation of acetoacetyl product. The cysteine-SH of the condensing enzyme is reset in the free thiol form. The acetoacetyl-*S*-pantetheine derivative is then processed as outlined in Figure 29, ultimately yielding palmitic acid. The essence of this mechanism is the involvement of the two subunits in the condensation reaction, in which the acyl group "see-saws" between the cysteine-SH and cysteamine-SH of the two subunits with each cycle adding C_2 units with the occurrence of two centers for the synthesis of palmitate within each active synthetase dimer, where each has its own complement of enzymes and perhaps functions independently. The multifunctional nature of the subunit and its organization and structural arrangement into the dimer form produce a highly efficient enzyme capable of carrying out sequentially and repetitively a total of 37 reactions in the synthesis of a molecule of palmitate from acetyl-CoA and malonyl-CoA.

B. Yeast

As previously stated, the $\alpha_6\beta_6$ structure is the oligomer active in palmitate synthesis. In this structure, a palmitate-synthesizing center consists of two complementary halves of two α-subunits and the arch β-subunit (Figure 27). In an $\alpha_6\beta_6$ structure, therefore, there are six sites for fatty acid synthesis, all of which function simultaneously (Stoops and Wakil, 1981b). Studies with dibromopropanone led Stoops and Wakil to propose that the active center of the β-ketoacyl synthetase consists of an acyl group attached to the cysteine-SH of one α-subunit (plate) and a malonyl group attached to the pantetheine-SH of an adjacent α-subunit (Figure 27). This arrangement appears to be necessary for the condensation reaction, yielding the β-ketoacyl-*S*-pantetheine derivative, which is analogous to the scheme proposed for the animal synthetase (Figure 29). The β-ketoacyl group is then reduced by NADPH to the β-hydroxy homologue at the β-keto reductase site of an α-subunit. Dehydration of the β-hydroxacyl derivative by the dehydratase of the arch β-subunit yields the α-β unsaturated acyl homologue, which is then reduced by NADPH through the FMN of the enoyl reductase of the β-subunit to the saturated acyl derivative. The latter is then transferred from the pantetheine-SH to the active cysteine-SH of the β-ketoacyl synthetase of the α-subunit, where the acyl group was bound prior to condensation. The free pantetheine-SH is then reloaded with another malonyl group and the sequence of reactions commences again. The sequential reactions are repeated until the acyl chain is elongated to 16 or 18 carbons, which are then transferred to CoA-SH by the palmitoyl transferase located on the arch β-subunit.

Condensation occurs by engaging the acyl-*S*-cysteine of one α-subunit and the malonyl-*S*-pantetheine of the second α-subunit; chain elongation occurs by transferring the acyl group back and forth between the pantetheine-SH and the cysteine-SH of the two complementary halves of the α-subunits. In essence, this mechanism is analogous to the one proposed for animal synthetase (Figure 29). This arrangement is a novel feature of our mechanism for the synthesis of fatty acids by the yeast FAS.

ACKNOWLEDGMENTS

Investigations reported from our laboratory were supported in part by grants from the National Institutes of Health (GM 19091), National Science Foundation (PCM 7-00969), and the Robert A. Welch Foundation (Q-587).

REFERENCES

Abdinejad, A., Fisher, A. M., and Kumar, S., 1981, Production and utilization of butyryl-CoA by fatty acid synthetase from mammalian tissues, *Arch. Biochem. Biophys.* **208:**135.

Agradi, E., Libertini, L., and Smith, S., 1976, Specific modification of fatty acid synthetase from lactating rat mammary gland by chymotrypsin and trypsin, *Biochem. Biophys. Res. Commun.* **68:**894.

Aprahamian, S., Arslanian, M. J., and Stoops, J. K., 1979, Effect of estrogen on fatty acid synthetase in the chicken oviduct and liver, *Lipids* **14:**1015.

Aprahamian, S. A., Arslanian, M. J., and Wakil, S. J., 1982, Comparative studies on the kinetic parameters and product analyses of chicken and rat liver and yeast fatty acid synthetase, *Comp. Biochem. Physiol.* **71B:**577.

Arslanian, M. J., Stoops, J. K., Oh, Y. H., and Wakil, S. J., 1976, On the 4′-phosphopantetheine content of chicken and rat liver fatty acid synthetase, *J. Biol. Chem.* **251:**3194.

Bedord, C. J., Kolattukudy, P. E., and Rogers, L., 1978, Isolation and characterization of a tryptic fragment containing the thioesterase segment of fatty acid synthetase from the uropygial gland of goose, *Arch. Biochem. Biophys.* **186:**139.

Bratcher, S. C., and Hsu, R. Y., 1975, Separation of active enzyme components from the fatty acid synthetase of chicken liver, *Biochim. Biophys. Acta* **410:**229.

Bressler, R., and Wakil, S. J., 1962, Studies on the mechanisms of fatty acid synthesis XI. The products of the reaction and role of sulfhydryl groups in the synthesis of fatty acids, *J. Biol. Chem.* **237:**1441.

Brindley, D. N., Matsumura, S., and Bloch, K., 1969, *Mycobacterium phlei* fatty acid synthetase—A bacterial multienzyme complex, *Nature (London)* **224:**666.

Brooks, J. L., and Stumpf, P. K., 1966, Fat metabolism in higher plants, *Arch. Biochim. Biophys.* **116:**108.

Buckner, J. S., and Kolattukudy, P. E., 1976, One-step purification and properties of a two-peptide fatty acid synthetase from the uropygial gland of the goose, *Biochemistry* **15:**1948.

Burton, D. W., Haavik, A. G., and Porter, J. W., 1968, Comparative studies of the rat and pigeon liver fatty acid synthetases, *Arch. Biochem. Biophys.* **126:**141.

Butterworth, P. H. W., and Bloch, K., 1970, Comparative aspects of fatty acid synthesis in *Bacillus subtilis* and *Escherichia coli, Eur. J. Biochem.* **12:**496.

Butterworth, P. H. W., Yang, P. C., Bock, R. M., and Porter, J. W., 1967, The partial dissociation and the reassociation of the pigeon liver fatty acid synthetase complex, *J. Biol. Chem.* **242:**3508.

Chalmers, J. H., Jr., and Hitzeman, R. A., 1980, Selection of clones expressing the fatty acid synthetase genes of yeast, *Fed. Proc.* **39:**1829.

Crisp, D., 1976, Cleavage of fatty acid synthetase by proteolytic enzymes, *Fed. Proc.* **35:**1500.

Crisp, D., and Wakil, S. J., 1982, Chicken liver fatty acid synthetase—proteolysis of synthetase by subtilisin and isolation and properties of palmitoyl thioesterase, *J. Protein Chem.* **1:**241.

D'Agnolo, G., Rosenfeld, I. S., and Vagelos, P. R., 1975, β-Ketoacyl-acyl carrier protein synthetase, *J. Biol. Chem.* **250:**5283.

Davidson, J. N., Rumsby, P. C., and Tamaren, J., 1981, Organization of a multifunctional protein in pyrimidine biosynthesis, *J. Biol. Chem.* **256:**5220.

Delo, J., Ernst-Fonberg, M. L., and Bloch, K., 1971, Fatty acid synthetases from *Euglena gracilis, Arch. Biochem. Biophys.* **143:**384.

Dimrother, P., Ringelmann, E., and Lynen, F., 1976, 6-Methylsalicylic acid synthetase from *Penicillium patulum, Eur. J. Biochem.* **68:**591.

Dodds, P. F., Guzman, M. G. F., Chalberg, S. C., Anderson, G. J., and Kumar, S., 1980, Acetoacetyl-CoA reductase activity of lactating bovine mammary fatty acid synthase, *J. Biol. Chem.* **256:**6282.

Dreuth, J., Jansonius, J. N., Koekoek, R., Swen, H. M., and Wothers, B. G., 1968, Structure of papain, *Nature (London)* **218:**929.

Dutler, H., Coon, M. J., Kull, A., Vogel, H., Waldvogel, G., and Prelog, V., 1971, Fatty acid synthetase from pig liver, *Eur. J. Biochem.* **22:**203.

Elovson, J., 1975, Purification and properties of the fatty acid synthetase complex from *Neurospora crassa* and the nature of the fas-mutation, *J. Bacteriol.* **124:**524.

Engeser, H., Hubner, K., Straub, J., and Lynen, F., 1979a, Identity of malonyl and palmitoyl transferase of fatty acid synthetase from yeast. 1. Functional interrelationships between the acyl transferases, *Eur. J. Biochem.* **101:**407.

Engeser, H., Hubner, K., Straub, J., and Lynen, F., 1979b, Identity of malonyl and palmitoyl transferase of fatty acid synthetase from yeast. 2. A comparison of active-site peptides, *Eur. J. Biochem.* **101:**413.

Ernst-Fonberg, M. L., and Bloch, K., 1971, A chloroplast-associated fatty acid synthetase system in *Euglena, Arch. Biochem. Biophys.* **143:**392.

Flick, P. K., Chen, J., Alberts, A. W., and Vagelos, P. R., 1978, Translation of rat liver fatty acid synthetase mRNA in a cell-free system derived from wheat germ, *Proc. Natl. Acad. Sci. USA* **75:**730.

Fox, J. L., and Lynen, F., 1980, Characterization of the flavoenzyme enoyl reductase of fatty acid synthetase from yeast, *Eur. J. Biochem.* **109:**417.

Goldberg, I., and Bloch, K., 1972, Fatty acid synthetases in *Euglena gracilis, J. Biol. Chem.* **247:**7349.

Goldman, P. W., Alberts, A. W., and Vagelos, P. R., 1963, The condensation reaction of fatty acid biosynthesis, *J. Biol. Chem.* **238:**1255.

Grunnet, I., and Knudsen, J., 1978, Molecular weight and subunit size of rabbit mammary-gland fatty acid synthetase, *Biochem. J.* **173:**929.

Guy, P., Law, S., and Hardie, G., 1978, Mammalian fatty acid synthetase: Evidence of subunit identity and specific removal of the thioesterase component using elastase digestion, *FEBS Lett.* **94:**33.

Holtermuller, K. H., Ringelmann, E., and Lynen, F., 1970, Reinigung und charakterislerung deter Fett saure—Synthetase aus *Penicillium patulum, Hoppe-Seyler's Z. Physiol. Chem.* **351:**1411.

Hsu, R. Y., and Yun, S., 1970, Stabilization and physicochemical properties of the fatty acid synthetase of chicken liver, *Biochemistry* **9:**239.

Hussain, S. S., and Lowe, G., 1968, Evidence for histidine in the active site of papain, *Biochim. J.* **108:**855.

Jacob, E. J., Butterworth, P. H. W., and Porter, J. W., 1968, Studies on the substrate binding sites of the pigeon liver fatty acid synthetase, *Arch. Biochem. Biophys.* **124:**392.

Kim, I., Unkefer, C. J., and Deal, W. C., Jr., 1977, Pig liver fatty acid synthetase: Purification and physicochemical properties, *Arch. Biochem. Biophys.* **178:**475.

Kinsella, J. E., Bruns, D., and Infante, J. P., 1975, Fatty acid synthetase of bovine mammary: Properties and products, *Lipids* **10:**227.

Kirschner, K., and Bisswanger, H., 1976, Multifunctional proteins, *Annu. Rev. Biochem.* **45:**143.

Knobling, A., Schiffman, D., Sickinger, H. S., and Schweizer, E., 1975, Malonyl and palmityl transferase-less mutants of the yeast fatty-acid-synthetase complex, *Eur. J. Biochem.* **56:**359.

Knoche, H. W., and Koths, K. E., 1973, Characterization of a fatty acid synthetase from *Cornyebacterium diphtheriae, J. Biol. Chem.* **248:**3517.

Kresze, G., Steber, L., Oesterhelt, D., and Lynen, F., 1977, Reaction of yeast fatty acid synthetase with iodoacetamide. 2. Identification of the amino acid residues reacting with iodoacetamide and primary structure of a peptide containing the peripheral sulfhydryl group, *Eur. J. Biochem.* **79:**181.

Kumar, S., Phillips, G. T., and Porter, J. W., 1972, Comparative biochemistry of fatty acid synthesizing enzyme systems: A review, *Int. J. Biochem.* **3:**15.

Kuziora, M. A., Chalmers, J. H., Jr., Hitzeman, R. A., Douglas, M. G., and Wakil, S. J., 1984, Molecular cloning of fatty acid synthetase genes from *Saccharomyces cerevisiae, J. Biol. Chem.,* in press.

Lennarz, W. J., Light, R. J., and Bloch, K., 1962, A fatty acid synthetase from *E. coli, Proc. Natl. Acad. Sci. USA* **48:**840.

Libertini, L. J., and Smith, S., 1979, Synthesis of long chain acyl-enzyme thioesters by modified fatty acid synthetases and their hydrolysis by a mammary gland thioesterase, *Arch. Biochem. Biophys.* **192:**47.

Linn, T. C., and Srere, P. A., 1980, Coenzyme A requirement for the termination reaction of rat liver fatty acid synthetase, *J. Biol. Chem.* **255:**10676.

Lornitzo, F. A., Qureshi, A. A., and Porter, J. W., 1974, Separation of the half-molecular weight nonidentical subunits of pigeon liver fatty acid synthetase by affinity chromatography, *J. Biol. Chem.* **249:** 1654.

Lynen, F., 1961, Biosynthesis of saturated fatty acids, *Fed. Proc.* **20:**941.

Lynen, F., 1964, Coordination of metabolic processes by multi-enzyme complexes, in: *New Perspectives in Biology,* Vol. 4 (M. Sela, ed.), Elsevier, Amsterdam, pp. 132–146.

Lynen, F., 1967a, The role of biotin-dependent carboxylations in biosynthetic reactions, *Biochem. J.* **102:**381.

Lynen, F., 1967b, Multienzyme complex of fatty acid synthetase, in: *Organizational Biosynthesis* (H. J. Vogel, J. O. Lampen, and V. Bryson, eds.), Academic Press, New York and London, pp. 243–266.

Lynen, F., 1980, On the structure of fatty acid synthetase of yeast, *Eur. J. Biochem.* **112:**431.

Maitra, S. K., and Kumar, S., 1974, Physicochemical properties of bovine mammary fatty acid synthetase, *J. Biol. Chem.* **249:**118.

Martin, D. B., Horning, M. G., and Vagelos, P. R., 1961, Fatty acid synthesis in adipose tissue, *J. Biol. Chem.* **236:**663.

Mattick, J. S., Zehner, Z. E., Calabro, M. A., and Wakil, S. J., 1981, The isolation and characterization of fatty-acid-synthetase mRNA from rat mammary gland, *Eur. J. Biochem.* **114:**643.

Mattick, J. S., Nickless, J., Mizugaki, M., Yang, C. Y., Uchiyama, S., and Wakil, S. J., 1984a, The architecture of the animal fatty acid synthetase. II. Separation of the core and thioesterase functions and determination of the N-C orientation of the subunit, *J. Biol. Chem.,* in press.

Mattick, J. S., Tsukamoto, Y., Nickless, J., and Wakil, S. J., 1984b, The architecture of the animal fatty acid synthetase. I. Proteolytic dissection and peptide mapping, *J. Biol. Chem.,* in press.

Morris, S. M., Jr., Nilson, J. H., Jenik, R. A., Winberry, L. K., McDevitt, M. A., and Goodridge, A. G., 1982, Molecular cloning of gene sequences for avian fatty acid synthase and evidence for nutritional regulation of fatty acid synthase mRNA concentration, *J. Biol. Chem.* **257:**3225.

Muesing, R. A., Lornitzo, F. A., Kumar, S., and Porter, J. W., 1975, Factors affecting the reassociation and reactivation of the half-molecular weight nonidentical subunits of pigeon liver fatty acid synthetase *J. Biol. Chem.* **250:**1814.

Nepokroeff, C. M., and Porter, J. W., 1978, Translation and characterization of the fatty acid synthetase messenger RNA, *J. Biol. Chem.* **253:**2279.

Oesterhelt, D., Bauer, H., Kresze, G., Steber, L., and Lynen, F., 1977, Reaction of yeast fatty acid synthetase with iodoacetamide. 1. Kinetics of inactivation and extent of carboxamidomethylation, *Eur. J. Biochem.* **79:**173.

Overath, P., and Stumpf, P. K., 1964, Fat metabolism in higher plants, *J. Biol. Chem.* **239:**4103.

Pabo, C. O., Sauer, R. T., Sturtevant, J. M., and Ptashne, M., 1979, The λ repressor contains two domains, *Proc. Natl. Acad. Sci. USA* **76:**1608.

Porter, R. R., 1959, The hydrolysis of rabbit γ-globulin and antibodies with crystalline papain, *Biochem. J.* **73:**119.

Poulose, A. J., and Kolattukudy, P. E., 1980a, Chemical modification of an essential lysine at the active site of enoyl-CoA reductase in fatty acid synthetase, *Arch. Biochem. Biophys.* **201:**313.

Poulose, A. J., and Kolattukudy, P. E., 1980b, Presence of one essential arginine that specifically binds the 2'phosphate of NADPH on each of the ketoacyl reductase and enoyl reductase active sites of fatty acid synthetase, *Arch. Biochem. Biophys.* **199:**457.

Poulose, A. J., and Kolattukudy, P. E., 1981, Role of the enoyl reductase domain in the regulation of fatty acid synthase activity by interdomain interaction, *J. Biol. Chem.* **256:**8379.

Poulose, A. J., and Kolattukudy, P. E., 1982, Evidence that the Coenzyme A requirement for avian fatty acid synthetase is not for the termination reaction, *Int. J. Biochem.* **14:**445.

Poulose, A. J., Rogers, L., and Kolattukudy, P. E., 1981, Primary structure of a chymotryptic peptide containing the "active serine" of the thioesterase domain of fatty acid synthetase, *Biochem. Biophys. Res. Commun.* **103:**377.

Pugh, E. L., Sauer, F., Waite, B. M., Toomey, R. E., and Wakil, S. J., 1966, Studies on the mechanism of fatty acid synthesis XIII. The role of β-hydroxyl acids in the synthesis of palmitate and *cis*-vaccenate by the *E. coli* enzyme system, *J. Biol. Chem.* **241:**2635.

Qureshi, A. A., Lornitzo, F. A., Hsu, R. Y., and Porter, J. W., 1976, Isolation, purification, and properties of mammalian and avian liver and yeast fatty acid synthetase acyl carrier proteins, *Arch. Biochem. Biophys.* **177:**379.

Roncari, D. A. K., 1974a, Dissociation of the acyl carrier protein subunit from dog liver fatty acid synthetase complex, *J. Biol. Chem.* **249:**7035.

Roncari, D. A. K., 1974b, Mammalian fatty acid synthetase. I. Purification and properties of human liver complex, *Can. J. Biochem.* **52:**221.

Roncari, D. A. K., Bradshaw, R. A., and Vagelos, P. R., 1972, Acyl carrier protein, *J. Biol. Chem.* **247:**6234.

Rossi, A., and Corcoron, J. W., 1973, Identification of a multienzyme complex synthesizing fatty acids in the actinomycete *Streptomyces erythreus, Biochem. Biophys. Res. Commun.* **50:**597.

Schreckenbach, T., Wobser, H., and Lynen, F., 1977, The palmityl binding sites of fatty acid synthetase from yeast, *Eur. J. Biochem.* **80:**13.

Schweizer, E., Piccinini, F., Duba, C., Gunter, S., Ritter, E., and Lynen, F., 1970, Die malonyl-bindengsstellen des Fettsaure-synthetase-komplexes aus Hefe, *Eur. J. Biochem.* **15:**483.

Schweizer, E., Kniep, B., Castorph, H., and Holzner, U., 1973, Pantetheine-free mutants of the yeast fatty-acid-synthetase complex, *Eur. J. Biochem.* **39:**353.

Schweizer, E., Dietlein, G., Gimmler, G., Knobling, A., Tahedl, H. W., Schwietz, H., and Schweizer, M., 1975, Yeast fatty acid synthetase comprising two multifunctional polypeptide chains, *Proc. 10th FEBS Meet.* **40:**85–97.

Setlow, P., Brutlag, D., and Kornberg, A., 1972, Deoxyribonucleic acid polymerase: Two distinct enzymes in one polypeptide, *J. Biol. Chem.* **247:**224.

Shimakata, T., and Stumpf, P. K., 1982a, The prokaryotic nature of the fatty acid synthetase of developing *Carthamus tinctorius* L. (Safflower) seeds, *Arch. Biochem. Biophys.* **217:**144.

Shimakata, T., and Stumpf, P. K., 1982b, Fatty acid synthetase of *Spinacea oleracea* leaves, *Plant Physiol.* **69:**1257.

Shoukry, S., Stoops, J. K., and Wakil, S. J., 1983, Inactivation of yeast fatty acid synthetase by modifying the β-ketoacyl reductase active lysine residue with pyridoxal 5'-phosphate, *Arch. Biochem. Biophys.* **226:**224.

Simons, S. S., Jr., and Johnson, D. F., 1978, Reaction of *o*-phthalaldehyde and thiols with primary amines: Formation of 1-alkyl (and aryl)thio-2-alkyliosoindoles, *J. Org. Chem.* **43:**2886.

Singh, N., Wakil, S. J., and Stoops, J. K., 1984, On the question of half- or full-site reactivity of animal fatty acid synthetase, *J. Biol. Chem.*, in press.

Sirevag, R., and Levine, R. P., 1972, Fatty acid synthetase from *Chlamydomonas reinhardii, J. Biol. Chem.* **247:**2586.

Smith, S., 1982, The effect of Coenzyme A and structurally related thiols on the mammalian fatty acid synthetase, *Arch. Biochem. Biophys.* **218:**249.

Smith, S., and Abraham, S., 1970, Fatty acid synthetase from lactating rat mammary gland, *J. Biol. Chem.* **245:**3209.

Smith, S., and Stern, A., 1979, Subunit structure of the mammalian fatty acid synthetase: Further evidence for a homodimer, *Arch. Biochem. Biophys.* **197:**379.

Smith, S., Agradi, E., Libertini, L., and Dileepan, K. N., 1976, Specific release of the thioesterase component of the fatty acid synthetase multienzyme complex by limited trypsinization, *Proc. Natl. Acad. Sci. USA* **73:**1184.

Steinberger, R., and Westheimer, F. H., 1951, Metal ion-catalyzed decarboxylation: A model for an enzyme system, *J. Am. Chem. Soc.* **73:**429.

Stern, A., Sedgwick, B., and Smith, S., 1982, The free coenzyme A requirement of animal fatty acid synthetase. Participation in the continuous exchange of acetyl and malonyl moieties between coenzyme A thioester and enzyme, *J. Biol. Chem.* **257:**799.

Stoops, J. K., and Wakil, S. J., 1978, The isolation of the two subunits of yeast fatty acid synthetase, *Biochem. Biophys. Res. Commun.* **84:**143.

Stoops, J. K., and Wakil, S. J., 1980, Yeast fatty acid synthetase: Structure–function relationship and the nature of the β-ketoacyl synthetase site, *Proc. Natl. Acad. Sci. USA* **77:**4544.

Stoops, J. K., and Wakil, S. J., 1981a, Animal fatty acid synthetase: A novel arrangement of the β-ketoacyl synthetase sites comprising domains of the two subunits, *J. Biol. Chem.* **256:**5128.

Stoops, J. K., and Wakil, S. J., 1981b, The yeast fatty acid synthetase: Structure–function relationship and the role of the active cysteine-SH and pantetheine-SH, *J. Biol. Chem.* **256:**8364.

Stoops, J. K., and Wakil, S. J., 1982a, Animal fatty acid synthetase: Identification of the residues comprising the novel arrangement of the β-ketoacyl synthetase site and their role in its cold inactivation, *J. Biol. Chem.* **257:**3230.

Stoops, J. K., and Wakil, S. J., 1982b, The reaction of chicken liver fatty acid synthetase with 5,5′-dithiobis(2-nitrobenzioc acid), *Biochem. Biophys. Res. Commun.* **104:**1018.

Stoops, J. K., Arslanian, M. J., Oh, Y. H., Aune, K. C., Vanaman, T. C., and Wakil, S. J., 1975, The presence of two polypeptide chains comprising the fatty acid synthetase, *Proc. Natl. Acad. Sci. USA* **72:**1940.

Stoops, J. K., Arslanian, M. J., Chalmers, J. H., Jr., Joshi, V. C., and Wakil, S. J., 1977, Fatty acid synthetase complexes, in: *Bioorganic Chemistry 1* (Van Tamelen, ed.), Academic Press, New York, pp. 339–370.

Stoops, J. K., Arslanian, M. J., Aune, K. C., and Wakil, S. J., 1978a, Further evidence for the multifunctional enzyme characteristic of the fatty acid synthetases of animal tissues: Physicochemical studies of the chicken liver fatty acid synthetase, *Arch. Biochem. Biophys.* **188:**348.

Stoops, J. K., Awad, E. S., Arslanian, M. J., Gunsberg, S., Wakil, S. J., and Oliver, R. M., 1978b, Studies on the yeast fatty acid synthetase: Subunit composition and structural organization of a large multifunctional enzyme complex, *J. Biol. Chem.* **253:**4464.

Stoops, J. K., Ross, P. R., Arslanian, M. J., Aune, K. C., and Wakil, S. J., 1979, Physicochemical studies of the rat liver and adipose fatty acid synthetases, *J. Biol. Chem.* **254:**7418.

Stoops, J. K., Henry, S. J., and Wakil, S. J., 1984, The arrangement and role of some of the amino acid residues in the β-ketoacyl synthetase site of chicken liver fatty acid synthetase, *J. Biol. Chem.*, in press.

Stram, K. A., and Kumar, S., 1979, Activation and inhibition of crotonyl-Coenzyme A reductase activity of bovine mammary fatty acid synthetase, *J. Biol. Chem.* **254:**8159.

Strong, C. R., and Dils, R., 1972, The fatty acid synthetase complex of lactating guinea-pig mammary gland, *Int. J. Biochem.* **3:**369.

Toomey, R. E., and Wakil, S. J., 1966, Studies on the mechanism of fatty acid synthesis XVI. Preparation and general properties of acyl-malonyl ACP condensing enzyme from *E. coli*, *J. Biol. Chem.* **241:**1159.

Tsukamoto, Y., Wong, H., Wakil, S. J., and Mattick, J. S., 1982, The architecture of animal fatty acid synthetase, *Fed. Proc.* **41:**1026.

Tsukamoto, Y., Mattick, J. S., Wong, H., and Wakil, S. J., 1984, The architecture of the animal fatty acid synthetase. IV. Active site determination study, *J. Biol. Chem.*, in press.

Vanaman, T. C., Wakil, S. J., and Hill, R. L., 1968, The complete amino acid sequence of the acyl carrier protein from *Escherichia coli*, *J. Biol. Chem.* **243:**6420.

Volpe, J. J., and Vagelos, P. R., 1977, Mechanisms and regulation of biosynthesis of saturated fatty acids, *Physiol. Rev.* **56:**339.

Wakil, S. J., 1971, Fatty acid metabolism, in: *Lipid Metabolism* (S. J. Wakil, ed.), Academic Press, New York, pp. 1–48.

Wakil, S. J., and Kuziora, M. A., 1983, *Manipulation of Expression of Genes in Eukaryotes* (A. W. Linnane, W. J. Peacock, and J. A. Pateman, eds.), Academic Press, Sydney, pp. 131–140.

Wakil, S. J., Stoops, J. K., and Mattick, J. S., 1981, The fatty acid synthetase—Structure–function relationship and mechanism of palmitate synthesis, *Cardiovasc. Res. Center Bull.* **20**(July–September):1–23.

Wakil, S. J., Stoops, J. K., and Joshi, V. C., 1983, Fatty acid synthesis and its regulation, *Annu. Rev. Biochem.* **52:**537.

Warren, S., Zerner, B., and Westheimer, F. H., 1966, Acetoacetate decarboxylase. Identification of lysine at the active site, *Biochemistry* **5:**817.

Wetlauter, D. B., 1973, Nucleation, rapid folding, and globular intrachain regions in proteins, *Proc. Natl. Acad. Sci. USA* **70:**697.

Wieland, F., Siess, E. A., Renner, L., Verfurth, C., and Lynen, F., 1978, Distribution of yeast fatty acid synthetase subunits: Three-dimensional model of the enzyme, *Proc. Natl. Acad. Sci. USA* **75:**5792.

Wieland, F., Renner, L., Verfurth, C., and Lynen, F., 1979, Studies on the multi-enzyme complex of yeast fatty-acid synthetase, *Eur. J. Biochem.* **94:**189.

Willecke, K., Ritter, E., and Lynen, F., 1969, Isolation of an acyl carrier protein component from the multienzyme complex of yeast fatty acid synthetase, *Eur. J. Biochem.* **8:**503.

Wong, H., Mattick, J. S., and Wakil, S. J., 1984, The architecture of the animal fatty acid synthetase. III. Isolation and characterization of β-ketoacyl reductase, *J. Biol. Chem.*, in press.

Wood, W. I., Peterson, D. O., and Bloch, K., 1978, Subunit structure of *Mycobacterium smegmatis* fatty acid, *J. Biol. Chem.* **253:**2650.

Yang, C. P., Butterworth, P. H. W., Bock, R. M., and Porter, J. W., 1967, Further studies on the properties of the pigeon liver fatty acid synthetase, *J. Biol. Chem.* **242:**3501.

Yung, S., and Hsu, R. Y., 1972, Fatty acid synthetase of chicken liver, *J. Biol. Chem.* **247:**2689.

Zehner, Z. E., Mattick, J. S., Stuart, R., and Wakil, S. J., 1980, Goose fatty acid synthetase mRNA, *J. Biol. Chem.* **255:**9519.

Ziegenhorn, J., Niedermeier, R., Nussler, C., and Lynen, F., 1972, Charakterisierung der acetyltransferase in der fettsauresynthetase aus hefe, *Eur. J. Biochem.* **30:**285.

16

Properties and Function of Phosphatidylcholine Transfer Proteins

Karel W. A. Wirtz, Tom Teerlink, and Rob Akeroyd

I. INTRODUCTION

When phospholipids extracted from biological membranes are dispersed in water, bilayer structures are spontaneously formed (Bangham *et al.*, 1965). This process of self-association is thermodynamically driven. It involves the extrusion of the hydrophobic acyl chains from the medium and the orientation of the polar head groups at the bilayer–water interface. In general, natural phospholipids have a very low critical micelle concentration. Consequently, the rate at which a monomer molecule dissociates from the bilayer is low, thus resulting in a slow approach to equilibrium (Reynolds, 1982). In agreement with this notion, spontaneous equilibration of phospholipid molecules between model membrane structures by diffusion of soluble monomers is a matter of days (Duckwitz-Peterlein *et al.*, 1977; McLean and Phillips, 1981). Interestingly, equilibration is greatly accelerated when membranes are prepared with phospholipids containing unnatural fatty acids like fluorescently labeled fatty acids (Roseman and Thompson, 1980; Nichols and Pagano, 1981) or myristic acid (Martin and MacDonald, 1976). This acceleration may be a reflection on the stability of these lipid molecules in the bilayer structure apparently resulting in an enhanced establishment of the bilayer–monomer equilibrium. Despite the great tendency for a phospholipid molecule to stay with the bilayer structure, numerous studies with intact cells have indicated that *in situ* phospholipid movement between subcellular organelles can be

Karel W. A. Wirtz, Tom Teerlink, and Rob Akeroyd • Laboratory of Biochemistry, State University of Utrecht, University Centre "De Uithof," NL-3584 CH Utrecht, The Netherlands.

very fast (Stein and Stein, 1969; Wirtz and Zilversmit, 1969; Jungalwala and Dawson, 1970). It is believed now that the phospholipid transfer proteins detected in the cytoplasm of mammalian and plant tissues are involved in this intermembranous transfer of phospholipids.

Phospholipid transfer activity was discovered when the 105,000 *g* supernatant from rat liver was found to greatly stimulate the transfer of phospholipids between rat liver mitochondria and microsomes (Wirtz and Zilversmit, 1968; McMurray and Dawson, 1969; Akiyama and Sakagami, 1969). Subsequently, similar transfer activity was detected in the cytosol of other mammalian tissues, plants (Abdelkader and Mazliak, 1970), yeast (Cobon *et al.*, 1976), and photosynthetic bacteria (Cohen *et al.*, 1979). Since the original observation in 1968, there has been a considerable effort in purifying the proteins responsible for this transfer activity. To date, phospholipid transfer proteins have been obtained in pure form from bovine liver (Kamp *et al.*, 1973), brain (Helmkamp *et al.*, 1974), and heart (DiCorleto *et al.*, 1979), and rat liver (Poorthuis *et al.*, 1980) and hepatoma (Dyatlovitskaya *et al.*, 1978), and maize seedlings (Dollady *et al.*, 1982). Several reports have appeared on partially purified transfer protein preparations, among them those from rat intestine (Yamada *et al.*, 1978) and lung (van Golde *et al.*, 1980), sheep lung (Robinson *et al.*, 1978), potato tuber (Kader, 1975), and castor bean endosperm (Tanaka and Yamada, 1979). In addition to these intracellular proteins, phospholipid transfer proteins have also been detected in human serum and partially purified from this source (Ihm *et al.*, 1980; Damen *et al.*, 1982). So far, three distinct classes of phospholipid transferring proteins have been identified. Thus, one distinguishes the phosphatidylcholine transfer protein (PC-TP*) which specifically catalyzes the transfer of PC (Kamp *et al.*, 1977), the phosphatidylinositol transfer protein (PI-TP) which preferentially transfers PI but is also capable of PC and PG transfer (Helmkamp *et al.*, 1976; Zborowski and Demel, 1982), and the nonspecific lipid transfer protein (nsL-TP) which catalyzes the transfer of all natural diacylphospholipids as well as cholesterol (Crain and Zilversmit, 1980).

This chapter will focus on the properties and characterization of PC-TP isolated from bovine and rat liver. Inasmuch as it serves to clarify this resume on PC-TP, other phospholipid transfer proteins will be mentioned. For further information on phospholipid transfer and the proteins involved, the interested reader be referred to some recent reviews (Wirtz, 1982; Kader *et al.*, 1983).

II. PHOSPHATIDYLCHOLINE TRANSFER PROTEIN FROM BOVINE LIVER

A. Introductory Remarks

Phospholipid transfer activity in crude 105,000 *g* supernatants is commonly determined by measuring the transfer of radiolabeled phospholipids from labeled donor

* Abbreviations used: PC-TP, phosphatidylcholine transfer protein; PI-TP, phosphatidylinositol transfer protein; nsL-TP, nonspecific lipid transfer protein; PC, phosphatidylcholine; PI, phosphatidylinositol; PE, phosphatidylethanolamine; PA, phosphatidic acid; PG, phosphatidylglycerol; PS, phosphatidylserine; ESR, electron spin resonance; IgG, immunoglobulin G.

membranes to unlabeled acceptor membranes (for a review, see Zilversmit and Hughes, 1976). These membranes can be of natural origin, e.g., mitochondria or microsomes, or be artificially prepared (single bilayer vesicles, multilamellar liposomes). In most assays, donor and acceptor membranes are sufficiently different in size and/or density to allow separation by centrifugation at the end of incubation. In some instances, donor phospholipid vesicles are sensitized with glycolipids so that prior to centrifugation these vesicles can be precipitated by addition of the appropriate agglutinin (Sasaki and Sakagami, 1978; Kasper and Helmkamp, 1981). Application of such assays has led to the detection of PC-transfer activity in a great variety of mammalian and plant tissues. According to what we currently know about the occurrence of phospholipid transfer proteins in mammalian tissues, the PC-transfer activity in the 105,000 *g* supernatant can be fully accounted for by the contributions of PC-TP, PI-TP, and nsL-TP. In an effort to understand the mode of action of these proteins, we have mostly concentrated on PC-TP from bovine liver which, by all criteria, is highly specific for PC.

For the purification of PC-TP, bovine liver was selected as this tissue is relatively rich in PC-transfer activity. The procedure originally described by Kamp *et al.* (1973) is still in use except for some minor modifications in the first steps (Westerman *et al.*, 1983a). Purification involves pH adjustment steps (pH 5.1 and pH 3.0, respectively), ammonium sulfate precipitation, fractionation by ion-exchange chromatography on DEAE- and CM-cellulose and molecular sieve chromatography on Sephadex G-50. Starting from 25,000 g of liver, the procedure renders a homogeneous protein (4200-fold purified) at a yield of 25%, i.e., 75 mg of PC-TP.

Here we will first describe the PC-TP-mediated transfer of PC between membranes and the regulation of this process by interfacial characteristics (Section II-B). Secondly, we will discuss various molecular aspects of PC-TP and relate this information to its mode of action (Section II-C).

B. Mode of Action

1. Exchange vs. Net Transfer

Extraction of purified bovine PC-TP with organic solvents has provided proof that the protein contains one molecule of noncovalently bound PC (Demel *et al.*, 1973). In order to gather further information on this lipid–protein complex, PC-TP was injected under a monolayer of [^{14}C]-PC spread at the air–water interface (Figure 1). Under these conditions, a steady decline of the surface radioactivity was observed, indicating that PC-TP removed [^{14}C]-PC from the interface. However, the surface pressure did not change. This shows that the actual number of PC molecules in the interface remained constant, which could only come about when PC-TP has exchanged its bound PC molecule for one present in the monolayer. By the same exchange mechanism PC-TP became radiolabeled when it was incubated with vesicles of [^{14}C]-PC (Kamp *et al.*, 1975a). This one-for-one molecular exchange reaction forms the basis for the PC-TP-mediated transfer of PC between membranes. In this process, PC-TP presumably functions as a freely diffusible carrier of PC. Inherent to this mechanism one would expect that PC-TP catalyzes only the exchange of PC between membranes.

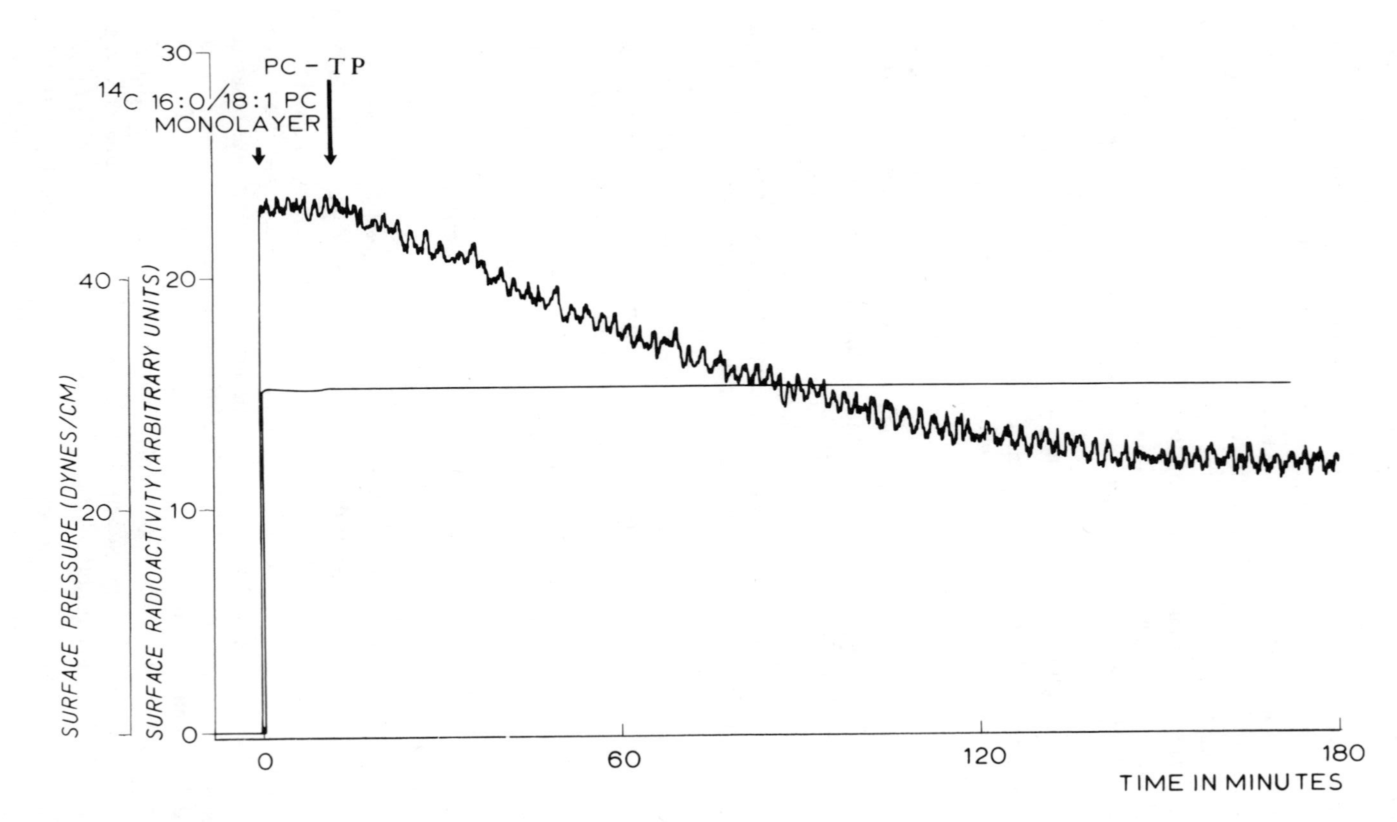

Figure 1. Exchange of PC between monolayer and PC-TP. The monolayer consisted of 20.6 nmoles 16 : 0/18 : 1 [^{14}C-methyl]PC; the subphase contained 22.7 nmoles PC-TP. From Demel *et al.* (1973).

This has recently been confirmed in a study where transfer was measured between two populations of single bilayer vesicles, one prepared with egg PC, the other with dimyristoyl-PC (Helmkamp, 1980). Irrespective of the difference in fatty acyl composition, as much egg PC was found to be transferred to the dimyristoyl-PC vesicles as dimyristoyl-PC to the egg PC vesicles.

These observations raise the question whether PC-TP can dissociate from a membrane without a PC molecule being bound. In other words, can PC-TP mediate a net transfer of PC between membranes? This question has been answered by incubating donor membranes that contain PC with acceptor membranes devoid of PC. Net transfer as part of a continuous process was detected by applying ESR spectroscopy to a mixture of donor vesicles consisting of PC, a nitroxide group on the *sn*-2-fatty acyl chain, and PA (75 : 25, mole%), and acceptor vesicles of PE and PA (81 : 19, mole%). In the absence of PC-TP, the spin–spin exchange-broadened spectrum (Figure 2, spectrum B) did not change with time, indicating a lack of spontaneous redistribution of spin-labeled PC between the vesicles. Addition of PC-TP, however, gave rise to the time-dependent appearance of a three-line spectrum superimposed on the exchange-broadened spectrum (Figure 2, spectrum A). This three-line spectrum resulted from the PC-TP-catalyzed insertion of spin-labeled PC from the donor into the acceptor vesicles. In the reverse experiment (the donor vesicles contained spin-labeled PE and the acceptor vesicles PC), PC-TP did not induce any spectral change. These two complementary experiments demonstrate that PC-TP incorporated PC into the PE-containing acceptor

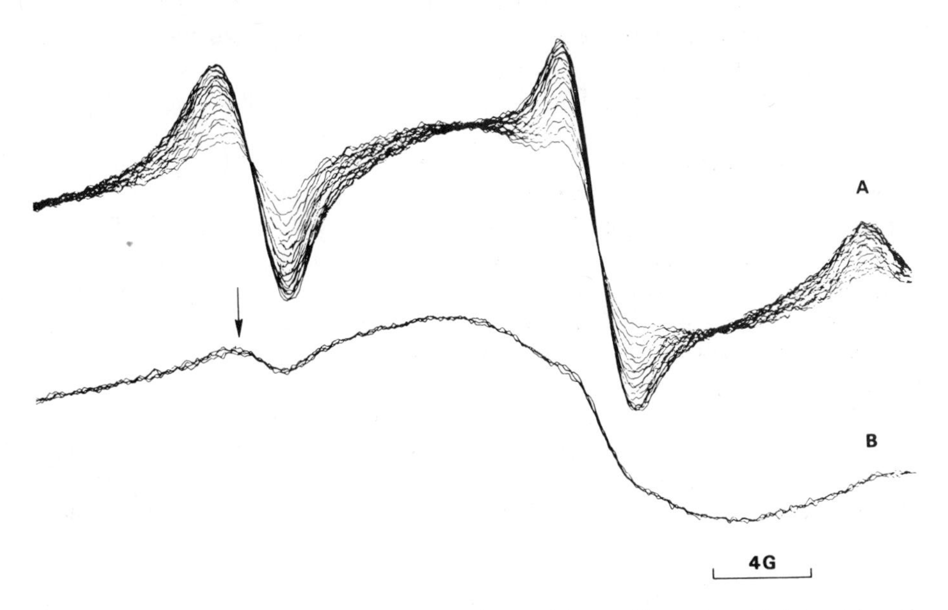

Figure 2. Change in the ESR spectra of spin-labeled PC vesicles on addition of nonlabeled vesicles and PC-TP. Spectra A: spin-labeled vesicles (6.3 nmoles spin-labeled PC-PA, 75 : 25 mole%) were mixed with nonlabeled vesicles (57 nmoles PE-PA, 81 : 19 mole%) and PC-TP (15 μg). Spectra B: as in A, without PC-TP. Spectra were recorded at a gain of 4×10^4 with scan rates of 8 min. From Wirtz *et al.* (1980a).

vesicles without transfer of PE in the opposite direction. This net transfer led to a buildup of PC in the acceptor up to an equilibrium concentration at which the PC-TP catalyzed transfer from donor to acceptor equaled the transfer from acceptor to donor vesicles, that is, exchange (Wirtz *et al.*, 1980a). It could be estimated that exchange occurred when about 2% of the acceptor phospholipid consisted of spin-labeled PC. Moreover, by varying the ratio of acceptor to donor phospholipid concentration, it was observed that there was a limit of about 20% to the depletion of PC from the donor vesicles.

Bovine PC-TP also facilitated a net transfer of PC from either plasma low density lipoproteins or vesicles of 1-palmitoyl-2-palmitoleoyl-PC to sphingomyelin-apo A-II complexes (Wilson *et al.*, 1980). Transfer of sphingomyelin in the opposite direction was not observed. From the data presented in this study, it can be calculated that at the end of the transfer reaction the sphingomyelin-apo A-II complex contained at least 10% PC. This is high in comparison to the 2% of spin-labeled PC incorporated by net transfer into the phospholipid vesicles (Wirtz *et al.*, 1980a). This difference may, in part, be due to the presence of apo A-II in the acceptor complex having a certain affinity for PC. At present, evidence is lacking that *in vivo* PC-TP is involved in net transfer of PC to intracellular sites in want of this phospholipid. On the other hand, the evidence *in vitro* indicates that, in principle, PC-TP can play such a role provided the receiving membranes effectively shield PC from interacting with PC-TP.

2. Regulation by the Interface

Phospholipid bilayers are minimum-energy structures. Extraction of individual phospholipid molecules from such structures by the phospholipid transfer proteins raises some fundamental questions about the energies involved. Or, as stated by Kasper and Helmkamp (1981), the problem is that "a rather substantial energy barrier must exist for the transfer of a PC molecule between an accommodating lipid environment and even the most hydrophobic protein environment." From past as well as more recent studies a picture emerges which becomes increasingly satisfactory as to what may happen at the interface. Among others, the kinetic model proposed by van den Besselaar *et al.* (1975) to describe the transfer process has substantially contributed to our present insight. This model is based on the following assumptions: (1) PC-TP as carrier forms a collision complex with a membrane, whereupon release of the endogenous PC molecule occurs, and (2) PC-TP does not dissociate from the interface without being associated with a PC molecule. These conditions of exchange are satisfied whenever PC-TP interacts with membranes that contain sufficient levels of PC.

The kinetic model for the transfer of PC between donor membrane L_1 and acceptor membrane L_2 is outlined in Figure 3. Application of steady-state approximations to the concentrations of the reaction components yields the following expression for the rate of transfer, V_O:

$$V_O = \frac{k_1[L_1]k_2[L_2]P_{tot}}{(k_1[L_1] + k_2[L_2])(1 + k_1/k_{-1}[L_1] + k_2/k_{-2}[L_2])} \tag{1}$$

where $[L_1]$ and $[L_2]$ are the concentrations of PC involved in the transfer reaction, P_{tot}

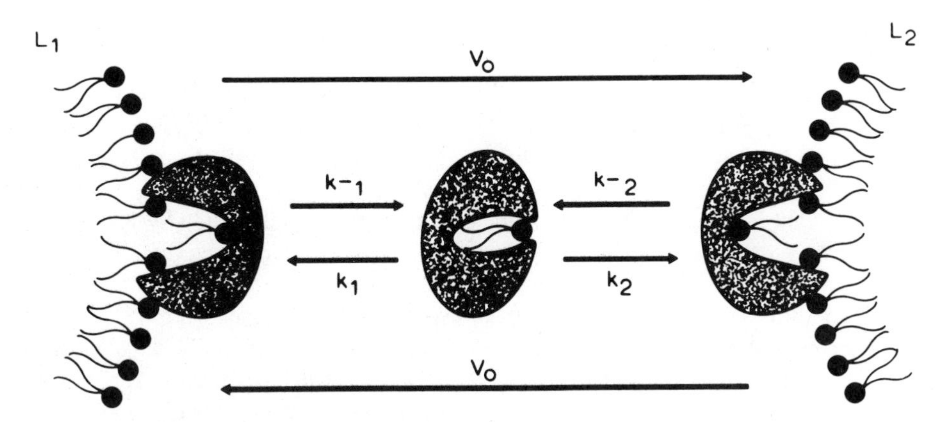

Figure 3. Kinetic model for the PC-TP-mediated transfer of PC between donor membrane L_1 and acceptor membrane L_2. From the model proposed by van den Besselaar *et al.* (1975).

is the total amount of PC-TP, and $k_1(k_2)$ and $k_{-1}(k_{-2})$ are the rate constants of association and dissociation, respectively. The values for the kinetic parameters (k_1, k_2, k_1/k_{-1}, k_2/k_{-2}) are estimated by fitting the experimentally determined rates of transfer, i.e., initial velocities, by computer with the theoretical rate equation (for details, see van den Besselaar *et al.*, 1975).

In a recent study (Bozzato and Tinker, 1982), application of this kinetic model has provided important information on how the "fluidity" of the membrane may regulate the PC-TP-mediated transfer reaction. Transfer of PC was measured between donor vesicles of egg PC and acceptor liposomes of dipalmitoyl-PC-dipalmitoyl-PG (95 : 5, mole%) both above and below the transition temperature of dipalmitoyl-PC, i.e., 41°C. For comparison, acceptor liposomes of egg PC-egg PG (95 : 5, mole%) were included as well. Fitting of the initial velocities to Eq. (1) gave the kinetic rate constants of the PC-TP-acceptor liposome complex (Table 1). For dipalmitoyl-PC, the rate constant of association (k_2) did not vary with the physical state of the acceptor interface. Very significantly, the rate constant of dissociation (k_{-2}), which represents the rate-limiting step in the overall transfer reaction, increased about five-fold above the phase transition temperature. This strongly suggests that in the liquid crystalline state, dissociation of the PC-TP-acceptor liposome complex is facilitated by the increase in fluidity. In

Table 1. Kinetic Rate Constants of the PC-TP-Acceptor Liposome Complex[a]

Acceptor liposome	Temperature of incubation	Association rate constant (k_2) (nmoles/min/μg/mM)	Dissociation rate constant (k_{-2}) (nmoles/min/μg)
Dipalmitoyl-PC	37°C	4.0 ± 1.7	0.12 ± 0.05
Dipalmitoyl-PC	45°C	5.2 ± 0.1	0.54 ± 0.03
Egg PC	37°C	7.5 ± 0.001	3.7 ± 0.8
Egg PC	45°C	10.9 ± 0.1	3.6 ± 0.5

[a] From Bozzato and Tinker (1982).

support of this conclusion, rate constants of dissociation were not influenced by temperature when the acceptor liposomes consisted of egg PC. Interestingly, one sees from Table 1 that in the liquid crystalline state the rate constant of dissociation is seven-fold higher for egg PC than dipalmitoyl-PC. In molecular terms this may mean that despite comparable physical conditions egg PC flips more easily from the interface onto PC-TP than dipalmitoyl-PC.

Relevant to the interpretation of the data in Table 1 is the study by Kasper and Helmkamp (1981) on the effect of gel to liquid crystalline transition on the PC-TP-mediated transfer of PC. Transfer was measured to vesicles prepared with egg PC, dimyristoyl-PC, and dipalmitoyl-PC, over a wide temperature range (11–45°C). From the transfer activity–temperature relationship, these investigators derived the apparent activation energies of transfer (Table 2). It is evident that for PC in the "fluid" state the activation energy is much lower than for PC in the gel state. This could mean that the crystalline arrangement of the PC molecules severely impairs the extraction of PC from the interface by PC-TP. Specifically, in the "fluid" state the activation energy for egg PC is 108 kJ/mole lower than for dipalmitoyl-PC in the gel state. This difference is presumably reflected in the 30-fold lower rate constant of dissociation for egg PC (see Table 1).

In addition to regulation by the physical state of the interface, PC-TP also "senses" whether PC is present in highly curved single bilayer vesicles or "flat" multilamellar liposomes. The rate of transfer of spin-labeled PC with 12-nitroxide stearate at the *sn*-2 position was more than 100 times larger between vesicles than between vesicles and liposomes (Machida and Ohnishi, 1980). Kinetic analyses involving interfaces of egg PC and PA (90 : 10, mole%) demonstrated that the rate constant of dissociation was about 100-fold lower for liposomes than vesicles (Wirtz *et al.*, 1979). This may reflect the less stable arrangement of PC in the vesicle interface facilitating the movement of PC from the interface onto PC-TP at the interface. Part of this instability follows from the sizable difference in average area per PC polar head group at the outer surface, i.e., 74 Å^2 for vesicles as compared to 46 Å^2 for liposomes (Huang and Mason, 1978). On the other hand we have to realize that at present we cannot be sure that the rate constant of dissociation only represents the desorption of a PC molecule from the interface by PC-TP (see also Kasper and Helmkamp, 1981; Bozzato and Tinker, 1982).

Table 2. Apparent Activation Energies for the PC-TP-Mediated Transfer of PC to Acceptor Vesicles above and below the Transition Temperature[a]

Acceptor vesicle	Physical state	Activation energy (kJ/mole)
Egg PC	Liquid crystalline	33
Dimyristoyl-PC	Crystalline	115
Dimyristoyl-PC	Liquid crystalline	35
Dipalmitoyl-PC	Crystalline	141

[a] From Kasper and Helmkamp (1981).

For example, there is the possibility that PC-TP penetrates into the bilayer. Desorption of PC-TP could then be a function of the physical state of the interface and thus contribute to the rate constant of dissociation.

At present, there is no evidence for the penetration of the membrane interface by PC-TP. Addition of PC-TP to the aqueous subphase of PC monolayers had no detectable effect on the surface pressure while PC-TP exchanged PC molecules up to surface pressures of 40 dynes/cm (Figure 1; Demel *et al.*, 1973). Significant transfer activity was observed with membranes in the gel state (Tables 1 and 2). It is commonly recognized that this condition does not favor penetration of proteins into membranes. It should be further noted that in the transfer between vesicles only PC in the outer monolayer is involved (Johnson *et al.*, 1975; Rothman and Dawidowicz, 1975; De Kruijff and Wirtz, 1977). Whatever interaction occurs at the interface PC-TP does not appear to disturb the bilayer to the extent that a transbilayer movement of phospholipid molecules is induced.

The specificity of bovine PC-TP for PC implies that the protein has a recognition site for the phosphorylcholine head group (Kamp *et al.*, 1977). Inherent to such a site, electrostatic interactions with PC-TP would be part of the mechanism of transfer. In support of this contention, cations inhibited the PC-TP-mediated transfer of PC between a monolayer and vesicles of pure PC (Wirtz *et al.*, 1976). Inhibition decreased in the order $La^{3+} > Mg^{2+} \geqslant Ca^{2+} > K^{+} = Na^{+}$, and was not related to the ionic strength. Interestingly, this order agrees with the declining affinity of these cations for the phosphodiester group of PC (Hauser *et al.*, 1977; Akutsu and Seelig, 1981). In case of electrostatic interactions one would expect the recognition site to contain one or more basic amino acid residues (see Section II-C.1). According to this interpretation, neutralization of the negative charge on the phosphorylcholine head group impairs the interaction with these positively charged residues.

PC-TP interacts strongly with negatively charged vesicles (Machida and Ohnishi, 1978; Wirtz *et al.*, 1979). This interaction becomes apparent when we measure the intrinsic tryptophanyl fluorescence (Wirtz and Moonen, 1977). Titration with negatively charged vesicles increased the fluorescence indicating that certain tryptophan residues of PC-TP "sensed" the different polarity of the membrane interface. This increase was more pronounced for vesicles with elevated levels of acidic phospholipids, e.g., PA (Figure 4). Lineweaver–Burk plots of the fluorescence intensity increase against the vesicle PC concentration have been used to estimate the apparent dissociation constants of the PC-TP-vesicle complexes. These constants decreased from 0.3 mM for vesicles with 5.6 mole% PA to 0.07 mM for vesicles with 15.1 mole% PA. Kinetic analyses based on the initial velocities of PC transfer to these vesicles have clearly shown that the rate constant of dissociation decreased and the rate constant of association increased with increasing vesicle PA content (van den Besselaar *et al.*, 1975). In summary, it is evident that PC-TP binds more strongly to the more negatively charged vesicles. This has been confirmed in a direct way by equilibrium gel chromatography (Wirtz *et al.*, 1979). Strong complex formation was found to occur with PC vesicles containing 10 and 20 mole% PA. Studies with vesicles of pure PS have clearly indicated that, in addition to negative charge, binding is also affected by the membrane curvature (Machida and Ohnishi, 1978, 1980). Strong binding occurred

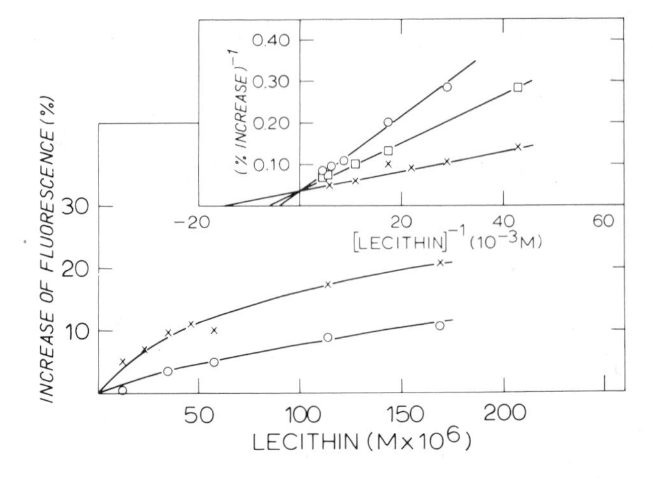

Figure 4. Fluorimetric titration of PC-TP with vesicles consisting of PC and various amounts of PA. (×) 5.6 mole% PA; (□) 9.4 mole% PA; (○) 15.1 mole% PA. Inset gives the Lineweaver–Burk plot of fluorescence increase against vesicle PC concentration. From Wirtz and Moonen (1977).

with vesicles of 17 nm but not with those of 22 nm. These studies suggest that very subtle changes in the molecular packing of phospholipid molecules may have a major effect on the interaction of PC-TP with the ensuing interface. Further studies along this line may explain why acidic phospholipids have been found to both inhibit and stimulate PC-TP-mediated transfer (Wirtz *et al.*, 1976; DiCorleto *et al.*, 1977).

C. Molecular Aspects

1. Primary Structure and Derived Characteristics

Inherent to its mode of action PC-TP is assumed to have a site that interacts with the membrane, e.g., the recognition site for the phosphorylcholine head group, and a lipid-binding site that accommodates the PC molecule. As a first step in gaining knowledge on these sites, we have recently elucidated the complete primary structure (Moonen *et al.*, 1980; Akeroyd *et al.*, 1981a). The protein (mol. wt. 24,681; isoelectric point of 5.8) consists of a single polypeptide chain of 213 amino acid residues of which *N*-acetyl methionine forms the blocked N-terminus and threonine the C-terminus (Figure 5). It contains two disulfide bridges at Cys^{17}–Cys^{63} and Cys^{93}–Cys^{207}. Further examination reveals several interesting features. The N-terminal half is particularly rich in charged amino acid residues. Virtually all arginine and glutamic acid residues are present in the segment up to residue 145. Only three out of a total of 12 aspartic acid residues are found in the C-terminal half (residues 124–213). Another remarkable point is that 22 out of the 213 residues are present in pairs, i.e, Glu^{10}-Glu^{11}, Ala^{28}-Ala^{29}, Leu^{32}-Leu^{33}, Leu^{44}-Leu^{45}, Gln^{47}-Gln^{48}, Leu^{67}-Leu^{68}, Val^{98}-Val^{99}, Tyr^{174}-Tyr^{175}, Gly^{180}-Gly^{181}, Ala^{191}-Ala^{192}, and Lys^{211}-Lys^{212}. Furthermore, PC-TP contains five tryptophan residues which upon excitation have an emission spectrum with a maximum at 327 nm (Wirtz and Moonen, 1977). This suggests that these residues are buried in

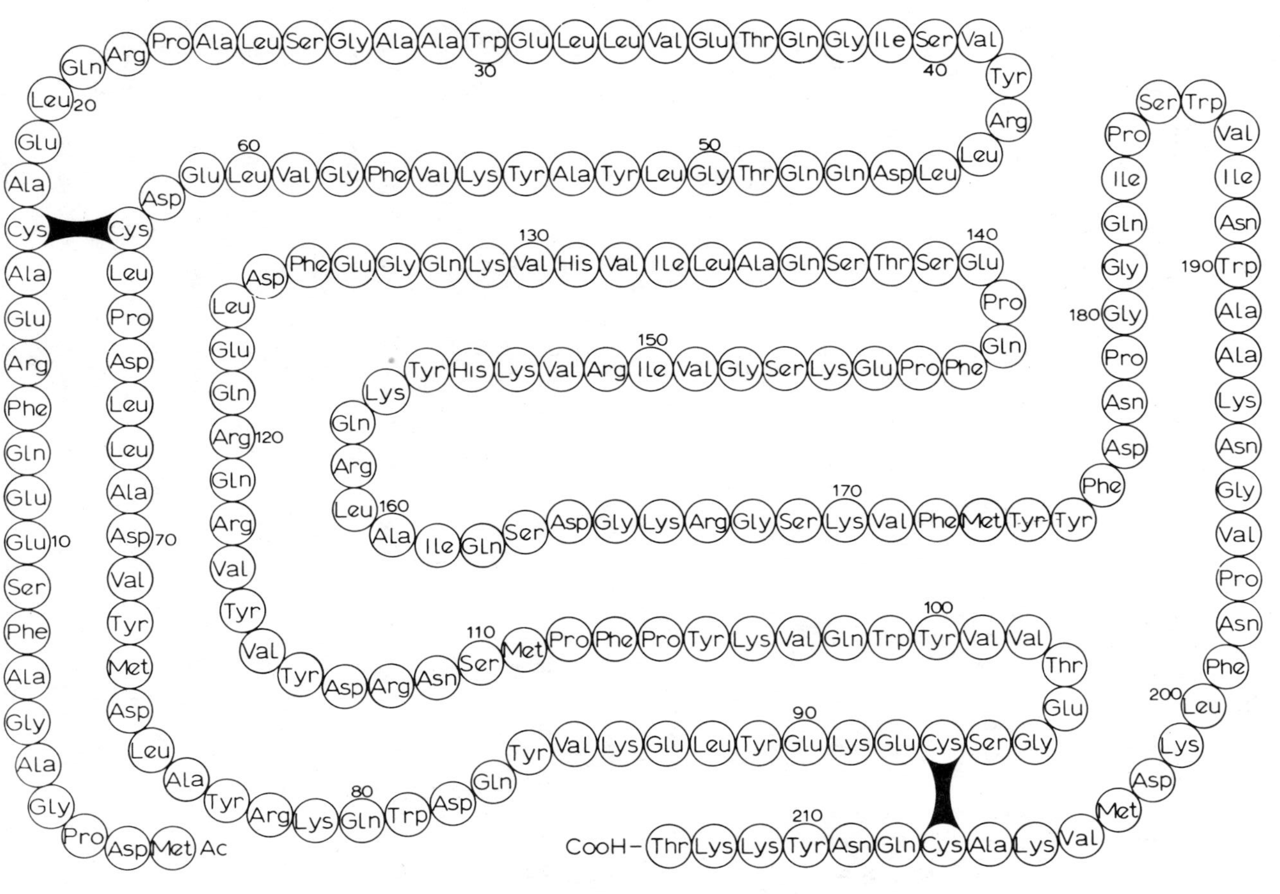

Figure 5. The primary structure of PC-TP from bovine liver. From Akeroyd *et al.* (1982).

the nonpolar region of the protein (Burstein *et al.*, 1973). In support of this suggestion it is seen from the primary structure that the four tryptophan residues at positions 30, 101, 186, and 190 are present in hydrophobic peptide segments (see also Figure 6). An exception is Trp^{81}, which is part of the peptide Arg^{78}-Lys-Gln-Trp-Asp-Gln^{83} and may be surface-oriented.

Another point of major interest is the presence of very hydrophobic peptide segments, of which the most prominent are Val^{98}-Val-Tyr-Trp-Gln-Val^{103}, Val^{171}-Phe-Met-Tyr-Tyr-Phe^{176}, and Trp^{186}-Val-Ile-Asn-Trp-Ala-Ala^{192}, with average hydrophobicities of 8402, 10,200, and 8611 J/residue as compared to 5680 J/residue for the total protein (Nozaki and Tanford, 1971; Manavalan and Ponnuswamy, 1978). A more systematic analysis by the method of Rose (1978) gives the distribution of hydrophobicity along the chain as shown in Figure 6A. In particular, the location of the maxima in the C-terminal half corresponding with the peptides Val^{171}–Phe^{176} and Trp^{186}–Ala^{122} is striking where these maxima are alternated by minima of zero hydrophobicity. In the method of Rose, the hydrophobicity profile is based on the hydrophobicity indices determined for each amino acid by measuring the free energy of transfer from water to an organic solvent (Nozaki and Tanford, 1971). Manavalan and Ponnuswamy (1978) have proposed to derive the hydrophobicity profile of a protein from the surrounding hydrophobicity of each amino acid. In this approach, the surrounding hydrophobicity of an amino acid is defined as the sum of the hydrophobic indices assigned to the various residues that appear within an 8 Å radius volume in the protein crystal. The values of these surrounding hydrophobicities are based on the crystallographic data on 14 protein molecules. From these indices, a surrounding hydrophobicity profile of PC-TP was derived (Figure 6B). The apparent maxima agree rather well with those in Figure 6A with the most distinct maxima being present in the C-terminal half. It is worth noting that in the latter analysis the peptide segments Val^{130}-His-Val-Ile-Leu-Ala^{135} and Gly^{148}-Val-Ile-Arg-Val^{152} are the most hydrophobic.

Hydrophobicity profiles are an index for the folding of the protein by which the maxima correspond to the peptide regions in the hydrophobic core and the minima to the peptide chain turns and solvent-exposed parts (Rose and Roy, 1980; Ponnuswamy *et al.*, 1980). Recently, hydration potentials of amino acid side chains were published (Wolfenden *et al.*, 1981). The hydration profile of PC-TP derived from these potentials is shown in Figure 6C. It is evident that the maxima of hydration correspond to the minima of hydrophobicity. It is striking that the most pronounced maximum of hydration representing the peptide Arg^{118}-Gln-Arg-Gln-Glu^{122} is very near the most pronounced maximum of surrounding hydrophobicity, i.e., the peptide Val^{130}–Ala^{135}.

As a first approximation, the occurrence of α-helix, β-strand and β-turn structures in PC-TP have been predicted by using the methods of Chou and Fasman (1978) and Lim (1974a,b). As shown in Figure 7, both methods yield predictions that agree rather well, with an estimated 33% of the residues having α-helix structure, 27% β-strand structure, and 6% β-turn. It is evident, particularly in the N-terminal, that half-extended segments of α-helix occur. The above-mentioned hydrophobic peptides, i.e., Val^{98}–Val^{103}, Val^{130}–Ala^{135}, Gly^{148}–Val^{152}, Val^{171}–Phe^{176}, and Trp^{186}–Ala^{192}, are all predicted to have a β-strand conformation. The method of Chou and Fasman predicts β-turns at positions 2–5, 165–168, and 178–181 of which the latter two are present on either

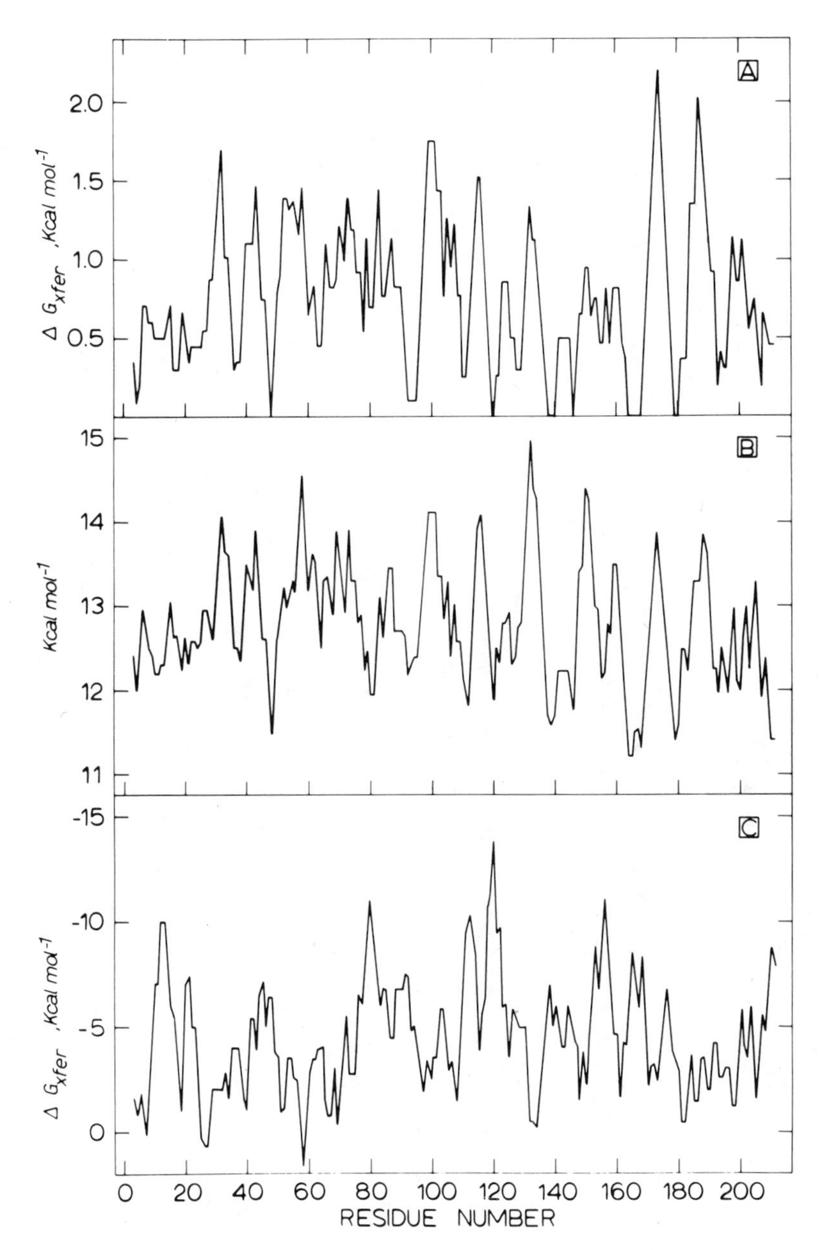

Figure 6. Hydrophobicity profile (A), surrounding hydrophobicity profile (B) and hydration profile (C) of PC-TP. The profiles were smoothed by taking a five-point moving average along the peptide chain Profile (A) was obtained by the method of Rose (1978), profile (B) by the method of Ponnuswamy *et al* (1980), and profile (C) by the method of Wolfenden *et al*. (1981).

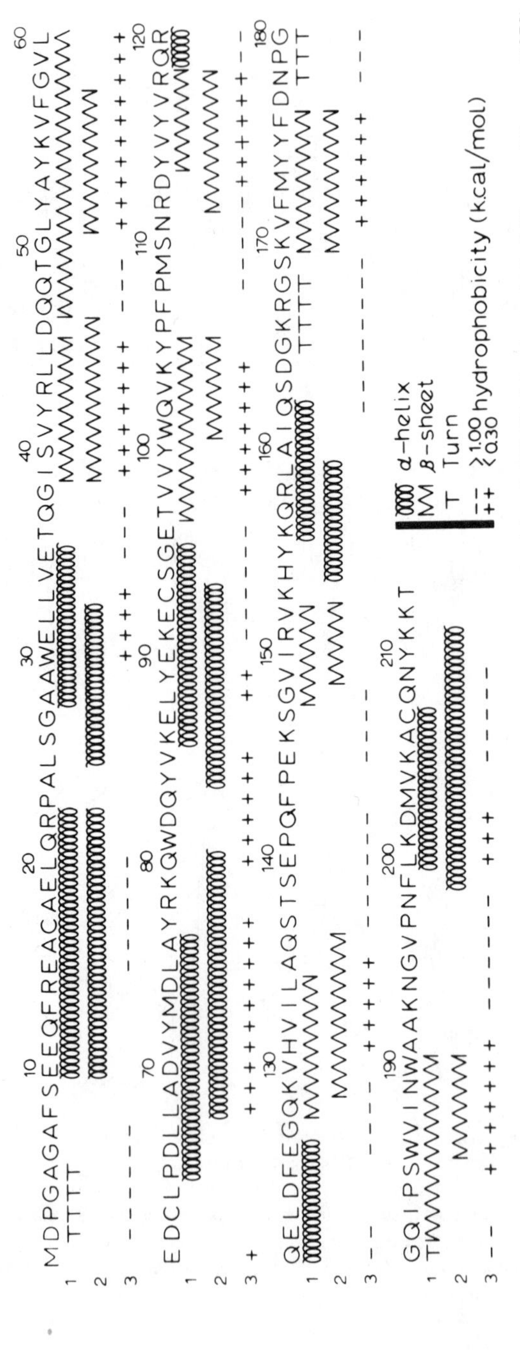

Figure 7. Predicted secondary structural elements and hydrophobicity maxima and minima in PC-TP. Line 1, method of Chou and Fasman (1978); line 2, method of Lim (1974a,b). Coils represent α-helix; zig-zags, β-strand; and T, β-turns. Line 3 indicates maxima (+ + +, 4.2 kJ/mole or more) and minima (---, 0-1.3 kJ/mole) of hydrophobicity. From Akeroyd *et al.* (1982).

side of the peptide Val[171]–Phe[176] that forms part of the lipid-binding site (see Section II-C.2). Regions with minimal (0-1.3 kJ/mole) and maximal (4.2 kJ/mole and up) hydrophobicity are indicated in Figure 7 as well. Virtually all maxima correspond with a predicted α-helix or β-strand structure.

The predicted secondary structural elements can be folded in a tentative model in which the information on hydrophobicity along the protein chain and the positions of the disulfide bonds are incorporated (Figure 8). In agreement with what is observed for numerous (α + β) proteins (Sternberg and Thornton, 1978; Janin and Chothia, 1980) the nonpolar core is formed by hydrophobic β-strands forming three antiparallel β-sheets (residues 39–46/51–57, 128–134/148–152, 170–176/182–190). The connecting peptide segments between these pairs of β-strands (residues 47–50, 135–147, 177–181, respectively) have a zero hydrophobicity suggesting that these segments are situated on the surface of the protein (Sternberg and Thornton, 1978; Rose and Roy, 1980). Another antiparallel β-sheet is thought to be formed by the hydrophobic β-strands of the sequences Tyr[114]-Val-Tyr-Val-Arg[118] and Val[99]-Tyr-Trp-Gln-Val-Lys-Tyr[105]. These sequences were predicted to have an amphiphilic character (Lim, 1974a). In view of both this character and the minimum hydrophobicity of the adjacent α-helix at positions 119–126, this β-sheet may be located more toward the surface of the protein.

The α-helical structures have been situated at the surface of the protein shielding the apolar core (Rose and Roy, 1980; Kanehisa and Tsong, 1980). This localization

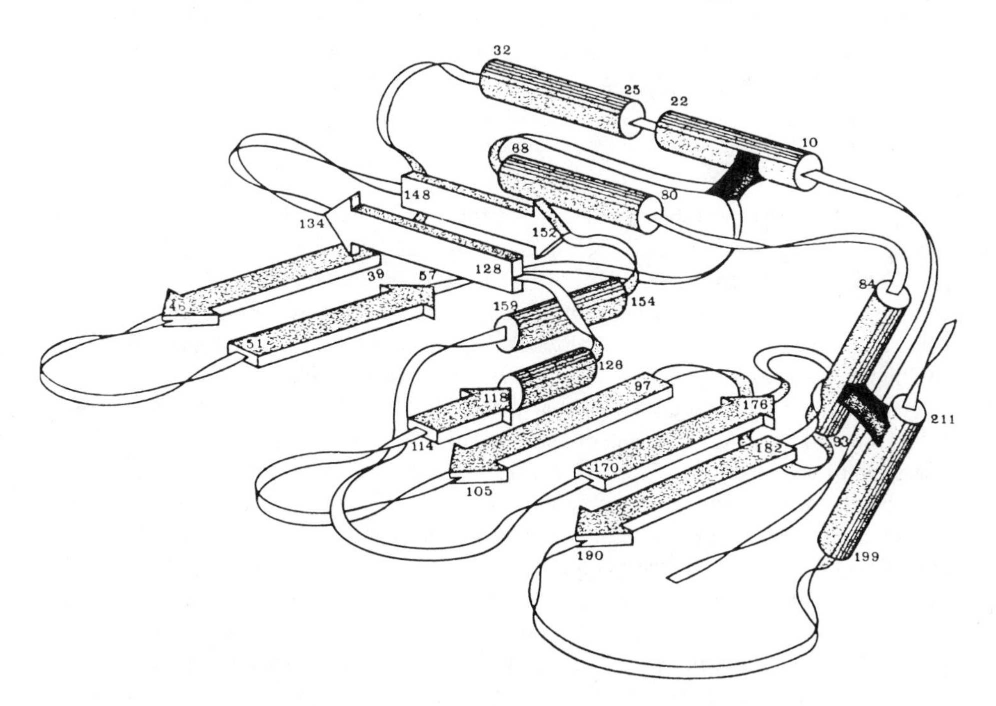

Figure 8. A folding model of the predicted secondary structural elements of PC-TP. Cylinders represent α-helix and arrows, β-strand. A slight modification of the model proposed by Akeroyd *et al.* (1982).

is in agreement with the very low hydrophobicity of the α-helices at positions 10–22, 84–93, 154–159, and 199–211 (see Figure 7). In spite of the relatively high hydrophobicity, the α-helices at positions 25–32 and 68–80 are also thought to occur at the interface as helical wheel constructions have indicated their amphiphilic character (Schulz and Schirmer, 1979).

The disulfide bonds at Cys^{17}–Cys^{63} and Cys^{93}–Cys^{207} are important for the stabilization of the protein. Reduction by dithiothreitol inhibited the transfer activity (Akeroyd *et al.*, 1982). It is evident that the bond at Cys^{17}–Cys^{63} restricts the arrangement of the secondary structures at the N-terminus. The bond at Cys^{93}–Cys^{207} may be important for the alignment of the strongly hydrophobic and amphipathic β-strand at positions 97–105 close to the β-sheet (residues 170–190) which forms part of the lipid-binding site (see Section II-C.2). At present, it is not known what segment of the protein interacts with the interface. In view of the position of the lipid-binding site in this tentative model, we suggest that the mixed β-sheet/α-helix at residues 97–126 may be involved in this interaction. This β-sheet contains two amphiphilic β-strands which would favor binding to interfacial phospholipid molecules. This interaction would be further stabilized by ionic bonds with the adjacent α-helix Arg^{118}-Gln-Arg-Gln-Glu-Leu-Asp-Phe-Glu^{126}. Recently, evidence was provided for the role of arginine residues in the PC-TP-mediated transfer reaction (Akeroyd *et al.*, 1981b). Specific modification with the α-dicarbonyl reagents butanedione and phenylglyoxal completely inhibited the transfer activity. Modification was limited to three arginine residues (out of a total of ten) of which one residue appeared to be essential. We have to await the identification of this essential residue to see whether it confirms our ideas on how PC-TP may be folded and interact with the membrane.

2. *Lipid-Binding Site*

As a carrier, PC-TP has a binding site for one PC molecule. In order to identify this lipid binding site, the photolabeling approach was tried with analogs of PC carrying photolabile 4-azido-2-nitrophenoxy- or m-diazirinophenoxy moieties on the *sn*-2-fatty acyl chain (Robson *et al.*, 1982). In a first study, PC was used which contained, at the glycerol-*sn*-2 position, the nitrene precursor 7-(4-azido-2-nitrophenoxy) [1-^{14}C]heptanoic acid (Moonen *et al.*, 1979). Incorporation into PC-TP occurred by exchange with vesicles prepared of this photolabile PC. Photolysis of the photolabile PC-TP complex resulted in covalent association of the reactive species with the protein in a yield of approximately 30%. The remaining 70% was not accounted for, but may have been lost by reaction of the generated nitrene with the *sn*-1-fatty acyl chain of the bound PC analog and/or water molecules present near the binding site. To determine the site(s) of crosslinking, the photolyzed complex was reduced with a thiol reagent, carboxymethylated, citraconylated, and then digested by a protease from *Staphylococcus aureus*. Fractionation by molecular sieve chromatography indicated that the ^{14}C label was primarily in a peptide fragment representing the C-terminal half (residues 146–213, see Figure 5). Further digestion with trypsin demonstrated that the coupling was mostly restricted to the peptide segment Gly^{168}-Ser-Lys-Val-Phe-Met-Tyr-Tyr^{175}. The point of interest is that this segment contains part of the extremely hydrophobic cluster of amino acids Val^{171}–Phe^{176} (Figure 6A). Application of PC carrying the carbene precursor 11-(m-diazirinophenoxy) [1-^{14}C]undecanoic acid in the *sn*-2 position (PC II; Figure 9) gave essentially the same information (Wirtz *et al.*, 1980b).

Recently, we have succeeded in gathering more specific information on the actual sites of crosslinking by the use of PC II and its shorter chain analog PC I which carried 6(m-diazirinophenoxy) [1-^{14}C]hexanoic acid in the *sn*-2-position (Figure 9). In this study (Westerman *et al.*, 1983b), PC-TP-containing covalently coupled PC I or PC II, was degraded by protease from *Staphylococcus aureus*, trypsin, and cyanogen bromide, and specific ^{14}C-labeled peptides were sequenced by automated Edman degradation. Major sites of coupling, as shown by release of radioactivity, were identified as Tyr^{54} and, in agreement with previous studies (Moonen *et al.*, 1979), the hydrophobic peptide segment Val^{171}-Phe-Met-Tyr-Tyr-Phe-Asp^{177}. Both PC I and PC II coupled extensively to Tyr^{54} (90% and 50% of total labeling, respectively). The remainder of the radioactivity was released from the peptide Val^{171}–Asp^{177} with a distinct preference in the pattern of release depending on whether PC I or PC II was used. Thus, coupling occurred preferentially to Tyr^{175} and Asp^{177} with PC I while Val^{171} and Met^{173} were labeled preferentially by PC II.

As pointed out in Section II-C.1 (Figure 8) the peptide segment Lys^{170}–Phe^{176} is predicted to form an antiparallel β-sheet with the segment Gln^{182}–Trp^{190}. A schematic drawing of this antiparallel β-sheet is presented in Figure 10. Also indicated are the *sn*-2-fatty acyl chains of PC I and II (drawn to scale) with the possible sites of coupling. The observed shift of Asp^{177}/Tyr^{175} for PC I to Met^{173}/Val^{171} for PC II is compatible with the difference in fatty acyl chain length of 0.6 nm, i.e., five carbons. From this shift, one may conclude that PC has one specific orientation in the lipid-binding site with a preference of the *sn*-2-fatty acyl chain for one phase of the β-strand as indicated by the alternating sites of coupling, i.e., Val^{171}, Met^{173}, Tyr^{175}, Asp^{177}. In this model, the polar head group points towards the β-turn and would be predicted to be accommodated in the surface of the protein.

This leaves the question as to how to interpret the extensive coupling of both PC I and PC II to Tyr^{54}. The fact that, in spite of the variation in *sn*-2-fatty acyl chain

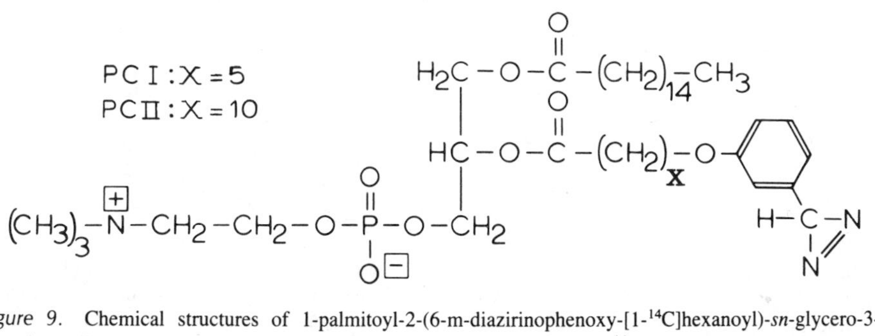

PC I : X = 5
PC II : X = 10

Figure 9. Chemical structures of 1-palmitoyl-2-(6-m-diazirinophenoxy-[1-^{14}C]hexanoyl)-*sn*-glycero-3-phosphocholine (x = 5) and 1-palmitoyl-2-(11-m-diazirinophenoxy-[1-^{14}C] undecanoyl)-*sn*-glycero-3-phosphocholine (x = 10).

length, Tyr^{54} and none of its neighboring amino acid residues was labeled, would argue against Tyr^{54} being a part of the actual binding site. On the other hand, this Tyr^{54} may favor a noncovalent association with the diazirinophenoxy group of the fatty acyl chain. Preferential labeling of specific tyrosine residues has previously been observed in photolabeling studies on the dinitrophenyl binding site of a mouse myeloma protein using 2,4-dinitrophenylazide (Yoshioka *et al.*, 1973; Lifter *et al.*, 1974).

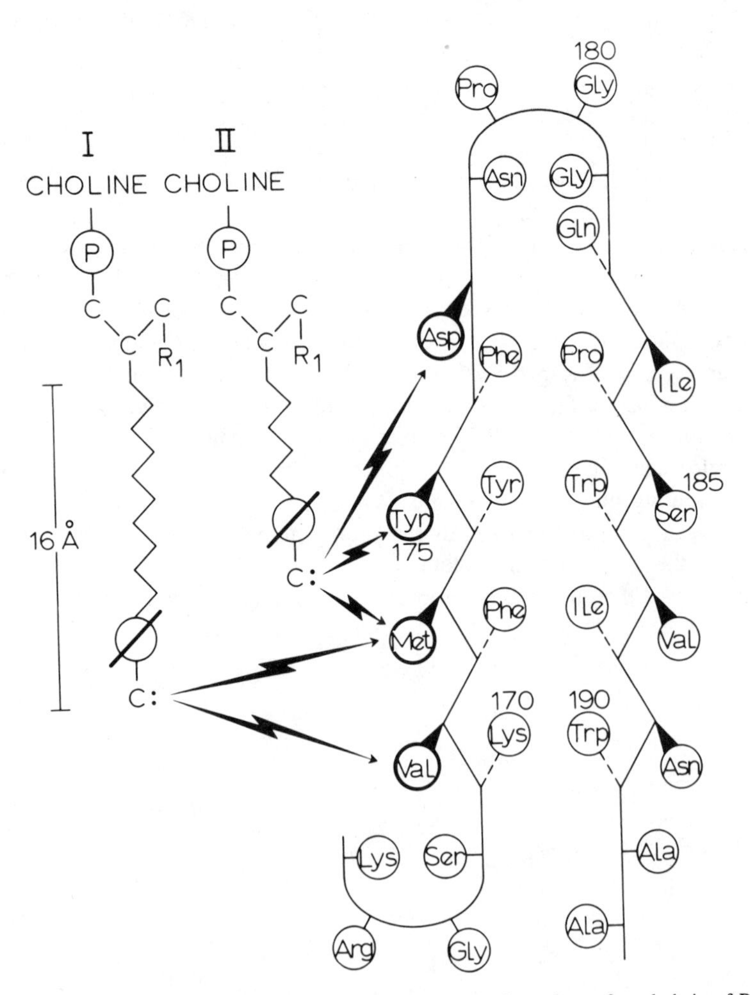

Figure 10. Tentative model of the peptide segment that accommodates the *sn*-2-acyl chain of PC in the lipid-binding site of PC-TP. Folding is based on the predictive analyses by Akeroyd *et al.* (1982). I and II represent the photogenerated carbenes of the PC analogs in Figure 9. Possible sites of crosslinking are indicated by the arrows.

In addition to photolabeling studies, knowledge about the organization of the lipid-binding site has been gained from studies with spin-labeled PC analogs and phospholipases. Incubation of PC-TP with vesicles containing PC with a nitroxide group on the *sn*-2-fatty acyl chain gave a spin-labeled PC-TP complex (Devaux *et al.*, 1977; Machida and Ohnishi, 1978). From the ESR spectrum it followed that the spin-labeled PC in the complex was strongly immobilized independent of whether the spin-label was close to the carboxy terminal (carbon-5) or methyl end (carbon-16) of the acyl chain. This suggests that the 2-acyl chain interacts strongly over its whole length, i.e., 16–18 carbon atoms, with the hydrophobic peptide segment Val^{171}–Phe^{176} (Figure 10). Very surprisingly, addition of ascorbate (2 mM at 0°C for 3 hr) had no effect on the intensity of the immobilized spectrum. This shows that, in the binding site, the nitroxide groups on the *sn*-2-fatty acyl chain are shielded from the medium. Studies with phospholipases have confirmed that the PC molecule in the lipid-binding site is not accessible from the outside (Kamp *et al.*, 1975b). Incubation of PC-TP containing $[^{14}C]$-PC, with a variety of phospholipases, demonstrated a lack of hydrolysis, suggesting that the fatty acid ester bonds and the polar head group were well protected.

D. Concluding Remarks

The available evidence strongly suggests that the lipid-binding site of PC-TP consists of a hydrophobic cleft which shields the bound PC molecule from the medium. This binding site is, in part, formed by the antiparallel β-sheet Lys^{170}–Phe^{176}/Gln^{182}–Trp^{190} noted for its high hydrophobicity (Figures 6 and 10). At present, we do not know how this binding site relates to the site that interacts with the membrane or, more specifically, to the site that recognizes the polar head group of PC. However, the interaction with the membrane appears to be governed by ionic bonds possibly involving the phosphorylcholine head group and arginine residue(s). From our current knowledge, it is tempting to propose the following tentative model for the mode of action of PC-TP.

As a carrier, PC-TP forms a transient collision complex with the interface. By entering an environment of a relatively low dielectric constant, the protein undergoes a conformational change. As a result, hydrophobic peptide segments (for example, the above hydrophobic antiparallel β-sheet) are exposed at the interface which leads to a local extrusion of water molecules. The ensuing neutralization of the "hydrophobic effect" will severely destabilize that part of the interface in contact with PC-TP. On the other hand, this apolar environment enhances the presumed ionic interaction between PC-TP and PC at the interface. As a consequence, PC bound to PC-TP will equilibrate with interfacial PC resulting in the exchange reaction. In this model, insertion and extraction of a PC molecule by PC-TP are not necessarily coupled events. If PC-TP forms an efficient collision complex with an interface devoid of PC, net insertion of PC may result. The fact that upon interaction PC-TP comes again off the membrane suggests a fine balance between hydrophobicity and hydrophilicity. However, it awaits X-ray analysis to make any further statements on the molecular organization of this protein.

III. PHOSPHATIDYLCHOLINE TRANSFER PROTEIN FROM RAT LIVER

A. Some Characteristics

From initial attempts to purify PC-TP from rat liver, it became clear that this protein differed substantially from bovine PC-TP. This led to the development of a new and specific purification method (Lumb *et al.*, 1976). Recently, a modified version of the original method was presented by which a homogeneous protein (5300-fold purified) was obtained at a yield of 17% (Poorthuis *et al.*, 1980). Characterization by sodium dodecyl sulfate gel electrophoresis demonstrated that rat PC-TP has a molecular weight of 28,000 which compares favorably with that of bovine PC-TP (Moonen *et al.*, 1980). On the other hand, isoelectric focusing showed a major difference in that rat PC-TP collected at a pH of 8.2 and bovine PC-TP at a pH of 5.8. Another remarkable difference was the lack of cross-reactivity for the antisera against rat and bovine PC-TP (Poorthuis *et al.*, 1980). Titration of rat PC-TP by anti-rat PC-TP-IgG completely inhibited the transfer activity, while the activity of bovine PC-TP was not affected (Figure 11). Moreover, anti-PC-TP-IgG was not cross-reactive with rat PI-TP and nsL-TP, confirming the high specificity for rat PC-TP (Teerlink *et al.*, 1981).

B. Levels in Various Tissues

1. Introductory Remarks

In eukaryotic cells, the biosynthesis of PC as well as other quantitatively important phospholipids, e.g., PE, PI, and PS, occurs exclusively in the endoplasmic reticulum (Wilgram and Kennedy, 1963; Jungalwala and Dawson, 1970; van Golde *et al.*, 1974). In view of this localization, PC-TP may be involved in the redistribution of PC from the site of biosynthesis to those intracellular sites in need of PC (see Section II-B.1). In a few studies, total PC-transfer activity in the 105,000 *g* supernatant has been correlated with certain aspects of membrane development and PC metabolism. Thus,

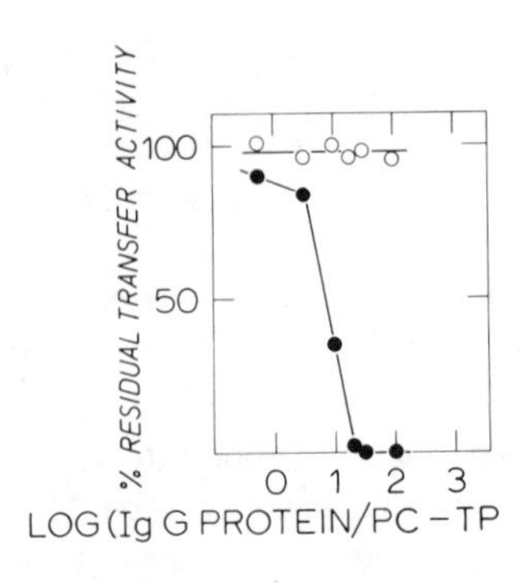

Figure 11. Effect of anti-rat liver PC-TP IgG on the activities of PC-TP from rat (●) and bovine (○) liver. PC-TP (3 μg) was incubated with increasing amounts of IgG and assayed for residual transfer activity. From Poorthuis *et al.* (1980).

in the case of maturation of the mouse lung surfactant system, PC-transfer activity and cholinephosphotransferase activity as the rate-limiting step in PC biosynthesis, changed in parallel (Engle *et al.*, 1978). In addition, for a great variety of animals, a significant correlation was found between PC-transfer activity and both the alveolar surface area and surface active material (Lumb *et al.*, 1980). As compared to normal and host liver PC-transfer activity was 2.6-fold higher in the fast growing Morris hepatoma 7777, while this activity was moderately (1.7-fold) or not increased in the slower growing 7787 and 9633 hepatomas, respectively (Poorthuis *et al.*, 1980). At this stage the validity of these correlations is difficult to judge. This is further complicated by the fact that the total PC-transfer activity is not a reliable measure for the levels of PC-TP. For example, titration of PC-transfer activity in rat liver cytosol by anti-PC-TP-IgG inhibited a maximum of 50% of the activity; the remaining activity was accounted for by PI-TP and nsL-TP (Poorthuis *et al.*, 1980). By similar titration experiments, the contribution of PC-TP to the PC-transfer activity of the 105,000 *g* supernatants from bovine brain, liver, and heart has been determined (Table 3). It is evident that the bulk of the PC-transfer activity in liver, i.e., 75%, is due to PC-TP; this protein accounts for only 15–20% of the activity in brain and heart. The remaining transfer activity which is virtually the same for all three tissues, was for the major part determined by PI-TP (Helmkamp *et al.*, 1976). In order to obtain a clear picture on what PC-TP amounts to in various rat tissues, a radioimmunoassay has been developed (Teerlink *et al.*, 1981, 1982).

2. *Radioimmunoassay*

The radioimmunoassay currently in use to measure levels of PC-TP makes use of ^{125}I-labeled PC-TP, specific anti-PC-TP-IgG isolated by antigen affinity chromatography, and *Staphylococcus aureus* cells to precipitate the antibody-bound ^{125}I-labeled PC-TP. A special feature is that the assay is performed in the presence of 1%

Table 3. Contribution of PC-TP to the Total PC-Transfer Activity in the 105,000 g Supernatant from Bovine Tissues[a]

Tissue	Transfer activity[b,c] (nmoles/min/μg)	Contribution of PC-TP (%)
Brain	0.74 (0.63)	15
Liver	2.38 (0.60)	75
Heart	0.82 (0.66)	20

[a] From Helmkamp *et al.* (1976).
[b] Activity was determined at 37°C by measuring transfer from rat liver microsomes specifically labeled with phosphatidyl [^{14}C]choline to vesicles.
[c] Numbers in parentheses represent the transfer activity corrected for the contribution of PC-TP.

(wt./vol.) Triton X-100. Application of this detergent made the assay suitable for determining levels of PC-TP bound to subcellular membranes (for further details, see Teerlink *et al.*, 1982).

For measuring the levels of PC-TP in various rat tissues, both the total homogenate and 105,000 *g* supernatant fraction were assayed (Table 4). As for the supernatants, the highest values were found in rat liver and intestinal mucosa in the range of 160–320 ng/mg protein. Kidney, spleen, and lung supernatant had intermediate values (30–90 ng/mg protein). Low levels of PC-TP were found in adrenals whereas brain and heart supernatant did not contain detectable amounts. The same trend was observed when levels of PC-TP were determined in total tissue homogenates, with the highest concentration in liver and the lowest concentration in heart. Subcellular distribution studies showed that in 10% (wt./vol.) homogenates of liver approximately 60% of PC-TP was present in the 105,000 x*g* supernatant fraction, the remainder being evenly distributed over the particulate fractions. Binding to the subcellular membranes was weak, as a single washing step released the protein almost completely. Subsequent studies with microsomes have confirmed the nonspecific nature of the binding (T. Teerlink, unpublished observation).

The high levels of PC-TP in liver and intestinal mucosa (Table 4) correlate with the very active PC metabolism of these tissues. However, the specific role of PC-TP in this metabolism remains to be established. Recently, Voelker and Kennedy (1982) suggested a role for PC-TP in the biosynthesis of sphingomyelin *in vivo*. This suggestion followed from their observations that in the plasma membrane PC was a substrate for sphingomyelin synthesis involving the transfer of the phosphocholine moiety to ceramide. With phosphatidyl[^{3}H]choline as substrate present in liposomes, the biosynthesis of [^{3}H]sphingomyelin *in vitro* was found to be almost completely dependent on PC-TP, presumably because of its ability to insert the labeled substrate in the plasma membrane.

Levels of PC-TP have also been determined in developing rat lung, together with the PC-transfer activity (Figure 12). In agreement with previous observations on de-

Table 4. Levels of PC-TP in Various Rat Tissues[a]

Tissue	PC-TP in 105,000 *g* supernatant (ng/mg protein)[b]	PC-TP in homogenate (μg/g wet weight)[b]
Liver	317 ± 65 (14)	30.3 ± 2.0 (9)
Intestinal mucosa	159 ± 16 (7)	Not determined
Kidney	92 ± 14 (4)	6.3 ± 1.3 (3)
Spleen	87 ± 8 (4)	7.8 ± 0.8 (3)
Lung	33 ± 4 (6)	1.6 ± 0.4 (3)
Adrenal	13 (2)	Not determined
Brain	Not detected	0.55 ± 0.04 (3)
Heart	Not detected	0.42 ± 0.05 (3)

[a] From Teerlink *et al.* (1982).
[b] Numbers in parentheses are the numbers of samples.

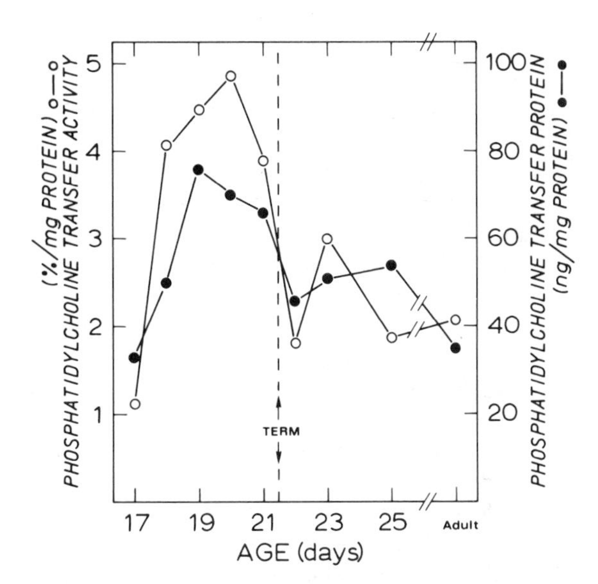

Figure 12. Level of PC-TP and PC-transfer activity in rat lung pH 5.1 supernatant during development. PC-TP (●) was determined by radioimmunoassay and transfer activity (○) by microsome-vesicle assay. From Teerlink *et al.* (1982).

veloping mouse lung (Engle *et al.*, 1978), PC-transfer activity was maximal at 2 days before term, concomitant with maximal levels of PC-TP. Available evidence suggests that there is a correlation between levels of PC-TP, rate of PC biosynthesis, and the production of lung surfactant. On the other hand, Post *et al.* (1980) failed to detect PC-TP in isolated alveolar epithelial type II cells which are considered to be the site of surfactant synthesis. The application of the radioimmunoassay has confirmed the absence of PC-TP from these types of cells (T. Teerlink, unpublished observation). Moreover, studies on rabbit lung have suggested that a specific transfer protein may exist for dipalmitoyl-PC, the major component of lung surfactant (Tsao, 1980). At present, more research is needed to clarify the specific role of PC-TP in the metabolism of lung phospholipids as well as other tissues. It is to be expected that the radioimmunoassay will help in achieving this goal.

IV. GENERAL CONCLUSIONS

This paper has given an overview of recent developments in the research on PC-TP from bovine and rat liver. This protein belongs to a series of lipid-transferring proteins that commonly occur in mammalian and plant tissues. The interest in these proteins stems from a number of considerations. First, the (phospho)lipid transfer proteins known to date have different specificities. This makes these proteins interesting models in the investigation of lipid–protein interactions. Second, these proteins have the unique property to insert and extract (phospho)lipid molecules from both biological and model membranes, that is, structures generally considered to be of a minimum energy conformation. This raises some fundamental questions about the forces which allow lipid molecules to shuttle between the accommodating lipid environment and presumed hydrophobic pockets on the proteins. Third, it seems inherent to their mode of action that the (phospho)lipid transfer proteins only "sense" the lipids in the outer monolayer, thereby not disturbing the membrane structure. This has made these proteins useful tools in the study of the phospholipid organization of membranes, providing information on phospholipid asymmetry and transbilayer movement. Last, intracellular lipid movement between membranes *in situ* occurs possibly integrated in processes like membrane biosynthesis and maintenance of membrane lipid composition. (Phospho)lipid transfer proteins may play an essential role in the regulation of this movement. The molecular characterization of bovine PC-TP has rendered insights in structure and mode of action, which may be relevant to other intracellular carriers of hydrophobic ligands, e.g., fatty acids, steroid hormones, vitamin A derivatives, and tocopherol. Further research may lead to some general principles characteristic for the functioning of those carrier proteins.

REFERENCES

Abdelkader, A. B., and Mazliak, P., 1970, Echange de lipides entre mitochondries, microsomes et surnageant cytoplasmique de cellules de pomme de terre ou de chou-fleur, *Eur. J. Biochem.* **15:**250–262.

Akeroyd, R., Moonen, P., Westerman, J., Puyk, W. C., and Wirtz, K. W. A., 1981a, The complete primary structure of the phosphatidylcholine-transfer protein from bovine liver. Isolation and characterization of the cyanogen bromide peptides, *Eur. J. Biochem.* **114:**385–391.

Akeroyd, R., Lange, L. G., Westerman, J., and Wirtz, K. W. A., 1981b, Modification of the phosphatidylcholine transfer protein from bovine liver with butanedione and phenylglyoxal. Evidence for one essential arginine residue, *Eur. J. Biochem.* **121:**77–81.

Akeroyd, R., Lenstra, J. A., Westerman, J., Vriend, G., Wirtz, K. W. A., and van Deenen, L. L. M., 1982, Prediction of secondary structural elements in the phosphatidylcholine transfer protein from bovine liver, *Eur. J. Biochem.* **121:**391–394.

Akiyama, M, and Sakagami, T., 1969, Exchange of mitochondria lecithin and cephalin with those in rat liver microsomes, *Biochim. Biophys. Acta* **187:**105–112.

Akutsu, H., and Seelig, J., 1981, Interaction of metal ions with phosphatidylcholine bilayer membranes, *Biochemistry* **20:**7366–7373.

Bangham, A. D., Standish, M. M., and Watkins, J. C., 1965, Diffusion of univalent ions across the lamellae of swollen phospholipids, *J. Mol. Biol.* **13:**238–252.

Bozzato, R. P., and Tinker, D. O., 1982, Effects of membrane fluidity and identification of the rate-limiting step in the protein-mediated phosphatidylcholine exchange reaction, *Can. J. Biochem.* **60:**409–418.

Burstein, E. A., Vedenkino, N. S., and Ivkova, M. N., 1973, Fluorescence and the location of the tryptophan residues in protein molecules, *Photochem. Photobiol.* **18:**263–279.

Chou, P. Y., and Fasman, G. D., 1978, Prediction of the secondary structure of proteins from their amino acid sequence, *Adv. Enzymol.* **47:**45–148.

Cobon, G. S., Crowfoot, P. D., Murphy, M., and Linnane, A. W., 1976, Exchange of phospholipids between mitochondria and microsomes *in vitro* stimulated by yeast cell cytosol, *Biochim. Biophys. Acta* **441:**255–259.

Cohen, L. K., Lueking, D. R., and Kaplan, S., 1979, Intermembrane phospholipid transfer mediated by cell-free extracts of *Rhodopseudomonas sphaeroides*, *J. Biol. Chem.* **254:**721–728.

Crain, R. C., and Zilversmit, D. B., 1980, Two nonspecific phospholipid exchange proteins from beef liver 1. Purification and characterization, *Biochemistry* **19:**1433–1439.

Damen, J., Regts, J., and Scherphof, G., 1982, Transfer of [^{14}C]phosphatidylcholine between liposomes and human plasma high density lipoprotein. Partial purification of a transfer-stimulating plasma factor using a rapid transfer assay, *Biochim. Biophys. Acta* **712:**444–452.

De Kruyff, B., and Wirtz, K. W. A., 1977, Induction of a relatively fast transbilayer movement of phosphatidylcholine in vesicles. A ^{13}C NMR study, *Biochim. Biophys. Acta* **468:**318–326.

Demel, R. A., Wirtz, K. W. A., Kamp, H. H., Geurts Van Kessel, W. S. M., and Van Deenen, L. L. M., 1973, Phosphatidylcholine exchange protein from beef liver, *Nature New Biol.* **246:**102–105.

Devaux, P. F., Moonen, P., Bienvenue, A., and Wirtz, K. W. A., 1977, Lipid-protein interaction in the phosphatidylcholine exchange protein, *Proc. Natl. Acad. Sci. USA* **74:**1807–1810.

DiCorleto, P. E., Fakharzadeh, F. F., Searles, L. L., and Zilversmit, D. B., 1977, Stimulation by acidic phospholipid of protein-catalyzed phosphatidylcholine transfer, *Biochim. Biophys. Acta* **468:**296–304.

DiCorleto, P. E., Warach, J. B., and Zilversmit, D. B., 1979, Purification and characterization of two phospholipid exchange proteins from bovine heart, *J. Biol. Chem.* **254:**7795–7802.

Dollady, D., Grosbois, M., Guerbette, F., and Kader, J. C., 1982, Purification of a basic phospholipid transfer protein from maize seedlings, *Biochim. Biophys. Acta* **710:**143–153.

Duckwitz-Peterlein, G., Eilenberger, G., and Overath, P., 1977, Phospholipid exchange between bilayer membranes, *Biochim. Biophys. Acta* **469:**311–325.

Dyatlovitskaya, E. V., Timofeeva, N. G., and Bergelson, L. D., 1978, A universal lipid exchange protein from rat hepatoma, *Eur. J. Biochem.* **82:**463–471.

Engle, M. J., Van Golde, L. M. G., and Wirtz, K. W. A., 1978, Transfer of phospholipids between subcellular fractions of the lung, *FEBS Lett.* **86:**277–281.

Hauser, H., Hinckley, C. C., Krebs, J., Levine, B. A., Phillips, M. C., and Williams, R. J. P., 1977, The interaction of ions with phosphatidylcholine bilayers, *Biochim. Biophys. Acta* **468:**364–377.

Helmkamp, G. M., 1980, Concerning the mechanism of action of bovine liver phospholipid exchange protein: Exchange or net transfer, *Biochem. Biophys. Res. Commun.,* **97:**1091–1096.

Helmkamp, G. M., Harvey, M. S., Wirtz, K. W. A., and Van Deenen, L. L. M., 1974, Phospholipid exchange between membranes. Purification of bovine brain proteins that preferentially catalyze the transfer of phosphatidylinositol, *J. Biol. Chem.* **249:**6382–6389.

Helmkamp, G. M., Nelemans, S. A., and Wirtz, K. W. A., 1976, Immunological comparison of phosphatidylinositol and phosphatidylcholine exchange proteins in bovine brain, liver and heart, *Biochim. Biophys. Acta* **424:**168–182.

Huang, C., and Mason, J. T., 1978, Geometric packing constraints in egg phosphatidylcholine vesicles, *Proc. Natl. Acad. Sci. USA* **75:**308–310.

Ihm, J., Harmony, J. A. K., Ellsworth, J. L., and Jackson, R. L., 1980, Simultaneous transfer of cholesteryl ester and phospholipid by protein(s) isolated from human lipoprotein-free plasma, *Biochem. Biophys. Res. Commun.* **93:**1114–1120.

Janin, J., and Chothia, C., 1980, Packing of α-helices onto β-pleated sheets and the anatomy of α/β proteins, *J. Mol. Biol.* **143:**95–128.

Johnson, L. W., Hughes, M. E., and Zilversmit, D. B., 1975, Use of phospholipid exchange protein to measure inside–outside transposition in phosphatidylcholine liposomes, *Biochim. Biophys. Acta* **375:**176–185.

Jungalwala, F. B., and Dawson, R. M. C., 1970, Phospholipid synthesis and exchange in isolated liver cells, *Biochem. J.* **117:**481–490.

Kader, J. C., 1975, Proteins and the intracellular exchange of lipids. 1. Stimulation of phospholipid exchange between mitochondria and microsomal fractions by proteins isolated from potato tuber, *Biochim. Biophys. Acta* **380:**31–44.

Kader, J. C., Dollady, D., and Mazlick, P., 1983, Phospholipid transfer proteins, in: *Phospholipids* (J. N. Hawthorne and G. B. Ansell, eds.), Elsevier Biomedical Press, Amsterdam, pp. 279–311.

Kamp, H. H., Wirtz, K. W. A., and Van Deenen, L. L. M., 1973, Some properties of phosphatidylcholine exchange protein purified from beef liver, *Biochim. Biophys. Acta* **318:**313–325.

Kamp, H. H., Wirtz, K. W. A., and Van Deenen, L. L. M., 1975a, Delipidation of the phosphatidylcholine exchange protein from beef liver by detergents, *Biochim. Biophys. Acta* **398:**401–414.

Kamp, H. H., Sprengers, E. D., Westerman, J., Wirtz, K. W. A., and Van Deenen, L. L. M., 1975b, Action of phospholipases on the phosphatidylcholine exchange protein from beef liver, *Biochim. Biophys. Acta* **398:**415–423.

Kamp, H. H., Wirtz, K. W. A., Baer, P. R., Slotboom, A. J., Rosenthal, A. F., Paltauf, F., and Van Deenen, L. L. M., 1977, Specificity of the phosphatidylcholine exchange protein from bovine liver, *Biochemistry* **16:**1310–1316.

Kanehisa, M. I., and Tsong, T. Y., 1980, Local hydrophobicity stabilizes secondary structures in proteins, *Biopolymers* **19:**1617–1628.

Kasper, A. M., and Helmkamp, G. M., 1981, Protein-catalyzed phospholipid exchange between gel and liquid-crystalline phospholipid vesicles, *Biochemistry* **20:**146–151.

Lifter, J., Hew, C. L., Yoshioka, M., Richards, F. F., and Konigsberg, W. H., 1974, Affinity-labeled peptides obtained from the combining region of myeloma protein 460, *Biochemistry* **13:**3567–3571.

Lim, V. I., 1974a, Structural principles of the globular organization of protein chains. A stereochemical theory of globular protein secondary structure, *J. Mol. Biol.* **88:**857–872.

Lim, V. I., 1974b, Algorithms for prediction of α-helical and β-structural regions in globular proteins, *J. Mol. Biol.* **88:**873–894.

Lumb, R. H., Kloosterman, A. D., Wirtz, K. W. A., and Van Deenen, L. L. M., 1976, Some properties of phospholipid exchange proteins from rat liver. *Eur. J. Biochem.* **69:**15–22.

Lumb, R. H., Cottle, D. A., White, L. C., Hoyle, S. N., Pool, G. L., and Brumley, G. W., 1980, Lung phosphotidylcholine transfer in six vertebrate species. Correlations with surfactant parameters, *Biochim. Biophys. Acta* **620:**172–175.

Machida, K., and Ohnishi, S., 1978, A spin-label study of phosphatidylcholine exchange protein. Regulation of the activity by phosphatidylserine and calcium ion, *Biochim. Biophys. Acta* **507:**156–164.

Machida, K., and Ohnishi, S., 1980, Effect of bilayer membrane curvature on activity of phosphatidylcholine exchange protein, *Biochim. Biophys. Acta* **596:**201–209.

Manavalan, P., and Ponnuswamy, P. K., 1978, Hydrophobic character of amino acid residues in globular proteins, *Nature (London)* **275:**673–674.

Martin, F. J., and MacDonald, R. C., 1976, Phospholiid exchange between bilayer membrane vesicles, *Biochemistry* **15:**321–327.

McLean, L. R., and Phillips, M. C., 1981, Mechanism of cholesterol and phosphatidylcholine exchange or transfer between unilamellar vesicles, *Biochemistry* **20:**2893–2900.

McMurray, W. C., and Dawson, R. M. C., 1969, Phospholipid exchange reactions within the liver cell, *Biochem. J.* **12:**91–108.

Moonen, P., Haagsman, H. P., Van Deenen, L. L. M., and Wirtz, K. W. A., 1979, Determination of the hydrophobic binding site of phosphatidylcholine exchange protein with photosensitive phosphatidylcholine, *Eur. J. Biochem.* **99:**439–445.

Moonen, P., Akeroyd, R., Westerman, J., Puyk, W. C., Smits, P., and Wirtz, K. W. A., 1980, The primary structure of the phosphatidylcholine exchange protein from bovine liver. Isolation and characterization of the staphylococcal protease peptides and the amino acid sequence of the N-terminal half (residues 11-122), *Eur. J. Biochem.* **106:**279–290.

Nichols, J. W., and Pagano, R. E., 1981, Kinetics of soluble lipid monomer diffusion between vesicles, *Biochemistry* **20:**2783–2789.

Nozaki, Y., and Tanford, C., 1971, The solubility of amino acids and two glycine peptides in aqueous ethanol and dioxane solutions, *J. Biol. Chem.* **246:**2211–2217.

Ponnuswamy, P. K., Prabhakaran, M., and Manavalan, P., 1980, Hydrophobic packing and spatial arrangement of amino acid residues in globular proteins, *Biochim. Biophys. Acta* **623:**301–316.

Poorthuis, B. J. H. M., Van der Krift, T. P., Teerlink, T., Akeroyd, R., Hostetler, K. Y., and Wirtz, K. W. A., 1980, Phospholipid transfer activities in Morris hepatomas and the specific contributions of the phosphatidylcholine exchange protein, *Biochim. Biophys. Acta* **600:**376–386.

Post, M., Batenburg, J. J., Schuurmans, E. A. J. M., and Van Golde, L. M. G., 1980, Phospholipid-transfer activity in type II cells isolated from adult rat lung, *Biochim. Biophys. Acta* **620:**317–321.

Reynolds, J. A., 1982, Interaction between proteins and amphiphiles, in: *Lipid-Protein Interactions,* Vol. 2 (P. C. Jost and O. H. Griffith, eds.), Wiley-Interscience, New York, pp. 193–224.

Robinson, M. E., Wu, L. N. Y., Brumley, G. W., and Lumb, R. H., 1978, A unique phosphatidylcholine exchange protein isolated from sheep lung, *FEBS Lett.* **87:**41–44.

Robson, R. J., Radhakrishnan, R., Ross, A. H., Takagaki, Y., and Khorana, H. G., 1982, Photochemical cross-linking in studies of lipid–protein interactions, in: *Lipid–Protein Interactions,* Vol. 2 (P. C. Jost and O. H. Griffith, eds.), Wiley-Interscience, New York, pp. 149–192.

Rose, G. D., 1978, Prediction of chain turns in globular proteins on a hydrophobic basis, *Nature (London)* **272:**586–590.

Rose, G. D., and Roy, S., 1980, Hydrophobic basis of packing in globular proteins, *Proc. Natl. Acad. Sci. USA* **77:**4643–4647.

Roseman, M. A., and Thompson, T. E., 1980, Mechanism of the spontaneous transfer of phospholipids between bilayers, *Biochemistry* **19:**439–444.

Rothman, J. E., and Dawidowicz, E. A., 1975, Asymmetric exchange of vesicle phospholipids catalyzed by the phosphatidylcholine exchange protein. Measurement of inside-outside transitions, *Biochemistry* **14:**2809–2816.

Sasaki, T., and Sakagami, T., 1978, A new assay system of phospholipid exchange activities using concanavalin A in the separation of donor and acceptor liposomes, *Biochim. Biophys. Acta* **512:**461–471.

Schulz, G. E., and Schirmer, R. H., 1979, Patterns of folding and association of polypeptide chains, in: *Principles of Protein Structure* (C. R. Cantor, ed.), Springer-Verlag, Heidelberg, pp. 66–165.

Stein, O., and Stein, Y., 1969, Lecithin synthesis, intracellular transport and secretion in rat liver, *J. Cell Biol.* **40:**461–483.

Sternberg, M. J. E., and Thornton, J. M., 1978, Prediction of protein structure from amino acid sequence, *Nature* **271:**15–20.

Tanaka, T., and Yamada, M., 1979, A phosphatidylcholine exchange protein isolated from germinated castor bean endosperms, *Plant Cell Physiol.* **20:**533–542.

Teerlink, T., Poorthuis, B. J. H. M., Van der Krift, T. P., and Wirtz, K. W. A., 1981, Measurement of phosphtidylcholine transfer protein in rat liver and hepatomas by radio-immunoassay, *Biochim. Biophys. Acta* **665:**74–80.

Teerlink, T., Van der Krift, T. P., Post, M., and Wirtz, K. W. A., 1982, Tissue distribution and subcellular localization of phosphatidylcholine transfer protein in rats as determined by radio-immunoassay, *Biochim. Biophys. Acta* **713:**61–67.

Tsao, F. H. C., 1980, Specific transfer of dipalmitoyl phosphatidylcholine in rabbit lung, *Biochim. Biophys. Acta* **601:**415–426.

Van den Besselaar, A. M. H. P., Helmkamp, G. M., and Wirtz, K. W. A., 1975, Kinetic model of the protein-mediated phosphatidylcholine exchange between single bilayer liposomes, *Biochemistry* **14:**1852–1858.

Van Golde, L. M. G., Raben, J., Batenburg, J. J., Fleischer, B., Zambrano, F., and Fleischer, S., 1974, Biosynthesis of lipids in Golgi complex and other subcellular fractions from rat liver, *Biochim. Biophys. Acta* **360:**179–192.

Van Golde, L. M. G., Oldenborg, V., Post, M., Batenburg, J. J., Poorthuis, B. J. H. M., and Wirtz, K. W. A., 1980, Phospholipid transfer proteins in rat lung: Identification of a protein specific for phosphatidylglycerol, *J. Biol. Chem.* **255:**6011–6013.

Voelker, D. R., and Kennedy, E. P., 1982, Cellular and enzymic synthesis of sphingomyelin, *J. Biol. Chem.* **21:**2753–2759.

Westerman, J., Kamp, H. H., and Wirtz, K. W. A., 1983a, Phosphatidylcholine transfer protein from bovine liver, in: *Methods in Enzymology,* Vol. 98. (S. Fleischer and B. Fleischer, eds.), Academic Press, New York, pp. 581–586.

Westerman, J., Wirtz, K. W. A., Berkhout, T., Van Deenen, L. L. M., Radhakrishnan, R., and Khorana, H. G., 1983b, Identification of the lipid binding site of phosphatidylcholine transfer protein with phosphatidylcholine analogs containing photoactivable carbene precursors, *Eur. J. Biochem.,* **132:** 441–449.

Wilgram, G. F., and Kennedy, E. P., 1963, Intracellular distribution of some enzymes catalyzing reactions in the biosynthesis of complex lipids, *J. Biol. Chem.* **238:**2615–2619.

Wilson, D. B., Ellsworth, J. L., and Jackson, R. L., 1980, Net transfer of phosphatidylcholine from plasma low density lipoproteins to sphingomyelin-apolipoprotein A-II complexes by bovine liver and human plasma phospholipid exchange proteins, *Biochim. Biophys. Acta* **620:**550–561.

Wirtz, K. W. A., 1982, Phospholipid transfer proteins, in: *Lipid–Protein Interactions,* Vol. 1 (P. C. Jost and O. H. Griffith, eds.), Wiley-Interscience, New York, pp. 151–231.

Wirtz, K. W. A., and Moonen, P., 1977, Interaction of the phosphatidylcholine exchange protein with phospholipids. A fluorescence and circular dichroism study, *Eur. J. Biochem.* **77:**437–443.

Wirtz, K. W. A., and Zilversmit, D. B., 1968, Exchange of phospholipids between liver mitochondria and microsomes *in vitro, J. Biol. Chem.* **243:**3596–3602.

Wirtz, K. W. A., and Zilversmit, D. B., 1969, The use of phenobarbital and carbon tetrachloride to examine liver phospholipid exchange in intact rats, *Biochim. Biophys. Acta* **187:**468–476.

Wirtz, K. W. A., Geurts van Kessel, W. S. M., Kamp, H. H., and Demel, R. A., 1976, The protein-mediated transfer of phosphatidylcholine between membranes. The effect of membrane lipid composition and ionic composition of the medium, *Eur. J. Biochem.* **61**:515–523.

Wirtz, K. W. A., Vriend, G., and Westerman, J., 1979, Kinetic analysis of the interaction of the phosphatidylcholine exchange protein with unilamellar vesicles and multi-lamellar liposomes, *Eur. J. Biochem.* **94**:215–221.

Wirtz, K. W. A., Devaux, P. F., and Bienvenue, A., 1980a, Phosphatidylcholine exchange protein catalyzes the net transfer of phosphatidylcholine to model membranes, *Biochemistry* **19**:3395–3399.

Wirtz, K. W. A., Moonen, P., Van Deenen, L. L. M., Radhakrishnan, R., and Khorana, H. G., 1980b, Identification of the lipid binding site of the phosphatidylcholine exchange protein with a photosensitive nitrene and carbene precursor of phosphatidylcholine, *Ann. N.Y. Acad. Sci.* **348**:244–255.

Wolfenden, R., Andersson, L., Cullis, P. M., and Southgate, C. C. B., 1981, Affinity of amino acid side chains for solvent water, *Biochemistry* **20**:849–855.

Yamada, K., Sasaki, T., and Sakagami, T., 1978, Separation and purification of phospholipid exchange proteins in rat small intestinal mucosa, *J. Biochem.* **84**:855–863.

Yoshioka, M., Lifter, J., Hew, C. L., Converse, C. A., Armstrong, M. Y. K., Konigsberg, W. H., and Richards, F. F., 1973, Studies on the combining region of protein 460, a mouse γA immunoglobulin which binds several haptens. Binding and reactivity of two types of photoaffinity labeling reagents, *Biochemistry* **12**:4679–4685.

Zborowski, J., and Demel, R. A., 1982, Transfer properties of the bovine brain phospholipid transfer protein. Effect of charged phospholipids and of phosphatidylcholine fatty acid composition, *Biochim. Biophys. Acta* **688**:381–387.

Zilversmit, D. B., and Hughes, M. E., 1976, Phospholipid exchange between membranes, in: *Methods in Membrane Biology,* Vol. 7 (E. D. Korn, ed.), Plenum Press, New York, pp. 211–259.

17

Carnitine Palmitoyltransferase and Transport of Fatty Acids

Charles L. Hoppel and Linda Brady

I. INTRODUCTION

This chapter will review the data describing the mitochondrial enzyme, carnitine palmitoyltransferase (hexadecanoyl-CoA:carnitine *O*-hexadecanoyltransferase, EC 2.3.1.23), and its role in the transport and metabolism of long-chain fatty acids.

A. Metabolic Fate of Long-Chain Fatty Acids

Long-chain fatty acids represent a major energy supply for the body (Fritz, 1961). These compounds are stored as triglycerides within the body. Following breakdown of the triglycerides, non-esterified or free fatty acids are released into the circulation where they are transported bound to albumin. After uptake by cells, free fatty acids are available for metabolism. The initial step in the metabolism of long-chain fatty acids is their activation to the long-chain acyl-CoA derivative. As shown in Figure 1, a number of metabolic pathways are available for these activated long-chain fatty acids. One pathway involves modification of the acyl-CoA, a second pathway is the formation of triglycerides or phospholipids, a third is the deacylation of long-chain acyl-CoA, which results in the formation of the free fatty acid and CoA, a fourth pathway is β-oxidation within peroxisomes which produces a chain-shortened fatty acyl-CoA and acetyl-CoA, and the fifth possibility is the β-oxidation within mitochondria of long-chain acyl-CoA involving carnitine acyltransferase.

Charles L. Hoppel • Medical Research Service, VA Medical Center, Cleveland, Ohio 44106. *Linda Brady* • Departments of Pharmacology and Medicine, Case Western Reserve University School of Medicine, Cleveland, Ohio 44106.

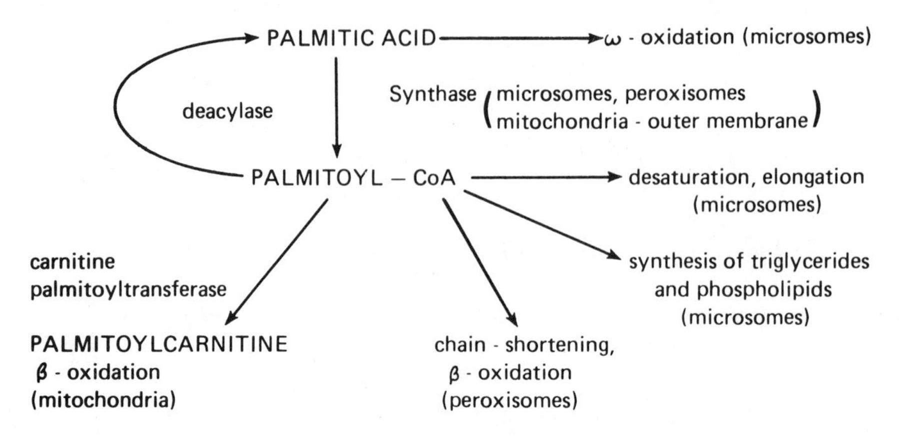

Figure 1. Scheme of long-chain fatty acid metabolism.

B. Role of Carnitine in Fatty Acid Oxidation

Fritz (1955, 1957, 1959) has demonstrated that addition of carnitine to liver preparations in the presence of radioactively labeled palmitic acid results in an increase in the incorporation of ^{14}C into $^{14}CO_2$. Detailed examination of this observation (Fritz and Yue, 1963; Bremer, 1962, 1963) has demonstrated that carnitine is a catalyst for the mitochondrial oxidation of long-chain acyl-CoA. Within the cell, long-chain acyl-CoAs are formed primarily by action of long-chain acyl-CoA synthetases present either in the endoplasmic reticulum or in the outer membrane of mitochondria. While long-chain acyl-CoA is formed in a compartment external to inner membrane of mitochondria, β-oxidation occurs within the mitochondrial matrix. Carnitine and palmitoyl-CoA are converted into palmitoyl-1-carnitine, which is involved in the transport of long-chain acyl groups across the inner membrane of the mitochondria. The reversible reaction is

$$\text{Palmitoyl-CoA} + \text{1-carnitine} \rightleftharpoons \text{palmitoyl-1-carnitine} + \text{CoA}$$

and is catalyzed by the enzyme, carnitine palmitoyltransferase.

In the past few years, the regulatory role of carnitine in this process has been explored. Although conditions associated with accelerated ketogenesis in the liver are accompanied by an increased intrahepatic content of carnitine (Brass and Hoppel, 1978a; McGarry *et al.*, 1975), this increase in liver carnitine alone is not sufficient to account for the accelerated ketogenesis.

C. Carnitine Acyltransferases

Besides carnitine palmitoyltransferase, other enzymes, carnitine acyltransferases, catalyze a reversible reaction involving acyl-CoA + carnitine leading to the formation of acylcarnitine + CoA. The reaction is

$$\text{Acyl-CoA} + \text{l-carnitine} \rightleftharpoons \text{acyl-l-carnitine} + \text{CoA}$$

Based on substrate specificity, four groups of carnitine acyltransferases have been described: (1) long-chain (carnitine palmitoyltransferase), (2) the intermediate chain (carnitine octanoyltransferase), (3) branched-chain, and (4) short-chain (carnitine acetyltransferase). The intermediate-chain carnitine acyltransferase and carnitine acetyltransferase activities have been demonstrated to be present not only in mitochondria, but also in peroxisomes and microsomes in liver (Markwell *et al.*, 1973). The rat liver microsomal and peroxisomal carnitine acetyltransferase activity has been partially purified (Markwell *et al.*, 1976) and recently, the peroxisomal carnitine octanoyltransferase has been purified and shown to be a separate enzyme from the other acyltransferases (Farrell and Bieber, 1983).

D. Carnitine Ester Hydrolase

In addition to carnitine acyltransferases, one other enzyme, carnitine ester hydrolase, has been described (Mahadevan and Sauer, 1969) that catalyzes the following reaction with acylcarnitines as substrate:

$$\text{Acyl-l-carnitine} + H_2O \rightarrow \text{fatty acid} + \text{l-carnitine}$$

and it is located in the microsomal fraction of rat hepatocytes (Mahadevan and Sauer, 1969; Hoppel and Tomec, 1972). In human liver, the enzyme is localized in mitochondria (Hagen *et al.*, 1979). The enzyme is specific for the l isomers of carnitine, has high K_ms for the carnitine esters, and does not catalyze the reverse reaction. In rats, the enzyme activity is present in liver, kidney, pancreas, small intestine, and the brain (Mahadevan and Sauer, 1969; Berge and Broch, 1981), and in guinea pigs, in brown adipose tissue mitochondria (Berge *et al.*, 1979). The physiological role of this enzyme has not been defined.

II. ASSAYS

A. General Comments

The measurement of the activity of carnitine palmitoyltransferase is complicated not only by specific problems, but also by general problems. Of the substrates, two, palmitoyl-CoA and palmitoylcarnitine, are surface-active agents (Zahler *et al.*, 1968; Yalkowsky and Zografi, 1970) and are potent detergents. The critical micelle concentration of palmitoyl-CoA is 3–4 μM (Zahler *et al.*, 1968), while that of the cationic form of palmitoylcarnitine is 15 μM (Yalkowsky and Zografi, 1970). Powell *et al.* (1981) have provided data that suggest the reported critical micelle concentration of acyl-CoAs is too low and the aggregation number reported earlier is too high. In addition, palmitoyl-CoA localizes at the liquid–air interface, thus reducing the actual aqueous concentration in dilute solutions (Barden and Cleland, 1969). Furthermore, the binding of palmitoyl-CoA to proteins is influenced by the ionic strength of the solution (Wood, 1973). These properties of the substrates must be taken into account in assaying carnitine palmitoyltransferase.

Three types of assay have been developed. Exchange activity is measured in one, whereas the other two assays directly measure activity. These direct assays may be classified as those favoring the measurement of the forward reaction and those favoring the measurement of the backward reaction.

B. Isotope-Exchange Method

The isotope-exchange method of assay was developed by Bremer (1963). Palmitoylcarnitine, carnitine, and CoA are incubated with the enzyme which is usually within mitochondria, and the exchange of radioactively labeled carnitine with palmitoylcarnitine is measured. Activity is expressed as counts per minute (cpm) of carnitine exchanged with palmitoylcarnitine because the specific activity of the carnitine is constantly changing during the assay. Objections have been raised to the use of this method in quantitating enzymatic activity (Bieber *et al.*, 1972; Saggerson, 1982). Bieber *et al.* (1972) showed that the presence of varying amounts of palmitoyl-CoA deacylase activity results in significant changes in exchange activity. Solberg (1974) modified the isotope-exchange assay to use either intramitochondrial (endogenous) CoA or extramitochondrial (exogenous) CoA + intramitochondrial CoA to allow comparison using these different pools of CoA.

C. Forward Reaction

The measurement of the forward reaction

$$\text{Palmitoyl-CoA} + \text{carnitine} \rightarrow \text{palmitoylcarnitine} + \text{CoA}$$

involves measuring either the disappearance of the substrate, palmitoyl-CoA, or the formation of one of the products, either palmitoylcarnitine or CoA. When palmitoyl-CoA disappearance or CoA release is measured, the presence of palmitoyl-CoA deacylase in the reaction

$$\text{Palmitoyl-CoA} + H_2O \rightarrow \text{palmitic acid} + \text{CoA}$$

must be assessed and corrections made to determine transferase activity. The thioester bond of acyl-CoA has an absorption peak at 232 nm. Measurement of the substrate disappearance at 232 nm is used primarily in purified enzymic fractions (Norum, 1964), but is not practical in crude preparations because mitochondria have high absorption in this area. The carnitine-dependent formation of CoA can be measured by coupling the release of CoA to the formation of colored complexes with various reagents such as Ellman's reagent (DTNB) (West *et al.*, 1971; Bieber *et al.*, 1972) or 4,4′-dipyridine disulfide (DPD) (Ramsay and Tubbs, 1975). These assays are sensitive because of the high extinction coefficient of the colored complexes with CoA.

The incorporation of radioactively-labeled carnitine into palmitoylcarnitine can be measured, providing a highly sensitive assay (Bremer, 1963; Bremer and Norum, 1967a; VanTol and Hulsmann, 1969; Hoppel and Tomec, 1972). This last assay is the most common assay used today with mitochondria. With the use of appropriate palmitoyl-CoA concentrations and short incubation times, the deacylase activity does not significantly change the substrate concentration. Particular attention needs to be placed on the concentration of carnitine used because of the differences in K_m between the two enzymatic activities in the mitochondria (Section IV-A).

D. *Backward Reaction*

Measurement of the backward reaction

$$\text{Palmitoylcarnitine} + \text{CoA} \rightarrow \text{palmitoyl-CoA} + \text{carnitine}$$

usually involves measuring the formation of the products, palmitoyl-CoA or carnitine. Palmitoyl-CoA formation can be measured either by the change in extinction at 232 nm in purified fractions (Norum, 1964), or reaction of palmitoyl-CoA with neutral hydroxylamine to form palmitoylhydroxymate (Fritz and Yue, 1963; Hoppel and Tomec, 1972). Hydroxylamine is a weak inhibitor of the transfer of palmitoylcarnitine and CoA (900 mM hydroxylamine produces about 50% inhibition), but the activity is sufficiently high that this poses no major problems (Hoppel and Tomec, 1972). Another approach is to measure the CoA-dependent release of carnitine from palmitoylcarnitine (Bremer and Norum, 1967a; Hoppel and Tomec, 1972); this allows measurement of palmitoylcarnitine hydrolase activity. The assay measuring the release of free carnitine using radioactively-labeled palmitoylcarnitine is both sensitive and reliable.

$$\text{Palmitoyl CoA} + \text{carnitine} \rightarrow \text{palmitoylcarnitine} + \text{coenzyme A}$$

III. MITOCHONDRIAL LOCALIZATION

A. *Intracellular Localization*

Carnitine palmitoyltransferase has been shown to be an exclusively mitochondrial enzyme when measured using the forward reaction (Hoppel and Tomec, 1972; Markwell *et al.*, 1973), the backward reaction (Hoppel and Tomec, 1972), or by the isotope-exchange method (Norum, 1965a; Norum and Bremer, 1967; VanTol and Hulsmann, 1969).

B. *Functional Studies in Mitochondria*

Using isolated mitochondria, the oxidation of palmitic acid supplemented with CoA and ATP is greatly stimulated by the addition of carnitine (Fritz, 1959; Fritz and Yue, 1963; Bremer, 1962). Under these circumstances, palmitoylcarnitine is formed during the incubation (Bremer, 1963; Fritz and Yue, 1963). Moreover, palmitoyl-CoA oxidation is stimulated by the addition of carnitine; again, palmitoylcarnitine is formed during the incubation (Fritz and Yue, 1963; Bremer and Norum, 1967c). With an isotope dilution experiment, Fritz and Yue (1963) have demonstrated that unlabeled palmitoylcarnitine decreased the radioactive incorporation from palmitoyl-[1-^{14}C]-CoA to $^{14}Co_2$. Palmitoylcarnitine is oxidized without the necessity of added cofactors. Acyl-CoA is the substrate for the first enzyme in β-oxidation, namely, acyl-CoA dehydrogenase; acylcarnitines are not substrates for the dehydrogenase. Fritz and Yue (1963) and Fritz (1967) have proposed a hypothesis to explain the role of carnitine in fatty acid oxidation in which mitochondria contain two carnitine palmitoyltransferases separated by a barrier for acyl-CoA. Two compartments are postulated involving the following reactions:

Compartment I:

Palmitoyl-CoA + carnitine → palmitoylcarnitine + coenzyme A

Compartment II:

Palmitoylcarnitine + coenzyme A → palmitoyl-CoA + carnitine.

C. Mitochondrial Membrane Fractionation

Further indirect evidence for a dual localization of transferase activity is presented by Yates and Garland (1966) who have demonstrated an overt (or soluble) as well as latent (membrane-bound) carnitine palmitoyltransferase in rat liver mitochondria. The overt carnitine palmitoyltransferase activity is inhibited by 2-bromostearoyl-CoA, while the latent activity is inhibited only if the mitochondria are disrupted before the addition of the inhibitor (Garland *et al.*, 1969; Yates and Garland, 1970). In contrast to this indirect evidence, mitochondrial fractionation studies by Norum *et al.* (1966), VanTol and Hulsmann (1969), and Haddock *et al.* (1970) do not show a dual distribution of carnitine palmitoyltransferase in rat liver mitochondria; instead, only an inner membrane localization is found. By combining mitochondrial membrane separation, kinetic properties, and the ability to oxidize palmitoyl-CoA + carnitine, Hoppel and Tomec (1972) have observed the release of a portion of mitochondrial carnitine palmitoyltransferase activity from the inner membrane-matrix fraction that coincided both with a change in the shape of the CoA-saturation curve and a loss in the ability of the resultant preparations to oxidize palmitoyl-CoA + carnitine. This evidence suggests that the released carnitine palmitoyltransferase activity is necessary for the oxidation of palmitoyl-CoA, but not for the oxidation of palmitoylcarnitine.

Based on these results, a scheme for the role of carnitine palmitoyltransferase in fatty acid oxidation is shown in Figure 2. An enzymatic activity catalyzing the formation of palmitoyl-1-carnitine from palmitoyl-CoA + 1-carnitine, designated carnitine palmitoyltransferase A, is loosely bound to the external surface of the inner membrane. Digitonin treatment releases this enzymatic activity from mitochondria, resulting in a preparation which has a severely limited ability for carnitine-dependent palmitoyl-CoA oxidation, but which retains the ability to oxidize palmitoyl-1-carnitine. The other

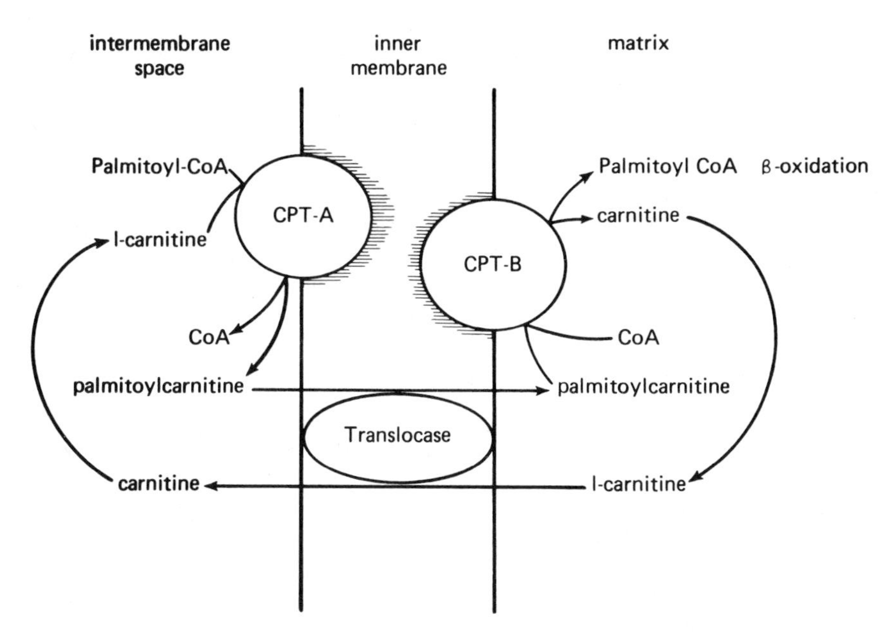

Figure 2. Scheme of the role of two carnitine palmitoyltransferases and carnitine–acylcarnitine translocase in long-chain fatty acid transport and oxidation in mitochondria. CPT-A = carnitine palmitoyltransferase A; CPT-B = carnitine palmitoyltransferase B; translocase = carnitine–acylcarnitine translocase.

mitochondrial enzymatic activity, carnitine palmitoyltransferase B, catalyzes the formation of intramitochondrial palmitoyl-CoA from exogenous palmitoylcarnitine and endogenous coenzyme A. This latter activity is firmly bound to the inner membrane and is not released by digitonin treatment.

Brosnan *et al.* (1973) have fractionated beef liver mitochondria using phospholipase A to remove the outer membrane. Using inhibition by antibodies prepared against their purified carnitine palmitoyltransferase (Kopec and Fritz, 1973; see Section IV), they have observed one carnitine palmitoyltransferase activity (CPT-1) in the inner membrane-matrix fraction (only external surface of the inner membrane exposed), and two activities (CPT-I and -II) in sonic preparations of the inner membrane (both external and internal surfaces of the inner membrane exposed).

D. Substrate-Specificity Studies in Mitochondria

Solberg (1974) has investigated the acyl-group specificity of the mitochondrial pools of carnitine acyltransferase activity. A method for assay of these pools has been developed using the "isotope-exchange method." The outer pool of transferase activity is measured in the presence of external CoA. The inner pool of carnitine palmitoyltransferase is measured in the absence of externally added CoA; any endogenous external CoA is removed by oxidation with tetrathionite. In rat liver, as well as in mouse and calf liver mitochondria, the inner transferase pool has a broad acyl-group specificity; maximum activity is observed in the region of the medium-chain acylcarnitines with the optimal substrate being at C_7. The outer transferase pool of activity in rat and mouse liver mitochondria shows little or no activity with short-chain acylcarnitines. Maximum activity of the outer pool is observed with C_9, and a lesser peak of activity at C_{14} and C_{15}. In contrast, the outer pool of transferase in calf liver mitochondria shows activity toward short-chain as well as toward medium- and long-chain acylcarnitines. Solberg (1974) has compared the specificity of the fractions obtained by treating rat liver mitochondria with digitonin. The specificity pattern in the soluble digitonin fraction (carnitine palmitoyltransferase A) resembles the pattern observed in the outer pool measured by the Solberg technique. The only difference noted is significant activity with short-chain acylcarnitines in the soluble digitonin fraction, an observation explained by partial extraction of the inner transferase from the mitochondria.

IV. PURIFICATION AND CHARACTERIZATION OF CARNITINE PALMITOYLTRANSFERASE

A. Purified Carnitine Palmitoyltransferase

As with most membrane-bound enzymes, the "solubilization" of the enzyme from the mitochondrial inner membrane has been a major obstacle in developing suitable purification schemes. Because the information available on purified carnitine palmitoyltransferases continues to be conflicting and fragmentary, this section will present first the data of investigators who have purified carnitine palmitoyltransferase(s) previously reviewed (Hoppel, 1976) and add the two new schemes for purification to this data base, and then our views on the correlation of this information with the scheme depicted in Figure 2. The first purification of the enzyme was reported by Norum (1964) and involves freeze-thawing mitochondria and washing in 1 mM glutathione. The resultant sediment in 1 mM glutathione is lyophilized and the residue extracted in buffer. This preparation serves as the partially purified enzyme for kinetic studies, while in some other studies the enzyme is further purified by DEAE-cellulose chromatography.

Using the former enzyme preparation, the K_m for carnitine is 0.25 mM and for palmitoyl-CoA is 10 μM. Palmitoylcarnitine has a K_m of 40 μM and CoA has a K_m of 50 μM (Bremer and Norum, 1967a). Palmitoyl-CoA is a competitive inhibitor (K_i = 3 μM) of carnitine (Bremer and Norum, 1967a). It is interesting that the K_i coincides with the critical micelle concentration for palmitoyl-CoA. Whether or not this reflects the surface-active properties is uncertain, although at low palmitoyl-CoA concentration, detergents are inhibitory without significantly altering the K_m for carnitine (Bremer and Norum, 1967b). However, at high palmitoyl-CoA concentrations, the addition of detergents produces a stimulation of activity with a resultant decrease in the K_m for carnitine.

Four groups have subsequently reported procedures for further purification of carnitine palmitoyltransferase. Table 1 summarizes the data on the purified carnitine

Table 1. Comparison of Purified Carnitine Palmitoyltransferases

				K_s				
Preparation[a]	Mitochondrial localization[b]	Isoelectric point	Molecular weight	Palmitoyl-carnitine	CoA	Palmitoyl CoA	K_m for carnitine	References
"Outer" CPT	CPT-A	6.0	59,000	12 μM		0.59 μM	0.14 mM	West *et al*. (1971), Edwards and Tubbs (personal communication)
"Inner" CPT	CPT-B	7.6	65,000	60 μM		9 μM	2.6 mM	West *et al*. (1971), Edwards and Tubbs (personal communication)
CPT	CPT-B			40 μM	50 μM	10 μM	0.25 mM	Norum (1964)
CPT-I	CPT-B		150,000	136 μM	5.5 μM	17.6 μM	0.45 mM	Kopec and Fritz (1971, 1973)
CPT-II			150,000					Kopec and Fritz (1971, 1973)
CPT_O	CPT-A		430,000	11 μM	35 μM	2.8 μM	0.28 mM	Bergstrom and Reitz (1980)
CPT_I	CPT-B		430,000	11 μM	34 μM	3.5 μM	0.30 mM	Bergstrom and Reitz (1980)
CPT[c]		8.05	67,000					Clarke and Bieber (1981a)

[a] CPT = carnitine palmitoyltransferase. The nomenclature is that used in the references.
[b] The assignment of mitochondrial localization is discussed in the text. This assignment is not given in the original references. The nomenclature of CPT-A and CPT-B is from Hoppel and Tomec (1972) and is shown in Figure 2.
[c] The kinetic constants are dependent on conditions for enzyme assay (Clarke and Bieber, 1981b).

palmitoyltransferases. West *et al.* (1971) start with frozen ox liver and separate enzyme activity into a soluble fraction and a membrane fraction. The soluble fraction, designated "outer," has been extensively purified. One of the criteria during purification is that 2-bromopalmitoyl-CoA + carnitine irreversibly inhibits this outer enzyme, whereas no effect is observed on the membrane or "inner" transferase. Similar patterns of inhibition are observed with 2-bromopalmitoylcarnitine + CoA. The purified "outer" CPT has an isoelectric point at 6.0, molecular weight of 59,000, and a specific activity of 16 μmoles (IU) product formed/min per mg (Edwards and Tubbs, personal communication). The K_m for carnitine is 140 μM in the presence of 50 μM palmitoyl-CoA. The K_m of 0.59 μM for palmitoyl-CoA with the "outer" transferase is below the critical micelle concentration. The distribution of acyl-chain-length specificity is shown in Table 2. Palmitoyl-CoA and myristoyl-CoA, the substrates of the forward reaction, have approximately the same V_{max}, whereas palmitoylcarnitine and hexanoylcarnitine, the substrates of the backward reaction, also have similar V_{max}.

The membrane-bound or "inner" transferase is purified from the pH 5.2 sediment of a water homogenate of frozen ox liver. A tenfold purification has been achieved. The resultant preparation has a specific activity of 1.33 IU/mg. This enzyme is inactive with acetyl-, propionyl-, or butyrylcarnitine. It is equally active with palmitoyl- and octanoylcarnitine and 150% more active with lauroyl(dodecanoyl)-carnitine. The molecular weight of the butanol-solubilized enzyme is 65,000. Relative rates with the CoA derivatives, palmitoyl-, lauroyl-, and octanoyl-CoA, are 100:70:15 (Edwards and Tubbs, personal communication).

The properties of the inner transferase depend on its state. For example, the K_m for palmitoyl-CoA is 2.2 μM in the prebutanol step, whereas after butanol treatment, it is 9 μM (Edwards and Tubbs, personal communication). The isoelectric point of a 0.5% Triton X-100 preparation of the high-speed pellet from ox liver homogenate is 5.0, but with isoelectric focusing of the butanol-soluble enzyme, it is 7.6. When ox liver mitochondria are suspended in 0.5% Triton X-100, two major regions of activity are observed, one at pH 4.8 (insensitive to 2-bromopalmitoyl-CoA) and the other at pH 5.7 (sensitive to 2-bromopalmitoyl-CoA). Therefore, the dissociation of the membrane-bound enzyme by butanol results in an increase in the K_m for palmitoyl-CoA and an increase in the isoelectric point (Edwards and Tubbs, personal communication).

Kopec and Fritz (1971) use the detergent Tween 20 to extract carnitine palmitoyltransferase activity from calf liver mitochondria. The enzymic activity in this extract has been purified (and designated CPT-I) to a specific activity of 12.3 to 23.4 IU/mg. The K_m for palmitoyl-CoA is 17.6 μM, whereas the K_m of the other substrates depends on the concentration of the companion substrate. The estimated K_m for carnitine is 0.45 mM, for palmitoylcarnitine, 0.136 mM, and for CoA, 5.5 μM. The V_{max} is greatest for myristoyl- and palmitoylcarnitine.

During purification, a portion of the enzyme protein is not absorbed by calcium phosphate gel. This enzyme fraction, designated CPT-II, shows equal enzymic rates with stearoyl- and palmitoylcarnitine as substrates, but has very low activity for myristoylcarnitine and shorter chain lengths (Kopec and Fritz, 1971). This substrate specificity is markedly different from that of the enzyme that adsorbs to the gel. Another characteristic of the CPT-II fraction is that it has to be preincubated with CoA before

Table 2. Acyl-Group Specificity of Carnitine Palmitoyltransferases

Derivatives	Preparation[c]	C_2	C_3	C_4	C_6	C_8	C_{10}	C_{12}	C_{14}	C_{16}[e]	References
Acyl-CoA[a]	"Outer" CPT	1	8	22	41	57	69	85	98	100	Edwards and Tubbs (personal communication)
	"Inner" CPT					15		70		100	
	CPT	0		1	43	113	226	166	100	100	Clarke and Bieber (1981a)
Acylcarnitine[b]	"Outer" pool	45	79	41	62	69	41	35	65	100	Solberg (1974)
	"Outer" CPT			40	45	45		76		100	Edwards and Tubbs (personal communication)
	"Inner" CPT	0	0	0	19	93		152		100	West *et al.* (1971)
	CPT	15		6		32	50	65	88	100	Norum (1964)
	CPT-I					2	26	31	99	100	Kopec and Fritz (1971)
	CPT	0		0	3	8	38	31	79	100	Clarke and Bieber (1981a)
	"Inner" pool	155	220	150	260	310	160	75	85	100	Solberg (1974)[d]

[a] Activity measuring forward reaction.
[b] Activity measuring back reaction.
[c] The sources of the preparations for enzyme purification are West *et al.* (1971) and Edwards and Tubbs (personal communication), ox liver; Norum (1964) and Kopec and Fritz (1971), calf liver mitochondria; Clarke and Bieber (1981a), beef heart mitochondria. Solberg (1974) uses intact calf liver mitochondria.
[d] These data represent the use of intramitochondrial CoA and extramitochondrial acylcarnitine as substrates. It includes activity of carnitine acetyltransferase.
[e] The velocity of the enzyme reaction with palmitoyl (C_{16}) group is taken as 100%.

activity can be measured, and the fraction is inactive with acyl-CoAs as substrate (Kopec and Fritz, 1973). If the enzynme or CoA is used to initiate the reaction, CPT-II is inactive. CPT-II is very unstable at 4°C or at liquid-nitrogen temperatures for more than 48 hr.

Kopec and Fritz (1973) injected CPT-I into three rabbits. Immunoglobulins from two of these animals (IG-I) produced 60% inhibition of CPT-I activity but did not inhibit the CPT-II fraction. However, the third rabbit (IG-II) had immunoglobulins which not only inhibited CPT-I to the same extent as did IG-I, but additionally produced 88% inhibition of CPT-II fraction activity.

Transformation of CPT-I into CPT-II has been suggested by Kopec and Fritz (1973). The dissociating agents, urea and guanidine, have been used to test this hypothesis. The criteria for conversion are the activity with myristoylcarnitine (poor substrate for CPT-II) and the necessity of CoA preincubation for CPT-II activity, but not for CPT-I. With incubation in urea, activity using myristoylcarnitine as substrate decreases, whereas activity with palmitoylcarnitine persists, although a variable recovery of total activity is observed. Although indirect, the data are consistent with the suspected transformation.

The approximate molecular weight of CPT-I is about 150,000 (Kopec and Fritz, 1973). After exposure to sodium dodecyl sulfate (SDS), CPT-I has a molecular weight of 75,000–76,000 (Kopec and Fritz, 1971, 1973) and is totally inactive. Following exposure to guanidine, the molecular weight of CPT is estimated to remain at 150,000 (Kopec and Fritz, 1973).

Using calf liver homogenized in phosphate buffer and centrifuged at low speed for 1 hr, the resultant supernatant is used by Solberg (1972) as the source of enzyme. This supernatant is fractionated with ammonium sulfate and then chromatographed on Sephadex G-200. Evaluation of results is based on chain-length specificity and recoveries through purification. Although the carnitine acyltransferase activity toward different chain lengths distributes differently through the purification, different chain-length activities could not be completely separated. These transferases have similar molecular weights and an isoelectric point of pH 6.0. This study probably deals with the "outer," easily solubilized carnitine palmitoyltransferase A activity, and with the carnitine palmitoyltransferase B ("inner") activity sedimenting during the long, low-speed centrifugation.

Bergstrom and Reitz (1980) have partially purified carnitine palmitoyltransferase from rat liver mitochondria. They initially separate the outer carnitine palmitoyltransferase activity by treatment of liver mitochondria with digitonin, and use the resultant supernatant for purification of the outer transferase. The mitochondrial membrane fraction is then disrupted and extracted with detergent; that extract is used for purification of the inner transferase. The investigators take special precautions to insure that each purification scheme was done in parallel. Therefore, using the scheme of Hoppel and Tomec (1972), they can separate the two forms based on localization within the mitochondria before purification. A 15-fold purification of the outer transferase and a 17-fold purification of the inner transferase have been achieved. The resultant preparations have specific activities of 6.8 and 36.7 nmoles/min per mg for the respective outer and inner enzymes. These specific activities are approximately

1000-fold less for the outer enzyme and 35-fold less in activity than those described by Edwards and Tubbs (personal communication). The molecular weight of both preparations is 430,000. SDS-disk gel electrophoresis of these partially purified materials shows 10–12 protein bands, but approximately 90% of the total protein is within one band. As shown in Table 1, the kinetic data are virtually identical for the two preparations. To test whether the differences *in situ* are due to the attachment of the inner carnitine palmitoyltransferase to the inner membrane, the investigators follow the reduction of acyl-CoA dehydrogenase flavoprotein to monitor movement of acyl groups across the mitochondrial inner membrane (Normann *et al.*, 1978; Norman and Flatmark, 1978). With this technique, the authors find a 450-fold greater activity of membrane-bound inner transferase than the activity of the outer transferase.

Clarke and Bieber (1981a) use beef heart mitochondria and have purified two fractions with carnitine octanoyltransferase activity, one identified as carnitine acetyltransferase and the other as carnitine palmitoyltransferase. The purification of the fraction containing carnitine palmitoyltransferase results in a single protein of greater than 95% purity. This migrates as part of a detergent micelle of apparent molecular weight 510,000; on sucrose density gradient isoelectric focusing, it has a pI of 8.05. The subunit molecular weight is 67,000 on SDS-gel electrophoresis. Kinetic data are not presented in Table 1 for this enzyme preparation because the authors find that the conditions for enzyme assay alter the kinetic determinations (Clarke and Bieber, 1981b). The distribution of acyl-chain length specificity is shown in Table 2. In the forward reaction, the activity profile is greatest with decanolycarnitine, and in the backward reaction, the acyl-group specificity is very similar to that of CPT-I described by Kopec and Fritz (1971).

B. Relationship of Purified Enzymes to Functional Localization in Mitochondria

The CPT-A activity, which we described as being located on the external surface of the inner membrane and being dissociated from the membrane by digitonin or sonication under hypotonic conditions, also appears to be dissociable under hypotonic conditions with freeze-thawing. Norum (1964) and Kopec and Fritz (1971) start with frozen mitochondria (calf liver) and, after thawing, wash the preparations under hypotonic conditions as preliminary steps. This probably results in the removal of most of the CPT-A activity from the inner membrane. Therefore, these two preparations of enzymes are probably derived mainly from the "inner" CPT-B activity. Norum (1964) uses calcium and palmitoylcarnitine, and Kopec and Fritz (1971) use Tween 20 as "solubilizing" agents. The butanol membrane fraction of West *et al.* (1971) also probably is derived from the CPT-B or inner activity. The difficulties inherent in working with membrane-bound enzymes are readily apparent from the information on the membrane-bound carnitine palmitoyltransferase. First is the change in isoelectric point, probably reflecting the removal of acidic components from the membrane, second is the change in K_m for palmitoyl-CoA, and last is the effect of urea on transforming CPT-I into CPT-II, resulting in activity with only acylcarnitine + CoA as substrates and the instability of CPT-II.

The recent studies (Bergstrom and Reitz, 1980; Clarke and Bieber, 1981a) suggest that mitochondria contain only one protein with carnitine palmitoyltransferase activity. The studies using either initial fractionation into the outer and inner form (Bergstrom and Reitz, 1980) or detergent solubilized mitochondria (Clark and Bieber, 1981a) support the concept that the same enzyme is different in functional properties by its localization in the membrane.

C. Properties of Membrane-Bound Carnitine Palmitoyltransferase

McGarry *et al.* (1978a) treat mitochondria with Tween 20, which releases a fraction of the transferase activity from the membrane. They find that this solubilized enzyme is insensitive to malonyl-CoA inhibition, while the enzyme that remains membrane-bound has increased sensitivity to malonyl-CoA. The solubilized fraction is presumably the CPT-A described by Hoppel and Tomec (1972) and the membrane-bound fraction, CPT-B (Hoppel and Tomec, 1972). While this appears contrary to malonyl-CoA inhibition of CPT-A, McGarry *et al.* (1978a) suggest that the results show that CPT-A(I) must be membrane-bound for malonyl-CoA inhibition to occur. Bremer (1981) has confirmed this observation.

Bergstrom and Reitz (1980) suggest that the catalytic properties of carnitine palmitoyltransferase are changed by their association either with the outer surface of the mitochondrial membrane or with the inner surface. The alteration produces an enhanced activity of the inner transferase. Clarke and Bieber (1981a) can purify only one protein and observe a marked effect of the state of the enzyme and substrates in measuring kinetic constants which suggests that the environment of the membrane is critical in determining the properties of the enzyme. The recent work primarily from Bremer's laboratory on the malonyl-CoA sensitivity and amount of outer carnitine palmitoyltransferase further supports this concept of the importance of the environment in determining the characteristics of the enzyme in the mitochondrial inner membrane. In particular, Bremer (1981) and Stakkestad and Bremer (1983) have shown that the total activity of carnitine palmitoyltransferase in mitochondria from fasted animals or animals who are hyperthyroid does not alter, but that the activity of the outer transferase is increased; the increase is associated with a decrease in sensitivity to malonyl-CoA. The sensitivity of CPT-A to malonyl-CoA may indicate two states of the enzyme exposed on the outer surface. Whether this is due to a binding site or protein that is different on the outer vs. the inner surface of the inner membrane or other local factors is uncertain.

D. Reversibility of Carnitine Palmitoyltransferase B

The studies of Kopec and Fritz (1971, 1973) suggest that CPT-II(B) is irreversible, catalyzing only the backward reaction. By using intramitochondrial-generated acyl-CoA, studies have shown that acyl groups can be removed from the mitochondria with the formation of an acylcarnitine; this is dependent on carnitine (Bremer and Wojtczak, 1972). We have examined the reversibility of CPT-B in isolated rat liver mitochondria when oxidizing palmitoylcarnitine at capacity (Brass and Hoppel, 1980). Under these conditions, CPT-B is reversible, while there is no change in oxygen consumption or acetoacetate production. However, the reaction leads to a decrease in the intramitochondrial acid-insoluble acyl-CoA (Brass and Hoppel, 1980). Therefore, the activity of CPT-B in the inner membrane appears to be reversible.

V. CARNITINE PALMITOYLTRANSFERASE AND MITOCHONDRIAL FATTY ACID OXIDATION

A. Is Carnitine Palmitoyltransferase Rate-Limiting for Fatty Acid Oxidation? The Role of Malonyl-CoA

The question of the rate-limiting role of carnitine palmitoyltransferase for hepatic mitochrondrial β-oxidation of fatty acids will be best resolved when specific measures of CPT-A are correlated with specific measures of β-oxidation and ketogenesis in the same preparations. At this point, this has not been done, and so there is still controversy over the role of carnitine palmitoyltransferase as the rate-limiting enzyme for mitochondrial β-oxidation of fatty acids. McGarry and Foster (1980) have proposed a tenable hypothesis of its rate-limiting function and its control. In the carbohydrate-fed state (low glucagon:insulin) hepatic fatty acid metabolism is directed toward synthesis, and the proposal suggests that fatty acid oxidation is inhibited under these conditions by high hepatic levels of malonyl-CoA (an inhibitor of CPT-A and product of acetyl-CoA carboxylase). Elevation of the glucagon:insulin ratio depresses fatty acid synthesis and malonyl-CoA production, and this lifts the inhibition of CPT-A and increases fatty acid oxidation. An increase in hepatic carnitine and acyl-CoA content under these conditions potentiates this effect, and β-oxidation and ketoacid production are further enhanced (McGarry *et al.*, 1975).

Support for the hypothesis that carnitine activates and malonyl-CoA inhibits CPT-A comes from a variety of sources (McGarry and Foster, 1979; McGarry *et al.*, 1973, 1977, 1978a,b). They have noted that any alterations which increase hepatic ketogenic capacity (AIS, glucagon, starvation alloxan diabetes) also increase hepatic carnitine. Adding carnitine to perfused livers increases ketoacid output with oleate as substrate. Early in their experiments, they realized that carnitine activation of CPT-A alone could not explain the regulation of β-oxidation since in lactating rats, hepatic carnitine was increased fourfold, with only a slight increase in ketoacid production over controls (Robles-Valdes *et al.*, 1976). These experiments led to the realization that the glycogen content of the liver is important. When depressed as in diabetes and starvation, carnitine increases hepatic ketogenesis, but when elevated as in lactation, the effect is not apparent. This could have to do with the levels of pyruvate derived from glycogen stores which become available for oxaloacetate synthesis, as will be discussed below. From these observations, McGarry and Foster have proposed a relationship between carbohydrate metabolism and fatty acid oxidation, but could find no metabolic effector among glycolytic intermediates. They propose that fatty acid synthesis and oxidation are reciprocal and note that malonyl-CoA, the first committed step in fatty acid synthesis, fluctuates in parallel with fatty acid synthesis, and inversely with fatty acid oxidation. Malonyl-CoA is proposed as a specific inhibitor of CPT-A, as it blocks oleate, palmitate, and palmitoyl-CoA oxidation, but not that of octanoate, octanoylcarnitine, or palmitoylcarnitine. They also find that the apparent K_i for malonyl-CoA (1–2 μM) is compatible with purported tissue levels. In experiments with isolated hepatocytes, malonyl-CoA levels correlate with fatty acid synthesis, and both are inverse to fatty acid oxidation. In addition, maximal fatty acid oxidation occurs in fed hepatocytes in the presence of glucagon which inhibits acetyl-CoA carboxylase (decreases malonyl-CoA) and is antiglycolytic (decreases oxaloacetate). In another series of experiments, glucose, lactate, and pyruvate were added to hepatocytes to obtain maximal fatty acid synthesis; glucagon and 5-tetradecloxy-2-furoate were added to maximize oxidation. Over a wide range of concentrations, lipogenesis correlates with malonyl-CoA levels and both are inverse to oxidation rates. Exposure of fed hepatocytes to glucagon and 5-tetradecyloxy-2-furoate increases fatty acid oxidation and ketogenesis five- and 12-fold compared to hepatocytes incubated with glucose, lactate, and pyruvate. Addition of carnitine has no effect on malonyl-CoA levels or fatty acid synthesis, but increases fatty acid oxidation to sevenfold and ketogenesis to 19-fold over the glucose, lactate, and pyruvate rates. These rates approximate fasted rates, and fasted hepatocytes do not respond to glucagon and carnitine.

There is evidence from other sources to both support and question the role of CPT-A as the primary regulator of fatty acid oxidation and ketogenesis. The increase in ketogenesis which occurs with carnitine addition to hepatocytes is not significant in fasted lean and obese Zucker rats, but that of glucagon is; addition of both produces additive results (Triscari *et al.*, 1982). Christiansen (1977) finds that glucagon stimulates palmitate oxidation from fed, but not fasted, hepatocytes, and the extent of stimulation is not dependent on intracellular carnitine. Faster, re-fed cells can be made to oxidize palmitate at the same rate as starved cells, with higher intracellular carnitine concentrations and increased concentration of palmitate. Glucagon increases long-chain acylcarnitine levels in re-fed cells, possibly due to increased long-chain acyl-CoA levels and increased carnitine levels. These data support a regulatory role for carnitine palmitoyltransferase in the increased oxidation. Harano *et al.* (1982) also find that carnitine and glucagon have an additive effect on ketogenesis in hepatocytes of chow-fed rats, but attribute the glucagon effect to a decreased K_m of CPT-A for palmitoyl-CoA. Brass and Hoppel (1978b) have administered 1-carnitine to fed rats *in vivo* and have measured hepatic long chain acylcarnitine content thereafter. The content increases to values found in fasted liver, but plasma ketone concentration remains constant in the fed carnitine-treated rats. However, they increase in fasted rats with equivalent hepatic acylcarnitine contents.

Some studies have compared carnitine palmitoyltransferase activity with measures of β-oxidation and ketogenesis. Pande (1971) argues that carnitine palmitoyltransferase cannot be rate-limiting, because its activity is higher than palmitoyl group utilization. Palmitoylcarnitine and palmitoyl-CoA oxidation rates are equivalent in his studies. DiMarco and Hoppel (1975) and Brady and Hoppel (1983a,b) have since confirmed the equivalent oxidation rates of these two substrates in rat liver mitochondria in starvation, diabetes, obesity, and with diet variation. In diabetes, Harano *et al.* (1972) have attempted to correlate CPT-1(A) with β-oxidation and find that both increase compared to controls. However, DiMarco and Hoppel (1975) have found that palmitoyl-CoA oxidation is unchanged with diabetes in isolated mitochondria, as is total carnitine palmitoyltransferase activity.

Data comparing lean and obese animals also reveal some discrepancies in the theory that CPT alone controls the rate of β-oxidation and ketogenesis. Triscari *et al.* (1982) find that in fasted obese hepatocytes, [^{14}C]palmitoylcarnitine is oxidized to ketones to the same extent as [^{14}C]palmitate, whereas in lean rats, it is oxidized at a higher rate. McCune *et al.* (1981) find that obese-rat hepatocytes are less sensitive in their ketogenic response than lean rat hepatocytes upon addition of dibutyryl-cAMP. They have suggested that increased malonyl-CoA levels in obese rats due to elevated insulin levels can blunt the dibutyryl-cAMP response. Nosadini *et al.* (1979) have perfused lean and obese fed livers with oleate and measured transferase activity with and without glucagon. In both lean and obese Zucker rats, carnitine palmitoyltransferase activity (measured by the isotope-exchange assay) is doubled with glucagon addition with no increase in ketone production. They suggest that insulin deficiency is necessary to see a correlation between increased carnitine palmitoyltransferase and increased ketogenesis. Their streptozotocin-treated lean Zucker rats have tripled the transferase activity and doubled the ketogenic rates. Brady and Hoppel (1983a) have measured total carnitine palmitoyltransferase activity (backward reaction) and β-oxidation in lean and obese Zucker rat liver mitochondria. While carnitine palmitoyltransferase is elevated in the fed obese rats, and in both lean and obese rats with starvation, β-oxidation is unchanged in every case.

Demaugre *et al.* (1983) also have done hepatocyte studies which suggest that carnitine palmitoyltransferase is not the sole regulator of β-oxidation and ketogenesis. They use α-cyanohydroxycinnamate and find that it increases ketogenesis in fed hepatocytes to fasting levels (this compound blocks pyruvate entry into the mitochondria and presumably depresses oxaloacetate levels). 5-Tetradecyloxy-2-furoate (inhibits acetyl-CoA carboxylase) increases fatty acid oxidation and ketogenesis from 0.35 mM oleate, but not from 1 mM oleate; the rates at 0.35 mM are not as high as starvation rates. When lactate, pyruvate, and mercaptopicolinate (PEP carboxykinase inhibitor) are included in the medium, ketone formation from oleate is decreased with similar effects on octanoate metabolism. The results taken together suggest a role for intracellular oxaloacetate in the regulation of fatty acid oxidation and ketogenesis. Benito and Williamson (1978) and Benito *et al.* (1979) also suggest that their data in hepatocytes cannot be explained completely by the malonyl-CoA hypothesis.

There is still dispute over the inhibitory role of malonyl-CoA on both carnitine palmitoyltransferase and β-oxidation and ketogenesis, particularly as sensitivity to effects of malonyl-CoA has been reported to decrease with fasting. Ontko and Johns (1980) and Cook *et al.* (1980) have found that oxidation of palmitate by hepatic mitochondria isolated from fed rats is inhibited to a greater degree by malonyl-CoA than in mitochondria isolated from starved rats. Bremer (1981) and Saggerson and Carpenter (1981a, 1983) have measured CPT activity and confirm that a decreased sensitivity to malonyl-CoA occurs in the enzyme of mitochondria isolated from starved rats [Bremer's notation is "outer transferase," and Saggerson and Carpenter call their enzyme CPT-1(A)].

McGarry and Foster (1981) have responded to these studies with a paper evaluating the effect of experimental conditions on the study of malonyl-CoA inhibition of CPT. They point out that the conditions used by others (Ontko and Johns, 1980; Cook *et al.*, 1980) are subject to a number of flaws which influence the interpretation of data, among them mitochondrial protein concentration, temperature of assay, time of assay, CoA concentration, and palmitate:albumin ratio. They present data which they interpret in the following way. If they set up conditions similar to Ontko and Johns (1980), they find that substrate depletion occurs, as well as malonyl-CoA hydrolysis. These conditions lead to an increased K_i of malonyl-CoA for CPT. If conditions are set up according to Cook *et al.* (1980), CoA is limiting and the palmitate:albumin ratio is too high which also leads to an increased K_i of malonyl-CoA for carnitine palmitoyltransferase.

They also present data on oleate oxidation in 18-hr-starved rats which show that shifting the malonyl-CoA concentration from 1 μM to 10 μM can produce a 75% inhibition of β-oxidation and a 95% inhibition of ketogenesis. They suggest that the sensitivity of carnitine palmitoyltransferase to malonyl-CoA is decreased by starvation, but that the decrease is not physiologically relevant. Saggerson (1982) has criticized the use of the isotope-exchange assay in these studies, but McGarry and Foster (1982) have responded that the isotope-exchange assay and the forward-reaction assay give essentially similar results (McGarry *et al.*, 1978a). However, McGarry and Foster use 0.1 mM and Saggerson uses 0.4 mM l-carnitine in their assay (K_m for CPT-A = 0.09 mM, Saggerson and Carpenter, 1981c; K_m for CPT-B = 2.1 mM, Norum, 1964; see also Section IV-A).

Robinson and Zammit (1982) have attempted to correlate isolated rat liver mitochondrial carnitine palmitoyltransferase activity to malonyl-CoA with *in vivo* levels of malonyl-CoA in tissues from which the mitochondria were derived. They find an inverse correlation between the amount of malonyl-CoA needed to produce 50% inhibition of transferase activity and the original tissue concentration of malonyl-CoA. The issue of conditions of assay is further complicated, as they use the assay of Saggerson and Carpenter (1981a), but with sixfold higher albumin; CPT-A activity is very low compared to Saggerson and Carpenter (1981a,b). They explain that low activity is due to the low palmitoyl-CoA concentration necessary to insure functional mitochondria as outlined by McGarry and Foster (1981).

The problems which exist in the complete recognition of the regulatory role of carnitine palmitoyltransferase in the control of β-oxidation and ketogenesis are CPT-A assay and the correlation of CPT-A and β-oxidation and ketogenesis, and the effect of malonyl-CoA on the enzyme as well as process.

B. Carnitine Palmitoyltransferase A

1. Assay

The importance of an intact mitochondrial membrane to see a malonyl-CoA effect has been stressed by McGarry *et al.* (1978a). To measure CPT-A and β-oxidation concurrently also required intact mitochondria. Thus, although kinetics are best done on pure enzymes, attempts to measure CPT-A in intact mitochondria are necessary. This necessitates verifying the integrity of the mitochondrial membrane, so that no CPT-B is exposed. The concentrations of carnitine and palmitoyl-CoA are also in dispute. A good kinetic assay will vary these concentrations over several-fold. The development of an assay that does not require albumin would be attractive, as the palmitate:albumin ratio will not be a question and the palmitate concentration will be known exactly.

2. *Correlation of CPT-A Activity and β-Oxidation*

CPT-A activity and β-oxidation rates must be measured under identical conditions, as must the effects of malonyl-CoA on the enzyme and on β-oxidation. It is possible that malonyl-CoA could have large inhibitory effects on CPT-A, but because its activity is higher than necessary for maximal mitochondrial β-oxidation, β-oxidation might not be affected.

VI. CARNITINE–ACYLCARNITINE TRANSLOCASE

If CPT-B is on the matrix surface of the mitochondrial inner membrane and its substrate, palmitoylcarnitine, is formed in the intermembrane space by the action of CPT-A, how does the palmitoylcarnitine traverse the inner membrane? Fritz and Yue (1963) have suggested that carnitine could move across the inner membrane and thus serve, with the transferases, as a system to carry fatty acyl groups from one side of the inner membrane to the other. Studies of the compartmentation of carnitine in mitochondria have shown that carnitine is present in the sucrose-permeable space and does not enter the matrix compartment (Yates and Garland, 1966; Brosnan and Fritz, 1971a). When incubated in an isoosmotic carnitine solution, mitochondria do not swell (Bremer, 1967; Levitsky and Skulachev, 1972). Furthermore, only a very low content of carnitine (0.1 nmole/mg protein) is observed in rat liver mitochondria (Yates and Garland, 1966). Thus, it appears that carnitine and acetylcarnitine cannot penetrate the mitochondrial inner membrane.

Levitsky and Skulachev (1972) have proposed that the acylcarnitine is protonated (the internal salt is converted to a cation) and the acylcarnitine cation moves across the mitochondrial inner membrane along a pH or electrochemical gradient. During CPT-B conversion of acylcarnitine to acyl-CoA + carnitine, the proton is lost and carnitine becomes an internal salt again and diffuses back across the membrane along a concentration gradient. Murray *et al.* (1980) have done conformational calculations on carnitine and acetylcarnitine. They propose that the "extended" conformation of carnitine interacts with the carnitine acyltransferase and the product, acylcarnitine, forms a "folded" (internal salt) conformation; this form is vectorially transferred through the inner membrane. Then the reverse reaction occurs and the "extended" conformer of carnitine is released. Therefore, their proposal is similar to that of Levitsky and Skulachev (1972), although the forms of carnitine and acylcarnitine are reversed.

Ramsay and Tubbs (1974, 1975) and Pande (1975) have provided an alternative solution. Ramsay and Tubbs (1974) find that ox heart mitochondria contain 2–4 nmoles carnitine/mg protein. When the mitochondria are incubated with [^{14}C]carnitine, there is uptake of the radioactivity into the mitochondria without an increase in the carnitine content. Pande (1975) finds that the carnitine-stimulated disposition of acetyl-CoA derived from pyruvate in rat heart mitochondria is inhibited by long-chain acyl-*d*-carnitines. This effect is not due to inhibition of carnitine acetyltransferase. Rat heart mitochondria also contain carnitine (at 2 nmoles/mg protein) and carnitine exchanges with carnitine or acylcarnitines. A series of studies (Ramsay and Tubbs, 1975, 1976; Pande and Parvin, 1976, 1980a; Parvin and Pande, 1979; Halperin and Pande, 1979; Idell-Wenger, 1981) have characterized the carnitine–acylcarnitine translocase. The translocase shows 1:1 exchange of carnitine and acylcarnitines. The K_m for carnitine in rat liver mitochondrial carnitine–carnitine exchange is 1.8 mM and the V_{max} is 1.2 nmoles/min per mg (Parvin and Pande, 1979). In the fasted rat, the liver mitochondrial exchange K_m does not change, but the V_{max} is increased (2.3 nmoles/min per mg). The rate of exchange of the translocase is rapid, and at physiological temperature, is not considered to be rate-limiting.

Carnitine is rapidly taken up by submitochondrial particles loaded with carnitine and this exchange is inhibited by mersalyl or *N*-ethylmaleimide (Schulz and Racker, 1979). Furthermore, facilitated diffusion occurs, but is slower than the carnitine–carnitine exchange. Proteoliposomes have been prepared using either submitochondrial particles or soluble proteins with octyl-β-glucoside from submitochondrial particles; both preparations exchange carnitine in the reconstituted vesicles (Schulz and Racker, 1979).

A series of compounds inhibit the exchange. Mersalyl produces a 50% decrease in exchange at 3 nmoles/mg mitochondrial protein (Pande and Parvin, 1976). Ramsay and Tubbs (1976) have synthesized 11-trimethylammonio-undecanoyl-1-carnitine, which is a competitive inhibitor and has a K_i of 56 μM. Both of the above inhibitors can be used as "stop inhibitors" in exchange assays. Acyl-*d*-carnitines are inhibitors, and as substrates, are less active than the corresponding acyl-1-carnitine (Pande, 1975). Sulfobetaines (*N*-alkyl-*N,N*-dimethyl-3-ammonio-1-propanesulfonates) have been found to be relatively specific and selective inhibitors of the translocase (Parvin *et al.*, 1980). (1-Pyrenebutyryl)carnitine is a potent inhibitor of the carnitine–acylcarnitine exchange (Wolkowicz *et al.*, 1982).

The majority of studies have examined carnitine–carnitine and carnitine–short-chain acylcarnitine exchanges. Is the carnitine–acylcarnitine translocase an integral part of the sequence of reactions in the mitochondrial inner membrane during long-chain fatty acid oxidation? As discussed above, mersalyl, acyl-*d*-carnitines and sulfobetaines are potent inhibitors of the translocase. They also inhibit palmitoylcarnitine oxidation, but the details of the inhibition have not been extensively studied. Mersalyl (10 μM) inhibits palmitoylcarnitine oxidation in uncoupled rat heart mitochondria. The timing of the addition of mersalyl is important. If added after palmitoylcarnitine, there is a lag before inhibition becomes apparent. This is explained by the uptake of palmitoylcarnitine into mitochondria occurring more rapidly than its further metabolism. Therefore, the inhibitor action will not be apparent until the accumulated palmitoylcarnitine is used. When mersalyl is examined as an inhibitor of CPT activity, the forward reaction has been measured. The amount of mersalyl required to decrease activity 50% is 100 nmol/mg mitochondrial protein. This compares with the 3 nmoles/mg for 50% inhibition of the translocase. The direction of the CPT-B activity in the mitochondria during palmitoylcarnitine oxidation is the backward reaction. The kinetics of the inhibition of the backward CPT-B activity are important, especially since the characteristics such as acyl-group specificity are different depending on the direction of the assay; this could be true for inhibitors also.

As discussed under the section on inhibitors (Section VIII-E), acyl-*d*-carnitines are competitive inhibitors of the carnitine-stimulated fatty acid oxidation in heart mitochondria (Fritz and Marquis, 1965). Pande (1975) has shown that acyl-*d*-carnitines inhibit the carnitine–acetylcarnitine exchange, and that increasing the carnitine concentration lessens the inhibition. Therefore, acyl-*d*-carnitine is proposed to decrease oxidation by inhibiting the translocase.

Sulfobetaines have been used to provide an argument for the role of the translocase in long-chain fatty acid oxidation. For the sulfobetaine, SB_{10} (the alkyl groups = decyl), the concentration to produce a 50% reduction in carnitine–carnitine exchange is 30 μM, whereas the concentration to decrease palmitoyl-1-carnitine oxidation 50% is approximately 1 mM. If the activity of the translocase is in excess, it will require achieving a concentration of inhibitor that will make the translocase rate-limiting for oxidation before inhibition will be observed. Thus, the lack of overlap in the concentration to achieve 50% inhibition is expected. The SB_{10} does not inhibit, and in fact leads to, a slight enhancement of carnitine palmitoyltransferase activity [when measured as the forward reaction (palmitoylcarnitine production)]. On the other hand, the carnitine palmitoyltransferase involved in palmitoylcarnitine oxidation is CPT-B, and the inhibitor may only affect the backward reaction (palmitoylcarnitine disapperance).

Trimethylammonio-acylcarnitines are competitive inhibitors of the translocase; trimethylammonio-hexadecanoylcarnitine has a K_i of 1 μM (Tubbs *et al.*, 1980). This derivative inhibits palmitoylcarnitine oxidation in rat liver mitochondria, but not oxidation of other substrates (50 μM produces almost complete inhibition).

(1-Pyrenebutyryl)carnitine is not a substrate for carnitine acyltransferase measured by the isotope-exchange assay (Wolkowicz *et al.*, 1982). Furthermore, it has no effect on either the forward or backward reaction of carnitine palmitoyltransferase. In contrast (1-pyrenebutyryl)carnitine produces a 50% inhibition of palmitoylcarnitine oxidation at 1.45 μM and of octanoylcarnitine oxidation at 40 nM. To test if the effect on acylcarnitine oxidation is due to inhibition of the carnitine–acylcarnitine translocase, the investigators have tested the effect on exchange of [^{14}C]carnitine-loaded mitochondria in the presence of either palmitoylcarnitine or carnitine. The K_i for (1-pyrenebutyryl)carnitine inhibition of palmitoylcarnitine–carnitine exchange is 0.58 μM and for carnitine–carnitine exchange is 23 nM (Wolkowicz *et al.*, 1982).

The inhibition of acylcarnitine oxidation by (1-pyrenebutyryl)carnitine and the inhibition of the acylcarnitine–carnitine exchange with similar K_is, but with no effect on carnitine acyltransferase activity, strongly support the concept that the carnitine–acylcarnitine translocase is an integral part of the transfer of acyl groups across the mitochondrial inner membrane.

VII. CARNITINE PALMITOYLTRANSFERASE: SUBSTRATE SPECIFICITY

A. Carnitine

The structural requirements for carnitine (Figure 3) to function in long-chain fatty acid oxidation and as a substrate for carnitine palmitoyltranferase are: (A) carboxyl groups, (B) 3-hydroxy (or 3-thio), and (C) 4-trimethylammonio (or dimethylammonio).

Norcarnitine (3-hydroxy-4-dimethylammoniobutanoate), a tertiary amino compound, is almost as effective as carnitine, although requiring higher concentrations, in stimulating fatty acid oxidation in heart muscle homogenates (Fritz *et al.*, 1962). Norcarnitine is a substrate for partially purified carnitine palmitoyltransferase with a K_m of 6 mM and a maximal velocity of 20% of that seen with carnitine (Norum, 1965b). It is also a competitive inhibitor (K_i = 20 mM) of carnitine.

When all three methyl groups are removed from the amino group (3-hydroxy-4-aminobutanoate), fatty acid oxidation is not stimulated (Fritz *et al.*, 1962). In addition, 3-hydroxy-4-aminobutanoate is not a substrate for carnitine palmitoyltransferase (Norum, 1965b). The monomethylamino derivative has not been tested. The 3-hydroxyl group can be replaced with a thiol (thiocarnitine) without loss of activity (Tubbs and Chase, 1970; Ferri *et al.*, 1980), but removal of the hydroxyl group (4-trimethylammoniobutanoate, butyrobetaine, deoxycarnitine) abolishes activity (Fritz *et al.*, 1962). Butyrobetaine is a competitive inhibitor (K_i = 0.2 M) of carnitine for carnitine palmitoyltransferase. Modifications and substitutions of the carboxyl group result in loss of activity to stimulate fatty acid oxidation or act as a substrate for the transferase.

Choline and choline derivatives cannot substitute for carnitine in supporting fatty acid oxidation. In the transferase reaction, choline is not a substrate, but the derivative palmitoylcholine stimulates transferase activity, probably as a result of a detergent effect. Therefore, there is a high degree of specificity for carnitine.

B. Acylcarnitine and Acyl-CoA

The specificity for the acyl group has been discussed in Section III-D for mitochondria and in Section IV for the various partially purified enzymes.

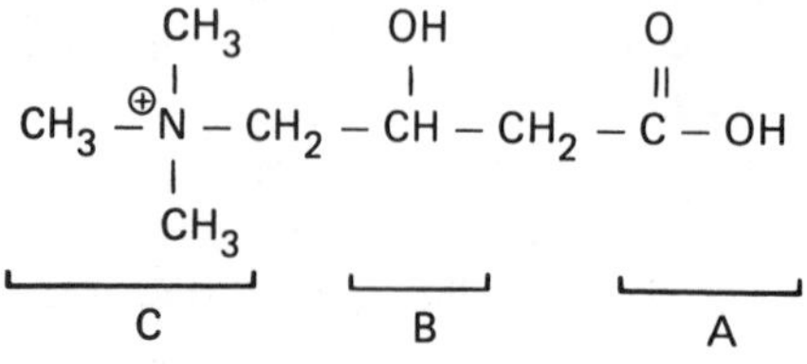

$$CH_3-\overset{\oplus}{N}(CH_3)_2-CH_2-CH(OH)-CH_2-C(=O)-OH$$

C B A

Figure 3. Chemical structure of carnitine. The substitutions at area A, B, or C are discussed in the text.

VIII. INHIBITORS OF CARNITINE PALMITOYLTRANSFERASE

A. 2-Tetradecylglycidic Acid

2-Tetradecylglycidic acid is an oral hypoglycemic agent that inhibits tissue long-chain fatty acid oxidation (Tutwiler *et al.*, 1981; Tutwiler and Dellevigne, 1979). The agent appears specific for long-chain fatty acids (Tutwiler and Dellevigne, 1979). In isolated rat liver mitochondria, 2-tetradecylglycidic acid is a potent irreversible inhibitor via the CoA ester (Tutwiler *et al.*, 1981). Following preincubation with 2-tetradecylglycidic acid, ATP, CoA, and Mg^{2+}, isolated rat liver mitochondria cannot oxidize palmitoyl-CoA + carnitine, but do oxidize palmitoylcarnitine (Tutwiler and Ryzlak, 1980). Thus, the inhibitor appears to inhibit CPT-A selectively under these conditions.

B. 2-Substituted Oxiran-2-carbonyl-CoA Esters

Other 2-substituted oxiran-2-carbonyl-CoA esters have been synthesized (besides 2-tetradecylglycidic acid) and are very potent inhibitors of CPT-A activity (Bartlett *et al.*, 1981). These inhibitors also do not inhibit the oxidation of palmitoylcarnitine.

C. 1-Pyrenebutyryl-CoA

1-Pyrenebutyryl-CoA is an inhibitor synthesized by Wolkowicz *et al.* (1982). 1-Pyrenebutyryl-CoA is not a substrate for carnitine acyltransferase, but is a competitive inhibitor (K_i = 2.1 μM) towards palmitoyl-CoA. It also is a competitive inhibitor (K_i = 14.5 μM) towards octanoyl-CoA.

D. Carba-analogue of Palmitoyl-CoA

Ciardelli *et al.* (1981) have synthesized the carba-analog of palmitoyl-CoA, heptadecan-2-onyldethio-CoA (thioester replaced by a methylene group) and find the isomeric mixture acts as a competitive inhibitor of CPT-B and has an apparent K_i of 1.4 μM. The designation of CPT-B is in contrast to their designation of CPT-I and is done here because they use the method of Kopec and Fritz (1971) to prepare the soluble partially purified preparation. They also have synthesized heptadecyl-CoA (carbonyl group is replaced by a methylene group), which is also a competitive inhibitor of CPT-B and has an apparent K_i of 3.3 μM. The effect of these inhibitors on fatty acid oxidation has not been examined. These analogues are potent inhibitors and should be useful tools to study carnitine palmitoyltransferase.

E. Acyl-d-carnitine

Palmitoyl-*d*-carnitine is a competitive inhibitor (K_i = 25–34 μM) of carnitine-induced fatty acid oxidation in heart muscle mitochondria (Fritz and Marquis, 1965). The derivative also inhibits palmitoyl-CoA + carnitine oxidation. Using a partially purified transferase from calf liver mitochondria, Fritz and Marquis (1965) have shown that palmitoyl-*d*-carnitine does not inhibit enzymic activity during either the forward or the backward reaction. When the transferase activity is examined using intact mitochondria with the isotope-exchange method, a biphasic response to palmitoyl-*d*-carnitine is observed that consists of an initial inhibition of activity and then, at higher concentrations of the inhibitor, a relief of inhibition with return of exchange activity to control level. However, in the presence of similar amounts of albumin, the amount of palmitoyl-*d*-carnitine necessary to produce 50% inhibition of the exchange activity in the intact mitochondria is ten times greater than the amount necessary to produce a comparable decrease in carnitine-dependent palmitic acid oxidation. Palmitoyl-*d*-carnitine has a detergent action but, under the conditions of these experiments, inhibition of fatty acid oxidation and the inhibition of the exchange activity occurs at lower concentrations of palmitoyl-*d*-carnitine than those producing the detergent effect. The relief of inhibition occurring at higher concentrations of palmitoyl-*d*-carnitine is associated with lysis of the mitochondria.

In agreement with the effect on exchange activity, Yates and Garland (1966) observe a K_i of 0.27 mM for palmitoyl-*d*-carnitine inhibition of the forward reaction (palmitoylcarnitine synthesis) in intact mitochondria. Presumably under these assay conditions, the "overt" ("outer") carnitine palmitoyltransferase A is measured. Following sonication of the mitochondria and centrifugation, the supernatant transferase activity is inhibited by palmitoyl-*d*-carnitine with a K_i of 0.2 mM, in agreement with the observation using intact mitochondria. On the other hand, the K_i for palmitoyl-*d*-carnitine is about ten times lower (25 μM) using the sedimented, membrane-bound transferase B activity. Although this suggests a relative specificity of acyl-*d*-carnitine for the inhibition of the two transferases, the lack of effect on the "solubilized" inner enzyme (Fritz and Marquis, 1965) has not been explained.

Metabolic actions of acyl-*d*-carnitine have been extensively explored. Its major effects are to decrease hepatic ketogenesis and gluconeogenesis (Delisle and Fritz, 1967; Williamson *et al.*, 1968; McGarry and Foster, 1973). It remains to be established whether or not these *in vivo* actions of the derivatives are due primarily to effects on carnitine palmitoyltransferase. Pande (1975) has proposed that the effects are due to inhibition of the translocase (Section VI).

F. 2-Bromoacyl Derivatives

Preincubation of rat liver mitochondria with long-chain 2-bromoacyl-CoA and l-carnitine abolishes the ability of the mitochondria to oxidize added palmitoyl-CoA, but not palmitoyl-l-carnitine (Chase and Tubbs, 1972). When rat liver mitochondria are pretreated with 2-bromostearoyl-CoA + carnitine, washed in bovine serum albumin, and then disrupted with sonic irradiation and the resultant suspension centrifuged, the greatest decrease in transferase activity is seen in the supernatant following centrifugation. However, if the mitochondria are first disrupted by sonication and then exposed to 2-bromostearoyl-CoA + carnitine, 90% of the carnitine palmitoyltransferase activity is inhibited (Garland *et al.*, 1969; Yates and Garland, 1970). Garland *et al.* (1969) and Yates and Garland (1970) conclude that both the overt and latent transferase activities are inhibited by the bromoacyl-CoA derivative. During the fractionation of frozen ox liver mitochondria, West *et al.* (1971) observe inhibition of the solubilized transferase activity in the water homogenate of the tissue, but the membrane-bound transferase activity is not inhibited when exposed to the bromoacyl-CoA derivative. Following the "solubilization" of the membrane-bound transferase activity by either butanol or Triton X-100, the preparations remain insensitive to the actions of the inhibitor. The factors responsible for these disparate results are unknown.

Preincubation of rat liver mitochondria with 2-bromopalmitoylcarnitine (Chase and Tubbs, 1972) or 2-bromomyristoylthiocarnitine (Tubbs and Chase, 1970) inhibits the oxidation of palmitoyl-CoA + carnitine, palmitoylcarnitine, pyruvate, α-ketoglutarate, and hexanoate, but not the oxidation of succinate. The addition of carnitine restores the oxidation of pyruvate, α-ketoglutarate, and hexanoate, but not of palmitoylcarnitine. In damaged mitochondria, 2-bromopalmitoyl-CoA (without added carnitine) inhibits palmitoylcarnitine oxidation but, under these conditions, does not affect pyruvate oxidation (Chase and Tubbs, 1972). The interpretation of these findings is that bromopalmitoyl-CoA inhibits long-chain acyl β-oxidation and, in addition, sequesters intramitochondrial CoA. Therefore, the inhibition of CoA-requiring substrates such as pyruvate, α-ketoglutarate, and hexanoate can be reversed by the addition of carnitine, which removes the sequestered intramitochondrial acyl groups, resulting in an increased availability of intramitochondrial CoA.

IX. CHANGES IN TISSUE ENZYMATIC ACTIVITY

A. Increases in Carnitine Palmitoyltransferase Activity

1. Inducing Agents in Rats

Clofibrate, a hypolipidemic agent, increases the activity of liver mitochondrial carnitine palmitoyltransferase during administration to rats (Daae and Aas, 1973; Kahonen, 1976). Solberg (1974) finds the inner transferase pool activity of rat liver mitochondria increases rather than the outer transferase pool. This has been confirmed by Stakkestad and Bremer (1983) who, using a different assay procedure from Solberg, observe no effect on the outer transferase, but an increase in total carnitine palmitoyltransferase activity. Clofibrate increases the mitochondrial mass in liver (Kurup *et al.*, 1970) and also increases carnitine acetyltransferase (Solberg *et al.*, 1972; Kahonen, 1976), acyl-CoA dehydrogenase, thiolase (Kahonen, 1979), and 2,4-dienoyl-CoA 4-reductase (Borrebaek *et al.*, 1980).

The oxidation of long-chain fatty acids and ketoacid production is increased in hepatocytes (Christiansen *et al.*, 1978; Mannaerts *et al.*, 1978) and homogenates (Mannaerts *et al.*, 1978; Pande and Parvin, 1980b) from clofibrate-fed rats. The liver mitochondrial oxidation of acylcarnitines is increased, whereas oxidation of succinate, glutamate, or pyruvate is not affected (Kahonen, 1979; Pande and Parvin, 1980b). The rate of oxidation of short-chain fatty acids is also increased (Christiansen *et al.*, 1978; Mannaerts *et al.*, 1978; Kahonen, 1979). At low doses of clofibrate, a selective increase in mitochondrial oxidation of polyunsaturated fatty acid occurs, while with high doses, a general stimulatory effect on fatty acid oxidation is observed (Osmundsen *et al.*, 1982).

2. *Obesity*

Hahn (1981) has observed higher liver mitochondrial carnitine palmitoyltransferase activity in young obese (ob/ob) mice than in their lean littermates. Brady and Hoppel (1983a) find higher liver mitochondrial carnitine palmitoyltransferase activity (measured by the backward reaction) in Zucker obese rats compared to their lean littermates. The mitochondrial oxidative capacity is identical in the Zucker obese and lean control rat liver mitochondria. In contrast to studies in other ketogenic situations, the mitochondrial carnitine palmitoyltransferase specific activity increases during fasting in both the lean and obese Zucker rats without a change in mitochondrial oxidative capacity for fatty acids. In addition, fatty acid oxidation in isolated hepatocytes from fed obese Zucker rats is reported to be decreased compared to lean controls; in rat hepatocytes from fasted obese Zucker rats, it is still depressed. This dissociation of carnitine palmitoyltransferase and both mitochondrial and hepatocyte fatty acid oxidation provides an intriguing model for future study.

Because the livers of obese Zucker rats are grossly fatty, we have examined the influence of dietary obesity (with fatty livers) in rats on mitochondrial function (Brady and Hoppel, 1983b). Neither carnitine palmitoyltransferase specific activity nor mitochondrial fatty acid oxidation capacity is different in the dietary obese rats compared to the lean controls. Also, during fasting, the mitochondrial specific activity of carnitine palmitoyltransferase does not change.

Nosadini *et al.* (1979) have found, using the isotope-exchange assay, that the activity of carnitine palmitoyltransferase is similar in obese Zucker rats compared to lean controls.

3. *Development of Carnitine Palmitoyltransferase and Fatty Acid Oxidation in Fetal and Neonatal Liver*

Blood ketoacids increase dramatically during the first 24 hr in suckling newborn rats (Foster and Bailey, 1976a; Yeh and Yee, 1976; Ferre *et al.*, 1978). Hepatic fatty acid oxidation increases at birth (Augenfeld and Fritz, 1970; Lockwood and Bailey, 1970; Warshaw, 1972; Foster and Bailey, 1976a) and, in addition, hepatic mitochondrial carnitine palmitoyltransferase increases (Augenfeld and Fritz, 1970; Warshaw, 1972; Foster and Bailey, 1976b; Yeh and Yee, 1979) and also the enzymes of β-oxidation (Foster and Bailey, 1976b). Saggerson and Carpenter (1982) find at birth that CPT-A increases sixfold with a decrease in sensitivity to inhibition by malonyl-CoA, whereas CPT-B activity is not affected.

In the liver of newborn piglets, carnitine palmitoyltransferase activity is about half that in 24-hr-old or 24-day-old animals (Bieber *et al.*, 1973). However, in liver mitochondria from 24-day-old animals, the rate of oxidation of palmitoyl-CoA is 2–3 times faster than in the 24-hr-old animals. Ketoacid production and hepatic fatty acid oxidation are low in the newborn guinea pig and then are markedly stimulated by 24 hr (Stanley *et al.*, 1983). The activity of CPT-A increases about 13-fold in the first 24 hr after birth. However, other sites for limitation of fatty acid oxidation are present besides CPT-A. Therefore, in all three models of suckling ketosis, a change in CPT-A activity occurs, but can only partially explain the developmental change.

B. Decreases in Carnitine Palmitoyltransferase Activity

1. Effects of Chronic Ethanol Ingestion

Parker *et al.* (1974) report a decrease in activity of liver carnitine palmitoyltransferase (20–60% of controls depending on the substrate) in rats that chronically ingest alcohol. In contrast, Cederbaum *et al.* (1975) report an increase in carnitine palmitoyltransferase, whereas hepatic mitochondrial fatty acid oxidation is decreased. Carnitine palmitoyltransferase activity is also decreased in the heart (Parker *et al.*, 1974), but Williams and Li (1977) observe no changes in heart mitochondrial carnitine palmitoyltransferase activity or fatty acid oxidation. Williams and Li (1977) use a liquid diet and then replace carbohydrate with ethanol (isocalorically) in the experimental group similar to Cederbaum *et al.* (1975). Therefore, the differences in treatment protocol and diets may be responsible for the decreases seen by Parker *et al.* (1974).

2. Chronic Myocardial Ischemia

Wood *et al.* (1973) observe decreased synthesis of palmitoylcarnitine in mitochondria isolated from ischemic canine ventricular muscle. The decrease in V_{max} is observed within 1 day of ischemia. A decrease in K_m for carnitine is observed during prolonged ischemia without a change in the K_m for palmitoyl-CoA. The change in the activation energy of transferase activity in ischemic heart mitochondria is believed to support a postulated change in the hydrophobic regions of the membrane carnitine palmitoyltransferase.

3. Fetal Heart Mitochondrial Carnitine Palmitoyltransferase

Newborn rat heart homogenates have a lower rate of oxidation of palmitate and palmitoyl-CoA than adult heart (Wittels and Bressler, 1965; Warshaw, 1972). Bovine fetal heart mitochondria oxidize palmitoyl-CoA + carnitine at greatly decreased rates compared to calf heart mitochondria (Warshaw and Terry, 1970; Tomec and Hoppel, 1975). This depressed rate of oxidation is only observed during the first 5 min of incubation. Preincubation of fetal heart mitochondria for 10 min with palmitoyl-CoA + carnitine results in oxidation rates comparable with those seen in calf heart (Tomec and Hoppel, 1975). The rate of oxidation of palmitoylcarnitine is similar in fetal and calf heart mitochondria. The total carnitine palmitoyltransferase activity of fetal heart mitochondria is not significantly different from calf heart. There is no deficiency of "overt" CPT-A in fetal heart mitochondria (Tomec and Hoppel, 1975), but an abnormal CoA-saturation curve for fetal heart transferase is observed. The relationship between this latter finding and the time-dependent oxidation of palmitoyl-CoA is not clear.

Although Brosnan and Fritz (1971b) have questioned the difference in capacity of fetal and calf heart mitochondria to oxidize palmitoyl-CoA, differences in the substrate concentration in the incubation medium probably account for their lack of observed differences (Tomec and Hoppel, 1975).

C. Changes in Distribution of CPT-A and CPT-B

1. Hypothyroidism and Hyperthyroidism in Rats

Thyroid hormone has been implicated to affect mitochondrial fatty acid oxidation by effects on carnitine palmitoyltransferase. In isolated rat hepatocytes or with the perfused liver from hyperthyroid rats, oxidation of long-chain fatty acids is increased (Keyes and Heimberg, 1979; Bartels and Sestoft, 1980; Keyes *et al.*, 1981; Laker and Mayes, 1981; Stakkestad and Bremer, 1982). Stakkestad and Bremer (1982) have found the liver mitochondrial outer CPT(A) activity is increased, but the total carnitine palmitoyltransferase activity is similar in hyperthyroid compared to control rats. Although the outer CPT(A) is inhibited by malonyl-CoA at low palmitoyl-CoA concentrations, the hyperthyroid liver mitochondria have a decreased sensitivity to malonyl-CoA. Fasting the hyperthyroid rat does not lead to any further changes than those seen in hyperthyroidism alone. Bremer's group (Stakkestad and Bremer, 1983) suggests that thyroid hormones exhibit their effects on fatty acid metabolism in the liver by an influence on the outer CPT(A) activity and that hyperthyroidism represents a maximal effect.

Hypothyroidism in rats results in a decreased fatty acid oxidation and decreased product formation (ketoacids and CO_2) in both isolated rat hepatocytes (Stakkestad and Bremer, 1983) and perfused liver (Keyes and Heimberg, 1979; Keyes *et al.*, 1981). Laker and Mayes (1981) have observed decreased CO_2 production compensated by an increased ketoacid production, so that total oxidation of oleate in perfused liver from hypothyroid rats is comparable to controls. Saggerson *et al.* (1982), using thyroidectomized rats, and Stakkestad and Bremer (1983), using propylthiouracil-treated rats, observe a decrease in the outer CPT(A) activity from hypothyroid rat liver. Stakkestad and Bremer (1983) find no change in the total carnitine palmitoyltransferase activity, whereas Saggerson *et al.* (1982) have observed a decrease in total carnitine palmitoyltransferase activity in hypothyroidism. The later total activity is less than half that observed by Stakkestad and Bremer (1983). The most obvious reason is that Saggerson *et al.* (1982) use 0.4 mM l-carnitine in their assay, whereas the K_m for carnitine of CPT-B is 2.0 mM and they do not have conditions to optimize the measurement of CPT-B. In both studies, fasting hypothyroid rats leads to an increase in liver mitochondrial outer CPT(A) activity and a decrease in sensitivity to malonyl-CoA. A normal thyroid hormone status does not appear to be necessary to see the changes produced by fasting [increased outer CPT(A) activity and decreased sensitivity to malonyl-CoA].

Stakkestad and Bremer (1983) have noted that the decrease in outer CPT(A) activity is not as striking as the decrease in palmitate oxidation in hypothyroid hepatocytes and suggested that additional factors must be operating in the hypothyroid liver to reduce fatty acid oxidation.

2. Metabolic Conditions

The specific activity of hepatic mitochondrial carnitine palmitoyltransferase activity measured by the back reaction (carnitine production) is not significantly different in fasting or diabetic animals compared to the controls (DiMarco and Hoppel, 1975). In addition, the capacity for palmitoyl-group oxidation in rat liver mitochondria isolated from fasted rats or during diabetic ketoacidosis is the same as the capacity in liver mitochondria isolated from control rats. In contrast, using the isotope-exchange assay method, Norum (1965a) has observed an increase in liver carnitine palmitoyltransferase activity in rats that were fat-fed, diabetic, or fasted. This increase is observed in both mitochondria and in the extramitochondrial compartment. Using fasted animals, the activity within the liver is not altered by treatment of the animals with ethionine, puromycin, or actinomycin. From these studies, it was suggested that the increased activity is not due to synthesis *de novo* but is caused by activation of preformed enzymes. The purported increase in activity of carnitine palmitoyltransferase during fasting also has been described by VanTol and Hulsmann (1969), Aas and Daae (1971), and VanTol (1974). However, the influence of palmitoyl-CoA deacylase activity on the exchange assay is not accounted for in these observations. We have observed, in unpublished experiments, that palmitoyl-CoA deacylase decreases during fasting, resulting in an increase in transferase activity measured by isotope exchange. Furthermore, when the data are expressed per gram wet weight of liver, this reflects the decrease in wet weight of liver resulting from dehydration, loss of glycogen, etc., with resultant increase in mitochondrial protein content per gram wet weight of liver.

Bremer (1981) has examined the distribution of transferase activity in rat liver mitochondria. He finds a near doubling of CPT-A activity after 24 hr of fasting, while the total activity increases only about 25%. The inhibition of CPT-A by malonyl-CoA is decreased in fasting rats. His studies suggest that during fasting, the increase in CPT-A activity is due to the addition of malonyl-CoA-insensitive activity accompanied by a decrease in the latent transferase (CPT-B?) activity.

D. Metabolic Myopathies Associated with Carnitine Palmitoyltransferase Deficiency

A number of patients have been described with episodic muscle pain and myoglobinuria; in skeletal muscle biopsies they have carnitine palmitoyltransferase deficiency (DiMauro and Trevisan, 1982). Some of these patients do not develop fasting ketosis and therefore are considered to have a deficiency in liver transferase as well as in skeletal muscle. Two patients have been described who have a deficiency of CPT-B (Patten *et al.*, 1979; Scholte *et al.*, 1979). In another patient with partial deficiency of skeletal muscle carnitine palmitoyltransferase with normal ketogenic response to fasting (Hostetler *et al.*, 1978), the isolated skeletal muscle mitochondria have normal respiratory control ratios and normal oxidation of nonlipid substrates. However, the mitochondria are unable to oxidize palmitoyl-CoA + carnitine, but retain the ability to oxidize palmitoylcarnitine. When examined kinetically, the enzymatic data are consistent with the loss of the enzyme activity (CPT-A) located on the external surface of the mitochondrial inner membrane, but with retention of CPT-B activity (Hoppel *et al.*, 1980). These data suggest that only one of the transferase activities is adversely affected by the development of this disease process. These mitochondria are very similar to rat liver mitochondria following treatment with selected amounts of digitonin to remove CPT-A activity. Both types of mitochondria have a functional loss of the ability to oxidize palmitoyl-CoA + carnitine, but retain the ability to oxidize palmitoylcarnitine. Although the data suggest that the two separate transferase activities are under different genetic control, if the enzyme is a single entity and the different properties are based on the attachment to the inner membrane, the attachment or binding site may be the pathogenic mechanism.

X. SUMMARY AND FUTURE

The sequence of reactions leading to the transport of long-chain acyl groups into the mitochondria involves CPT-A, located on the outer surface of the inner membrane, the carnitine–acylcarnitine translocase and CPT-B, located in the inner membrane (Figure 2). A number of unanswered questions still exist in this pathway.

First, the question of the existence of one or two carnitine palmitoyltransferase(s) is not answered. If there is only one enzyme, then the environment of the inner membrane is critical to the function of the two activities. The metabolic myopathies with either CPT-A or CPT-B deficiency may be due to the absence of a "binding" protein that anchors the enzyme into the inner membrane. It seems reasonable to suggest that one protein for CPT-A is located on the outer surface of the inner membrane, while another protein for CPT-B is located on the inner surface. On the other hand, CPT-A and CPT-B may be isoenzymes as originally suggested. There are no data on where in the cell the enzymes are made and, if in the endoplasmic reticulum, how and in what form they enter the mitochondrial inner membrane.

The carnitine–acylcarnitine translocase has been shown to be involved in the transport of acylcarnitines into the mitochondria. The activity changes under physiological stress, but it does not appear to be rate-limiting for fatty acid oxidation.

A very stimulating hypothesis has been proposed by McGarry and Foster that malonyl-CoA concentration in liver controls long-chain fatty acid oxidation by inhibiting CPT-A. From this, a number of experimental approaches have been developed to evaluate and challenge the role of the transferases. Is malonyl-CoA the major effector of long-chain fatty acid oxidation and ketogenesis? However this is answered in the future, the hypothesis itself has been important in the development of new approaches to control mechanisms in fatty acid metabolism.

The recent development of specific inhibitors of CPT-A, CPT-B, and the translocase provides the opportunity to further clarify the relationships between these steps in fatty acid oxidation.

Finally, the existence of clinical syndromes with apparent CPT-A or CPT-B deficiency provides unique opportunities for further dissection of the transferase. Is carnitine palmitoyltransferase mobile in the inner membrane and can it change from one type of activity to the other? How does this relate to the disease syndromes? Will the newer oral hypoglycemic agents that inhibit CPT-A become clinically useful agents and help to delineate the role of the transferase in clinical syndromes?

ACKNOWLEDGMENTS

The work done in the authors' laboratories has been supported by grants from the National Institutes of Health (AM 15804) and the Medical Research Service of the Veterans Administration. Dr. Brady was supported by a fellowship from the American Heart Association (Northeastern Ohio Chapter). We thank Ms. D. Collins for typing the manuscript.

REFERENCES

Aas, M., and Daae, L. N. W., 1971, Fatty acid activation and acyl transfer in organs from rats in different nutritional states, *Biochim. Biophys. Acta* **239:**208–216.

Augenfeld, J., and Fritz, I. B., 1970, Carnitine palmitoyltransferase activity and fatty acid oxidation by livers from fetal and neonatal rats, *Can. J. Biochem.* **48:**288–294.

Barden, R. E., and Cleland, W. W., 1969, Alteration of the concentrations of dilute palmitoyl-CoA solutions by surface adsorption, *Biochem. Biophys. Res. Commun.* **34:**555–559.

Bartels, P. D., and Sestoft, L., 1980, Thyroid hormone induced changes in gluconeogenesis and ketogenesis in perfused rat liver, *Biochim. Biophys. Acta* **633:**56–57.

Bartlett, K., Bone, A. J., Koundajian, P., Meredith, P., Turnbull, D. M., and Sherratt, H. S. A., 1981, Inhibition of mitochondrial β-oxidation at the stage of carnitine palmitoyltransferase I by the coenzyme A esters of some substituted hypoglycaemic oxiran-2-carboxylic acids, *Biochem. Soc. Trans.* **9:**574–575.

Benito, M., and Williamson, D. H., 1978, Evidence for a reciprocal relationship between lipogenesis and ketogenesis in hepatocytes from fed virgin and lactating rats, *Biochem. J.* **176:**331–334.

Benito, M., Whitelaw, E., and Williamson, D. H., 1979, Regulation of ketogenesis during the suckling-weanling transition in the rat, *Biochem. J.* **180:**137–144.

Berge, R. K., and Broch, O. J., 1981, Regional and subcellular distribution of acyl-CoA hydrolase and acyl-1-carnitine hydrolase in young and adult rat brain, *Int. J. Biochem.* **13:**1157–1162.

Berge, R. K., Slinde, E., and Farstad, M., 1979, Intracellular localization of long-chain acyl-coenzyme A hydrolase and acyl-1-carnitine hydrolase in brown adipose tissue from guinea pigs, *Biochim. J.* **182:**347–351.

Bergstrom, J. D., and Reitz, R. C., 1980, Studies on carnitine palmitoyltransferase: The similar nature of CPT_i (inner form) and CPT_o (outer form), *Arch. Biochem. Biophys.* **204:**71–79.

Bieber, L. L., Abraham, T., and Helmrath, T., 1972, A rapid spectrophotometric assay for carnitine palmitoyltransferase, *Anal. Biochem.* **50:**509–518.

Bieber, L. L., Markwell, M. A. K., Blair, M., and Helmrath, T. A., 1973, Studies on the development of carnitine palmitoyltransferase and fatty acid oxidation in liver mitochondria of neonatal pigs, *Biochem. Biophys. Acta* **326:**145–154.

Borrebaek, B., Osmundsen, H., and Bremer, J., 1980, *In vivo* induction of 4-enoyl-CoA reductase by clofibrate in liver mitochondria and its effect on pent-4-enoate metabolism, *Biochem. Biophys. Res. Commun.* **93:**1173–1180.

Brady, L. J., and Hoppel, C. L., 1983a, Hepatic mitochondrial function in lean and obese Zucker rats, *Am. J. Physiol.,* **245:**E239–E245.

Brady, L. J., and Hoppel, C. L., 1983b, Effect of diet and starvation on hepatic mitochondrial function in the rat, *J. Nutr.,* **113:**2129–2137.

Brass, E. P., and Hoppel, C. L., 1978a, Carnitine metabolism in the fasting rat, *J. Biol. Chem.* **253:**2688–2693.

Brass, E. P., and Hoppel, C. L., 1978b, Disassociation between acid insoluble acylcarnitines and ketogenesis following carnitine administration *in vivo, J. Biol. Chem.* **253:**5274–5276.

Brass, E. P., and Hoppel, C. L., 1980, Effect of carnitine on mitochondrial oxidation of palmitoylcarnitine, *Biochem. J.* **188:**451–458.

Bremer, J., 1962, Carnitine in intermediary metabolism. The metabolism of fatty acid esters of carnitine by mitochondria, *J. Biol. Chem.* **237:**3628–3632.

Bremer, J., 1963, Carnitine in intermediary metabolism. The biosynthesis of palmityl-carnitine by cell subfractions, *J. Biol. Chem.* **238:**2774–2779.

Bremer, J., 1967, The function of palmityl-CoA:carnitine palmityltransferase in the oxidation of fatty acids, in: *Protides of the Biological Fluids,* Vol. 15 (H. Peeters, ed.), Elsevier, Amsterdam, pp. 185–189.

Bremer, J., 1981, The effect of fasting on the activity of liver carnitine palmitoyltransferase and its inhibition by malonyl-CoA, *Biochim. Biophys. Acta* **665:**628–631.

Bremer, J., and Norum, K. R., 1967a, The mechanism of substrate inhibition of palmityl coenzyme A:carnitine palmityltransferase by palmityl-coenzyme A, *J. Biol. Chem.* **242:**1744–1748.

Bremer, J., and Norum, K. R., 1967b, The effects of detergents on palmityl coenzyme A:carnitine palmityltransferase, *J. Biol. Chem.* **242:**1749–1755.

Bremer, J., and Norum, K. R., 1967c, Palmityl-CoA:carnitine-*o*-palmityltransferase in the mitochondrial oxidation of palmityl-CoA, *Eur. J. Biochem.* **1:**427–433.

Bremer, J., and Wojtczak, A. B., 1972, Factors controlling the rate of fatty acid beta-oxidation in rat liver mitochondria, *Biochim. Biophys. Acta* **280:**515–530.

Brosnan, J. T., and Fritz, I. B., 1971a, The permeability of mitochondria to carnitine and acetylcarnitine, *Biochem. J.* **125:**94P.

Brosnan, J. T., and Fritz, I. B., 1971b, The oxidation of fatty-acyl derivatives by mitochondria from bovine fetal and calf hearts, *Can. J. Biochem.* **49:**1296–1300.

Brosnan, J. T., Kopec, B., and Fritz, I. B., 1973, The localization of carnitine palmitoyltransferase on the inner membrane of bovine liver mitochondria, *J. Biol. Chem.* **248:**4075–4082.

Cederbaum, A. I., Lieber, C. S., Beattie, D. S., and Rubin, E., 1975, Effect of chronic ethanol ingestion on fatty acid oxidation by hepatic mitochondria, *J. Biol. Chem.* **250:**5122–5129.

Chase, J. F. A., and Tubbs, P. K., 1972, Specific inhibition of mitochondrial fatty acid oxidation by 2-bromopalmitate and its coenzyme A and carnitine esters, *Biochem. J.* **129:**55–65.

Christiansen, R. Z., 1977, Regulation of palmitate metabolism by carnitine and glucagon in hepatocytes isolated from fasted and carbohydrate refed rats, *Biochim. Biophys. Acta* **488:**249–262.

Christiansen, R. Z., 1978, The effect of clofibrate-feeding on hepatic fatty acid metabolism, *Biochim. Biophys. Acta* **530:**314–324.

Christiansen, R. Z., Osmundsen, H., Borrebaeck, B., and Bremer, J., 1978, The effects of clofibrate feeding on the metabolism of palmitate and erucate in isolated hepatocytes, *Lipids* **13:**487–491.

Ciardelli, T., Stewart, C. J., Seeliger, A., and Wieland, T., 1981, Synthesis of a carba-analog of *S*-palmitoyl-coenzyme-A, heptadecan-2-onyldethio-CoA, and of 5-heptadecyl-CoA; effective inhibitors of citrate synthase and carnitine palmitoyltransferase, *Liebigs Ann. Chem.* **1981:**828–841.

Clarke, P. R. H., and Bieber, L. L., 1981a, Isolation and purification of mitochondrial carnitine octanoyltransferase activities from beef heart, *J. Biol. Chem.* **256:**9861–9868.

Clarke, P. R. H., and Bieber, L. L., 1981b, Effect of micelles on the kinetics of purified beef heart mitochondrial carnitine palmitoyltransferase, *J. Biol. Chem.* **256:**9869–9873.

Cook, G. A., Otto, D. A., and Cornell, N. W., 1980, Differential inhibition of ketogenesis by malonyl-CoA in mitochondria from fed and starved rats, *Biochem. J.* **192:**955–958.

Daae, L. N. W., and Aas, M., 1973, Fatty acid activation and acyl transfer in rat liver during clofibrate feeding, *Atherosclerosis* **17:**389–400.

Delisle, G., and Fritz, I. B., 1967, Interrelations between hepatic fatty acid oxidation and gluconeogenesis: A possible regulatory role of carnitine palmityltransferase, *Proc. Natl. Acad. Sci. USA* **58:**790–797.

Demaugre, F., Buc, H., Girard, J., and Leroux, J. P., 1983, Role of the mitochondrial metabolism of pyruvate on the regulation of ketogenesis in rat hepatocytes, *Metabolism* **32:**40–48.

DiMarco, J. P., and Hoppel, C., 1975, Hepatic mitochondrial function in ketogenic states. Diabetes, starvation, and after growth hormone administration, *J. Clin. Invest.* **55:**1237–1244.

DiMauro, S., and Trevisan, C., 1982, Carnitine palmitoyltransferase (CPT) deficiency: A review, in: *Disorders of the Motor Unit* (D. L. Schotland, ed.), John Wiley and Sons, New York, pp. 657–666.

Farrell, S. O., and Bieber, L. L., 1983, Carnitine octanoyltransferase of mouse liver peroxisomes: Properties and effect of hypolipidemic drugs, *Arch. Biochem. Biophys.* **222:**123–132.

Ferre, P., Pegorier, J. P., Williamson, D. H., and Girard, J. R., 1978, The development of ketogenesis at birth in the rat, *Biochem. J.* **176:**759–765.

Ferri, L., Jocelyn, P. C., and Siliprandi, N., 1980, The mitochondrial handling of D,L-thiocarnitine and its *s*-acetyl derivative, *FEBS Lett.* **121:**19–22.

Foster, P. C., and Bailey, E., 1976a, Changes in hepatic fatty acid degradation and blood lipid and ketone body content during development of the rat, *Enzyme* **21**:397–407.

Foster, P. C., and Bailey, E., 1976b, Changes in the activities of the enzymes of hepatic fatty acid oxidation during development of the rat, *Biochem. J.* **154**:49–56.

Fritz, I. B., 1955, The effects of muscle extracts on the oxidation of palmitic acid by liver slices and homogenates, *Acta Physiol. Scand.* **34**:367–385.

Fritz, I. B., 1957, Effects of choline deficiency and carnitine on palmitic acid oxidation by rat liver homogenates, *Am. J. Physiol.* **190**:449–452.

Fritz, I. B., 1959, Action of carnitine on long chain fatty acid oxidation by liver, *Am. J. Physiol.* **197**:297–304.

Fritz, I. B., 1961, Factors influencing the rates of long-chain fatty acid oxidation and synthesis in mammalian systems, *Physiol. Rev.* **41**:52–129.

Fritz, I. B., 1967, An hypothesis concerning the role of carnitine in the control of intermediates between fatty acid and carbohydrate metabolism, *Perspect. Biol. Med.* **10**:643–677.

Fritz, I. B., and Marquis, N. R., 1965, The role of acylcarnitine esters and carnitine palmityltransferase in the transport of fatty acyl groups across mitochondrial membranes, *Proc. Natl. Acad. Sci. USA* **54**:1226–1233.

Fritz, I. B., and Yue, K. T. N., 1963, Long-chain carnitine acyltransferase and the role of acylcarnitine derivatives in the catalytic increase of fatty acid oxidation induced by carnitine, *J. Lipid Res.* **4**:279–288.

Fritz, I. B., Kaplan, E., and Yue, K. T. N., 1962, Specificity of carnitine action on fatty acid oxidation by heart muscle, *Am. J. Physiol.* **202**:117–121.

Garland, P. B., Haddock, B. A., and Yates, D. W., 1969, Components and compartments of mitochondrial fatty acid oxidation, in: *FEBS Symp. 17* (L. Ernster and Z. Drahota, eds.), Academic Press, London, pp. 111–126.

Haddock, B. A., Yates, D. W., and Garland, P. D., 1970, The localization of some coenzyme A-dependent enzymes in rat liver mitochondria, *Biochem. J.* **119**:565–573.

Hagen, L. E., Berge, R. K., and Farstad, M., 1979, Different subcellular localization of palmitoyl-1-carnitine hydrolysis in human and rat liver, *FEBS Lett.* **104**:297–299.

Hahn, P., 1981, Serum carnitine levels and hepatic and adipose tissue carnitine transferases in obese mice, *Nutr. Res.* **1**:93–99.

Halperin, M. L., and Pande, S. V., 1979, Fatty acyl group transport into mitochondria: Carnitine palmitoyltransferase EC 2.3.1.23 and the carnitine-acylcarnitine translocase, *Meth. Enzymol.* **56**:368–378.

Harano, Y., Kowal, J., Yamazaki, R., Lavine, L., and Miller, M., 1972, Carnitine palmitoyltransferase activities (1 and 2) and the rate of palmitate oxidation in liver mitochondria from diabetic rats, *Arch. Biochim. Biophys.* **153**:426–437.

Harano, Y., Kosugi, K., Kashiwagi, A., Nakano, T., Hidaka, H., and Shigeta, Y., 1982, Regulatory mechanism of ketogenesis by glucagon and insulin in isolated and cultured hepatocytes, *J. Biochem.* **91**:1739–1748.

Hoppel, C. L., 1976, Carnitine palmitoyltransferase and transport of fatty acids, in: *The Enzymes of Biological Membranes,* Vol. 2 (A. Martonosi, ed.), Plenum Press, New York, pp. 119–143.

Hoppel, C. L., and Tomec, R. J., 1972, Carnitine palmityltransferase. Location of two enzymatic activities in rat liver mitochondria, *J. Biol. Chem.* **247**:832–844.

Hoppel, C. L., Genuth, S., Brass, E., Fuller, R., and Hostetler, K., 1980, Carnitine and carnitine palmitoyltransferase in metabolic studies, in: *Carnitine Biosynthesis, Metabolism, and Functions* (R. A. Frenkel and J. D. McGarry, eds.), Academic Press, New York, pp. 287–305.

Hostetler, K. Y., Hoppel, C. L., Romine, J. S., Sipe, J. C., Gross, S. R., and Higginbottom, P. A., 1978, Partial deficiency of muscle carnitine palmitoyltransferase with normal ketone production, *New Engl. J. Med.* **298**:553–557.

Idell-Wenger, J. A., 1981, Carnitine:acetylcarnitine translocase of rat heart mitochondria. Competition for carnitine uptake by carnitine esters, *J. Biol. Chem.* **256**:5597–5603.

Kahonen, M. T., 1976, Effect of clofibrate treatment on carnitine acyltransferases in different subcellular fractions of rat liver, *Biochim. Biophys. Acta* **428**:690–701.

Kahonen, M. T., 1979, Effect of clofibrate treatment on acylcarnitine oxidation in isolated rat liver mitochondria, *Med. Biol.* **67**:58–65.

Keyes, W. G., and Heimberg, M., 1979, Influence of thyroid status on lipid metabolism in the perfused rat liver, *J. Clin. Invest.* **64:**182–190.

Keyes, W. G., Wilcox, H. G., and Heimberg, M., 1981, Formation of the very low density lipoproteins and metabolism of [1-^{14}C]oleate by perfused livers from rats treated with triiodothyronine or propylthiouracil, *Metabolism* **30:**135–146.

Kopec, B., and Fritz, I. B., 1971, Properties of a purified carnitine palmitoyltransferase, and evidence for the existence of other carnitine acyltransferases, *Can. J. Biochem.* **49:**941–948.

Kopec, B., and Fritz, I. B., 1973, Comparison of properties of carnitine palmitoyltransferase I with those of carnitine palmitoyltransferase II, and preparation of antibodies to carnitine palmitoyltransferases, *J. Biol. Chem.* **248:**4069–4074.

Kurup, C. K. R., Aithal, H. N., and Ramasarma, T., 1970, Increase in hepatic mitochondria on administration of ethyl-α-*p*-chlorophenoxy isobutyrate to the rat, *Biochem. J.* **116:**773–779.

Laker, M. E., and Mayes, P. A., 1981, Effect of hyperthroidism and hypothyroidism on lipid and carbohydrate metabolism of the perfused rat liver, *Biochem. J.* **196:**247–255.

Levitsky, D. O., and Skulachev, V. P., 1972, Carnitine: The carrier transporting fatty acids into mitochondria by means of an electrochemical gradient of H^+, *Biochim. Biophys. Acta* **275:**33–50.

Lockwood, E. A., and Bailey, E., 1970, Fatty acid utilization during development of the rat, *Biochem. J.* **120:**49–54.

Mahadevan, S., and Sauer, F., 1969, Carnitine ester hydrolase of rat liver, *J. Biol. Chem.* **244:**4448–4453.

Mannaerts, G. P., Thomas, J., Debeer, L. J., McGarry, J. D., and Foster, D. W., 1978, Hepatic fatty acid oxidation and ketogenesis after clofibrate treatment, *Biochim. Biophys. Acta* **529:**201–211.

Markwell, M. A. K., McGroarty, E. J., Bieber, L. L., and Tolbert, N. E., 1973, The subcellular distribution of carnitine acyltransferases in mammalian liver and kidney. A new peroxisomal enzyme, *J. Biol. Chem.* **248:**3426–3432.

Markwell, M. A. K., Tolbert, N. E., and Bieber, L. L., 1976, Comparison of the carnitine acyltransferase activities from rat liver peroxisomes and microsomes, *Arch. Biochem. Biophys.* **176:**479–488.

McCune, S. A., Durant, P. J., Jenkins, P. A., and Harris, R. A., 1981, Comparative studies on fatty acid synthesis, glycogen metabolism, and gluconeogenesis by hepatocytes isolated from lean and obese Zucker rats, *Metabolism* **30:**1170–1178.

McGarry, J. D., and Foster, D. W., 1973, Acute reversal of experimental diabetic ketoacidosis in the rat with (+)-decanoylcarnitine, *J. Clin. Invest.* **52:**877–884.

McGarry, J. D., and Foster, D. W., 1979, In support of the roles of malonyl-CoA and carnitine acyltransferase I in the regulation of hepatic fatty acid oxidation and ketogenesis, *J. Biol. Chem.* **254:**8163–8168.

McGarry, J. D., and Foster, D. W., 1980, Regulation of hepatic fatty acid oxidation and ketone production, *Annu. Rev. Biochem.* **49:**395–420.

McGarry, J. D., and Foster, D. W., 1981, Importance of experimental conditions in evaluating the malonyl-CoA sensitivity of liver carnitine acyltransferase, *Biochem. J.* **200:**217–223.

McGarry, J. D., and Foster, D. W., 1982, Malonyl-CoA inhibition of carnitine palmitoyltransferase I: Response to Dr. Saggerson's letter, *Biochem. J.* **208:**527–528.

McGarry, J. D., Meier, J. M., and Foster, D. W., 1973, The effects of starvation and refeeding on carbohydrate and lipid metabolism *in vivo* and in the perfused rat liver. The relationship between fatty acid oxidation and esterification in the regulation of ketogenesis, *J. Biol. Chem.* **248:**270–278.

McGarry, J. D., Robles-Valdes,C., and Foster, D. W., 1975, Role of carnitine in hepatic ketogenesis, *Proc. Natl. Acad. Sci. USA* **72:**4385–4388.

McGarry, J. D., Mannaerts, G. P., and Foster, D. W., 1977, A possible role for malonyl-CoA in the regulation of hepatic fatty acid oxidation and ketogenesis, *J. Clin. Invest.* **60:**265–270.

McGarry, J. D., Leatherman, G. F., and Foster, D. W., 1978a, Carnitine palmitoyltransferase I. The site of inhibition of hepatic fatty acid oxidation by malonyl-CoA, *J. Biol. Chem.* **253:**4128–4136.

McGarry, J. D., Takabayashi, Y., and Foster, D. W., 1978b, The role of malonyl-CoA in the coordination of fatty acid synthesis and oxidation in isolated rat hepatocytes, *J. Biol. Chem.* **253:**8294–8300.

Murray, W. J., Reed, K. W., and Roche, E. B., 1980, Conformations of carnitine and acetylcarnitine and the relationship of mitochondrial transport of fatty acids, *J. Theoret. Biol.* **82:**559–572.

Normann, P. T., and Flatmark, T., 1978, Long-chain acyl-CoA synthetase and "outer" carnitine long-chain acyltransferase activities of intact brown adipose tissue mitochondria, *Biochim. Biophys. Acta* **530:**461–473.

Normann, P. T., Ingebretsen, O. C., and Flatmark, T., 1978, On the rate-limiting step in the transfer of long-chain acyl groups across the inner membrane of brown adipose tissue mitochondria, *Biochim. Biophys. Acta* **501**:286–295.

Norum, K. R., 1964, Palmityl-CoA:carnitine palmityltransferase purification from calf-liver mitochondria and some properties of the enzyme, *Biochim. Biophys. Acta* **89**:95–108.

Norum, K. R., 1965a, Activation of palmityl-CoA:carnitine palmityltransferase in livers from fasted, fat-fed, or diabetic rats, *Biochim. Biophys. Acta* **98**:652–654.

Norum, K. R., 1965b, Palmityl-CoA:carnitine palmityltransferase. Studies on the substrate specificity of the enzyme, *Biochim. Biophys. Acta* **99**:511–522.

Norum, K. R., and Bremer, J., 1967, The localization of acyl coenzyme A-carnitine acyltransferases in rat liver cells, *J. Biol. Chem.* **242**:407–411.

Norum, K. R., Farstad, M., and Bremer, J., 1966, The submitochondrial distribution of acid:CoA ligase (AMP) and palmityl-CoA:carnitine palmityltransferase in rat liver mitochondria, *Biochem. Biophys. Res. Commun.* **24**:797–804.

Nosadini, R., Ursini, F., Tessari, P., Giengo, A., and Gregolin, C., 1979, Perfused liver carnitine palmitoyltransferase activity and ketogenesis in streptozotocin treated and genetic hyperinsulinemic rats. Effect of glucagon, *Horm. Metabol. Res.* **11**:661–664.

Ontko, J. A., and Johns, M. L., 1980, Evaluation of malonyl-CoA in the regulation of long-chain fatty acid oxidation in the liver. Evidence for an unidentified regulatory component of the system, *Biochem. J.* **192**:959–962.

Osmundsen, H., Cervenka, J., and Bremer, J., 1982, A role for 2,4-enoyl-CoA reductase in mitochondrial β-oxidation of polyunsaturated fatty acids. Effects of treatment with clofibrate on oxidation of polyunsaturated acylcarnitines by isolated rat liver mitochondria, *Biochem. J.* **208**:749–757.

Pande, S. V., 1971, On rate controlling factors of long chain fatty acid oxidation, *J. Biol. Chem.* **246**:5384–5390.

Pande, S. V., 1975, A mitochondrial carnitine acylcarnitine translocase system, *Proc. Natl. Acad. Sci. USA* **72**:883–887.

Pande, S. V., and Parvin, R., 1976, Characterization of carnitine acylcarnitine translocase system of heart mitochondria, *J. Biol. Chem.* **251**:6683–6691.

Pande, S. V., and Parvin, R., 1980a, Carnitine-acylcarnitine translocase catalyzes an equilibrating unidirectional transport as well, *J. Biol. Chem.* **255**:2994–3001.

Pande, S. V., and Parvin, R., 1980b, Clofibrate enhancement of mitochondrial carnitine transport system of rat liver and augmentation of liver carnitine and γ-butyrobetaine hydroxylase activity by thyroxine, *Biochim. Biophys. Acta* **617**:363–370.

Parker, S. L., Thompson, J. A., and Reitz, R. C., 1974, Effects of chronic ethanol ingestion upon acyl-CoA:carnitine acyltransferase in liver and heart, *Lipids* **9**:520–525.

Parvin, R., and Pande, S. V., 1979, Enhancement of mitochondrial carnitine and carnitine acylcarnitine translocase-mediated transport of fatty acids into liver mitochondria under ketogenic conditions, *J. Biol. Chem.* **254**:5423–5429.

Parvin, R., Goswami, T., and Pande, S. V., 1980, Inhibition of mitochondrial carnitine-acylcarnitine translocase by sulfobetaines, *Can. J. Biochem.* **58**:822–830.

Patten, B. M., Wood, J. M., Harati, Y., Hefferan, P., and Howell, R. R., 1979, Familial recurrent rhabdomyolysis due to carnitine palmityltransferase deficiency, *Am. J. Med.* **67**:167–171.

Powell, G. L., Grothusen, J. R., Zimmerman, J. K., Evans, C. A., and Fish, W. W., 1981, A re-examination of some properties of fatty acyl-CoA micelles, *J. Biol. Chem.* **256**:12740–12747.

Ramsay, R. R., and Tubbs, P. K., 1974, Exchange of the endogenous carnitine of ox heart mitochondria with external carnitine and its possible relevance to the mechanism of fatty-acyl transport into mitochondria, *Biochem. Soc. Trans.* **2**:1285–1286.

Ramsay, R. R., and Tubbs, P. K., 1975, The mechanism of fatty acid uptake by heart mitochondria: An acylcarnitine–carnitine exchange, *FEBS Lett.* **54**:21–25.

Ramsay, R. R., and Tubbs, P. K., 1976, The effects of temperature and some inhibitors on the carnitine exchange system of heart mitochondria, *Eur. J. Biochem.* **69**:299–303.

Robinson, L. N., and Zammit, V. A., 1982, Sensitivity of carnitine acyltransferase I to malonyl-CoA inhibition in isolated rat liver mitochondria is quantitatively related to hepatic malonyl-CoA concentration *in vivo*, *Biochem. J.* **206**:177–179.

Robles-Valdes, C., McGarry, J. D., and Foster, D. W., 1976, Maternal–fetal carnitine relationships and neonatal ketosis in the rat, *J. Biol. Chem.* **251**:6007–6012.

Saggerson, E. D., 1982, Does fasting decrease the inhibitory effect of malonyl-CoA on hepatic β-oxidation? *Biochem. J.* **208**:525–526.

Saggerson, E. D., and Carpenter, C. A., 1981a, Effects of fasting, adrenalectomy and streptozotocin-diabetes on sensitivity of hepatic carnitine acyltransferase to malonyl-CoA, *FEBS Lett.* **129**:225–228.

Saggerson, E. D., and Carpenter, C. A., 1981b, Carnitine palmitoyltransferase and carnitine octanoyltransferase activities in liver, kidney cortex, adipocyte, lactating mammary gland, skeletal muscle and heart; relative activities, latency, and effect of malonyl-CoA, *FEBS Lett.* **129**:229–232.

Saggerson, E. D., and Carpenter, C. A., 1981c, Effects of fasting and malonyl-CoA on the kinetics of carnitine palmitoyltransferase and carnitine octanoyltransferase in intact rat liver mitochondria, *FEBS Lett.* **132**:166–168.

Saggerson, E. D., and Carpenter, C. A., 1982, Regulation of hepatic carnitine palmitoyltransferase activity during the foetal-neonatal transition, *FEBS Lett.* **150**:177–180.

Saggerson, E. D., and Carpenter, C. A., 1983, The effect of malonyl-CoA on overt and latent carnitine acyltransferase activities in rat liver and adipocyte mitochondria, *Biochem. J.* **210**:591–597.

Saggerson, E. D., Carpenter, C. A., and Tselentis, B. S., 1982, Effects of thyroidectomy and starvation on the activity and properties of hepatic carnitine palmitoyltransferase, *Biochem. J.* **208**:667–672.

Scholte, H. R., Jennekens, F. G. I., and Bouvy, J. J. B. J., 1979, Carnitine palmitoyltransferase II deficiency with normal carnitine palmitoyltransferase I in skeletal muscle and leukocytes, *J. Neurol. Sci.* **40**:39–51.

Schulz, H., and Racker, E., 1979, Carnitine transport in submitochondrial particles and reconstituted proteoliposomes, *Biochem. Biophys. Res. Commun.* **89**:134–140.

Solberg, H. E., 1972, Different carnitine acyltransferases in calf liver, *Biochim. Biophys. Acta* **280**:422–433.

Solberg, H. E., 1974, Acyl group specificity of mitochondrial pool of carnitine acyltransferases, *Biochim. Biophys. Acta* **360**:101–112.

Solberg, H. E., Aas, M., and Daae, L. N. W., 1972, The activity of the different carnitine acyltransferases in the liver of clofibrate-fed rats, *Biochim. Biophys. Acta* **280**:434–439.

Stakkestad, J. A., and Bremer, J., 1982, The metabolism of fatty acids in hepatocytes isolated from triiodothyronine-treated rats, *Biochim. Biophys. Acta* **711**:90–100.

Stakkestad, J. A., and Bremer, J., 1983, The outer carnitine palmitoyltransferase and regulation of fatty acid metabolism in rat liver in different thyroid states, *Biochem. Biophys. Acta,* **750**:244–252.

Stanley, C. A., Gonzales, E., and Baker, L., 1983, Development of hepatic fatty acid oxidation and ketogenesis in the newborn guinea pig, *Pediat. Res.* **17**:224–229.

Tomec, R. J., and Hoppel, C. L., 1975, Carnitine palmitoyltransferase in bovine fetal heart mitochondria, *Arch. Biochem. Biophys.* **170**:716–723.

Triscari, J., Greenwood, M. R. C., and Sullivan, A. C., 1982, Oxidation and ketogenesis in hepatocytes of lean and obese Zucker rats, *Metabolism* **31**:223–228.

Tubbs, P. K., and Chase, J. F. A., 1970, Effects of an acylcarnitine analogue, 2-bromomyristoyl-thiocarnitine, on mitochondrial respiration, *Biochem. J.* **116**:34p.

Tubbs, P. K., Ramsay, R. R., and Edwards, M. R., 1980, Inhibitors of carnitine transport and metabolism, in: *Carnitine Biosynthesis, Metabolism, and Functions* (R. A. Frenkel and J. D. McGarry, eds.), Academic Press, New York, pp. 207–217.

Tutwiler, G. F., and Dellevigne, P., 1979, Action of the oral hypoglycemic agent 2-tetradecylglycidic acid on hepatic fatty acid oxidation and gluconeogenesis, *J. Biol. Chem.* **254**:2935–2941.

Tutwiler, G. F., and Ryzlak, M. T., 1980, Inhibition of mitochondrial carnitine palmitoyltransferase by 2-tetradecylglycidic acid (McN-3802), *Life Sci.* **26**:393–397.

Tutwiler, G. F., Ho, W., and Mahrbrachr, R. J., 1981, Tetradecylglycidic acid, *Meth. Enzymol.* **72**:533–551.

VanTol, A., 1974, The effect of fasting on the acylation of carnitine and glycerophosphate in rat liver subcellular fractions, *Biochim. Biophys. Acta* **357**:14–23.

VanTol, A., and Hulsmann, W. C., 1969, The localization of palmitoyl -CoA:carnitine palmitoyltransferase in rat liver, *Biochim. Biophys. Acta* **189**:342–353.

Warshaw, J. B., 1972, Cellular energy metabolism during fetal development. IV. Fatty acid activation, acyl transfer and fatty acid oxidation during development of the chick and rat, *Dev. Biol.* **28**:537–544.

Warshaw, J. B., and Terry, M. L., 1970, Cellular energy metabolism during fetal development. II. Fatty acid oxidation by the developing heart. *J. Cell. Biol.* **44:**354–360.

West, D. W., Chase, J. F. A., and Tubbs, P. K., 1971, The separation and properties of two forms of carnitine palmitoyltransferase from ox liver mitochondria, *Biochem. Biophys. Res. Commun.* **42:**912–918.

Williams, E. S., and Li, T.-K., 1977, The effect of chronic alcohol administration on fatty acid metabolism and pyruvate oxidation of heart mitochondria, *J. Mol. Cell. Cardiol.* **9:**1003–1011.

Williamson, J. R., Browning, E. T., Scholz, R., Kreisberg, R. A., and Fritz, I. B., 1968, Inhibition of fatty acid stimulation of gluconeogenesis by (+)-decanoylcarnitine in perfused rat liver, *Diabetes* **17:**194–208.

Wittels, B., and Bressler, R., 1965, Lipid metabolism in the newborn rat, *J. Clin. Invest.* **44:**1639–1646.

Wolkowicz, P. E., Pownall, H. J., and McMillin-Wood, J. B., 1982, (1-Pyrenebutyryl)carnitine and 1-pyrenebutyryl coenzyme A: Fluorescent probes for lipid metabolite studies in artificial and natural membranes, *Biochemistry* **21:**2990–2996.

Wood, J. McM., 1973, Effect of ionic strength on the activity of carnitine palmityltransferase I, *Biochemistry* **12:**5268–5273.

Wood, J. McM., Sordahl, L. A., Lewis, R. M., and Schwartz, A., 1973, Effect of chronic myocardial ischemia on the activity of carnitine palmitylcoenzyme A transferase of isolated canine heart mitochondria, *Circ. Res.* **32:**340–347.

Yalkowsky, S. H., and Zografi, G., 1970, Potentiometric titration of monomeric and micellar acylcarnitines, *J. Pharm. Sci.* **59:**798–802.

Yates, D. W., and Garland, P. B., 1966, The partial latency and intramitochondrial distribution of carnitine palmitoyltransferase (EC 2.3.1.-), and the CoASH and carnitine permeable space of rat liver mitochondria, *Biochem. Biophys. Res. Commun.* **23:**460–465.

Yates, D. W., and Garland, P. B., 1970, Carnitine palmitoyltransferase activities (EC 2.3.1.-) of rat liver mitochondria, *Biochem. J.* **119:**547–552.

Yeh, Y. Y., and Yee, P., 1976, Insulin, a possible regulator of ketosis in newborn and suckling rats, *Pediat. Res.* **10:**192–197.

Yeh, Y.-Y., and Yee, P., 1979, Fatty acid oxidation in isolated rat liver mitochondria. Developmental changes and their relation to hepatic levels of carnitine and glycogen and to carnitine acyltransferase activity, *Arch. Biochem. Biophys.* **199:**560–569.

Zahler, W. L., Barden, R. E., and Cleland, W. W., 1968, Some physical properties of palmitoyl-coenzyme A micelles, *Biochim. Biophys. Acta* **164:**1–11.

18

Membrane-Bound Enzymes of Cholesterol Biosynthesis: Resolution and Identification of the Components Required for Cholesterol Synthesis from Squalene

James M. Trzaskos and James L. Gaylor

I. INTRODUCTION

A. Scope

All 22 enzymic steps in the synthesis of cholesterol (Figure 1) from squalene are catalyzed by membrane-bound enzymes of the endoplasmic reticulum which is isolated as the microsomal fraction of cell-free tissue preparations. In addition, 3-hydroxy-3-methylglutaryl coenzyme A reductase (HMG-CoA reductase), the rate-limiting enzyme for the overall cholesterol biosynthetic pathway, is also membrane-bound and found associated with microsomes. These observations are significant since they suggest that the membrane may play important roles in both the catalysis and regulation of synthesis of cellular sterols.

James M. Trzaskos and James L. Gaylor • Central Research and Development Department, E. I. du Pont de Nemours and Company, Glenolden, Pennsylvania 19036.

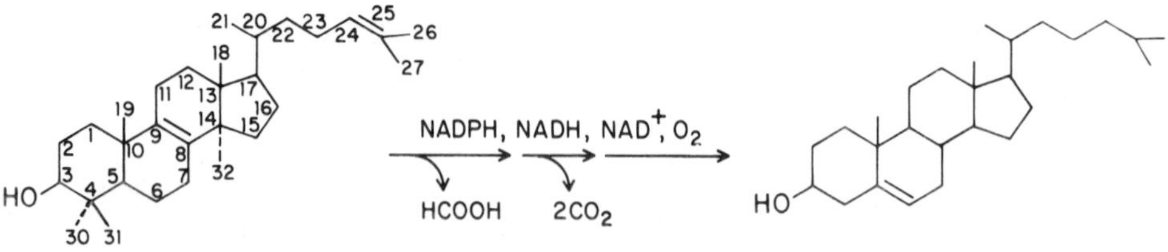

NADPH, NADH, NAD^+, O_2

HCOOH $2CO_2$

LANOSTEROL

(4,4,14α–trimethyl–5α–cholesta–8,24–dien–3β–ol)

CHOLESTEROL

(cholest–5–en–3β–ol)

Figure 1. Microsomal formation of cholesterol from lanosterol. All of the enzymes are membrane-bound, and in the presence of reactants shown, lanosterol is converted by rat liver microsomes to the mature end product, cholesterol. (Conventional numbering of the steroid nucleus is shown for the reader.)

Several functions which can be envisioned for the membrane in this regard are (1) as a common framework for the enzymes in the pathway providing a close physical association between consecutive enzyme reactions that operate efficiently in sequence, (2) as a regulator of enzymic activity through changes in lipid and/or protein microenvironments which may affect either individual enzymic reaction rates or transfers between enzymes, (3) as an essential component of various enzyme-catalyzed reactions which are incapable of catalysis in the absence of lipid, (4) as a medium in which the lipid-soluble intermediates generated along the pathway can remain in a hydrophobic domain provided by the membrane lipid with each metabolite being the product of one reaction and the substrate for the next without diffusion into the aqueous cytosol, and (5) as a supporting medium for the obligatory electron-transfer reactions that are needed for maintenance of both NADH- and NADPH-dependent monooxygenases. Our knowledge of specific functions is fragmentary, and documentation of these possible functions will necessarily await purification of the components and reincorporation into membranes.

Our discussion will consider only those membrane-bound enzymes in the squalene-to-cholesterol sequence. The rate-limiting enzyme of sterol biosynthesis, HMG-CoA reductase, will not be considered except peripherally in the section on coordinate control mechanisms. [An excellent monograph dealing with numerous aspects of HMG-CoA reductase has recently appeared (Sabine, 1983)].

B. Objectives

Although the membrane environment of the endoplasmic reticulum presumably provides supportive functions for the synthesis of lipophilic sterols that are necessary for cellular proliferation, the membrane provides the biochemist with numerous physical and chemical problems that are associated with attempts to study the metabolic processes involved in the cholesterol synthetic pathway. The lipid nature of the membrane and the insolubility of the constitutive synthetic enzymes require purification methods which must not only resolve various protein species by conventional separation techniques, but also maintain the enzymes in a soluble, active form throughout the purification process primarily through the use of detergents and stabilizing agents. To observe activity, it may be necessary to perform laborious reconstitution studies by adding all essential membrane components that may have been removed during purification. The ultimate objective for each enzyme in this study is to resolve the pure enzyme, electron carriers, lipid, and ancillary protein that may be necessary to observe each catalytic activity.

Second, with a sequence of enzyme-catalyzed reactions associated on a membrane in which the product of one reaction becomes the substrate for the next enzyme without diffusion from the membrane, the biochemist must devise necessary separations or interruptions to study individual steps in the sequence. This development of an appropriate assay is followed by isolation and purification of the enzyme with characterization of the enzyme that includes identification of the sterol metabolites that are formed. Interruption is especially necessary in the study of the cholesterol synthetic pathway where numerous isolated and synthetic sterols have been proposed as possible intermediates (Schroepfer, 1982); yet no confirmed, established pathway can emerge because the evidence is permissive rather than compelling for any given sterol that has not been shown to be formed. Gaylor (1981, 1982) has presented experimental avenues to address these problems by investigating details of each enzyme step. The criteria necessary to establish validity will not be argued in this review. Instead, we have chosen to describe the more practical aspects of solubilization and purification methods that have been successful in the resolution of component reactions of cholesterol biosynthesis.

The discussion of enzymes is presented in the order of expected reaction sequence (Gaylor, 1981). This order should not be construed as being established since, as mentioned in the preceding paragraph, many of the enzymes exhibit a broad substrate specificity and many of the alternatives, although narrowed, are still open to question (Schroepfer, 1982). Indeed, there may be no single pathway; even very few broader substrate specificities may allow several sequential options. Also, unless otherwise indicated, the processes described in this review are those known for mammalian microsomes. Basically, the microsomal enzymes of yeast, plants, and animals do not differ markedly with the exception of those discrete steps that impart species individuality, e.g., C-24 substitution (see Nes and McKean, 1977).

II. SQUALENE SYNTHETASE

A. Purification Methods

After HMG-CoA reductase, squalene synthetase is the next membrane-bound enzyme in the cholesterol synthetic pathway (Figure 2). The enzyme-catalyzed condensation of two molecules of farnesyl pyrophosphate to form squalene requires NADPH as a cofactor and a divalent cation for activity (Lynen *et al.*, 1958; Popjak, 1959). Omission of NADPH from a yeast microsomal preparation that was incubated with farnesyl pyrophosphate first showed that a C_{30}-pyrophosphate intermediate (presqual-

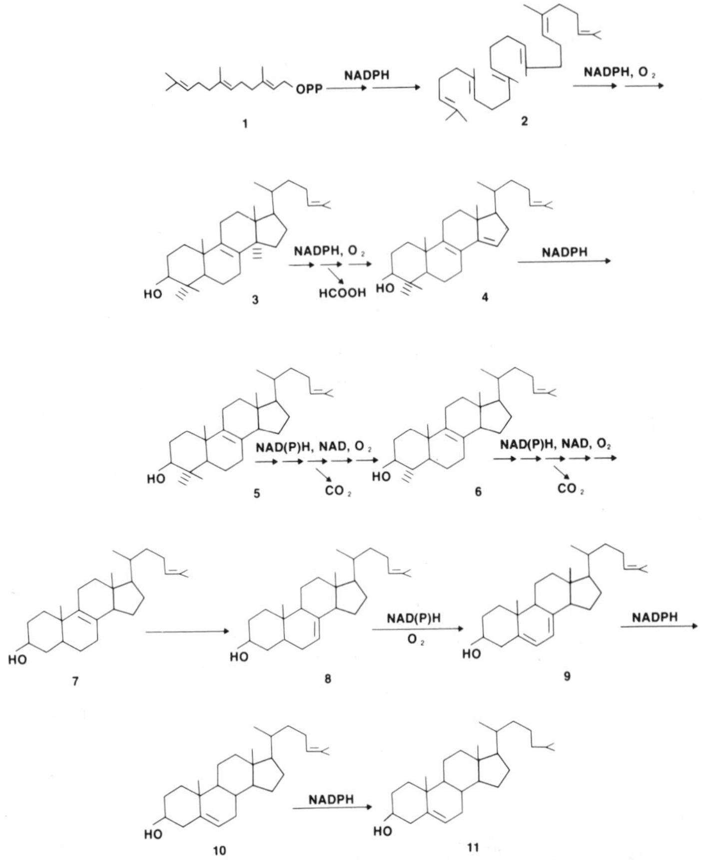

Figure 2. Intermediates of cholesterol biosynthesis which are generated and/or metabolized by membrane-bound enzymes. Metabolism of all steroidal intermediates (3–11) in the pathway in addition to (1) farnesyl pyrophosphate, (2) squalene, and hydroxy-methylglutaryl-coenzyme A (not shown) is performed by membrane-bound enzymes. The intermediates shown here are not all-inclusive and elaboration of the more involved reactions are displayed in detail in other figures. Numbered compounds include: (1) farnesyl pyrophosphate, (2) squalene, (3) lanosterol, (4) 4,4-dimethyl-5α-cholesta-8,14,24-trien-3-βol, (5) 4,4-dimethyl-5α-cholesta-8,24-dien-3β-ol, (6) 4α-monomethyl-5α-cholesta-8,24-dien-3β-ol, (7) zymosterol, (8) 5α-cholesta-7,24-dien-3β-ol, (9) cholesta-5,7,24-trien-3β-ol, (10) desmosterol, and (11) cholesterol.

ene pyrophosphate) convertable to squalene in the presence of NADPH is the immediate condensation product of the synthetase reaction (Rilling, 1966). Thus, two separate steps in the squalene synthetase reaction were suggested, condensation and reduction. The involvement of two enzymic activities has now been shown by enzyme solubilization, purification, and reconstitution experiments (see Popjak and Agnew, 1979).

Squalene synthetase activity has been obtained in soluble form by treating microsomal preparations with either ultrasound (Dugan and Porter, 1972), high-density sucrose solution (Qureshi *et al.*, 1973), or detergents (Shechter and Bloch, 1971;

Agnew and Popjak, 1978). The most successful of these techniques, detergent solubilition, employs sodium deoxycholate as the solubilizing agent. Shechter and Bloch (1971) have shown that the resulting solubilized enzyme preparations can be further purified, after detergent removal, by ion-exchange chromatography. The resulting labile enzyme displays a molecular weight of 426,000 and exhibits no requirement for ancillary proteins or phospholipids. Alternatively, Agnew and Popjak (1978) have characterized the enzyme in the presence of detergent by sucrose density-gradient sedimentation and Sephadex G-200 gel filtration column chromatography. These conditions produce an enzyme preparation of molecular weight 54,000 which is thermally labile in the presence of detergent (50% inactivation in 5 min at 26°C), but the stability and activity of which are restored upon detergent removal. In addition, stability and activity restoration of the synthetase enzyme can be accomplished by treatment with either phosphatidyl choline or phosphatidyl ethanolamine; neither phosphatidylserine nor phosphatidyl-inositol is effective.

Attempts to resolve the two component activities of squalene synthesis from farnesyl pyrophosphate in detergent-solubilized preparations have not been successful (Popjak and Agnew, 1979). Thus, a single protein species carrying both activities is suggested. In contrast, Qureshi *et al.* (1973), using high-density sucrose solubilized preparations, report that a purified polymeric enzyme is capable of squalene synthesis from farnesyl pyrophosphate without any requirement for phospholipid. Treatment of the isolated enzyme of molecular weight 450,000 with decreasing glycerol concentrations in buffer leads to a preparation that catalyzes only formation of presqualene pyrophosphate from farnesyl pyrophosphate. The basis for the observed differences between the various preparations in different laboratories will not be clarified until complete resolution and purification of the protein is accomplished.

B. Mechanism

Condensation of two farnesyl pyrophosphate molecules appears to involve enzyme-directed nucleophilic attack at C-3 of the allylic double bond of one molecule of farnesyl pyrophosphate (Figure 3). The result is C-2 to C-1 bond formation with the other farnesyl pyrophosphate molecule accompanied by elimination of one pyrophosphate anion (Popjak and Agnew, 1979). Completion of the initial phase of catalysis is accomplished by proton loss and elimination of the enzyme nucleophile resulting in formation of a cyclopropane ring containing C-3 and C-2 originating from the initial farnesyl pyrophosphate molecule. Pyridine nucleotide dependence is observed in the subsequent conversion of presqualene pyrophosphate to squalene which proceeds by ring expansion accompanied by either pyrophosphate migration or ion-pair formation. Pyridine-nucleotide-dependent ring cleavage and double-bond formation complete the reaction sequence (Popjak and Agnew, 1979). Kinetically, a ping-pong bi-bi mechanism has been proposed by Dugan and Porter (1972) for the enzyme reaction in particulate pig liver preparations. Precise kinetic interpretations, however, seem premature and clouded by both substrate micellar formation and the use of an impure enzyme preparation.

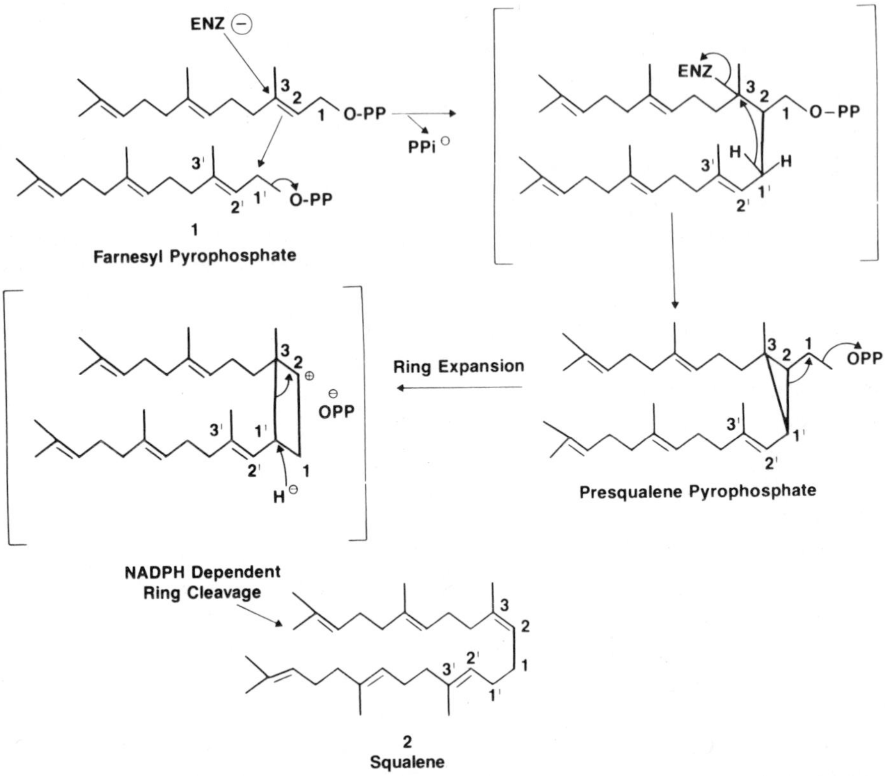

Figure 3. Squalene-synthetase-catalyzed condensation reaction mechanism of two farnesyl pyrophosphate molecules to form squalene. Enzyme-catalyzed condensation proceeds through a proposed enzyme-bound intermediate which generates presqualene pyrophosphate by formation of the cyclopropane ring system. Ring expansion is proposed with loss of pyrophosphate followed by NADPH-dependent ring cleavage resulting in formation of the squalene molecule. The numbers underneath structures refer to Figure 2. Carbon atoms 1, 2, and 3 in each farnesyl pyrophosphate molecule are numbered to assist the reader in following the condensation reaction as described in the text.

III. SQUALENE OXIDASE

A. Squalene Epoxidase

Epoxidation of squalene to squalene 2,3-oxide is the first oxygen-requiring reaction of the microsomal enzymes in the overall conversion of squalene to cholesterol (Tchen and Bloch, 1957). The epoxidase-catalyzed reaction has been shown by Yamamoto and Bloch (1970) to require NADPH, molecular oxygen, and soluble cellular components which include phospholipid and protein (Tai and Bloch, 1972; Ferguson and Bloch, 1977). Triton X-100 can replace the soluble components necessary for squalene epoxidase activity, and Triton X-100 has also proven to be a useful detergent for obtaining the enzyme in soluble form (Ono and Bloch, 1975; Ono *et al.*, 1980).

DEAE-cellulose chromatography of the Triton X-100-solubilized enzyme resolves epoxidase activity into two components (Ono and Bloch, 1975). One component has been identified as NADPH-cytochrome *c* reductase (Ono and Bloch, 1975; Ono *et al.*, 1977) whose involvement as an electron carrier in numerous microsomal oxidase reactions is well established (King *et al.*, 1973). This reductase can be obtained in purified form by several procedures (Yasukochi and Masters, 1976; Vermilion and Coon, 1974). Resolution of the second membrane component, the terminal oxidase, has been obtained by affinity adsorption chromatography on Cibacron Blue Sepharose 4B (Ono *et al.*, 1980). The resolved oxidase preparation is composed of two proteins of molecular weight 47,000 and 27,000 in a ratio of 3 : 1 respectively, with a protoheme content of 0.9 nmoles/mg protein. In addition, the cytochrome P-450 content of this preparation, as determined by CO difference spectroscopy of the ferrohemoprotein, is only 0.1 nmole. With these data, in addition to lack of characteristic inhibition of epoxidase activity by either carbon monoxide or cyanide (Ono and Bloch, 1975), and in spite of the participation of cytochrome P-450 reductase, it seems improbable that cytochrome P-450 is involved in the squalene epoxidase reaction, and a distinct terminal oxidase is likely responsible for catalysis.

Reconstitution of epoxidase activity is dependent upon purified oxidase fraction, NADPH-cytochrome P-450 reductase, NADPH, O_2, and FAD (Ono *et al.*, 1980). The requirement for FAD suggests that the purified oxidase has a dissociable FAD prosthetic group and that the purified enzyme may be an apoflavoprotein. When reconstituted with 0.035% Triton X-100, neither phospholipid nor a soluble protein is required. Although such evidence may lead to questioning the role of phospholipid in the microsomal epoxidase reaction (see Section III-C) other microsomal monooxygenases have been shown to be supported by detergents in the absence of the natural lipids derived from the membrane.

B. Squalene 2,3-Oxide-Lanosterol Cyclase

The second step in the two-step conversion of squalene to lanosterol is cyclization of squalene 2,3-oxide. The reaction proceeds anaerobically in the absence of added cofactors and without detection of stable intermediates in the cyclization process (Corey *et al.*, 1966). Characterization of the pig liver enzyme by Dean *et al.* (1966) and Yamamoto *et al.* (1969) showed that cyclase activity may be readily solubilized with sodium deoxycholate and purified about 20-fold by salt fractionation. Sephadex G-200 column chromatography of the cyclase in high-ionic-strength buffers results in an active enzyme preparation with an estimated molecular weight of 90,000. Use of low-ionic-strength buffers in the absence of sodium deoxycholate results in aggregation, insolubility, and loss of enzyme activity. In contrast, Shechter *et al.* (1970) have shown that the enzyme of yeast is found in the soluble fraction of cell homogenates and is active in low-ionic-strength buffers. Triton X-100 activates the yeast enzyme, but not liver cyclase, which responds to deoxycholate treatment in contrast to the yeast preparation. Despite the ease of solubilization and response to detergent treatment, no acceptable preparation of the cyclase has been reported to date that can be used in reconstitution experiments. Further work on this enzyme is needed.

C. Cytosolic Protein Effectors

Various preparations of soluble cellular proteins have been described which enhance the rate of squalene metabolism in liver microsome preparations (Tai and Bloch, 1972; Scallen *et al.*, 1971; Ritter and Dempsey, 1971). Purification from two of these preparations by Ferguson and Bloch (1977) and Srikantaiah *et al.* (1976) yields a protein of molecular weight 47,000 designated SPF and SCP_1,* respectively. Recently, the proteins have been shown to be identical (Noland *et al.*, 1980), thus ending several years of debate over possible multiplicity of soluble protein involvement in squalene-metabolizing reactions. Both enzymes in the squalene-to-lanosterol conversion, i.e., oxidase and cyclase, are affected by the cytosolic protein when the enzymes are bound to membrane preparations (Saat and Bloch, 1976; Caras and Bloch, 1979; Garvey and Scallen, 1978), but detergent solubilization of enzymes eliminates the requirement for cytosolic protein (Yamamoto *et al.*, 1969; Ono and Bloch, 1975). Such observations argue strongly against an absolute requirement for the protein in enzyme catalysis, but these observations provide interesting insight for speculation and experimentation of membrane-associated reactions. For example, both Garvey and Scallen (1978) and Caras and Bloch (1979) have shown that SCP_1 and SPF enhance conversion of endogenously generated and/or microsomal-bound squalene to sterol product. Each laboratory implies an *intra*membrane function for the protein, perhaps as a site-specific director of substrate within the membrane. In addition, Kojima *et al.* (1981) have demonstrated that *inter*membrane squalene transfers can be facilitated by the protein possibly functioning as a "carrier" of squalene between membrane populations, although direct binding of squalene to the protein does not appear to occur (Caras *et al.*, 1980). A "carrier" function was initially suggested for proteins of this type when studied with exogenously supplied squalene as substrate (Scallen *et al.*, 1971). And although a "carrier" role for such proteins may now be questioned as an obligatory step of the biosynthetic processes in intact cells (Trzaskos and Gaylor, 1983b), their involvement in inter- and intramembrane communication plus cellular transport of sterols is of particular interest and quite worthy of continued investigation.

D. Mechanism

Squalene cyclization is a two-step process of aerobic activation and anaerobic cyclization (Figure 4). The initial NADPH-requiring monooxygenase-dependent formation of squalene 2,3-oxide has not been studied in much mechanistic detail since, only recently, a distinct oxidase has been found to function in this reaction (Ono *et al.*, 1980). Cyclization of squalene oxide is thought to be initiated by protonation of the oxygen center followed by a series of nucleophilic rearrangements terminating in generation of a protosteroid cation (Nes and McKean, 1977). Intramolecular rearrangement and deprotonation yields the neutral cyclization product, lanosterol. Cyclase enzyme imparts the specificity for product formation, and, as pointed out by Nes and

* SPF, supernatant protein factor; SCP_1, sterol carrier $protein_1$.

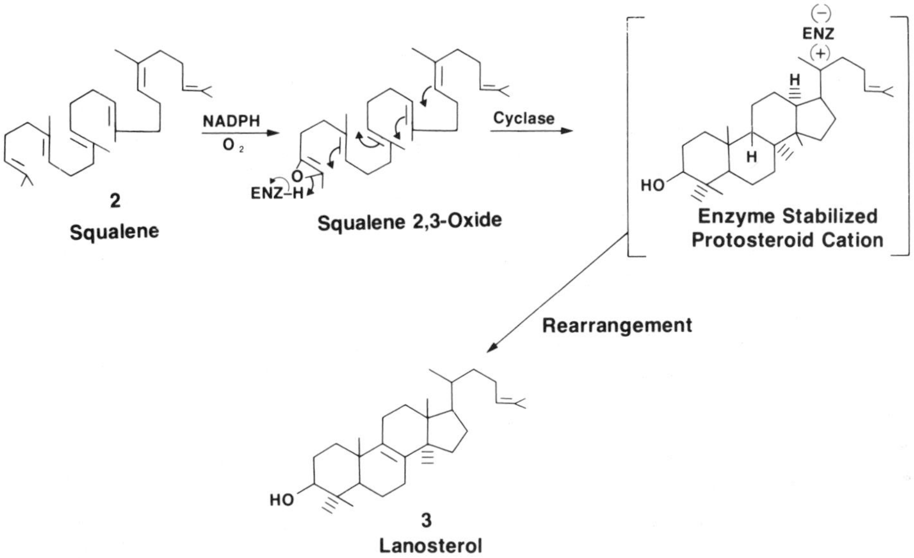

Figure 4. Lanosterol formation from squalene catalyzed by squalene epoxidase and lanosterol cyclase. Squalene oxidation results in formation of the stable, isolable squalene 2,3-oxide. Cyclization of oxidized intermediate to lanosterol is dependent upon lanosterol cyclase which provides stabilization of the protosteroid cation and directs rearrangement resulting in the first steroidal intermediate in the cholesterol biosynthetic pathway. Numbers underneath structures refer to Figure 2.

McKean (1977), the cyclase is responsible for cation stabilization by a mechanism which remains to be elucidated. Further much-needed work on both the oxidase and cyclase must await rigorous purification and reconstitution.

IV. OXIDATIVE DEMETHYLATION OF C-32 FROM LANOSTEROL

A. Metabolite Identification

Any proposed reaction mechanism and sequence involving microsomal enzymes and other membrane components which catalyze the demethylation of lanosterol (or dihydrolanosterol) must be consistent with several observations that attempt to describe the initial reactions in the transformation of this first steroid intermediate to cholesterol. These observations include (1) loss of C-32 at the aldehyde stage of oxidation of the steroid with release of C-32 as formic acid (Alexander *et al.*, 1972; Akhtar *et al.*, 1978; Gibbons *et al.*, 1979), (2) labilization of the 15α-hydrogen atom with formation of a conjugated $\Delta^{14(15)}$-double bond (Gibbons *et al.*, 1968a,b; Schroepfer *et al.*, 1972), (3) dependence upon NADPH and oxygen (Alexander *et al.*, 1972), and (4) obligatory participation of the $\Delta^{8(9)}$-double bond [4,4,14α-trimethylcholestan-3β-ol is not metabolized under conditions which support conversion of the natural, unsaturated sterol to cholesterol (Sharpless *et al.*, 1968)]. In addition, based upon conversion studies with synthetic sterols, it is assumed (Figure 5) that the oxidized intermediates of C-32 oxidative demethylation include 5α-lanost-8-ene-3β,32-diol and 3β-hydroxy-5α-lanost-8-en-32-al followed by deformylation to 4,4-dimethyl-5α-cholesta-8,14-dien-3β-ol (Nes and McKean, 1977; Schroepfer, 1982). Although the C-32 aldehyde has recently been isolated from tissue sources (Tabacik *et al.*, 1981), the C-32 alcohol has not. Both have been made synthetically and shown to be converted to both the proposed demethylation product, $\Delta^{8,14}$-diene (Akhtar *et al.*, 1978), and cholesterol (Gibbons *et al.*, 1979).

The actual deformylation reaction mechanism of the 32-aldehyde and the generation of the $\Delta^{8,14}$-diene is not known, and encompassing all of the above observations without first having the benefit of studies with resolved enzymes has produced considerable debate (Akhtar *et al.*, 1978; Schroepfer, 1982). Interruption of the reaction sequence from 32-ol to $\Delta^{8,14}$-diene through manipulation of microsomal incubation conditions has not as yet been successful; thus, isolation of biosynthetic intermediates has not been forthcoming. Resolution of the reaction sequences, isolation of the intermediates involved, and detailed discussion of mechanisms should be tempered until after purification and resolution of the component enzymes.

B. Cytochrome P-450 Involvement

Evidence for cytochrome P-450 involvement in the oxidative demethylation of C-32 of lanosterol has been obtained with rat liver homogenates employing radiolabeled mevalonate which is converted into lanosterol and 24,25-dihydrolanosterol when the synthetic sequence is interrupted with CO (Gibbons and Mitropoulos, 1972, 1973). Similarly, Alexander *et al.* (1974) demonstrated partial CO inhibition of lanosterol conversion to demethylated products in cell-free yeast preparations, thus suggesting a common cytochrome P-450 involvement in the demethylation reactions of yeast as well as liver. The site of cytochrome P-450 participation appears to be at the level of initial oxidation of C-32 in the lanosterol demethylation sequence because CO does not inhibit conversion of either the C-32 alcohol or C-32 aldehyde to cholesterol, whereas substantial CO inhibition of 24,25-dihydrolanosterol, i.e, C-32 methyl group, conversion to cholesterol is consistently observed (Gibbons *et al.*, 1979).

The CO inhibition data described above prompted Gibbons *et al.* (1979) to suggest that two distinct oxidases may be involved in C-32 demethylation. Such a suggestion conflicts with the proposal by Akhtar *et al.* (1978) that the initial and subsequent oxidations of C-32 demethylation are catalyzed by the same enzyme system. In addition, Aoyama and Yoshida (1978) have shown, in a reconstituted system of partially purified yeast microsomal cytochrome P-450 and NADPH-cytochrome P-450 reductase, that C-32 demethylation of lanosterol (M_r 426) results in production of a metabolite with a molecular weight of 410 which is consistent with loss of a methyl group and double-bond insertion. Only one cytochrome P-450 oxidase appears to be responsible for oxidation of C-32 by the reconstituted yeast system with demethylation yielding the $\Delta^{8,14}$-diene intermediate. However, the cytochrome P-450 employed in this study was not homogeneous; therefore, the possibility still remains that other proteins or perhaps more than one isozyme of cytochrome P-450 present in the preparation may have been responsible for further metabolism of the initial oxidation

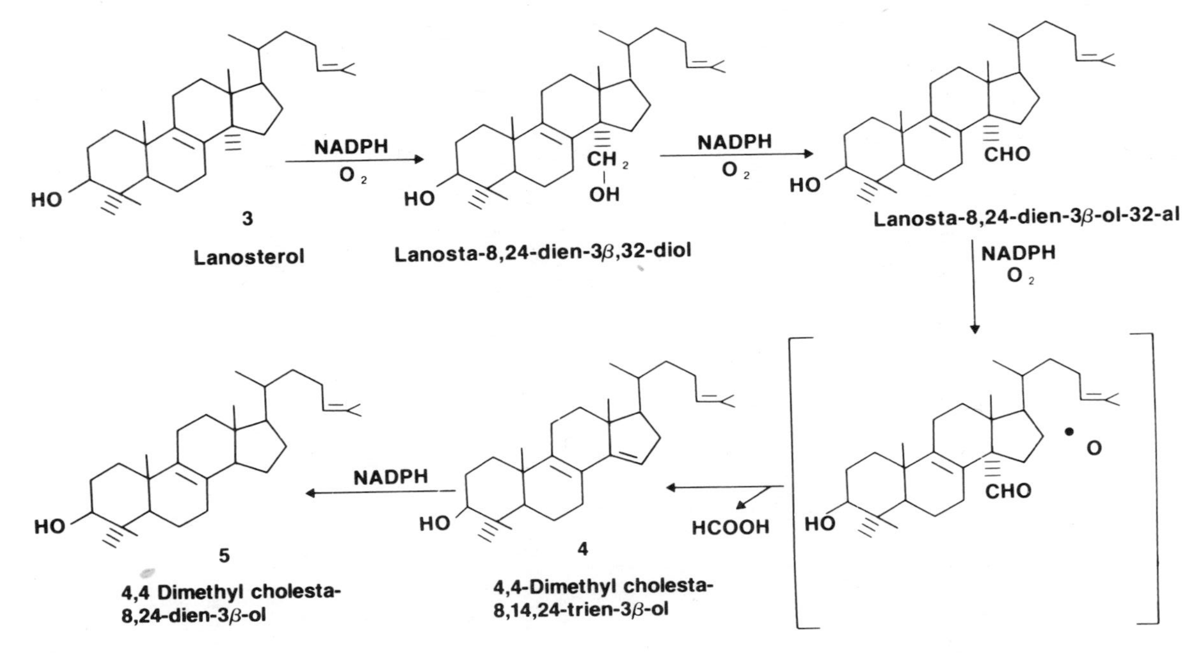

Figure 5. Proposed lanosterol 14α-demethylation reaction sequence with presumed oxygenated intermediates. Lanosterol 14α-demethylation proceeds by mixed-function-oxidase-dependent oxidation to the aldehyde state of oxidation. Formate release is dependent upon a third, yet unidentified oxidation step which results in the $\Delta^{14(15)}$-conjugated diene system. Reduction of the $\Delta^{14(15)}$-double bond completes the 14-demethylation reaction sequence. Numbers underneath structures refer to Figure 2.

product yielding the isolated diene. It now appears increasingly apparent that resolution of problems associated with interpretation of the number and type of oxidases will occur only with investigation of the individual steps of C-32 oxidation after purification of the oxidase(s) and the required electron carriers.

V. STEROID 14-REDUCTASE

A. Description of Enzymatic Activity

The existence of an enzyme capable of reducing the $\Delta^{14(15)}$-double bond of 5α-cholesta-8,14-dien-3β-ol has been described by Schroepfer *et al.* (1972). Subsequently, Gibbons and Mitropoulos (1975) have shown that 4,4-dimethyl-5α-cholesta-8,14-dien-3β-ol accumulates in a cell-free liver preparation in the presence of a reductase inhibitor, AY-9944 [*trans*-1,4-*bis*(2-chlorobenzyl-aminomethyl) cyclohexane dihydrochloride], when either mevalonic acid or 24,25-dihydrolanosterol is used as a substrate for cholesterol synthesis. In the absence of inhibitor, the 4,4-dimethyl-5α-cholesta-8,14-dien-3β-ol is converted to cholesterol. Thus, liver appears to contain an enzyme capable of reduction of the $\Delta^{14(15)}$-double bond of the $\Delta^{8,14}$-diene intermediate, and because a 4,4-dimethyl intermediate accumulates with inhibition, an obligatory reduction appears to occur immediately after C-32 demethylation.

Early mechanistic studies by Watkinson *et al.* (1971a) have shown that $\Delta^{14(15)}$-double bond reduction is dependent upon NADPH and proceeds via a *trans*-reduction with apparent inversion at C-15. Other studies by Lutsky *et al.* (1971) indicated that substrate specificity for $\Delta^{14(15)}$-double bond reduction is not limited to the $\Delta^{8,14}$-diene series since cholesta-7,14-dien-3β-ol is efficiently converted to the Δ^{7}-sterol product by rat liver preparations. Studies in our laboratory (Paik and Gaylor, 1982) have also employed 4,4-dimethyl-5α-cholesta-7,14-dien-3β-ol for the anaerobic assay of this enzymic activity, thus confirming the indifference of the enzyme to both the position of the second nuclear double bond and methyl group substitution at C-4, in addition to the microsomal location of the enzyme. The anaerobic assay employed by Paik and Gaylor (1982) also demonstrates a novel application of a method of interruption of the overall cholesterol metabolic sequence in microsomes since further conversion of the resulting 4,4-dimethyl-Δ^{7}-sterol product is fully dependent upon O_2 (see Section VI). Furthermore, by employing $\Delta^{7,14}$-diene sterol substrate, steroid 8 → 7 isomerization (also active under anaerobic conditions) is obviated. Thus, a specific assay can be employed to study this single microsomal enzyme activity without interference from either sequential conversion of the sterol product or concomitant metabolism by other microsomal enzymes of broad substrate specificity. Characterization, solubilization, and partial purification of this enzyme have been achieved with the use of this assay (Paik and Gaylor 1982; Paik *et al.*, 1983).

B. Solubilization

Detergent solubilization of the steroid 14-reductase activity has been achieved with the combination of detergents, octylglucoside and taurodeoxycholic acid (Paik and Gaylor, 1982; Paik *et al.*, 1983). This detergent combination has proved to be generally useful as several nonoxidative enzymes of the cholesterol synthetic pathway (see Sections IX and X) along with an oxidative enzymic activity (see Section VIII) have been successfully obtained in soluble, active form. Also, these nonpolymeric detergents are readily removed from solubilized proteins. The solubilized steroid 14-reductase activity does not require other additions, e.g., phospholipid, for activity and is optimal under anaerobic conditions with NADPH as cofactor. All properties observed to date for solubilized enzyme are similar to the membrane-associated enzyme (Paik and Gaylor, 1982). Thus, solubilization has not drastically altered the catalytic properties of the enzyme, and further characterization and reconstitution of the steroid 14-reductase activity will be forthcoming shortly.

VI. OXIDATIVE DEMETHYLATION OF 4,4-DIMETHYL STEROLS

Demethylation of C-30 and C-31 of 4,4-dimethyl sterols in cholesterol biosynthesis is a complex, ten-step reaction sequence composed of three separate enzymes with differing cofactor requirements, supported by various membrane-bound electron carriers, and modulated by cytosolic, soluble ancillary proteins (Gaylor, 1981; Gaylor and Delwiche, 1973). This process is characteristic of the multienzymic nature of the enzymes of cholesterol synthesis where close association of membrane-bound enzymes results in an efficient catalytic conversion of substrate to product without dissociation of intermediates from the membrane environment. The 4α-demethylation process will be addressed below in the sequence of the catalytic reactions that result in loss of the 4-gem dimethyl substituent as two molecules of CO_2 (Figure 6).

A. Methyl Sterol Oxidase

1. Description of Enzymatic Activity

4-Methyl sterol oxidase, the first enzyme in sequence in the 4α-demethylation process, is rate-limiting for 4-demethylation (Williams *et al.*, 1977). Methyl sterol oxidase in three successive oxidative cycles catalyzes the conversion of 4α-methyl sterols to 4α-carboxylic acid sterols in the presence of molecular oxygen and reduced pyridine nucleotide (Miller *et al.*, 1971). In contrast to 14α-demethylation, 4α-demethylation is not dependent upon microsomal cytochrome P-450 (Gaylor and Mason, 1968), but requires a different electron transport chain to carry out the observed oxidations. Evidence for cytochrome b_5 dependence in 4-methyl sterol oxidase has been obtained and is discussed below. Cofactor requirement, cytochrome b_5 dependence, and cyanide sensitivity of 4-methyl sterol oxidase resembles other lipid-oxidizing enzymes of liver microsomes such as acyl-CoA desaturases, Δ^7-sterol 5-desaturase, and alkanyl ether oxidations.

2. Cytochrome b_5 Dependence

Immunoglobulin titration of membrane-bound cytochrome b_5 with anticytochrome b_5 antibody in both liver and yeast microsomes results in a significant (60–70%)

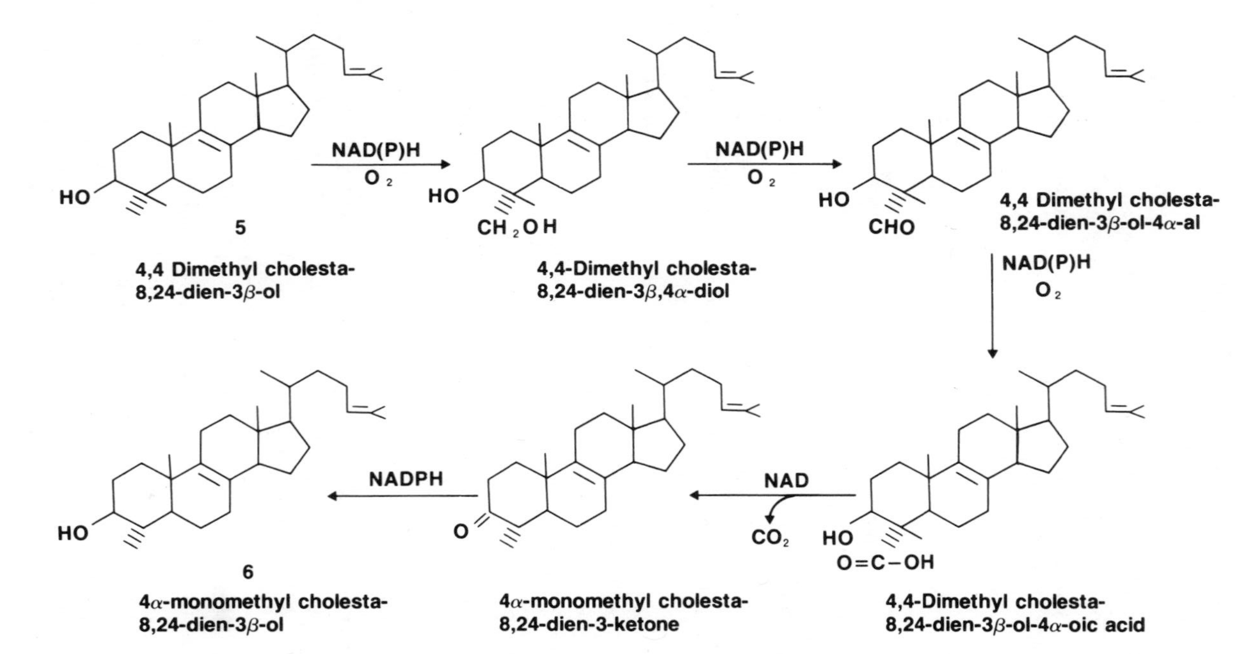

Figure 6. 4α-Methyl sterol demethylase reaction sequence showing isolated and characterized reaction intermediates. 4α-Methyl sterols are oxidized to the corresponding carboxylic acid by three successive oxidations catalyzed by 4-methyl sterol oxidase. Decarboxylation catalyzed by the NAD-dependent 4α-oic decarboxylase results in CO_2 release and formation of the corresponding 3-keto steroid. NADPH-dependent 3-keto steroid reductase completes the sequence with generation of the steroidal-3β-ol. When 4,4-dimethyl sterols are metabolized, epimerization of the 4β-methyl to the 4α-position occurs and the cycle is repeated resulting in loss of both methyl groups from the 4α-position as CO_2.

inhibition of 4-methyl sterol oxidase activity (Fukushima *et al.*, 1981; Aoyama *et al.*, 1981). In addition, Fukushima *et al.* (1981) have shown that after proteolytic digestion of microsomal preparation with trypsin, both cytochrome b_5 and 4-methyl sterol oxidase are below levels of detectability. Addition of purified, detergent-solubilized cytochrome b_5 restores 4-methyl sterol oxidase activity in proportion to the amount of cytochrome b_5 reincorporated into the trypsin-treated microsomes. Thus, microsomal electron transfer to the terminal oxidase in 4-demethylation requires the participation of cytochrome b_5.

3. *Oxidase Purification*

The demonstrated dependence of 4-methyl sterol oxidase upon cytochrome b_5 in intact microsomal preparations indicated that the electron carrier would be an obligatory component needed during solubilization and purification to assay 4-methyl sterol oxidase activity. Indeed, solubilization of the membrane-bound oxidase with Renex 690 and chromatography on DEAE-cellulose results in isolation of an oxidase preparation which is fully dependent upon cytochrome b_5 (Fukushima *et al.*, 1981). In addition, the chromatographic fraction that contains 4-methyl sterol oxidase also contains substantial amounts of NADH-cytochrome b_5 reductase (flavoprotein 1) but very low levels of NADPH-cytochrome P-450 reductase (flavoprotein 2). Addition of flavoprotein 2 is needed for maximal oxidase activity when NADPH is used as a source of reducing equivalents rather than NADH in a similar microsomal oxidase reaction (Grinstead and Gaylor, 1982). Further resolution of the same oxidase preparation by affinity chromatography on ADP-agarose results in a fraction which, in addition to cytochrome b_5, requires NADH-cytochrome b_5 reductase and phospholipid for maximal oxidase activity (Grinstead and Gaylor, 1982). Thus, 4-methyl sterol oxidase, when freed from the membrane by detergent solubilization, can be resolved into a terminal oxidase which has an obligatory requirement for cytochrome b_5 and is maximally active in the presence of both flavoproteins 1 and 2.

Purification of the terminal oxidase of C-30 and C-31 demethylation to homogeneity has been reported by Maitra *et al.* (1982) from Triton CF-10-solubilized rat liver microsomes by employing polyethylene glycol precipitation and chromatography on Sepharose 6B, DEAE-52 cellulose and hydroxylapatite. The preparation exhibits properties similar to those observed earlier by Gaylor and Mason (1968) who solubilized the same oxidase from microsomes with deoxycholate and carried purification to greater than 100-fold enrichment over microsomal methyl sterol oxidase. The deoxycholate-solubilized oxidase proved too labile, however, for more extensive studies. Neither preparation of "purified oxidase" curiously is dependent upon cytochrome b_5 nor inhibitable by anticytochrome P-450 reductase immunoglobulin, as has been shown for the membrane-bound enzyme, but each remains characteristically cyanide-sensitive. The exploration of the apparent contradiction in electron-carrier dependence between these soluble oxidase preparations and the components utilized by intact microsomes probably lies in both the effects of detergent disruption and the extremely slow rates of oxidase achieved upon reconstitutions to date.

B. Decarboxylase

Elimination of C-30 and C-31 as CO_2 with obligatory formation of 3-keto steroid products is NAD^+-dependent and catalyzed by membrane-bound steroid 4α-oic acid decarboxylase (Swindell and Gaylor, 1968). Partially purified preparations of the decarboxylase can be obtained by membrane solubilization with deoxycholate and chromatography on DEAE Sephadex A-50 (Rahimtula and Gaylor, 1972). The solubilized enzyme does not require phospholipid, electron carriers, or metal ions for activity, and the resolved decarboxylase is specific for β-NAD^+ as the required cofactor.

Detailed mechanistic studies have not been performed on the decarboxylase-catalyzed reaction despite the ease of purification and stability of the decarboxylase. However, it can be envisioned that in a two-step process, hydride ion abstraction at C-3 results in the formation of an unstable 3-keto steroid with subsequent decarboxylation of the labile β-keto acid intermediate. Formation of the obligatory 3-keto steroid probably facilitates epimerization of the 4β-methyl group of 4,4-disubstituted sterol substrate to the 4α-position, which is required for continued oxidative attack by methyl sterol oxidase in a repeat of the same 4α-methyl demethylation sequence. The precise mechanism of the two-step decarboxylation plus epimerization reactions will have to await further study with pure enzyme.

C. 3-Ketosteroid Reductase

The final enzyme of the 4α-demethylation sequence catalyzes the NADPH-dependent reduction of the 3-keto steroid that is generated during the decarboxylation reaction (Swindell and Gaylor, 1968). The 3-keto steroid reductase has been obtained in a soluble, partially purified form from rat liver microsomes (Billheimer *et al.*, 1981). The enzyme, solubilized by a 1 : 1 (wt./wt.) mixture of Lubrol WX and cholic acid, is highly labile, and stability during subsequent chromatographic purification steps is enhanced only modestly by the inclusion of 20% ethylene glycol in all buffers. Unlike the other enzymes of the 4-demethylation sequence, the 3-keto steroid reductase is retained upon chromatography on DEAE-cellulose, and the 3-ketosteroid reductase can be obtained free of methyl sterol oxidase and steroid 4α-oic acid decarboxylase activity in this single chromatographic step. The partially purified reductase preparation retains the catalytic properties of the membrane-bound enzyme in regard to substrate specificity and cofactor requirement (Billheimer *et al.*, 1981). Thus, the enzyme obtained in this partially purified form is suitable for reconstitution studies designed to investigate the complex membrane relationship of the ten steps of 4,4-dimethyl sterol demethylation.

D. Cytosolic Protein Effectors

The rate of 4α-methyl sterol demethylation observed in intact microsomes is enhanced by the addition of a cytosolic protein preparation obtained from the soluble fraction of liver homogenates (Gaylor and Delwiche, 1976; Spence and Gaylor, 1977).

The protein preparation, when resolved into components by isoelectric focusing, is composed of two protein species which have been further characterized as the highly abundant Z-protein of cytosol and a very active, but grossly less abundant nonspecific lipid-transfer protein (Billheimer and Gaylor, 1980; Trzaskos and Gaylor, 1983a). The mechanisms through which these (and possibly other) cytosolic proteins enhance 4-demethylation activity are quite complex (Trzaskos and Gaylor, 1983b). *With membrane-bound enzymes,* Z-protein reverses fatty acyl-CoA inhibition of several microsomal enzymes (Grinstead *et al.,* 1983) and enhances availability of exchangeable hematin to oxidase enzymes (Billheimer and Gaylor, 1980). The lipid-transfer protein enhances substrate availability *only with intact microsomes* as a result of nonspecific transfer of artificially introduced substrate in the *in vitro* assay systems (Trzaskos and Gaylor, 1983a). The effects of these proteins are not limited to methyl sterol oxidase of the demethylation sequence because membrane-bound 3-ketosteroid reductase and steroid 4α-oic acid decarboxylase activities are also enhanced by the same isolated fraction containing both proteins (Spence and Gaylor, 1977; Billheimer *et al.,* 1981). Although these proteins can be shown to enhance membrane-bound enzymic activities, in our hands, enzyme solubilization or disruption of membrane integrity by detergent treatment eliminates all stimulatory effects (Billheimer *et al.,* 1981; Trzaskos and Gaylor, 1983a). Thus, in intact cells with membrane-bound enzymes, these proteins may not be obligatory components of the cholesterol synthetic pathway particularly as "sterol carrier proteins" since full enzymic activity can be demonstrated in their absence. Although their biochemical properties are becoming better characterized, the significance of their involvement as true participants rather than ancillary modulators in sterol synthesis is open to question. However, the modulating role of Z-protein can be documented, e.g., reversal of acyl-CoA inhibition, and the transfer protein(s) may be important in relocation of cholesterol *from* the site of synthesis in the endoplasmic reticulum to other membranes for either further transformation as substrate for bile acids and steroid hormones (Chanderbhan *et al.,* 1982) or incorporation into membranes structurally.

VII. $\Delta^8 \rightarrow \Delta^7$ ISOMERASE

A. Purification Data

Following removal of the three "extra" methyl groups present in lanosterol, the next anticipated reaction in the cholesterol synthetic pathway is isomerization of the $\Delta^{8(9)}$-double bond to the $\Delta^{7(8)}$-position in the B-ring of the steroid nucleus. Maximal enzymic activity is detected under anaerobic conditions (Paik and Gaylor, 1982) employing 5α-cholesta-8,24-dien-3β-ol (zymosterol) as substrate (Gaylor *et al.,* 1966). Attempted solubilizations of the steroid 8-ene isomerase activity had not been successful owing to intrinsic lability of the enzyme and apparent dependence upon phospholipid for activity (Yamaga and Gaylor, 1978). However, Paik and Gaylor (1982) have recently obtained a solubilized preparation of the steroid 8-ene isomerase from rat liver microsomes by employing a combination of octylglucoside and taurodeoxycholate as solubilizing agents. The solubilized preparation retains the catalytic properties of the membrane-bound enzyme, and further purification and reconstitution studies should prove interesting.

B. Mechanism

Catalytically, the Δ^8-steroid isomerase reaction is one of the fastest in the cholesterol synthetic pathway (Gaylor, 1981). The preferred substrate appears to be the fully demethylated cholesta-8,24-dien-3β-ol, whereas (in decreasing order), cholest-8-en-3β-ol, 4α-monomethyl- and then 4,4-dimethyl-cholest-8-en-3β-ols are metabolized with either the 4,4,14α-trimethyl cholest-8-en- or $\Delta^{8,24}$-dien-3β-ols being unreactive (Gaylor *et al.*, 1966). Thus, the presence of a 14α-methyl substituent appears to prevent isomerization. The enzyme has no cofactor or metal ion requirements. During isomerization, a hydrogen atom is taken up from the medium at C-9 (Lee *et al.*, 1969; Wilton *et al.*, 1969) with loss of the 7α-hydrogen or 7β-hydrogen atom occurring in yeast and animal preparations, respectively (Caspi and Ramm, 1969). Reversibility of the isomerase reaction has been demonstrated (Wilton *et al.*, 1969; Scala *et al.*, 1974), and Pascal and Schroepfer (1980) have shown that even the 14α-alkyl-substituted Δ^7-sterol can be converted to a Δ^8-counterpart in the reverse reaction. Thus, although the $\Delta^8 \rightarrow \Delta^7$ isomerization normally does not seem to occur in the expected cholesterol synthetic pathway with a 14α-methyl substituent on the sterol nucleus (Gaylor *et al.*, 1966), the reverse reaction may occur which accounts, in part, for the naturally occurring 14α-methyl sterols of both the Δ^7- and Δ^8-families. In some organisms, 4-methyl sterol oxidase also appears to be active without prior loss of the 14α-methyl group thus adding to the complexity of a wide variety of naturally occurring sterols.

VIII. Δ^5-Desaturase

A. Purification and Reconstitution

Introduction of the Δ^5-bond in cholesterol synthesis by microsomal preparations requires both molecular oxygen and reduced pyridine nucleotide (Dempsey *et al.*, 1964; Reddy and Caspi, 1976). In addition, the oxidase reaction is inhibitable by cyanide, but is not affected by CO (Dempsey *et al.*, 1968; Gaylor *et al.*, 1973). Thus, although evidence against cytochrome P-450 involvement in Δ^7-sterol 5-desaturase exists, and indirect evidence for cytochrome b_5 participation in the microsomal desaturase reaction in both rat liver and yeast has been presented (Reddy *et al.*, 1977; Osumi *et al.*, 1979), demonstration of obligate electron carriers required for desaturase activity had not been forthcoming until recently for the membrane-bound enzyme. Grinstead and Gaylor (1982) obtained a detergent-solubilized enzyme preparation employing octylglucoside and sodium taurodeoxycholate. Upon chromatography on DEAE-cellulose, Sepharose 4B, and ADP-Agarose, the oxidase preparation was shown to be dependent upon cytochrome b_5 and phospholipid, plus maximal activity was observed only in the presence of both flavoproteins 1 and 2. This work clearly defines the necessary components of the desaturase reaction through solubilization, purification, and reconstitution. Interestingly, the same oxidase preparation under identical reconstitution conditions also displays 4-methyl sterol oxidase activity (see Section VI-A). Activities of the two oxidases remain in the same proportion during resolution that suggests that a single oxidase may catalyze both reactions. If this suggestion proves valid, the mechanisms of the oxidase-catalyzed reactions after the cytochrome P-450-dependent C-32 elimination would be of considerable interest both mechanistically and biologically. The reconstitution of 5-desaturase with purified components eliminates the earlier suggestion of an $NAD(P)^+$-dependent dehydrogenase mechanism.

B. Mechanism

In animals and yeast, the Δ^5-desaturase reaction occurs by *cis* elimination of the 5α and 6α-hydrogens (Atkin *et al.*, 1972; Akhtar and Parvez, 1968). An oxygen requirement had been interpreted as evidence for hydroxylation followed by dehydroxylation (Akhtar and Parvez, 1968), but as indicated by Nes and McKean (1977) little success has been obtained with synthetic oxygenated substrates in the liver system, and isolation of possible intermediates has not been fruitful. However, a 5α-hydroxy sterol has been isolated from yeast cultured under aerobic conditions (Fryberg *et al.*, 1973), and further metabolism of the hydroxy sterol can be observed under both aerobic and anaerobic atmospheres (Fryberg *et al.*, 1973; Topham and Gaylor, 1970, 1972). The use of purified oxidase enzyme and electron transport components in both the yeast and animal systems should help in both the isolation of intermediates, if they exist, and resolution of the 5-desaturase mechanism.

C. Cytosolic Protein Effectors

Soluble proteins have been shown to enhance microsomal Δ^7-sterol 5-desaturase activity and the stimulation of enzyme activity has been used to purify one of these proteins, squalene and sterol carrier protein, SCP, from liver cytosol (Dempsey *et al.*, 1981). Trzaskos and Gaylor (1983a) have shown that a second distinct, nonspecific lipid transfer protein identical with SCP_2* and purified by Noland *et al.* (1980) is also active in stimulating this enzyme-catalyzed reaction. The possible mechanism(s) of these proteins has been addressed (Trzaskos and Gaylor, 1983b), and as in the cytosolic protein activation of other membrane-bound enzymes of cholesterol biosynthesis, possible functions should be considered when reconstitution experiments are conducted.

* SCP_2, sterol carrier $protein_2$.

IX. 7-DEHYDROCHOLESTEROL Δ^7-REDUCTASE

A. Purification Data

Reduction of the Δ^7-double bond in the microsomal conversion of 7-dehydrocholesterol to cholesterol has been shown by Schroepfer and Frantz (1961) and Dempsey (1965) to proceed in the absence of oxygen and to require NADPH as cofactor. Study of the substrate specificity of the enzyme has been aided by the use of the sterol 24-reductase inhibitor, triparanol, showing that either $\Delta^{5,7,24}$-cholestatrien-3β-ol or $\Delta^{5,7}$-cholestadien-3β-ol can act as substrate for the Δ^7-reductase (Dempsey, 1965). Limited purification of the NADPH-dependent Δ^7-reductase of rat liver microsomes has been reported by Dempsey (1969). Employing sodium deoxycholate for solubilization, ammonium sulfate precipitation, and Sephadex G-25 column chromatography, Dempsey (1969) obtained enzyme preparations enriched about ten-fold over microsomal preparations. The isolated fractions were extremely labile to either freezing and thawing or brief exposure to heat, but activity was maintained when the fractions were stored at 4°C for approximately 48 hr. Paik and Gaylor (1982) have also obtained a soluble preparation of Δ^7-reductase using the detergent combination of sodium taurodeoxycholate and octylglucoside. The isolated enzyme fraction is maximally active under anaerobic conditions and stable to storage at −80°C for several weeks. Unlike the preparation described by Dempsey (1969) which requires a soluble activator protein for enzyme detection, that of Paik and Gaylor (1982) is fully active in the absence of added soluble protein. Sensitivity to soluble protein suggests that the earlier preparation of Dempsey is an aggregate containing phospholipid and proteins.

B. Cytosolic Protein Effectors

Enhanced rates of conversion of 7-dehydrocholesterol to cholesterol by rat liver microsomes in the presence of soluble protein preparations have led to attempted purification of the protein(s) responsible for the activation. Ritter and Dempsey (1970) and Scallen *et al.* (1974) have employed activation of the Δ^7-reductase reaction to obtain protein fractions designated "sterol carrier proteins" which apparently function by enhancing substrate availability to the membrane-bound enzyme, hence the name, "carrier." Purification of the isolated protein fractions to apparent homogeneity has shown that SCP_2 obtained by Noland *et al.* (1980) and SCP purified by Dempsey *et al.* (1981) are distinctly different molecular entities. Controversy exists over which protein species may be responsible for the sterol "carrier role" in the activation of membrane-bound sterol synthetic enzymes. A lipid-transfer protein purified by Trzaskos and Gaylor (1983a) which is identical to SCP_2 does enhance microsomal Δ^7-reductase activity. The actual function of these proteins, their mechanism of action, and demonstration of any obligate requirement for synthetic activity in intact cells and membranes will have to await total reconstitution of the enzyme processes which they are described to enhance, since solubilized microsomal enzyme preparations do not require the proteins for activity.

C. Mechanism

Stereochemically, reduction of 7-dehydrocholesterol by Δ^7-reductase during cholesterol synthesis proceeds by a 1,2-*trans*-diaxial addition of hydrogen to the carbon–carbon double bond (Wilton *et al.*, 1968). In this same communication, it was shown that the 7α- and 8β-hydrogen atoms were derived from NADPH and a proton source most likely enzyme, respectively. Isotope effects observed during the reduction process suggested that protonation yielding the 8β-hydrogen is rate-limiting for the reaction and that hydride transfer to the stabilized carbonium ion at C-7 is quite rapid (Wilton *et al.*, 1968). Bjorkhem and Holmberg (1973) confirmed these suggestions on reaction rates, thus establishing protonation followed by hydride transfer as the reduction mechanism.

X. Δ^{24}-STEROL 24-REDUCTASE

A. Purification Data

Reduction of the Δ^{24}-side chain double bond of cholesta-5,24-dien-3β-ol appears to be the final enzyme-catalyzed reaction in the formation of cholesterol from lanosterol in animal systems (Gaylor, 1981). Despite the seeming simplicity of the NADPH-dependent 24-reductase reaction, little information is available on solubilization and purification procedures. Paik and Gaylor (1982) have now obtained the 24-reductase in soluble form using the octylglucoside–taurodeoxycholate detergent combination, and these workers have had some success on further purification of the solubilized enzyme. Successful purification of the 24-reductase should be attainable now that solubilization and stabilization have been achieved.

B. Mechanism

Reduction of the $\Delta^{24(25)}$-double bond of desmosterol may proceed similarly to the reduction of the $\Delta^{7(8)}$-double bond of 7-dehydrocholesterol with protonation at C-24 followed by hydride transfer from NADPH to the resulting carbonium ion at C-25 (Watkinson *et al.*, 1971b). Kienle *et al.* (1973) showed that stereochemically, however, the reaction involves *cis* rather than *trans* addition of the two hydrogen atoms to the $\Delta^{24(25)}$-double bond, thus establishing the overall reaction mechanism. In spite of the elegant chemistry which has established the reduction mechanism, lack of substrate specificity for the Δ^{24}-reductase has not permitted assignment of a most favored substrate or a definitive sequence location for the enzymic activity, since side-chain reduction could occur early, late, or, for that matter, at any point along the cholesterol synthetic pathway (Steinberg and Avigan, 1969). However, as pointed out by Gaylor (1981), a specific pathway may be an unnecessary preoccupation since the primary process of the nuclear transformations of lanosterol to cholesterol can proceed by several different routes regardless of side-chain saturation status.

XI. COORDINATE CONTROL MECHANISMS OF STEROL SYNTHETIC ENZYMES

A. HMG-CoA Reductase

The accepted rate-limiting enzyme governing overall carbon flux through the cholesterol synthetic pathway is the membrane-bound microsomal enzyme HMG-CoA reductase (Rodwell *et al.*, 1973, 1976; Sabine, 1983). Variations in HMG-CoA reductase activity have been shown to parallel those of overall cholesterol synthesis under a variety of conditions associated with either an increased demand or a decreased need for *de novo* synthesized cellular cholesterol. Considerable effort has been expended in the study of the regulatory mechanism(s) responsible for the alterations in reductase activity, and our understanding of this key regulatory enzyme has expanded substantially in the last 20 years (Rodwell *et al.*, 1976). However, as evident in the recent monograph on HMG-CoA reductase (Sabine, 1983), numerous agents or metabolic states can alter reductase activity, and the alterations are multifaceted with changes in catalytic efficiency, enzyme amount, and even ancillary regulatory protein levels modulating observed activity changes. Extension of the parallel between overall cholesterol synthetic rate and enzyme activity to other enzymes in the cholesterol synthetic pathway has been made (Gaylor, 1981; Chang and Limanek, 1980), and although the mechanisms governing these enzymes have not been as exhaustively investigated as those of HMG-CoA reductase, the suggestion for multilevel control seems reasonable. Furthermore, because mevalonic acid, the product of reduction of HMG-CoA, is converted to other metabolites, regulation in concert between mevalonate formation and metabolism appears essential as described below.

B. Multilevel Control

Considerable evidence exists for multilevel control in the cholesterol synthetic pathway for several membrane-bound sterol synthetic enzymes in response to cholesterol feeding, diurnal variation, and cholestyramine feeding. Activities of the three enzymes responsible for 4-demethylation and the $\Delta^8 \rightarrow \Delta^7$ isomerase have been shown by Gaylor (1981) to parallel those of HMG-CoA reductase and cholesterol synthesis under the conditions listed above. In addition, squalene epoxidase (Kojima and Sakurada, 1982) and the soluble enzymes, HMG-CoA synthetase, acetoacetyl coenzyme A thiolase, and mevalonate kinase (Chang and Limanek, 1980) appear to be under similar metabolic control with enzymic activity paralleling that of the overall cholesterol synthetic rate. Further possible sites for regulation have also been proposed by Havel *et al.* (1979) to exist at the level of lanosterol demethylation with an endogenous sterol synthesis inhibitor being produced at this part in the pathway being recently suggested (Tabacik *et al.*, 1981). Thus, the control not only of HMG-CoA reductase, but of other membrane-bound and even soluble enzymes appears to be regulated in concert with activity fluctuations consistent with multilevel control. The possible direct inhibition by a synthetic intermediate other than end product cholesterol for the pathway would be consistent with regulation seen in pathways of this type.

The close membrane association of several enzymes associated with squalene synthesis, squalene cyclization, 4-demethylation, and 14-demethylation have made it difficult to isolate individual protein species and the generated metabolites presumably due to the physical proximity within the microsomal membrane. It might also be that the proximity and membrane association of the various activities found later in the pathways are for regulatory purposes and that the coincidental location of HMG-CoA reductase in the same membrane is not an oddity of the pathway, but a physical coupling for activity regulation. The membrane, thus, may play an intermediary or communicative role in the regulatory process allowing both major adjustments in carbon flux as well as fine tuning of metabolic rates under the same general control mechanism. Continued study of all of the membrane-associated cholesterol synthetic enzymes through purification and characterization will ultimately determine whether the same mechanisms of altered catalytic efficiency and altered enzyme amount are involved in control.

C. Soluble-Protein Involvement

A final point consistent with a coordinate control mechanism is the observed enhancement and regulation of a number of cholesterol synthetic enzymes by soluble noncatalytic proteins (Scallen *et al.*, 1971; Ritter and Dempsey, 1970). As indicated in our discussions of several of the enzymes in the pathway, soluble proteins seem to alter membrane-bound catalytic activity of the enzyme while solubilized preparations of microsomes do not respond. A review of soluble-protein involvement in conjunction with HMG-CoA reductase activity has been presented (Trzaskos and Gaylor, 1983b), and it is our feeling that these proteins do not function as substrate "carriers" in sterol synthesis, but the cytosolic protein may be instrumental in communication both inter- and intramembranously allowing a rapid change in synthetic rate to occur with only a minor change in steady-state metabolite status. Definitive answers to the mode of soluble-protein involvement in sterol synthesis and an understanding of the regulatory mechanism observed will come only as enzyme purification and reconstitution studies progress to the point that all components in the system are identified, characterized, and reconstituted to relate potential function to requirements for catalytic activity. Subsequently, an equally difficult challenge requires that studies also must be carried out in intact cells.

REFERENCES

Agnew, W. S., and Popjak, G., 1978, Squalene synthetase. Stoichiometry and kinetics of presqualene pyrophosphate and squalene synthesis by yeast microsomes, *J. Biol. Chem.* **253**:4566–4573.

Akhtar, M., and Parvez, M. A., 1968, The mechanism of the elaboration of ring b in ergosterol biosynthesis, *Biochem. J.* **108**:527–531.

Akhtar, M., Alexander, K., Boar, R. B., McGhie, J. F., and Barton, D. H. R., 1978, Chemical and enzymic studies on the characterization of intermediates during the removal of the 14α-methyl group in cholesterol biosynthesis. The use of 32-functionalized lanosterol derivatives, *Biochem. J.* **169**:449–463.

Alexander, K., Akhtar, M., Boar, R. B., McGhie, J. F., and Barton, D. H. R., 1972, The removal of the 32-carbon atom as formic acid in cholesterol biosynthesis, *J. Chem. Soc. Chem. Commun.* **7**:383–385.

Alexander, K. T. W., Mitropoulos, K. A., and Gibbons, G. F., 1974, A possible role for cytochrome P-450 during the biosynthesis of zymosterol from lanosterol by *Saccharomyces cerevisiae, Biochem. Biophys. Res. Commun.* **60**:460–467.

Aoyama, Y., and Yoshida, Y., 1978, The 14α-demethylation of lanosterol by a reconstituted cytochrome P-450 system from yeast microsomes, *Biochem. Biophys. Res. Commun.* **85**:28–34.

Aoyama, Y., Yoshida, Y., Sato, R., Susani, M., and Ruis, H., 1981, Involvement of cytochrome b_5 and a cyanide-sensitive monooxygenase in the 4-demethylation of 4,4-dimethylzymosterol by yeast microsomes, *Biochim. Biophys. Acta* **663**:194–202.

Atkin, S. D., Palmer, E. D., English, P. D., Morgan, B., Cawthorne, M. A., and Green, J., 1972, Role of cytochrome P-450 in cholesterol biogenesis and catabolism, *Biochem. J.* **128**:237–242.

Billheimer, J. T., and Gaylor, J. L., 1980, Cytosolic modulators of activities of microsomal enzymes of cholesterol biosynthesis. Role of a cytosolic protein with properties similar to Z-protein (fatty acid-binding protein), *J. Biol. Chem.* **255**:8128–8135.

Billheimer, J. T., Alcorn, M., and Gaylor, J. L., 1981, Solubilization and partial purification of a microsomal 3-ketosteroid reductase of cholesterol biosynthesis, *Arch. Biochem. Biophys.* **211**:430–438.

Bjorkhem, I., and Holmberg, I., 1973, Mechanism of enzymatic reduction of steroid double bonds, *Eur. J. Biochem.* **33**:364–367.

Caras, I. W., and Bloch, K., 1979, Effects of a supernatant protein activator on microsomal squalene-2,3-oxide-lanosterol cyclase, *J. Biol. Chem.* **254**:11816–11821.

Caras, I. W., Friedlander, E. J., and Bloch, K., 1980, Interactions of supernatant protein factor with components of the microsomal squalene epoxidase system, *J. Biol. Chem.* **255**:3575–3580.

Caspi, E., and Ramm, P. J., 1969, Stereochemical differences in the biosynthesis of C^{27}-Δ^7-steroidal intermediates, *Tetrahedron Lett.* **3**:181–185.

Chanderbhan, R., Noland, B. J., Scallen T. L., and Vahouny, G. V., 1982, Sterol carrier protein$_2$. Delivery of cholesterol from adrenal lipid droplets to mitochondria for pregnenolone synthesis, *J. Biol. Chem.* **257**:8928–8934.

Chang, T.-Y., and Limanek, J. S., 1980, Regulation of cytosolic acetoacetyl coenzyme A thiolase, 3-hydroxy-3-methylglutaryl coenzyme A reductase, and mevalonate kinase by low density lipoprotein and by 25-hydroxycholesterol in Chinese hamster ovary cells, *J. Biol. Chem.* **255**:7787–7795.

Corey, E. J., Russey, W. E., and Ortiz de Montellano, P. R., 1966, 2,3-Oxidosqualene, an intermediate in the biological synthesis of sterols from squalene, *J. Am. Chem. Soc.* **88**:4750–4751.

Dean, P. D. G., Ortiz de Montellano, P. R., Bloch, K., and Corey, E. J., 1966, A soluble 2,3-oxidosqualene sterol cyclase, *J. Biol. Chem.* **242**:3014–3015.

Dempsey, M. E., 1965, Pathways of enzymic synthesis and conversion to cholesterol of $\Delta^{5,7,24}$-cholestatrien-3β-ol and other naturally occurring sterols, *J. Biol. Chem.* **240**:4176–4188.

Dempsey, M. E., 1969, Δ^7-Sterol Δ^5-dehydrogenase and $\Delta^{5,7}$-sterol Δ^7-reductase of rat liver, *Meth. Enzymol.* **15**:501–514.

Dempsey, M. E., Seaton, J. D., Schroepfer, G. J., and Trockman, R. W., 1964, Intermediary role of 5,7-cholestadien-3β-ol in cholesterol biosynthesis, *J. Biol. Chem.* **239**:1381–1387.

Dempsey, M. E., Bissett, K. J., and Ritter, M. C., 1968, Enzymic formation of the Δ^5-bond of cholesterol, *Circulation* **38**(Suppl.):VI-5.

Dempsey, M. E., McCoy, K. E., Baker, H. N., Damitriadou-Vafiadou, A., Lorsbach, T., and Howard, J. B., 1981, Large-scale purification and structural characterization of squalene and sterol carrier protein, *J. Biol. Chem.* **256**:1867–1873.

Dugan, R. E., and Porter, J. W., 1972, Hog liver squalene synthetase: The partial purification of the particulate enzyme and kinetic analysis of the reaction, *Arch. Biochem. Biophys.* **152**:28–35.

Ferguson, J. B., and Bloch, K., 1977, Purification and properties of a soluble protein activator of rat liver squalene epoxidase, *J. Biol. Chem.* **252**:5381–5385.

Fryberg, M., Oehlschlager, A. C., and Unrau, A. M., 1973, Formation of a 5α-hydroxy sterol by Saccharomyces cerevisiae, *Biochem. Biophys. Res. Commun.* **51**:219–222.

Fukushima, H., Grinstead, G. F., and Gaylor, J. L., 1981, Total enzymic synthesis of cholesterol from lanosterol. Cytochrome b_5-dependence of 4-methyl sterol oxidase, *J. Biol. Chem.* **256**:4822–4826.

Garvey, K. L., and Scallen, T. J., 1978, Studies on the conversion of enzymatically generated, microsomal-bound squalene to sterol, *J. Biol. Chem.* **253:**5476–5483.

Gaylor, J. L., 1981, Formation of sterols in animals, in: *Biochemistry of Isoprenoids,* Vol. 1 (J. W. Porter and S. L. Springer, eds.), John Wiley & Sons, New York, pp. 482–543.

Gaylor, J. L., 1982, Membrane-bound enzymes of cholesterol biosynthesis from lanosterol, in: *Membranes and Transport,* Vol. 1 (A. N. Martonosi, ed.), Plenum, New York, pp. 249–253.

Gaylor, J. L., and Delwiche, C. V., 1973, Investigation of the multienzymic system of microsomal cholesterol biosynthesis, in: *Annals of the New York Academy of Science,* Vol. 212 (D. Y. Cooper and H. A. Salhanick, eds.), Academy Publishing Co., New York, pp. 122–138.

Gaylor, J. L., and Delwiche, C. V., 1976, Purification of a soluble rat liver protein that stimulates microsomal 4-methyl sterol oxidase activity, *J. Biol. Chem.* **251:**6638–6645.

Gaylor, J. L., and Mason, H. S., 1968, Investigation of the component reactions of sterol demethylation. Evidence against participation of cytochrome P-450, *J. Biol. Chem.* **243:**5546–5555.

Gaylor, J. L., Delwiche, C. V., and Swindell, A. C., 1966, Enzymatic isomerization ($\Delta^8 \rightarrow \Delta^7$) of intermediates of sterol biosynthesis, *Steroids* **8:**353–363.

Gaylor, J. L., Hsu, S. T., Delwiche, C. V., Comai, K., and Seifried, H. E., 1973, Noncytochrome P-450-dependent oxidase of liver microsomes. Oxidation of methyl sterols and stearoyl coenzyme A, in: *Oxidases and Related Redox Systems* (T. E. King, H. S. Mason, and M. Morrison, eds.), University Park Press, Baltimore, pp. 431–444.

Gibbons, G. F., and Mitropoulos, K. A., 1972, Inhibition of cholesterol biosynthesis by carbon monoxide: Accumulation of lanosterol and 24,25 dihydrolanosterol, *Biochem. J.* **127:**315–317.

Gibbons, G. F., and Mitropoulos, K. A., 1973, The effect of carbon monoxide on the nature of the accumulated 4,4-dimethyl sterol precursors of cholesterol during biosynthesis from [2-^{14}C]mevalonic acid *in vitro, Biochem. J.* **132:**439–448.

Gibbons, G. F., and Mitropoulos, K. A., 1975, Effect of *trans*-1,4-bis (2-chlorobenzylaminomethyl) cyclohexane dihydrochloride and carbon monoxide on hepatic cholesterol biosynthesis from 4,4-dimethyl sterols *in vitro, Biochim. Biophys. Acta* **380:**270–281.

Gibbons, G. F., Goad, L. J., and Goodwin, T. W., 1968a, Stereochemistry of hydrogen elimination from C-15 during cholesterol biosynthesis, *Chem. Commun.* **22:**1458–1460.

Gibbons, G. F., Goad, L. J., and Goodwin, T. W., 1968b, The stereochemistry of hydrogen elimination at C-7 during cholesterol biosynthesis in rat liver, *Chem. Commun.* **20:**1212–1214.

Gibbons, G. F., Pullinger, C. R., and Mitropoulos, K. A., 1979, Studies on the mechanism of lanosterol 14α-demethylation. A requirement for two distinct types of mixed-function oxidase systems, *Biochem. J.* **183:**309–315.

Grinstead, G. F., and Gaylor, J. L., 1982, Total enzymic synthesis of cholesterol from 4,4,14α-trimethyl-5α-cholesta-8,24-dien-3β-ol: Solubilization, resolution, and reconstitution of Δ^7-sterol 5-desaturase, *J. Biol. Chem.* **257:**13937–13944.

Grinstead, G., Trzaskos, J. M., Billheimer, J. T., and Gaylor, J. L., 1983, Cytosolic modulators of activities of microsomal enzymes of cholesterol biosynthesis. Effects of acyl-CoA inhibition and cytosolic Z-protein, *Biochim. Biophys. Acta,* **751:**41–51.

Havel, C., Hansbury, E., Scallen, T. J., and Watson, J. A., 1979, Regulation of cholesterol synthesis in primary rat hepatocyte culture cells. Possible regulatory site at sterol demethylation, *J. Biol. Chem.* **254:**9573–9582.

Kienle, M. G., Varma, R. K., Mulheirn, L. J., Yagen, B., and Caspi, E., 1973, Reduction of Δ^{24} of lanosterol in the biosynthesis of cholesterol by rat liver enzymes. II. Stereochemistry of addition of the C-25 proton, *J. Am. Chem. Soc.* **95:**1996–2001.

King, H. S., Mason, H. S., and Morrison, M. (eds.), 1973, *Oxidases and Related Redox Systems,* Vol. 2, University Park Press, Baltimore.

Kojima, Y., and Sakurada, T., 1982, Regulation of mouse liver squalene epoxidase, XII International Congress of Biochemistry, p. 349.

Kojima, Y., Friedlander, E. J., and Bloch, K., 1981, Protein-facilitated intermembrane transfer of squalene. Demonstration by density gradient centrifugation, *J. Biol. Chem.* **256:**7235–7239.

Lee, W.-H., Kammereck, R., Lutsky, B. N., McCloskey, J. A., and Schroepfer, G. J., Jr., 1969, Mechanism of the enzymic conversion of cholest-8-en-3β-ol to cholest-7-en-3β-ol, *J. Biol. Chem.* **244:**2033–2040.

Lutsky, B. N., Martin, J. A., and Schroepfer, G. J. Jr., 1971, Studies of the metabolism of 5α-cholesta-8,14-dien-3β-ol and 5α-cholesta-7,14-diene-3β-ol in rat liver homogenate preparations, *J. Biol. Chem.* **246:**6437–6744.

Lynen, F., Eggerer, H., Henning, U., and Kessel, I., 1958, Biosynthesis of the terpenes. III. Farnesyl pyrophosphate and 3-methyl-3-buten-l-yl pyrophosphate, the biological precursors of squalene, *Angew. Chem.* **70:**738–742.

Maitra, V. S., Mohau, V. P., Kochi, H., Shankar, V., Adlersberg, M., Liu, K.-P., Ponticorvo, L., and Sprinson, D. B., 1982, Purification of a terminal oxygenase in demethylation of C-30 of lanosterol, *Biochem. Biophys. Res. Commun.* **108:**517–525.

Miller, W. L., Brady, D. R., and Gaylor, J. L., 1971, Investigation of the component reactions of oxidative demethylation of sterols. Metabolism of 4α-hydroxymethyl steroids, *J. Biol. Chem.* **246:**5147–5153.

Nes, W. R., and McKean, M. L., 1977, *Biochemistry of Steroids and Other Isopentenoids*, 1st Ed., University Park Press, Baltimore, Maryland.

Noland, B. J., Arebalo, R. E., Hansbury, E., and Scallen, T. J., 1980, Purification and properties of sterol carrier protein$_2$, *J. Biol. Chem.* **255:**4282–4289.

Ono, T., and Bloch, K., 1975, Solubilization and partial purification of rat liver squalene epoxidase, *J. Biol. Chem.* **250:**1571–1579.

Ono, T., Ozasa, S., Hasegawa, F., and Imai, Y., 1977, Involvement of NADPH-cytochrome *c* reductase in the rat liver squalene epoxidase system, *Biochim. Biophys. Acta* **486:**401–407.

Ono, T., Takahashi, K., Odani, S., Konno, H., and Imai, Y., 1980, Purification of squalene epoxidase from rat liver microsomes, *Biochem. Biophys. Res. Commun.* **96:**522–528.

Osumi, T., Nishino, T., and Katsuki, H., 1979, Studies on the delta 5-desaturation in ergosterol biosynthesis in yeast, *J. Biochem. (Tokyo)* **85:**819–826.

Paik, Y. K., and Gaylor, J. L., 1982, Complete resolution and reconstitution of cholesterol synthesis from lanosterol: Nonoxidative enzymes, 184th ACS meeting, Kansas City, Missouri, Abstract No. 39.

Paik, Y. K., Shafiee, A., and Gaylor, J. L., 1984, Microsomal enzymes of cholesterol biosynthesis from lanosterol: Characterization of the $\Delta^{8,14}$-steroid 14-reductase, *J. Biol. Chem.*, in press.

Pascal, R. A., and Schroepfer, G. J., 1980, Enzymatic isomerization ($\Delta^7 \rightarrow \Delta^8$) of the nuclear double bond of 14α-methyl substituted sterol precursors of cholesterol, *Biochem. Biophys. Res. Commun.* **94:**932–939.

Popjak, G., 1959, Biosynthesis of derivatives of allylic alcohols from mevalonate-2-C^{14} in liver enzyme preparations and their relation to the synthesis of squalene, *Tetrahedron Lett.* **19:**19–28.

Popjak, G., and Agnew, W. S., 1979, Squalene synthetase, *Mol. Cell. Biochem.* **27:**97–116.

Qureshi, A. A., Beytia, E., and Porter, J. W., 1973, Squalene synthetase. II. Purification and properties of baker's yeast enzyme, *J. Biol. Chem.* **248:**1848–1855.

Rahimtula, A. D., and Gaylor, J. L., 1972, Investigation of the component reactions of oxidative sterol demethylation. Partial purification of a microsomal sterol 4α-carboxylic acid decarboxylase, *J. Biol. Chem.* **247:**9–15.

Reddy, V. V. R., and Caspi, E., 1976, The mechanism of C-5(6) double-bond introduction in the biosynthesis of cholesterol by rat liver microsomes. Consideration of a mechanism similar to the oxidation of *o*-diphenols, *Eur. J. Biochem.* **69:**577–582.

Reddy, V. V. R., Kupfer, D., and Caspi, E., 1977, Mechanism of C-5 double-bond introduction in the biosynthesis of cholesterol by rat liver microsomes. Evidence for the participation of microsomal cytochrome b_5, *J. Biol. Chem.* **252:**2797–2801.

Rilling, H. C., 1966, A new intermediate in the biosynthesis of squalene, *J. Biol. Chem.* **241:**3233–3236.

Ritter, M. C., and Dempsey, M. E., 1970, Purification and characterization of a naturally occurring activator of cholesterol biosynthesis from $\Delta^{5,7}$-cholestadienol and other precursors, *Biochem. Biophys. Res. Commun.* **38:**921–929.

Ritter, M. C., and Dempsey, M. E., 1971, Specificity and role in cholesterol biosynthesis of a squalene and sterol carrier protein, *J. Biol. Chem.* **246:**1536–1539.

Rodwell, V. W., McNamara, D. J., and Shapiro, D. J., 1973, Regulation of hepatic 3-hydroxy-3-methylglutaryl coenzyme A reductase, *Adv. Enzymol.* **38:**373–412.

Rodwell, V. W., Nordstrom, J. L., and Mitschelen, J. J., 1976, Regulation of HMG-CoA reductase, *Adv. Lipid Res.* **14:**1–74.

Saat, Y. A., and Bloch, K. E., 1976, Effect of a supernatant protein on microsomal squalene epoxidase and 2,3-oxidosqualene-lanosterol cyclase, *J. Biol. Chem.* **251:**5155–5160.

Sabine, J. R. (ed.), 1983, *3-Hydroxy-3-methylglutaryl Coenzyme A Reductase,* in: *The Uniscience Series in Enzyme Biology,* CRC Press, Boca Raton, in press.

Scala, A., Galli-Kienie, M., Anastasia, M., and Galli, G., 1974, The reversibility of the isomerization of the Δ^8 to Δ^7 bond in cholesterol biosynthesis, *Eur. J. Biochem.* **48:**263–269.

Scallen, T. J., Schuster, M. W., and Dhar, A. K., 1971, Evidence for a noncatalytic carrier protein in cholesterol biosynthesis, *J. Biol. Chem.* **246:**224–230.

Scallen, T. J., Srikantaiah, M. V., Seetharam, B., Hansbury, E., and Gavey, K. L., 1974, Sterol carrier protein hypothesis, *Fed. Proc.* **33:**1733–1746.

Schroepfer, G. J., 1982, Sterol biosynthesis, *Annu. Rev. Biochem.* **51:**555–585.

Schroepfer, G. J., and Frantz, I. D., Jr., 1961, Conversion of Δ^7-cholestenol-4-C^{14} and 7-dehydrocholesterol-4-C^{14} to cholesterol, *J. Biol. Chem.* **236:**3137–3140.

Schroepfer, G. J., Jr., Lutsky, B. N., Martin, J. A., Huntoon, S., Fourcans, B., Lee, W.-H., and Vermilion, J. L., 1972, Recent investigation on the nature of sterol intermediates in the biosynthesis of cholesterol, *Proc. R. Soc. Lond. B.* **180:**125–146.

Sharpless, K. B., Snyder, T. E., Spencer, T. A., Maheshwari, K. K., Guhn, G., and Clayton, R. B., 1968, Biological demethylation of 4,4-dimethyl sterols. Initial removal of the 4α-methyl group, *J. Am. Chem. Soc.* **90:**6874–6875.

Shechter, I., and Bloch, K., 1971, Solubilization and purification of *trans*-farnesyl pyrophosphate-squalene synthetase, *J. Biol. Chem.* **246:**7690–7696.

Shechter, I., Sweat, F. W., and Bloch, K., 1970, Comparative properties of 2,3-oxidosqualene-lanosterol cyclase from yeast and liver, *Biochim. Biophys. Acta* **220:**463–468.

Spence, J. T., and Gaylor, J. L., 1977, Investigation of regulation of microsomal hydroxymethylglutaryl coenzyme A reductase and methyl sterol oxidase of cholesterol biosynthesis, *J. Biol. Chem.* **252:**5852–5858.

Srikantaiah, M. V., Hansbury, E., Loughran, E. D., and Scallen, T. J., 1976, Purification and properties of sterol carrier protein$_1$, *J. Biol. Chem.* **251:**5496–5504.

Steinberg, D., and Avigan, J., 1969, Rat liver sterol Δ^{24}-reductase, *Meth. Enzymol.* **15:**514–522.

Swindell, A. C., and Gaylor, J. L., 1968, Investigation of the component reactions of oxidative sterol demethylation. Formation and metabolism of 3-ketosteroid intermediates, *J. Biol. Chem.* **243:**5546–5555.

Tabacik, C., Aliau, S., Serrou, B., and de Paulet, A. C., 1981, Post-HMG-CoA reductase regulation of cholesterol biosynthesis in normal lymphocytes: Lanosten-3β-ol-32-al, a natural inhibitor, *Biochem. Biophys. Res. Commun.* **101:**1087–1095.

Tai, H.-H., and Bloch, K., 1972, Squalene epoxidase of rat liver, *J. Biol. Chem.* **247:**3767–3773.

Tchen, T. T., and Bloch, K., 1957, Conversion of squalene to lanosterol *in vitro, J. Biol. Chem.* **226:** 921–930.

Topham, R. W., and Gaylor, J. L., 1970, Isolation and purification of a 5α-hydroxy sterol dehydrase of yeast, *J. Biol. Chem.* **245:**2319–2327.

Topham, R. W., and Gaylor, J. L., 1972, Further characterization of the 5α-hydroxy sterol dehydrase of yeast, *Biochem. Biophys. Res. Commun.* **47:**180–186.

Trzaskos, J. M., and Gaylor, J. L., 1983a, Cytosolic modulations of activities of microsomal enzymes of cholesterol biosynthesis. Purification and characterization of a nonspecific lipid transfer protein. *Biochim. Biophys. Acta,* **751:**52–65.

Trzaskos, J. M., and Gaylor, J. L., 1983b, Molecular control of HMG-CoA reductase: The role of cytosolic proteins, in: *3-Hydroxy-3-methylglutaryl Coenzyme A Reductase,* in: *The Uniscience Series in Enzyme Biology* (J. R. Sabine, ed.), CRC Press, Boca Raton.

Vermilion, J. L., and Coon, M. J., 1974, Highly purified detergent-solubilized NADPH-cytochrome P-450 reductase from phenobarbital-induced rat liver microsomes, *Biochem. Biophys. Res. Commun.* **60:**1315–1322.

Watkinson, I. A., Wilton, D. C., Munday, K. A., and Akhtar, M., 1971a, Formation and reduction of the 14,15-double-bond in cholesterol biosynthesis, *Biochem. J.* **121:**131–137.

Watkinson, I. A., Wilton, D. C., Rahimtula, A. D., and Akhtar, M. M., 1971b, The substrate activation in some pyridine nucleotide linked enzymic reactions. The conversion of desmosterol into cholesterol, *Eur. J. Biochem.* **23:**1–6.

Williams, M. T., Gaylor, J. L., and Morris, H. P., 1977, Characterization of microsomal sterol demethylase in two Morris hepatomas, *Cancer Res.* **36:**291–297.

Wilton, D. C., Munday, K. A., Skinner, S. J. M., and Akhtar, M., 1968, The biological conversion of 7-dehydro-cholesterol into cholesterol and comments on the reduction of double bonds, *Biochem. J.* **106:**803–810.

Wilton, D. C., Rahimtula, A. D., and Akhtar, M., 1969, The reversibility of the Δ^8-cholestenol-Δ^7-cholestenol isomerase reaction in cholesterol biosynthesis, *Biochem. J.* **114:**71–73.

Yamaga, N., and Gaylor, J. L., 1978, Characterization of the microsomal steroid-8-ene isomerase of cholesterol biosynthesis, *J. Lipid Res.* **19:**375–382.

Yamamoto, S., and Bloch, K., 1970, Studies on squalene epoxidase of rat liver, *J. Biol. Chem.* **245:**1670–1674.

Yamamoto, S., Lin, K., and Bloch, K., 1969, Some properties of the microsomal 2,3-oxidosqualene sterol cyclase, *Proc. Natl. Acad. Sci. USA* **63:**110–117.

Yasukochi, Y., and Masters, B. S. S., 1976, Some properties of a detergent-solubilized NADPH-cytochrome *c* (cytochrome P-450) reductase purified by biospecific affinity chromatography, *J. Biol. Chem.* **251:**5337–5344.

19

Membrane-Bound Enzymes in Plant Sterol Biosynthesis

Trevor W. Goodwin, C.B.E., F.R.S.

I. INTRODUCTION

The basic pattern of sterol biosynthesis in plants is similar to that involved in cholesterol biosynthesis in mammals, but there are important differences of detail. There are also additional reactions in plants such as alkylation at C-24, glucosylation at C-3, and side-chain desaturation, most frequently at C-2. Sterol biosynthesis in mammals has been studied at the enzyme level to a much greater extent than has phytosterol biosynthesis. Part of the reason is that plant enzymes, particularly those in higher plants, are notoriously difficult to deal with (Loomis, 1973). However, reliable information is accumulating, and in presenting the evidence for the involvement of membrane-bound enzymes in sterol biosynthesis in plants one realizes that the foundations for future developments, not only in enzymology but in studies on the control of synthesis, are now reasonably well established.

The pathway in animals from the basic C_2 unit (acetyl-CoA) to the C_{15} unit (farnesyl pyrophosphate) is carried out by cytoplasmic enzymes with the exception of hydroxymethylglutaryl-CoA reductase (HMG-CoA reductase) which reduces HMG-CoA to mevalonate; this is a microsomal enzyme (see Nes and McKean, 1977). This is also the case with the enzyme from maize seedlings (Brooker and Russell, 1975a,b, 1979, Wong *et al.*, 1982); but in yeast (Shimizu *et al.*, 1973) and the fungus *Cochliobolus heterostrophus* (Kawaguchi *et al.*, 1973b), it is located in the mitochondria. In contrast the enzyme from a *Pseudomonas* sp., which has been purified 21-fold, is a soluble protein (Bensch and Rodwell, 1970). The steps involved in the conversion

Trevor W. Goodwin • Department of Biochemistry, The University of Liverpool, Liverpool L69 3BX, England.

of farnesyl pyrophosphate into squalene and then squalene eventually into cholesterol are, in mammals, all catalyzed by microsomal enzymes, although a soluble carrier protein may be necessary for some steps (Nes and McKean, 1977). The situation is likely to be the same in plants and protista, but this has not yet been fully documented.

II. INITIAL STAGES

A. HMG-CoA Reductase (EC 1.1.1.34)

This enzyme catalyzes the reduction of HMG-CoA to mevalonic acid (Figure 1) in the presence of NADH; coenzyme A is liberated in the reaction. The yeast enzyme has been studied in detail. It was first purified 200-fold by Kirtley and Rudney (1967) but later Qureshi *et al.* (1976a) achieved a 5000-fold purification, the resulting protein having an activity of 9–11 units/mg. The enzyme (mol. wt. 260,000) a tetramer, and each subunit has a molecular weight of 60,000–65,000 (Qureshi *et al.*, 1976a). The early view that the mechanism of the reduction is ping-pong with no ternary complexes involved (Kirtley and Rudney, 1967) has not been substantiated. Later work indicated that each reductive step proceeds sequentially, and that binding of substrates to the enzyme and the release of products therefrom has a considerable degree of order. A key intermediate in the proposed mechanism is a hemithioacetal of mevaldic acid (Scheme 1) and coenzyme A (Qureshi *et al.*, 1976b). Similar results were observed

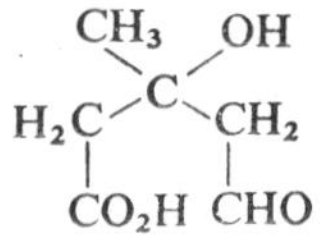

$$\begin{array}{ccc} & CH_3 \quad OH & \\ & \diagdown \; \diagup & \\ H_2C & C & CH_2 \\ | & & | \\ CO_2H & & CHO \end{array}$$

Scheme 1

with the *Pseudomonas* enzymes (Bensch and Rodwell, 1970). The first demonstration of a particulate HMG-CoA reductase in plants came from studies on the latex rubber of *Hevea brasiliensis* (Hepper and Audley, 1969), and the enzyme has recently been examined in detail (Sipat, 1982). It requires NADPH specifically, has an optimum pH of 6.6–6.9 in 0.1 M phosphate buffer, and it is difficult to solubilize. In peas, the enzyme exists in the cytosol (microsomes), mitochondria, and plastids (Brooker and Russell, 1975a,b, 1979; Wong *et al.*, 1982). The plastid enzyme has an affinity for its substrate some 200 times greater than that of the extra-plastidic enzymes; it also has distinctive kinetics and is under phytochrome control (Wong *et al.*, 1982).

It is not known whether HMG-CoA reductase plays a central role in regulation of sterol synthesis in plants as it does in animals (Dempsey, 1974). However, in yeast,

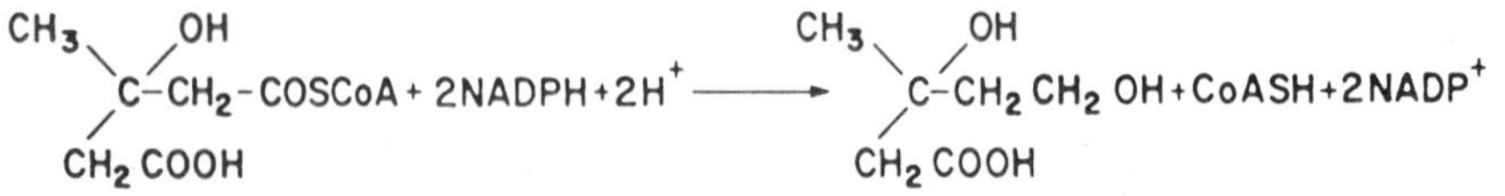

$$\begin{array}{l} CH_3 \quad OH \\ \quad \diagdown \; \diagup \\ \quad\; C{-}CH_2{-}COSCoA + 2NADPH + 2H^+ \\ \quad \diagup \\ CH_2COOH \end{array} \longrightarrow \begin{array}{l} CH_3 \quad OH \\ \quad \diagdown \; \diagup \\ \quad\; C{-}CH_2CH_2OH + CoASH + 2NADP^+ \\ \quad \diagup \\ CH_2COOH \end{array}$$

Figure 1. The reaction catalyzed by HMG-CoA reductase.

stimulation of sterol synthesis is paralleled by an increase in HMG-CoA reductase activity (Berndt *et al.*, 1973; Boll *et al.*, 1975).

B. *Farnesyl Pyrophosphate (EC 2.5.1.1): Squalene Synthesis*

This enzyme converts two molecules of farnesyl pyrophosphate into squalene (Scheme 2) via presqualene pyrophosphate (Scheme 3). The enzyme requires a divalent

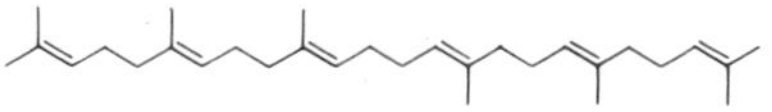

Scheme 2

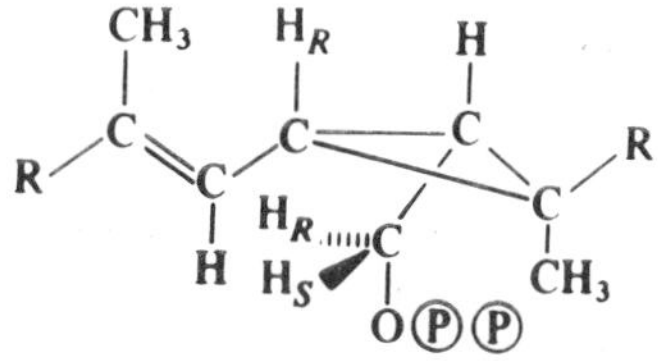

Scheme 3

cation, Mg^{2+} or Mn^{2+}, and either NADH or NADPH as co-factors. Absence of the reductant results in the accumulation of presqualene pyrophosphate; indeed, that was how the intermediate was discovered (Rilling, 1966). Its structure was later fully established (Rilling and Epstein, 1969; Epstein and Rilling, 1970), and its absolute stereochemistry (1*R*, 2*R*, 3*R*) (Scheme 3) elucidated (Popják *et al.*, 1973). Thus, the reactions involve two discrete steps: the formation of presqualene pyrophosphate and the reductive rearrangement of presqualene pyrophosphate with the concomitant ejection of pyrophosphate. Although, as will be explained later, there are doubts about whether the enzyme has been purified to homogeneity; there are no biochemical indications that more than one enzyme is involved. Rather, it is probably an enzyme with two active centers. However, three of a number of temperature-sensitive mutants of yeast defective in ergosterol biosynthesis have been assigned by indirect methods to the farnesyl pyrophosphate → squalene step (Karst and Lacroute, 1977). This suggests that three proteins might be involved, but not all need be enzymes; one might be a carrier protein.

The mechanism involved in the formation of squalene has been discussed very thoroughly over the years (Cornforth, 1973; Rees and Goodwin, 1975; Popják and Agnew, 1979; Poulter and Rilling, 1981) and need not be considered here. The yeast enzyme was first solubilized with deoxycholate by Shechter *et al.* (1970) who partly purified it (45-fold purification) in the presence of glycerol to maintain activity. A different procedure involving high concentrations of glucose and glycerol which was said to lead to a nearly homogeneous protein (Qureshi *et al.*, 1972, 1973) was not so successful in other hands (Popják and Agnew, 1979; Rilling, quoted by Poulter and Rilling, 1981).

The enzyme from etiolated leaves is associated mainly with smooth and rough endoplasmic reticulum and not with etioplasts or mitochondria (Hartmann *et al.*, 1973). This suggests that the enzyme may not be present in chloroplasts, which develop from etioplasts. A cell-free system from germinating pea seedlings has been obtained which will convert synthetic presqualene pyrophosphate with the same configuration as the natural material (Scheme 3) into squalene; the 1*S*, 2*S*, 3*S* enantiomer is not converted (Barton *et al.*, 1972).

III. CYCLIZATION OF SQUALENE

A. Squalene Monooxidase (EC 1.14.99.7)

This enzyme converts squalene into *S*-squalene 2,3-oxide in the presence of O_2 and NADPH (Figure 2). Nothing is known about the plant enzyme although the labeled oxide has been detected after adding $[1-{}^{14}C]$ acetate to tissue cultures of *Nicotiana tabacum* (Benveniste and Massy-Westropp,1967) and to *Euphorbia cyparissias* latex (Ponsinet and Ourisson, 1967). It also accumulates in various plant systems in the presence of the inhibitors SKF-7989 (Reid, 1968; Heintz and Benveniste, 1974), iminosqualene (Corey and Ortiz De Montellano, 1967), AMO 1618, and CCC (Douglas and Paleg, 1974). The mammalian enzyme has been examined in detail but has not been solubilized and purified (Yamamoto and Bloch, 1970).

B. Oxidosqualene Cyclases

Animals content themselves with just one cyclase (2,3-oxidosqualene:lanosterol cyclase) which forms lanosterol (Scheme 4), the first stable cyclic intermediate in the

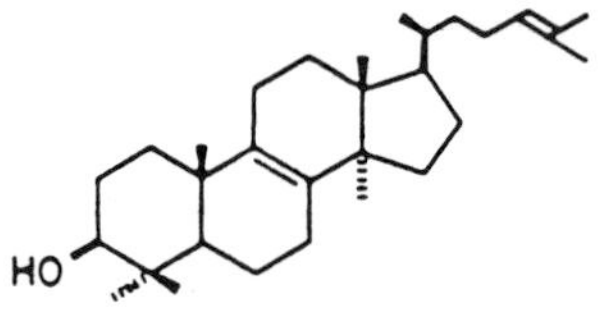

HO

Scheme 4

biosynthesis of cholesterol. Plants, on the other hand, synthesize a number of different cyclases although all use the same antipodal form of the substrate, that is, *S*-squalene 2,3-oxide (Barton *et al.*, 1974). Fungi synthesize the lanosterol cyclase, but higher plants and algae, i.e., photosynthetic organisms, form instead an enzyme which produces not lanosterol but an isomer, cycloartenol (Scheme 5), as the first stable cyclic product. However, in some fungi and many plants, there clearly exist enzymes in which the first cyclic product is not a sterol precursor. A typical example is β-amyrin (Scheme 6; Corey and Ortiz De Montellano, 1967). Enzymes also exist which will cyclize squalene 2,3,22,23 diepoxide as in the formation of α-onocerin (Scheme 7; Rowan *et al.*, 1971; Rowan and Dean, 1972), and bramble microsomes will convert

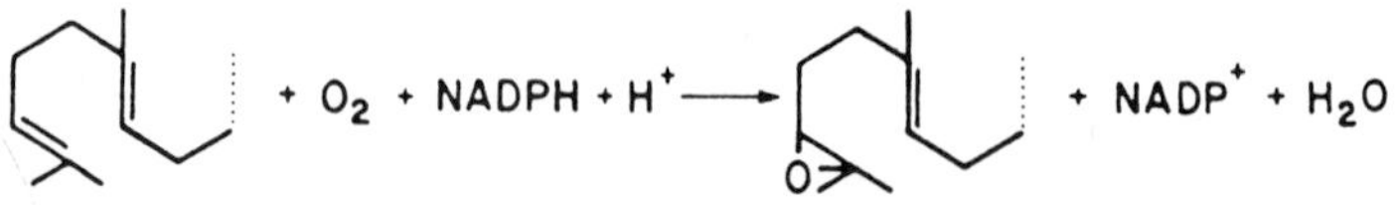

+ O_2 + NADPH + H^+ ⟶ O + $NADP^+$ + H_2O

Figure 2. The action of squalene oxidase.

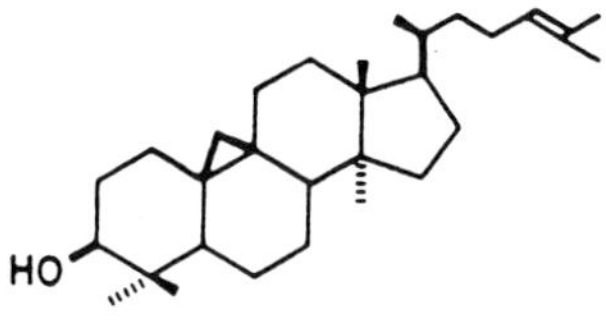

Scheme 5

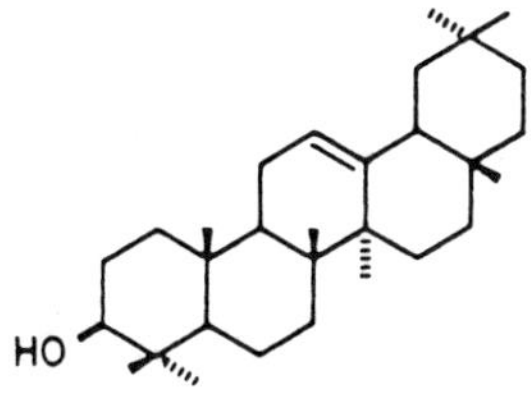

Scheme 6

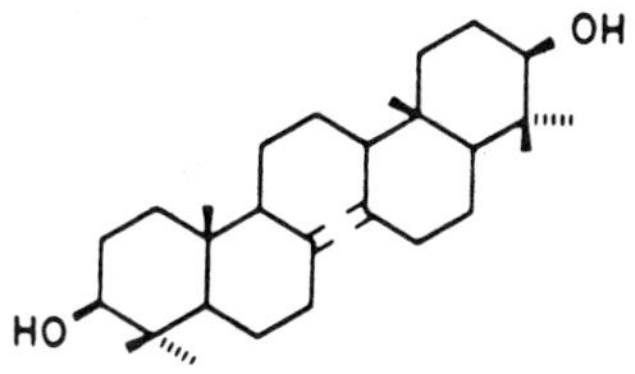

Scheme 7

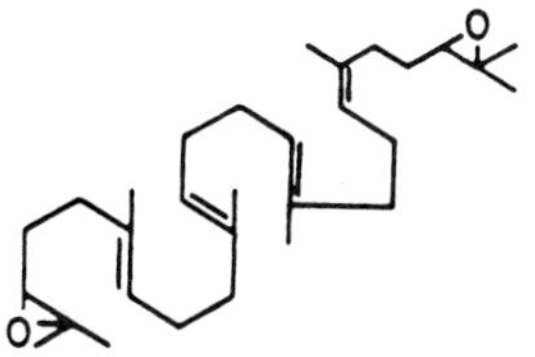

Scheme 8

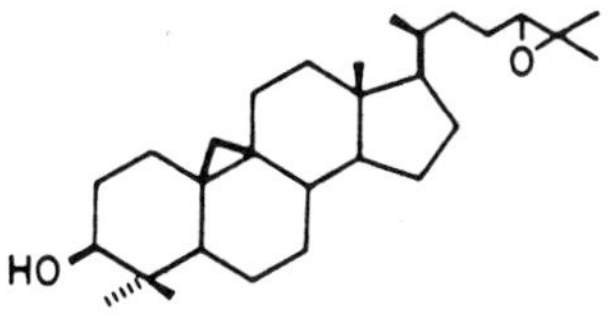

Scheme 9

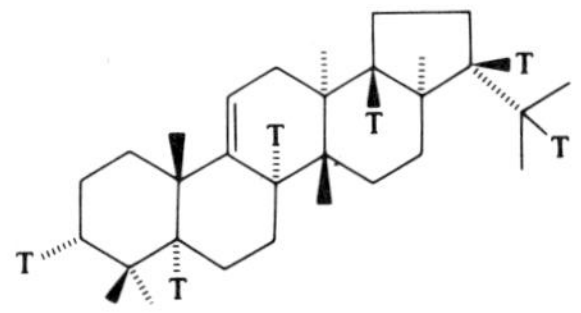

Scheme 10

squalene diepoxide (Scheme 8) into 24,25-epoxycycloartanol (Scheme 9; Heintz *et al.*, 1970). In the latter case, this is probably the result of the normal cyclase acting on an abnormal substrate. *Fusidium coccineum* converts squalene 2,3-oxide into fusidic acid probably by way of the protosterol 3-hydroxyprotosta-17(20),24-diene (Figure 3; Gotfredson *et al.*, 1968). The latter is interesting in that it is formed from the C-20 carbonium ion, the postulated intermediate in lanosterol synthesis, without the backbone rearrangement necessary for lanosterol formation (Caspi and Mulheirn, 1971).

Cyclization of squalene itself, rather than its epoxide, must occur to produce cyclic hydrocarbons such as fernene (Scheme 10) which are found in ferns (Barton *et*

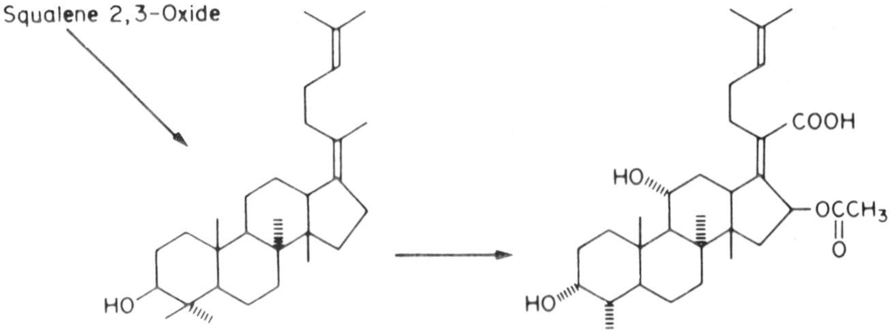

Squalene 2,3-Oxide

HO HO HO COOH OCCH3 O

Figure 3. A mechanism for the formation of fusidic acid from squalene 2,3-oxide without backbone rearrangement (reproduced with permission from Porter and Spurgeon, 1981).

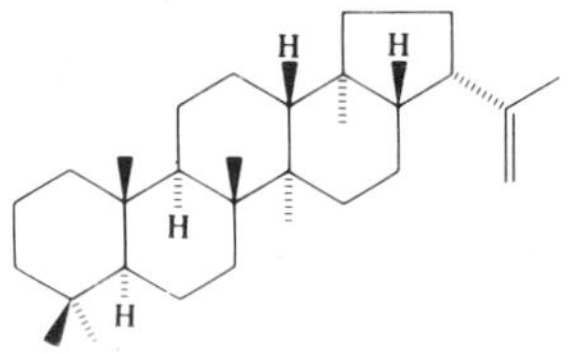

H H H H

Scheme 11

al., 1969, 1971; Ghisalberti *et al.*, 1970). Cell-free systems from *Acetobacter pasteurianum* and *Methylococcus capsulatus* have recently been reported which convert squalene into hopane derivatives such as diploptene (Scheme 11; Rohmer *et al.*, 1980).

1. 2,3-Oxidosqualene:Lanosterol Cyclase (EC 5.4.99.7)

The enzyme from yeast occurs in the microsomes as it does in liver, but it differs from the liver enzyme in that it is "soluble" in the sense that it can be extracted from the microsomes without the aid of surface-active agents, and that it remains in solution after centrifugation at 100,000*g* for up to 3 hr. It also differs from the liver enzyme in that deoxycholate does not enhance its catalytic activity; on the other hand, Triton X-100 is stimulatory. Whereas the activity of the liver enzyme is maximal in the presence of 0.4M KCl and deoxycholate, the activity of the yeast enzyme in Triton X-100 is inhibited in the presence of KCl. In spite of the "solubility" of the enzyme, all attempts to purify it by the usual procedures applied to soluble enzymes have so far failed (Shechter *et al.*, 1970). Microsomal preparations catalyzing the same reaction have been prepared from *Phycomyces blakeleeanus* (Mercer and Johnson, 1969) and *Cephalosporium caerulens* (Kawaguchi *et al.*, 1973a). The mechanism involved in the cyclization is now well established. The first step is a forward cyclization initiated by proton attack which leads to an electron deficiency at C-20 with the production of a transient carbonium ion. A backward rearrangement as indicated in (Figure 4a) results in the expulsion of a proton from C-9 with the formation of lanosterol.

2. 2,3-Oxidosqualene : Cycloartenol Cyclase (EC 5.4.99.8)

Formation of cycloartenol from 2,3-oxidosqualene occurs in a 40,000 *g* supernatant from bean leaf homogenates (Rees *et al.*, 1968b), in microsomes from tissue

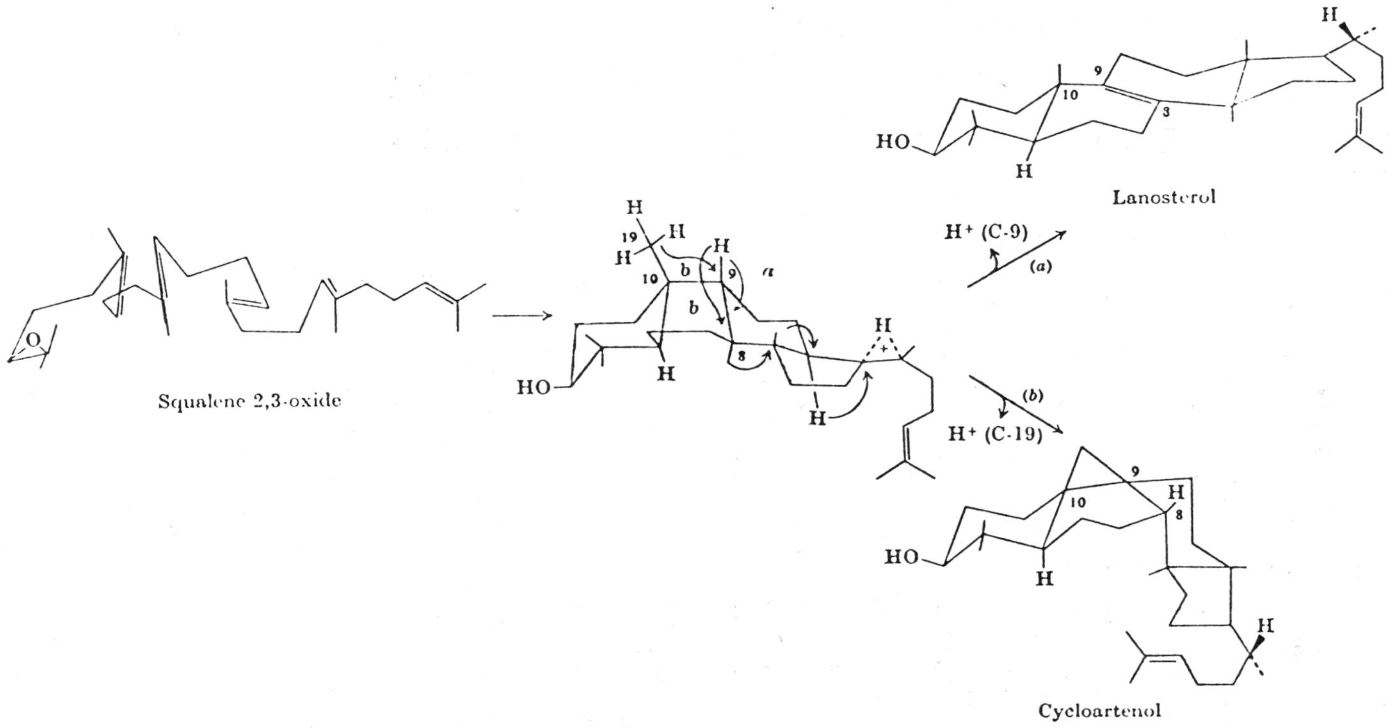

Figure 4. Mechanism of formation of lanosterol and cycloartenol from squalene 2,3-oxide (Rees *et al.*, 1968b).

cultures of tobacco (Heintz and Benveniste, 1970), bramble (Heintz, 1973), and maize embryos (Cattel *et al.*, 1976), as well as in barley coleoptiles (Heintz, 1973). The chrysophyte alga *Ochromonas malhamensis* is a good source of the enzyme (Rees *et al.*, 1969; Beastall *et al.*, 1971). It has been purified some 25-fold from a microsomal pellet, and it is optimally activated by 0.1% deoxycholate by 0.35 M KCl. In this respect, it resembles the liver lanosterol cyclase rather than the yeast enzyme. The overall mechanism for the formation of cycloartenol is superficially very similar to that for lanosterol formation; the difference being that stabilization is achieved by the loss of a proton from the methyl group at C-10 with the formation of the 9,19 methylene group of cycloartenol (Figure 4b). However, the mechanism indicated in Figure 4 would result in the final migration of the hydrogen from C-19 being *cis* to the C-9 hydrogen transfer, which is contrary to the biogenetic isoprene rule (Ruzicka, 1959). It was therefore postulated that a nucleophilic intermediate X^-(?enzyme) is involved. The neutralization of the charge generated at C-9 during the backward rearrangement of the C-20 cation by X^- would allow subsequent removal of X^- with the formation of the cyclopropane ring in a *trans* manner (Rees *et al.*, 1968b; Figure 5).

3. 2,3-Oxidosqualene:β-Amyrin Cyclase

A particulate homogenate from peas (*Pisum sativum*) will convert 2,3-oxidosqualene into β-amyrin (Capstack *et al.*, 1965; Corey and Ortiz De Montellano, 1967; Horan *et al.*, 1973). The mechanism involved (Figure 6) was demonstrated by experiments with stereospecifically labeled mevalonic acids (Rees *et al.*, 1968a).

IV. FORMATION OF STEROLS FROM CYCLOARTENOL

A. S-Adenosylmethionine:Δ^{24}-Triterpene Methyl Transferase

A characteristic of plant sterols is the presence of additional carbon atoms, one or two at C-24, as for example in ergosterol (Scheme 12) and poriferasterol (Scheme 13). These carbon atoms are added to a Δ^{24} precursor by a single or double transmethylation involving *S*-adenosylmethionine as methyl donor. The mechanism involved varies according to the organism under investigation and the final chirality at C-24 (Goad *et al.*, 1974; Goodwin, 1979, 1980, 1981, 1982).

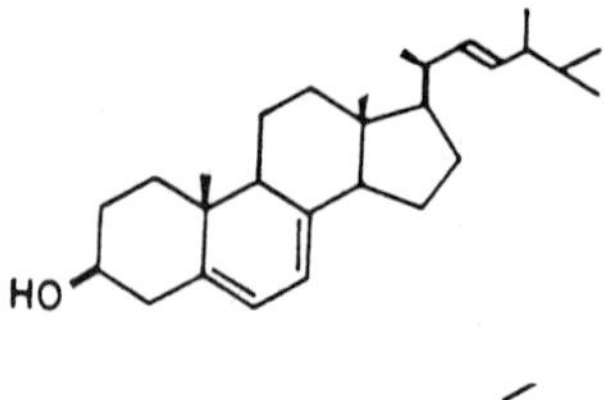

HO

Scheme 12

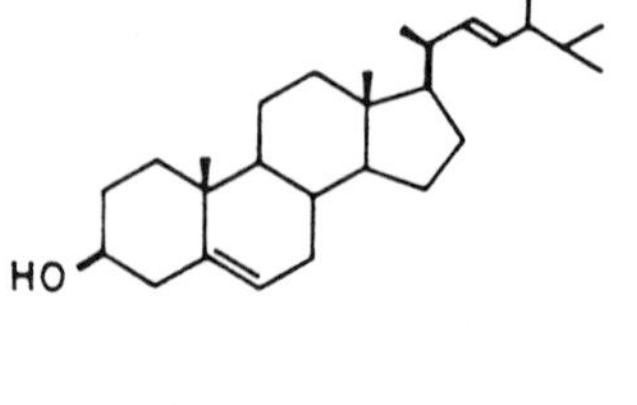

HO

Scheme 13

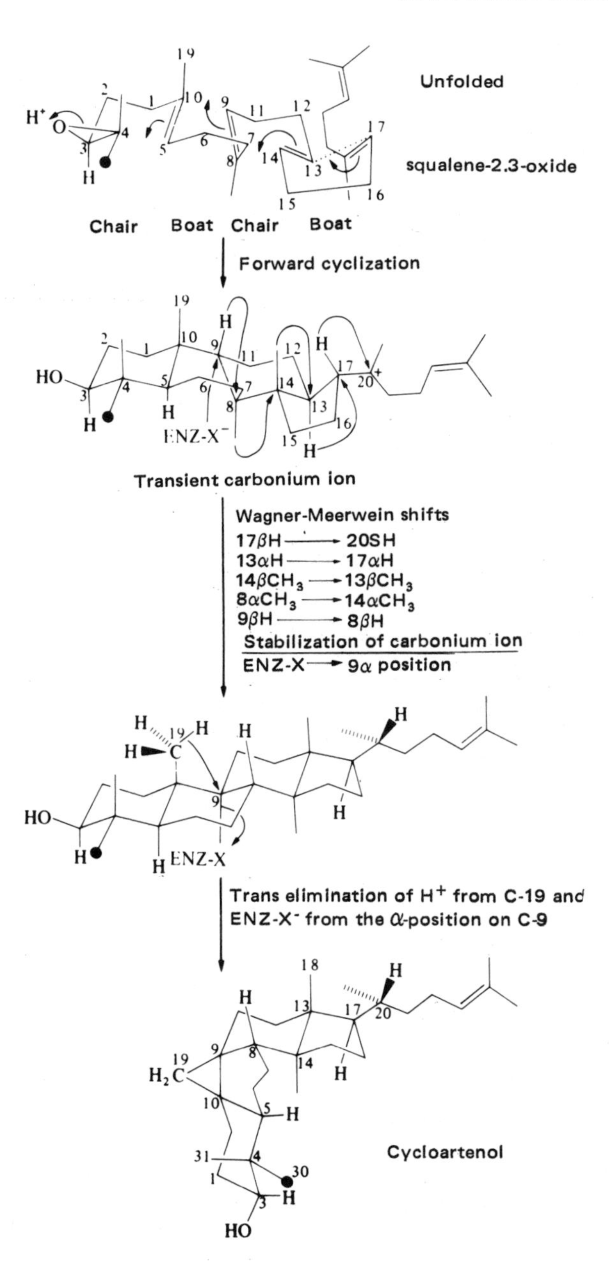

Figure 5. Mechanism for the cyclization of squalene 2,3-oxide to cycloartenol (reproduced with permission from Goodwin and Mercer, 1983.

1. S-Adenosylmethionine:Cycloartenol Methyl Transferase from Algae

Cell preparations from the algae *Trebouxia* sp. 213/3 and *Scenedesmus obliquus* will methylate cycloartenol (Scheme 5) with the production of 24-methylene cycloartanol (Scheme 14) and cyclolaudenol (Scheme 15; Wojciechowski *et al.*, 1973). The

chair–chair–chair–boat-Squalene 2,3-oxide

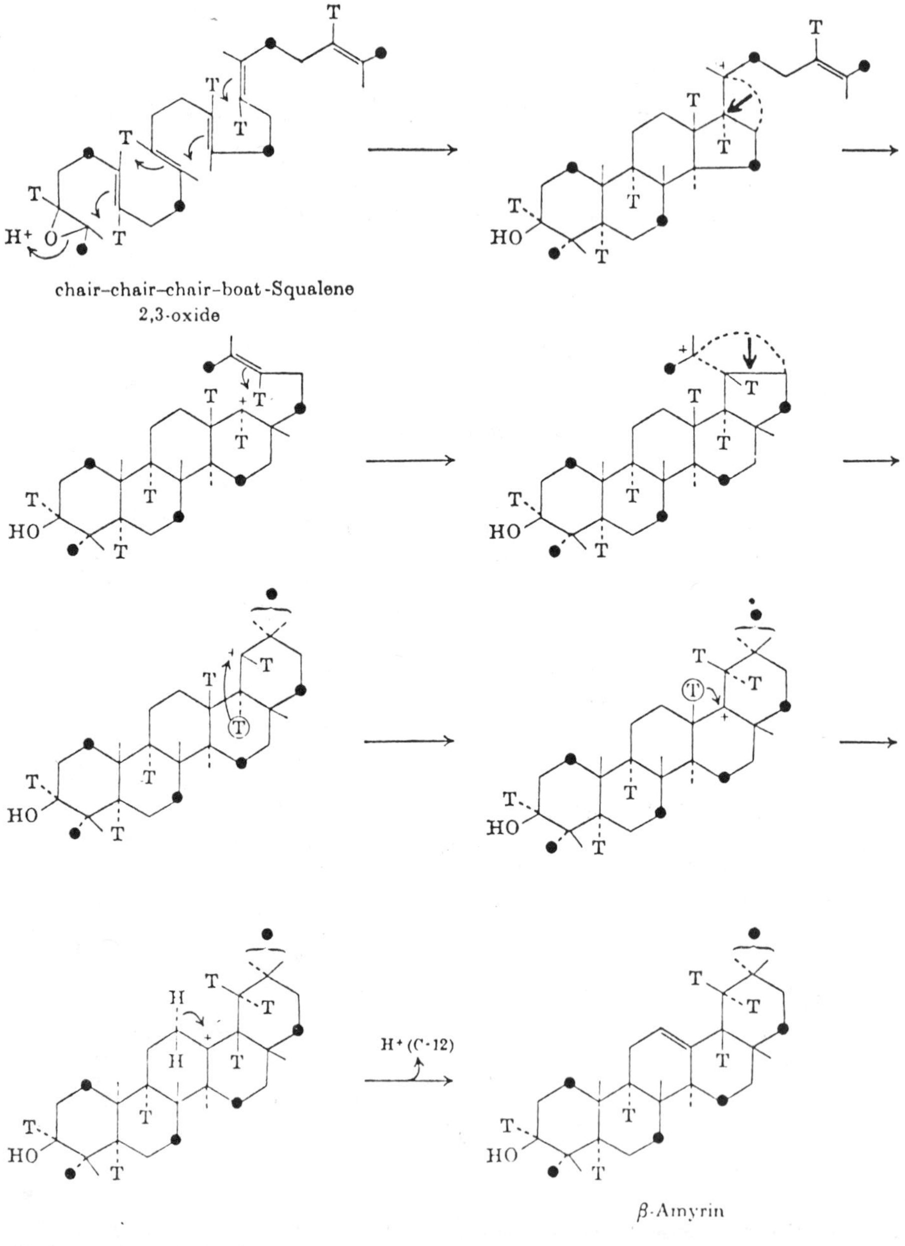

H⁺ (C-12)

β-Amyrin

Figure 6. The mechanism involved in converting squalene 2,3-epoxide into β-amyrin (Rees *et al.*, 1968a). ● = Carbons from C-2 of mevalonic acid; T = tritium from [4*R*,4T] mevalonic acid.

enzyme requires glutathione for maximum activity and will not alkylate compounds which do not possess a C-24(25) double bond. The enzyme is probably microsomal, although in preliminary experiments on *Trebouxia*, activity was observed in the soluble fraction. Cycloartenol is the best substrate, although some activity was observed with lanosterol (Scheme 4) or with cycloeucalenol (30-norcycloartenol; Scheme 16). This means that methylation at C-24 is the first step in the conversion of cycloartenol into algal sterols. The enzyme preparation produces 24-methylene cycloartanol (Scheme 14) and cyclolaudenol (Scheme 15). The mechanism involved in forming these two compounds is outlined in Figure 7. It is important to note that the 24-methylene cycloartanol is further metabolized to 24β-ethyl sterols, whereas cyclolaudenol is reduced to give 24β-methyl sterols. Thus, in these algae, there are two separate

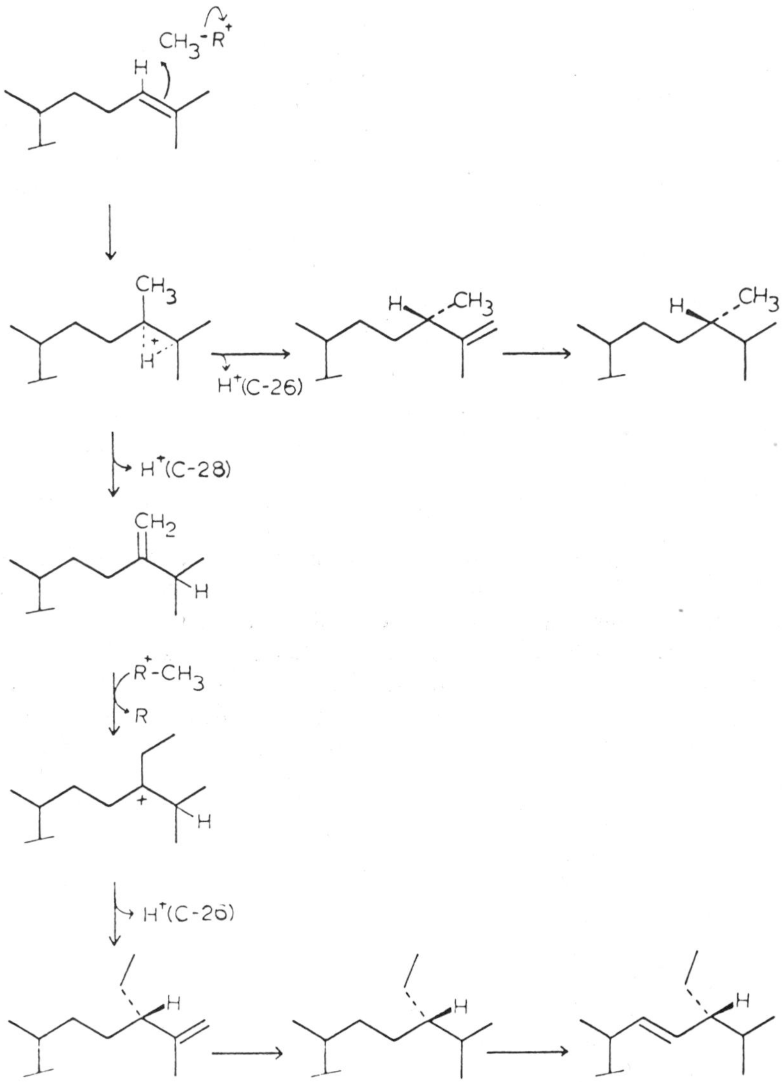

Figure 7. Formation of 24-methylene cycloartanol and cyclolaudenol by *Trebouxia* and *Scenedesmus* (Wojciechowski *et al.*, 1973).

pathways for the formation of C_{28} and C_{29} sterols, the former arising from cyclolaudenol and the latter from 24-methylene cycloartanol. It will be important to decide whether one or two enzymes are involved in these methylations.

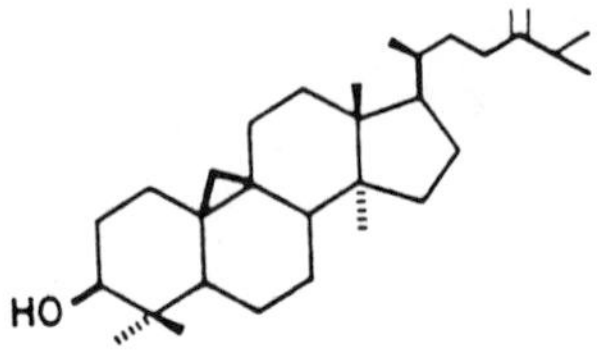

Scheme 14

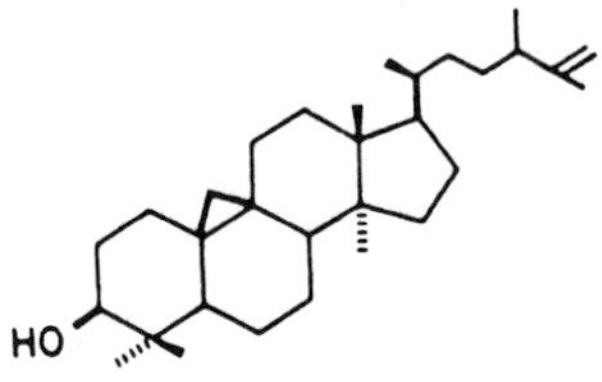

Scheme 15

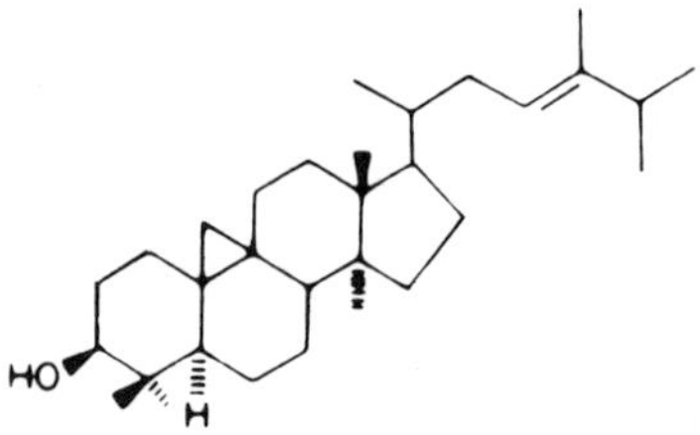

Scheme 16

2. *S-Adenosylmethionine:Cycloartenol Methyl Transferase from Higher Plants*

This enzyme preparation from maize seedlings are, as with the algal preparation, associated with the microsomal fraction and produces both 24-methylene cycloartanol (Scheme 14; Hartmann *et al.*, 1977; M. Zakelj and L. J. Goad, unpublished experiments quoted by Goodwin, 1979) and cyclolaudenol (Zakelj and Goad, 1983). However, the preparation also synthesizes cyclosadol (Scheme 16; Scheid *et al.*, 1982; Misso and Goad, 1983). 24-Methylene cycloartanol was the major product of the reaction, and cycloartenol was the best substrate tested (Scheid *et al.*, 1982). Figure 8 indicates how

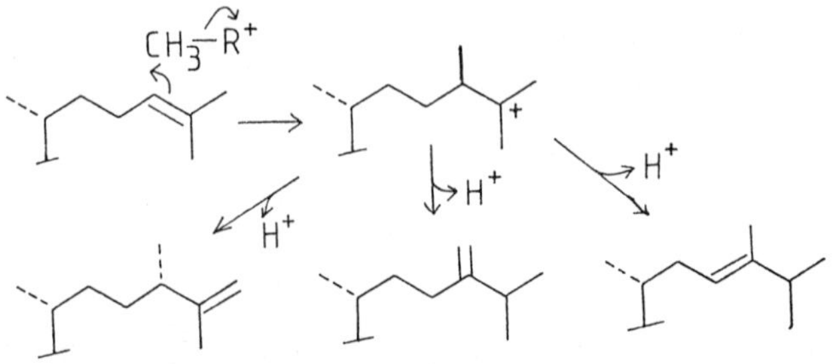

Figure 8. Mechanism for formation of 24-methylene cycloartanol, cyclolaudenol, and cyclosadol from cycloartenol.

these three compounds can be formed from a common carbonium ion precursor by the elimination of three different protons.

3. *S-Adenosylmethionine: Δ^{24}-Zymosterol Methyl Transferase (EC 2.1.1.41) from Yeast*

The enzyme is tightly bound to microsomes from which it can be obtained in a soluble form by extracting an acetone powder of microsomes with 0.1 M Tris-HCl buffer (pH 7.5) and centrifuging at 10,000 *g* for 20 min. An ammonium sulphate precipitate (45–55% saturation) is stable at $-25°C$. Glutathione, Mg^{2+}, and pH 7.0 are required for optimal activity (Moore and Gaylor, 1969). In contrast to the higher plant and algal enzymes, the enzyme is not active on the first product of squalene cyclization, in this case lanosterol, but is most active in the C-28 compounds, zymosterol (Scheme 17) which is almost at the end of the line in the conversion of lanosterol into cholesterol (Scheme 18; Moore and Gaylor, 1970). Therefore, *in vivo* demethylation at C-4 and C-14 must occur before alkylation can take place.

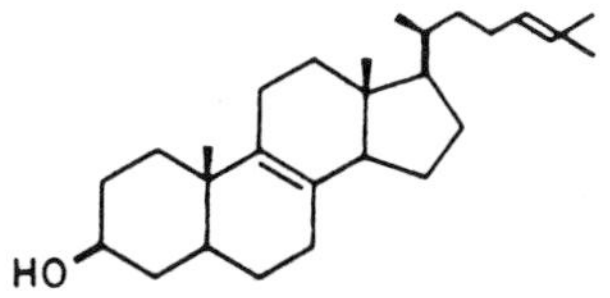

HO

Scheme 17

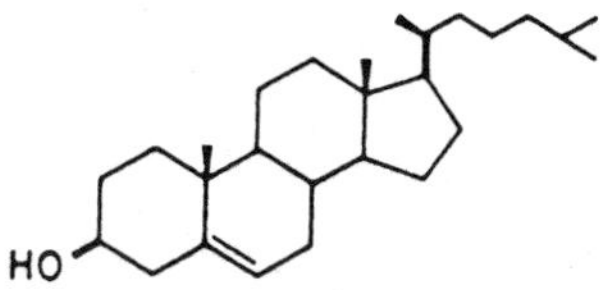

HO

Scheme 18

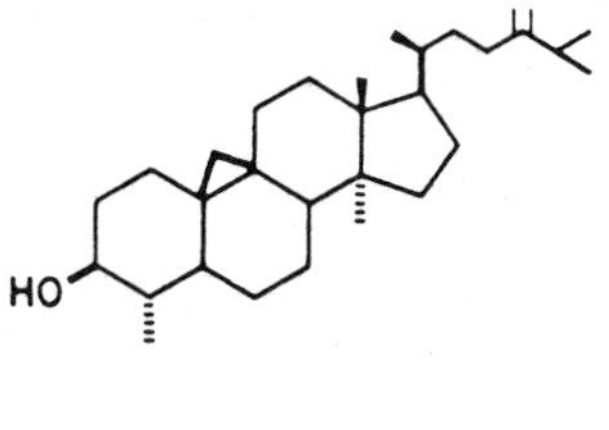

HO

Scheme 19

B. *Cycloeucalenol:Obtusifoliol Isomerase*

24-Methylene cycloartanol, formed by the methylation of cycloartenol is demethylated at C-4 to form cycloeucalenol (Scheme 19). This is the preferred substrate for cycloeucalenol, obtusifoliol isomerase (Figure 9). A microsomal preparation from bramble tissue cultures very effectively carries out this isomerization which involves the opening of 9β,19β-cyclopropane ring (Heintz *et al.*, 1972a,b). The enzyme is also present in maize microsomes (Rahier *et al.*, 1982). The substrate specificity of the enzyme is high for it is over 100 times less active with cycloartenol or 24-methylene

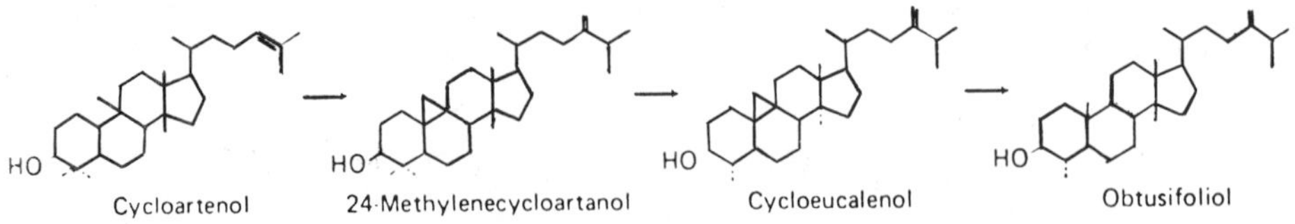

Cycloartenol → 24-Methylenecycloartanol → Cycloeucalenol → Obtusifoliol

Figure 9. The biosynthetic pathway from cycloartenol to obtusifoliol in higher plants.

cycloartanol as substrate than it is with cycloeucalenol. This specificity explains the absence of lanosterol from plants and the fact that it has never been detected as a biosynthetic intermediate in plant sterol biosynthesis (Goad and Goodwin, 1972). The opening of the cyclopropane ring as indicated in Figure 9 requires the uptake of a proton from water. The reality of this has been demonstrated by germinating pea seedlings in D_2O. The resulting sterols contained a deuterium atom attached to C-19 (Caspi and Sliwowski, 1975). The enzyme is inhibited by 7-oxo-24(28)-dihydrocycloeucalenol (Rahier *et al.*, 1982).

C. *S-Adenosylmethionine:24-Methylenelophenol Transferase*

After the formation of obtusifoliol (Figure 9) the next steps in the biosynthetic sequence are the removal of the methyl group from C-14 and the isomerization of the double bond from C-8 to C-7 to form 24-methylenelophenol (Scheme 20). This compound is the substrate for the second methylation in the sequence with the production in higher plants of 24-ethylidene lophenol (Scheme 21). The enzyme has been demonstrated in the microsomal fraction from bramble cell cultures (Fonteneau *et al.*, 1977).

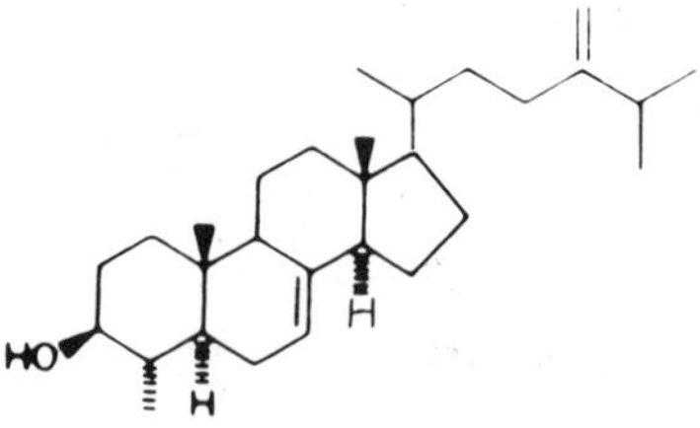

Scheme 20

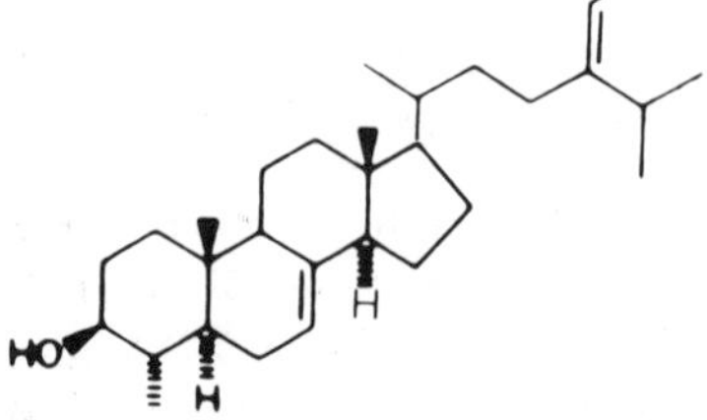

Scheme 21

D. *Δ^{25}-Sterol Reductase*

The major ethyl sterols in higher plants have the 24β configuration so that the eventual reduction of the $\Delta^{24(28)}$-double bond must give rise to this configuration. In green algae however, the major sterols have the 24β configuration (Figure 7), and this appears to be achieved during the alkylation reaction itself. The product in the case of the methyl derivative is cyclolaudenol (Scheme 15) which has the 24β configuration. The immediate product of the second methylation has not yet been detected but one would expect it to be the 24β-ethyl analogue of cyclolaudenol. However, Δ^{25} compounds with 24β-ethyl substituents such as clerosterol (Scheme 22), which are further on the pathway, have been detected in green algae (Rubinstein and Goad, 1974) and are converted in *Trebouxia* into clionasterol (Scheme 23) and poriferasterol (Scheme

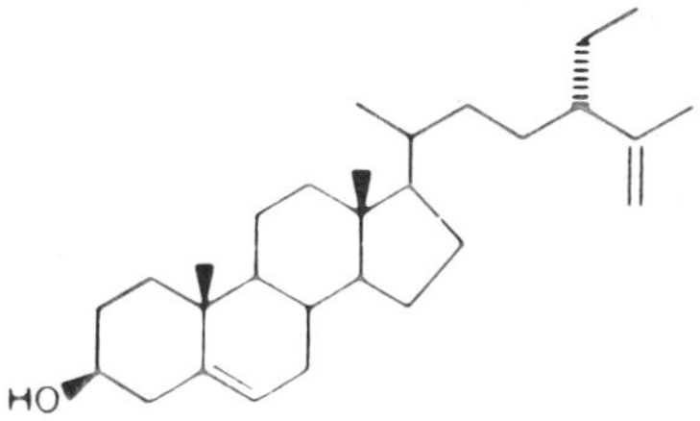

HO

Scheme 22

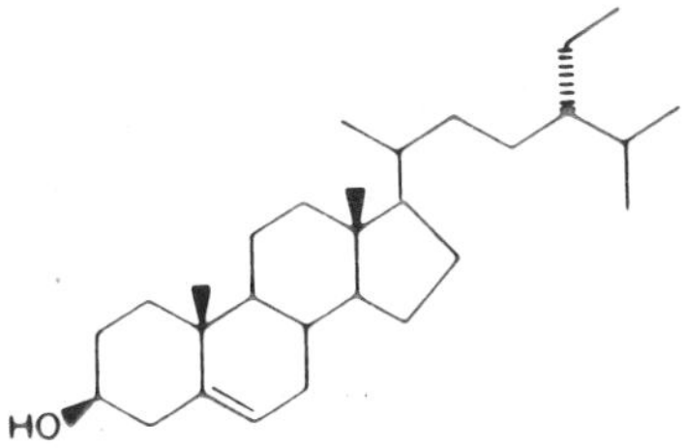

HO

Scheme 23

13; Largeau *et al.*, 1977). Recently, an active microsomal preparation has been obtained from *Trebouxia* which will convert clerosterol (Scheme 22) into poriferasterol (Scheme 13) in the presence of NADPH; it has an optimum pH of 7.4. The enzyme is very specific for the Δ^{25} double bond; very little reduction of the double bond at Δ^{22} was recorded (Wilkomirski and Goad, 1983). This observation, combined with the demonstration that in intact cells clionasterol is very poorly incorporated into poriferasterol (Largeau *et al.*, 1977), indicates that desaturation at C-22 must occur at an earlier stage in the biosynthetic pathway than previously thought.

V. GLYCOSYLATION OF PHYTOSTEROLS AND ACYLATION OF STEROL GLYCOSIDES

Sterol glycosides are widely distributed in higher plants and they are frequently accompanied by their 6-*O*-acyl derivatives (Lepage, 1964; Kiribuchi *et al.*, 1966; Laine and Elbein, 1971; Mudd, 1980; Wojciechowski, 1983).

A. Uridine Diphosphate Glucose:Sterol Transglucosylase

The existence of UDPG:sterol transglucosylase activity in particulate preparations from higher plants is well established from experiments with soya bean seeds (Hou *et al.*, 1968), mung bean shoots (Kauss, 1968), various leaves (Eichenberger and

Newman, 1968), cauliflower inflorescences (Ongun and Mudd, 1970), corn starch grains (Lavintman and Cardini, 1970), wheat shoots (Peaud-Lenoël and Axelos, 1972), marigold leaves (Wojciechowski, 1972, 1974; Lercher and Wojciechowski, 1976), tobacco seedlings (Bush and Grunwald, 1972, 1974), cotton fibers (Forsee *et al.*, 1974), onion stems (Lercher and Wojciechowski, 1976), potato tubers (Lavintman *et al.*, 1977), daffodil coronas (Liedvogel and Kleinig, 1977), pea seeds (Bowles *et al.*, 1977; Baisted, 1978), and maize leaves (Hartmann *et al.*, 1977). These preparations carry out the reaction

$$\text{UDP-glucose} + \text{sterol} \rightarrow \text{sterol glucoside} + \text{UDP}$$

In cotton fibers (Forsee *et al.*, 1974), onion and marigold preparations (Lercher and Wojciechowski, 1976), and pea preparations (Bowles *et al.*, 1977), the enzyme is located in the Golgi complex and this may be so in maize leaf preparations (Hartmann *et al.*, 1977). The activity is associated with starch grains (amyloplastids of potato tubers; Lavintman *et al.*, 1977) and also with chromoplasts of daffodil coronas (Liedvogel and Kleinig, 1977). The latter preparations will also utilize UDP-galactose as substrate to form the corresponding sterol galactoside.

The reported requirement for metal ions varies with the preparation. Mung bean preparations needed no addition for maximum activity (Kauss, 1968), whereas, a 2- to 3-fold increase was observed in maize (Lavintman and Cardini, 1970) and pea seed preparations (Baisted, 1978). Reported pH optima varied from 6.3–6.6 to 9.0 (see Mudd, 1980), with 8.5–9.0 recorded for a purified preparation (Forsee *et al.*, 1974).

The enzyme can be solubilized by treatment of the particulate preparations with Triton X-100 (Mudd, 1980). Although the enzyme is specific for UDP-glucose (Forsee *et al.*, 1974) it is catholic in its taste for sterols and will glucosylate all the plant

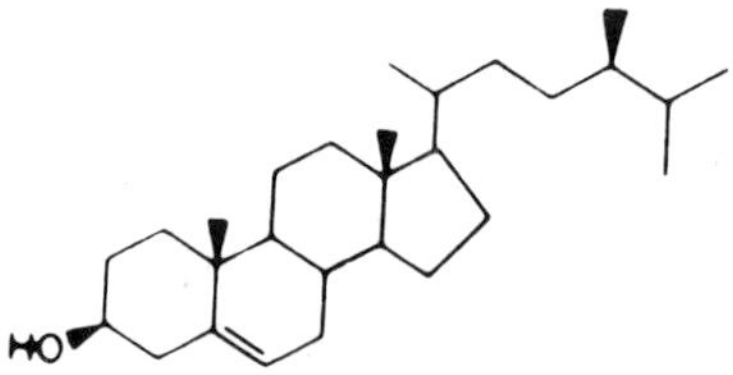

HO

Scheme 24

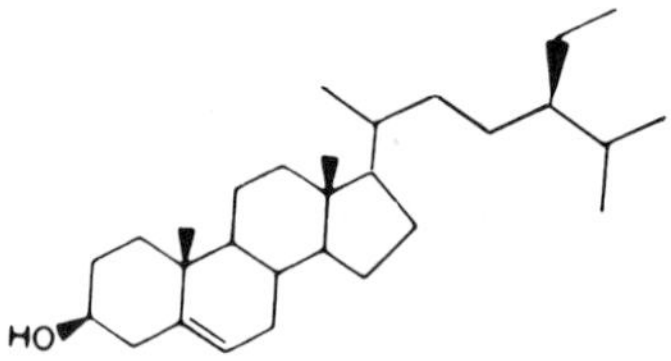

HO

Scheme 25

HO

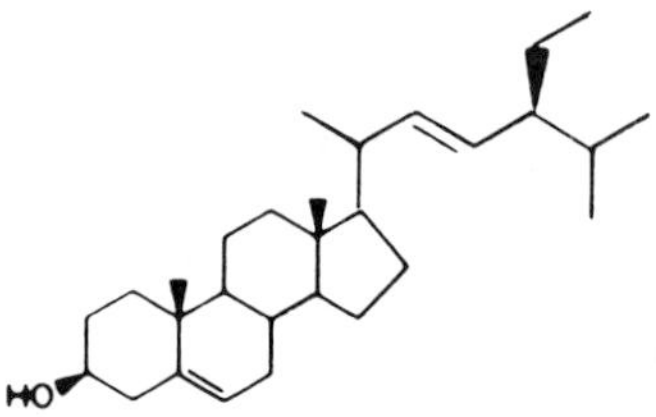

Scheme 26

sterols tested, viz., cholesterol (Scheme 18), campesterol (Scheme 24), stigmasterol (Scheme 25), and sitosterol (Scheme 26; Bush and Grunwald, 1974).

B. Phosphatidylethanolamine:Sterol Transacylase

Particulate preparations which acylate sterol glucosides have been obtained from wheat roots (Peaud-Lenoël and Axelos, 1972), cotton fibers (Forsee *et al.*, 1976), and marigold seedlings (Wojciechowski and Zimowski, 1975). In the latter source, the activity was associated with the Golgi complex. A 30-fold purification of the cotton enzyme was obtained by standard methods after solubilization with Triton X-100; this purified preparation had a pH optimum of 7.5 (Forsee *et al.*, 1976). The preferred acyl donor in all cases is phosphatidyl ethanolamine, although other phospholipids are active to some extent, and the transfer is to C-6 of the glucose residue. The overall reaction is thus:

Sterol glucose + phosphatidylethanolamine →
steryl-3 – (6-acyl-D-glucose) + lysophosphatidyl ethanolamine

It should be pointed out that soluble transacylases have been found in carrot and radish roots (Eichenberger and Siegrist, 1975) and bean leaves (Heinze *et al.*, 1975). In this case, the best acyl donor is digalactosyldiglyceride.

VI. ESTERIFICATION OF STEROLS

A. Diacylglycerol:Sterol Acyltransferase

Particulate preparations which will esterify plant sterols have been obtained by Garcia and Mudd (see Mudd, 1980, for a summary). The most effective acyl donor was diacyl glycerol. In the fungus *Phycomyces blakesleeanus* the acyl donor is phosphatidylcholine (Bartlett *et al.*, 1974) as it is in animals (see Mudd 1980).

VII. HYDROLYSIS OF STEROL ESTERS

Sterol ester hydrolase activity is located in the cell membrane of white mustard (*Sinapis alba*) seedlings and is solubilized with 0.1% Triton X-100. Activity is highest with esters of C_{14}–C_{18} fatty acids (Kalinowska and Wojciechowski, 1983).

VIII. SUMMARY

The membrane-bound enzymes involved in the biosynthesis of plant sterols have been much less extensively studied than those concerned with animal sterologenesis. The important regulatory enzyme HMG-CoA reductase is particulate in yeast and higher plants but soluble in a *Pseudomonas* sp. Plants produce a number of cyclases involving squalene 2,3-oxide as substrate, whereas animals produce only one (oxidosqualene-lanosterol cyclase). However, only two plant enzymes have been examined in detail, that synthesizing lanosterol in yeast and that synthesizing cycloartenol in algae and higher plants; both are microsomal. In the conversion of cycloartenol into plant sterols, the following enzymes are particulate: *S*-adenosylmethionine: cycloartenol methyl transferase; cycloeucalenol: obtusifoliol isomerase; *S*-adenosyl methionine: 24-methylene lophenol transferase; and Δ^{25}-sterol reductase. Glycosylation of sterols and acylation of sterol glycosides are also brought about by particulate enzymes, which are probably located in the Golgi apparatus. The enzyme involved in esterification of sterols is probably particulate.

This summary reveals the large gaps in our knowledge of the enzymology of plant sterol biosynthesis compared with the detailed information available on reaction mechanisms and their stereochemistry (see Goodwin, 1979, 1980, 1981). One hopes that with the eventual accumulation of precise information on the enzymes involved it will be possible to explain the differences in the active site of enzymes which result in the formation of lanosterol as the first stabilized cyclic product of sterol biosynthesis in animals and fungi compared with the formation of its isomer, cycloartenol, in higher plants and algae. Similarly, one would expect the unraveling of the detailed mechanism of the actions of other cyclases, not only on squalene 2,3-oxide, but on the hydrocarbon squalene itself.

REFERENCES

Baisted, D. J., 1978, Steryl glucoside and acyl glucoside biosynthesis in maturing pea seeds, *Phytochemistry* **17:**435–438.

Bartlett, K., Keat, M. J., and Mercer, E. I., 1974, Biosynthesis of sterol esters in *Phycomyces blakesleeanus, Phytochemistry* **13:**1107–1113.

Barton, D. H. R., Gosden, A. F., Mellows, G., and Widdowson, D. A., 1969, Biosynthesis of fern-9-one in *Polypodium vulgare* Linn., *Chem. Commun.* 184–186.

Barton, D. H. R., Mellows, G., and Widdowson, D. A., 1971, Biosynthesis of terpenes and steroids III. Squalene cyclization in the biosynthesis of triterpenoids; the biosynthesis of fern-9-one in *Polypodium vulgare* Linn., *J. Chem. Soc.* **C:**110–116.

Barton, D. H. R., Jarman, R. R., Watson, K. G., Widdowson, D. A., Boar, R. B., and Damps, K., 1974, Assimilation of the antipodal forms of squalene 2,3-oxide by mammalian, yeast and plant systems, *J. Chem. Soc.***D:**861–862.

Beastall, G. H., Rees, H. H., and Goodwin, T. W., 1971, Properties of a 2,3-oxidosqualene-cycloartenol cyclase from *Ochromonas malhamensis, FEBS Lett.* **18:**175–178.

Beastall, G. H., Rees, H. H., and Goodwin, T. W., 1972, The conversion of presqualene pyrophosphate into squalene by a cell-free preparation of *Pisum sativum, FEBS Lett.* **28:**243–256.

Bensch, W. R., and Rodwell, V. W., 1970, Purification and properties of 3-hydroxy-3-methylglutaryl coenzyme A reductase from *Pseudomonas, J. Biol. Chem.* **245:**3755–3762.

Benveniste, P., and Massy-Westropp, R. A., 1967, Mise en evidence de l'epoxyde-2,3 de squalene dans les tissus de tabac, *Tetrahedron Lett.* **37:**3553–3556.

Berndt, J., Boll, M., Lowel, M., and Gaument, R., 1973, Regulation of sterol biosynthesis in yeast: Induction of 3-hydroxy-3-methylglutaryl-CoA reductase by glucose, *Biochem. Biophys. Res. Commun.* **51:**843–848.

Boll, M., Lowel, M., Still, J., and Berndt, J., 1975, Sterol biosynthesis in yeast 3-hydroxy-3-methylglutaryl-coenzyme A reductase as a regulatory enzyme, *Eur. J. Biochem.* **54:**435–444.

Bowles, D. J., Lehle, L., and Kauss, H., 1977, Glucosylation of sterols and polyprenol phosphate in the Golgi apparatus of *Phaseolus aureus, Planta* **134:** 177–181.

Brooker, J. D., and Russell, D. W., 1975a, Properties of microsomal 3-hydroxy-3-methylglutaryl coenzyme A reductase from *Pisum sativum* seedlings *Arch. Biochem. Biophys.***167:**732–739.

Brooker, J. D., and Russell, D. W., 1975b, Subcellular localization of 3-hydroxy-3-methylglutaryl coenzyme A reductase in *Pisum sativum* seedlings, *Arch. Biochem. Biophys.* **167:**730–737.

Brooker, J. D., and Russell, D. W., 1979, Regulation of microsomal 3-hydroxy-3-methylglutaryl coenzyme A reductase from pea seedlings; rapid post-translational phytochrome-mediated decrease in activity and *in vivo* regulation by isoprenoid products, *Arch. Biochem. Biophys.* **198:**232–334.

Bush, P. B., and Grunwald, C., 1972, Sterol changes during germination of *Nicotiana tabacum, Plant Physiol.* **50:**69–72.

Bush, P. B., and Grunwald, C., 1974, Steryl glycoside formation in seedlings of *Nicotiana tabacum* L., *Plant Physiol.***53:**131–135.

Capstack, E., Rosin, N., Blondin, G. A., and Nes, W. R., 1965, Squalene in *Pisum sativum,* its cyclization to β-amyrin and labelling pattern, *J. Biol. Chem.***240:**3258.

Caspi, E., and Mulheirn, L. J., 1971, Mechanism of squalene cyclization. The biosynthesis of fusidic acid, *J. Biol. Chem.* **246:**2494–2501.

Caspi, E., and Sliwowski, J., 1975, On the role of cycloartenol in the formation of phytosterols. Biosynthesis of [19^{2}H] sitosterol in deuterium oxide germinated peas, *J. Am. Chem. Soc.* **97:**5032–5034.

Cattel, L., Anding, C., and Benveniste, P., 1976, Cyclisation of 1-*trans*-1′-norsqualene-2,3-epoxide and 1-*cis*-1′-norsqualene-2-3-epoxide by a cell-free system of corn embryos, *Phytochemistry* **15:**931–938.

Corey, E. J., and Ortiz De Montellano, P. R., 1967, Enzymic synthesis of β-amyrin from 2,3-oxidosqualene, *J. Am. Chem. Soc.* **89:**3362–3363.

Cornforth, J. W., 1973, The logic of working with enzymes, *Chem. Soc. Rev.* **2:**1–20.

Dempsey, M. E., 1974, Regulation of steroid biosynthesis, *Annu. Rev. Biochem.* **43:**967–990.

Douglas, T. J., and Paleg, L. G., 1974, Plant growth retardants as inhibitors of sterol biosynthesis in tobacco seedlings, *Plant Physiol.***54:**238–245.

Eichenberger, W., and Newman, D. W., 1968, Hexose transfer from UDP hexose in formation of sterol glycosides and esterified sterol glycosides in leaves, *Biochem. Biophys. Res. Commun.* **32:**336–374.

Eichenberger, W., and Siegrist, H. P., 1975, Steryl glycoside acyl transferase from carrots, *FEBS Lett.* **52:**153–156.

Epstein, W. W., and Rilling, H. C., 1970, Studies on the mechanism of squalene biosynthesis, the structure of presqualene pyrophosphate, *J. Biol. Chem.* **245:**4597–4605.

Fonteneau, P., Hartmann-Bouillon, M. A., and Benveniste, P., 1977, A 24-methylene lophenol C-28 methyl transferase from suspension cultures of bramble cells, *Plant Sci. Lett.* **10:**147–155.

Forsee, W. T., Laime, R. A., and Elbein, A. D., 1974, Solubilization of a particulate UDP-sucrose: sterol β-glucosyl transferase in developing cotton fibres and seeds and characterization of steryl β-acyl-D-glucosides, *Arch. Biochem. Biophys.* **161:**248–259.

Forsee, W. T., Valkovich, G., and Elbein, A. D., 1976, Acylation of steryl glucosides by phospholipids, *Arch. Biochem. Biophys.* **172:**410–418.

Ghisalberti, E. L., De Souza, N. J., Rees, H. H., and Goodwin, T. W., 1970, Biosynthesis of the triterpene hydrocarbons of *Polypodium vulgare, Phytochemistry* **9:**1817–1823.

Goad, L. J., and Goodwin, T. W., 1972, The biosynthesis of plant sterols, *Prog. Phytochem.* **3:**113–298.

Goad, L. J., Lenton, J. R., Knapp, F. F., and Goodwin, T. W., 1974, Phytosterol side chain biosynthesis, *Lipids* **9:**582–595.

Goodwin, T. W., 1979, Biosynthesis of terpenoids, *Annu. Rev. Plant Physiol.* **30:**369–404.

Goodwin, T. W., 1980, Biosynthesis of sterols, in: *The Biochemistry of Plants,* Vol. 4 (P. K. Stumpf, ed.), Academic Press, New York.

Goodwin, T. W., 1981, Biosynthesis of plant sterols and other plant terpenoids, in: *Biosynthesis of Isoprenoid Compounds,* Vol. 1 (J. W. Porter and S. L. Spurgeon, eds.), Wiley, New York.

Goodwin, T. W., 1982, Aspects of sterol biosynthesis in plants, Roussel Prize Lecture.

Goodwin, T. W., and Mercer, E. I., 1983, *An Introduction to Plant Biochemistry,* 2nd Ed., Pergamon Press, London.

Gotfredson, W. O., Loreh, H., Van Tamelen, E. E., Willet, J. D., and Clayton, R. B., 1968, Biosynthesis of fusidic acid from squalene 2,3-oxide, *J. Am. Chem. Soc.* **90:**208–209.

Hartmann, M. A., Ferne, M., Gigot, C., Brandt, R., and Benveniste, P., 1973, Isolement, caracterisation et composition en sterols de fractions subcellulaires de feuilles etoilées de Haricot, *Physiol. Veg.* **11:**209–230.

Hartmann, M. A., Fonteneau, P., and Benveniste, P., 1977, Subcellular localization of sterol synthesizing enzymes in maize coleoptiles, *Plant Sci. Lett.* **10:**147.

Heintz, R., 1973, Utilisation de fractions subcellulaires pour l'étude de la biosynthese des sterols de végetaux supèrieurs, Thesis, University of Strasbourg, France.

Heintz, R., and Benveniste, P., 1970, Cyclization de l'epoxide-2,3 de squalene par des microsomes extraits de tisus de tabac cultivés *in vitro, Phytochemistry* **9:**1499–1503.

Heintz, R., and Benveniste, P., 1974, Plant sterol metabolism. Enzymatic cleavage of the 9β 19β cyclopropane ring of cyclopropylsterols in bramble tissue cultures, *J. Biol. Chem.* **249:**4267–4274.

Heintz, R., Schaeffer, P. C., and Benveniste, P., 1970, Cyclization of squalene 2,3,22,23-diepoxide by microsomes from bramble (*Rubus fruticosa*) tissues grown *in vitro, Chem. Commun.* 946–947.

Heintz, R., Benveniste, P., and Bimpson, T., 1972a, Plant sterol metabolism. Evidence for the presence of an enzyme capable of opening the cyclopropane ring of cycloeucalenol, *Biochem. Biophys. Res. Commun.* **46:**766–772.

Heintz, R., Bimpson, T., and Benveniste, P., 1972b, Plant sterol metabolism. Studies on the substrate specificity of an enzyme capable of opening the cyclopropane ring of cycloeucalenol, *Biochem. Biophys. Res. Commun.* **49:**820–826.

Heinze, E., Dieler, H. P., and Rullkötter, J., 1975, Enzymatic acylation of steryl glycoside, *J. Plant Physiol.* **75:**78–87.

Hepper, C. M., and Audley, B. G., 1969, The biosynthesis of rubber from β-hydroxy-β-methylglutaryl coenzyme A in *Hevea brasiliensis* latex, *Biochem. J.* **114:**379–386.

Horan, H., McCormick, J. P., and Arigoni, D., 1973, Enzyme-catalyzed formation of β-amyrin from a bicyclic isomer of 2,3-epoxy-squalene, *J. Chem. Soc.* **D:** 73–74.

Hou, C. T., Umemura, Y., Nakamura, M., and Funahashi, S., 1968, Enzymic synthesis of steryl glucosides by a particulate preparation from immature soyabean seeds, *J. Biochem.* **63:**351–360.

Kalinowska, M., and Wojciechowski, A., 1983, The occurrence of sterol ester hydrolase activity in roots of white mustard seedlings, *Phytochemistry* **22:**59–63.

Karst, F., and Lacroute, F., 1977, Ergosterol biosynthesis in *Saccharomyces cerevisiae,* mutants deficient in the early sites of the pathway, *Mol. Gen. Genet.* **154:**269–277.

Kauss, H., 1968, Enzymatische glucosylierung von pflanzlichen sterinen, *Z. Naturforsch.* **23b:**1522–1526.

Kawaguchi, A., Kobayashi, H., and Okuda, S., 1973a, Cyclization of 2,3-oxidosqualene with microsomal fraction of *Cephalosporium caerulens, Chem. Pharm. Bull.* **21:** 577–583.

Kawaguchi, A., Nozoe, S., and Okuda, S., 1973b, Subcellular distribution of sesterterpene- and sterol-biosynthesis activities in *Cochliobolus heterostrophus, Biochim. Biophys. Acta* **196:**615–623.

Kiribuchi, T., Mizumaga, T., and Funahashi, S., 1966, Separation of soyabean sterols by fluorisil chromatography and characterization of acylated sterol glycosides, *Agric. Biol. Chem.* **30:**770–778.

Kirtley, M. E., and Rudney, H., 1967, Some properties and mechanism of action of the β-hydroxy-β-methylglutaryl coenzyme A reductase of yeast, *Biochemistry* **6:**230–238.

Laine, R. A., and Elbein, A. D., 1971, Sterol glucosides in *Phaseolus aureus.,* Use of GLC and MS for structural identification, *Biochemistry* **10:**2547–2553.

Langeau, C., Goad, L. J., and Goodwin, T. W., 1977, Conversion of a 24-β-ethyl-25-methylene intermediate into poriferasterol by *Trebouxia* species, *Phytochemistry* **16:**1931–1933.

Lavintman, N., and Cardini, C. E., 1970, Biosynthesis of a glycolipid in starch grains from sweetcorn, *Biochim. Biophys. Acta.* **201:**508–510.

Lavintman, N., Tandecarz, J., and Cardini, C. E., 1977, Enzymatic glycosylation of steroid alkaloids in potato tuber, *Plant Sci. Lett.* **8:**65–70.

Lepage, M., 1964, Isolation and characterization of an esterified form of sterol glucoside, *J. Lipid Res.* **5:**587–592.

Lercher, M., and Wojciechowski, Z. A., 1976, Localization of plant UDP-glucose: Sterol glycosyltransferase in the Golgi membranes, *Plant Sci. Lett.* **7:**227–340.

Liedvogel, B., and Kleinig, H., 1977, Lipid metabolism in chromoplast membranes from the daffodil: Glycosylation and acylation. *Planta* **133:**249–253.

Loomis, W. D., 1973, Overcoming problems of phenolics and quinones in the isolation of plant enzymes and organelles, in: *Methods in Enzymology,* Vol. 31, Academic Press, New York, pp. 528–544.

Mercer, E. I., and Johnson, M. W., 1969, Cyclization of squalene-2,3-oxide to lanosterol in a cell-free system from *Phycomyces blakesleeanus, Phytochemistry* **8:**2329–2331.

Misso, N. L. A., and Goad, L. J., 1983, Cyclolaudenol production by a microsomal preparation from *Zea mays,* shoots, *Phytochemistry,* **22:**2473–2479.

Moore, J. T., Jr., and Gaylor, J. L., 1969, Isolation and purification of an S-adenosylmethionine: Δ^{24}-sterol methyltransferase from yeast, *J. Biol. Chem.* **233:**6334–6340.

Moore, J. T., Jr., and Gaylor, J. L., 1970, Investigation of an S-adenosylmethionine: Δ^{24}-sterol methyltransferase in ergosterol biosynthesis in yeast, *J. Biol. Chem.* **245:**4684–4688.

Mudd, J. B., 1980, Sterol interconversions, in: *Biochemistry of Plants,* Vol. 4 (P. K. Stumpf, ed.), Academic Press, New York.

Nes, W. R., and McKean, M. L., 1977, *Biochemistry of Steroids and Other Isoprenoids,* University Park Press, Baltimore.

Ongun, A., and Mudd, J. B., 1970, The biosynthesis of steryl glucosides in plants, *Plant Physiol.* **45:** 255–262.

Peaud-Lenoël, C., and Axelos, M., 1972, D-Glucosylation des phytosterols et acylation des steryl-D glucosides en presence d'enzymes de plantes, *Carbohydr. Res.* **24:**247–262.

Ponsinet, G., and Ourisson, G., 1967, Biosynthèse *in vitro* des triterpenes dans le latex d'Euphorbia, *Phytochemistry* **6:**1235–1243.

Popják, G., and Agnew, W. S., 1979, Squalene synthetase, *Mol. Cell. Biochem.* **27:**97–116.

Popják, G., Edmond, J., and Wong, S. W., 1973, Absolute configuration of presqualene alcohol, *J. Am. Chem. Soc.* **95:**2713–2714.

Porter, J. W., and Spurgeon, S. L. (eds.), 1981, *Biosynthesis of Isoprenoid Compounds,* Vol. 1, Wiley-Interscience, New York.

Poulter, C. D., and Rilling, H. C., 1981, in: *Biosynthesis of Isoprenoid Compounds,* Vol. 1 (J. W. Porter and S. L. Spurgeon, eds.) Wiley, New York, pp. 161–224.

Qureshi, A. A., Beytia, E. D., and Porter, J. W., 1972, Squalene synthetase I. Dissociation and reassociation of enzyme complex, *Biochem. Biophys. Res. Commun.,* **48:**1123.

Qureshi, A. A., Beytia, E., and Porter, J. W., 1973, Squalene synthase II. Purification and properties of bakers' yeast enzyme, *J. Biol. Chem.* **248:**1848–1853.

Qureshi, N., Dugan, R. E., Nimmannit, S., Wu, W. H., and Porter, J. W., 1976a, Purification of β-hydroxy-β-methylglutaryl coenzyme A reductase from yeast, *Biochemistry* **15:**4185–4190.

Qureshi, N., Dugan, R. E., Cleland, W. W., and Porter, J. W., 1976b, Kinetic analysis of the individual reductive steps catalysed by β-hydroxy-β-methylglutaryl coenzyme A reductase obtained from yeast, *Biochemistry* **15:**4191–4197.

Rahier, A., Schmitt, P., and Benveniste, P., 1982, 7-oxo-24(28)-dehydrocycloeucalenol, a potent inhibitor of plant sterol biosynthesis, *Phytochemistry* **21:**1969–1974.

Rees, H. H., and Goodwin, T. W., 1975, Biosynthesis of triterpenes, steroids and carotenoids, in: *Biosynthesis,* Vol. 3 (Specialist Periodical Reports) (T. A. Geissman, ed.), *The Chemical Society,* London, pp. 14–88.

Rees, H. H., Britton, G., and Goodwin, T. W., 1968a, The biosynthesis of β-amyrin: Mechanism of squalene cyclization, *Biochem. J.* **106:**659–665.

Rees, H. H., Goad, L. J., and Goodwin, T. W., 1968b, Cyclization of 2,3-oxidosqualene to cycloartenol in a cell-free system from higher plants, *Tetrahedron Lett.* **6:**723–725.

Rees, H. H., Goad, L. J., and Goodwin, T. W., 1969, 2,3-oxidosqualene cycloartenol cyclase from *Ochromonas malhamensis, Biochim. Biophys. Acta* **176:**892–894.

Reid, W. W., 1968, Accumulation of squalene 2,3-oxide during inhibition of phytosterol biosynthesis in *Nicotiana tabacum, Phytochemistry* **7:**451–452.

Rilling, H. C., 1966, A new intermediate in the biosynthesis of squalene, *J. Biol. Chem.* **241:**3233–3236.

Rilling, H. C., and Epstein, W. W., 1969, Studies in the mechanism of squalene biosynthesis. Presqualene, a phosphorylated precursor to squalene, *J. Am. Chem. Soc.* **91:**1041–1042.

Rohmer, M., Anding, C., and Ourisson, G., 1980, Non-specific biosynthesis of hopane triterpenes by a cell-free system from *Acetobacter pasteurianum, Eur. J. Biochem.* **112:**541–547.

Rowan, M. G., and Dean, P. D. G., 1972, Properties of squalene-2(3),22(23)-diepoxide-α-onocerin cyclase from *Ononis spinosa* root, *Phytochemistry* **11**:2111–2118.

Rowan, M. G., Dean, P. D. G., and Goodwin, T. W., 1971, The enzymic conversion of squalene 2,(3),22-(23)-diepoxide to α-onocerin by a cell-free extract of *Ononis spinosa, FEBS Lett.* **12**:229–232.

Rubinstein, I., and Goad, L. J., 1974, Occurrence of (24*S*)-24-methylcholesta-5,22E-dier-3 β-ol in the diatom *Phaeodactylum triconutum, Phytochemistry* **13**:455–487.

Ruzicka, L., 1959, History of the isoprene rule, *Proc. Chem. Soc.* 341–360.

Scheid, G., Rohmer, M., and Benveniste, P., 1982, Biosynthesis of $\Delta^{5,23}$ sterols in etiolated coleoptiles from *Zea mays, Phytochemistry* **21**:1959–1968.

Shechter, I., Sweet, F. W., and Bloch, K., 1970, Comparative properties of 2,3-oxidosqualene-lanosterol cyclase from yeast and liver, *Biochim. Biophys. Acta* **220**:463–468.

Shimizu, I., Nagai, J., Hatanoka, H., and Katsuki, H., 1973, Mevalonate synthesis in the mitochondria of yeast, *Biochim. Biophys. Acta* **296**:310–320.

Sipat, A. B., 1982, Hydroxymethylglutaryl CoA reductase (NADPH) in the latex of *Hevea brasiliensis, Phytochemistry* **21**:2613–2618.

Wilkomirski, B., and Goad, L. J., 1983, The conversion of (24*S*)-24-ethylcholesta-5,22,25-trien-3β-ol into poriferasterol both *in vivo* and with a cell-free homogenate of the alga *Trebouxia* sp, *Phytochemistry* **22**:929–932.

Wojciechowski, Z. A., 1972, Biosynthesis of sterol glycosides in cell-free preparations from *Calendula officinalis, Acta Biochem. Pol.* **19**:43–49.

Wojciechowski, Z. A., 1974, Changes in UDPG-sterol glycosyl transferase activity in *Calendula officinalis, Phytochemistry* **13**:2091–2094.

Wojciechowski, Z. A., 1983, The biosynthesis of plant steryl glycosides and saponins, Biochem. Soc. Trans. **11**:565–568.

Wojciechowski, Z. A., Goad, L. J., and Goodwin, T. W., 1973, S-Adenosyl-L-methionine-cycloartenol methyltransferase activity in cell-free systems from *Trebouxia* sp. and *Scenedesmus obliquus, Biochem. J.* **136**:405–412.

Wong, R. J., McCormack, D. K., and Russell, D. W., 1982, Plastid 3-hydroxy-3-methylglutaryl coenzyme A reductase; distinctive kinetic and regulatory features, *Arch. Biochem. Biophys.* **261**:631–638.

Yamamoto, S., and Bloch, K., 1970, Enzymic studies on the oxidative cyclization of squalene, *Biochem. Soc. Symp.* **29**:35–43.

Zakalj, M., and Goad, L. J., 1983, Observations on the biosynthesis of 24-methylcholesterol and 24-ethylcholesterol by Zea mais, *Phytochemistry* **22**:1931–1936.

20

Glycosyltransferases Involved in the Biosynthesis of Protein-Bound Oligosaccharides of the Asparagine-N-Acetyl-D-Glucosamine and Serine(Threonine)-N-Acetyl-D-Galactosamine Types

Harry Schachter, Saroja Narasimhan, Paul Gleeson, George Vella, and Inka Brockhausen

I. INTRODUCTION

Glycoproteins are a diverse group of biopolymers containing one or more carbohydrate chains linked covalently to a polypeptide backbone. The carbohydrate chains are classified according to the linkage between sugar and amino acid. This chapter will deal with the biosynthesis of mammalian and avian protein-bound oligosaccharides linked to polypeptide via asparagine-*N*-acetyl-D-glucosamine (Asn-GlcNAc, *N*-glycosidic) and serine(threonine)-*N*-acetyl-D-galactosamine [Ser(Thr)-GalNAc, *O*-glycosidic] linkages. Several recent reviews have considered various aspects of this topic

---

Harry Schachter, Saroja Narasimhan, Paul Gleeson, George Vella, and Inka Brockhausen • Department of Biochemistry, University of Toronto, and Division of Biochemistry Research, Hospital for Sick Children, Toronto, Ontario, Canada M5G 1X8.

(Schachter, 1978; Schachter and Roseman, 1980; Schachter and Williams, 1982; Beyer *et al.*, 1981).

Most glycoproteins contain only oligosaccharides of the Asn-GlcNAc type. These glycoproteins carry out many diverse functions. They can be secreted proteins such as enzymes (amylases, ribonuclease B), plasma proteins (α_1-acid glycoprotein, transferrin, fibrinogen), immunoglobulins, hormones (thyroglobulin), and storage proteins (ovalbumin), or, they can be membranous proteins such as transport proteins (erythrocyte band 3), mammalian lectins, and receptors for hormones, toxins, viruses, and other molecules. Certain glycoproteins, e.g., human erythrocyte glycophorin A, human chorionic gonadotropin, human IgA, and calf fetuin, contain oligosaccharides with both *O*-glycosidic and *N*-glycosidic linkages. Glycoproteins such as the antifreeze glycoproteins found in Antarctic fish and mammalian mucins contain only *O*-glycosidic oligosaccharides. Mucins (mucous glycoproteins) are responsible for the gel-forming properties of mucus, the viscous fluid lining the epithelium of the gastrointestinal, respiratory and genitourinary tracts. The function of the mucous gel is to protect and lubricate the mucous epithelium.

Protein-bound oligosaccharides range in size from a single monosaccharide to as many as 18 or more residues. Glycoproteins with *N*-glycosidically linked oligosaccharides usually have from one to five oligosaccharides per polypeptide, whereas mucins may have as many as several hundred oligosaccharide chains per polypeptide. Recent advances in the purification of glycopeptides and oligosaccharides, and in the determination of oligosaccharide fine structure have been applied to many *N*- and *O*-glycosidically linked oligosaccharides. As will be discussed in this chapter, this new structural information is an essential prerequisite to the understanding of the biosynthesis of these molecules.

Glycosylation of proteins can be studied in two general ways (Schachter, 1978). Radiolabeled monosaccharides, amino acids, and sulfate can be administered to whole animals, tissue incubations, or cell cultures, and the incorporation of radioactivity into glycoproteins can be followed either by autoradiography or by chemical analyses of tissue fluids and subcellular fractions. Alternatively, cell-free preparations can be monitored for glycosyltransferase activities. The autoradiographic approach has two serious disadvantages. First, the glycoprotein being labeled cannot be identified with certainty. Second, many radiolabeled monosaccharide precursors undergo extensive conversions to other monosaccharides. The intact-cell approach using either autoradiography or biochemical analyses has, nevertheless, shown that the polypeptide backbones of glycoproteins are assembled on membrane-bound polyribosomes in the rough endoplasmic reticulum, and that the addition of carbohydrate occurs as the glycoprotein moves through the endomembrane system of the cell from rough endoplasmic reticulum to the Golgi apparatus. The site of formation of the amino acid–carbohydrate bond appears to be the rough endoplasmic reticulum for Asn-GlcNAc and the smooth endoplasmic reticulum–Golgi apparatus for Ser(Thr)-GalNAc. Thus, for example, Hanover *et al.* (1982) used a human choriocarcinoma cell line to study the incorporation of radiolabeled amino acids and glucosamine into human chorionic gonadotropin (hCG), a glycoprotein containing both *N*- and *O*-linked oligosaccharides. They found that *N*-glycosylation of the polypeptide backbone of this protein occurred cotranslationally or shortly after translation in the rough endoplasmic reticulum, while *O*-glycosylation of the same protein occurred shortly before secretion, presumably in the Golgi apparatus.

The intact-cell approach will not be considered further in this chapter. We will deal only with the glycosyltransferases involved in the assembly of Asn-GlcNAc and Ser(Thr)-GalNAc oligosaccharides.

The glycosyltransferases are a large group of enzymes catalyzing the following general reaction:

$$\text{Nucleotide sugar} + \text{acceptor} \rightarrow \text{sugar-acceptor} + \text{nucleotide}$$

The glycosyltransferases have been extensively reviewed (Schachter, 1978; Schachter and Roseman, 1980; Beyer *et al.*, 1981). They are membrane-bound enzymes and require detergent for optimum activity. The recent application of affinity chromatography to detergent-solubilized enzyme preparations has resulted in the purification of many of these enzymes (Beyer *et al.*, 1981). One of their most interesting features is the ability to discriminate between different oligosaccharide structures. This substrate specificity directs the synthetic pathway along various routes and is a major factor in the control of oligosaccharide synthesis. The study of highly purified glycosyltransferases has provided evidence for the hypothesis that every sugar–sugar linkage requires a separate transferase for its synthesis. This one linkage–one glycosyltransferase hypothesis appears to hold true in most instances, although at least one exception is now known (the Lewis blood group-dependent fucosyltransferase catalyzes the synthesis of both α1-3 and α1-4 linkages; Prieels *et al.*, 1981). Also, it has long been known that identical linkages in different complex carbohydrates may be synthesized by the same glycosyltransferase, e.g., the determinants for the human blood groups A, B, H, Le^a, and Le^b are synthesized by the same transferases whether they occur in glycosphingolipids, *N*-glycosyl oligosaccharides, *O*-glycosyl oligosaccharides, or milk oligosaccharides with lactose at the reducing terminus. Examples have appeared in the literature in which a particular linkage is made by more than one glycosyltransferase.

II. Asn-GlcNAc OLIGOSACCHARIDE STRUCTURE

Protein-bound oligosaccharides are classified according to the covalent linkage between amino acid and carbohydrate. The major linkages in avian and mammalian glycoproteins are the *N*-glycosidic linkage between Asn and GlcNAc and three types of *O*-glycosidic linkage: Ser(Thr)-GalNAc, Ser-xylose, and hydroxylysine-Gal (Kornfeld and Kornfeld, 1980). Ser-xylose oligosaccharides occur in glycosaminoglycans and hydroxylysine-Gal oligosaccharides are found in collagens and basement membranes; these structures are not considered in this chapter. The Ser(Thr)-GalNAc oligosaccharides are considered in Section VII. The synthesis of the Asn-GlcNAc linkage is presently believed to be carried out by an oligosaccharide transferase in the rough endoplasmic reticulum which catalyzes the transfer of oligosaccharide from

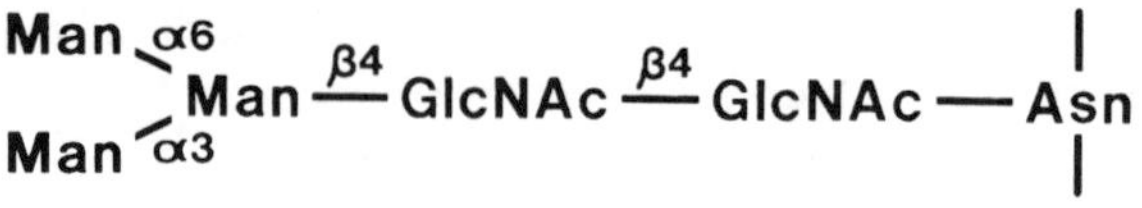

$$\begin{array}{l} \text{Man}\,\alpha 6 \\ \quad\quad\quad\ \text{Man} \xrightarrow{\beta 4} \text{GlcNAc} \xrightarrow{\beta 4} \text{GlcNAc} — \text{Asn} \\ \text{Man}\,\alpha 3 \end{array}$$

Figure 1. The structure of the core common to most *N*-glycosyl oligosaccharides. This structure is named MM to indicate that both arms are terminated in Man (M) residues.

dolichol pyrophosphate oligosaccharide to either nascent ribosome-bound peptide or to postribosomal peptide (Struck and Lennarz, 1980). In several tissues, it has been shown that the initial peptide-bound oligosaccharide contains three D-glucose (Glc), nine D-mannose (Man), and two *N*-acetyl-D-glucosamine (GlcNAc) residues, but this structure has not been established for all tissues. The protein-bound oligosaccharide is believed to be the precursor of all *N*-glycosyl oligosaccharides. These oligosaccharides all share a common core containing three Man and two GlcNAc residues (Figure 1).

The large protein-bound oligosaccharide, $(Glc)_3(Man)_9(GlcNAc)_2$, undergoes a series of processing reactions to form the various *N*-glycosyl oligosaccharides shown in Figure 2. The three Glc residues are removed by at least two α-glucosidases in the rough endoplasmic reticulum, and zero to six of the nine Man residues are removed primarily in the Golgi apparatus (Tabas and Kornfeld, 1982). The number of Man residues removed varies from one oligosaccharide to the other. If four or less of the peripheral α2-linked Man residues are removed by α2-mannosidase(s) I, the result is a high-mannose type of oligosaccharide (Figure 2). A subclass of high-mannose structure with mannose-6-phosphate residues has been found in some lysosomal hydrolases (Kornfeld, 1982). Further oligosaccharide processing followed by a series of elongation reactions leads the synthetic pathway from M_5 (the product of α2-mannosidase(s) I action, Figure 2) towards at least 11 other classes of *N*-glycosyl oligosaccharides (Figure 2; Carver and Grey, 1981).

Bi-, tri-, and tetra-antennary complex (also known as *N*-acetyllactosamine type) oligosaccharides carry two, three, or four branches or antennae attached to the common core (Figure 2). A branch can be incomplete (either a single GlcNAc, or a Gal-GlcNAc disaccharide) or complete (a sialyl-Gal-GlcNAc trisaccharide). The linkage between Gal and GlcNAc is usually β1-4 (hence the name *N*-acetyllactosamine) but is occasionally β1-3. The linkage between sialic acid and Gal is either α2-3 or α2-6. Bisected

→

Figure 2. *N*-Glycosyl oligosaccharides can be classified into at least 12 types of structures. Examples of these types are shown in the figure. High-mannose oligosaccharides are named according to the number of Man (M) residues, e.g., M_5 and M_9. Complex oligosaccharides are named according to the terminal sugars present on the arms [M, Man; Gn, GlcNAc; G, Gal; S, sialic acid; (Gn) represents a bisecting GlcNAc residue]. The *X* can be either H, Galβ1-4(or 3), or sialylα2-6(or 3)Galβ1-4(or 3). The hybrids shown in the figure have only incomplete antennae (GlcNAc) and are named according to the positions of these GlcNAc residues using the code shown in Figure 3.

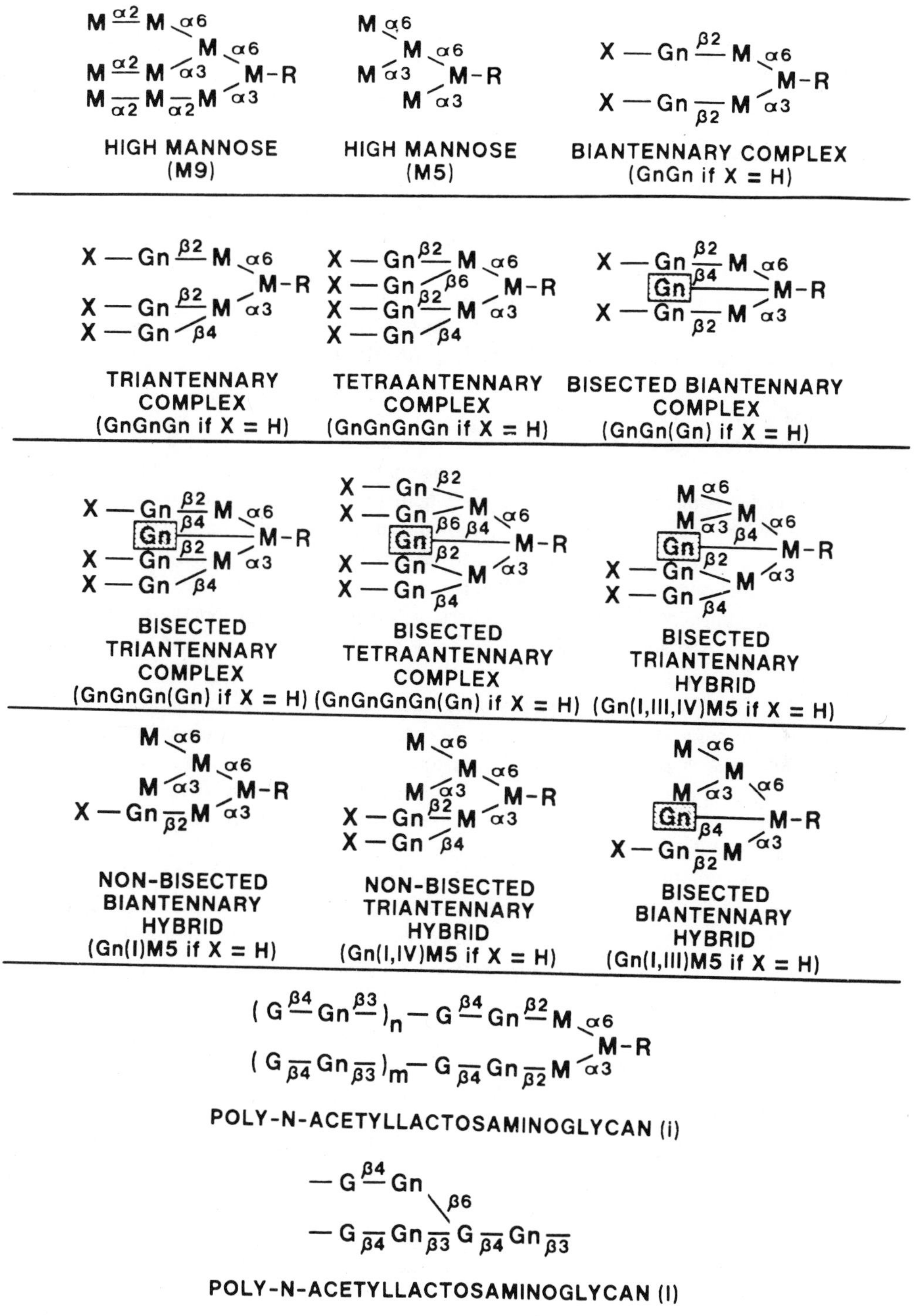

M α2 M α6 M α6 M–R; M α2 M α3; M α2 M α2 M α3

HIGH MANNOSE
(M9)

M α6 M α6 M–R; M α3; M α3

HIGH MANNOSE
(M5)

X — Gn β2 M α6 M–R; X — Gn β2 M α3

BIANTENNARY COMPLEX
(GnGn if X = H)

X — Gn β2 M α6 M–R; X — Gn β2 M α3; X — Gn β4

TRIANTENNARY
COMPLEX
(GnGnGn if X = H)

X — Gn β2 M α6; X — Gn β6; M–R; X — Gn β2 M α3; X — Gn β4

TETRAANTENNARY
COMPLEX
(GnGnGnGn if X = H)

X — Gn β2 M α6; Gn β4 M–R; X — Gn β2 M α3

BISECTED BIANTENNARY
COMPLEX
(GnGn(Gn) if X = H)

X — Gn β2 M α6; Gn β4 M–R; X — Gn β2 M α3; X — Gn β4

BISECTED
TRIANTENNARY
COMPLEX
(GnGnGn(Gn) if X = H)

X — Gn β2; X — Gn β6 M α6; Gn β4 M–R; X — Gn β2 M α3; X — Gn β4

BISECTED
TETRAANTENNARY
COMPLEX
(GnGnGnGn(Gn) if X = H)

M α6; M α3 M α6; Gn β4 M–R; X — Gn β2 M α3; X — Gn β4

BISECTED
TRIANTENNARY
HYBRID
(Gn(I,III,IV)M5 if X = H)

M α6 M α6 M–R; M α3; X — Gn β2 M α3

NON-BISECTED
BIANTENNARY
HYBRID
(Gn(I)M5 if X = H)

M α6 M α6 M–R; M α3; X — Gn β2 M α3; X — Gn β4

NON-BISECTED
TRIANTENNARY
HYBRID
(Gn(I,IV)M5 if X = H)

M α6 M α6; M α3; Gn β4 M–R; X — Gn β2 M α3

BISECTED
BIANTENNARY
HYBRID
(Gn(I,III)M5 if X = H)

(G β4 Gn β3)n — G β4 Gn β2 M α6 M–R
(G β4 Gn β3)m — G β4 Gn β2 M α3

POLY-N-ACETYLLACTOSAMINOGLYCAN (i)

— G β4 Gn β6
— G β4 Gn β3 G β4 Gn β3

POLY-N-ACETYLLACTOSAMINOGLYCAN (I)

complex oligosaccharides are similar except for an additional GlcNAc residue linked β1-4 to the β-linked Man residue of the core; this GlcNAc bisects the branches on the two arms of the core (Figure 2), e.g., ovotransferrin (Dorland *et al.*, 1979). Hybrid oligosaccharides contain either one or two Man residues attached to the Man-α1-6 arm of the core and one or two complex-type branches (usually incomplete) on the Man-α1-3 arm of the core. The structure is thus a hybrid in the sense of having a high-mannose structure on one arm of the core and a complex structure on the other arm. Hybrid oligosaccharides are usually isolated as bisected structures, e.g., from ovalbumin (Tai *et al.*, 1977). Nonbisected hybrid oligosaccharides have, however, been isolated from rhodopsin (Liang *et al.*, 1979; Fukuda *et al.*, 1979c). Poly-*N*-acetyllactosaminoglycan structures carrying i and I blood group antigenic determinants have also been found attached to the common *N*-glycosyl core in human erythrocyte band 3 (Fukuda *et al.*, 1979a,b; Jarnefelt *et al.*, 1978), K-562 cells derived from human chronic myelogenous leukemia (Turco *et al.*, 1980), and Chinese hamster ovary cell membranes (Li *et al.*, 1980).

Some glycoproteins contain high-mannose oligosaccharides, others contain complex oligosaccharides, and some glycoproteins contain both, e.g., calf thyroglobulin. The biosynthetic factors which determine whether an oligosaccharide remains high mannose or is processed further are not known, but it is probable that the amino acid sequence near the glycosylation site is involved. The high-mannose $(Man)_5(GlcNAc)_2$-Asn structure shown in Figure 2 (M_5) is the starting point for the biosynthesis of all complex and hybrid *N*-glycosyl oligosaccharides. A series of Golgi apparatus-localized glycosyltransferases and α-mannosidases carry out these processing and elongation reactions.

Although the function of most complex carbohydrates remains unknown, there is evidence to suggest that the branching of *N*-glycosyl oligosaccharides on the cell surface may play a role in oncogenic transformation. For example, Takasaki *et al.* (1980) found that transformation of baby hamster kidney cells with polyoma virus led to a reduction in bi-antennary oligosaccharides, an increase in tetra-antennary oligosaccharides, and the appearance of novel penta- and hexa-antennary oligosaccharides. Branches are usually initiated by the incorporation of a GlcNAc residue. In fact, branches can be initiated on the *N*-glycosyl core in at least seven different ways (Figure 3), and if one assumes the one linkage–one enzyme hypothesis (Schachter and Roden, 1973; Schachter and Roseman, 1980), there should be at least seven different initiating

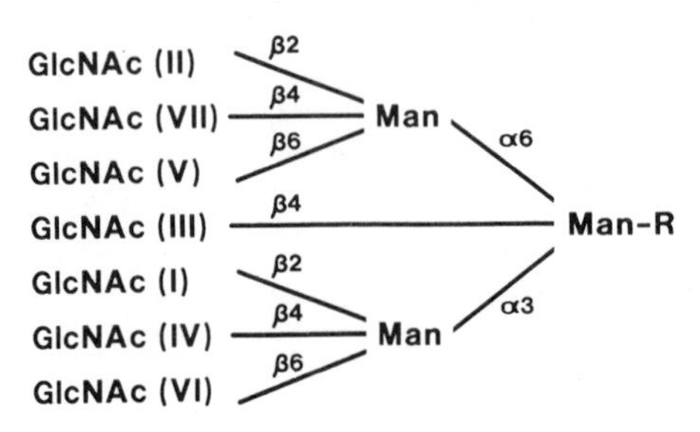

Figure 3. A hypothetical structure showing all the antennary GlcNAc residues described to date. GlcNAc linked 1-3 to either arm of the core, 1-2 to the β-linked Man, and in α-linkages are all theoretically possible but have not yet been described. We have numbered these GlcNAc residues as indicated. This numbering system is used to name compounds (see hybrids in Figure 2) and GlcNAc-transferases. Abbreviations as for Figure 2.

GlcNAc-transferases. Five of these transferases have, in fact, been demonstrated *in vitro* to date. These enzymes incorporate the GlcNAc residues labeled as I, II, III, IV, and probably V in Figure 3, and the transferases are numbered accordingly.

III. Asn-GlcNAc OLIGOSACCHARIDE INITIATION AND PROCESSING

A detailed discussion of Asn-GlcNAc oligosaccharide initiation and processing is beyond the scope of this chapter. The first stage in the process is the synthesis of dolichol pyrophosphate oligosaccharide (Figure 4; Struck and Lennarz, 1980). It has now been established that the first five Man residues to be incorporated into dolichol pyrophosphate oligosaccharide are derived from GDP-Man, and the next four Man residues are derived from dolichol monophosphate mannose. A mouse lymphoma mutant cell line (Thy-1-negative, class E) lacking Thy-1 antigen on its cell surface (Chapman *et al.*, 1979, 1980) is unable to make dolichol monophosphate mannose and therefore makes dolichol pyrophosphate oligosaccharide containing five instead of nine Man residues (Figure 4). Normal cells deprived of glucose at low cell densities also make truncated oligosaccharides with five instead of nine Man residues (Gershman and Robbins, 1981; Rearick *et al.*, 1981a).

The second stage is the transfer of oligosaccharide from dolichol pyrophosphate oligosaccharide to an Asn residue in the polypeptide chain (Struck and Lennarz, 1980). The oligosaccharide transferase carrying out this reaction is located primarily in the

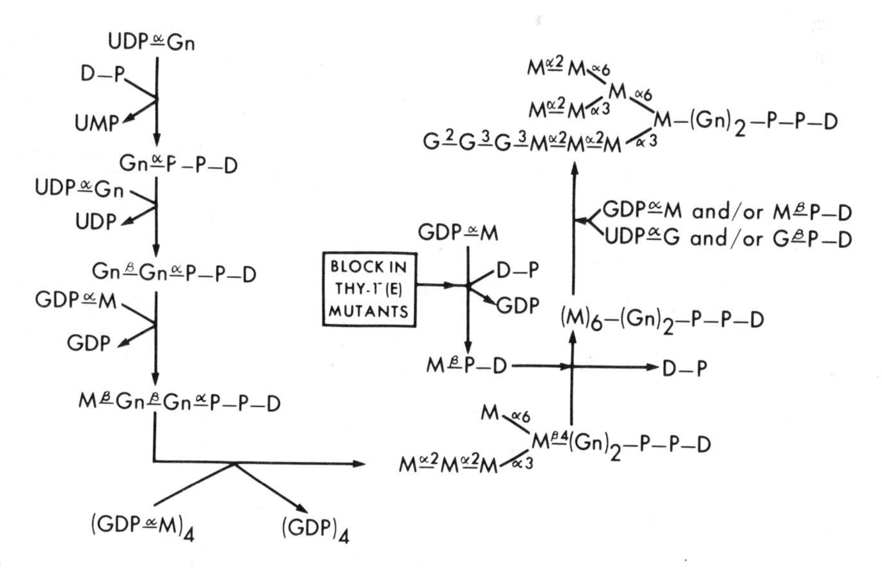

Figure 4. Biosynthesis of dolichol pyrophosphate oligosaccharide. Abbreviations: D, dolichol; P, phosphate; UDP, uridine diphosphate; UMP, uridine monophosphate; GDP, guanosine diphosphate; Gn, GlcNAc; M, Man; G, Glc.

rough endoplasmic reticulum. The enzyme acts most efficiently on the complete $(Glc)_3(Man)_9(GlcNAc)_2$ oligosaccharide and removal of even one of the three Glc residues causes a significant reduction in transfer rate (Turco *et al.*, 1977). The oligosaccharide transferase will, however, transfer smaller oligosaccharides, even Manβ1-4GlcNAcβ1-4GlcNAc and GlcNAcβ1-4GlcNAc, but at appreciably slower rates. The physiological importance of these slower transfer reactions has not been established. The transferase also catalyzes transfer to polypeptide of the $(Glc)_3(Man)_5(GlcNAc)_2$ oligosaccharide formed in the absence of dolichol monophosphate mannose (see above and Figure 4). Many glycoproteins containing this truncated oligosaccharide can be processed by an alternative pathway (Kornfeld, 1982) but some glycoproteins, e.g., Thy-1, cannot be processed normally in the truncated form. The oligosaccharide transferase probably acts primarily on nascent polypeptides still bound to the polyribosome, but posttranslational glycosylation may occur in some situations (Jamieson, 1977). The transferase acts on polypeptides containing an Asn-*X*-Ser(Thr) sequence and even the tripeptide Asn-*X*-Ser(Thr) will act as an acceptor if the amino and carboxy termini

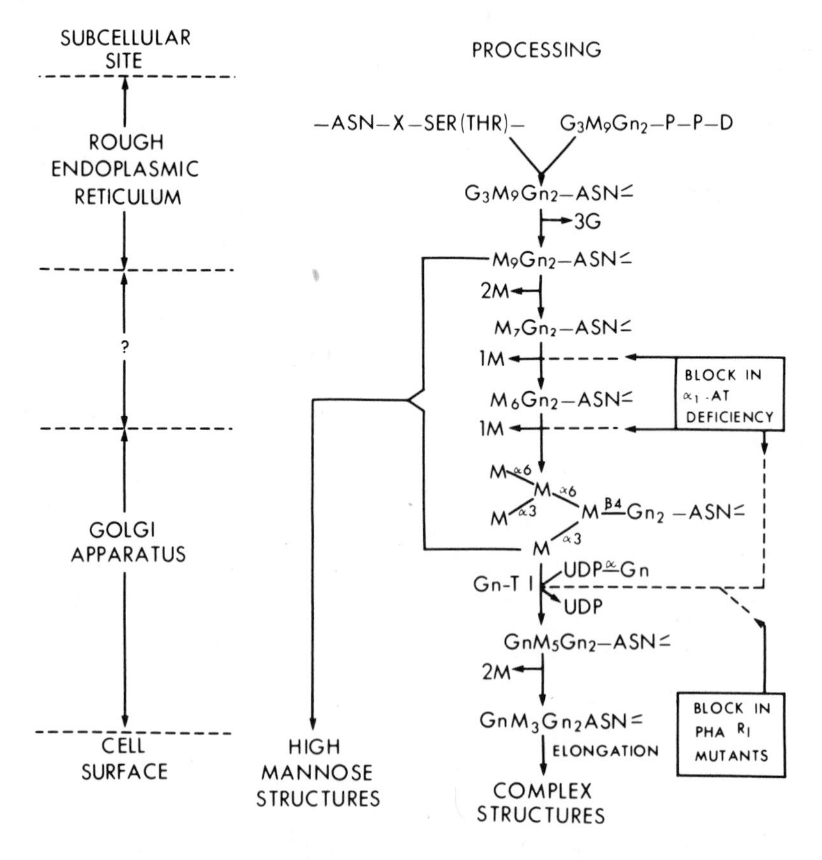

Figure 5. Oligosaccharide processing of *N*-glycosyl oligosaccharides. Abbreviations: T, transferase; others as in legend to Figure 4. See text for an explanation of this figure.

are both blocked. However, certain proteins with the Asn-*X*-Ser(Thr) sequence are not acceptors even after partial unfolding. It is now evident that the tripeptide sequence must be sterically available if the oligosaccharide transferase is to act efficiently. The three-dimensional folding of the polypeptide chain appears to control not only glycosylation of the Asn residue but also the degree of subsequent oligosaccharide processing.

The third stage in the synthesis of Asn-GlcNAc oligosaccharides is the processing of the protein-bound $(Glc)_3(Man)_9(GlcNAc)_2$ oligosaccharide (Figure 5). Oligosaccharide processing has been elucidated by work in the laboratories of D. F. Summers, P. W. Robbins, S. Kornfeld, R. G. Spiro, and others, and has been reviewed recently by Tabas and Kornfeld (1982). The three Glc residues are first removed by two α-glucosidases in the rough endoplasmic reticulum. The result is a high-mannose oligosaccharide containing nine Man residues. An enzyme called α2-mannosidase I then removes from one to four α2-linked Man residues to yield high-mannose oligosaccharides of varying sizes. Mannosidase I is believed to be mainly a Golgi-localized enzyme but some Man is probably also removed within the endoplasmic reticulum. For example, the abnormal α_1-antitrypsin which accumulates in the rough endoplasmic reticulum of the livers of humans with *ZZ*-genotype antitrypsin deficiency contains high-mannose oligosaccharides with five, six, and seven Man residues implying removal of at least some Man in the rough endoplasmic reticulum (Hercz *et al.*, 1978; Hercz and Harpaz, 1980).

The final product of mannosidase I is the 5-mannose compound (M_5) shown in Figure 2 and 5 and at the top of Figure 6. This important intermediate is the precursor of all complex and hybrid Asn-GlcNAc oligosaccharides. Lectin-resistant somatic cell mutants lacking the next enzyme in the scheme, UDP-GlcNAc:α-D-mannoside β2-GlcNAc-transferase I, have been isolated in various laboratories, e.g., the Chinese hamster ovary cell line Pha^{R1} resistant to phytohemagglutinin (Figure 5; Narasimhan *et al.*, 1977). GlcNAc-transferase I-deficient cells accumulate glycoproteins carrying the M_5 structure (Li and Kornfeld, 1978; Robertson *et al.*, 1978). When GlcNAc-transferase I has incorporated a GlcNAc in β1-2 linkage into the Manα1-3-terminus of M_5 (Figure 6), an enzyme called GlcNAc-transferase I-dependent α3/6-mannosidase (Harpaz and Schachter, 1980b) or mannosidase II (Tabas and Kornfeld, 1978, 1982; Tulsiani *et al.*, 1982a,b) removes the last two Man residues to give the core structure common to all complex *N*-glycosyl oligosaccharides (Figure 1). Mannosidase II cannot act on M_5 nor on the bisected structure shown in Figure 2 as Gn(I,III)M_5.

IV. SYNTHESIS OF NONBISECTED BI-ANTENNARY COMPLEX N-GLYCOSYL OLIGOSACCHARIDES

The 5-Man structure shown at the top of Figure 6 is the precursor of all complex and hybrid *N*-glycosyl oligosaccharides. Nonbisected bi-antennary complex oligosaccharides (Figure 2) are the simplest of the complex oligosaccharides and are of widespread occurrence, e.g., they have been found in immunoglobulins, plasma glycoproteins, and cell surface glycoproteins. The glycosyltransferases involved in the

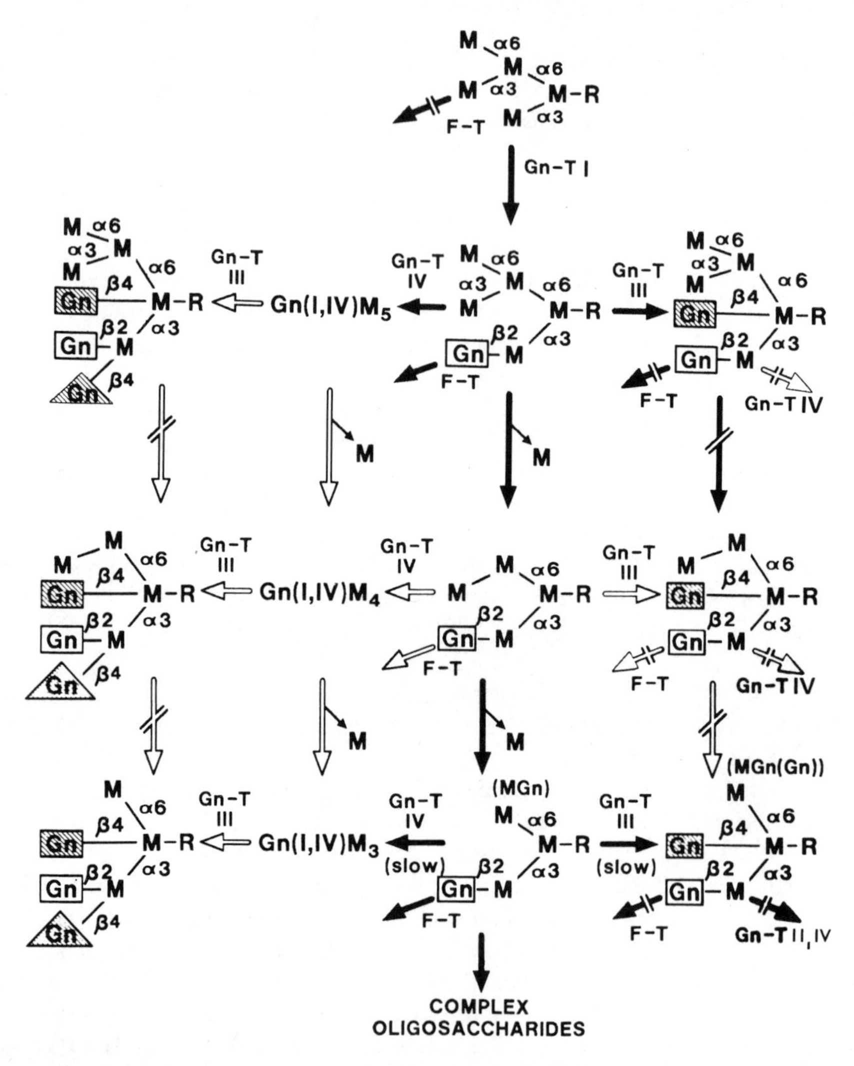

Figure 6. Pathway showing the action of GlcNAc-transferase I, GlcNAc-transferase I-dependent α3/6-mannosidase (mannosidase II, Tulsiani *et al.*, 1982a,b), and the synthesis of hybrid oligosaccharides. Abbreviations: M, Man; Gn, GlcNAc; F, Fuc; T, transferase. Proven or highly likely reactions are shown by solid arrows, unproven reactions by hollow arrows, reactions proven not to function *in vitro* by blocked solid arrows, and reactions which probably do not function *in vitro* but have not yet been tested are shown in blocked hollow arrows. Data to support the action of GlcNAc-transferases III and IV at the 5-Man stage (Allen *et al.*, 1984) and of GlcNAc-transferase IV on MGn (Gleeson and Schachter, 1983; Table 3) and data in support of the solid arrows representing mannosidase II action (Harpaz and Schachter, 1980b), GlcNAc-transferase I action (Table 1; Harpaz and Schachter, 1980a), GlcNAc-transferase III action on MGn (Narasimhan, 1982; Table 3), and Fuc-transferase action (Longmore and Schachter, 1982) have been published.

conversion of M_5 to nonbisected bi-antennary complex oligosaccharides will now be considered.

A. UDP-GlcNAc:α-D-Mannoside (GlcNAc to Manα1-3) β1-2-GlcNAc-Transferase I

Johnston *et al.* (1966) first reported the presence in goat colostrum of GlcNAc-transferase acting on α_1-acid glycoprotein pretreated with neuraminidase, β-galactosidase, and β-*N*-acetylglucosaminidase. A similar enzyme has been found in many tissues (Schachter and Roseman, 1980; Schachter, 1978) of the rat, pig, guinea pig, and human and has been localized in the Golgi apparatus of rat liver (Munro *et al.*, 1975). Studies on lectin-resistant mutants of Chinese hamster ovary cells (Stanley *et al.*, 1975; Narasimhan *et al.*, 1977; Gottlieb *et al.*, 1975) and baby hamster kidney cells (Vischer and Hughes, 1981) indicated that the high molecular weight acceptor detects at least two different GlcNAc-transferases. One of these enzymes, UDP-GlcNAc : α-D-mannoside β1-2-GlcNAc-transferase I (GlcNAc-transferase I), attaches GlcNAc in β1-2 linkage to the Manα1-3-terminus of the tri-mannosyl core of glycopeptide M_5, as shown in Figure 6. This enzyme has been purified from bovine colostrum (Harpaz and Schachter, 1980a), rabbit liver (Oppenheimer and Hill, 1981), pig liver (Oppenheimer *et al.*, 1981), and pig tracheal mucosa (Oppenheimer *et al.*, 1981; Mendicino *et al.*, 1981).

The substrate specificity of GlcNAc-transferase I is summarized in Table 1. The enzyme appears to be specific for the terminal Manα1-3 residue of the core since it does not act on the other more peripheral terminal Manα1-3 residue of M_5 (Figures 2 and 6). The physiological substrate is probably the 5-mannose oligosaccharide M_5 since Chinese hamster ovary cell mutants lacking GlcNAc-transferase I contain glycoproteins with the M_5 structure and fail to make complex or hybrid oligosaccharides (Tabas *et al.*, 1978; Li and Kornfeld, 1978). However, Kornfeld (1982) has pointed out that normal cells possess an alternate pathway involving the action of GlcNAc-transferase I on the 3-mannose structure MM shown in Figure 1. This alternate pathway is usually minor in normal cells but becomes the major pathway under certain conditions in normal cells, e.g., glucose deprivation, or in cell mutants with an inability to make dolichol monophosphate mannose. The result is a truncated dolichol pyrophosphate oligosaccharide containing 5 instead of 9 mannose residues (Figure 4), which on oligosaccharide processing leads to the MM structure shown in Figure 1.

B. GlcNAc-Transferase-I-Dependent α3/6-Mannosidase(s) (Mannosidase II)

The product of GlcNAc-transferase I, Gn(I)M_5, sits at an important crossroads (Figure 6) in that it can be acted on by at least three glycosyltransferases and one highly specific α-mannosidase. The glycosyltransferases lead towards the synthesis of hybrid oligosaccharides (discussed in Section V). The "main-line" pathway via α-mannosidase leads towards the various complex oligosaccharide structures shown in Figure 2. These structures all have only 3 Man residues. The enzyme which removes

Table 1. Substrate Specificity of GlcNAc-Transferases I and II

Substrate[a]	GlcNAc-transferase I K_m (mM)	GlcNAc-transferase II K_m (mM)
Mβ4Gn /α3 M	7.4	Not active
M \α6 Mβ4Gn-*R* /α3 M	0.20	Not active
Mα3M \α6 Mβ4Gn-*R* /α3 M	Active	Not tested
M \α6 Mα3M \α6 Mβ4Gn-*R* /α3 M	0.12	Not active
M \α6 Mβ4Gn-*R* /α3 Gnβ2M	10[b]	0.1
M \α6 Gnβ4Mβ4Gn-*R* /α3 Gnβ2M	Not active	Not active

[a] Abbreviations: M, Man; Gn, GlcNAc; *R*, β4(+/-Fucα6)Gn-Asn-.
[b] Oppenheimer and Hill (1981) report that their highly purified GlcNAc-transferase I is totally inactive with this substrate.

2 Man residues from Gn(I)M_5 is an α-mannosidase absolutely dependent on the prior action of GlcNAc-transferase I (Tabas and Kornfeld, 1978; Harpaz and Schachter, 1980b); the enzyme will not work on M_5 but only on Gn(I)M_5 (Figures 2 and 6). This GlcNAc-transferase I-dependent α3/6-mannosidase is the first committed step towards the synthesis of complex oligosaccharides. It appears to be the same enzyme as the α-mannosidase II purified from rat liver Golgi-rich membranes by Tulsiani *et al.* (1982b). The highly purified enzyme catalyzes the release of both the α3- and α6-

Man residues from Gn(I)M_5; it is uncertain which Man residue is released preferentially (Harpaz and Schachter, 1980b). Tulsiani *et al.* (1982a) have recently shown that swainsonine causes the accumulation of hybrid intermediates by inhibiting the action of α-mannosidase II.

C. *UDP-GlcNAc:α-D-Mannoside (GlcNAc to Manα1-6) β1-2-GlcNAc-Transferase II*

The study of lectin-resistant mutants of Chinese hamster ovary cells (Stanley *et al.*, 1975; Narasimhan *et al.*, 1977; Gottlieb *et al.*, 1975) and baby hamster kidney cells (Vischer and Hughes, 1981) deficient in GlcNAc-transferase I indicated the presence of a second UDP-GlcNAc:α-D-mannoside β1-2-GlcNAc-transferase designated GlcNAc-transferase II. The function of GlcNAc-transferase II is the initiation of the second antenna or branch by addition of GlcNAc in β1-2 linkage to the Manα1-6 terminus of glycopeptide MGn (Figure 7). The enzyme has not been purified to homogeneity but partially purified preparations have been obtained from bovine colostrum (Harpaz and Schachter, 1980a), pig liver (Oppenheimer *et al.*, 1981), and pig trachea (Oppenheimer *et al.*, 1981; Mendicino *et al.*, 1981).

The only effective substrate for GlcNAc-transferase II is MGn (Table 1), the product of GlcNAc-transferase I-dependent α3/6-mannosidase. GlcNAc-transferase II does not act on MM (Figure 1) indicating that GlcNAc-transferase I action must precede GlcNAc-transferase II action. The product of GlcNAc-transferase II is GnGn (Figure 2). GnGn is the major precursor for all complex oligosaccharides.

D. *UDP-Gal:GlcNAc β1-4-Galactosyltransferase*

The structure GnGn is at another crossroads in the biosynthetic pathway and can be acted on by at least four glycosyltransferases (Figure 7). If UDP-Gal : GlcNAc β1-4-Gal-transferase acts on GnGn before the other enzymes to form GnG, the pathway is committed to synthesis of nonbisected bi-antennary oligosaccharides. Enzyme activities capable of transferring Gal from UDP-Gal to GlcNAc, or to oligosaccharides, glycopeptides, or glycoproteins with GlcNAc at their nonreducing termini, have been widely described in many species and tissues (Schachter and Roseman, 1980; Schachter, 1978; Beyer *et al.*, 1981). However, the identification of the product as Galβ1-4GlcNAc has been carried out only with free GlcNAc as acceptor and only with a limited number of enzyme sources. The Gal-transferase activity occurs in most tissues in a membrane-bound form localized primarily to the Golgi apparatus and in a soluble form in various tissue fluids. The reaction catalyzed is as follows:

$$\text{UDP-Gal} + \text{GlcNAc-R} \rightarrow \text{Gal}\beta\text{1-4GlcNAc-R} + \text{UDP}$$

Brodbeck and Ebner (1966) first resolved bovine milk lactose synthetase (UDP-Gal : Glc β1-4-galactosyltransferase) into 2 proteins, called A and B, both of which are needed for lactose synthetase activity. The B protein was later shown to be α-lactalbumin, a protein totally devoid of any catalytic activity. The A protein was shown

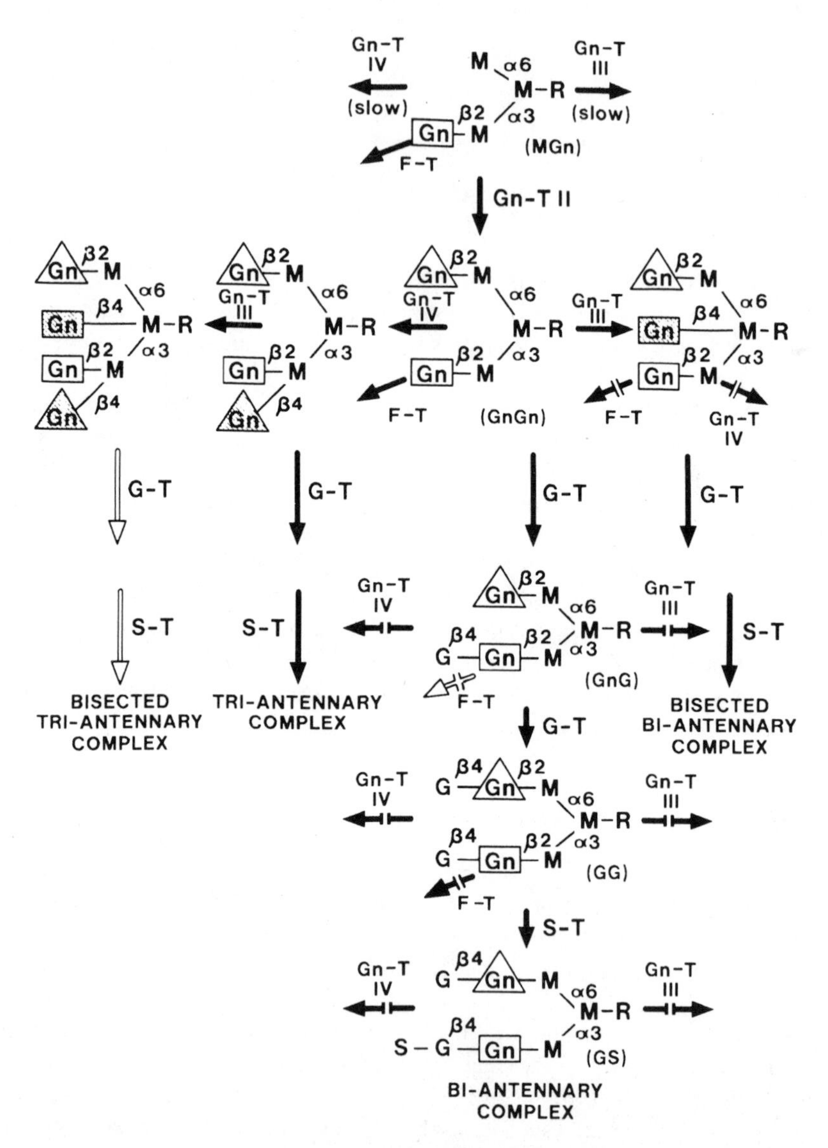

Figure 7. Pathway showing the synthesis of complex oligosaccharides. Abbreviations and definitions for arrows are as for Figure 6. Oligosaccharides are named according to the terminal sugars with the sugar on the Manα1-6 arm being named first. Solid arrows representing Gal-transferase action are based on M. Paquet, S. Narasimhan, H. Schachter, and M. A. Moscarello, J. Biol. Chem., in press, and data in the literature (Beyer and Hill, 1982). Sialyltransferase arrows are based on data in the literature (Beyer and Hill, 1982; Beyer *et al.*, 1981; Schachter and Roseman, 1980; Van den Eijnden *et al.*, 1980); the presence of a bisecting GlcNAc residue probably has no effect on sialyltransferase action but this has not been proven definitively. Data for GlcNAc-transferase II are shown in Table 1 (Harpaz and Schacter, 1980a). Data for GlcNAc-transferases III (Narasimhan, 1982) and IV (Gleeson and Schachter, 1983) and for Fuc-transferase (Longmore and Schachter, 1982) are shown in Table 3. Solid arrows indicate the lack of action of GlcNAc-transferases III and IV on GS because neither GG nor SS are substrates (Table 3); GS, however, has not yet been tested directly.

to be a β1-4-galactosyltransferase which in the absence of α-lactalbumin could transfer Gal to GlcNAc to make *N*-acetyllactosamine but had minimal ability to synthesize lactose (Brew *et al.*, 1968).

The A protein (UDP-Gal:GlcNAc β1-4-galactosyltransferase) has been purified (Barker *et al.*, 1972; Geren *et al.*, 1976; Schachter and Roseman, 1980) from bovine and human milk, bovine colostrum, rat, calf, and human serum, human malignant effusions, ovine mammary gland, and swine mesentary lymph nodes, using classical techniques as well as affinity chromatography on α-lactalbumin-agarose, UDP-hexanolamine-agarose, GlcNAc-agarose, *p*-aminophenyl-β-GlcNAc-agarose, and UDP-glucuronyl-6-aminohexyl-agarose.

The kinetic parameters are listed in Table 2. Many kinetic studies have been

Table 2. Kinetic Parameters of Glycosyltransferases Acting on N-*Glycosyl Oligosaccharides*[a]

Enzyme	Source	Substrate	K_m (mM)	pH Optimum	Optimum concentration of Mn^{2+} (mM)
GlcNAc-*T* I	Bovine colostrum	UDP-GlcNAc	0.1	6.0	18–26
		M_5	0.12		
	Rabbit liver	M_5	0.45	6.3	ND[b]
	Pig liver	M_5	0.39	ND	ND
GlcNAc-*T* II	Bovine colostrum	MGn	0.1	ND	ND
	Pig liver	MGn	0.2	6.0–6.5	ND
GlcNAc-*T* III	Hen oviduct	UDP-GlcNAc	1.1	6.0–7.0	12
		GnGn	0.23		
β1-4-Gal-*T*	Bovine milk	UDP-Gal	0.1	7.5–9.0	13
		GlcNAc	6–8		
α2-6-sialyl-*T*	Bovine colostrum	CMP-NeuNAc	0.17	6.4–7.2	None[d]
		Asialo-AGP[c]	1.6		
	Rat liver	Galβ1-4GlcNAc	1.6	ND	None
		Galβ1-4Glc	129		
		Asialo-AGP[c]	0.18		
α2-3-sialyl-*T*	Rat liver	Galβ1-3GlcNAc	0.64	ND	None
		Galβ1-4GlcNAc	2.7		
		Galβ1-4Glc	9.4		
		Asialo-AGP[c]	0.29		
		Galβ1-3GlcNAc-protein	0.006		
α1-6-Fuc-*T*	Pig liver	GnGn	0.08	ND	ND
		MGn	0.6		
		Asialo, agalacto-AGP[c]	17.0		
α1-3/4-Fuc-*T*	Human milk	GDP-Fuc	0.005–0.013	7.0–7.8	5
		Galβ1-3GlcNAc	1.9		
		Galβ1-4GlcNAc	1.6		
		Galβ1-4Glc	59		

[a] Glycopeptide structures are shown in Figures 2 and 7, T, transferase.
[b] ND, not done or not reported.
[c] AGP, α_1-acid glycoprotein.
[d] Metal is not essential for activity.

published and various enzyme mechanisms have been proposed. There appear to be 2 Mn^{2+} sites, a high-affinity site I and a low-affinity site II. The mechanism is an ordered addition in which the enzyme first binds Mn^{2+} at site I, followed by a branched pathway in which either Mn^{2+}-UDP-Gal binds to site II or UDP-Gal alone binds to the enzyme. This is followed by binding of GlcNAc and eventual release of products. The K_m for glucose is 1.4 M but the addition of α-lactalbumin lowers this value to 5 mM. The ordered enzyme mechanism explains the behavior of the enzyme on affinity chromatography. The enzyme adheres strongly to α-lactalbumin-Sepharose only in the presence of acceptor (GlcNAc or Glc), to UDP-hexanolamine-Sepharose only in the presence of Mn^{2+}, and to GlcNAc-Sepharose only in the presence of UDP or UMP and Mn^{2+}.

The purified enzyme transfers Gal to either free GlcNAc or to oligosaccharides, glycopeptides, and glycoproteins with a β-linked GlcNAc at the nonreducing terminus. Glucose, glucosamine, and oligosaccharides with terminal α- or β-linked Glc residues are poor acceptors. The addition of α-lactalbumin lowers the K_m for Glc about 300-fold, lowers transfer to GlcNAc by over 70%, but has only a small inhibitory effect on Gal transfer to larger acceptors. The physiological function of the enzyme in most tissues is the elongation of *N*-glycosyl oligosaccharides and it probably also functions in the synthesis of *O*-glycosyl oligosaccharides and glycolipids. In mammary gland during lactation, it combines with α-lactalbumin to make lactose.

Recent data (Paquet, *et al.*, 1984) indicate that the UDP-Gal:GlcNAc β1-4-Gal-transferase from both bovine milk and rat liver shows a marked preference for the terminal GlcNAc on the Manα1-3 arm when GnGn is used as substrate. It is likely therefore that GnG rather than GGn is the physiological route towards bi-antennary structures, as indicated in Figure 7.

E. CMP-NeuNAc:Galβ1-4GlcNAc α2-6-Sialyltransferase

Several different sialyltransferases acting on *N*- and *O*-glycosyl oligosaccharides and on glycosphingolipids have been described (Paulson *et al.*, 1977a,b, 1982; Weinstein *et al.*, 1982a,b; Van den Eijnden and Schiphorst, 1981; Schachter and Roseman, 1980; Schachter, 1978; Beyer *et al.*, 1981; Sadler *et al.*, 1982). Two of these sialyltransferases (the α2-6-sialyltransferase discussed in this section and the α2-3-sialyltransferase discussed in Section F) act on the disaccharide Galβ1-4GlcNAc and on *N*-glycosyl oligosaccharides terminated at their reducing ends by this disaccharide. Many tissues are capable of transferring NeuNAc from CMP-NeuNAc to Galβ1-4GlcNAc or to asialo-glycoproteins, glycopeptides, or oligosaccharides terminating with this disaccharide, but very few studies have determined the linkage formed between sialic acid and Gal. When in fact such a study was carried out with rat liver as an enzyme source (Van den Eijnden *et al.*, 1977), the only linkage found was α2-6. However, the α2-3-sialyltransferase was later shown to be also present in rat liver (see Section F). The α2-6-sialyltransferase occurs in a soluble form in goat, cow, and human colostrum and in human, pig, and rat serum, and has been demonstrated as a Golgi membrane-bound enzyme in a variety of tissues, particularly in the livers of the rat, cow, pig, guinea pig, and human (Schachter and Roseman, 1980; Beyer *et al.*, 1981). It has been purified 440,000-fold to homogeneity from bovine colostrum (Paulson *et al.*, 1977a,b) and 22,800-fold to homogeneity from detergent extracts of rat liver membranes (Weinstein *et al.*, 1982a). The reaction catalyzed is

$$\text{CMP-NeuNAc} + \text{Gal}\beta\text{1-4GlcNAc-R} \rightarrow$$
$$\text{NeuNAc}\alpha\text{2-6Gal}\beta\text{1-4GlcNAc-R} + \text{CMP}$$

where *R* is H, oligosaccharide, or glycoprotein. The kinetic parameters are summarized in Table 2. The enzyme appears to have an equilibrium random order mechanism. Galβ1-4GlcNAc and oligosaccharides, glycopeptides, and glycoproteins with Galβ1-4GlcNAc termini are the only good acceptors. Galβ1-3GlcNAc, Galβ1-6GlcNAc, Galβ1-4Glc (lactose), and other β-galactosides have much higher K_m values (140–1780 mM) than *N*-acetyllactosamine (K_m = 12 mM). Substitution of the terminal Gal of the acceptor with sialic acid or removal of the terminal Gal prevents enzyme action. Galβ1-3GalNAc-terminal compounds, which are excellent acceptors for two sialyltransferases (α2-3 to Gal and α2-6 to GalNAc) acting on Ser(Thr)-GalNAc-type glycoproteins, are not acceptors for either the pure bovine colostrum or rat liver enzymes. It is interesting that the enzyme prefers to attach sialic acid to the Gal on the Manα1-3 arm of the core (Figure 7; Van den Eijnden *et al.*, 1980). The physiological function of this sialyltransferase is almost certainly the synthesis of the NeuNAcα2-6Galβ1-4GlcNAc antennae of complex *N*-glycosyl oligosaccharides.

F. *CMP-NeuNAc:Galβ1-3(4)GlcNAc α2-3-Sialyltransferase*

The sialylα2-3Gal linkage is found in both *N*- and *O*-glycosyl oligosaccharides, in milk oligosaccharides, and in gangliosides. At least three different and distinct sialyltransferases are now known to be involved in the synthesis of this linkage (Schachter and Roseman, 1980; Beyer *et al.*, 1981; Sadler *et al.*, 1982), i.e., enzymes which respectively convert (1) lactosyl-ceramide to sialylα2-3Galβ1-4Glc-ceramide (ganglioside G_{M3}), (2) Galβ1-3GalNAc-R (where R is H, oligosaccharide, mucin-type polypeptide backbone, or glycosphingolipid) to sialylα2-3Galβ1-3GalNAc-R, and (3) Galβ1-3(4)GlcNAc-R (where R is H, oligosaccharide, or a glycoprotein) to sialylα2-3Galβ1-3(4)GlcNAc-R. The latter enzyme, CMP-NeuNAc : Galβ1-3(4)GlcNAc α2-3-sialyltransferase, has been found in rat liver, fetal calf liver, embryonic chicken brain, rat testis, rabbit spleen and kidney, and human placenta (Van den Eijnden and Schiphorst, 1981; Van den Eijnden *et al.*, 1977; Paulson *et al.*, 1982; Weinstein *et al.*, 1982a,b). It has been purified 859,000-fold to homogeneity from detergent extracts of rat liver membranes using affinity chromatography on CDP-Sepharose and asialo-prothrombin-Sepharose (Weinstein *et al.*, 1982a). The reactions catalyzed by the pure enzyme are

$$\text{CMP-NeuNAc} + \text{Gal}\beta\text{1-3GlcNAc-}R \rightarrow \text{NeuNAc}\alpha\text{2-3Gal}\beta\text{1-3GlcNAc-R} + \text{CMP}$$

$$\text{CMP-NeuNAc} + \text{Gal}\beta\text{1-4GlcNAc-}R \rightarrow \text{NeuNAc}\alpha\text{2-3Gal}\beta\text{1-4GlcNAc-R} + \text{CMP}$$

where R is H, oligosaccharide, or glycoprotein. The kinetic parameters are summarized in Table 2. Whereas the α2-6-sialyltransferase discussed in Section E shows a strict specificity for Galβ1-4GlcNAc-terminated compounds, the α2-3-sialyltransferase can act on both Galβ1-4GlcNAc and Galβ1-3GlcNAc-terminated acceptors, although the latter have lower K_m values (Table 2). Lactose can be used as an acceptor but has a high K_m (Table 2). Galβ1-6GlcNAc, Galβ1-6GalNAc, Galβ1-3GalNAc, glycoproteins with the Galβ1-3GalNAc-Ser(Thr) and GalNAc-Ser(Thr) moieties are all ineffective as substrates. The enzyme is therefore different from the α2-3- and α2-6-sialyltransferases acting on Ser(Thr)-GalNAc oligosaccharides and appears to be involved in the synthesis of antennae for complex *N*-glycosyl oligosaccharides. Substitution or removal of the terminal Gal prevents enzyme action.

N-Glycosyl oligosaccharides have been described with sialic acid in α2-4 linkage to Gal, α2-6 linkage to GlcNAc, and α2-8 linkage to sialic acid (Weinstein *et al.*, 1982a,b), but the relevant sialyltransferases have not as yet been described.

G. GDP-Fuc:β-N-Acetylglucosaminide (Fuc to Asn-Linked GlcNAc) α1-6-Fucosyltransferase

Fucose has been reported to occur in α1-6 or α1-3 linkage to the Asn-linked GlcNAc of *N*-glycosyl oligosaccharides (Schachter and Roseman, 1980; Schachter, 1978; Beyer *et al.*, 1981). Fucose also occurs in a variety of antigenic determinants (Table 4) at the nonreducing termini of milk and urinary oligosaccharides, *N*- and *O*-linked protein-bound oligosaccharides and glycosphingolipids (Watkins, 1980; Schachter and Tilley, 1978). Golgi apparatus-enriched membrane preparations from pig and rat liver and rat and mouse testis, crude extracts from HeLa cells, and human serum, can all transfer Fuc from GDP-Fuc to *N*-glycosyl oligosaccharides with terminal GlcNAc residues. The rat liver enzyme has been shown to transfer Fuc to the Asn-linked GlcNAc residue but the linkage has not been determined (Wilson *et al.*, 1976). The pork liver enzyme has been shown to catalyze the following reaction (Longmore and Schachter, 1982):

GDP-β-L-Fuc + GlcNAcβ1-2Manα1-3(R-Manα1-6)Manβ1-4GlcNAcβ1-4GlcNAc-Asn-X
↓
GlcNAcβ1-2Manα1-3(R-Manα1-6)Manβ1-4GlcNAcβ1-4(Fucα1-6)GlcNAc-Asn-X + GDP

where R is H (glycopeptide MGn), GlcNAcβ1-2 (glycopeptide GnGn), or Manα1-6(Manα1-3). No product was detected with Fuc in α1-3 linkage to the Asn-linked GlcNAc. The kinetic parameters are indicated in Table 2. The substrate specificity of the α1-6-fucosyltransferase is summarized in Table 3. The enzyme does not act on substrates which lack the GlcNAcβ1-2Manα1-3 sequence or on any substrate carrying a bisecting GlcNAc residue (GlcNAc attached β1-4 to the β-linked Man). Elongation of antennae by addition of Galβ1-4 or sialyl-Galβ1-4 sequences also prevents enzyme action. Glycopeptides with only a single Gal residue (GGn and GnG) have, however, not yet been tested as substrates for this enzyme. These studies predict that high-mannose *N*-glycosyl oligosaccharides should not have a Fuc in the core and, indeed,

Table 3. Glycosyltransferase Substrate Specificities

Substrate[a]	Specific activity (μunits/mg)			
	Gn-T III pH 5.7	Gn-T III pH 7.0	Gn-T IV pH 7.0	Fuc-T pH 8.0
GnGn (+Fuc)	85	73	83	ND[b]
GnGn (−Fuc)		75	95	1800
MGn (+Fuc)	≤13	10	18	0
MGn (−Fuc)				3100
MM (+Fuc)	≤7	≤16	≤16	0
MM (−Fuc)				0
GGn (+Fuc)		48	30	
GnG (+Fuc)		≤5	≤5	
GG (−Fuc)		≤8	≤8	0[c]
SS (−Fuc)		0	0	0
GnGn(Gn) (−Fuc)			≤3	0
MGn(Gn) (−Fuc)				0

[a] Hen oviduct membranes were used for studies with GlcNAc-transferase III (Gn-T III) at pH 5.7 and 7.0, and GlcNAc-transferase IV (Gn-T IV) at pH 7.0. Golgi-enriched membranes from pig liver were used for the studies on GDP-Fuc:β-*N*-acetylglucosaminide (Fuc to Asn-linked GlcNAc) α1-6-fucosyltransferase (Fuc-T). These membranes are enriched 30-fold in Fuc-T relative to homogenate. Glycopeptide MM is shown in Figure 1. Structures of glycopeptides GnGn, GG, and GnGn(Gn) are shown in Figures 2 and 7. Glycopeptides MGn and GnG are shown in Figure 7. GGn is similar to GnG except that the Gal residue is on the Manα1-6 arm instead of the Manα1-3 arm. SS is fully sialylated bi-antennary complex glycopeptide with both sialyl residues linked α2-6. The glycopeptides are named according to the terminal sugar residues. M, Man; Gn, GlcNAc; G, Gal; S, sialyl; (Gn), bisecting GlcNAc. The sugar on the Manα1-6 arm is named first.

[b] Not done.

[c] Based on work with rat liver Golgi membranes which lack fucosyltransferase activities towards Galβ1-4GlcNAc-terminated acceptors (Munro *et al.*, 1975).

this prediction has thus far not been violated. Fucose is probably added to the core at the MGn and GnGn stages of synthesis (Figure 7). All the complex oligosaccharides in some glycoproteins have a core fucose residue, e.g., immunoglobulins, while the same oligosaccharides in other glycoproteins are totally devoid of core fucose residues, e.g., α_1-acid glycoprotein and transferrin. The reasons for these differences are not clear but may reflect the availability of the oligosaccharides to the α6-fucosyltransferase.

H. GDP-Fuc:β-Galactoside α1-2,α1-3, and α1-4-Fucosyltransferases

There are at least three separate fucosyltransferases acting on acceptors with a β-galactoside nonreducing terminus (Prieels *et al.*, 1981; Johnson *et al.*, 1981; Schachter and Roseman, 1980; Beyer *et al.*, 1981; Watkins, 1980): (1) the human blood group gene *H*-dependent GDP-L-Fuc:β-galactoside α1-2-fucosyltransferase, (2) the human blood group gene *Le*(Lewis)-dependent GDP-L-Fuc:Galβ1-3GlcNAc-R (Fuc to GlcNAc) α1-4-fucosyltransferase, and (3) GDP-L-Fuc:Galβ1-4GlcNAc-R (Fuc to GlcNAc) α1-3-fucosyltransferase. All three enzymes have an absolute requirement for a terminal Gal residue and are therefore different from the α1-6-fucosyltransferase discussed in Section G above.

The blood group H enzyme attaches Fuc in α1-2 linkage to the terminal Gal residue of any β-D-galactoside irrespective of the subterminal sugars and, indeed, the blood group H determinant occurs on both Type 1 and Type 2 chains (Table 4) and on the Galβ1-3GalNAc sequence. Phenyl β-D-galactoside is a useful specific substrate for the enzyme since it is inactive with the other known fucosyltransferases. Structures like Galβ1-3(Fucα1-4)GlcNAc-R or Galβ1-4(Fucα1-3)GlcNAc-*R* are not acceptors since the second fucose residue inhibits enzyme action. The enzyme occurs in the serum and bone marrow of all humans with the *H* gene irrespective of secretor status. The enzyme also occurs in human milk, submaxillary glands, and gastric mucosa but only in individuals who are secretors, i.e., have the *Se* gene. A similar enzyme occurs in other species, e.g., dog, pig, rat, and cow. Such an enzyme has been purified 124,000-fold to homogeneity from pig submaxillary glands (Beyer *et al.*, 1981).

The blood group Lewis enzyme attaches Fuc in α1-4 linkage to the subterminal GlcNAc of Galβ1-3GlcNAc-R (Type 1) structures (Table 4). The enzyme has been found in the submaxillary glands, gastric mucosa, and milk of individuals with the Lewis (*Le*) gene. The enzyme is not present in human serum irrespective of the Lewis genotype. The best substrates for specific assay of the enzyme are milk oligosaccharides like lacto-*N*-fucopentaose I (Fucα1-2Galβ1-3GlcNAcβ1-3Galβ1-4Glc) which are not acceptors for the other known fucosyltransferases. The enzyme has been purified over

Table 4. Antigenic Determinants

Structure of determinant	Name of antigen
Type 1[a]	
Fucαl-2Galβ1-3GlcNAc-	Blood group H, Type 1 (Le^{dH})
GalNAcα1-3(Fucα1-2)Galβ1-3GlcNAc-	Blood group A, Type 1 (ALe^{d})
Galα1-3(Fucα1-2)Galβ1-3GlcNAc-	Blood group B, Type 1 (BLe^{d})
Galβ1-3(Fucα1-4)GlcNAc-	Blood group Lewis[a] (Le^{a})
Fucα1-2Galβ1-3(Fucα1-4)GlcNAc-	Blood group Lewis[b] (Le^{bH})
Type 2[b]	
Fucα1-2Galβ1-4GlcNAc-	Blood group H, Type 2
GalNAcα1-3(Fucα1-2)Galβ1-4GlcNAc-	Blood group A, Type 2
Galα1-3(Fucα1-2)Galβ1-4GlcNAc-	Blood group B, Type 2
Galβ1-4(Fucα1-3)GlcNAc-	Determinant *X*, Stage-Specific Embryonic Antigen-1 (SSEA-1)
Fucα1-2Galβ1-4(Fucα1-3)GlcNAc-	Determinant *Y*
Polylactosaminoglycan	
Galβ1-4GlcNAcβ1-3Galβ1-4GlcNAcβ1-3-R	Blood group i
Galβ1-4GlcNAcβ1-6 \ Galβ1-4GlcNAc-R / Galβ1-4GlcNAcβ1-3	Blood group I

[a] Determinants containing the sequence Galβ1-3GlcNAc are called Type 1.
[b] Determinants containing the sequence Galβ1-4GlcNAc are called Type 2.

500,000-fold to homogeneity from human milk (Prieels *et al.*, 1981; Beyer *et al.*, 1981). The highly purified enzyme carries out all three of the following reactions:

GDP-Fuc + Galβ1-3GlcNAc-R → Galβ1-3(Fucα1-4)GlcNAc-R + GDP

GDP-Fuc + Galβ1-4GlcNAc-R → Galβ1-4(Fucα1-3)GlcNAc-*R* + GDP

GDP-Fuc + Galβ1-4Glc → Galβ1-4(Fucα1-3)Glc + GDP

The kinetic parameters for these substrates are shown in Table 2. The α1-3- and α1-4-fucosyltransferase activities copurify, are activated to the same extent by various cations, and are inactivated at identical rates by various procedures. Kinetic analysis indicates that a single active site is involved in both α1-3- and α1-4-Fuc incorporation. This is the first example of a single glycosyltransferase catalyzing sugar attachment to two different carbon atoms at appreciable rates, and is therefore the first example of an exception to the "one linkage–one glycosyltransferase" rule. Enzyme activity is enhanced by the presence of a Fucα1-2 residue on the terminal Gal of a substrate. The enzyme is inactive towards substrates with the structure Galβ1-3GalNAc-R commonly found in *O*-glycosyl oligosaccharides.

It has recently been demonstrated (Johnson *et al.*, 1981) that saliva from individuals with the *Le* gene can catalyze the three reactions shown above, in agreement with the finding that the purified *Le*-dependent fucosyltransferase catalyzes these reactions. Saliva from *Le*-negative individuals, however, can only catalyze the attachment of Fuc in α1-3 linkage to GlcNAc. There is, therefore, a third enzyme, GDP-L-Fuc:Galβ1-4GlcNAc (Fuc to GlcNAc) α1-3-fucosyltransferase, which is different from both the *H*- and *Le*-dependent fucosyltransferases. This enzyme and the *Le*-dependent enzyme can, in fact, be separated by isoelectric focusing (Johnson *et al.*, 1981) of human saliva from *Le*-positive individuals. Also, human serum contains the non-*Le*-dependent α1-3-fucosyltransferase but lacks completely the *Le*-dependent α1-3/4-fucosyltransferase even in *Le*-positive individuals. Neither the serum enzyme nor the *Le*-negative salivary enzyme can add Fuc in α1-4 linkage to Galβ1-3GlcNAc or in α1-3 linkage to Galβ1-4Glc.

Pig liver contains a GDP-L-Fuc:β-galactoside fucosyltransferase which is highly active towards Galβ1-4GlcNAc but not towards Galβ1-3GlcNAc, Galβ1-6GlcNAc, or Galβ1-4Glc (Jabbal and Schachter, 1971). It thus resembles the human non-*Le*-dependent α1-3-fucosyltransferase in substrate specificity. The structure Galβ1-4(Fucα1-3)GlcNAc occurs in complex *N*-glycosyl oligosaccharides, e.g., in human serum α_1-acid glycoprotein (Fournet *et al.*, 1978), and is probably synthesized by the non-*Le*-dependent α1-3-fucosyltransferase. This enzyme has not as yet been purified.

Table 4 lists some of the antigenic determinants which contain fucose. These determinants occur at the nonreducing ends of *N*-glycosyl oligosaccharides of the poly-*N*-acetyllactosaminoglycan type (Figure 2 and Table 4), *O*-glycosyl oligosaccharides, and glycosphingolipids. Some of these determinants have been attached to the nonreducing ends of complex *N*-glycosyl oligosaccharides by *in vitro* enzymatic methods (Beyer *et al.*, 1981).

V. SYNTHESIS OF BISECTED OLIGOSACCHARIDES

Bisected *N*-glycosyl oligosaccharides occur in both complex and hybrid forms (Figure 2). Bisected complex oligosaccharides (containing three Man residues) are relatively uncommon but have been found in human immunoglobulins, hen ovotransferrin and ovomucoid, human erythrocyte glycophorin A and band 3, and in a few other glycoproteins (Narasimhan, 1982). Such structures cannot be made by the incorporation of a bisecting GlcNAc residue (GlcNAc linked β1-4 to the β-linked Man residue of the core) at either the 5-Man or 4-Man stages since mannosidase II cannot act if a bisecting GlcNAc is present (Figure 6). A third point of entry into bisected oligosaccharides at the 3-Man stage was therefore predicted by Harpaz and Schachter (1980b) and has recently been established by Narasimhan (1982).

A. UDP-GlcNAc:GnGn (GlcNAc to Manβ1-4) β1-4-GlcNAc-Transferase III

GnGn, the product of GlcNAc-transferase II, sits at an important crossroads in the synthetic pathway since it is a substrate for at least four glycosyltransferases (Figure 7). One of these enzymes is UDP-GlcNAc:GnGn (GlcNAc to Manβ1-4) β1-4-GlcNAc-transferase III, the action of which leads towards bisected bi-antennary oligosaccharides (Figure 7). The enzyme has recently been reported in hen oviduct membranes (Narasimhan *et al.*, 1981; Narasimhan, 1982). It attaches a bisecting GlcNAc residue to glycopeptide GnGn. A similar reaction (discussed in Section V-B) also occurs in hen oviduct membranes at the 5-Man, and possibly the 4-Man stages (Figure 6), but it is not yet known whether these various activities are catalyzed by the same enzyme. GlcNAc-transferase III has been partially purified from hen oviduct (G. Vella and H. Schachter, unpublished data). The product of the reaction has been identified as GnGn(Gn) (Figures 2 and 7) by methylation analysis and high-resolution proton NMR spectrometry at 360 MHz (Narasimhan, 1982).

The substrate specificity of GlcNAc-transferase III is shown in Table 3. The activity of crude hen oviduct membranes towards glycopeptide GnGn is about 85 μunits/mg protein under standard assay conditions. GnGn preparations both with and without a Fucα1-6 residue attached to the asparagine-linked GlcNAc give the same activity. Activities with MGn (Figure 7) and MM (Figure 1) relative to GnGn are less than 15% and 8%, respectively, at pH 5.7. Activity with glycopeptide GGn (defined in Table 3) is about 60% of that with GnGn. However, the enzyme is essentially inactive with glycopeptide GnG and with the other glycopeptides tested.

The substrate specificity studies shown in Table 3 indicate that GlcNAc-transferase III has a narrow window of action at the 3-Man stage (Figure 7). MGn is a relatively poor substrate, GGn is probably not formed in large amounts *in vivo* (see Section IV-D) and GnG, GG, and the sialylated compounds are not substrates. Thus, GnGn is the major entry point for bisected bi-antennary complex oligosaccharides. GlcNAc-transferase IV has a similarly narrow window of action (see Section VI and Table 3) and therefore GnGn is also the main point of entry for both bisected and nonbisected tri-antennary oligosaccharides (Figure 7).

B. Hybrid Oligosaccharide Synthesis

Hybrid oligosaccharides can occur either in a bisected or nonbisected form (Figure 2). It is not clear why a glycoprotein like rhodopsin has nonbisected hybrid structures (Liang *et al.*, 1979; Fukuda *et al.*, 1979c). The scheme in Figure 6 suggests that there may be interference with either α-mannosidase II or GlcNAc-transferase II. Ovalbumin is rich in bisected hybrid oligosaccharides (Tai *et al.*, 1977). Hen oviduct membranes are very rich in GlcNAc-transferase III and relatively deficient in GlcNAc-transferase I-dependent α3/6-mannosidase (mannosidase II) (unpublished data). Narasimhan *et al.* (1981) have shown that GlcNAc-transferase III′ can incorporate a bisecting GlcNAc into Gn(I)M_5 (Figure 6). Harpaz and Schachter (1980b) have observed that α-mannosidase II does not act if the substrate has a bisecting GlcNAc residue. These observations suggest strongly that hen oviduct membranes shunt the synthetic pathway from Gn(I)M_5 to bisected bi- and tri-antennary hybrid structures (Figure 6). Once GlcNAc-transferase III′ has acted, further Man removal cannot occur. The only available paths are further elongation by Gal- and sialyl-transferases. Since the presence of a bisecting GlcNAc prevents GlcNAc-transferase IV action (Table 3), it is suggested that bisected bi-antennary hybrids result if GlcNAc-transferase III′ acts before GlcNAc-transferase IV, and bisected tri-antennary hybrids result if transferase IV acts before transferase III′ (Figure 6). The existence of hybrids with only four Man residues suggests that GlcNAc-transferases III and IV can act at both the 5-Man and 4-Man stages but the 4-Man compounds have not been tested with these enzymes.

C. Elongation Reactions on Bisected and Hybrid Oligosaccharides

Bisected complex oligosaccharides have been isolated with incomplete antennae consisting of a single GlcNAc residue, e.g., from ovotransferrin, or with complete antennae (sialyl-Gal-GlcNAc) and a Fuc-linked α1-6 to the asparagine-linked GlcNAc, e.g., from immunoglobulins. The α1-6-fucosyltransferase (Section IV-G) cannot act on bisected oligosaccharides so that Fuc incorporation must precede action of GlcNAc-transferase III. We have shown (unpublished data) that bovine milk UDP-Gal:GlcNAc β1-4-Gal-transferase can introduce a Gal residue into GnGn(Gn), but a systematic study of bisected structures as substrates for the Gal-, Fuc-, and sialyl-transferases discussed in Sections IV-D–IV-H remains to be done.

Hybrid oligosaccharides are usually isolated with incomplete antennae (GlcNAc or Gal-GlcNAc) but sialylated hybrid structures have been reported recently. Elongation reactions with hybrid substrates have not been studied *in vitro*. The α6-fucosyltransferase discussed in Section IV-G above has been shown to act on Gn(I)M_5 (Longmore and Schachter, 1982). This suggests that fucosylated hybrids are theoretically possible but such structures have not yet been described.

VI. SYNTHESIS OF TRI-ANTENNARY N-GLYCOSYL OLIGOSACCHARIDES

A. UDP-GlcNAc:GnGn (GlcNAc to Manα1-3) β1-4-GlcNAc-Transferase IV

An enzyme, UDP-GlcNAc:GnGn (GlcNAc to Manα1-3) β1-4-GlcNAc-transferase IV, has recently been described in hen oviduct membranes (Gleeson *et al.*, 1982; Gleeson and Schachter, 1983) which catalyzes the attachment of a GlcNAc residue in β1-4 linkage to the Manα1-3 arm of glycopeptide GnGn thereby initiating the third antenna or branch (see Figure 7 for the reaction catalyzed). The enzyme therefore routes the synthetic pathway towards tri-antennary, bisected tri-antennary, and tetra-antennary complex oligosaccharides.

The assay of both GlcNAc-transferases III and IV is complicated by the fact that they act on a common substrate, GnGn (Figure 7). It is therefore necessary to separate the products [GnGn(Gn) and GnGnGn, respectively] when crude enzyme sources containing both activities are being assayed. This can be accomplished by either concanavalin A (Con A)–Sepharose or pea lectin–agarose chromatography. GnGn adheres to Con A–Sepharose, and GlcNAc-transferase III product [GnGn(Gn)] is retarded on the column while any product with a GlcNAc linked β1-4 or β1-6 to either the Manα1-3 or Manα1-6 arm passes through Con A–Sepharose unretarded. The products of GlcNAc-transferases IV, V, VI, or VII (see Figure 3 for nomenclature) will therefore all pass through the lectin column unretarded. The unretarded product produced by hen oviduct has been identified as GnGnGn (Figure 2) containing GlcNAc residues I, II, and IV (Figure 3). Identification was by methylation analysis and high resolution proton NMR spectrometry at 360 MHz. Small amounts of product containing GlcNAc in β1-4 linkage to Manα1-6 or in β1-6 linkage to either Manα1-3 or Manα1-6 may be produced but have not been detected. Con A–Sepharose columns can therefore be used to identify GlcNAc-transferase III product, but the linkage position of GlcNAc in unretarded product requires further identification work such as methylation analysis, NMR, and chromatography on pea lectin–Sepharose or lentil lectin–Sepharose columns (Cummings *et al.*, 1982).

The substrate specificity of GlcNAc-transferase IV is shown in Figures 6 and 7 and in Table 3; it is very similar to that of GlcNAc-transferase III. Substrates are Gn(I)M_5, GnGn (with or without a Fucα1-6 attached to the asparagine-linked GlcNAc), and GGn. MGn is a poor substrate and MM, GnG, GG, and SS are ineffective. An interesting finding is that the presence of a bisecting GlcNAc on glycopeptide GnGn completely inhibits GlcNAc-transferase IV action (Table 3). Thus, GlcNAc-transferase IV must act before GlcNAc-transferase III to form bisected tri-antennary complex oligosaccharides (Figure 7). This sequence is probably also required for the synthesis of bisected tri-antennary hybrid oligosaccharides (Figure 6). It should be pointed out that the ability of M_5 to act as a substrate for either GlcNAc-transferase III or IV has not yet been tested due to the difficulties involved using crude enzyme preparations. More detailed substrate specificity studies must await the availability of pure enzymes.

Substrate specificity studies on crude enzyme preparations (Table 3) are often difficult and subject to artefacts. For example, not only is GnGn a substrate for both GlcNAc-transferases III and IV but the presence of β-*N*-acetylglucosaminidase in crude preparations can convert GnGn to MGn and MM, respective substrates for GlcNAc-transferases II and I. This problem was approached in two ways. First, a β-*N*-acetylglucosaminidase inhibitor, e.g., GlcNAc, was always present in the incubation mixture. Second, the radioactive products of GlcNAc-transferases I and II (MGn and GnGn, respectively) adhere to Con A–Sepharose and are thus separated from the products of GlcNAc-transferases III and IV. Under standard assay conditions, no more than about 2% of the radioactivity adheres to the lectin column. This is due to a relatively low rate of GnGn breakdown and to a relatively low level of GlcNAc-transferase I in hen oviduct membranes (G. Vella and H. Schachter, unpublished). However, some tissues, e.g., the mononuclear fraction from normal human white cells, are very rich in glycosidases, and the standard assay produces a very large radioactive peak adhering to Con A–Sepharose. This destruction of GnGn may interfere with the accurate assay of GlcNAc-transferases III and IV.

Table 3 shows that GnG is a very poor substrate for either GlcNAc-transferase III or IV, whereas GGn is a good substrate for both enzymes. Since it is believed that the GlcNAcβ1-2Manα1-3 arm must be sterically available for both GlcNAc-transferase III and IV action, it is perhaps not surprising that substitution of the GlcNAcβ1-2Manα1-3 arm by a Gal residue to give GnG results in an inactive substrate, whereas a Gal residue on the GlcNAcβ1-2Manα1-6 arm (GGn) has no effect. Recent data (Paquet *et al.*, 1984) have shown that UDP-Gal:GlcNAc β4-Gal-transferase prefers the GlcNAc on the Manα1-3 arm of the core to form GnG (Figure 7) suggesting that, *in vivo*, the incorporation of even a single Gal residue into GnGn prevents further branching by either GlcNAc-transferase III or IV. Thus, the action of the β4-Gal-transferase is the first committed step towards synthesis of nonbisected bi-antennary oligosaccharides.

Since GGn is a good substrate for both transferases III and IV (Table 3), it is possible that the relatively small amount of GGn that is formed is an alternate and probably minor path towards bisected bi-antennary and highly-branched oligosaccharides.

B. UDP-GlcNAc:GnGn (GlcNAc to Man) β1-6-GlcNAc-Transferase V

An enzyme activity has been described in a mouse lymphoma cell line (Cummings *et al.*, 1982) which appears to incorporate GlcNAc in β1-6 linkage into the subterminal Man of either the Manα1-3 or Manα1-6 arm of glycopeptide GnGn. The activity was of very low level and product identification was based on methylation analyses of small amounts of radioactive product and on behavior on various lectin columns. A lectin-resistant cell line was isolated which lacked this enzyme activity. It is interesting that GlcNAc-transferase V was unable to act on either GG (Figure 7) or fully sialylated bi-antennary complex oligosaccharide (SS), a substrate specificity identical to that of GlcNAc-transferases III and IV (Table 3).

C. Elongation of Tri- and Tetra-Antennary Oligosaccharides

Tri- and tetra-antennary complex *N*-glycosyl oligosaccharides have been isolated which carry complete antennae (sialyl-Gal-GlcNAc), e.g., from α_1-acid glycoprotein (Fournet *et al.*, 1978). CMP-NeuNAc:Galβ1-4GlcNAc α2-6-sialyltransferase from bovine colostrum (Section IV-E) has been shown to act on tri-antennary glycopeptide (Van den Eijnden *et al.*, 1980), but a detailed study of the action of the elongation glycosyltransferases on tri- and tetra-antennary oligosaccharides has not been carried out.

VII. Ser(Thr)-GalNAc OLIGOSACCHARIDE STRUCTURES

Structural information is an essential prerequisite to an understanding of oligosaccharide biosynthesis. The sugars commonly found in mucin-type oligosaccharides are GalNAc, Gal, GlcNAc, sialic acid, and fucose, but mannose and glucose have also been reported (Dutta and Rao, 1982). Gal, GlcNAc, and GalNAc may occur in a sulfated form and thus contribute with sialic acid to the acidity of some mucin-type oligosaccharides. It is convenient to classify *O*-glycosyl oligosaccharides according to their core structures (Schachter and Williams, 1982; Hounsell *et al.*, 1982).

1. Monosaccharide Chain

The simplest oligosaccharide of the Ser(Thr)-GalNAc class is obviously GalNAc. GalNAc is found in ovine submaxillary mucin (OSM) and in other salivary mucins.

2. Disaccharide Chains

Four types of Ser(Thr)-GalNAc disaccharide chains have been described to date: (1) sialylα2-6GalNAc-*X*, (2) Galβ1-3GalNAc-*X*, (3) GlcNAcβ1-3GalNAc-*X*, and (4) GalNAcα1-3GalNAc-*X*.

3. Chains with More than Two Residues

These chains vary in size from trisaccharides to chains with 18 or more residues. The larger chains contain three distinct regions, a core, a backbone, and a nonreducing terminus.

a. Core Classes. There are at least four established core classes of the Ser(Thr)-GalNAc type although preliminary reports indicate that other classes may also occur. We have named these classes as follows (Schachter and Williams, 1982): (1) core class 1: Galβ1-3GalNAc-*X*, (2) core class 2: Galβ1-3(GlcNAcβ1-6)GalNAc-*X*, (3) core class 3: GlcNAcβ1-3GalNAc-*X*, (4) core class 4: GlcNAcβ1-3(GlcNAcβ1-6)GalNAc-*X*.

b. Backbone Structures. The backbone of the Ser(Thr)-GalNAc oligosaccharide is formed by elongation of the core. Elongation usually involves addition of Gal and GlcNAc residues in β-linkages. The following moieties (Table 4) are commonly found in the backbone structure (Hounsell *et al.*, 1982; Feizi, 1982): Galβ1-3GlcNAc, usually referred to as a Type 1 structure; Galβ1-4GlcNAc, usually referred to as a Type 2 structure; and linear and branched sequences of repeating Type 2 disaccharides, representing respectively the i and I antigenic determinants. Other branched structures have also been described as backbone components in Ser(Thr)-GalNAc oligosaccharides (Hounsell *et al.*, 1982).

c. Nonreducing Termini. The larger Ser(Thr)-GalNAc oligosaccharides are frequently terminated by antigenic determinants such as the human blood group ABH, Le^a and Le^b determinants, and the mouse stage-specific embryonic antigen SSEA-1 (Table 4). These determinants are all due to α-linked sugars. Other α-linked sugars have also been found in terminal nonreducing positions, e.g., GlcNAcα1-4. Feizi (1982) has pointed out that the incorporation of a branch into the blood group i determinant is essential for blood group I activity but blocks expression of blood group i activity. Similarly, the presence of the H determinant blocks expression of determinants i, I, and SSEA-1, and the presence of either the blood group A or B determinant blocks the H determinant. Thus, internal antigenic determinants are blocked as new antigenic determinants are added to the growing oligosaccharide chain.

VIII. INITIATION OF Ser(Thr)-GalNAc OLIGOSACCHARIDE SYNTHESIS

All Ser(Thr)-GalNAc oligosaccharides appear to be initiated by the incorporation of a single GalNAc residue from UDP-GalNAc into the polypeptide chain. This mechanism of initiation differs markedly from the complex process required for synthesis of the Asn-GlcNAc linkage (Section III) which involves transfer of oligosaccharide from dolichol pyrophosphate oligosaccharide to the polypeptide chain. Further, *N*-glycosylation by the oligosaccharide transferase occurs in the rough endoplasmic reticulum while *O*-glycosylation appears to occur in smooth membranes (probably the Golgi apparatus).

UDP-GalNAc:polypeptide α-*N*-acetylgalactosaminyltransferase (polypeptide GalNAc-transferase) was first described by McGuire and Roseman (1967) in sheep submaxillary glands. It has subsequently been found in many other tissues (Schachter, 1978) and has been purified (48,100-fold, 4% yield) from ascites hepatoma AH 66 cells (Sugiura *et al.*, 1982). Purification was achieved by affinity chromatography on a column of carbohydrate-free bovine submaxillary apomucin coupled to cyanogen bromide-activated Sepharose 4B.

The ability of various compounds to accept GalNAc has been studied by many laboratories with both crude and purified enzyme preparations (McGuire and Roseman, 1967; Schachter, 1978; Hill *et al.*, 1977b; Hagopian *et al.*, 1971; Young *et al.*, 1979; Briand *et al.*, 1981; Sugiura *et al.*, 1982). Effective acceptors are apomucins made from ovine or bovine submaxillary glands, κ-casein, bovine myelin A1 protein (basic myelin protein), and carbohydrate-free antifreeze glycoprotein from Antarctic fish. Native mucins, asialo-mucins, or pronase-digested apomucins are ineffective acceptors. Large tryptic peptides from apomucin were found to be effective acceptors (Hill *et al.*, 1977b), but no unique primary sequences were identified adjacent to the *O*-glycosidically-substituted Ser and Thr residues. More recently (Young *et al.*, 1979; Briand *et al.*, 1981), synthetic polypeptides have been tested as acceptors for the GalNAc-transferase. The smallest effective peptide is Thr-Pro-Pro-Pro with a blocked amino terminus, and all peptides with this sequence are effective acceptors. At least three Pro residues are essential. It is interesting that 20 of the 28 amino acid sequences adjacent to *O*-glycosidically-substituted Ser or Thr residues in ovine submaxillary mucin have at least one Pro residue within 4 residues of the hydroxylamino acid (Hill *et al.*, 1977a,b). The string of Pro residues apparently required for polypeptide GalNAc-transferase action may either be adjacent to the *O*-glycosidically-substituted amino acid in the primary sequence, or one or more Pro residues may be brought into the region of the enzyme active site by three-dimensional factors.

Hanover *et al.* (1980) found that the action of polypeptide GalNAc-transferase from both porcine submaxillary glands and hen oviduct is not mediated by a dolichol intermediate. Babczinski (1980) similarly concluded that bovine submaxillary gland membrane preparations could transfer GalNAc to endogenous polypeptide without the formation of GalNAc-containing lipid intermediates.

Autoradiographic evidence has indicated that the Golgi apparatus is the primary site of *O*-glycosylation. Various attempts have also been made to determine the subcellular location of the polypeptide GalNAc-transferase. The evidence presently available indicates that the bulk of *O*-glycosylation occurs after release of nascent peptide from the polyribosomes and when the peptide has moved to the smooth endoplasmic reticulum–Golgi apparatus region of the cell (Kim *et al.*, 1971; Ko and Raghupathy, 1972; Hanover *et al.*, 1980; Hagopian *et al.*, 1968).

IX. SYNTHESIS OF Ser(Thr)-GalNAc OLIGOSACCHARIDES

The following discussion will deal with the glycosyltransferases that have been shown to be involved in Ser(Thr)-GalNAc oligosaccharide synthesis. Hypothetical synthesis schemes will also be presented, based on the *O*-glycosyl oligosaccharide structures that have been reported in the literature to date.

A. Assembly of Oligosaccharides with Core Class 1

1. UDP-Gal:GalNAc-X β3-Galactosyltransferase: Synthesis of the Core

The enzyme which routes the biosynthetic pathway towards synthesis of core class 1 oligosaccharides is UDP-Gal:GalNAc-*X* β3-galactosyltransferase. The enzyme catalyzes the following reaction:

$$\text{UDP-Gal} + \text{GalNAc-}X \rightarrow \text{Gal}\beta1\text{-3GalNAc-}X + \text{UDP}$$

The β3-Gal-transferase was first described in porcine and ovine submaxillary glands (Schachter *et al.*, 1971), and has since been reported in rat pancreas, liver, stomach and intestine, human erythrocytes, platelets, serum and trachea, porcine Cowper's gland, and canine trachea (Schachter, 1978; Van den Eijnden *et al.*, 1979a; Carlson *et al.*, 1973a; Shier and Roloson, 1977; Andersson and Eriksson, 1979, 1980, 1981; Hesford *et al.*, 1981; Hesford and Berger, 1981; Berger and Kozdrowski, 1978; Cheng and Bona, 1982; Mendicino *et al.*, 1982; Cartron and Nurden, 1979; Strous *et al.*, 1980; Baker and Munro, 1971; Ronzio, 1973a,b). The enzyme is membrane-bound and requires detergent and Mn^{2+} for activity.

The enzyme has recently been purified (2000-fold, 20% yield) from detergent extracts of swine trachea (Mendicino *et al.*, 1982). The authors prepared an affinity column by using ionic forces to attach asialo Cowper's gland mucin to DEAE-cellulose. The β3-Gal-transferase adhered to the affinity column in the presence of 10 mM Mg^{2+} and 1 mM UMP and was eluted with EDTA. The most effective acceptors are mucins containing terminal GalNAc residues, e.g., asialo ovine submaxillary, asialo porcine submaxillary, or asialo porcine Cowper's gland mucins. The enzyme can transfer Gal to free GalNAc or to glycosides of GalNAc, but the K_m for these acceptors (100–180 mM) is much higher than for mucins (90 μM for the pure enzyme from pig trachea). Mucins in which the Ser(Thr)-linked GalNAc residue is substituted by one or more sugar residues and glycoproteins lacking the Ser(Thr)-GalNAc moiety are not effective as acceptors. It is interesting to note that substitution of the Ser(Thr)-linked GalNAc by a sialylα2-6 residue, e.g., native ovine submaxillary, porcine submaxillary, or porcine Cowper's gland mucins, prevents enzyme action. This serves as an important control in the biosynthesis of mucins (Schachter *et al.*, 1971; McGuire, 1970). If the CMP-NAN:GalNAc-*X* α6-sialyltransferase (see below) acts first, the synthetic pathway stops at the disaccharide sialyl-α2-6-GalNAc-*X*. Mucins rich in this disaccharide (such as ovine submaxillary mucin or porcine Cowper's gland mucin) are made in tissues which have a high α6-sialyltransferase relative to β3-Gal-transferase. On the other hand, mucins which have larger oligosaccharides with core class 1 (such as porcine submaxillary mucin) are made in tissues in which the β3-Gal-transferase level is high relative to the α6-sialyltransferase level. Prior action of the β3-Gal-transferase does not prevent subsequent action of the α6-sialyltransferase.

The β3-Gal-transferase acts not only on mucins but is required for the synthesis of core class 1 oligosaccharides on glycoproteins such as human erythrocyte glycophorin A, human IgA, calf fetuin, and Antarctic fish antifreeze glycoprotein. Derivatives of antifreeze glycoprotein (Shier and Roloson, 1977) and glycophorin (Hesford *et al.*, 1981) do in fact act as acceptors for the β3-Gal-transferase.

The subcellular location of the β3-Gal-transferase has been studied in rat pancreas (Ronzio, 1973a,b), rat liver (Andersson and Eriksson, 1979, 1980, 1981), and porcine Cowper's gland (Mendicino *et al.*, 1982). The consensus is that the enzyme is greatly enriched in the Golgi apparatus.

2. *Elongation of the Core: Synthesis of the Blood Group Ii Backbone*

When the Galβ1-3GalNAc core is synthesized (Figure 8), elongation can proceed along a variety of different pathways (Figure 9). Elongation almost always involves

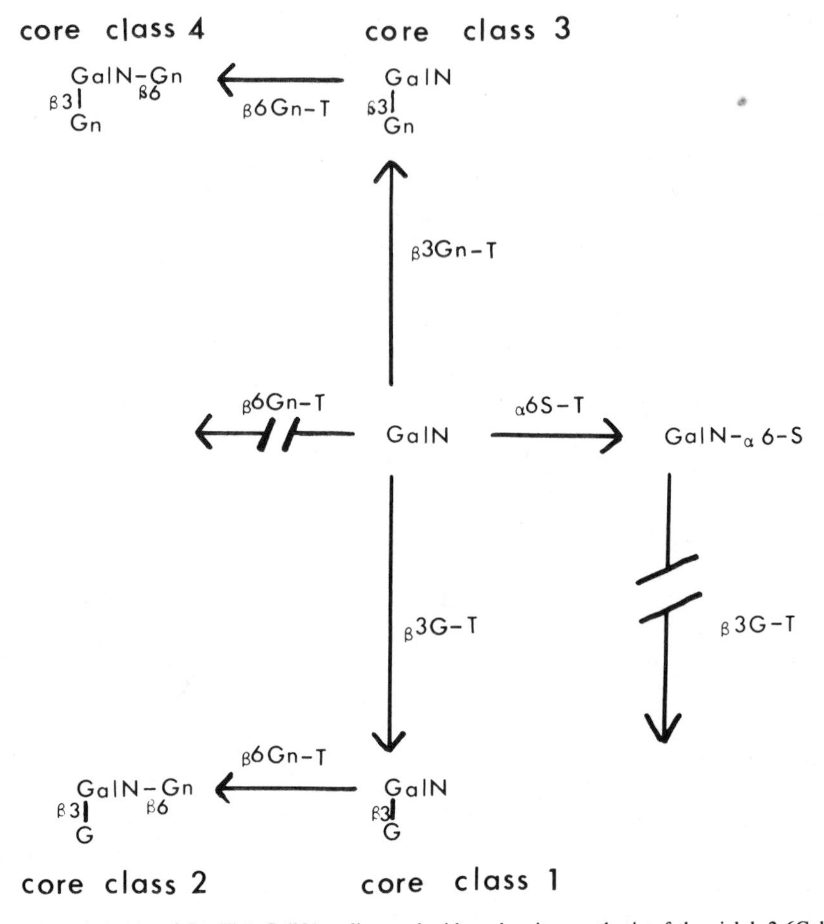

Figure 8. Assembly of Ser(Thr)-GalNAc oligosaccharides, showing synthesis of the sialylα2-6GalNAc disaccharide and of the four core classes (see text). Abbreviations: GalN, Ser(Thr)-linked GalNAc residue; S, sialic acid; G, D-galactose (Gal); Gn, *N*-acetyl-D-glucosamine (GlcNAc); T, transferase. All the reactions in the figure have been detected in cell-free preparations. References to the structural data and abbreviations for the various mucins and glycoproteins are as follows: human ovarian cyst mucins, HOCM (Rovis *et al.*, 1973; Wu *et al.*, 1982; Tanaka *et al.*, 1982), human bronchial mucin, HBM (Van Halbeek *et al.*, 1982a), human salivary mucin, HSM (Reddy *et al.*, 1979, 1982), human gastric mucin, HGM (Oates *et al.*, 1974), human cervical mucin (Yurewicz and Moghissi, 1981; Yurewicz *et al.*, 1982a,b), human melanoma plasma membrane (Umemoto *et al.*, 1981), porcine submaxillary mucin, PSM (Carlson and Blackwell, 1968; Van Halbeek *et al.*, 1981a), porcine gastric mucin, PGM (Van Halbeek *et al.*, 1982b), porcine Cowper's gland mucin (Mendicino *et al.*, 1982), ovine submaxillary mucin, OSM (Hill *et al.*, 1977a,b), ovine gastric mucin, OGM (Hounsell *et al.*, 1980, 1981; Wood *et al.*, 1980, 1981), equine gastric mucin, EGM (Newman and Kabat, 1976), canine submaxillary mucin, CSM (Lombart and Winzler, 1974), canine gastric mucin, CGM (Slomiany *et al.*, 1982), bovine submaxillary mucin, BSM (Herp *et al.*, 1979), goat submaxillary mucin (Dutta and Rao, 1982), armadillo submandibular mucin (Wu *et al.*, 1979), rat salivary mucins, RSM (Tabak *et al.*, 1982; Slomiany and Slomiany, 1978), rat small intestinal mucin (Carlsson *et al.*, 1978), rat colonic mucin (Slomiany *et al.*, 1980), monkey cervical mucin (Hatcher *et al.*, 1977), human glycophorin (Lisowska *et al.*, 1980; Thomas and Winzler, 1969), bovine glycophorin (Fukuda *et al.*, 1981, 1982), equine glycophorin (Fukuda *et al.*, 1980), porcine glycophorin (Kawashima *et al.*, 1982), human plasminogen (Hayes and Castellino, 1979), fetuin (Spiro and Bhoyroo, 1974), immunoglobulin A, IgA (Pierce-

the addition of Gal and GlcNAc residues in β1-3, β1-4, and β1-6 linkages. The repeating sequence (Galβ1-4GlcNAcβ1-3-)$_n$ often serves in *O*-glycosyl oligosaccharides as an intermediary sequence between core and nonreducing terminus. This repeating sequence is also found in some glycosphingolipids and *N*-glycosyl oligosaccharides, and has been identified as the antigenic determinant for an interesting developmentally regulated blood group antigen on human erythrocytes named i (Feizi, 1982). GlcNAc residues can be inserted in β1-6 linkage to a Gal residue in this repeating sequence to create a branched structure Galβ1-4GlcNAcβ1-3(Galβ1-4GlcNAcβ1-6)Gal. This branched structure has been associated with human blood group I antigenic activity (Feizi, 1982). Elongation of the core can lead to linear repeating sequences characteristic of the human blood group i, or to various branched structures reactive with antibodies to the human blood group I, or to other structures (Table 4 and Figure 9). Most of the enzymes involved in elongation have not yet been demonstrated *in vitro* and the scheme shown in Figure 9 is based primarily on structural data in the literature.

For example, GlcNAc addition in β1-3 linkage to the core to form GlcNAcβ1-3Galβ1-3GalNAc-*X* is indicated in Figure 9 mainly because formation of this linkage must occur during synthesis of bovine and porcine glycophorin, porcine and ovine gastric mucin, monkey cervical mucin, epiglycanin, human ovarian cyst mucin, bronchial mucin, and IgA. *In vitro* data to support this enzymatic step have appeared in the literature but are very preliminary. Ziderman *et al.* (1967) showed the ability of rabbit, human, and baboon stomach mucosal extracts to catalyze the transfer of GlcNAc from UDP-GlcNAc to β-methylgalactoside and β-Gal-terminal disaccharides such as Galβ1-4GlcNAc. The linkage was identified as a β-linkage but was not further characterized. Preliminary data from our laboratory indicate that porcine gastric membrane preparations can attach GlcNAc in β1-3 linkage to the Gal residue of Galβ1-3GalNAc-mucin. Several reports have appeared suggesting the existence of a UDP-GlcNAc:Galβ1-4GlcNAc-*R* β1-3-GlcNAc-transferase (Rearick *et al.*, 1981b; Humphreys-Beher and Carlson, 1982; Russin *et al.*, 1981; Basu *et al.*, 1982) which incorporates GlcNAc

---

Cretel *et al.*, 1981), human chorionic gonadotropin, hCG (Bedi *et al.*, 1979), κ-casein (Saito *et al.*, 1981; Van Halbeek *et al.*, 1980, 1981b; Soulier *et al.*, 1980), swarm rat chondrosarcoma (Nilsson *et al.*, 1982), epiglycanin (Van den Eijnden *et al.*, 1979b), ascites hepatoma AH 66 plasma membrane (Funakoshi and Yamashina, 1982), bovine adrenal medulla chromaffin granules (Kiang *et al.*, 1982). The sialylα2-6GalNAc moiety, either alone or as part of a larger oligosaccharide, is found in PSM, OSM, BSM, CSM, RSM, rat colonic mucin, human and monkey cervical mucins, porcine Cowper's gland mucin, adrenal medulla chromaffin granules, casein, human, equine, and porcine glycophorins, hCG, fetuin, human plasminogen, swarm rat chondrosarcoma, and epiglycanin. The class 1 core has been found in PSM, OSM, CSM, RSM, HSM, goat submaxillary mucin, armadillo submandibular mucin, HBM, PGM, EGM, OGM, CGM, HOCM, human and monkey cervical mucins, porcine Cowper's gland mucin, rat small intestinal mucin, hCG, fetuin, human, equine, porcine, and bovine glycophorins, adrenal medulla chromaffin granules, human plasminogen, casein, swarm rat chondrosarcoma, human melanoma plasma membranes, IgA, epiglycanin, and antifreeze glycoprotein from Antarctic fish. The class 2 core has been found in HSM, HBM, HOCM, human cervical mucin, EGM, PGM, OGM, HGM, casein, IgA, ascites hepatoma AH 66 plasma membranes, swarm rat chondrosarcoma, and bovine glycophorin. The class 3 core has been found in rat sublingual, colonic, and small intestinal mucins, EGM, HBM, and monkey cervical mucin. The class 4 core has been found only in OGM and HBM.

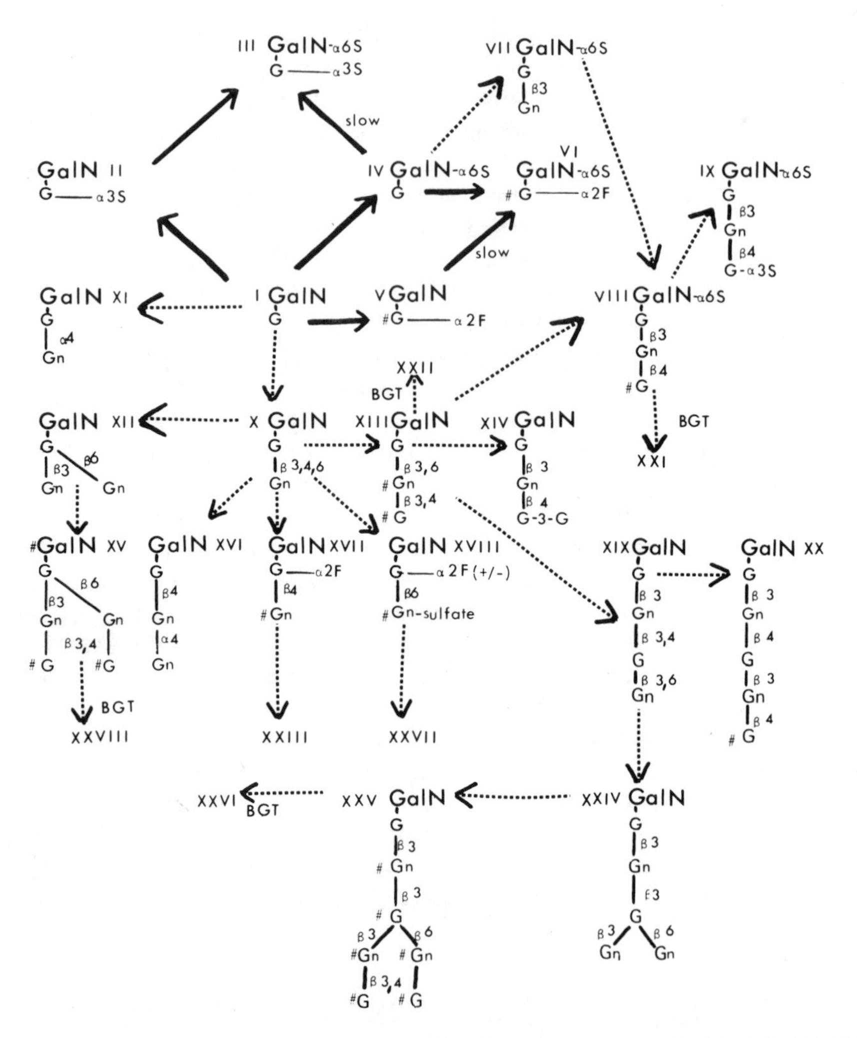

Figure 9. Synthesis of oligosaccharides with class 1 core. Abbreviations: GalN, Ser(Thr)-linked GalNAc; S, sialic acid; G, Gal; Gn, GlcNAc; F, L-fucose (Fuc); BGT, blood group transferases; #, point of attachment of various terminating α-linked sugars (see text). Continuous arrows indicate reactions established in cell-free systems. Discontinuous arrows indicate hypothetical reactions based on known structures. The occurrence of the various structures is as follows (see legend to Figure 8 for references and for abbreviations of mucins and glycoproteins): I, anti-freeze glycoprotein; II, HSM, human and horse glycophorins, casein, IgA, epiglycanin, AH 66 plasma membranes, adrenal medulla chromaffin granules, swarm rat chondrosarcoma; III, porcine Cowper's gland mucin, human cervical mucin, human and equine glycophorins, hCG, fetuin, adrenal medulla chromaffin granules, human plasminogen, casein, swarm rat chondrosarcoma; IV, PSM, CSM, porcine glycophorin, adrenal medulla chromaffin granules, casein; V, PSM, OSM, HSM, RSM, HBM, HOCM, EGM, rat small intestinal mucin, human erythrocytes; VI, PSM, CSM, RSM, human erythrocytes; VII, casein; VIII, monkey cervical mucin; IX, epiglycanin; X, CSM; XI, PGM; XIII, CSM, HBM, HOCM, OGM; XIV, bovine and porcine glycophorins; XV, HOCM, bovine glycophorin; XVI, EGM; XVII, goat submaxillary mucin; XVIII, CSM; XX, OGM, bovine glycophorin; XXI, monkey cervical mucin; XXII, HOCM, HBM; XXIII, goat submaxillary mucin; XXV, PGM, HOCM, IgA; XXVI, PGM, HOCM; XXVII, CSM; XXVIII, PGM, CGM, HOCM, bovine glycophorin.

into various Galβ1-4GlcNAc-terminated acceptors to form the i structure, GlcNAcβ1-3Galβ1-4GlcNAc, although product identification in these studies was preliminary.

The enzyme UDP-Gal:GlcNAc β4-galactosyltransferase (Section IV-D) has been purified from bovine milk and from various other sources and is one of the most thoroughly studied glycosyltransferases (Schachter and Roseman, 1980; Beyer *et al.*, 1981). Its main physiological role appears to be synthesis of the Galβ1-4GlcNAc moiety of complex *N*-glycosyl oligosaccharides. This same enzyme appears to be capable of the synthesis of the Galβ1-4GlcNAc moiety found in the poly-*N*-acetyllactosaminoglycans with blood group i antigenic activity (Kaur *et al.*, 1982; Schenkel-Brunner *et al.*, 1979; Schenkel-Brunner, 1973). In an interesting recent study, Blanken *et al.* (1982) showed that the pure bovine colostrum UDP-Gal:GlcNAc β4-Gal-transferase could act on the synthetic branched trisaccharide GlcNAcβ1-3(GlcNAcβ1-6)Gal to form the pentasaccharide Galβ1-4GlcNAcβ1-3(Galβ1-4GlcNAcβ1-6)Gal. This pentasaccharide shows blood group I activity with some anti-I antibodies. The Gal-transferase showed a marked preference for the GlcNAc linked β1-6 over the GlcNAc linked β1-3, although with time both GlcNAc residues were galactosylated.

The poly-*N*-acetyllactosaminoglycan structures also occur as extensions of the *N*-glycosyl oligosaccharide core (Figure 2). Galβ1-4GlcNAc antennae may be capped by a sialic acid residue or by a repeating poly-*N*-acetyllactosaminoglycan. The controls are not known but probably involve compartmentation or competition between sialyltransferases and UDP-GlcNAc:Galβ1-4GlcNAc-R β1-3-GlcNAc-transferase (see above).

More recently, a UDP-Gal:GlcNAc β3-Gal-transferase has been highly purified from pig trachea (Sheares and Carlson, 1982; Sheares *et al.*, 1982). This enzyme is clearly different from the previously studied UDP-Gal:GlcNAc β4-Gal-transferase and UDP-Gal:GalNAc-*R* β3-Gal-transferase. The physiological function of this enzyme appears to be the elongation of Ser(Thr)-GalNAc oligosaccharides since the enzyme does not work on *N*-glycosyl oligosaccharides.

3. *Terminal Glycosylation: The Addition of α-Linked Sugar Residues*

A common feature of all Ser(Thr)-GalNAc oligosaccharides is the presence of nonreducing terminal α-linked sugars. The incorporation of a sugar in α-linkage usually prevents further sugar additions to that residue although elongation of other parts of the oligosaccharide may occur. For example, elongation of the antennae or branches of a complex *N*-glycosyl oligosaccharide can occur after incorporation of a fucose in α1-6 linkage to the Asn-linked GlcNAc but further addition of sugars to the fucose does not occur (Longmore and Schachter, 1982). Similarly, addition of a sialic acid in α2-6 linkage to the GalNAc residue of Galβ1-3GalNAc-Ser(Thr)-*X* does not prevent further sugar additions to the Gal residue although the sialyl residue usually remains unsubstituted. However, sialyl residues can be substituted by another sialyl residue in α2-8 linkage in both gangliosides and glycoproteins (Finne *et al.*, 1977).

The most common α-linked terminal sugars are sialic acid and L-fucose, but terminal Gal, GalNAc, and GlcNAc residues also occur in α-linkage. Frequently, the α-linked sugars, either alone (blood group H determinant) or in combination (blood group Le^b determinant), constitute an antigenic determinant (Table 4). The synthesis of the human ABO and Lewis determinants has been thoroughly reviewed (Watkins, 1980; Schachter and Tilley, 1978; Beyer *et al.*, 1981).

a. CMP-Sialic Acid:GalNAc-X *α6-Sialyltransferase*.

It was pointed out above that UDP-Gal:GalNAc-*X* β3-Gal-transferase could not act after the α6-sialyltransferase (Figure 8), i.e., the incorporation of a sialic acid in α2-6 linkage into GalNAc-mucin prevented further sugar addition and accounted for the relatively large amounts of sialylα2-6GalNAc disaccharides in many mucins such as ovine submaxillary and procine Cowper's gland mucins. Mucins with larger oligosaccharides containing the sialylα2-6GalNAc moiety, such as porcine submaxillary mucin, are made in tissues containing relatively more of the β3-Gal-transferase than the α6-sialyltransferase. The α6-sialyltransferase catalyzes the following reaction:

$$\text{CMP-sialic acid} + Y\text{-GalNAc-Ser(Thr)-}X \rightarrow$$
$$Y\text{-(sialyl}\alpha\text{2-6)GalNAc-Ser(Thr)-}X + \text{CMP}$$

where *Y* can be H or Galβ1-3, and where the sialyl residue can be either *N*-acetylneuraminic acid or *N*-glycollylneuraminic acid. The β3-Gal-transferase and the β6-sialyltransferase both compete for the same substrate, GalNAc-*X* (Figure 8). The sialyltransferase was first described in sheep submaxillary glands (Carlson *et al.*, 1973b), has also been found in porcine, bovine, and canine submaxillary glands, and has been highly purified from porcine submaxillary glands (Sadler *et al.*, 1979a,b; Beyer *et al.*, 1981). Structural considerations indicate that the enzyme is also present in rat salivary glands, human, porcine and equine bone marrow, porcine Cowper's gland, human cervix, and various other tissues (Figure 9).

Purification of the α6-sialyltransferase (Sadler *et al.*, 1979a,b) was complicated by the presence of another enzyme, CMP-sialic acid:Galβ1-3GalNAc-*X* α3-sialyltransferase (see below), in porcine submaxillary gland extracts. Both enzymes were purified to homogeneity from Triton X-100 extracts of submaxillary gland by repeated adsorptions to CDP-hexanolamine-Sepharose columns. The final preparation of the α6-sialyltransferase was obtained in 2% yield with a purification factor of 117,000, had a specific activity of 44.6 units/mg and was free of the α3-sialyltransferase. The enzyme has no requirement for cation.

As indicated in Figure 9, the α6-sialyltransferase acts poorly on Fucα1-2Galβ1-3GalNAc-*X*, and therefore the preferential pathway towards Fucα1-2Galβ1-3(sialylα2-6)GalNAc-*X* involves the prior action of the sialyltransferase followed by the fucosyltransferase (Beyer *et al.*, 1979, 1981). Pigs and other species with the appropriate genotype make mucins carrying the blood group A determinant (Table 4). Thus, porcine submaxillary mucin from A^+-pigs contains appreciable amounts of GalNAcα1-3(Fucα1-2)Galβ1-3GalNAc-*X* and GalNAcα1-3(Fucα1-2)Galβ1-3(sialylα2-6)GalNAc-*X*. The α6-sialyltransferase is totally inactive towards GalNAcα1-3(Fucα1-2)Galβ1-3GalNAc-*X*, and therefore the blood group A-active pentasaccharide must be synthesized by the action of the blood group A-dependent α3-GalNAc-transferase (see below) on Fucα1-2Galβ1-3(sialylα2-6)GalNAc-*X* (Beyer *et al.*, 1979, 1981).

b. CMP-Sialic Acid:Galβ1-3GalNAc-X α3-Sialyltransferase. Porcine submaxillary glands contain two distinct sialyltransferases acting on Ser(Thr)-GalNAc oligosaccharides, the α6-sialyltransferase discussed above, and an α3-sialyltransferase catalyzing the following reaction:

$$\text{CMP-sialic acid} + \text{Gal}\beta\text{1-3GalNAc-}X \rightarrow \text{sialyl}\alpha\text{2-3Gal}\beta\text{1-3GalNAc-}X + \text{CMP}$$

Both enzymes have been purified to homogeneity. The final preparation of the α3-sialyltransferase (Sadler *et al.*, 1979a,b; Rearick *et al.*, 1979) was obtained in 5% yield, had a purification factor of 92,000, and a specific activity of 10.6 units per mg protein. The purified enzyme can form the sialylα2-3Gal bond with a variety of Gal-terminal acceptors, namely, Galβ1-3GalNAc-Ser(Thr)-proteins (various asialo-mucins, antifreeze glycoprotein, asialo-fetuin), Galβ1-3GalNAc (K_m = 0.2 mM), Galβ1-3GlcNAc (K_m = 85 mM), Galβ1-4Glc (K_m = 130 mM), and ganglioside G_{M1} (to form ganglioside G_{D1a}).

These two enzymes are different from the two sialyltransferases which act on *N*-glycosyl oligosaccharides (Sections IV-E and IV-F). The α6-sialyltransferases acting on *N*- and *O*-glycosyl oligosaccharides synthesize different linkages, i.e., sialylα2-6Gal and sialylα2-6GalNAc, respectively, and it is therefore not surprising that the enzymes are different. The α3-sialyltransferases, however, both synthesize the sialylα2-3Gal moiety and yet are quite distinct enzymes; evidently, these enzymes recognize not only the terminal Gal residue of the acceptor but also the penultimate residue.

The action of the highly purified CMP-sialic acid:Galβ1-3GalNAc-*X* α3-sialyltransferase on ganglioside G_{M1} implies that the same enzyme acts on both Ser(Thr)-GalNAc oligosaccharides and glycolipids. The sialyltransferase does not, however, act on lactosyl-ceramide implying the existence of a third α3-sialyltransferase responsible for synthesis of sialylα2-3Galβ1-4Glc-ceramide.

The α3-sialyltransferase has been reported in ovine submaxillary glands (Bergh *et al.*, 1982) and must be present in human salivary glands, human and equine bone marrow, in certain cancer lines, and various other tissues to account for the synthesis of sialylα2-3Galβ1-3GalNAc-*X* (Figure 9), and in human bone marrow and cervix, porcine Cowper's gland, calf liver, certain cancer cells, and other tissues to account for the synthesis of sialylα2-3Galβ1-3(sialylα2-6)GalNAc-*X* (Figure 9). The synthesis of the latter tetrasaccharide probably proceeds by addition of sialic acid in α2-6 linkage to the GalNAc of sialylα2-3Galβ1-3GalNAc-*X* (Figure 9) since the α2-3-sialyltransferase acts poorly on Galβ1-3(sialylα2-6)GalNAc-*X* (Beyer *et al.*, 1979, 1981; Figure 9). Since porcine submaxillary glands are rich in both the α3- and α6-sialyltransferases acting on *O*-glycosyl oligosaccharides, it is not clear why porcine submaxillary mucin contains the sialylα2-6GalNAc moiety but not the sialylα2-3Gal structure.

The sialylα2-3Galβ1-3GalNAc moiety may be important at cell surfaces since it is found in several plasma membrane components, i.e., human and equine glycophorin, epiglycanin, ascites cell AH 66 plasma membrane, and several gangliosides (Figure 9).

Porcine liver, and probably also rat, human, and canine liver, contain a CMP-NAN:Galβ1-3GalNAc-*X* α3-sialyltransferase which transfers sialic acid in α2-3 linkage to the Gal residue of various asialo-mucins and ganglioside G_{M1} (Bergh *et al.*, 1981a,b; Wetmore *et al.*, 1974; Van den Eijnden *et al.*, 1981). The synthesis of the tetrasaccharide sialylα2-3Galβ1-3(sialylα2-6)GalNAc-*X* both in fetal calf liver (Van den Eijnden *et al.*, 1982) and in salivary glands (Beyer *et al.*, 1979, 1981) proceeds by α3-sialyltransferase acting before α6-sialyltransferase (Figure 9). However, whereas the salivary gland α6-sialyltransferase can act on GalNAc-mucin (Figure 8), Galβ1-3GalNAc-mucin (Figure 9), and sialylα2-3Galβ1-3GalNAc-mucin (Figure 9), the fetal calf liver α6-sialyltransferase acts only on the latter acceptor and is therefore a different enzyme from the salivary gland enzyme. This interesting finding suggests that the generalized schemes presented in this chapter, based on findings in many species and tissues, may not hold true for all situations.

c. GDP-Fuc:β-Galactoside α1-2,α1-3, and α1-4-Fucosyltransferases. Table 4 lists some of the antigenic determinants which contain fucose. These determinants occur at the nonreducing ends of *N*-glycosyl oligosaccharides of the poly-*N*-acetyllactosaminoglycan type, *O*-glycosyl oligosaccharides, and glycosphingolipids. The three fucosyltransferases acting on acceptors with a β-galactoside nonreducing terminus were discussed in Section IV-H.

The blood group H enzyme attaches Fuc in α1-2 linkage to the terminal Gal residue of any β-D-galactoside, both large and small, irrespective of the subterminal sugars. The blood group H determinant occurs on both Type 1 and Type 2 chains (Table 4) at the termini of *N*- and *O*-glycosyl oligosaccharides and glycosphingolipids, and on the Galβ1-3GalNAc sequence itself, e.g., in porcine, ovine, canine, and goat submaxillary mucins, human ovarian cyst, salivary, and bronchial mucins, human erythrocytes, monkey cervical mucin, rat intestinal and salivary mucins, and porcine and equine gastric mucins (Figure 9).

The blood group Lewis enzyme attaches Fuc in α1-4 linkage to the subterminal GlcNAc of Galβ1-3GlcNAc-R (Type 1) structures (Table 4). The enzyme incorporates Fuc in both α1-4 and α1-3 linkages (Section IV-H and Table 2). The enzyme is inactive towards substrates with the structure Galβ1-3GalNAc-Ser(Thr), but *O*-glycosyl oligosaccharides with core class 1 do carry the Lewis determinant at Galβ1-3GlcNAc termini of backbone structures, e.g., in pig gastric mucin and human ovarian cyst mucin.

The structure Galβ1-4(Fucα1-3)GlcNAc is the antigenic determinant for mouse stage-specific embryonic antigen (Table 4) and occurs in complex *N*-glycosyl oligosaccharides, e.g., in human serum α_1-acid glycoprotein, as well as in *O*-glycosyl oligosaccharides with core class 1, e.g., pig gastric mucin and human ovarian cyst and bronchial mucins. Non-Lewis-dependent GDP-Fuc:β-galactoside α3-fucosyltransferase probably carries out synthesis of this determinant. This enzyme can incorporate Fuc in α1-3 linkage to the GlcNAc of Galβ1-4GlcNAc but not in α1-3 linkage to the Glc of Galβ1-4Glc or in α1-4 linkage to the GlcNAc of Galβ1-3GlcNAc.

d. Blood-Group-A- and B-Dependent α3-GalNAc- and α3-Gal-Transferases. The determinants for the human blood groups A and B are shown in Table 4. These determinants have also been found in other species. The enzymes responsible for the synthesis of these determinants are, respectively, UDP-GalNAc:Fucα1-2Gal-*X* α3-*N*-acetylgalactosaminyltransferase and UDP-Gal:Fucα1-2Gal-*X* α3-galactosyltransferase (Watkins, 1980; Schachter and Tilley, 1978; Beyer *et al.*, 1981). Both enzymes have a very similar substrate specificity; i.e., they require the H determinant (Table 4) at the terminus of the substrate. The enzymes can, however, act on the disaccharide Fucα1-2Gal, as well as on larger oligosaccharides, *N*- and *O*-glycosyl protein-bound oligosaccharides, and glycosphingolipids, provided there is Fucα1-2Gal at the terminus. In some species, e.g., rabbit bone marrow (Basu and Basu, 1973), the blood group-A- or B-dependent glycosyltransferases do not require the presence of the α2-linked fucose residue. The A-dependent GalNAc-transferase has been purified from human serum and from porcine submaxillary glands. The porcine submaxillary gland enzyme was purified 38,000-fold to homogeneity by repeated chromatography of Triton X-100 extracts on UDP-hexanolamine-agarose (Schwyzer and Hill, 1977a,b). The purified enzyme was obtained in 13% yield and had a specific activity of 30 units per mg protein.

The A and B determinants occur on *N*- and *O*-glycosidically-linked oligosaccharides and on glycosphingolipids. Much of the early structural work on the human blood group antigens was carried out on the water-soluble ovarian cyst mucins (Watkins, 1980; Rovis *et al.*, 1973; Wu *et al.*, 1982). Other glycoproteins with core class 1 *O*-glycosyl oligosaccharides carrying either A or B blood group determinants are porcine submaxillary and gastric mucins, rat intestinal and salivary mucins, and equine gastric mucin (Figure 9).

B. Synthesis of Oligosaccharides with Core Class 2

1. Synthesis of the Core: UDP-GlcNAc:Galβ1-3GalNAc-X (GlcNAc to GalNAc) β6-N-Acetylglucosaminyltransferase

The Galβ1-3(GlcNAcβ1-6)GalNAc-*X* core is synthesized from Galβ1-3GalNAc-*X* by an enzyme recently described in canine submaxillary glands (Williams and Schachter, 1980; Williams *et. al.*, 1980), rabbit intestine (Wingert and Cheng, 1982), and porcine gastric mucosa (unpublished data from this laboratory), which catalyzes the following reaction:

$$\text{UDP-GlcNAc} + \text{Gal}\beta 1\text{-3GalNAc-}X \rightarrow \text{Gal}\beta 1\text{-3(GlcNAc}\beta 1\text{-6)GalNAc-}X + \text{UDP.}$$

O-Glycosyl oligosaccharides with this core class 2 have been described in bovine κ-casein and glycophorin, human IgA, human gastric, ovarian cyst, cervical, salivary, and bronchial mucins, ovine, equine, and porcine gastric mucins, and certain tumor cells (Figure 10). It is not clear why canine submaxillary mucin does not appear to have core class 2 structures (Lombart and Winzler, 1974) since canine submaxillary glands are very rich in the β6-GlcNAc-transferase.

The β6-GlcNAc-transferase has an absolute requirement for acceptors with the Galβ1-3GalNAc terminus. The enzyme will work on the disaccharide Galβ1-3GalNAc-*X* (*X* = H, benzyl, *o*- or *p*-nitrophenyl, methyl) as well as on Galβ1-3GalNAc-Ser(Thr)-mucin and antifreeze glycoprotein. The enzyme is totally inactive on GalNAc-Ser(Thr)-

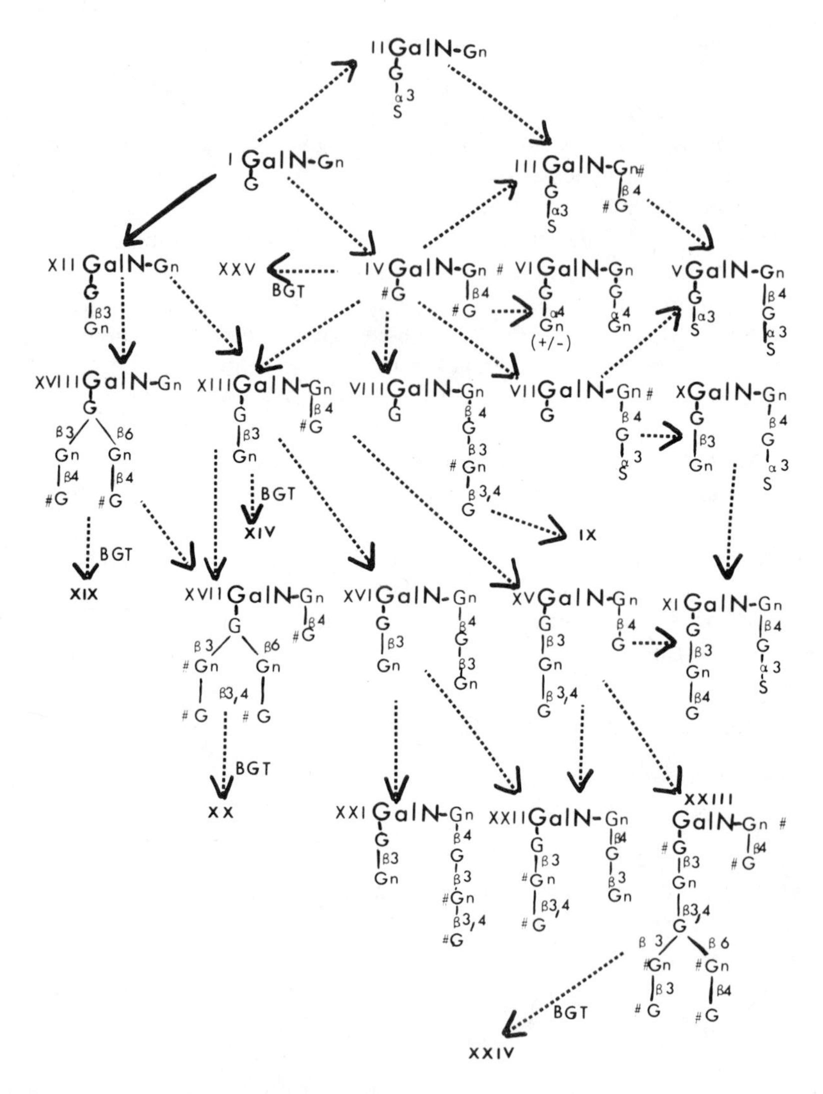

Figure 10. Synthesis of oligosaccharides with class 2 core. See the legends to Figures 8 and 9 for abbreviations and references. The occurrence of the various structures is as follows: II, ascites hepatoma AH 66 plasma membranes; III, HSM, human cervical mucin, casein, AH 66 plasma membranes; IV, HBM, HOCM, human cervical mucin, casein, IgA, AH 66 plasma membranes; V, swarm rat chondrosarcoma, AH 66 plasma membranes; VI, HOCM, PGM; VII, human cervical mucin, AH 66 plasma membranes; VIII and IX, HOCM, OGM; XI, bovine glycophorin; XIV, OGM; XV, OGM, PGM, HGM; XIX, PGM; XX, HOCM, PGM; XXI and XXII, OGM; XXIV, HOCM; XXV, HSM, HOCM, human cervical mucin, HBM, EGM.

mucin indicating that the UDP-Gal:GalNAc-mucin β3-Gal-transferase must act before the β6-GlcNAc-transferase as indicated in Figure 8. The presence of a fucose residue, i.e., Fucα1-2Galβ1-3GalNAc-*X*, does not prevent β6-GlcNAc-transferase action but makes it less favorable. In fact, the tetrasaccharide structure Fucα1-2Galβ1-3(GlcNAcβ1-6)GalNAc-*X* has not yet been reported. However, the pentasaccharide Fucα1-2Galβ1-

3(Galβ1-4GlcNAcβ1-6)GalNAc-*X* occurs in equine gastric mucin and human ovarian cyst mucin, suggesting the sequence of synthesis shown in Figure 10, i.e., β4-Gal on the GlcNAc of the core followed by α2-Fuc on the Gal of the core.

2. *Elongation of the Core: Synthesis of the Ii Backbone*

The elongation of the class 2 core is very similar to the process described above for class 1 core, i.e., the process involves addition of GlcNAc and Gal residues in β1-3, β1-4, and β1-6 linkages (Figure 10). The process is more complex than the scheme shown in Figure 9 because elongation can occur both at the Gal residue of the core as well as at the GlcNAc residue.

We have found in porcine gastric mucosa (Brockhausen *et al.*, 1984) an elongation enzyme, UDP-GlcNAc:Galβ1-3(GlcNAcβ1-6)GalNAc-*X* (GlcNAc to Gal) β3-*N*-acetylglucosaminyltransferase which catalyzes the reaction:

UDP-GlcNAc + Galβ1-3(GlcNAcβ1-6)GalNAc-α-*X*
↓
GlcNAcβ1-3Galβ1-3(GlcNAcβ1-6)GalNAc-α-*X* + UDP

where *X* is *o*-nitrophenyl or benzyl. The enzyme shows a strict requirement for divalent cation (Mn^{2+} or Co^{2+}) and is markedly stimulated by addition of Triton X-100. Since canine submaxillary glands are rich in the UDP-GlcNAc:Galβ1-3GalNAc-*X* (GlcNAc to GalNAc) β6-GlcNAc-transferase responsible for core synthesis (Williams and Schachter, 1980; Williams *et al.*, 1980) but lack the β3-GlcNAc-transferase activity, the β3- and β6-enzymes must be different.

The β3-GlcNAc-transferase does not appear to act on Galβ1-3GalNAc-*X* (*X* = *o*-nitrophenyl or benzyl) acceptors suggesting that the prior synthesis of Galβ1-3(GlcNAcβ1-6)GalNAc-*X* (class 2 core) is essential for elongation. However, there must exist a mechanism for elongation of the Gal residue of class 1 core (Galβ1-3GalNAc-*X*) by addition of a GlcNAc residue in β1-3 linkage. It is possible that there are two elongation β3-GlcNAc-transferases.

The tetrasaccharide GlcNAcβ1-3Galβ1-3(GlcNAcβ1-6)GalNAc-*X* (Figure 10) has been found as a component of various Ser(Thr)-GalNAc oligosaccharides, e.g., in gastric mucins from pigs (Van Halbeek *et al.*, 1982b), humans (Oates *et al.*, 1974), and sheep (Hounsell *et al.*, 1981), in human submaxillary mucin (Reddy *et al.*, 1979), and in human ovarian cyst mucin (Rovis *et al.*, 1973; Wu *et al.*, 1982).

3. *Addition of α-Linked Sugars*

The termination of core class 2 oligosaccharides follows the same rules as outlined for class 1 oligosaccharides. Addition of α-linked sialyl, fucosyl, GalNAc, and Gal residues can occur on oligosaccharides attached both to the Gal and GlcNAc residues of the core (Figure 10). Sialylated class 2 oligosaccharides have been described in bovine κ-casein and glycophorin, ascites hepatoma AH 66 plasma membranes, and human salivary and cervical mucins. Fucosylated oligosaccharides have been described in ovine, equine, and porcine gastric mucins, and human ovarian cyst, cervical, salivary, and bronchial mucins. Core class 2 oligosaccharides terminated with blood group A or B determinants have been described in ovine gastric, equine gastric, and human ovarian cyst mucins. Finally, there have also been reports of termination by α-linked GlcNAc residues in porcine gastric and human ovarian cyst mucins.

C. *Synthesis of Oligosaccharides with Core Classes 3 and 4*

Preliminary data from our laboratory have indicated that rat colon mucosal extracts can catalyze the following two reactions in the presence of Triton X-100 and Mn^{2+}:

$$\text{UDP-GlcNAc} + \text{GalNAc-}X \rightarrow \text{GlcNAc}\beta\text{1-3GalNAc-}X$$

$$\text{UDP-GlcNAc} + \text{GlcNAc}\beta\text{1-3GalNAc-}X \rightarrow \text{GlcNAc}\beta\text{1-3(GlcNAc}\beta\text{1-6)GalNAc-}X + \text{UDP}.$$

The second reaction proceeds at an appreciably faster rate than the first. Core class 3 oligosaccharides (Figure 11) have been found in rat small intestinal, colonic, and salivary mucins, human bronchial mucin, equine gastric mucin, and monkey cevical mucin. Core class 4 oligosaccharides (Figure 11) have to date been described only in ovine gastric and human bronchial mucins.

Elongation and termination reactions for core class 3 and 4 oligosaccharides have not been studied. Hypothetical schemes based on published structures are indicated in Figure 11. Sialylated core class 3 oligosaccharides have been found in rat sublingual

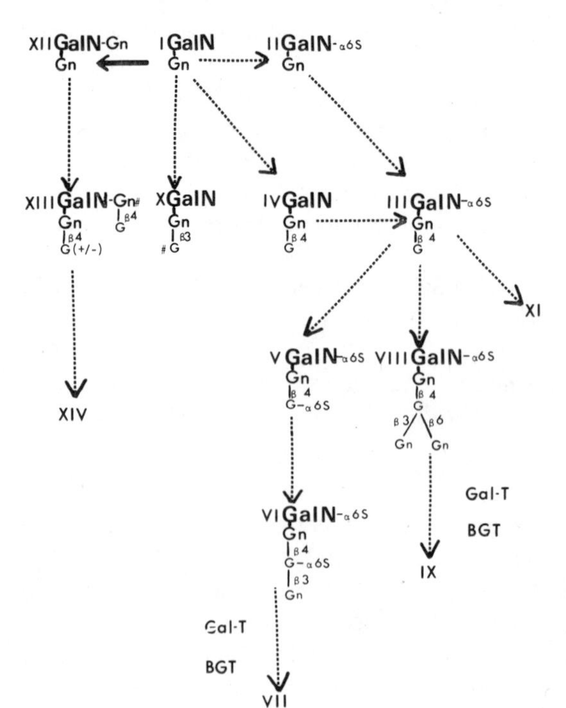

Figure 11. Synthesis of oligosaccharides with core classes 3 and 4. See the legends to Figures 8 and 9 for abbreviations and references. The occurrence of the various structures is as follows: II and III, rat colonic mucin; IV, HBM, EGM, rat colonic mucin; V, RSM; VI, RSM, rat colonic mucin; VII, rat colonic mucin; VIII and IX, rat colonic mucin; X, HBM; XI, monkey cervical mucin; XIII, HBM; XIV, OGM.

and colonic mucins and in monkey cervical mucin. Fucosylated core class 3 and 4 oligosaccharides occur in rat colonic and human bronchial mucins. Blood group A determinants are found in rat colonic mucin.

X. CONCLUDING REMARKS: CONTROL OF SYNTHESIS

The control of protein and nucleic acid synthesis depends primarily on the assembly of exact copies on templates. A template biosynthetic mechanism is only feasible, however, with linear structures like proteins and nucleic acids. One of the most characteristic and potentially very important features of complex carbohydrates is their branched structure. The control of the synthesis of such a structure cannot be via a template mechanism. Rather, control is exerted through a variety of factors. Obviously, the genes must control complex carbohydrate synthesis but they do so by being structural genes for the glycosyltransferases and for the protein backbones of the glycoproteins, and by determining the structure of the endomembrane system on which glycosylation takes place.

There are many critical points during oligosaccharide assembly at which control can be exerted. The first is the initiation process which is highly complex in *N*-glycosyl oligosaccharide assembly (Figures 4 and 5 and Section III) and relatively more simple in *O*-glycosyl oligosaccharide synthesis (Section VIII). Once initiation has been achieved, the growing glycoprotein passes through the endomembrane system where glycosidase processing, in the case of *N*-glycosyl oligosaccharides, and various multiglycosyltransferase systems determine the final oligosaccharide structure. Many factors can come into play in this process, e.g., compartmentation of different molecules in different intracellular compartments or in different cell types, availability of cations, cofactors, and nucleotide-sugars, transit rate through the endomembrane system, steric availability of the protein-bound oligosaccharide to glycosidases and glycosyltransferases, and presence of inhibitors, modifiers, and activators. We have frequently taken the view that it is the presence of batteries of glycosyltransferases capable of highly specific discrimination between different oligosaccharides which plays a major role in the oligosaccharide assembly process. We have tried in this review to point out the many examples of glycosyltransferase specificity.

There are three general patterns which repeat themselves in the synthetic schemes for both *N*- and *O*-glycosyl oligosaccharides, as follows:

(1) There are many points at which two or more glycosyltransferases compete for a common substrate, e.g., competition occurs for Gn(I)M_5 (Figure 6), for GnGn (Figure 7), and for GalNAc-Ser(Thr)-*X* (Figure 8). The dominant transferase in such a competition decides the synthetic fate, e.g., whether synthesis proceeds into hybrid or complex oligosaccharides at Gn(I)M_5, into nonbisected bi-antennary, bisected bi-antennary, or more highly branched oligosaccharides at GnGn, into sialylα2-6GalNAc-Ser(Thr)-*X*, or into larger *O*-glycosyl oligosaccharides at GalNAc-Ser(Thr)-*X*. The reasons why one transferase dominates over another are not clear. It may be simply a matter of differences in activity, e.g., ovine submaxillary glands have a higher α2-6-sialyltransferase activity than β1-3-Gal-transferase activity (Figure 8) whereas por-

cine submaxillary glands have the reverse situation. Or there may be more subtle controls that require elucidation.

(2) There are other points at which a certain glycosidase or glycosyltransferase is unable to act until a key glycosyl residue is introduced into the oligosaccharide. The GlcNAc introduced in β1-2 linkage into the Manα1-3 terminus of M_5 by GlcNAc-transferase I (Figure 6) is such a key residue. Until this residue is present, the following five enzymes cannot act: GlcNAc-transferases II, III, and IV, (Fuc to Asn-linked GlcNAc) α1-6-fucosyltransferase, and mannosidase II. Similarly, the β1-6-GlcNAc-transferase which synthesizes *O*-glycosyl oligosaccharide core class 2 (Figure 8) requires the prior insertion of a Gal residue.

(3) Other key glycosyl residues do exactly the reverse, i.e., they prevent enzyme action. A good example is the bisecting GlcNAc residue introduced by GlcNAc-transferase III. The insertion of this residue prevents the action of four enzymes: GlcNAc-transferases II and IV, (Fuc to Asn-linked GlcNAc) α1-6-fucosyltransferase, and mannosidase II. The presence of a Gal on the GlcNAcβ1-2Manα1-3 arm prevents the action of G1cNAc-transferases II, III and IV, mannosidase II and α6-fucosyltransferase. Recent work in our group using high resolution proton NMR spectroscopy to study three-dimensional oligosaccharide structures has suggested an explanation for these findings. It appears that the GlcNAcβ1-2Manα1-3 structure is essential for all five of the above enzymes. A bisecting GlcNAc (linked β1-4 to the β-linked Man of the core) or substitution by a Gal residue prevents access of these enzymes to the GlcNAcβ1-2Manα1-3 structure.

Substrate specificity studies on the glycosyltransferase thus shed light not only on the control of glycoprotein synthesis but provide models for the problem of oligosaccharide–protein interaction and, in this way, may help elucidate the biological functions of complex carbohydrates.

ACKNOWLEDGMENTS

This research has been supported by the Medical Research Council of Canada and by NIH grant RO1-HD-07889.

REFERENCES

Andersson, G. N., and Eriksson, L. C., 1979, Characterization of UDP-galactosyl:asialo-mucin transferase activity in the Golgi system of rat liver, *Biochim. Biophys. Acta* **570**:239–247.

Andersson, G. N., and Eriksson, L. C., 1980, Studies on the latency of UDP-galactose-asialo-mucin galactosyltransferase activity in microsomal and Golgi subfractions from rat liver, *Biochim. Biophys. Acta* **600**:571–576.

Andersson, G. N., and Eriksson, L. C., 1981, Endogenous localization of UDP-galactose:asialomucin galactosyltransferase activity in rat liver endoplasmic reticulum and Golgi apparatus, *J. Biol. Chem.* **256**:9633–9639.

Babczinski, P., 1980, Evidence against the participation of lipid intermediates in the *in vitro* biosynthesis of serine(threonine)-*N*-acetyl-D-galactosamine linkages in submaxillary mucin, *FEBS Lett.* **117**:207–211.

Baker, A. P., and Munro, J. R., 1971, Multiglycosyltransferase system of canine respiratory tissue. Uridine diphosphate galactose:mucin galactosyltransferase, *J. Biol. Chem.* **246**:4358–4362.

Barker, R., Olsen, K. W., Shaper, J. H., and Hill, R. L., 1972, Agarose derivatives of uridine diphosphate and *N*-acetylglucosamine for the purification of a galactosyltransferase, *J. Biol. Chem.* **247**:7135–7147.

Basu, M., and Basu, S., 1973, Enzymatic synthesis of a blood group B-related pentaglycosylceramide by an α-galactosyltransferase from rabbit bone marrow, *J. Biol. Chem.* **248:**1700–1706.

Basu, M., Chon, H-C., and Basu, S., 1982, Biosynthesis *in vitro* of core structure LM1 and Ii glycosphingolipids in mouse lymphomas, Abst. Soc. Complex Carbohydrates, No. 63, Hershey, Pennsylvania.

Bedi, G. S., Reddy, M. S., Shah, R. H., and Bahl, O. P., 1979, Carbohydrate structure of gonadotropins, Abst. Soc. Complex Carbohydrates, No. 2, Toronto, Canada.

Berger, E. G., and Kozdrowski, I., 1978, Permanent mixed-field polyagglutinable erythrocytes lack galactosyltransferase activity, *FEBS Lett.* **93:**105–108.

Bergh, M. L. E., Hooghwinkel, G. J. M., and Van den Eijnden, D. H., 1981a, Specificity of porcine liver Galβ(1-3)GalNAc-*R* α(2-3)sialyltransferase. Sialylation of mucin-type acceptors and ganglioside G_{M1} *in vitro, Biochim. Biophys. Acta* **660:**161–169.

Bergh, M. L. E., Koppen, P., and Van den Eijnden, D. H., 1981b, High pressure liquid chromatography of sialic acid-containing oligosaccharides, *Carbohydrate Res.* **94:**225–229.

Bergh, M. L. E., Koppen, P. L., and Van den Eijnden, D. H., 1982, Specificity of ovine submaxillary-gland sialyltransferase. Application of high-pressure liquid chromatography in the identification of asialo-oligosaccharide products, *Biochem. J.* **201:**411–415.

Beyer, T. A., and Hill, R. L., 1982, Glycosylation pathways in the biosynthesis of nonreducing terminal sequences in oligosaccharides of glycoproteins, in: *The Glycoconjugates,* Vol. III (M. I. Horowitz, ed.), Academic Press, New York, pp. 25–45.

Beyer, T. A., Rearick, J. I., Paulson, J. C., Prieels, J. P., Sadler, J. E., and Hill, R. L., 1979, Biosynthesis of mammalian glycoproteins. Glycosylation pathways in the synthesis of the non-reducing terminal sequences, *J. Biol. Chem.* **254:**12531–12541.

Beyer, T. A., Sadler, J. E., Rearick, J. I., Paulson, J. C., and Hill, R. L., 1981, Glycosyltransferases and their use in assessing oligosaccharide structure and structure–function relationships, in: *Advances in Enzymology,* Vol. 52 (A. Meister, ed.), John Wiley and Sons, New York, pp. 23–175.

Blanken, W. M., Hooghwinkel, G. J. M., and Van den Eijnden, D. H., 1982, Biosynthesis of blood-group I and i substances. Specificity of bovine colostrum β-*N*-acetyl-D-glucosaminide β1-4 galactosyltransferase, *Eur. J. Biochem.* **127:**547–552.

Brew, K., Vanaman, T. C., and Hill, R. L., 1968, The role of α-lactalbumin and the A protein in lactose synthetase: A unique mechanism for the control of a biological reaction, *Proc. Natl. Acad. Sci. USA* **59:**491–497.

Briand, J. P., Andrews, S. P., Cahill, E., Conway, N. A., and Young, J. D., 1981, Investigation of the requirements for *O*-glycosylation by bovine submaxillary gland UDP-*N*-acetylgalactosamine:polypeptide *N*-acetylgalactosamine transferase using synthetic peptide substrates, *J. Biol. Chem.* **256:**12205–12207.

Brodbeck, U., and Ebner, K. E., 1966, Resolution of a soluble lactose synthetase into two protein components and solubilization of microsomal lactose synthetase, *J. Biol. Chem.* **241:**762–764.

Carlson, D. M., and Blackwell, C., 1968, Structures and immunochemical properties of oligosaccharides isolated from pig submaxillary mucins, *J. Biol. Chem.* **243:**616–626.

Carlson, D. M., David, J., and Rutter, W. J., 1973a, Galactosyltransferase activities in pancreas, liver, and gut of the developing rat, *Arch. Biochem. Biophys.* **157:**605–612.

Carlson, D. M., McGuire, E. J., Jourdian, G. W., and Roseman, S., 1973b, The sialic acids. XVI. Isolation of a mucin sialyltransferase from sheep submaxillary gland, *J. Biol. Chem.* **248:**5763–5773.

Carlsson, H. E., Sundblad, G., Hammarstrom, A., and Lonngren, J., 1978, Structure of some oligosaccharides derived from rat-intestinal glycoproteins, *Carbohydrate Res.* **64:**181–188.

Cartron, J. P., and Nurden, A. T., 1979, Galactosyltransferase and membrane glycoprotein abnormality in human platelets from T_n-syndrome donors, *Nature* **282:**621–623.

Carver, J. P., and Grey, A. A., 1981, Determination of glycopeptide primary structure by 360-MHz proton magnetic resonance spectroscopy, *Biochemistry* **20:**6607–6616.

Chapman, A., Trowbridge, I. S., Hyman, R., and Kornfeld, S., 1979, Structure of the lipid-linked oligosaccharides that accumulate in class E Thy-1-negative mutant lymphomas, *Cell* **17:**509–515.

Chapman, A., Fujimoto, K., and Kornfeld, S., 1980, The primary glycosylation defect in class E Thy-1-negative mutant mouse lymphoma cells is an inability to synthesize dolichol-P-mannose, *J. Biol. Chem.* **255:**4441–4446.

Cheng, P. W., and Bona, S. J., 1982, Mucin biosynthesis. Characterization of UDP-galactose: α-*N*-acetylgalactosaminide β3-galactosyltransferase from human tracheal epithelium, *J. Biol. Chem.* **257**:6251–6258.

Cummings, R. D., Trowbridge, I. S., and Kornfeld, S., 1982, A mouse lymphoma cell line resistant to the leukoagglutinating lectin from Phaseolus vulgaris is deficient in UDP-GlcNAc:α-D-mannoside β1,6 *N*-acetylglucosaminyltransferase, *J. Biol. Chem.* **257**:13421–13427.

Dorland, L., Haverkamp, J., Vliegenthart, J. F. G., Spik, G., Fournet, B., and Montreuil, J., 1979, Investigation by 360-MHz ^{1}H-nuclear-magnetic-resonance spectroscopy and methylation analysis of the single glycan chain of chicken ovotransferrin, *Eur. J. Biochem.* **100**:569–574.

Dutta, B., and Rao, C. V. N., 1982, Structures of carbohydrate chains of glycoprotein isolated from goat submaxillary mucin, *Biochim. Biophys. Acta* **701**:72–85.

Feizi, T., 1982, Antigenicities of mucins—their relevance to tumour associated and stage specific embryonic antigens, in: *Mucus in Health and Disease,* Vol. II (E. N. Chantler, J. B. Elder, and M. Elstein, eds.), Plenum Press, New York, pp. 29–37.

Finne, J., Krusius, T., and Rauvala, H., 1977, Occurrence of disialosyl groups in glycoproteins, *Biochem. Biophys. Res. Commun.* **74**:405–410.

Fournet, B., Montreuil, J., Strecker, G., Dorland, L., Haverkamp, J., Vliegenthart, J. F. G., Binette, J. P., and Schmid, K., 1978, Determination of the primary structures of 16 asialo-carbohydrate units derived from human plasma α_1-acid glycoprotein by 360-MHz ^{1}H NMR spectroscopy and permethylation analysis, *Biochemistry* **17**:5206–5214.

Fukuda, M., Fukuda, M. N., and Hakomori, S., 1979a, Developmental change and genetic defect in the carbohydrate structure of Band 3 glycoprotein of human erythrocyte membrane, *J. Biol. Chem.* **254**:3700–3703.

Fukuda, M. N., Fukuda, M., and Hakomori, S., 1979b, Cell surface modification by endo-β-galactosidase. Change of blood group activities and release of oligosaccharides from glycoproteins and glycosphingolipids of human erythrocytes, *J. Biol. Chem.* **254**:5458–5465.

Fukuda, M. N., Papermaster, D. S., and Hargrave, P. A., 1979c, Rhodopsin carbohydrate. Structure of small oligosaccharides attached at two sites near the amino terminus, *J. Biol. Chem.* **254**: 8201–8207.

Fukuda, K., Tomita, M., and Hamada, A., 1980, Isolation and characterization of alkali-labile oligosaccharide units from horse glycophorin, *J. Biochem. (Tokyo)* **87**:687–693.

Fukuda, K., Tomita, M., and Hamada, A., 1981, Isolation and structural studies of the neutral oligosaccharide units from bovine glycophorin, *Biochim. Biophys. Acta* **677**:462–470.

Fukuda, K., Kawashima, I., Tomita, M., and Hamada, A., 1982, Structural studies of the acidic oligosaccharide units from bovine glycophorin, *Biochim. Biophys. Acta* **717**:278–288.

Funakoshi, I., and Yamashina, I., 1982, Structure of *O*-glycosidically linked sugar units from plasma membranes of an ascites hepatoma, AH 66, *J. Biol. Chem.* **257**:3782–3787.

Geren, C. R., Magee, S. C., and Ebner, K. E., 1976, Hydrophobic chromatography of galactosyltransferase, *Arch. Biochem. Biophys.* **172**:149–155.

Gershman, H., and Robbins, P. W., 1981, Transitory effects of glucose starvation on the synthesis of dolichol-linked oligosaccharides in mammalian cells, *J. Biol. Chem.* **256**:7774–7780.

Gleeson, P. A., and Schachter, H., 1983, Control of glycoprotein synthesis, *J. Biol. Chem.* **258**: 6162–6173.

Gleeson, P., Vella, G., Narasimhan, S., and Schachter, H., 1982, Hen oviduct *N*-acetylglucosaminyltransferase IV (Gn-T IV) responsible for branching of complex *N*-glycosyl oligosaccharides, *Fed. Proc.* (Abst. No. 5127) **41**:1147.

Gottlieb, C., Baenziger, J., and Kornfeld, S., 1975, Deficient uridine diphosphate-*N*-acetylglucosamine:glycoprotein *N*-acetylglucosaminyltransferase activity in a clone of Chinese hamster ovary cells with altered surface glycoproteins, *J. Biol. Chem.* **250**:3303–3309.

Hagopian, A., Bosmann, H. B., and Eylar, E. H., 1968, Glycoprotein biosynthesis: The localization of polypeptidyl:*N*-acetylgalactosaminyl, collagen:glucosyl, and glycoprotein:galactosyl transferases in HeLa cell membrane fractions, *Arch. Biochem. Biophys.* **128**:387–396.

Hagopian, A., Westall, F. C., Whitehead, J. S., and Eylar, E. H., 1971, Glycosylation of the A1 protein from myelin by a polypeptide *N*-acetylgalactosaminyltransferase. Identification of the receptor sequence, *J. Biol. Chem.* **246**:2519–2523.

Hanover, J. A., Lennarz, W. J., and Young, J. D., 1980, Synthesis of *N*- and *O*-linked glycopeptides in oviduct membrane preparations, *J. Biol. Chem.* **255:**6713–6716.

Hanover, J. A., Elting, J., Mintz, G. R., and Lennarz, W. J., 1982, Temporal aspects of the *N*- and *O*-glycosylation of human chorionic gonadotropin, *J. Biol. Chem.* **257:**10172–10177.

Harpaz, N., and Schachter, H., 1980a, Control of glycoprotein synthesis. Bovine colostrum UDP-*N*-acetylglucosamine:α-D-mannoside β-2-*N*-acetylglucosaminyltransferase I. Separation from UDP-*N*-acetylglucosamine:α-D-mannoside β-2-*N*-acetylglucosaminyltransferase II, partial purification and substrate specificity, *J. Biol. Chem.* **255:**4885–4893.

Harpaz, N., and Schachter, H., 1980b, Control of glycoprotein synthesis. Processing of asparagine-linked oligosaccharides by one or more rat liver Golgi α-D-mannosidases dependent on the prior action of UDP-*N*-acetylglucosamine:α-D-mannoside β-2-*N*-acetylglucosaminyltransferase I. *J. Biol. Chem.* **255:**4894–4902.

Hatcher, V. B., Schwarzmann, G. O. H., Jeanloz, R. W., and McArthur, J. W., 1977, Purification, properties, and partial structure elucidation of a high-molecular-weight glycoprotein from cervical mucus of the bonnet monkey (Macaca radiata), *Biochemistry* **16:**1518–1524.

Hayes, M. L., and Castellino, F. J., 1979, Carbohydrate of the human plasminogen variants. III. Structure of the *O*-glycosidically linked oligosaccharide units, *J. Biol. Chem.* **254:**8777–8780.

Hercz, A., and Harpaz, N., 1980, Characterization of the oligosaccharides of liver Z variant α_1-antitrypsin, *Can. J. Biochem.* **58:**644–648.

Hercz, A., Katona, E., Cutz, E., Wilson, J. R., and Barton, M., 1978, α_1-Antitrypsin: The presence of excess mannose in the Z variant isolated from liver, *Science* **201:**1229–1232.

Herp, A., Wu, A. M., and Moschera, J., 1979, Current concepts of the structure and nature of mammalian salivary mucous glycoproteins, *Mol. Cell. Biochem.* **23:**27–44.

Hesford, F. J., and Berger, E. G., 1981, Human erythrocyte galactosyltransferase. Characterization, membrane association and sidedness of active site, *Biochim. Biophys. Acta* **649:**709–716.

Hesford, F. J., Berger, E. G., and Van den Eijnden, D. H., 1981, Identification of the product formed by human erythrocyte galactosyltransferase, *Biochim. Biophys. Acta* **659:**302–311.

Hill, H. D., Reynolds, J. A., and Hill, R. L., 1977a, Purification, composition, molecular weight and subunit structure of ovine submaxillary mucin, *J. Biol. Chem.* **252:**3791–3798.

Hill, H. D., Schwyzer, M., Steinman, H. M., and Hill, R. L., 1977b, Ovine submaxillary mucin: Primary structure and peptide substrates of UDP-*N*-acetylgalactosamine:mucin transferase, *J. Biol. Chem.* **252:**3799–3804.

Hounsell, E. F., Fukuda, M., Powell, M. E., Feizi, T., and Hakomori, S., 1980, A new *O*-glycosidically linked tri-hexosamine core structure in sheep gastric mucin: A preliminary note, *Biochem. Biophys. Res. Commun.* **92:**1143–1150.

Hounsell, E. F., Wood, E., Feizi, T., Fukuda, M., Powell, M. E., and Hakomori, S., 1981, Structural analysis of hexa- to octa-saccharide fractions isolated from sheep gastric glycoproteins having blood-group I and i activities, *Carbohydrate Res.* **90:**283–307.

Hounsell, E. F., Lawson, A. M., and Feizi, T., 1982, Structural and antigenic diversity in mucin carbohydrate chains, in: *Mucus in Health and Disease,* Vol. II (E. N. Chantler, J. B. Elder, and M. Elstein, eds.), Plenum Press, New York, pp. 39–41.

Humphreys-Beher, M., and Carlson, D. M., 1982, Synthesis of a GlcNAc-Gal linkage on asialo-orosomucoid. Abst. Soc. Complex Carbohydrates, No. 25, Hershey, Pennsylvania.

Jabbal, I., and Schachter, H., 1971, Pork liver guanosine diphosphate-L-fucose glycoprotein fucosyltransferases, *J. Biol. Chem.* **246:**5154–5161.

Jamieson, J. C., 1977, Studies on the site of addition of sialic acid and glucosamine to rat α_1-acid glycoprotein, *Can. J. Biochem.* **55:**408–414.

Jarnefelt, J., Rush, J., Li, Y-T., and Laine, R. A., 1978, Erythroglycan, a high molecular weight glycopeptide with the repeating structure [galactosyl-(1-4)-2-deoxy-2-acetamidoglucosyl(1-3)] comprising more than one-third of the protein-bound carbohydrate of human erythrocyte stroma, *J. Biol. Chem.* **253:**8006–8009.

Johnson, P. H., Yates, A. D., and Watkins, W. M., 1981, Human salivary fucosyltransferases: Evidence for two distinct α-3-L-fucosyltransferase activities one of which is associated with the Lewis blood group *Le* gene, *Biochem. Biophys. Res. Commun.* **100:**1611–1618.

Johnston, I. R., McGuire, E. J., Jourdian, G. W., and Roseman, S., 1966, Incorporation of *N*-acetyl-D-glucosamine into glycoproteins, *J. Biol. Chem.* **241:**5735–5737.

Kaur, K. J., Turco, S. J., and Laine, R. A., 1982, Erythroglycan can be elongated by bovine milk UDP-galactose:D-glucose 4-β-galactosyltransferase, *Biochem. Int.* **4:**345–351.

Kawashima, I., Fukuda, K., Tomita, M., and Hamada, A., 1982, Isolation and characterization of alkali-labile oligosaccharide units from porcine erythrocyte glycophorin, *J. Biochem. (Tokyo)* **91:**865–872.

Kiang, W. L., Krusius, T., Finne, J., Margolis, R. U., and Margolis, R. K., 1982, Glycoproteins and proteoglycans of the chromaffin granule matrix, *J. Biol. Chem.* **257:**1651–1659.

Kim, Y. S., Perdomo, J., and Nordberg, J., 1971, Glycoprotein biosynthesis in small intestinal mucosa. I. A study of glycosyltransferases in microsomal subfractions, *J. Biol. Chem.* **246:**5466–5476.

Ko, G. K. W., and Raghupathy, E., 1972, Glycoprotein biosynthesis in the developing rat brain. II. Microsomal galactosaminyltransferase utilizing endogenous and exogenous protein acceptors, *Biochim. Biophys. Acta* **264:**129–143.

Kornfeld, R., and Kornfeld, S., 1980, Structure of glycoproteins and their oligosaccharide units, in: *The Biochemistry of Glycoproteins and Proteoglycans* (W. J. Lennarz, ed.), Plenum Press, New York, pp. 1–34.

Kornfeld, S., 1982, Oligosaccharide processing during glycoprotein biosynthesis, in: *The Glycoconjugates*, Vol. III (M. I. Horowitz, ed.), Academic Press, New York, pp. 3–23.

Li, E., and Kornfeld, S., 1978, Structure of the altered oligosaccharide present in glycoproteins from a clone of Chinese hamster ovary cells deficient in *N*-acetylglucosaminyltransferase activity, *J. Biol. Chem.* **253:**6426–6431.

Li, E., Gibson, R., and Kornfeld, S., 1980, Structure of an unusual complex-type oligosaccharide isolated from Chinese hamster ovary cells, *Arch. Biochem. Biophys.* **199:**393–399.

Liang, C-J., Yamashita, K., Muellenberg, C. G., Shichi, H., and Kobata, A., 1979, Structure of the carbohydrate moieties of bovine rhodopsin, *J. Biol. Chem.* **254:**6414–6418.

Lisowska, E., Duk, M., and Dahr, W., 1980, Comparison of alkali-labile oligosaccharide chains of M and N blood-group glycopeptides from human erythrocyte membrane, *Carbohydrate Res.* **79:**103–113.

Lombart, C. G., and Winzler, R. J., 1974, Isolation and characterization of oligosaccharides from canine submaxillary mucin, *Eur. J. Biochem.* **49:**77–86.

Longmore, G. D., and Schachter, H., 1982, Product identification and substrate specificity studies of the GDP-L-fucose:2-acetamido-2-deoxy-β-D-glucoside (Fuc to Asn-linked GlcNAc) 6-α-L-fucosyltransferase in a Golgi-rich fraction from porcine liver, *Carbohydrate Res.* **100:**365–392.

McGuire, E. J., 1970, Biosynthesis of submaxillary mucins, in: *Blood and Tissue Antigens* (D. Aminoff, ed.), Academic Press, New York, pp. 461–478.

McGuire, E. J., and Roseman, S., 1967, Enzymatic synthesis of the protein-hexosamine linkage in sheep submaxillary mucin, *J. Biol. Chem.* **242:**3745–3747.

Mendicino, J., Chandrasekaran, E. V., Anumula, K. R., and Davila, M., 1981, Isolation and properties of α-D-mannose:β-1,2-*N*-acetylglucosaminyltransferase from trachea mucosa, *Biochemistry* **20:**967–976.

Mendicino, J., Sivakami, S., Davila, M., and Chandrasekaran, E. V., 1982, Purification and properties of UDP-Gal:*N*-acetylgalactosaminide mucin:β1,3-galactosyltransferase from swine trachea mucosa, *J. Biol. Chem.* **257:**3987–3994.

Munro, J. R., Narasimhan, S., Wetmore, S., Riordan, J. R., and Schachter, H., 1975, Intracellular localization of GDP-L-fucose:glycoprotein and CMP-sialic acid:apolipoprotein glycosyltransferases in rat and pork liver, *Arch. Biochem. Biophys.* **169:**269–277.

Narasimhan, S., 1982, Control of glycoprotein synthesis. UDP-GlcNAc:glycopeptide β4-*N*-acetylglucosaminyltransferase III, an enzyme in hen oviduct which adds GlcNAc in β1-4 linkage to the β-linked mannose of the tri-mannosyl core of *N*-glycosyl oligosaccharides, *J. Biol. Chem.* **257:**10235–10242.

Narasimhan, S., Stanley, P., and Schachter, H., 1977, Control of glycoprotein synthesis. Lectin-resistant mutant containing only one of two distinct *N*-acetylglucosaminyltransferase activities present in wild type Chinese hamster ovary cells, *J. Biol. Chem.* **252:**3926–3933.

Narasimhan, S., Tsai, D., and Schachter, H., 1981, An *N*-acetylglucosaminyltransferase (Gn-T) in hen oviduct which adds GlcNAc to the β-linked Man of the tri-mannosyl core of Asn-linked oligosaccharides, *Fed. Proc.* (Abst. No. 329) **40:**1597.

Newman, W., and Kabat, E. A., 1976, Immunochemical studies on blood groups. Structures and immunochemical properties of nine oligosaccharides from B-active and non-B-active blood group substances of horse gastric mucosae, *Arch. Biochem. Biophys.* **172:**535–550.

Nilsson, B., De Luca, S., Lohmander, S., and Hascall, V. C., 1982, Structures of *N*-linked and *O*-linked oligosaccharides on proteoglycan monomer isolated from the swarm rat chondrosarcoma, *J. Biol. Chem.* **257:**10920–10927.

Oates, M. D. G., Rosbottom, A. C., and Schrager, J., 1974, Further investigations into the structure of human gastric mucin: The structural configuration of the oligosaccharide chains, *Carbohydrate Res.* **34:**115–137.

Oppenheimer, C. L., and Hill, R. L., 1981, Purification and characterization of a rabbit liver α1-3 mannoside β1-2 *N*-acetylglucosaminyltransferase, *J. Biol. Chem.* **256:**799–804.

Oppenheimer, C. L., Eckhardt, A. E., and Hill, R. L., 1981, The non-identity of porcine *N*-acetylglucosaminyltransferases I and II, *J. Biol. Chem.* **256:**11477–11482.

Paquet, M., Narasimhan, S., Schachter, H., and Moscarello, M. A., 1984, Branch specificities of purified rat liver golgi UDP-Gal:GlcNAc β1-4-Gal-transferase, *J. Biol. Chem.* **259:**4716–4721.

Paulson, J. C., Beranek, W. E., and Hill, R. L., 1977a, Purification of a sialyltransferase from bovine colostrum by affinity chromatography on CDP-agarose, *J. Biol. Chem.* **252:**2356–2362.

Paulson, J. C., Rearick, J. I., and Hill, R. L., 1977b, Enzymatic properties of β-D-galactoside α2-6-sialyltransferase from bovine colostrum, *J. Biol. Chem.* **252:**2363–2371.

Paulson, J. C., Weinstein, J., and de Souza-e-Silva, U., 1982, Identification of a Galβ1-3GlcNAc α2-3-sialyltransferase in rat liver, *J. Biol. Chem.* **257:**4034–4037.

Pierce-Cretel, A., Pamblanco, M., Strecker, G., Montreuil, J., and Spik, G., 1981, Heterogeneity of the glycans *O*-glycosidically linked to the hinge region of the secretory immunoglobulins from human milk, *Eur. J. Biochem.* **114:**169–178.

Prieels, J-P., Monnom, D., Dolmans, M., Beyer, T. A., and Hill, R. L., 1981, Co-purification of the Lewis blood group *N*-acetylglucosaminide α1-4 fucosyltransferase and *N*-acetylglucosaminide α1-3 fucosyltransferase from human milk, *J. Biol. Chem.* **256:**10456–10463.

Rearick, J. I., Sadler, J. E., Paulson, J. C., and Hill, R. L., 1979, Enzymatic characterization of β-D-galactoside α2-3-sialyltransferase from porcine submaxillary gland, *J. Biol. Chem.* **254:**4444–4451.

Rearick, J. I., Chapman, A., and Kornfeld, S., 1981a, Glucose starvation alters lipid-linked oligosaccharide biosynthesis in Chinese hamster ovary cells, *J. Biol. Chem.* **256:**6255–6261.

Rearick, J. I., Cummings, R., and Kornfeld, S., 1981b, Specific assay for UDP-GlcNAc:β-galactoside *N*-acetylglucosaminyltransferase, Abst. Soc. Complex Carbohydrates, No. 63, Philadelphia, Pennsylvania.

Reddy, M. S., Shah, R. H., and Bahl, O. P., 1979, Structures of the carbohydrate units of mucins from normal and fibrocystic human submaxillary secretions, *Abst. Soc. Complex Carbohydrates, No. 40, Toronto, Canada.*

Reddy, M. S., Prakobphol, A., Levine, M. J., and Tabak, L. A., 1982, Structures of the *O*-glycosidic units of a lower molecular weight salivary mucin, Abst. Soc. Complex Carbohydrates, No. 29, Hershey, Pennsylvania.

Robertson, M. A., Etchison, J. R., Robertson, J. S., Summers, D. F., and Stanley, P., 1978, Specific changes in the oligosaccharide moieties of VSV grown in different lectin-resistant CHO cells, *Cell* **13:**515–526.

Ronzio, R. A., 1973a, Glycoprotein synthesis in the adult rat pancreas. I. Subcellular distributions of uridine diphosphate galactose:glycoprotein galactosyltransferase and thiamine pyrophosphate phosphohydrolase, *Biochim. Biophys. Acta* **313:**286–295.

Ronzio, R. A., 1973b, Glycoprotein synthesis in the adult rat pancreas. II. Characterization of Golgi-rich fractions, *Arch. Biochem. Biophys.* **159:**777–784.

Rovis, L., Anderson, B., Kabat, E. A., Gruezo, F., and Liao, J., 1973, Structures of oligosaccharides produced by base-borohydride degradation of human ovarian cyst blood group H, Le^b and Le^a active glycoproteins, *Biochemistry* **12:**5340–5354.

Russin, T. Z., Laine, R. A., and Turco, S. J., 1981, Cell-free biosynthesis of erythroglycan in a microsomal fraction from K-562 cells, *Biochem. J.* **197:**327–332.

Sadler, J. E., Rearick, J. I., Paulson, J. C., and Hill, R. L., 1979a, Purification to homogeneity of a β-galactoside α2-3-sialyltransferase and partial purification of an α-*N*-acetylgalactosaminide α2-6-sialyltransferase from porcine submaxillary glands, *J. Biol. Chem.* **254:**4434–4443.

Sadler, J. E., Rearick, J. I., and Hill, R. L., 1979b, Purification to homogeneity and enzymatic characterization of an α-*N*-acetylgalactosaminide α2-6-sialyltransferase from porcine submaxillary glands, *J. Biol. Chem.* **254:**5934–5941.

Sadler, J. E., Beyer, T. A., Oppenheimer, C. L., Paulson, J. C., Prieels, J-P., Rearick, J. I., and Hill, R. L., 1982, Purification of mammalian glycosyltransferases, *Meth. Enzymol.* **83:**458–514.

Saito, T., Itoh, T., and Adachi, S., 1981, The chemical structure of a tetrasaccharide containing *N*-acetylglucosamine obtained from bovine colostrum kappa-casein, *Biochim. Biophys. Acta* **673:**487–494.

Schachter, H., 1978, Glycoprotein biosynthesis, in: *The GLycoconjugates,* Vol. 2 (W. Pigman and M. I. Horowitz, eds.), Academic Press, New York, pp. 87–181.

Schachter, H., and Roden, L., 1973, The biosynthesis of animal glycoproteins, in: *Metabolic Conjugation and Metabolic Hydrolysis,* Vol. III (W. H. Fishman, ed.), Academic Press, New York, pp. 1–149.

Schachter, H., and Roseman, S., 1980, Mammalian glycosyltransferases: Their role in the synthesis and function of complex carbohydrates and glycolipids, in: *Biochemistry of Glycoproteins and Proteoglycans* (W. J. Lennarz, ed.), Plenum Press, New York, pp. 85–160.

Schachter, H., and Tilley, C. A., 1978, The biosynthesis of human blood group substances, in: *International Review of Biochemistry,* Vol. 16, *Biochemistry of Carbohydrates, Series II* (D. J. Manners, ed.), University Park Press, Baltimore, pp. 209–246.

Schachter, H., and Williams, D., 1982, Biosynthesis of mucus glycoproteins, in: *Mucus in Health and Disease,* Vol. II (E. N. Chantler, J. B. Elder, and M. Elstein, eds.), *Adv. Exp. Med. Biol.* **144:**3–28, Plenum Press, New York and London.

Schachter, H., McGuire, E. J., and Roseman, S., 1971, Sialic acids. XIII. A uridine diphosphate D-galactose:mucin galactosyltransferase from porcine submaxillary gland, *J. Biol. Chem.* **246:**5321–5328.

Schenkel-Brunner, H., 1973, Incorporation of galactose into blood-groups (ABH) precursor substance by lactose synthetase from human milk. Effects on cross reactivity with anti-Type-14 pneumococcus serum, *Eur. J. Biochem.* **33:**30–35.

Schenkel-Brunner, H., Kabat, E. A., and Liao, J., 1979, Biosynthesis of a blood-group-I determinant reacting with anti-I Ma serum (Group 1), *Eur. J. Biochem.* **98:**573–575.

Schwyzer, M., and Hill, R. L., 1977a, Porcine A blood group-specific *N*-acetylgalactosaminyltransferase. I. Purification from porcine submaxillary glands, *J. Biol. Chem.* **252:**2338–2345.

Schwyzer, M., and Hill, R. L., 1977b, Porcine A blood group-specific *N*-acetylgalactosaminyltransferase. II. Enzymatic properties, *J. Biol. Chem.* **252:**2346–2355.

Sheares, B. T., and Carlson, D. M., 1982, Isolation of UDP-Gal:GlcNAc 3β-galactosyltransferase from pig trachea, Abst. Soc. Complex Carbohydrate, No. 62, Hershey, Pennsylvania.

Sheares, B. T., Lau, J. T. Y., and Carlson, D. M., 1982, Biosynthesis of galactosyl-β1,3-*N*-acetylglucosamine, *J. Biol. Chem.* **257:**599–602.

Shier, W. T., and Roloson, G. J., 1977, Preparation and galactosyltransferase acceptor activity of derivatives of antifreeze glycoproteins of an Antarctic fish, *Can. J. Biochem.* **55:**886–893.

Slomiany, A., and Slomiany, B. L., 1978, Structures of the acidic oligosaccharides isolated from rat sublingual glycoprotein, *J. Biol. Chem.* **253:**7301–7306.

Slomiany, B. L., Murty, V. L. N., and Slomiany, A., 1980, Isolation and characterization of oligosaccharides from rat colonic mucus glycoprotein, *J. Biol. Chem.* **255:**9719–9723.

Slomiany, B. L., Banas-Gruszka, Z., Zdebska, E., and Slomiany, A., 1982, Characterization of the Forssman-active oligosaccharides from dog gastric mucus glycoprotein isolated with the use of a monoclonal antibody, *J. Biol. Chem.* **257:**9561–9565.

Soulier, S., Sarfati, R. S., and Szabo, L., 1980, Structure of the asialyl oligosaccharide chains of casein isolated from ovine colostrum, *Eur. J. Biochem.* **108:**465–472.

Spiro, R. G., and Bhoyroo, V. D., 1974, Structure of the *O*-glycosidically linked carbohydrate units of fetuin, *J. Biol. Chem.* **249:**5704–5717.

Stanley, P., Narasimhan, S., Siminovitch, L., and Schachter, H., 1975, Chinese hamster ovary cells selected for resistance to the cytotoxicity of phytohemagglutinin are deficient in a UDP-*N*-acetylglucosamine:glycoprotein *N*-acetylglucosaminyltransferase activity, *Proc. Natl. Acad. Sci. USA* **72:**3323–3327.

Strous, G. J. A. M., Hendriks, H. G. C. J. M., and Kramer, M. F., 1980, Role of galactosyl-transferases in rat gastric epithelial glycoprotein synthesis, *Biochim. Biophys. Acta* **613:**381–391.

Struck, D. K., and Lennarz, W. J., 1980, The function of saccharide-lipids in synthesis of glycoproteins, in: *The Biochemistry of Glycoproteins and Proteoglycans* (W. J. Lennarz, ed.), Plenum Press, New York, pp. 35–83.

Sugiura, M., Kawasaki, T., and Yamashina, I., 1982, Purification and characterization of UDP-GalNAc:polypeptide *N*-acetylgalactosamine transferase from an ascites hepatoma, AH 66, *J. Biol. Chem.* **257:**9501–9507.

Tabak, L. A., Dickson, L., Reddy, M. S., Levine, M. J., Kuatt, B. L., and Baum, B. J., 1982, Isolation and partial characterization of a mucin-glycoprotein from rat submandibular glands, Abst. Soc. Complex Carbohydrates, No. 27, Hershey, Pennsylvania.

Tabas, I., and Kornfeld, S., 1978, The synthesis of complex type oligosaccharides. III. Identification of an α-D-mannosidase activity involved in a late stage of processing of complex-type oligosaccharides, *J. Biol. Chem.* **253:**7779–7786.

Tabas, I., and Kornfeld, S., 1982, *N*-asparagine-linked oligosaccharides: Processing, *Meth. Enzymol.* **83:**416–429.

Tabas, I., Schlesinger, S., and Kornfeld, S., 1978, Processing of high mannose oligosaccharides to form complex type oligosaccharides on the newly synthesized polypeptides of the vesicular stomatitis virus G protein and the IgG heavy chain, *J. Biol. Chem.* **253:**716–722.

Tai, T., Yamashita, K., Ito, S., and Kobata, A., 1977, Structures of the carbohydrate moiety of ovalbumin glycopeptide III and the difference in specificity of endo-β-*N*-acetylglucosaminidases C_{II} and H, *J. Biol. Chem.* **252:**6687–6694.

Takasaki, S., Ikehira, H., and Kobata, A., 1980, Increase of asparagine-linked oligosaccharides with branched outer chains caused by cell transformation, *Biochem. Biophys. Res. Commun.* **92:** 735–742.

Tanaka, M., Anderson, B., and Dube, V. E., 1982, A new I-active sequence in oligosaccharides of ovarian cyst glycoprotein, Abst. Soc. Complex Carbohydrates, No. 10, Hershey, Pennsylvania.

Thomas, D. B., and Winzler, R. J., 1969, Structural studies on human erythrocyte glycoproteins: Alkali-labile oligosaccharides, *J. Biol. Chem.* **244:**5943–5946.

Tulsiani, D. R. P., Harris, T. M., and Touster, O., 1982a, Swainsonine inhibits the biosynthesis of complex glycoproteins by inhibition of Golgi mannosidase II. *J. Biol. Chem.* **257:**7936–7939.

Tulsiani, D. R. P., Hubbard, S. C., Robbins, P. W., and Touster, O., 1982b, α-D-Mannosidases of rat liver Golgi membranes. Mannosidase II is the GlcNAcMan$_5$-cleaving enzyme in glycoprotein biosynthesis and mannosidases IA and IB are the enzymes converting Man$_9$ precursors to Man$_5$ intermediates, *J. Biol. Chem.* **257:**3660–3668.

Turco, S. J., Stetson, B., and Robbins, P. W., 1977, Comparative rates of transfer of lipid-linked oligosaccharides to endogenous glycoprotein acceptors *in vitro, Proc. Natl. Acad. Sci. USA* **74:**4411–4414.

Turco, S. J., Rush, J. S., and Laine, R. A., 1980, Presence of erythroglycan on human K-562 chronic myelogenous leukemia-derived cells, *J. Biol. Chem.* **255:**3266–3269.

Umemoto, J., Bhavanandan, V. P., and Davidson, E. A., 1981, Isolation and partial characterization of a mucin-type glycoprotein from plasma membranes of human melanoma cells, *Biochim. Biophys. Acta* **646:**402–410.

Van den Eijnden, D. H., and Schiphorst, W. E. C. M., 1981, Detection of β-galactosyl(1-4)*N*-acetylglucosaminide α(2-3)-sialyltransferase activity in fetal calf liver and other tissues, *J. Biol. Chem.* **256:**3159–3162.

Van den Eijnden, D. H., Stoffyn, P., Stoffyn, A., and Schiphorst, W. E. C. M., 1977, Specificity of sialyltransferase: Structure of α_1-acid glycoprotein sialylated *in vitro, Eur. J. Biochem.* **81:**1–7.

Van den Eijnden, D. H., Barneveld, R. A., and Schiphorst, W. E. C. M., 1979a, Structure of the disaccharide chain of galactosyl-*N*-acetylgalactosaminyl-protein synthesized *in vitro, Eur. J. Biochem.* **95:**629–637.

Van den Eijnden, D. H., Evans, N. A., Codington, J. F., Reinhold, V., Silber, C., and Jeanloz, R. W., 1979b, Chemical structure of epiglycanin, the major glycoprotein of the TA3-Ha ascites cell. The carbohydrate chains, *J. Biol. Chem.* **254:**12153–12159.

Van den Eijnden, D. H., Joziasse, D. H., Dorland, L., Van Halbeek, H., Vliegenthart, J. F. G., and Schmid, K., 1980, Specificity in the enzymic transfer of sialic acid to the oligosaccharide branches of bi- and triantennary glycopeptides of α_1-acid glycoprotein, *Biochem. Biophys. Res. Commun.* **92:**839–845.

Van den Eijnden, D. H., Bergh, M. L. E., Dieleman, B., and Schiphorst, W. E. C. M., 1981, Specificity of sialyltransferase: Sialylation of ovine submaxillary mucin *in vitro, Hoppe-Seyler's Z. Physiol. Chem.* **362:**113–124.

Van den Eijnden, D. H., Bergh, M. L. E., Joziasse, D. H., Blanken, W. M., and Koppen, P. L., 1982, Application of high-pressure liquid chromatography (HPLC) in the study of glycosyltransferases. Abst. XIth Int. Carbohydrate Symposium, No. IV-21, Vancouver, Canada.

Van Halbeek, H., Dorland, L., Vliegenthart, J. F. G., Fiat, A. M., and Jolles, P., 1980, A 360-MHz ^{1}H-NMR study of three oligosaccharides isolated from cow kappa-casein, *Biochim. Biophys. Acta* **623:**295–300.

Van Halbeek, H., Dorland, L., Haverkamp, J., Veldink, G. A., Vliegenthart, J. F. G., Fournet, B., Ricart, G., Montreuil, J., Gathmann, W. D., and Aminoff, D., 1981a, Structure determination of oligosaccharides isolated from A^+, H^+ and A-H-hog-submaxillary-gland mucin glycoproteins, by 360-MHz ^{1}H-NMR spectroscopy, permethylation analysis and mass spectrometry, *Eur. J. Biochem.* **118:**487–495.

Van Halbeek, H., Dorland, L., Vliegenthart, J. F. G., Fiat, A. M., and Jolles, P., 1981b, Structural characterization of a novel acidic oligosaccharide unit derived from cow colostrum kappa-casein, *FEBS Lett.* **133:**45–50.

Van Halbeek, H., Dorland, L., Vliegenthart, J. F. G., Hull, W. E., Lamblin, G., Lhermitte, M., Boersma, A., and Roussel, P., 1982a, Primary-structure determination of fourteen neutral oligosaccharides derived from bronchial-mucus glycoproteins of patients suffering from Cystic Fibrosis, employing 500-MHz ^{1}H-NMR spectroscopy, *Eur. J. Biochem.* **127:**7–20.

Van Halbeek, H., Dorland, L., Vliegenthart, J. F. G., Kochetkov, N. K., Arbatsky, N. P., and Derevitskaya, V. A., 1982b, Characterization of the primary structure and the microheterogeneity of the carbohydrate chains of porcine blood-group H substance by 500-MHz ^{1}H-NMR spectroscopy, *Eur. J. Biochem.* **127:**21–29.

Vischer, P., and Hughes, R. C., 1981, Glycosyl transferases of baby-hamster-kidney (BHK) cells and ricin-resistant mutants. *N*-glycan biosynthesis, *Eur. J. Biochem.* **117:**275–284.

Watkins, W. M., 1980, Biochemistry and genetics of the ABO, Lewis and P blood group systems, in: *Advances in Human Genetics,* Vol. 10 (H. Harris and K. Hirschhorn, eds.), Plenum Press, New York, pp. 1–136, 379–385.

Weinstein, J., De Souza-e-Silva, U., and Paulson, J. C., 1982a, Purification of a Galβ1-4GlcNAc α2-6-sialyltransferase and a Galβ1-3(4)GlcNAc α2-3-sialyltransferase to homogeneity from rat liver, *J. Biol. Chem.* **257:**13835–13844.

Weinstein, J., DeSouza-e-Silva, U., and Paulson, J. C., 1982b, Sialylation of glycoprotein oligosaccharides *N*-linked to asparagine. Enzymatic characterization of a Galβ1-3(4)GlcNAc α2-3-sialyltransferase and a Galβ1-4GlcNAc α2-6-sialyltransferase from rat liver, *J. Biol. Chem.* **257:**13845–13853.

Wetmore, S., Mahley, R. W., Brown, W. V., and Schachter, H., 1974, Incorporation of sialic acid into sialidase-treated apolipoprotein of human very low density lipoprotein by a pork liver sialyltransferase, *Can. J. Biochem.* **52:**655–664.

Williams, D., and Schachter, H., 1980, Mucin synthesis. I. Detection in canine submaxillary glands of an *N*-acetylglucosaminyltransferase which acts on mucin substrates, *J. Biol. Chem.* **255:**11247–11252.

Williams, D., Longmore, G. D., Matta, K. L., and Schachter, H., 1980, Mucin synthesis. II. Substrate specificity and product identification studies on canine submaxillary gland UDP-GlcNAc:Galβ1-3GalNAc (GlcNAc to GalNAc) β6-*N*-acetylglucosaminyltransferase, *J. Biol. Chem.* **255:**11253–11261.

Wilson, J. R., Williams, D., and Schachter, H., 1976, The control of glycoprotein synthesis: *N*-acetylglucosamine linkage to mannose residue as a signal for the attachment of L-fucose to the asparagine-linked *N*-acetylglucosamine residue of glycopeptide from α_1-acid glycoprotein, *Biochem. Biophys. Res. Commun.* **72:**909–916.

Wingert, W. E., and Cheng, P-W., 1982, Characterization of rabbit intestinal UDP-GlcNAc:Galβ3GalNAc (GlcNAc to GalNAc) β6-*N*-acetylglucosaminyltransferase, Abst. Soc. Complex Carbohydrates, No. 49, Hershey, Pennsylvania.

Wood, E., Hounsell, E. F., Langhorne, J., and Feizi, T., 1980, Sheep gastric mucins as a source of blood-group-I and -i antigens, *Biochem. J.* **187:**711–718.

Wood, E., Hounsell, E. F., and Feizi, T., 1981, Preparative affinity chromatography of sheep gastric mucins having blood-group Ii activity, and release of antigenically active oligosaccharides by alkaline-borohydride degradation, *Carbohydrate Res.* **90:**269–282.

Wu, A. M., Slomiany, A., Herp, A., and Slomiany, B. L., 1979, Structural studies on the carbohydrate units of armadillo submandibular glycoprotein, *Biochim. Biophys. Acta* **578:**297–304.

Wu, A. M., Kabat, E. A., Pereira, M. E. A., Gruezo, F. G., and Liao, J., 1982, Immunochemical studies on blood groups: The internal structure and immunological properties of water-soluble human blood group A substance studied by Smith degradation, liberation, and fractionation of oligosaccharides and reaction with lectins, *Arch. Biochem. Biophys.* **215:**390–404.

Young, J. D., Tsuchiya, D., Sandlin, D. E., and Holroyde, M. J., 1979, Enzymic *O*-glycosylation of synthetic peptides from sequences in basic myelin protein, *Biochemistry* **18:**4444–4448.

Yurewicz, E. C., and Moghissi, K. S., 1981, Purification of human midcycle cervical mucin and characterization of its oligosaccharides with respect to size, composition, and microheterogeneity, *J. Biol. Chem.* **256:**11895–11904.

Yurewicz, E. C., Matsuura, F., and Moghissi, K. S., 1982a, Structural characterization of neutral oligosaccharides of human midcycle cervical mucin, *J. Biol. Chem.* **257:**2314–2322.

Yurewicz, E. C., Matsuura, F., and Moghissi, K. S., 1982b, Sialylated oligosaccharides of human cervical mucin, Abst. Soc. Complex Carbohydrates, No. 30, Hershey, Pennsylvania.

Ziderman, D., Gompertz, S., Smith, Z. G., and Watkins, W. M., 1967, Glycosyltransferases in mammalian gastric mucosal linings, *Biochem. Biophys. Res. Commun.* **29:**56–61.

21

Biosynthesis of the Bacterial Envelope Polymers Teichoic Acid and Teichuronic Acid

Ian C. Hancock and James Baddiley

I. INTRODUCTION

Bacteria are surrounded by a cell wall which is responsible for the shape and osmotic stability of the cell. Electron microscopy of thin sections of Gram-positive bacteria reveals a single layer of wall, in which little fine structures can be determined lying outside the cytoplasmic membrane. This contrasts with the picture of a Gram-negative cell envelope, in which an outer membrane surrounds a much thinner, structureless layer of peptidoglycan outside the cytoplasmic membrane.

The difference in morphology is reflected in the wall chemistry. Although environmental conditions may influence the precise composition (Ellwood and Tempest, 1972), in general, cell walls of Gram-positive bacteria contain large proportions of teichoic acids, teichuronic acids, and polysaccharides, as well as peptidoglycan. These polymers are covalently linked to the peptidoglycan layer throughout its thickness. On the other hand, the additional components of the Gram-negative envelope lie in the outer membrane and, apart from lipoprotein, are not covalently linked to peptidoglycan.

The way these polymers of the Gram-positive wall, external to the cytoplasmic membrane, are synthesized by membrane-bound enzymes from cytoplasmic nucleotide precursors and become assembled into the complex structure of the wall will be the principal subject of this review.

Ian C. Hancock • Department of Microbiology, University of Newcastle upon Tyne, Newcastle upon Tyne NE1 7RU, England. *James Baddiley* • Department of Biochemistry, University of Cambridge, Cambridge CB1 1QW, England.

II. TEICHOIC ACID STRUCTURE

The teichoic acids are a group of phosphate-containing polymers found in many Gram-positive bacterial cell walls and membranes. (The capsular antigen of one strain of *E. coli* and the lipoteichoic acids of representatives of the genus of rumen bacteria *Butyrivibrio,* are the only known occurrences of teichoic acids in Gram-negative bacteria). The name was originally applied to polymers of ribitol phosphate and glycerol phosphate but is now more generally used for related polymers which also contain sugar residues as an integral part of the polymer chain. Teichoic acids can represent between 20 and 60% of the weight of the wall, while nearly all Gram-positive bacteria also contain membrane teichoic acids (lipoteichoic acids, LTA) which are bound to the cytoplasmic membrane by a covalently linked glycolipid moiety. The structure and function of teichoic acids have been previously reviewed by Baddiley (1972) and Archibald (1974), where complete references can be found.

A. Poly(Alditol Phosphate)

These are polymers of glycerol phosphate or ribitol phosphate in which the alditol residues are joined together by phosphodiester linkages.

1. Ribitol Teichoic Acids

These are located exclusively in bacterial cell walls. The phosphodiester linkages extend between positions 1 and 5 on adjacent ribitol residues (Figure 1a). The ribitol teichoic acids of strains of *Staphylococcus aureus* have been most thoroughly investigated (Baddiley *et al.*, 1962). They have been shown to carry *N*-acetylglucosamine (GlcNAc) and D-alanine substituents on the ribitol residues. GlcNAc may be in α- or β-glycosidic linkage on the D-4 position of ribitol, with D-alanine in ester linkage at D-2. Both types of linkage may be present in the teichoic acid from a single strain, although in any one molecule only one configuration is found. Alpha- and β-linked glucosyl residues are found in other species, and more than one substituent may be present, as for example in the teichoic acid from *Lactobacillus arabinosus* 17-5, where glucosyl groups are found at both D-3 and D-4 of some ribitol residues.

2. Glycerol Teichoic Acids

These occur as both wall and membrane polymers and have the general structure shown in Figure 1b (Shaw and Baddiley, 1964). Phosphodiester linkages join glycerol residues between the hydroxyl groups at C-1 and C-3 in most examples, but a few cases of linkage between 2 and 3 have been reported, and the detailed structure of such a polymer from *Streptomyces antibioticus* 39 has recently been described (Shashkova *et al.*, 1979). As in ribitol teichoic acids, glycosyl and D-alanyl substituents are usually present. Cases of partial glycosylation of individual polymer chains have been reported, but in other cases, mixtures of fully glycosylated and unglycosylated chains are found.

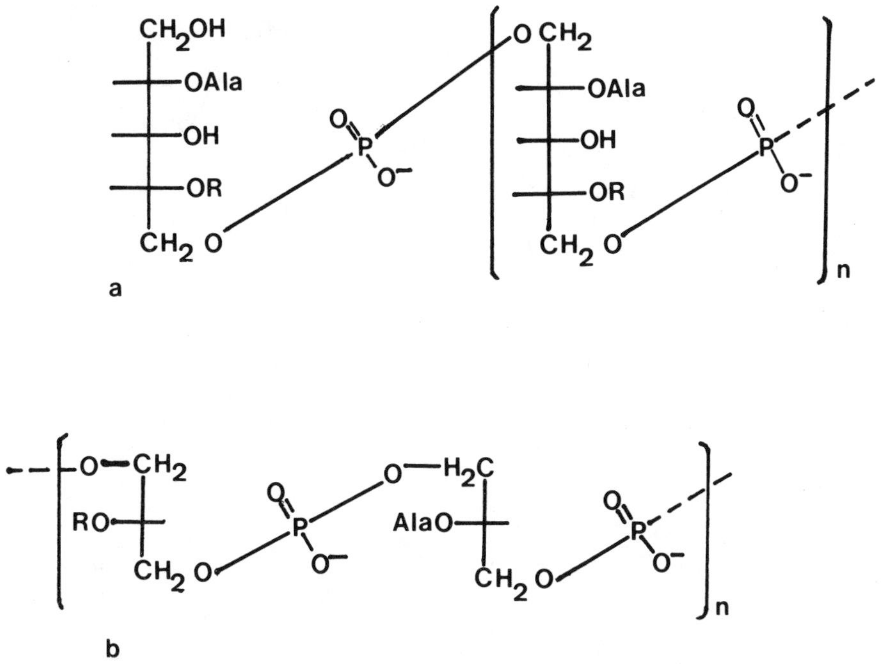

Figure 1. Poly (alditol phosphate) teichoic acids. (a) Poly(ribitol phosphate), (b) poly(glycerol phosphate). R, glycosyl; Ala, D-alanine.

3. Mannitol Teichoic Acids

There have been two recent reports of poly(mannitol phosphate) in the walls of coryneform bacteria. Anderton and Wilkinson (1980) found such a polymer in the unclassified species N.C.T.C. 9742, and Fiedler *et al.* (1981) observed a polymer with the same backbone but with different sugar substituents in *Brevibacterium linens*. A detailed structure is not available for either polymer.

B. Polymers with Glycosyl Residues as Part of the Chain

Poly(glycosyl glycerol phosphate) and poly(galactosylglycerol phosphate) have been identified in the cell wall of *B. licheniformis* A.T.C.C. 9945 (Burger and Glaser, 1966). A mixture of related teichoic acids also occurs in *L. plantarum* C106, including two polymers of isomeric diglucosylglycerol phosphate (Figure 2). Similar polymers have been found in a number of other bacteria, perhaps the most interesting being the K2 capsular antigen of *E. coli,* one of the rare examples of a teichoic acid in a Gram-negative bacterium. Fischer *et al.* (1982) have shown that it consists of repeating units of galactofuranosylglycerol phosphate and galactopyranosyl glycerol phosphate.

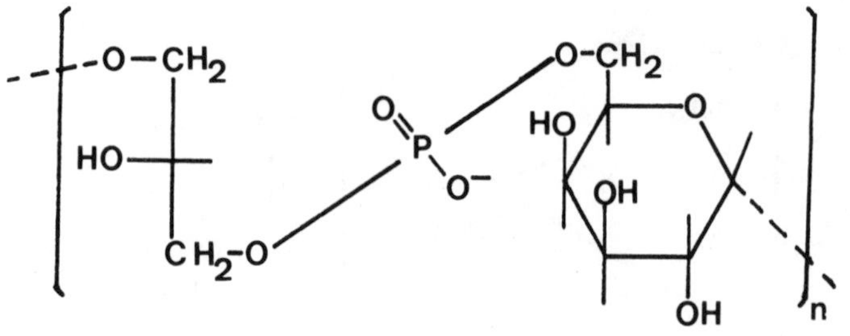

Figure 2. Teichoic acids with sugars in the chain. (a) Poly(glycerophosphory glucose) from *Bacillus licheniformis* ATCC 9945, (b) poly(glycerophosphoryldiglucose) from *Lactobacillus plantarum* C106.

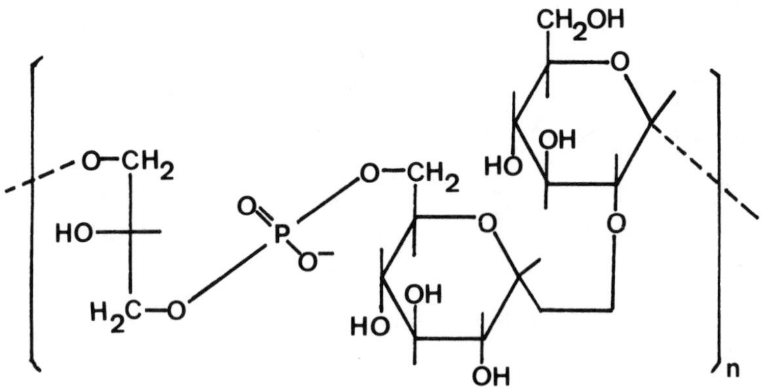

C. Sugar 1-Phosphate Polymers

These polymers contain sugar 1-phosphate linkages in the main polymer chains, which are therefore very acid-labile. Figure 3 shows the best characterized polymers of this type, those from *Staphylococcus lactis* I3 and *Micrococcus* sp. 2102 (Archibald *et al.*, 1968).

D. Attachment of Wall Teichoic Acids to Peptidoglycan

It was known for many years that the covalent linkage of teichoic acids to peptidoglycan in the cell wall was very susceptible to hydrolysis under both acid and alkaline conditions. The identification of muramic acid 6-phosphate in acid hydrolysates of the walls of many Gram-positive bacteria (Liu and Gottschlick, 1967) led to the general acceptance that phosphodiester groups were involved in the attachment of secondary polymers, but the precise nature of the linkage was only recognized following the discovery by Heckels *et al.* (1975) that in *S. aureus,* poly(ribitol phosphate) was attached by a linkage unit containing tri(glycerol phosphate). The subsequent discovery that biosynthesis of this linkage unit (LU) required UDP-GlcNAc as well as CDP-glycerol (Section III-B.2) led to the identification of a single *N*-acetylglucosamine 1-phosphate residue as part of the LU, intervening between the tri(glycerol phosphate) part and C-6 of muramic acid in the peptidoglycan (Coley *et al.*, 1977). The complete structure is shown in Figure 4, which indicates the sites of acid and alkali hydrolysis. Determination of linkage unit structure has been described by Coley *et al.* (1978). More recently, there has been a report of a slightly different linkage unit, containing

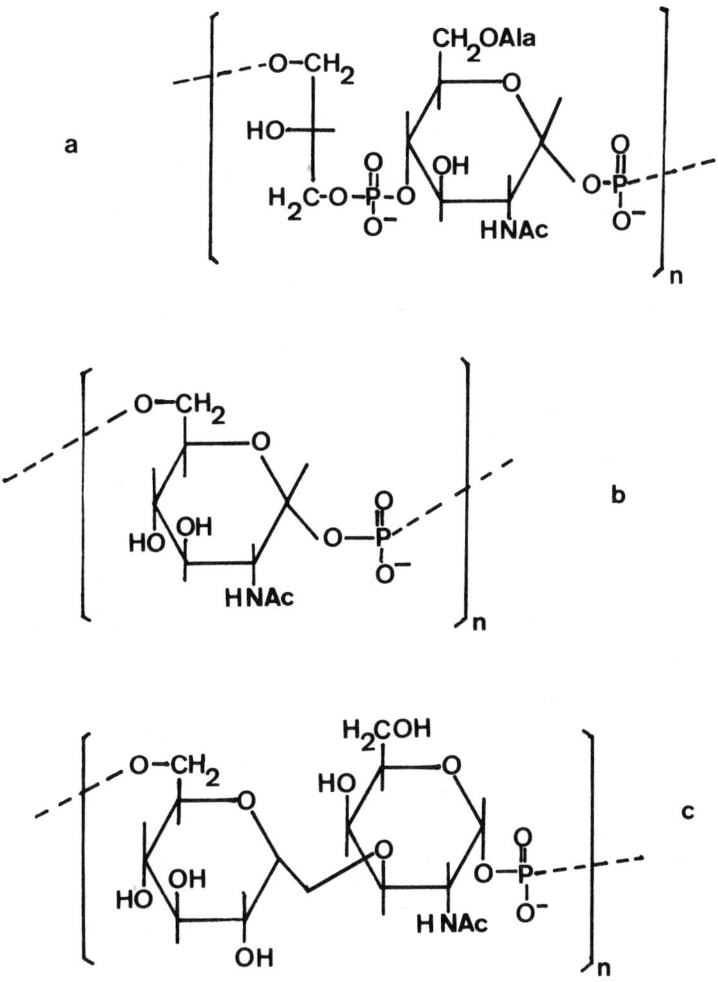

Figure 3. Teichoic acids containing sugar 1-phosphate residues. (a) From *Staphylococcus lactis* I3, (b) from *Micrococcus varians* ATCC29750 (*S. lactis* 2102), (c) from *Micrococcus* sp. A1.

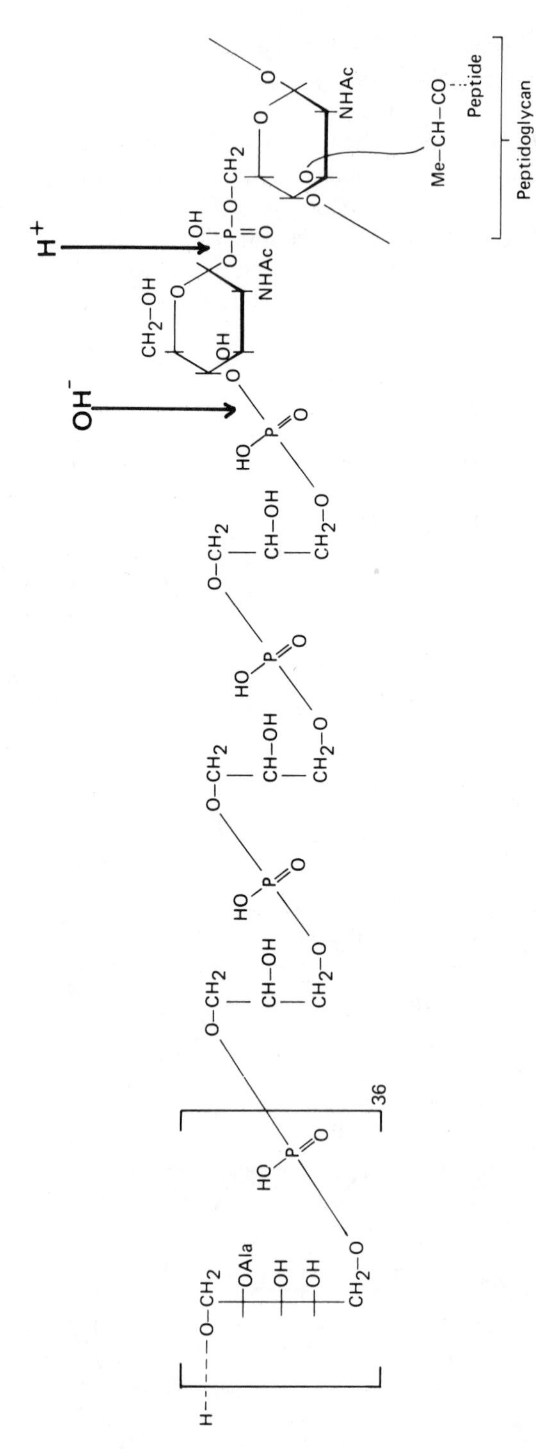

Figure 4. Structure of the linkage unit that joins poly(ribitol phosphate) to peptidoglycan in *Staphylococcus aureus*. The arrows indicate the bonds that are particularly labile to acid (H^+) and alkali (OH^-).

N-acetylmannosaminyl *N*-acetylglucosamine 1-phosphate, attaching poly(glycerol phosphate) to peptidoglycan in *B. cereus* (Saski *et al.*, 1980).

III. BIOSYNTHESIS OF TEICHOIC ACIDS

The biosynthetic pathways can be envisaged in three stages: synthesis of nucleotide precursors, assembly of the polymer on a membrane-bound carrier, and finally, attachment of the peptidoglycan.

A. Biosynthesis of Nucleotide Precursors

The donors of glycerol and ribitol phosphate residues for main chain synthesis are CDP-glycerol and CDP-ribitol, whose discovery preceded the recognition of teichoic acids (Baddiley *et al.*, 1956). These are synthesized from CTP and D-glycerol 1-phosphate or D-ribitol 5-phosphate, in reactions catalyzed by the appropriate pyrophosphorylases. These enzymes are generally considered to be soluble cytoplasmic proteins, having been isolated from the cytoplasmic fractions of a number of bacteria (Shaw, 1962). However, there is evidence that under some circumstances they may be associated with the cytoplasmic membrane *in vivo*. CDP-glycerol pyrophosphorylase activity has been demonstrated in washed membranes from *Lactobacillus buchneri* and *S. lactis* I3 after mechanical cell disruption. Douglas (1968, cited in Archibald *et al.*, 1968) and Hurly *et al.* (1982) have recently obtained evidence that this enzyme is membrane-bound during outgrowth of germinated spores of *B. subtilis*. UDP-GlcNAc pyrophosphorylase has also been found associated with membrane preparations from *S. lactis* I3 (Douglas, 1968) and *Micrococcus* sp. 2102 (Brooks and Baddiley, 1969), as well as in the soluble fraction of the cell. The functional significance of these observations is not known, but it has been suggested that UDP-glucose, the substrate for glucose incorporation into teichoic acids, may be synthesized by a membrane-associated complex (Maino and Young, 1974).

B. Biosynthesis of Poly(Alditol Phosphate) Polymers

The biosynthesis of both poly(glycerol phosphate) and poly(ribitol phosphate) was first demonstrated *in vitro* using isolated membrane preparations. Early investigations established requirements for the appropriate nucleotide precursors, CDP-glycerol and CDP-ribitol, and high concentrations of bivalent cations, usually magnesium ions. Some properties of the enzyme systems that have been investigated are listed in Table 1, which includes the appropriate references. The reactions catalyzed may be presented as follows:

$$n\text{CDP-alditol} + \text{acceptor} \xrightarrow{Mg^{2+}} n\text{CMP} + (\text{alditol phosphate})_n\text{-acceptor}$$

Table 1. Properties of Some Enzymes Involved in Teichoic Acid Synthesis

Organism	State	Nucleotide K_m	Poly(alditol phosphate)K_m	pH optimum	Metal ion requirement	Reference
Poly(ribitol phosphate)synthetase						
B. subtilis W23	Membrane-bound					Chin *et al.*, (1966)
S. aureus H	Solubilized with Triton X-100, purified 475-fold	CDP-ribitol		8.0	Mg^{2+} (spermidine)	Fiedler and Glaser (1974a)
S. aureus Copenhagen	Membrane-bound	CDP-ribitol 1×10^{-4} M				
L. plantarum ATCC 8014	Membrane-bound	CDP-ribitol 1×10^{-5} M		8.3	Mg^{2+} $8 \times 10^{-3}M^b$ Ca^{2+} $4 \times 10^{-3}M^b$	Glaser and Burger (1964)
Poly(glycerol phosphate) synthetase						
B. licheniformis ATCC 9945	Membrane-bound	CDP-glycerol 8.3×10^{-4} M		7.2 and 9.0	Mg^{2+} $4 \times 10^{-2}M^b$ Ca^{2+} $1 \times 10^{-2}M^b$	Burger and Glaser (1966)
B. subtilis Marburg ATCC 6051	Solubilized with Triton X-100	CDP-glycerol 2×10^{-4} M		7.5	Mg^{2+}	Mauck and Glaser (1972a)
UDP-glucose:poly(ribitol phosphate) glucosyl transferase						
B. subtilis	Membrane-bound	UDP-glucose 1.2×10^{-4} M	1×10^{-6} M	7.0	Mg^{2+} $5 \times 10^{-2}M^b$ Ca^{2+}, Mn^{2+} } $1 \times 10^{-2}M^b$	Chin *et al.* (1966)
UDP-*N*-acetylglucosamine:poly(ribitol phosphate) *N*-acetylglucosamine transferase						
S. aureus Copenhagen	Membrane-bound	UDP-GlcNAc 7×10^{-4} M	6×10^{-5} M	8–8.3	Mg^{2+} $9 \times 10^{-3}M^b$	Nathenson and Strominger (1963)

Organism	Enzyme system	State	Nucleotide precursor K_m	Metal ion requirement optimum concentration	pH optimum	Reference
Synthesis of polymers with sugars within the polymer chains						
B. licheniformis ATCC 9945	Poly(galactosyl-glycerol phosphate) synthetase	Membrane-bound	UDP-galactose 2.5×10^{-4} M, CDP-glycerol 2.5×10^{-4} M			
B. licheniformis ATCC 9945	Poly(glucosyl-glycerol phosphate) synthetase	Membrane-bound	UDP-glucose 3.3×10^{-4} M, CDP-glycerol 2.2×10^{-4} M	Mg^{2+}, 15 mM, 10 mM[a]	7.5	Burger and Glaser (1966) Hancock and Baddiley (1972)
S. lactis I3	Poly(glycerol phosphate *N*-acetylglucosamine phosphate synthetase)	Membrane-bound	UDP-*N*-acetylglucosamine 1.25×10^{-3} M, CDP-glycerol	Mg^{2+} or Mn^{2+}, 20 mM	8.0	Baddiley *et al.* (1968)
S. lactis 2102	Poly(*N*-acetylglucosamine phosphate) synthetase	Membrane-bound	UDP-*N*-acetyl-glucosamine	Mg^{2+} or Mn^{2+}, 70 mM	8.5	Brooks and Baddiley (1969)
Linkage unit synthesis						
B. subtilis W23	UDP-*N*-acetylglucosamine:lipid phosphate GlcNAc 1-phosphate transferase	Membrane-bound	UDP-*N*-acetyl-glucosamine	Mg^{2+}	8.0	Hancock (1981)
B. subtilis W23	CDP-glycerol:lipid pyrophosphate GlcNAc glycerophosphate		CDP-glycerol 5×10^{-6} M	Mg^{2+}, 10 mM	8.0	Hancock (1981)

Organism	State	Nucleotide K_m	Poly(alditol phosphate) K_m	pH optimum	Metal ion requirement	Reference
UDP-glucose:poly(glycerol phosphate) glucosyl transferase						
B. subtilis NCTC 3610	Membrane-bound	4×10^{-5} M	1×10^{-4} M	8.0	Mg^{2+}, Ca^{2+} } 30 mM[a]	Glaser and Burger (1964)
B. subtilis 168	(1) Membrane-bound	1.3×10^{-5} M	0.5 mg/ml	7.5	Mg^{2+} 3.3×10^{-2} M[b]	Brooks *et al.* (1971)
	(2) Solubilized with chaotrope	1.7×10^{-5} M	0.1 mg/ml	6.8	Mg^{2+} 3.3×10^{-2} M[b]	

[a] Optimum concentration.
[b] K_m.

1. Lipoteichoic Acid Carriers (LTC) in Poly(Alditolphosphate Synthesis

Poly(glycerol phosphate) polymerase and poly(ribitol phosphate) polymerase have been solubilized and partially purified from membranes of *B. subtilis* and *S. aureus* (Mauck and Glaser, 1972a; Fiedler and Glaser, 1974a). Solubilization was achieved by extraction with Tris buffer containing Triton X-100 with spermidine and dithiothreitol as stabilizing agents. The enzymes were partially purified by ion-exchange chromatography followed, in the case of poly(RP) polymerase, by gel-permeation chromatography. The latter enzyme was purified 475-fold. Both preparations were activated by "lipoteichoic acid carrier" in the presence of Triton X-100. In addition, the more highly purified enzyme required phospholipid. The acceptor LTC could be extracted from the membranes of a wide range of Gram-positive bacteria into Triton X-100 (Fiedler and Glaser, 1974b). Examination of the products of the reaction by polyacrylamide gel electrophoresis clearly showed that poly(ribitol phosphate) synthesized by the solubilized *S. aureus* enzyme became covalently linked to the LTC. The kinetics of synthesis and the size-distribution of the product were consistent with a single-chain mechanism of synthesis (Fiedler and Glaser, 1974b). The chain length of the product was consistently close to 30 repeating units, even in the presence of a large excess of CDP-ribitol.

Analysis of the LTC acceptor, after purification by ion exchange chromatography, showed that it closely resembled membrane lipoteichoic acid (Section VI) in containing a poly(glycerol phosphate) chain carrying a glycolipid moiety at its phosphate terminus. Fiedler and Glaser (1974c) originally proposed a structure in which a chain of 12–14 glycerolphosphate units was linked to monoglucosyldiglyceride but their results, combined with those of Duckworth *et al.* (1975), were reinterpreted to show a polymer chain of 28–30 repeating units carrying a diglucosyldiglyceride lipid moiety identical in these respects to the lipoteichoic acid from the membrane of *S. aureus* (Duckworth *et al.*, 1975). However, Lambert *et al.* (1977a) found that material active as LTC made up only a small proportion of the total lipoteichoic acid isolated from *S. aureus* membrane, and could be separated from the bulk of the material by ion-exchange chromatography on DEAE-cellulose in the presence of Triton X-100.

More recently, the discovery of the structure of the linkage unit that attaches teichoic acid to peptidoglycan (Coley *et al.*, 1977), and the demonstration that the lipid-bound precursor of linkage unit can act as acceptor for poly(ribitol phosphate) polymerase from *S. aureus*, (Bracha *et al.*, 1978) have led Fischer *et al.* (1980a,b) to make a thorough reappraisal of the properties and structure of LTC. They have confirmed that lipoteichoic acids from a variety of sources, prepared by conventional techniques involving phenol extraction and gel permeation chromatography, will all act as acceptors with the poly(ribitol phosphate) polymerase of *S. aureus*. Acceptor activity was dependent on the length of the poly(glycerol phosphate) chain but was not entirely lost until the average chain length fell below 4.6 glycerophosphate units (Fischer *et al.*, 1980a). On the other hand, acceptor activity was entirely dependent on the presence of the lipophilic portion of the LTA. Most importantly, they found that the presence of alanyl and glycosyl substituents on the poly(glycerol phosphate) chain severely reduced the LTC activity of the LTA. It had been known for several years that native LTA was substituted with D-alanyl residues (Keleman and Baddiley, 1961), but previous workers on its function in biosynthesis had largely overlooked the fact that the conventional extraction technique for LTA, using buffers at pH8, would cause the hydrolysis of the ester linkage of the D-alanine due to the influence of the neighboring phosphate group (Shabarova *et al.*, 1962). By careful purification of LTA at pH 4, Fischer and Rosel (1980) isolated material, in which half the glycerol phosphate units were alanylated, that was entirely inactive as LTC until the alanyl residues were chemically removed (Fischer *et al.*, 1980b). Likewise, LTA from *S. lactis,* in which 21% of the residues carried alanine and 39% carried α-galactosyl substituents, was inactive unless alanine or galactose was removed. Detailed examination of the "native" LTA from *S. aureus* revealed heterogeneity in the degree of alanylation of molecules of the polymer, but under no conditions could polymer with a mean degree of substitution of less than 20% be detected. It was therefore concluded that no LTA that could act as LTC was present in *S. aureus in vivo,* and that the material identified as LTC was entirely artefactual.

2. *Linkage Unit and Polymer Synthesis*

Following the observation (Heckels *et al.*, 1975) that *S. aureus* teichoic acid was attached to peptidoglycan by a linkage unit (LU) containing three glycerol phosphate residues, the pathway for the synthesis of LU was quickly identified. Hancock and Baddiley (1976) demonstrated that the unit was synthesized in a membrane-bound form from CDP-glycerol and UDP-*N*-acetylglucosamine before attachment to it of teichoic acid, while Bracha and Glaser (1976a) showed that the same two nucleotides were essential for the attachment of newly synthesized teichoic acid to peptidoglycan in wall-membrane preparations from *S. aureus*. Subsequently, the membrane-bound form of LU was found to contain a polyisoprenyl phosphate lipid of the type involved in peptidoglycan synthesis and the various stages in its assembly were elucidated (McArthur *et al.*, 1978, 1980a). The antibiotic tunicamycin was found to inhibit the first step of the pathway, transfer of *N*-acetylglucosamine 1-phosphate from UDPGlcNAc to the lipid phosphate (Hancock *et al.*, 1976), and the incorporation of teichoic acid into the cell wall (Bracha and Glaser, 1976b), but not the synthesis of teichoic acid on endogenous LTC in the membrane. The pathway of linkage unit synthesis and incorporation of teichoic acid into the wall of *S. aureus* is shown in Figure 5. The identification of the carrier lipid as polyisoprenyl phosphate is based on the chemical properties of the pyrophosphate diester linkage between it and *N*-acetylglucosamine, and on the inhibition of synthesis of lipid I by tunicamycin, which is known to inhibit transfer of GlcNAc 1-phosphate to polyisoprenyl phosphate in other systems (Tkacz, 1982).

Essentially the same pathway has been demonstrated in *Micrococcus varians* (*Micrococcus* sp. 2102), *Bacillus subtilis* W23, *Staphylococcus lactis* I3 and *Lactobacillus plantarum*. In *B. subtilis* 168, which has a poly(glycerol phosphate) teichoic acid, it would be difficult to identify a separate role for CDP-glycerol in synthesis of linkage unit but UDP-*N*-acetylglucosamine stimulates polymer synthesis from CDP-glycerol in cells made permeable to nucleotides by treatment with toluene. Hancock

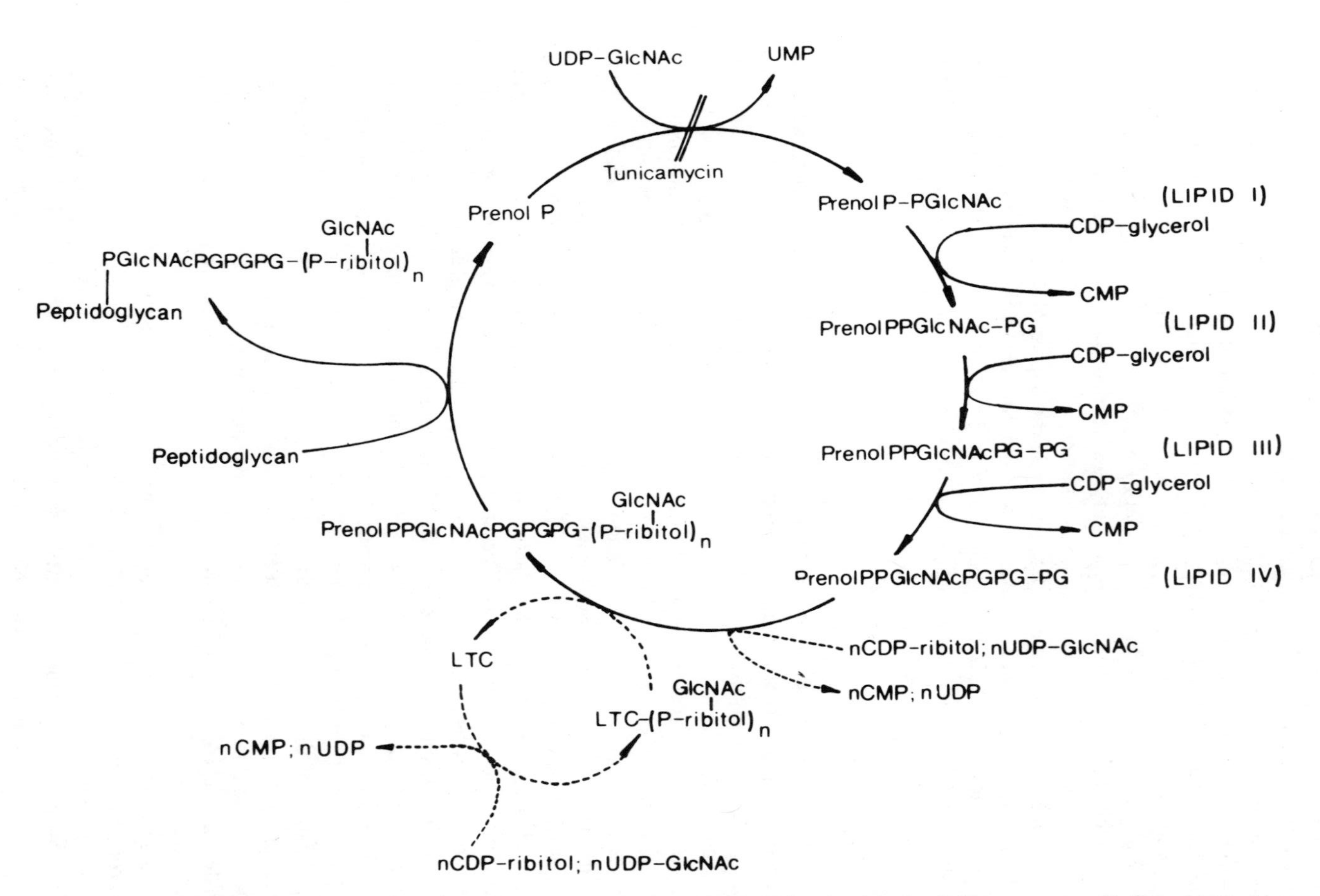

Figure 5. The pathway of synthesis of linkage unit and attachment of poly(ribitol phosphate) in *Staphylococcus aureus* H. The questionable participation of lipoteichoic acid carrier (LTC) is indicated by the dotted arrows.

(1981) and McArthur *et al.* (1981) have demonstrated that linkage unit lipid intermediates (mainly lipid III) synthesized from CDP-glycerol and UDP-*N*-acetylglucosamine in *S. aureus* can act as acceptors of poly(glycerol phosphate) when supplied to membranes of *B. subtilis* 168 in the presence of Triton X-100. It is interesting that linkage unit lipid containing only two glycerol phosphate units (lipid III) can act as acceptor without the participation of additional CDP-glycerol.

As mentioned above, the discovery of lipid-bound linkage unit threw into question the role of LTC in teichoic acid synthesis and strong evidence has since been presented that LTC is an artefact. Bracha *et al.* (1978) have demonstrated that lipid-bound linkage unit intermediates from *S. aureus* can act as acceptor for poly(ribitol phosphate) synthesis from CDP-ribitol, with the solubilized polymerase from *S. aureus* in the absence of LTC. It has been proposed that LTA lacking alanine, LTC, is a fortuitous analogue of lipid-bound linkage unit. For these reasons, involvement of LTC has been shown only as a dotted line in Figure 5. Nevertheless, it might be unwise to disregard LTC completely. It has been shown to be a derivative of LTA, which is synthesized by polymerization of glycerol phosphate moieties donated by the membrane phospholipid phosphatidyl glycerol, (see Section VI) whereas the glycerophosphate residues of linkage unit arise exclusively in CDP-glycerol (Roberts *et al.*, 1979). Thus, the terminal glycerol phosphate of LTC must have the opposite stereochemical configuration to that in linkage unit, and it is somewhat surprising if the polymerase does not distinguish between them. Moreover, McArthur *et al.* (1981) have demonstrated that poly(ribitol phosphate) attached to LTC in *S. aureus* membrane can be transferred to membrane-bound linkage unit synthesized in a separate batch of membrane, when the two membrane samples are mixed in the presence of Triton X-100. Recently, Arakawa *et al.* 1981) have reported the isolation from cells of *S. aureus* of poly(ribitol phosphate) attached to lipoteichoic acid.

C. Biosynthesis of Teichoic Acids with Sugar Residues in the Main Chain

Studies of the pathways for the synthesis of poly(*N*-acetylglucosamine phosphate) in *Micrococcus* sp. 2102 (Brooks and Baddiley, 1969) and of poly(*N*-acetylglucosamine phosphate glycerol phosphate) in *S. lactis* I3 (Hussey and Baddiley, 1972) revealed membrane-bound lipid intermediates that were believed to be involved in the polymerization reactions. Considerable evidence was obtained that these were polyisoprenyl phosphate lipids. The later discovery of the linkage unit and its synthesis via lipid-bound intermediates, from CDP-glycerol and UDP-*N*-acetylglucosamine, provides a possible alternative explanation for the earlier results. Thus, the lipid found to turn over into poly(*N*-acetylglucosamine phosphate) in *Micrococcus* 2102 has the same structure as lipid I of linkage unit synthesis, while the major lipid formed in the pathway to teichoic acid in *S. lactis* I3 is identical with lipid II of linkage unit synthesis (see Figure 5). It would therefore be very difficult to distinguish whether polymerization in these strains proceeds through the observed lipid intermediates, or whether the lipids are precursors of lipid-bound linkage unit on which the polymer chain is built up directly from sugar nucleotides. This latter explanation is the most likely, since in this case polymer synthesis itself is not inhibited by tunicamycin.

D. Addition of Glycosyl and Alanyl Substitutents

In vitro systems have been shown to catalyze transfer of glycosyl residues from nucleotide precursors to poly(alditol phosphate) polymers. The transfer of glucose from UDP-glucose (or TDP-glucose) to poly(glycerol phosphate) has been demonstrated in a particulate preparation from *B. subtilis* ATCC 6051 (Glaser and Burger, 1964). The glucose acceptors in the reaction are partially glucosylated, membrane-bound teichoic acid, or poly(glycerol phosphate) synthesized *in vitro*. The product is a fully glucosylated polymer in which every glycerol residue is substituted at position C-2 by an α-D-glucopyranosyl residue. In more recent studies, UDP-glucose:poly(glycerol phosphate) glucosyl transferase (TAG transferase) has been solubilized from the membrane of *B. subtilis* 168 (Brooks *et al.*, 1971. TAG transferase was shown to be partly soluble *in vivo* in the early stages of growth. As growth proceeds, the amount of soluble enzyme decreases while the amount and specific activity of the membrane-bound enzyme increase until they reach a maximum in late exponential phase of growth. The membrane-bound activity falls sharply in stationary phase. The membrane-bound enzyme was solubilized using the chaotrope, sodium perchlorate. Once solubilized, the enzyme is not readily re-bound to the depleted membrane. Some properties of the solubilized and membrane-bound enzymes are indicated in Table 1. These indicate that the enzyme is not greatly changed on solubilization.

The synthesis of the glucosylated 2,3-linked poly(glycerol phosphate) of *B. stearothermophilus* appears to depend on the simultaneous presence of CDP-glycerol and UDP-glucose (Kennedy, 1974). In the absence of UDP-glucose 1,3-linked poly(glycerol phosphate) is synthesized, and this cannot be glucosylated in the presence of UDP-glucose by the *B. stearothermophilus* enzyme although it is a substrate for the enzyme from *B. subtilis* ATCC 6051. In the presence of UDP-glucose glucosylated 2,3-linked polymer is synthesized as well as unglucosylated 1,3-linked polymer, and the proportion of glucosylated polymer is increased with increasing UDP-glucose:CDP-glycerol ratio. The glycosyl transferase and polymerase in this organism are closely interdependent even *in vitro*, and this enzyme system may more truly reflect the *in vivo* situation in all organisms.

The teichoic acid of *B. subtilis* W23 consists of a mixture of glucosylated and unglucosylated poly(ribitol phosphate). Membrane preparations from this organism will transfer β-D-glucopyranosyl residues from UDP-glucose to endogenous acceptor and also to added poly(ribitol phosphate) obtained from the cell wall. It will not transfer glucose to poly(glycerol phosphate). *In vitro* poly(ribitol phosphate) extracted from the wall can be fully glucosylated, and the polymer in intact walls can also be glucosylated to some extent by isolated membranes (Chin *et al.*, 1966) and in toluenized bacteria (Hancock, 1981). It follows, therefore, that *in vivo,* two polymerase systems must exist, one which contains the glucosyl transferase and one which does not. Unlike TAG transferase of *B. subtilis* 168, there is no change in enzyme activity during growth and no decay in stationary phase (Chin *et al.*, 1966).

The *N*-acetylglucosamine substituents on the poly(ribitol phosphate) of *S. aureus* are in α-, β-, or α- and β-linkage depending on the strain. The teichoic acid of *S.*

aureus Copenhagen contains approximately 85% β- and 15% α-*N*-acetylglucosaminyl substituents (Sanderson *et al.*, 1962). The UDP-*N*-acetylglucosamine : poly(ribitol phosphate) *N*-acetylglucosaminyl transferase of this organism will transfer *N*-acetylglucosamine residues to α-*N*-acetylglucosamine-containing poly(ribitol phosphate) obtained by treatment of extracted wall teichoic acid with β-*N*-acetylglucosaminidase (Nathenson and Strominger, 1963). The enzymatic product contained 71% β-linked and 29% α-linked *N*-acetylglucosamine residues. As the enzyme preparation aged, the proportion of β-linked residues in the product fell. Two enzymes therefore are probably responsible for *N*-acetylglucosamine transfer, one catalyzing formation of α-linked residues and the other β-linked residues. Other strains of *S. aureus* which contain different proportions of α- and β-linked *N*-acetylglucosamine residues in their teichoic acids have been investigated, and it is notable that enzyme preparations from each of them have the capacity to synthesize both α- and β-linkages irrespective of the type of linkage found *in vivo* (Nathenson *et al.*, 1966). While the transferase from *S. aureus* would utilize added poly(ribitol phosphate) and teichoic acid synthesized *in vitro* as glycosyl acceptors, the velocity and extent of incorporation of *N*-acetylglucosamine was much greater when transfer was coupled with polymer synthesis from CDP-ribitol. Thus, addition of *N*-acetylglucosamine residues to ribitol phosphate might occur sequentially, concomitant with chain elongation (Ishimoto and Strominger, 1966).

The addition of D-alanine to membrane lipoteichoic acid is described in section VI. It is not known whether the mechanism of alanylation of wall teichoic acid is similar.

IV. TEICHURONIC ACID STRUCTURE AND SYNTHESIS

The walls of a wide range of Gram-positive bacteria contain, either in place of teichoic acid or in addition to it, anionic polymers whose negative charges are provided by the carboxyl groups of glycuronic acid residues rather than by phosphate. Representative examples are shown in Figure 6. These polymers have been less well studied

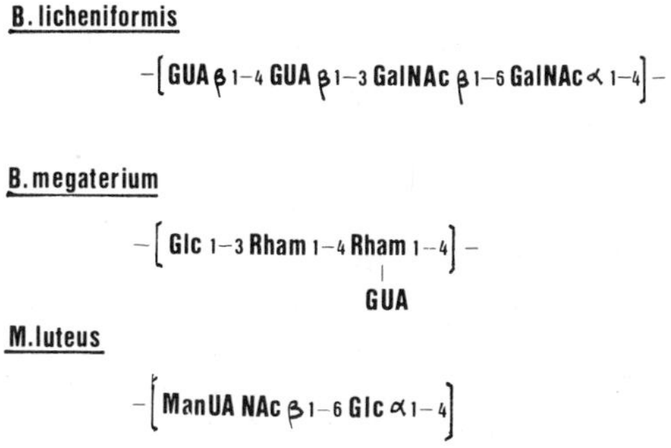

B. licheniformis

−[GUA β1–4 GUA β1–3 GalNAc β1–6 GalNAc α1–4]−

B. megaterium

−[Glc 1–3 Rham 1–4 Rham 1–4]−
GUA

M. luteus

−[ManUA NAc β1–6 Glc α1–4]

Figure 6. Teichuronic acids whose structures have been thoroughly investigated. GUA, D-glucuronic acid; ManUANAc, *N*-acetylmannosamineuronic acid.

than the teichoic acids and the biosynthesis of only two types, those from *Micrococcus luteus* and from *Bacillus licheniformis* have been investigated in detail.

As in the case of teichoic acid, the polymers are covalently linked through phosphate diesters to muramic acid residues in the peptidoglycan of the wall. In *M. luteus*, the attachment involves a structure in many ways analogous to the linkage unit of teichoic acids (Hase and Matsushima, 1977) in which an *N*-acetylglucosamine 1-phosphate residue bridges between the muramic acid and two additional *N*-acetylmannosamine uronic acid residues before the first repeating unit of the main polymer chain is attached. In *B. licheniformis*, however, no evidence for a linkage unit other than the single-phosphate group between the terminal *N*-acetylgalactosamine residue of the polymer and muramic acid has been obtained, either from chemical (Lifely *et al.*, 1980) or biosynthetic studies (Ward and Curtis, 1982).

Anderson and his co-workers have elucidated the pathway for the synthesis of the teichuronic acid of *M. luteus* by membrane-bound enzymes. Early work indicated that although the polymer chain consisted of repeating units of glucose and *N*-acetylmannosamine uronic acid synthesized from the corresponding UDP-linked sugars, the presence of UDP-*N*-acetylglucosamine was required for maximum polymer synthesis in a cell-free system. More recently (Stark *et al.*, 1977), they have shown that the UDP-*N*-acetylglucosamine is required for the first step in the synthesis of the polymer, the transfer of *N*-acetylglucosamine to a lipid, presumably a polyisoprenyl phosphate. The rest of the linkage unit, and subsequently the polymer chain, is built up on this membrane-bound lipid. The direction of chain extension is therefore the same as that in teichoic acid synthesis, in which the last repeating unit to be added is that most distant from the membrane linkage site. The pathway in *B. licheniformis*, studied by Ward and Curtis (1982), is very different. Here, no membrane-bound linkage unit is involved. Instead, the repeating units of *N*-acetylgalactosamine and glucuronic acid are built up individually on a lipid carrier, again presumed to be a polyisoprenyl phosphate, and the reducing end of the growing polysaccharide chain is transferred from one lipid carrier to the nonreducing glucuronic acid of the next disaccharide repeating unit, attached to its own lipid carrier through the *N*-acetylgalactosamine 1-phosphate. Thus, the latest unit to be added is always adjacent to the lipid carrier in the membrane. This is the type of chain extension mechanism found for peptidoglycan and the *O*-antigenic polysaccharide chains of lipopolysaccharide, but is the reverse of the direction observed for teichoic acids (Kennedy and Shaw, 1968; Hussey *et al.*, 1969) in which the most recently added unit is at the "glycerol end" of the chain. As in the case of peptidoglycan synthesis, transfer of the growing teichuronic acid chain from its lipid carrier leaves a polyisoprenyl pyrophosphate molecule which can only be recycled following removal of the terminal phosphate residue. This reaction is inhibited by bacitracin, which therefore prevents chain growth. However, tunicamycin has no effect on the *N*-acetylgalactosamine 1-phosphate transferase that catalyzes the first step of the synthetic pathway.

The structure of the teichuronic acid obtained *in vitro* by biosynthesis in *B. licheniformis* 94 by Ward and Curtis (1982) was apparently different from that of the polymer from the walls of strain ATCC 9945 determined chemically by Lifely *et al.* (1980). Biosynthesis has not been studied in the latter strain.

V. INCORPORATION OF TEICHOIC ACID AND TEICHURONIC ACID INTO THE CELL WALL

Studies of the biochemistry of attachment of teichoic acid and teichuronic acid, synthesized on the membrane in conjunction with their linkage units to peptidoglycan and hence to the wall, require cellular or subcellular preparations in which the interaction between the membrane and the cell wall is maintained. Two types of preparation have been used: wall-membrane, in which large wall fragments with membrane adhering to them are isolated following careful cell breakage by mechanical grinding with alumina or glass beads, and toluenized cells, rendered permeable to small molecules such as the nucleotide precursors of wall synthesis and washed free of internal nucleotides. Wall membranes have rather low specific enzyme activities compared with growing bacteria or toluenized cells, probably due to the presence of membrane not adequately coordinated with the wall; but Bracha *et al.* (1978) have described a method for purifying functional fragments by density gradient centrifugation.

Experiments with whole cells have shown convincingly that during normal wall synthesis, teichoic acid and teichuronic acid molecules become attached to molecules of peptidoglycan synthesized concomitantly (Mauck and Glaser, 1972b). Moreover, peptidoglycan synthesized in the presence of penicillin, and therefore not crosslinked to the cell wall, is released into the growth medium unattached to other polymers (Tynecka and Ward, 1975), and this observation has led to the suggestion that peptidoglycan must be linked to the wall by transpeptidation before attachment of teichoic or teichuronic acids can occur. Neither of these features is consistently reproduced by *in vitro* preparations that catalyze linkage of polymers. Wall-membrane preparation from *B. licheniformis* most closely reflect the *in vivo* observation. These preparations accomplish the synthesis of teichoic acid or teichuronic acid from the appropriate nucleotide precursors, but only attach it to the wall in the presence of concomitant synthesis of peptidoglycan from its nucleotide precursors. Penicillin does not prevent the synthesis of these polymers but does prevent their attachment to the wall. Moreover, the unattached peptidoglycan does not have teichoic acid attached to it (Wyke and Ward, 1977a). In contrast to this, wall-membrane preparations (Wyke and Ward, 1977b) and toluene-treated cells (Hancock, 1981) of *B. subtilis* W23 attach newly synthesized teichoic acid to the wall in the absence of concomitant peptidoglycan synthesis, while wall-membrane preparations from *S. aureus* use both endogenous peptidoglycan and concomitantly synthesized polymer as acceptor for teichoic acid (Bracha and Glaser, 1976). The attachment of newly synthesised teichuronic acid to peptidoglycan in wall-membrane preparations of *M. luteus* resembles *B. subtilis* W23 in being independent of new peptidoglycan synthesis (Weston and Perkins, 1977). These results suggest that *in vivo* attachment of teichoic acid takes place to peptidoglycan chains that are still in the process of elongation by transglycosylation and are therefore attached to the biosynthetic sites on the membrane, but are already cross-linked, by transpeptidation, to the inner surface of the wall. Whether attachment sites on these molecules of peptidoglycan remain accessible to newly synthesized secondary polymers once the cell has been broken or made permeable may depend on the stability of the complex of enzymes, nascent peptidoglycan, and wall at the biosynthetic sites.

If the organization remained intact, nascent peptidoglycan would be available for attachment of polymers synthesized *in vitro,* but if the nascent peptidoglycan became detached from the sites during cell treatment, initiation of new acceptor peptidoglycan molecules would be necessary. The hypothesis that it is the geometry of the biosynthetic complex rather than enzyme specificity that governs the requirement for prior cross-linking of peptidoglycan *in vivo* is supported by the observation that a wall-free fragmented membrane preparation from *M. varians* was able to link together newly synthesized peptidoglycans and teichoic acid under conditions where transpeptidation of the peptidoglycan was impossible (McArthur *et al.*, 1980b).

VI. SYNTHESIS OF MEMBRANE LIPOTEICHOIC ACID

The observation that the group antigen of Group D Streptococci was a glycerol teichoic acid that was not covalently linked to the cell wall (Whicken *et al.*, 1963) led to the discovery of the widespread occurrence of membrane-bound teichoic acids in Gram-positive bacteria (Lambert *et al.*, 1977b). Although the type and extent of glycosyl and alanyl substitution of these polymers varied between bacterial strains, the polymer backbone was invariably poly(glycerol phosphate). Whicken and Knox (1970) first reported that the membrane teichoic acid could be extracted from cells in a tight association with a glycolipid, and Toon *et al.* (1972) showed that there was a covalent phosphodiester link between the terminal glycerol phosphate residue of the polymer chain and one of the sugar residues of the glycolipid. Because of their potential medical importance, membrane teichoic acids, now usually called lipoteichoic acids (LTA), have attracted interest and their structures have been investigated in detail (Whicken and Knox, 1975). They have also been found in one genus of Gram -negative bacteria, *Butyrivibrio* (Hewett *et al.*, 1976).

In 1974, independent experiments by Emdur and Chiu (1974) and Glaser and Lindsay (1974) showed that the origin of the glycerol phosphate residues in LTA from *Strep. sanguis* and *S. aureus* was not CDP-glycerol as in wall teichoic acid synthesis, but the membrane phospholipid phosphatidylglycerol. This lipid is also known to donate glycerol phosphate residues during the synthesis of the membrane-derived glucose oligosaccharides in Gram-negative bacteria (Schulman and Kennedy, 1977, 1979) and outer membrane lipoprotein in *E. coli* (Chattopadhyay and Wu, 1977). Radioactivity from glycerol in the culture medium of both *Strep. sanguis* and *S. aureus* was incorporated into phosphatidylglycerol and subsequently turned over into LTA, about half of which, when isolated, was active as LTC in *S. aureus*. Subsequent *in vitro* experiments with *Strep. sanguis* membrane preparation, in which radioactively-labeled phosphatidylglycerol in sonicated aqueous dispersion was mixed with membrane fragments, demonstrated the incorporation of both the glycerol phosphate and diglyceride moieties of the phospholipid into a mixture of polymers, one of which resembled native LTA (Emdur and Chiu, 1975). UDP-glucose appeared to be the donor of the glycosyl residues (Mancuso *et al.*, 1979). The precise structure of the lipid part of the LTA of *Strep. sanguis* is not known (Chiu *et al.*, 1974), but in *Strep. faecium* it consists of a diglucosyl (kojibiosyl) diglyceride (Toon *et al.*, 1972) substituted on one of the glucose residues with a phosphatidyl group (Pieringer, 1972). This phosphoglycolipid is synthesized from kojibiosyl diglyceride by the transfer of the phosphatidyl group from phosphatidylglycerol (Pieringer, 1972), and acts as acceptor for the growing polymer chain. Pieringer and his co-workers have shown that when the phosphoglycolipid is added to isolated membranes from *Strep. faecium* in the presence of Triton X-100, it accepts glycerol phosphate residues from phosphatidylglycerol, giving LTA and lipophilic short chain precursors (Ganfield and Pieringer, 1980). Cabacungan and Pieringer (1981) have recently demonstrated that the glycerol phosphate residues are added to the growing chain at the end distal from the glycolipid portion of the molecule.

The process of acylation of LTA by alanine has been studied by Neuhaus and his co-workers. The reaction requires a soluble D-alanine activating enzyme, acting in the presence of ATP and an additional membrane-bound "membrane acceptor ligase" (Neuhaus *et al.*, 1974). In early work, the membrane-bound macromolecular acceptor of the alanyl residues was not thoroughly characterized, but more recent work has clearly identified it as nascent LTA (Childs and Neuhaus, 1980). Further studies, using isolated membranes and toluene-treated whole cells of *Lactobacillus casei* have revealed that addition of alanine residues takes place during elongation of the poly(glycerol phosphate) chain. Brautigan *et al.* (1981) have isolated from incubation mixtures containing ATP, radioactive alanine, and membrane fragments or toluene-treated cells (containing endogenous phosphatidylglycerol), lipophilic compounds containing alanine that turn over into LTA on further incubation. The lipophilic intermediates could be degraded by a phosphodiesterase from *Aspergillus niger* to give 2-alanylglycerol, and they resembled the lipophilic products obtained by shortening the chain of authentic LTA with the same enzyme. However, alanylphosphatidylglycerol could not be detected, and it therefore appears that alanyl residues are added to each glycerol phosphate residue following its transfer to the growing chain. It is interesting that the membrane from a stable L-form of *Streptococcus pyogenes,* which contains alanine-deficient LTA, would not act as acceptor for D-alanine, although both activating enzyme and ligase were present. This may indicate a requirement for an unrecognized factor in the alanyl transfer reaction or may be related to the requirement of the L-form for a high concentration of salt in the growth medium, since alanyl transfer is inhibited at high ionic strengths (Neuhaus *et al.*, 1974).

VII. MODE OF ACTION OF THE ENZYMES IN THE MEMBRANE

Since teichoic acid-LU becomes attached to peptidoglycan already linked to the cell wall in wall membrane and toluene-treated cell preparations, it must be assumed that this reaction takes place on the outer surface of the cytoplasmic membrane. The precursors of teichoic acid and LU, the sugar nucleotides and alditol nucleotides, are synthesized in the cytoplasm by the action of pyrophosphorylases that are believed to be soluble enzymes. However, the locations of the membrane-bound enzymes that catalyze the intervening reactions in the synthesis of linkage unit and teichoic acid are unknown, and none of the enzymes has so far been purified to homogeneity. Despite this paucity of information, there is indirect evidence that the enzymes comprise a tightly organized transmembrane complex which may in turn work in a coordinated way with the enzymes of peptidoglycan synthesis. This concept was first put forward by Watkinson *et al.* (1971) who obtained evidence that teichoic acid synthesis and peptidoglycan synthesis shared access to a pool of prenyl phosphate carrier lipid in *S. lactis* I3. The antibiotic bacitracin did not inhibit the synthesis of teichoic acid catalyzed by membrane preparations in the absence of peptidoglycan synthesis. If, however, peptidoglycan precursors were also present, bacitracin reduced the rate of teichoic acid synthesis. The antibiotic is known to sequester prenyl pyrophosphate formed during peptidoglycan synthesis and thus prevent its recycling for further rounds of polymer synthesis. It was concluded that, since prenyl pyrophosphate was not formed during teichoic acid synthesis, the effect of bacitracin must have been to reduce the size of a prenyl monophosphate pool common to the two pathways. Similar results were later obtained with *B. licheniformis* and *B. subtilis* (Anderson *et al.*, 1972). The work of Mauck and Glaser (1972b) clearly showed that molecules of teichoic acid and peptidoglycan found linked together in the wall of *B. subtilis* were synthesized concomitantly, and the same conclusion was reached by Tomasz *et al.* (1975) in a study of *Streptococcus pneumoniae*. Moreover, Ward and Curtis (1982) have demonstrated that in walls of *B. licheniformis* that contain covalently linked polymers in addition to teichoic acid, individual peptidoglycan chains always carry only one type of polymer or teichoic acid. All these observations support the view that sites of synthesis of peptidoglycan in the membrane are in close contact with the sites of synthesis of the other wall polymers.

The first information about the organization of the teichoic-acid-synthesizing enzymes came from an investigation of the recovery of teichoic acid synthesis by synchronously germinating spores of *B. subtilis* W23 (Chin *et al.*, 1968). The first activity to appear produced fully glucosylated poly(ribitol phosphate) but additonal activity gained later in the germination process synthesized unglucosylated polymer. Apparently, only a proportion of the biosynthetic sites acquired glucosyl transferase, and indeed, walls of vegetative cells of this strain contain a mixture of fully glucosylated and unglucosylated teichoic acids (Torii *et al.*, 1964). In some strains of *S. aureus*, each biosynthetic site apparently contains either an α- or a β-*N*-acetylglucosamine transferase but not both; the teichoic acid chains can be separated immunologically into those substituted only with the α-anomer and those carrying only the β-anomer.

Since the teichoic acid linkage unit is assembled on a prenyl phosphate lipid, it is attractive to suppose that this lipid forms a center around which the enzymes cluster. Leaver *et al.* (1981) have solubilized enzymes that synthesize teichoic acid and linkage unit from membranes of *M. varians* in Triton X-100. Fractionation of the extract on a sucrose density gradient yielded a protein fraction that catalyzed the synthesis of the main polymer chain in the absence of LU-precursors and another much less dense fraction that catalyzed the synthesis and accumulation of LU-lipid intermediates I to III but not polymer synthesis. The latter fraction, which had a distinctive polypeptide composition, could not be fractionated further without loss of activity and therefore appeared to consist of a complex of the enzymes associated with prenyl lipid acceptor.

As previously shown by Glaser and co-workers (Mauck and Glaser, 1972a; Fiedler and Glaser, 1974a) for other species of bacteria, the polymerase could be separated from the membrane without loss of its activity, as can glycosyl transferases (Brooks *et al.*, 1971). The core of the biosynthetic complex is thus the group of enzymes and lipid involved in linkage unit synthesis. The first reaction in this complex is the transfer of *N*-acetylglucosamine 1-phosphate from UDP-*N*-acetylglucosamine to the prenyl phosphate. This reaction is analogous to the initial reaction of peptidoglycan synthesis, the transfer of *N*-acetylmuramylpeptide phosphate to prenyl phosphate, which Neuhaus and his colleagues have termed the translocase reaction. It should be emphasized, however, that there is no evidence that such reactions involve translocation of the sugar phosphate group across the membrane while attached to the lipid carrier. All the physical studies of prenyl phosphate carrier lipids in membranes have failed to detect any potential for transmembrane "flip-flop" motion. The balance of probabilities (Hanover and Lennartz, 1979; McCloskey and Troy, 1980; Weppner and Neuhaus, 1977) favors a transmembrane protein complex as the mediator of group translocation in the synthesis of cell wall polymers. In this connection, the observation (Bertram *et al.*, 1981) that *B. subtilis* W23 cells suspended in isotonic buffer rapidly acquire the ability to synthesize teichoic acid and peptidoglycan from externally added nucleotide precursors, as their walls begin to autolyze, is particularly interesting. No membrane permeability to nucleotides can be detected in these protoplasting cells, or in free protoplasts which are also active. The polymeric products are extracellular and their synthesis can be blocked reversibly by extracellular trypsin; activity is regained on the addition of trypsin inhibitor. When the protoplasts are gently burst, the total synthetic activity approximately doubles. It therefore appears that either a proportion of the biosynthetic sites is permanently exposed on the outer surface of the membrane, but is shielded from the external environment by close contact with the wall, or that the protein complex at each site is temporarily exposed to the outside of the membrane at some stage in the biosynthetic cycle. The latter hypothesis implies some type of reorientation of the complex in the membrane during biosynthesis of the polymers. Lack of fluidity in the surrounding bilayer below the transition point might hinder this movement and account for the inefficiency of membranes of *M. varians* in attaching main polymer chain to LU-lipid at temperatures below 35°, although LU-lipids continue to be synthesized rapidly (Roberts *et al.*, 1979).

The studies on teichoic acid synthesis with protoplasts have recently been extended to peptidoglycan by Harrington and Baddiley (1983). Although the presence of lysozyme and autolysins on the surface of the protoplasts prevents the accumulation of polymer, the lipid intermediates are formed from externally supplied nucleotide precursors, and their formation is inhibited by bacitracin and vancomycin, thus showing that the biosynthetic cycle is complete. In similar experiments with partly autolyzed cells, uncrosslinked peptidoglycan is produced. An incompletely characterized polymer is formed from UDP-GlcNAc alone in both cases. It seems likely then that there is a close similarity between the organization of enzymes in the complexes for teichoic acid and peptidoglycan synthesis.

An alternative to the rotating or reorienting enzyme complex hypothesis is that the respective complexes, which normally span the membrane, are readily removed from this position to one in which they occupy an approximately transverse position on the outer surface of the membrane. The membrane itself reseals and the complex is now able to utilize externally supplied nucleotides and form polymer outside the protoplast or partly autolyzed cell. The prenyl phosphate would accompany the enzymes in the complexes.

On present evidence it is not possible to decide between these two hypotheses. The ease of removal of the complexes from a transmembrane location in an active site is consistent with their removal by freeze-thawing (Thompson *et al.*, 1980), and their ability to remain intact during fractionation (Leaver *et al.*, 1981). However, ease of removal implies relatively loose association with the membrane and so possibly ease of rotation or reorientation. Thus, a complex which is not firmly held in position might require relatively little energy to rotate. There are several well-known sources of energy available in a membrane in addition to thermal energy.

VIII. REGULATION OF SYNTHESIS OF TEICHOIC AND TEICHURONIC ACIDS

The amount of wall teichoic acid in live bacteria responds to changes in the physiological state of the cells (Ellwood and Tempest, 1972) and the responses to phosphate-limitation of growth in a chemostat and to phosphate-starvation of batch cultures have received particular attention. When the growth limitation of *B. licheniformis* (Forsberg *et al.*, 1973; Hussey *et al.*, 1978) or *B. subtilis* (Ellwood and Tempest, 1972; Archibald and Coapes, 1976) in a chemostat is changed from carbon or potassium limitation to phosphate limitation, the bacteria terminate the synthesis of wall teichoic acid and begin to introduce another anionic wall polymer teichuronic acid (Ellwood and Tempest, 1972). Phosphate-starved batch cultures also lose the ability to synthesize teichoic acid and acquire the enzymes of teichuronic acid synthesis, but the change is not, in this case, reflected by a gross change in wall composition since the bacteria hardly grow (Mauck and Glaser, 1972b). The amount of teichoic acid in walls of *B. subtilis* W23 in balanced growth in a chemostat depends on the concentration of inorganic orthophosphate in the medium, and ranges from 50–300 nmols of teichoic acid per milligram of bacteria in media containing between 1.5–4 mM phosphate (Anderson *et al.*, 1978), with a reciprocal change in the amount of teichuronic acid. Similar changes occur in *B. subtilis* 168, which has a glycerol teichoic acid (Lang *et al.*, 1982).

Investigation of the enzymic basis for these changes is recent, but preliminary results indicate that both genetic and enzymic regulation of the pathway occur. Hussey *et al.* (1978) found that about one cell generation time after the attainment of phosphate-limitation in a chemostat culture of *B. licheniformis* previously limited by the supply of glucose, synthesis of CDP-glycerol pyrophosphorylase, and poly(glycerol phosphate) polymerase, ceased and at the same time the amount of teichoic acid in the walls began to decrease. However, whereas the teichoic acid was lost from the walls at the rate expected on the basis of wall turnover and wash-out from the chemostat, both enzyme activities decayed at a much higher rate. The pyrophosphorylase disappeared within half a cell generation and 50% of the polymerase was lost over the same period. Thus, the enzymes are subject to specific inactivation during phosphate-limited growth. A similar study of *B. subtilis* W23 has revealed intriguing differences from *B. licheniformis* (Cheah *et al.*, 1981). In the latter species, the product of CDP-glycerol pyrophosphorylase, CDP-glycerol, is the substrate for the teichoic acid polymerase as well as for linkage synthesis. In strain W23, CDP-ribitol pyrophosphorylase provided CDP-ribitol for the equivalent polymerase, poly(ribitol phosphate) polymerase. Synthesis of this enzyme, too, was repressed during phosphate-limited growth, but it was not subject to the rapid inactivation observed with CDP-glycerol pyrophosphorylase in *B. licheniformis* or with the polymerase in *B. subtilis* W23. On the other hand, in strain W23 CDP-glycerol pyrophosphorylase, whose only role is the production of CDP-glycerol for linkage-unit synthesis, was rapidly inactivated.

The reappearance of teichoic acid synthesis on addition of excess of phosphate to cultures that have been phosphate-limited in a chemostat is largely dependent on protein synthesis in both *B. licheniformis* (Hussey *et al.*, 1978) and *B. subtilis* (Rosenberger, 1976; Glaser and Loewy, 1979a). There is evidence, however, that a mechanism other than induction and repression of enzyme synthesis may also be involved in these physiological changes, although until recently this evidence has been obtained with bacteria starved of phosphate in batch culture. Rosenberger (1976), using *B. subtilis* 168 and Glaser and Loewy (1979a), using *B. subtilis* B42, a derivative of the Marburg strain, both found that transfer of bacteria to phosphate-free medium rapidly led to a decrease in rate of incorporation of teichoic acid and peptidoglycan into cell wall, the effect being much more pronounced on teichoic acid. However, whereas inhibition of protein synthesis during phosphate starvation further reduced the rate of peptidoglycan incorporation, it relieved some of the inhibition of teichoic acid incorporation, although this did not regain the activity observed in phosphate replete bacteria. Thus, a protein not directly involved in the teichoic acid-synthetic pathway, or the enzyme product of such a protein, might also be involved in the inhibition. Rosenberger (1976) found that chemostat phosphate-limited bacteria transferred to phosphate-rich medium did not lose teichuronic acid-synthesizing activity, or recover teichoic acid synthesis if protein synthesis was inhibited. He suggested that the two pathways might be regulated by each others' metabolic intermediates. It is difficult to make direct comparisons between these experiments and those with chemostat cultures (Thompson *et al.*, 1980). In the latter case, the bacteria are in balanced growth, whereas phosphate-starved bacteria have much depleted nucleotide pools (Lopez *et al.*, 1979) incapable of supporting high rates of wall polymer synthesis from externally added glycerol or glucose. This probably accounts for the stimulation of peptidoglycan synthesis by added inorganic phosphate in the absence of protein synthesis observed in both the above experiments. In phosphate-starved cells of strain B42, addition of inorganic phosphate in the presence of chloramphenicol stimulated teichoic acid synthesis more than peptidoglycan synthesis. Glaser and Loewy (1979a) suggested that this may reflect a nongenetic control of teichoic acid synthesis by P_i or by a phosphorylated metabolite, although it was possible that the addition of phosphate simply increased the size of nucleotide pools available to residual levels of teichoic acid-synthesizing enzymes.

Measurement of enzyme activities in toluenized cells of *B. subtilis* W23 undergoing recovery from phosphate limitation has recently confirmed that regulation of the membrane-bound enzymes of polymer synthesis is entirely independent of protein synthesis during short-term changes of phosphate concentration and that the effect of inhibition of protein synthesis on the changes *in vivo* is due to inhibition of synthesis of the soluble enzyme CDP-glycerol pyrophosphorylase (Hancock, 1983). Work with phosphate-starved and chemostat phosphate-limited cultures of *B. subtilis* has shown that the activities of the enzymes of linkage unit synthesis control the role of addition of teichoic acid into the wall (Glaser and Loewy, 1979b; Cheah *et al.*, 1982), but it is not yet known whether one of these enzymes is the one that is activated and inactivated independently of protein synthesis. The activity of the first enzyme of linkage unit synthesis is inhibited in isolated membranes from phosphate-starved cells, and both this enzyme and that responsible for addition of glycerol phosphate units to linkage unit, as well as, to a lesser extent, the polymerase, are reduced in activity in phosphate-limited chemostat-grown bacteria in balanced growth. The mechanism of regulation of the membrane-bound enzymes is unknown.

In balanced growth under any particular condition, the composition of the wall remains constant. There is, again, evidence that the activities of the enzymes of linkage unit synthesis control the rate of teichoic acid incorporation under such conditions (Cheah *et al.*, 1982), but it is not known how the rate of teichoic acid synthesis is related to that of peptidoglycan synthesis in order to maintain a constant proportion of the two polymers. One hypothesis proposed by Watkinson *et al.* (1971) and Anderson *et al.* (1972), that in isolated membrane preparations of *S. lactis* and *B. licheniformis* there is competition for a limited pool of polyisoprenyl phosphate between the pathways of teichoic acid and peptidoglycan synthesis, has already been described (see Section VII). Since it is now known that synthesis of teichuronic acid also requires a prenyl phosphate carrier lipid (Ward and Curtis, 1982), a similar mechanism might regulate the balance between teichoic acid and teichuronic acid synthesis in bacilli.

REFERENCES

Anderson, A. J., Green, R. S., and Archibald, A. R., 1978, Wall composition and phage-binding properties of *Bacillus subtilis* W23 grown in chemostat culture in media containing varied concentrations of phosphate, *FEMS Microbiol. Lett.* **4:**129–132.

Anderson, R. G., Hussey, H., and Baddiley, J., 1972, The mechanism of wall synthesis in bacteria, *Biochem. J.* **127:**11–25.

Anderton, W. J., and Wilkinson, S. G., 1980, Evidence for the presence of a new class of teichoic acid in the cell wall of bacterium NCTC 9742, *J. Gen. Microbiol.* **118:**343–351.

Arakawa, H., Shimada, A., Ishimoto, N., and Ito, E., 1981, Occurrence of ribitol-containing lipoteichoic acid in *Staphylococcus aureus* H and its glycosylation, *J. Biochem. (Japan)* **89:**1555–1563.

Archibald, A. R., 1974, The structure, biosynthesis and function of teichoic acids, *Adv. Microbial Physiol.* **10:**53–95.

Archibald, A. R., and Coapes, H. E., 1976, Bacteriophage SP50 as a marker for cell wall growth in *Bacillus subtilis, J. Bacteriol.* **125:**1195–1206.

Archibald, A. R., Baddiley, J., Button, D., Heptinstall, S., and Stafford, G. H., 1968, Occurrence of polymers containing *N*-acetylglucosamine 1-phosphate in bacterial walls, *Nature (London)* **219:**855–856.

Baddiley, J., 1972, Teichoic acids in cell walls and membranes of bacteria, *Essays Biochem.* **8:**35–77.

Baddiley, J., Blumson, N. L., and Douglas, L. J., 1968, The biosynthesis of wall teichoic acid in *Staphylococcus lactis* I3, *Biochem. J.* **110:**565–571.

Baddiley, J., Buchanan, J. G., Carss, B., Mathias, A. P., and Sanderson, A. R., 1956, The isolation of CDP-glycerol, CDP-ribitol and mannitol 1-phosphate from *Lactobacillus arabinosus, Biochem. J.* **64:**599–603.

Baddiley, J., Buchanan, J. G., Martin, R. O., and Rajbhandary, U. L., 1962, Teichoic acid from the walls of *Staphylococcus aureus* H, *Biochem. J.* **85:**49–56.

Bertram, K., Hancock, I. C., and Baddiley, J., 1981, Synthesis of teichoic acid by protoplasts of *Bacillus subtilis, J. Bacteriol.* **148:**406–412.

Bracha, R., and Glaser, L., 1976a, *In vitro* synthesis of teichoic acid linked to peptidoglycan, *J. Bacteriol.* **125:**872–879.

Bracha, R., and Glaser, L., 1976b, An intermediate in teichoic acid biosynthesis, *Biochem. Biophys. Res. Commun.* **72:**1091–1098.

Bracha, R., Chang, M., Fiedler, F., and Glaser, L., 1978, Biosynthesis of teichoic acids, *Meth. Enzymol.* **50:**387–402.

Brautigan, V. M., Childs, W. C., and Neuhaus, F. C., 1981, Biosynthesis of D-alanyl lipoteichoic acid in *Lactobacillus casei*: D-alanyl lipophylic compounds as intermediates, *J. Bacteriol.* **146:**239–250.

Brooks, D., and Baddiley, J., 1969, A lipid intermediate in the synthesis of poly (*N*-acetylglucosamine 1-phosphate) from the wall of *Staphylococcus lactis* 2102, *Biochem. J.* **115:**307–314.

Brooks, D., Mays, L. L., Hatefi, Y., and Young, F. E., 1971, Glucosylation of teichoic acid: Solubilisation and partial characterisation of the uridine diphosphoglucose : polyglycerolphosphate teichoic acid glucosyl transferase from membranes of *Bacillus subtilis, J. Bacteriol.* **107:**223–229.

Burger, M. M., and Glaser, L., 1966, The synthesis of teichoic acids V: Polyglucosyl glycerolphosphate and polygalactosylglycerolphosphate, *J. Biol. Chem.* **241:**494–506.

Cabacungan, E., and Pieringer, R. A., 1981, Mode of elongation of the glycerol phosphate polymer of membrane lipoteichoic acid in *Streptococcus faecium* ATCC 9790, *J. Bacteriol.* **147:**75–79.

Chattopadhyay, P. K., and Wu, H. C., 1977, Biosynthesis of the covalent diglyceride in murein lipoprotein of *Escherichia coli, Proc. Natl. Acad. Sci. USA* **74:**5318–5322.

Cheah, S-C., Hussey, H., and Baddiley, J., 1981, Control of synthesis of wall teichoic acid in phosphate-starved cultures of *Bacillus subtilis* W23, *Eur. J. Biochem.* **118:**497–500.

Cheah, S-C., Hussey, H., Hancock, I. C., and Baddiley, J., 1982, Control of synthesis of wall teichoic acid during balanced growth of *Bacillus subtilis* W23, *J. Gen. Microbiol.* **128:**593–599.

Childs, W. C., and Neuhaus, F. C., 1980, Biosynthesis of D-alanyl lipoteichoic acid: Characterisation of ester-linked D-alanine in the *in vitro* synthesised product, *J. Bacteriol.* **143:**293–301.

Chin, T., Burger, M. M., and Glaser, L., 1966, Synthesis of teichoic acids VI: The formation of multiple wall polymers in *Bacillus subtilis* W23, *Arch. Biochem. Biophys.* **116:**358–367.

Chin, T., Younger, J., and Glaser, L., 1968, Synthesis of teichoic acids during spore germination, *J. Bacteriol.* **95:**2044–2050.

Chiu, T. H., Emdur, L. I., and Platt, D., 1974, Lipoteichoic acids from *Streptococcus sanguis, J. Bacteriol.* **118:**471–479.

Coley, J., Archibald, A. R., and Baddiley, J., 1977, The presence of *N*-acetylglucosamine 1-phosphate in the linkage unit that connects teichoic acid to peptidoglycan in *Staphylococcus aureus, FEBS Lett.* **80:**405–407.

Coley, J., Tarelli, E., Archibald, A. R., and Baddiley, J., 1978, The linkage between teichoic acid and peptidoglycan in bacterial cell walls, *FEBS Lett.* **88:**1–9.

Douglas, L. J., 1968, Studies on the biosynthesis of the wall teichoic acid from *Staphylococcus lactis* I3, Ph.D. thesis, University of Newcastle upon Tyne, United Kingdom.

Duckworth, M., Archibald, A. R., and Baddiley, J., 1975, Lipoteichoic acid and lipoteichoic acid carrier in *Staphylococcus aureus* H, *FEBS Lett.* **53:**176–179.

Ellwood, D. C., and Tempest, D. W., 1972, Effects of environment on bacterial wall content and composition, *Adv. Microbial Physiol.* **7:**83–116.

Emdur, L. I., and Chiu, T. H., 1974, Turnover of phosphatidyl glycerol in *Streptococcus sanguis, Biochem. Biophys. Res. Commun.* **59:**1137–1144.

Emdur, L. I., and Chiu, T. H., 1975, The role of phosphatidyl glycerol in the *in vitro* biosynthesis of lipoteichoic acid in *Streptococcus sanguis, FEBS Lett.* **55:**216–219.

Fiedler, F., and Glaser, L., 1974a, The synthesis of poly(ribitolphosphate) I. Purification of Poly(ribitolphosphate) polymerase and lipoteichoic acid carrier, *J. Biol. Chem.* **249:**2684–2689.

Fiedler, F., and Glaser, L., 1974b, The synthesis of poly(ribitolphosphate) II. On the mechanism of poly(ribitolphosphate) polymerase, *J. Biol. Chem.* **249:**2690–2695.

Fiedler, F., and Glaser, L., 1974c, The attachment of poly(ribitolphosphate) to lipoteichoic acid carrier, *Carbohydrate Res.* **37:**37–46.

Fiedler, F., Stackebrandt, E., and Schaffler, M. J., 1981, Biochemical and nucleic acid hybridisation studies on *Brevibacterium linens* and related strains, *Arch. Microbiol.* **129:**85–93.

Fischer, W., and Rosel, P., 1980, The alanine ester substitution of lipoteichoic acid in *Staphylococcus aureus, FEBS Lett.* **119:**224–226.

Fischer, W., Koch, H. U., Rosel, P., and Fiedler, F., 1980a, Alanine ester-containing native lipoteichoic acids do not act as lipoteichoic acid carrier, *J. Biol. Chem.* **255:**4557–4562.

Fischer, W., Koch, H. U., Rosel, P., Fiedler, F., and Schmuck, L., 1980b, Structural requirements for lipoteichoic acid carrier for recognition by the poly(ribitolphosphate) polymerase of *Staphylococcus aureus* H, *J. Biol. Chem.* **255:**4550–4556.

Fischer, W., Schmidt, M. A., Jann, B., and Jann, K., 1982, Structure of the *E. coli* K2 capsular antigen. Stereochemical configuration of the glycerophosphate and distribution of galactopyranosyl and galactofuranosyl residues, *Biochemistry* **21:**1279–1284.

Forsberg, C. W., Wyrick, P. B., Ward, J. B., and Rogers, H. J., 1973, The effect of phosphate limitation on the morphology and wall composition of *Bacillus licheniformis* and its phosphoglucomutase-deficient mutants, *J. Bacteriol.* **113:**969–984.

Ganfield, M-C. W., and Pieringer, R. A., 1980, The biosynthesis of nascent membrane teichoic acid of *Streptococcus faecium* ATCC 9790 from phosphatidylkojibiosyl diglyceride and phosphatidyl glycerol, *J. Biol. Chem.* **255:**5164–5169.

Glaser, L., and Burger, M. M., 1964, The synthesis of teichoic acids III. Glycosylation of polyglycerophosphate, *J. Biol. Chem.* **239:**3187–3191.

Glaser, L., and Lindsay, B., 1974, The synthesis of lipoteichoic acid carrier, *Biochem. Biophys. Res. Commun.* **59:**1131–1136.

Glaser, L., and Loewy, A., 1979a, Control of teichoic acid synthesis during phosphate limitation, *J. Bacteriol.* **137:**327–331.

Glaser, L., and Loewy, A., 1979b, Regulation of teichoic acid synthesis during phosphate limitation, *J. Biol. Chem.* **254:**2184–2186.

Hancock, I. C., 1981, The biosynthesis of wall teichoic acid by toluenised cells of *Bacillus subtilis* W23, *Eur. J. Biochem.* **119:**85–90.

Hancock, I. C., 1983, Activation and inactivation of synthesis of secondary wall polymers in *Bacillus subtilis* W23, *Arch. Microbiol.* **134:**222–226.

Hancock, I. C., and Baddiley, J., 1972, Biosynthesis of the wall teichoic acid in *Bacillus licheniformis, Biochem. J.* **127:**27–37.

Hancock, I. C., and Baddiley, J., 1976, *In vitro* synthesis of the unit that links teichoic acid to peptidoglycan, *J. Bacteriol.* **125:**880–886.

Hancock, I. C., Wiseman, G., and Baddiley, J., 1976, Biosynthesis of the unit that links teichoic acid to the bacterial wall: inhibition by tunicamycin, *FEBS Letts.* **69:**75–80.

Hanover, J. A., and Lennartz, W. J., 1979, The topological orientation of *N,N'*-diacetylchitobiosyl pyrophosphoryl dolichol in artificial and natural membranes, *J. Biol. Chem.* **254:**9237–9246.

Harrington, C. R., and Baddiley, J., 1983, Peptidoglycan synthesis by partly autolysed cells of *Bacillus subtilis* W23, *J. Bacteriol.* **155:**776–792.

Hase, S., and Matsushima, Y., 1977, The structure of the branching point between acidic polysaccharide and peptidoglycan in *Micrococcus lysodeikticus* cell wall, *J. Biochem. (Japan)* **81:**1181–1186.

Heckels, J. E., Archibald, A. R., and Baddiley, J., 1975, Studies on the linkage between teichoic acid and peptidoglycan in a bacteriophage-resistant strain of *Staphylococcus aureus* H, *Biochem. J.* **149:**637–647.

Hewett, M. J., Whicken, A. J., Knox, K. W., and Sharpe, M. E., 1976, Isolation of lipoteichoic acids from *Butyrivibrio fibrisolvens, J. Gen. Microbiol.* **94:**126–130.

Hurly, B. J., Hale, S. G., and Reusch, V. M., 1982, Differential expression of cell wall enzymes in *Bacillus subtilis* 168, *Proc. Annu. Meeting A.S.M.*, cc11–9.

Hussey, H., and Baddiley, J., 1972, Lipid intermediates in the biosynthesis of wall teichoic acid in *Staphylococcus lactis* I3, *Biochem. J.* **127:**39–50.

Hussey, H., Brooks, D., and Baddiley, J., 1969, Direction of chain extension during the synthesis of teichoic acids in bacterial cell walls, *Nature (London)* **221:**665–666.

Hussey, H., Sueda, S., Cheah, S-C., and Baddiley, J., 1978, Control of teichoic acid synthesis in *Bacillus licheniformis* ATCC 9945, *Eur. J. Biochem.* **82:**169–174.

Ishimoto, N., and Strominger, J. L., 1966, Polyribitolphosphate synthetase of *Staphylococcus aureus* strain Copenhagen, *J. Biol. Chem.* **241:**639–650.

Keleman, M. V., and Baddiley, J., 1961, Structure of the intracellular glycerol teichoic acid from *Lactobacillus casei* ATCC 7469, *Biochem. J.* **80:**246–254.

Kennedy, L. D., 1974, Teichoic acid synthesis in *Bacillus stearothermophilus, Biochem. J.* **138:**525–535.

Kennedy, L. D., and Shaw, D. R. D., 1968, Direction of polyglycerophosphate chain growth in *Bacillus subtilis, Biochem. Biophys. Res. Commun.* **32:**861–865.

Lambert, P. A., Coley, J., and Baddiley, J., 1977a, The nature of lipoteichoic acid carrier (LTC) in *Staphylococcus aureus* H, *FEBS Lett.* **79:**327–330.

Lambert, P. A., Hancock, I. C., and Baddiley, J., 1977b, Occurrence and function of membrane teichoic acids, *Biochim. Biophys. Acta* **472:**1–12.

Lang, W. K., Glassey, K., and Archibald, A. R., 1982, Influence of phosphate supply on teichoic acid and teichuronic acid content of *Bacillus subtilis* cell walls, *J. Bacteriol.* **151:**367–375.

Leaver, J., Hancock, I. C., and Baddiley, J., 1981, Fractionation studies of the enzyme complex involved in teichoic acid synthesis, *J. Bacteriol.* **146:**847–852.

Lifely, M. R., Tarelli, E., and Baddiley, J., 1980, The teichuronic acid from the walls of *Bacillus licheniformis* ATCC 9945, *Biochem. J.* **191:**305–318.

Liu, T. Y., and Gottschlick, E. C., 1967, Muramic acid phosphate as a component of the mucopeptide of Gram-positive bacteria, *J. Biol. Chem.* **242:**471–476.

Lopez, J. M., Marks, C. L., and Freese, E., 1979, The decrease of guanine nucleotides initiates sporulation of *Bacillus subtilis, Biochim. Biophys. Acta* **587:**238–252.

Maino, V. C., and Young, F. E., 1974, Regulation of glucosylation of teichoic acid II. Partial characterisation of phosphoglucomutase in *Bacillus subtilis* 168, *J. Biol. Chem.* **249:**5176–5181.

Mancuso, D. D., Junker, D. D., Hsu, S. C., and Chiu, T-H., 1979, Biosynthesis of glycosylated glycerolphosphate polymers in *Streptococcus sanguis, J. Bacteriol.* **140:**547–554.

Mauck, J., and Glaser, L., 1972a, An acceptor-dependent polyglycerophosphate polymerase, *Proc. Natl. Acad. Sci. USA* **69:**2386–2390.

Mauck, J., and Glaser, L., 1972b, On the *in vivo* assembly of the cell wall of *Bacillus subtilis, J. Biol. Chem.* **247:**1180–1187.

McArthur, H. A. I., Roberts, F. M., Hancock, I. C., and Baddiley, J., 1978, Lipid intermediates in the biosynthesis of the linkage unit between teichoic acids and peptidoglycan, *FEBS Lett.* **86:**193–200.

McArthur, H. A. I., Hancock, I. C., Roberts, F. M., and Baddiley, J., 1980a, Biosynthesis of teichoic acid in *Micrococcus varians* ATCC 29750. Characterisation of a further lipid intermediate, *FEBS Lett.* **111:**317–323.

McArthur, H. A. I., Roberts, F. M., Hancock, I. C., and Baddiley, J., 1980b, Concomitant synthesis and attachment of cell wall polymers by a membrane preparation from *Micrococcus varians* ATCC 29750, *Bioorg. Chem.* **9:**55–62.

McArthur, H. A. I., Hancock, I. C., and Baddiley, J., 1981, Attachment of the main chain to linkage unit in the synthesis of teichoic acids, *J. Bacteriol.* **145:**1222–1231.

McCloskey, M. A., and Troy, F. A., 1980, Paramagnetic isoprenoid carrier lipids. 2. Dispersion and dynamics in lipid membranes, *Biochemistry* **19:**2061–2066.

Nathenson, S. G., and Strominger, J. L., 1963, Enzymatic synthesis of *N*-acetyl-glucosaminyl ribitol linkages in teichoic acid from *Staphylococcus aureus* strain Copenhagen, *J. Biol. Chem.* **238:**3161–3169.

Nathenson, S. G., Ishimoto, N., Anderson, J. S., and Strominger, J. L., 1966, Enzymatic synthesis and immunochemistry of α- and β-*N*-acetylglucosaminyl ribitol linkages in teichoic acids from several strains of *Staphylococcus aureus, J. Biol. Chem.* **241:**651–658.

Neuhaus, F. C., Linzer, R., and Reusch, V. M., 1974, Biosynthesis of membrane teichoic acid: Role of D-alanine : membrane acceptor ligase, *Ann. N.Y. Acad. Sci.* **235:**502–518.

Pieringer, R. A., 1972, Biosynthesis of phosphatidyl diglucosyl diglyceride of Streptococcus faecalis ATCC 9790, *Biochem. Biophys. Res. Commun.* **49:**502–508.

Reusch, V. M., and Neuhaus, F. C., 1971, D-alanine : membrane acceptor ligase from *Lactobacillus casei, J. Biol. Chem.* **246:**6136–6143.

Roberts, F. M., McArthur, H. A. I., Hancock, I. C., and Baddiley, J., 1979, Biosynthesis of the linkage unit joining teichoic acid to peptidoglycan in walls of *Micrococcus varians* ATCC 29750, *FEBS Lett.* **97:**211–216.

Rosenberger, R. F., 1976, Control of teichoic and teichuronic acid synthesis in *Bacillus subtilis* 168 trp$^-$, *Biochim. Biophys. Acta* **428:**516–524.

Sanderson, A. R., Strominger, J. L., and Nathenson, S. G., 1962, Chemical structure of teichoic acid from *Staphylococcus aureus* strain Copenhagen, *J. Biol. Chem.* **237:**3603–3613.

Saski, Y., Araki, Y., and Ito, E., 1980, Structure of the linkage region between teichoic acid and peptidoglycan in *Bacillus cereus* AHU 1030, *Biochem. Biophys. Res. Commun.* **96:**529–534.

Schulman, H., and Kennedy, E. P., 1977, Relation of turnover of membrane phospholipids to synthesis of membrane-derived oligosaccharides of *Escherichia coli, J. Biol. Chem.* **252:**4250–4255.

Schulman, H., and Kennedy, E. P., 1979, Localisation of membrane-derived oligosaccharides in the outer envelope of *Escherichia coli* and their occurrence in other Gram-negative bacteria, *J. Bacteriol.* **137:**686–688.

Shabarova, Z. A. Hughes, N. A., and Baddiley, J., 1962, The influence of adjacent phosphate and hydroxyl groups on amino acid esters, *Biochem. J.* **83:**216–219.

Shashkova, A. S., Zaretskaya, M. S., Yarotsky, S. V., Naumova, I. B., Chizov, O. S., and Shabarova, Z. A., 1979, On the structure of the teichoic acid from the cell wall of *Streptomyces antibioticus* 39, *Eur. J. Biochem.* **102:**477–481.

Shaw, N., and Baddiley, J., 1964, The teichoic acid from the walls of *Lactobacillus buchneri* NCIB 8007, *Biochem. J.* **93:**317–321.

Shaw, R. D., 1962, Pyrophosphorolysis and enzymatic synthesis of cytidine diphosphate glycerol and cytidine diphosphate ribitol, *Biochem. J.* **82:**297–312.

Stark, N. J., Levy, G. N., Rohr, T. E., and Anderson, J. S., 1977, Initial reactions of teichuronic acid synthesis in *Micrococcus lysodeikticus, J. Biol. Chem.* **252:**3466–3472.

Thompson, S., Hancock, I. C., and Baddiley, J., 1980, Identification of polypeptides synthesised during the acquisition of teichoic acid synthesis by *Bacillus licheniformis, Biochim. Biophys. Acta* **630:**537–544.

Tkacz, J. S., 1982, Tunicamycin and related antibiotics, in: *Antibiotics VI, Modes and Mechanisms of Microbial Growth Inhibitors,* Springer-Verlag, New York.

Tomasz, A., McDonnell, M., Westphal, M., and Zanati, E., 1975, Coordinated incorporation of nascent peptidoglycan and teichoic acid into Pneumococcal cell walls and conservation of peptidoglycan during growth, *J. Biol. Chem.* **250:**337–341.

Toon, P., Brown, P. E., and Baddiley, J., 1972, The lipid-teichoic acid complex in the cytoplasmic membrane of *Streptococcus faecalis* NCIB 8191, *Biochem. J.* **127:**399–409.

Torii, M., Kabat, E. A., and Bezer, A. E., 1964, Separation of teichoic acid of *Staphylococcus aureus* into two immunologically distinct polysaccharides with α- and β -*N*-acetylglucosaminyl linkages respectively, *J. Exp. Med.* **120:**13–15.

Tynecka, Z., and Ward, J. B., 1975, Peptidoglycan synthesis in *Bacillus licheniformis, Biochem. J.* **146:**253–267.

Ward, J. B., and Curtis, C. A. M., 1982, The biosynthesis and linkage of teichuronic acid to peptidoglycan in *Bacillus licheniformis, Eur. J. Biochem.* **122:**125–132.

Watkinson, B. J., Hussey, H., and Baddiley, J., 1971, Shared lipid phosphate carrier in the biosynthesis of teichoic acid and peptidoglycan, *Nature (London) New Biol.* **229:**57–59.

Weppner, L., and Neuhaus, F. C., 1977, A fluorescent substrate for peptidoglycan biosynthesis, *J. Biol. Chem.* **252:**2296–2303.

Weston, A., and Perkins, H. R., 1977, Biosynthesis of wall-linked teichuronic acid by a wall-plus-membrane preparation from *Micrococcus luteus, FEBS Lett.* **76:**195–198.

Whicken, A. J., and Knox, K. W., 1970, Studies on the Group F antigen of lactobacilli: Isolation of a teichoic acid-lipid complex from *Lactobacillus fermenti* NCTC 6991, *J. Gen. Microbiol.* **60:**293–301.

Whicken, A. J., and Knox, K. W., 1975, Lipoteichoic acids—a new class of bacterial antigen, *Science* **187:**1161–1167.

Whicken, A. J., Elliott, S. D., and Baddiley, J., 1963, The identity of streptococcal Group D antigen with teichoic acid, *J. Gen. Microbiol.* **31:**231–239.

Wyke, A. W., and Ward, J. B., 1977a, The biosynthesis of muramic acid phosphate in *Bacillus licheniformis, FEBS Lett.* **73:**159–163.

Wyke, A. W., and Ward, J. B., 1977b, Biosynthesis of wall polymers in *Bacillus subtilis J. Bacteriol.* **130:**1055–1063.

22

The Major Outer Membrane Lipoprotein of Escherichia coli: Secretion, Modification, and Processing

George P. Vlasuk, John Ghrayeb, and Masayori Inouye

I. INTRODUCTION

The outer membrane of gram-negative bacteria has proven to be a valuable model system in which to study the many complexities of biomembrane structure, function, and biogenesis which are common not only to prokaryotic but also eukaryotic cells. The advantages in utilizing the *Escherichia coli* outer membrane to examine the various aspects of membrane biology are (1) the outer membrane consists of relatively few major proteins, many of which have been purified, (2) a finely detailed genetic map for *E. coli* as well as the structure of the genes coding for many of these proteins have been established, and (3) all of the outer membrane proteins have been shown to be synthesized in the cytoplasmic compartment and subsequently secreted across the cytoplasmic membrane prior to assembly in the outer membrane. Therefore, by utilizing these aspects of the *E. coli* outer membrane, one can directly address many of the fundamental questions concerning biomembranes including the functions and structure of integral membrane proteins, molecular mechanisms of biosynthesis and assembly of membrane proteins, interactions between membrane proteins themselves and with

George P. Vlasuk, John Ghrayeb, and Masayori Inouye • Department of Biochemistry, State University of New York at Stony Brook, Stony Brook, New York 11794.

other membrane components, regulation of membrane biogenesis, and mechanisms of protein translocation across bilayer membranes.

The following brief description of the *E. coli* membrane systems will familiarize the reader with the various components that constitute this structure. A more detailed treatment of this topic can be found in a comprehensive monograph on the subject (Inouye, 1979a).

E. coli as well as other gram-negative bacteria are surrounded by an envelope consisting of two distinct membranes, the outer membrane and the inner (cytoplasmic) membrane (Figure 1). The peptidoglycan or cell wall which is located between these two membranes provides the cell with structural rigidity allowing the bacteria to survive in osmotically hypotonic environments. Both the outer and cytoplasmic membranes

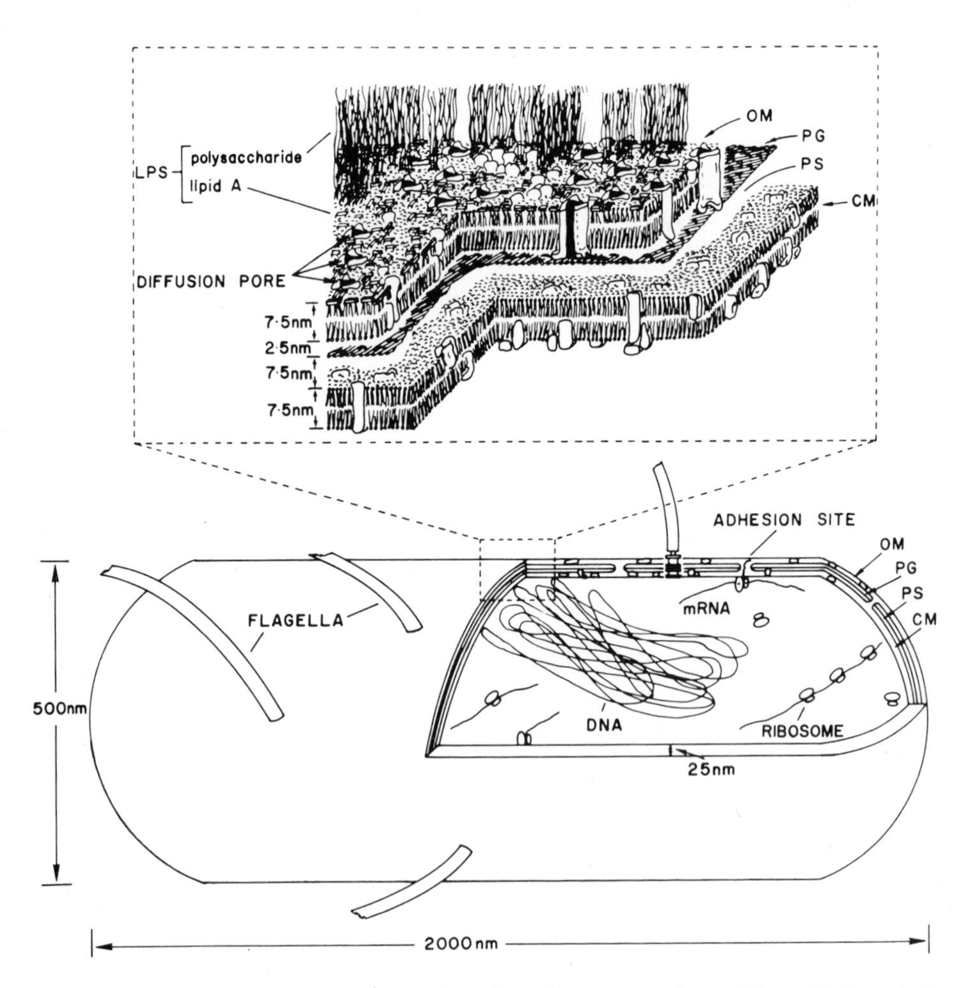

Figure 1. Molecular architecture of the *E. coli* envelope. OM, outer membrane; PG, peptidoglycan; PS, periplasmic space; CM, cytoplasmic membrane. From Inouye (1979c).

show typical features of a unit membrane having a thickness of about 75 Å and consisting of phospholipids and proteins. In addition, the outer membrane also has an associated lipopolysaccharide moiety which, among other functions, has been postulated to be associated with certain outer membrane proteins as well as being intimately associated with their synthesis and localization (Rick *et al.*, 1983).

The cytoplasmic membrane of *E. coli* displays many important functions such as energy metabolism, active transport, synthesis of lipids, peptidoglycan, and polysaccharides. In contrast, the functions of the outer membrane remain somewhat obscure. It is known that the outer membrane serves to keep the periplasmic enzymes and other proteins in the periplasmic space which resides between the cytoplasmic and outer membrane. At the same time, the outer membrane serves as a selective barrier to the cells' exterior preventing toxic substances such as certain antibiotics from entering the cells' interior. On the other hand, the outer membrane appears to have passive diffusion pores for those materials required for cell growth, since the active transport systems for these substances are located exclusively in the cytoplasmic membrane.

The major outer membrane lipoprotein is by far the most thoroughly investigated of all the outer membrane proteins with respect to its biosynthesis and assembly. This form of lipoprotein is the most abundant protein in *E. coli* (approximately 7.5×10^5 molecules per wild-type cell) and its complete protein as well as genomic structure have been determined. In addition to the major form of lipoprotein, there also exists minor forms (Ichihara *et al.*, 1981). Although little information regarding the structure of these minor forms exists, it is believed that their assembly into the outer membrane may be similar to the major form (Ichihara *et al.*, 1981).

Many lipoprotein mutants have been isolated which has helped delineate the mechanisms involved in the secretion and unique modifications of this protein prior to its assembly in the outer membrane. It is the purpose of this review to describe the latest progress dealing with the secretion, modification, and processing of this protein in *E. coli*.

II. THE STRUCTURE OF LIPOPROTEIN AND ITS GENE

The major lipoprotein of *E. coli* consists of 58 amino-acid residues and lacks histidine, tryptophan, glutamic acid, glycine, proline, and phenylalanine as shown in Figure 2. The amino-terminal structure of lipoprotein is unique, consisting of a glycerylcysteine [*S*-(propane-2′-3′-diol)-3-thioaminopropionic acid] to which two fatty acids are linked by two ester linkages and one fatty acid by an amide linkage. The composition of the amide-linked fatty acid has been determined to be palmitic acid (65%), palmitoleic acid (11%), and *cis*-vaccinic acid (11%). The glycerylcysteine-linked fatty acid composition was found to be similar to the fatty acid composition of the same cells, indicating that this diglyceride moiety is derived from a stage in the pathway of phospholipid metabolism (Hantke and Braun, 1973). One-third of the lipoprotein molecules are covalently bound to the peptidoglycan via the ε-amino group

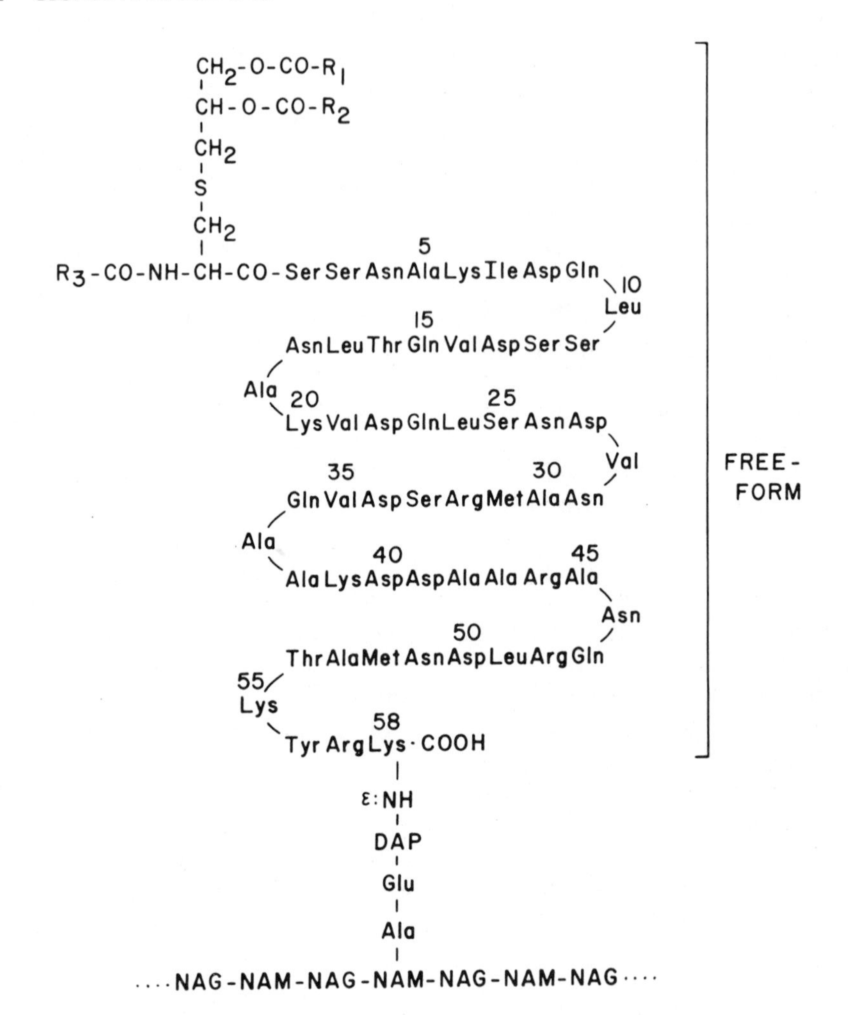

Figure 2. The complete chemical structure of the bound form of the *E. coli* lipoprotein (Braun and Bosch, 1972b; Hantke and Braun, 1973; Nakamura and Inouye, 1979). DAP, diaminopilmelic acid; NAM, *N*-acetylmuramic acid; NAG, *N*-acetylglucosamine. R_1, R_2, and R_3 represent hydrocarbon chains of fatty acids. From Inouye, (1982).

of the carboxy-terminal lysine to the carboxy group of every tenth to twelfth *meso*-diaminopimelic acid residue (Braun and Bosch, 1972a; Inouye *et al.*, 1972).

By using purified lipoprotein mRNA as a probe, the entire lipoprotein gene (*lpp*) of *E. coli* as well as several other Enterobacteriaieae have been isolated, cloned, and their entire DNA sequences determined. A detailed discussion of these sequences has been published elsewhere (Inouye, 1982) and will not be repeated here. The availability of the cloned *lpp* gene has allowed the construction of a variety of plasmid expression vectors in our laboratory (Nakamura *et al.*, 1982; Nakamura and Inouye, 1982). Utilizing these vectors, we have created a variety of mutations within the *lpp* gene to determine the function of these various regions in the biosynthesis and assembly of lipoprotein in *E. coli*. The analysis of many of these mutations with respect to the secretion modification and processing of lipoprotein are discussed below.

III. SECRETION OF LIPOPROTEIN ACROSS THE CYTOPLASMIC MEMBRANE

A. Characterization of Prolipoprotein, the Secretory Precursor of Lipoprotein

Lipoprotein has been shown both *in vivo* (Halegoua and Inouye, 1979; DiRenzo and Inouye, 1979) and *in vitro* (Inouye *et al.*, 1977b) to be initially synthesized as a higher molecular weight precursor, prolipoprotein, which contains an additional 20 amino-acid residues at its amino-terminal end. The discovery and characterization of this precursor has been described in some detail previously (Inouye, 1979b). The structure of the *E. coli* prolipoprotein leader peptide has been shown to be extremely well conserved in prolipoproteins from other gram-negative bacteria (see Table I in Vlasuk *et al.*, 1983).

The fact that mature lipoprotein does not have the leader peptide suggested that an endoproteolytic cleavage of this sequence occurred during the assembly of lipoprotein in the outer membrane. In addition, since lipoprotein must be synthesized in the cytoplasm and translocated across the cytoplasmic membrane in order to reach the outer membrane, the leader peptide may be involved in the translocation process. This was supported by the finding that all secreted proteins, either from the outer membrane or periplasm studied thus far, are also initially synthesized as higher molecular weight precursors containing an amino-terminal extension similar to the prolipoprotein leader peptide (see Table I in Vlasuk *et al.*, 1983). Therefore, the function of the leader peptide (or as it will be designated hence forth, the signal peptide) in the secretion of proteins in *E. coli* may be analagous to the function of the signal peptide of eukaryotic secretory proteins as described in the signal hypothesis (Blobel and Dobberstein, 1975).

From the amino acid sequences of the signal peptide extension of prolipoprotein as well as other *E. coli* secretory precursors, several common structural features, which possibly play important roles in the translocation of these proteins, can be described. These features include (1) 1–3 basic amino-acid residues at the amino terminus, (2) a sequence of 10–15 hydrophobic amino acids directly following the positively-charged amino terminus, (3) in most signal peptides, a proline and/or glycine residue(s) located within the hydrophobic domain (4) a serine or threonine residue(s) following the hydrophobic core and located close to the carboxy-terminus, and (5) an alanine or glycine (prolipoproteins) residue at the carboxy-terminal end (cleavage site).

B. Role of the Prolipoprotein Signal Peptide in Translocation across the Cytoplasmic Membrane

How are the general structural features of the prolipoprotein signal peptide, as well as other prokaryotic signal peptides mentioned above, important in the translocation of these proteins across the cytoplasmic membrane? A model describing the possible functions of these regions has been described by Inouye and Halegoua (1980).

In this proposal, termed the loop model, the positively charged amino terminus of the signal peptide facilitates the initial attachment of the emerging nascent signal peptide, and consequently the active polysome, to the negatively charged inner surface of the cytoplasmic membrane via ionic interactions. Alternatively, there may be a specific receptor residing in the cytoplasmic membrane which interacts with this portion of the signal peptide. This association allows the emerging hydrophobic core of the signal peptide to interact with the lipid bilayer of the cytoplasmic membrane. Further protein elongation allows the signal peptide to form a loop structure which bends in regions of low α-helicity at the glycine or proline residues. During or following protein elongation, the signal peptide is cleaved from the remainder of the protein catalyzed by a specific endopeptidase termed the signal peptidase.

The serine or threonine residues, in the case of *E. coli* lipoprotein, may be serving as specific recognition sites for either the signal peptidase, modification enzymes, or possibly some, as of yet unidentified, secretory component. Alternatively, these amino acids may be involved in the maintenance of a particular conformation of the signal peptide which may be required for efficient secretion to occur.

In support of this model, it is known that certain prokaryotic secretory proteins, and presumably lipoprotein, are synthesized on membrane-bound polysomes which appear to be anchored only by their nascent chains (Smith *et al.*, 1978). In addition, it has been reported that wild-type prolipoprotein cannot be released from the cytoplasmic membrane without cleavage of the signal peptide (Ichihara *et al.*, 1982). However, to definitely test the various proposals made in the loop model in comparison with other models of translocation (Wickner, 1979), we have created specific mutations within the various segments of the prolipoprotein signal peptide followed by characterization of their effects on its secretion. The results obtained from these studied are summarized below.

C. Analysis of Mutations Involving the Prolipoprotein Signal Peptide

Mutational analysis has proved to be a valuable tool in the study of secretion in *E. coli*. The analysis of random mutations within the signal peptide in secretory deficient mutants has provided evidence for the absolute requirement of the signal peptide in the secretion of several other membrane proteins (see the review by Michealis and Beckwith, 1982).

The determination of the entire DNA sequence of lipoprotein (Nakamura and Inouye, 1979) as well as cloning of this DNA into a variety of plasmid vectors (Nakamura and Inouye, 1982) has provided us with the opportunity to manipulate the secretion of lipoprotein using recombinant DNA technology. Experiments of this type, however, require that an inducible system be used to express the mutant lipoprotein genes (*lpp*) since it is expected that many of the mutations could be lethal to the cell. Therefore the constitutive *lpp* gene was made inducible by inserting a *lac* promoter-operator fragment between the *lpp* promoter and the *lpp* structural gene so that the *lpp* gene is expressed only in the presence of the *lac* inducer, isopropyl-β-D-1 thio-galactopyranoside (IPTG); (Nakamura and Inouye, 1982).

To obtain specific mutations within the prolipoprotein signal sequence, we have employed oligodeoxyribonucleotide-directed mutagenesis of the cloned wild-type *lpp* gene (Vlasuk and Inouye, 1983). With this method, we have obtained mutations ranging from single-base changes to three base deletions at specific regions within the prolipoprotein signal sequence (Inouye *et al.*, 1982, 1983a,b, 1984; Vlasuk *et al.*, 1983, 1984). In addition, by utilizing various restriction endonuclease sites within the *lpp* gene, we have also created a variety of mutants in which extra amino acids have been inserted into specific regions of the lipoprotein sequence (Inukai *et al.*, 1983b).

The results of these studies with respect to the role of the various regions of the signal peptide in the secretion of lipoprotein are outlined below.

1. The Amino Terminus

To determine if the positive charges at the amino terminus of the prolipoprotein signal peptide are required for translocation as proposed in the loop model (see above), we have constructed a series of seven unique mutations which systematically altered the charge in this region from +2 (wild type) to −2 (Figure 3). Mutants I-1, I-2, I-

	NH_2-MET	LYS	ALA	THR	LYS	LEU	VAL	LEU	GLY	ALA	VAL	ILE	LEU	GLY	SER	THR	LEU	LEU	ALA	GLY ▼	CYS
I-1			ASP																		
I-2		DEL																			
I-3		DEL	ASP																		
I-4		GLU	ASP																		
I-5					ASN																
I-6		DEL			ASN																
I-7		GLU	ASP		ASN																
A-1									VAL												
A-2									DEL												
B-1														VAL							
B-2														DEL							
A_1B_1									VAL					VAL							
A_1B_2									VAL					DEL							
A_2B_1									DEL					VAL							
A_2B_2									DEL					DEL							
H_1															ALA						
H_2																ALA					
H_3															ALA	ALA					
C_1																				ALA	
C_2																					DEL
C_3																				DEL	

Figure 3. Site-directed mutations made in the *E. coli* prolipoprotein signal peptide. DEL, deletion; (▼) denotes prolipoprotein cleavage site.

3 (Inouye *et al.*, 1982), I-5, and I-6 (Vlasuk *et al.*, 1983) were found to export and process their respective prolipoproteins as efficiently as wild type, albeit at a lower rate of synthesis (see below). In mutants, I-1, I-2, I-3, and I-5, this result was not surprising, since there was one remaining positive charge at the amino terminus. However, the results for mutant I-6 were quite surprising since this mutant lipoprotein is devoid of any positively-charged amino acids at its amino terminus. Therefore, the results for mutant I-6 clearly demonstrates that the basic amino acids present at the amino terminus of prolipoprotein are *not* absolutely required for translocation and processing of this precursor. However, because of its small size, the effects of the mutation on prolipoprotein secretion may not be apparent under the experimental conditions employed in these studies. Therefore, we cannot definitively rule out the involvement of basic amino acids present at the amino terminus of other prokaryotic secretory proteins in the export process until similar studies on these proteins become available.

In contrast to the mutants described above, the synthesis and export of prolipoprotein in mutants I-4 (Inouye *et al.*, 1982) and I-7 (Vlasuk *et al.*, 1983) were strikingly different. Both of these mutant prolipoproteins have a net negatively charged amino terminus in contrast to the other amino-terminal mutants described above. In both of these mutants, there was a significant initial accumulation of unmodified prolipoprotein in the cytoplasmic fraction. Such an accumulation could not be detected in any of the other amino-terminal mutants or wild-type. Both mutant prolipoproteins, i.e., I-4 and I-7, were found to be posttranslationally translocated across the cytoplasmic membrane and processed at a rate which was dependent on the number of positive charges present, i.e., the rate of translocation for mutant I-7 (two negative charges) was slower by a factor of two than that measured for mutant I-4 (two negative charges, one positive charge). Moreover, the posttranslational translocation of cytoplasmic prolipoprotein in mutant I-7 was inhibited by carbonylcyanide-*m*-chlorophenylhydrazone (CCCP) indicating that this process may require a transmembrane electrochemical potential similar to the posttranslational insertion of phage M13 coat protein into the cytoplasmic membrane (Date *et al.*, 1980).

At present, it is not known how the positively charged amino terminus of prolipoprotein participates in the initial stages of secretion. The results obtained from these seven amino-terminal mutants clearly indicate that there is not an absolute requirement of a positive charge at the amino terminus of prolipoprotein for its secretion. However, whether the presence of such a charge in this region facilitates this process could not be determined from the present experiments. In contrast, the presence of a net negative charge in this region severely affects the rate of translocation. It is not known whether this effect on translocation is directly due to the charge at the amino terminus as predicted by the loop model or instead to a secondary effect of these mutations on the level of protein synthesis. In this regard, it appears that the positively charged amino terminus is required for maximal prolipoprotein synthesis since the rate of prolipoprotein synthesis was reduced in all of the amino-terminal mutants studied thus far. This reduced synthesis does not appear to be due to the formation of more stable secondary structures within the region of the mRNA corresponding to the amino terminus as was recently shown for a pro-*lamB* amino-terminal mutant (Schwartz *et al.*, 1981). Instead, the decreased level of prolipoprotein synthesis appears to be related to the net charge present at the amino terminus with mutants having a net negative charge, i.e., mutants I-4 and I-7, being the lowest. The exact influence of the amino-terminal charge on the translational efficiency of these mutants is not known at this time, although a possible coupling between translation and translocation, mediated through systems similar to those described for eucaryotic secretory protein biosynthesis (Walter and Blobel, 1981) or those described in a model proposed by Hall and Schwartz (1982) and Hall *et al.* (1983), may be involved. However, it is clear that the amino-terminal region of prolipoprotein plays an important role in initiating the secretion of this protein in *E. coli*.

2. The Glycine Residues

The two glycine residues at positions 9 and 14 in the prolipoprotein signal sequence have been implicated as being important in the processing of this precursor, since replacement of Gly^{14} with aspartic acid prevented the modification with glycerol or fatty acids and cleavage of the signal peptide (Wu *et al.*, 1977; see Section IV). However, this mutation did not affect the secretion of the mutant prolipoprotein which was found primarily in the outer membrane fraction (Wu *et al.*, 1977; Lee *et al.*, 1983). We have chosen to study in more detail the importance of these residues not only in the processing but also in the secretion of prolipoprotein by analyzing eight systematic site-directed mutations as shown in Figure 3 (Inouye *et al.*, 1984).

The results on the individual mutants, i.e., A-1, A-2, B-1, and B-2, indicated that only the deletion of Gly^{14} (mutant B-2) resulted in an accumulation of membrane-bound prolipoprotein which has been modified with a diacyl glycerol-linked fatty acid moiety (prolipoprotein III; see Figure 5; Section IV). The other individual mutants could not be distinguished from wild type with respect to their secretion and processing. The precursor seen in mutant B-2 has also been observed in cells treated with the antibiotic globomycin which specifically inhibits the lipoprotein signal peptidase (see Section IV). However, in contrast to the precursor accumulated in the presence of globomycin, the accumulated prolipoprotein in mutant B-2 could be "chased," in the presence of excess nonradioactive methionine, to mature lipoprotein indicating that the deletion of Gly^{14} alters the efficiency of signal peptide cleavage by the signal peptidase, possibly through a conformational change in this region of the signal peptide.

The same phenotype observed in mutant B-2 was also observed in the double mutant A_1B_2 in which Gly^9 was changed to Val and Gly^{14} was deleted (Figure 3). However, the effect of deleting Gly^{14} could be compensated for by the deletion of Gly^9 as evidenced by the normal secretion of mutant A_2B_2. By eliminating both glycine residues without changing the length of the hydrophobic core as in mutant A_1B_1 (Figure 3), we have definitively shown that the glycine residues *per se* are not required for normal secretion since the expression of this mutant prolipoprotein could not be distinguished from wild type.

These results indicate that the "helix-bending" function of the glycine residues within the lipoprotein signal peptide is not an important parameter in the secretion of this protein in *E. coli*. Moreover, the effect of the various mutations on the secondary structure of the signal peptide, as predicted from the rules developed by Chou and

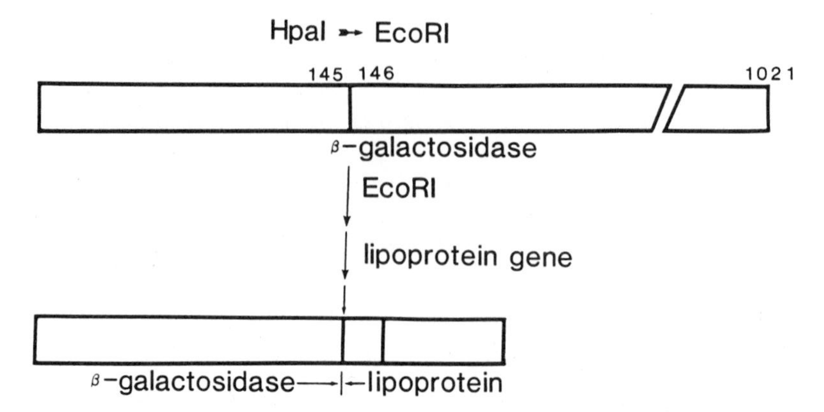

Figure 4. Construction of fusion protein containing an internalized prolipoprotein signal peptide. The β-galactosidase portion of the fusion protein consists of the first 145 amino acids from its amino terminus. The entire prolipoprotein sequence is fused at this point. Numbers correspond to amino acid residues, not base pairs.

Fasman (1977), do not correlate well with the expression studies outlined above. The results obtained from these mutants may, however, indicate that the distance of the cleavage site from a putative recognition site within the signal peptide is important in the rapid and efficient cleavage of the signal peptide. Further experiments will be required to definitively assign a role, if any, for the glycine residues in the prolipoprotein signal peptide and to determine whether a recognition site does indeed exist.

3. *The Serine and Threonine Residues*

The lipoprotein signal sequence has serine and threonine residues located at positions 15 and 16, respectively (Figure 3). To investigate their importance in the secretion of prolipoprotein, we have systematically changed these residues to alanine as shown in Figure 3.

When these mutants were analyzed, mutant H_1 and, to a greater extent, H_3 (Figure 3) were found to accumulate prolipoprotein in the membrane fraction, while mutant H_2 and wild type did not. This effect was more pronounced at 42°C than at 37°C (Vlasvk *et al.*, 1984). Subsequent analysis indicated that the accumulated prolipoprotein in both mutants H_1 and H_3 was unmodified prolipoprotein I (see Figure 5; Section IV). This accumulated prolipoprotein in both mutants was found to be slowly converted to mature lipoprotein. In addition, glyceride-modified prolipoprotein (prolipoprotein III) was not detected in these mutants indicating that the glyceride modification step in the conversion to mature lipoprotein, was slow followed by a rapid cleavage of the signal peptide.

Another interesting aspect that was observed was the higher mobilities of the three mutant lipoproteins compared to the wild type species. This was in contrast to the migration of mature lipoprotein which was the same as wild type in all cases. To determine whether the observed differences in electrophoretic mobilities was due to conformational differences caused by these mutations or an alternative processing event, the first 17 amino-terminal amino acids of the mutant H3 prolipoprotein were sequenced. The results indicated that the mutant prolipoprotein was full length thereby eliminating an alternative processing event. Therefore, these changes in the amino acid composition of the signal peptide in the three mutants appeared to have affected the SDS-binding capacity of the mutant prolipoproteins and/or their conformation, which was reflected in their abnormal electrophoretic mobility.

The results obtained for mutants H1 and H3 are in contrast to another replacement mutation within the same region of the signal peptide which has been shown to completely inhibit fatty acid modification (Wu *et al.*, 1977). In this mutant, where Gly^{14} was replaced by aspartic acid, the unmodified prolipoprotein was secreted across the cytoplasmic membrane and assembled primarily in the outer membrane fraction.

This effect appeared to be specific to replacement of Gly^{14} with aspartic acid since replacement with valine had no effect on modification or processing (see above and Inouye *et al.*, 1984). Therefore the effects of mutants H1 and H3 were unique among the series of prolipoprotein signal peptide mutants studied thus far.

The membrane bound unmodified precursor accumulated in mutant H3 is believed to be in the cytoplasmic membrane fraction anchored via its signal peptide. The fact that this precursor is slowly converted to mature lipoprotein could reflect the fact that it is not yet accessible to the modification and/or processing enzymes, (i.e., not translocated across the membrane bilayer). This may explain why in both wild type and mutant H3, this membrane bound precursor persists in the presence of the ionophore CCCP which has been proposed to inhibit the translocation event (Wickner, 1983). Alternatively, the mutation in mutant H3, like CCCP, may be inhibiting the enzymes responsible for the lipid modification of prolipoprotein which would also result in the slow processing of the unmodified bound precursor. Which of these two processes is actually occurring is currently being investigated.

If one analyzes the possible secondary structures of the prolipoprotein signal peptide region which can be predicted utilizing the rules developed by Chou and Fassman (1977) an interesting correlation between the secondary structure and prolipoprotein maturation is seen. The predicted secondary structure of the wild-type signal peptide indicates an α-helical conformation from residues 1-13 with the region from residue 14 through 17 exhibiting a higher probability for a β-turn structure. In contrast to this, in both mutants H1 and H3, the probability of the β-turn structure between residues 14 and 17 is less than that for an α-helical structure, whereas in mutant H2 this structure remains. These data correlate extremely well with the observed expression results obtained for mutants H1 and H3 in which the accumulated unmodified, membrane bound prolipoprotein was slowly matured whereas mutant H2 maturation was similar to wild type.

Although a strong correlation exists between the predicted secondary structure of the signal peptide and its rate of maturation, we cannot definitively say whether these structures are important in membrane translocation or fatty acid modification. However, since serine and/or threonine occur within roughly the same region in virtually every prokaryotic signal peptide, it is tempting to speculate that their function may be more related to the process of membrane translocation rather than simply fatty acid modification.

From the analysis of the site-specific mutations described above, it is clear that this powerful method of mutagenesis can provide novel and important information regarding the role of the signal peptide in secretion. Many more mutations are currently being pursued and, hopefully, from the analysis of these mutants an overall picture of signal peptide function will be obtained.

4. Internalization of the Prolipoprotein Signal Peptide

The loop model for protein secretion predicts that the initiation of secretion, mediated by the signal peptide, can occur even if this sequence is *not* present at the amino terminus of the precursor protein. In addition, this model also predicts that the amino-terminal portion of the signal peptide would remain on the cytoplasmic face of the cytoplasmic membrane serving as an anchor of sorts. This role of the signal peptide is in contrast to the linear model of cotranslational translocation originally proposed by Blobel and Dobberstein (1975), which predicts that the amino-terminal end of the signal peptide is vectorially translocated across the bilayer membrane.

To determine which of the two models is correct, we have constructed a variety of fusion proteins in which the prolipoprotein signal peptide has been internalized, i.e., not at the amino terminus. In one of these fusion proteins, all but three amino-terminal amino acids of prolipoprotein were fused to a position in the cytoplasmic protein, β-galactosidase, 145 amino acids from its amino terminus (see Figure 4). Lipoprotein-deficient cells in which this protein was expressed, were found to have a full size fusion protein in the cytoplasmic fraction following a 2 min pulse with [^{35}S]methionine. In addition, a small amount of mature lipoprotein was also detected during this time in the membrane fraction. The cytoplasmic hybrid protein was efficiently "chased" in the presence of excess nonradioactive methionine, into mature, membrane-bound lipoprotein. Long-term labeling of these cells in the presence of globomycin, which inhibits the prolipoprotein signal peptidase (see below), resulted in the cytoplasmic precursor becoming bound to the membrane (J. Coleman and M. Inouye, manuscript in preparation).

These results support the loop model in that the prolipoprotein signal sequence can still be functional in initiating secretion even though it is far removed from the amino terminus. The only requirement for an internalized signal peptide to be functional in secretion is that it be exposed to the inner surface of the cytoplasmic membrane.

IV. MODIFICATION AND PROCESSING OF LIPOPROTEIN

As discussed earlier, the *E. coli* lipoprotein is initially produced as a secretory precursor (prolipoprotein I) containing a peptide extension (signal peptide) consisting of 20 amino acid residues at the amino terminal end. The precursor lipoprotein then undergoes a complex set of modification and processing steps which result in the production of mature lipoprotein which is found in the outer membrane. The steps involved in these processes are summarized in Figure 5 and detailed below.

A. Formation of Glyceryl Prolipoprotein

The first modification step is the transfer of a nonacylated glycerol group from phosphatidyl glycerol to form glycerylcysteine of prolipoprotein II (Chattopadhyay

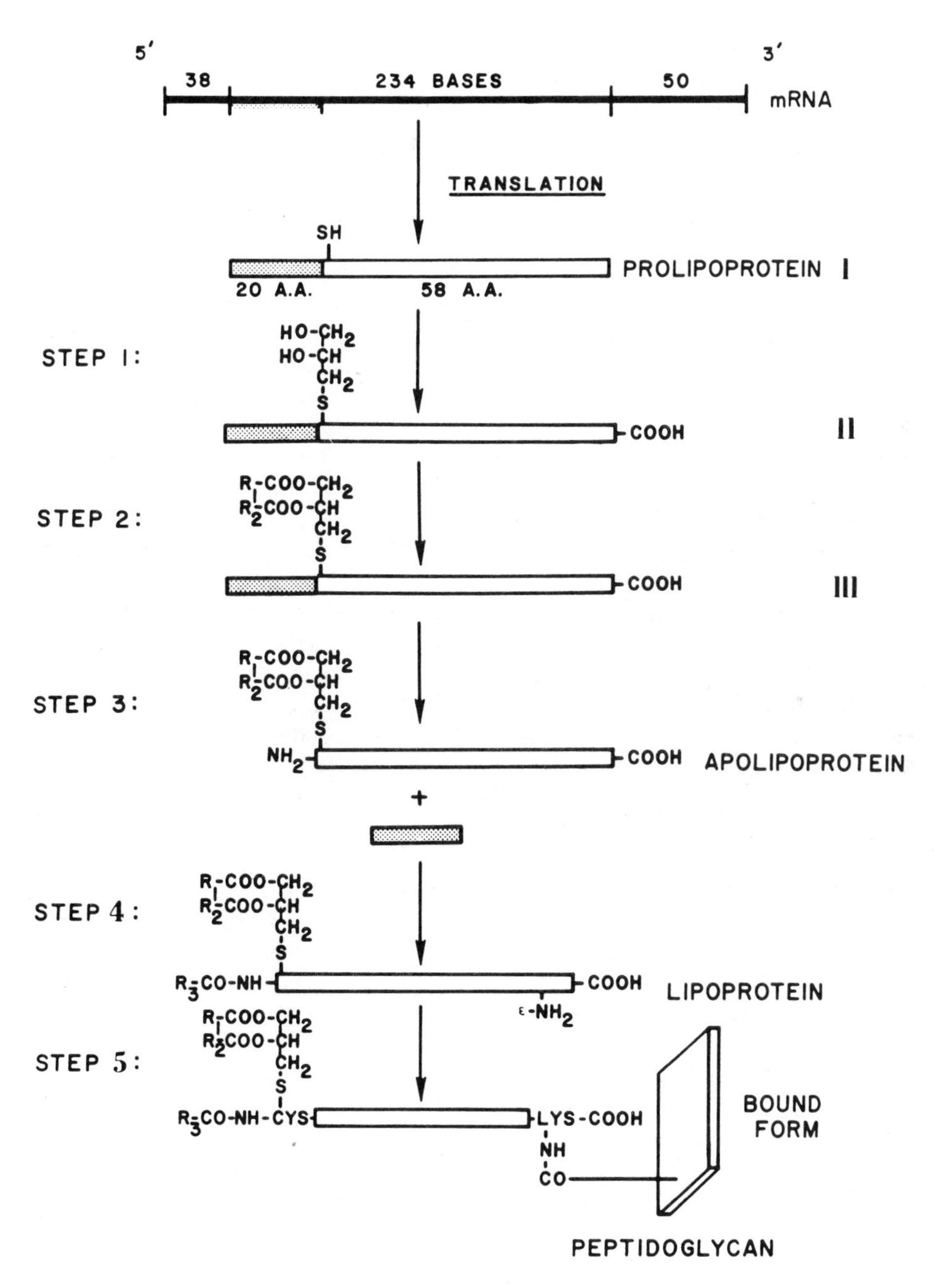

Figure 5. Proposed pathway involved in the modification of the major outer membrane lipoprotein of *E. coli*.

and Wu, 1977; Chattopadhyay *et al.*, 1979; Tokunaga *et al.*, 1982a). Little is known about the transferase enzyme catalyzing this step, except that it is inactive *in vitro* at pH 5.0 (Tokunaga *et al.*, 1982a). In wild-type cells, the glyceryl prolipoprotein intermediate (prolipoprotein II) has never been detected, persumably due to the rapid reaction of the subsequent acylation step (see below). However, in a lipoprotein mutant in which the arginine at position 57 is replaced by cysteine, the lipoprotein purified in our laboratory was found to contain only two moles of fatty acid per mole of lipoprotein compared to three moles of fatty acid per mole of wild-type lipoprotein. In addition, one of the fatty acids in the mutant lipoprotein was bound via an ester linkage (to glycerol) and the other via an amide linkage to the amino group of the amino-terminal cysteine (Inouye *et al.*, 1977a). Almost at the same time, Braun and his group examined the same mutant and concluded that only one amide-linked fatty acid was present in the mutant lipoprotein (Rotering and Braun, 1977). Recently, we have re-examined the same mutant and found that [^{3}H]palmitic acid was incorporated into the prolipoprotein in the presence of the signal peptidase inhibitor, globomycin (see below; J. Ghrayeb and M. Inouye, unpublished data). This result clearly indicates that the glyceryl group of this mutant lipoprotein is acylated. Therefore, results obtained by Braun are most likely due to deacylation during the purification procedure.

Several mutants have been described in which the unmodified mutant prolipoprotein I accumulates in the cell. Wu and his associates as well as our group, independently isolated a mutant lipoprotein in which the glycine at the fourteenth position in the signal peptide of prolipoprotein was changed to an aspartic acid (Lin *et al.*, 1978; Lee *et al.*, 1983). The results from both laboratories clearly indicate that this mutant accumulates the unmodified prolipoprotein in the membrane fraction. More recently, it has been shown in our laboratory that a lipoprotein mutant in which the glycine at the cleavage site was deleted (mutant C-3 in Figure 3), cannot be modified with glycerol and the accumulated prolipoprotein I is found in the outer membrane and is lethal to the cell. This lethality may be due to an abnormal association of the precursor to the outer membrane. In another mutant in which glycine at the cleavage site was replaced by alanine (mutant C-1 in Figure 3), modification and processing occur normally (Inouye *et al.*, 1983b). Therefore, it appears that the glyceryl transferase enzyme may require a recognition site within the signal peptide at a fixed distance from the amino-terminal cysteine. The fact that deletion of Gly^{14}, which also shortens the signal peptide by one amino acid, has no effect on glycerol modification (see mutant B_2 above), indicates that this putative recognition site may lie somewhere between Gly^{14} and the amino-terminal cysteine. Experiments aimed at introducing deletions or insertions within this region of the signal peptide should clarify this hypothesis.

B. Formation of Glyceride Prolipoprotein

In the second modification step, two fatty acid moieties are transferred from cellular phospholipid intermediates to form prolipoprotein III (Figure 5; Lai and Wu, 1980; Lai *et al.*, 1980). This acyl transferase activity can be reconstituted *in vitro* by using cell envelope containing prolipoprotein I and cell envelope from a lipoprotein-deficient strain (Tokunaga *et al.*, 1982a). The glyceride prolipoprotein can be readily detected when the signal peptidase is inhibited as discussed below.

C. Proteolytic Cleavage of the Prolipoprotein Signal Peptide

The next step involves the cleavage of the signal peptide to yield apolipoprotein. The lipoprotein signal peptidase has been shown to be functionally and immunologically distinct from the leader peptidase responsible for the processing of phage M13 coat protein and several *E. coli* secretory proteins (Wolfe *et al.*, 1982; Ito, 1982; Zimmermann and Wickner, 1983; Tokunaga *et al.*, 1982b). Unlike leader peptidase, the lipoprotein signal peptidase has not been isolated or characterized. In 1978, however, a group led by Arai in Japan, reported the isolation of a naturally occurring cyclic peptide which was found to be lethal to a variety of bacteria. It was postulated that globomycin is a specific inhibitor of prolipoprotein signal peptide cleavage (Inukai *et al.*, 1978a). It was observed that prolipoprotein III was accumulated in globomycin-treated cells (Hussain *et al.*, 1980; Inukai *et al.*, 1978b; Ichihara *et al.*, 1981), and that mutants which produced only unmodified prolipoprotein were more resistant to globomycin (Lai *et al.*, 1981; Wu *et al.*, 1982).

From results with globomycin inhibition, one can infer that a fully modified cysteine at the cleavage site may be required by the lipoprotein signal peptidase. This was shown recently by Wu and his group using an *in vitro* assay (Tokunaga *et al.*, 1982a). In their *in vitro* system, they incubated cell envelope containing unmodified prolipoprotein in the presence of detergent with cell envelope from a strain lacking lipoprotein which was labeled either with [^{3}H]glycerol or [^{3}H]palmitic acid. They found that at pH 5.0, where the glyceryl transferase activity is inhibited (see above), no lipid modification or signal peptide cleavage occurred. However, if the substrate used was the fully modified prolipoprotein III, the signal peptide was cleaved even at pH 5.0.

The requirement for a modified cysteine for cleavage of the signal peptide was shown more directly by the construction of a lipoprotein mutant in which the cysteine was replaced by a glycine (mutant C-2 in Figure 3; Inouye *et al.*, 1983a). It was found that the mutant prolipoprotein accumulated in the outer membrane and no mature-like lipoprotein was detected. In addition, the accumulated prolipoprotein was lethal to the cell. It seems clear, therefore, from this result and the results discussed above, that the lipoprotein signal peptidase has an absolute requirement for a glycerol-modified cysteine. Whether the acylation of the glycerol moiety by fatty acids is required for signal peptide cleavage remains to be seen. Further studies on the lipoprotein mutant lacking one or two fatty acids (see above) are required to determine the exact requirement for signal peptide cleavage.

It seems clear from the above results that translocation of prolipoprotein from the cytoplasm to the outer membrane can occur without cleavage of the signal peptide. Thus, the accumulated prolipoprotein III, due to treatment of cells with globomycin, is almost exclusively found in the outer membrane fraction when the cytoplasmic membrane is removed by solubilization with detergent. However, when the outer membrane was separated from the cytoplasmic membrane by sucrose density gradient centrifugation, the prolipoprotein was found in the cytoplasmic membrane (Inukai and Inouye, 1983). These results indicate that the prolipoprotein accumulated in the presence of globomycin, interacts with both the cytoplasmic and the outer membranes. In this interaction, the lipoprotein part of the prolipoprotein is most likely translocated across the cytoplasmic membrane and is interacting with the outer membrane while it is still associated with the cytoplasmic membrane, presumably via the signal peptide. This conclusion is also substantiated by the fact that a fraction of the prolipoprotein has been found to be linked to the peptidoglycan (Inukai *et al.*, 1979; Ichihara *et al.*, 1982) since the bound form cannot be formed without the carboxyl terminal end of the lipoprotein being translocated across the cytoplasmic membrane. In both cleavage, site mutants C-2 and C-3 discussed above, the unmodified mutant prolipoprotein was found exclusively in the outer membrane using both membrane fractionation techniques (Inouye *et al.*, 1983a,b). Mutants C-2 and C-3 lack the glyceride modification which may weaken the association of the prolipoprotein with the cytoplasmic membrane. This may explain why the fully modified prolipoprotein is found in the cytoplasmic membrane when the outer and cytoplasmic membranes are separated by sucrose density gradient centrifugation which is a more physical means of separation. In any event, it is clear that the prolipoprotein can be pulled towards the outer membrane due to its strong affinity for that membrane. It is not known which part of the mature lipoprotein structure is responsible for its strong affinity to the outer membrane. In this regard, we have constructed a hybrid protein in which the first nine amino acids of the mature lipoprotein were fused to β-lactamase. The hybrid product was found to contain a lipid-modified cysteine, is processed in a globomycin-sensitive manner, and was localized in the outer membrane (Ghrayeb and Inouye, 1984). In another lipoprotein mutant, eight amino acids were inserted between the glutamine at the ninth position and the leucine at the tenth position of the mature lipoprotein. This mutant lipoprotein was normally modified, processed, and assembled in the outer membrane but its production was lethal to the cell (Inukai *et al.*, 1983b). These results may indicate that the first nine amino acids are sufficient to "pull" the lipoprotein towards the outer membrane, but the correct insertion into the outer membrane requires additional features found in the mature lipoprotein structure.

The characterization of the lipoprotein signal peptidase has not been achieved to date. However, we have found that the lipoprotein signal peptidase can be inhibited by benzyloxycarbonylalanine chloromethyl ketone (Z-Ala-CH_2Cl) resulting in the accumulation of prolipoprotein III. Furthermore, by using Z-[^3H]AlaCH_2Cl, it was found that two cytoplasmic membrane proteins with M_r = 78,000 and 46,000 were predominantly labeled (Maeda *et al.*, 1981). More recently, an *E. coli* mutant was isolated containing a temperature-sensitive mutation in the signal peptidase (Yamagata *et al.*, 1982). In addition, the isolation of an *E. coli* clone which overproduces the prolipoprotein signal peptidase from the cloned gene has been described (Tokunaga *et al.*, 1983). Further work on this mutant is needed in order to identify the gene coding for the signal peptidase and the eventual characterization of the enzyme.

D. Acylation of Apolipoprotein

After the removal of the signal peptide, the apolipoprotein is acylated at the amino terminal with a fatty acid to form the mature lipoprotein (Figure 5). In contrast to the temperature-resistant property of the signal peptidase (Hirota *et al.*, 1977; Hussain *et al.*, 1982), the acyl transferase responsible for modifying the apolipoprotein, appears to be inactive at high temperatures. This was shown by incubating cell envelope containing prolipoprotein III and observing its conversion to apolipoprotein at 60°C or 80°C (Inukai *et al.*, 1983b; Hussain *et al.*, 1982). The apolipoprotein has also been detected *in vivo* in the membrane fraction of an *E. coli* strain carrying an inducible plasmid containing two tandemly repeated lipoprotein genes. When this strain is pulse labeled with [^{35}S]methionine at 42°C, the apolipoprotein is clearly detected in the outer membrane and can be chased into mature lipoprotein by adding excess unlabeled methionine (Inukai *et al.*, 1984).

E. Peptidoglycan Attachment of Mature Lipoprotein

The final step in the assembly of the lipoprotein is the covalent attachment of the mature lipoprotein, via its carboxy-terminal lysine residue, to the peptidoglycan layer thus giving the bound form of the lipoprotein (Braun and Bosch, 1972b). In addition, it was found that only one-third of the total lipoprotein found in the cell is bound to the peptidoglycan and that the free form is a precursor of, and is in dynamic equilibrium with, the bound form of lipoprotein (Inouye *et al.*, 1972; Weinsink and Witholt, 1981).

The binding of lipoprotein may be its major function in the cell. A class of mutants of *Salmonella typhimurium* (*lky* D mutants) were described by Rothfield's group. These mutants are characterized by a defect in the morphogenesis of the division septum which leads to the formation of blebs of outer membrane over the defective septa which contain peptidoglycan and cytoplasmic membrane but not outer membrane. The outer membrane in these mutants showed a marked reduction in the bound form of lipoprotein and an increase in the free form and this is thought to cause the abnormalities in the outer membrane which leads to the formation of defective division septa (Weigand *et al.*, 1976; Fung *et al.*, 1980). Thus, it appears that the main function of lipoprotein may be, as earlier suggested by Braun, to put the intermediate peptidoglycan layer together with the outer membrane by using the lipoprotein as an anchor in the outer membrane (Braun and Sieglin, 1970). The reversibility of the binding of the lipoprotein (Inouye *et al.*, 1972) may be important in adjusting local differences in the biosynthesis rate of each layer in the envelope.

Another well-conserved structural feature of lipoprotein is a unique tyrosine residue near the carboxy terminus. This tyrosine may be thought to be involved in the reactions resulting in the covalent binding of lipoprotein to the peptidoglycan. However, preliminary results with an *E. coli* lipoprotein mutant in which the tyrosine was replaced by phenylalanine, indicate that such a mutant lipoprotein is still capable of binding to the peptidoglycan (G. P. Vlasuk and M. Inouye, unpublished data). It may still be possible that it is not tyrosine *per se* that is required for binding but rather that an aromatic side chain is needed. Work is currently under way to construct mutants in which the tyrosine will be changed to an aliphatic, neutral, or charged, residue and to examine the effects of these changes on the binding of lipoprotein to the peptidoglycan.

V. CONCLUSIONS

A substantial amount of information has been accumulated in the past 3 years with respect to the biosynthesis and assembly of the major outer membrane lipoprotein of *E. coli*. The elucidation of the lipoprotein gene structure and its sequence as well as its cloning into versatile plasmid expression vectors, has opened up new opportunities in which to manipulate the processes involved in the biosynthesis of lipoprotein via the powerful methods available through recombinant DNA technology.

Some of these studies, which have been described above, have just begun to shed light on the molecular details involved in the translocation of proteins across bilayer membranes. It is clear that the various regions of the prolipoprotein signal peptide play different roles in the secretion and processing of this precursor. Some of the predictions made in the various models describing these functions do not seem to be consistent with the results obtained thus far with prolipoprotein. Therefore, an overall view of the mechanistic components involved in procaryotic secretion must include similar studies on other secretory proteins.

As reviewed above, the pathway for the complex modification steps involved in the maturation of lipoprotein has been worked out in some detail. What remains however, is the elucidation of the mechanistic details involved in these reactions and their overall importance in the function of lipoprotein in *E. coli*.

By pursuing these studies on *E. coli* lipoprotein, we may be able to understand, at the molecular level, the processes of membrane translocation as well as the assembly and association of proteins within the membrane structure which are processes common to all living cells.

REFERENCES

Blobel, G., and Dobberstein, B., 1975, Transfer of proteins across membranes, I. Presence of proteolytically processed and unprocessed nascent immunoglobulin light chains on membrane-bound ribosomes of murine myeloma, *J. Cell Biol.* **67:**835–851.

Braun, V., and Bosch, V., 1972a, Sequence of the murein lipoprotein and the attachment site of the lipid, *Eur. J. Biochem.* **28:**51–69.

Braun, V., and Bosch, V., 1972b, Repetitive sequences in the murein-lipoprotein of the cell wall of *Escherichia coli, Proc. Natl. Acad. Sci. USA* **69:**970–974.

Braun, V., and Sieglin, U., 1970, The covalent murein-lipoprotein structure of the *Escherichia coli* cell wall, *Eur. J. Biochem.* **13:**336–346.

Chattopadhyay, P. K., and Wu, H. C., 1977, Biosynthesis of the covalently-linked diglyceride in murein lipoprotein of *Escherichia coli, Proc. Natl. Acad. Sci. USA* **74:**5318–5322.

Chattopadhyay, P. K., Lai, J. S., and Wu, H. C., 1979, Incorporation of phosphatidylglycerol into murein lipoprotein in intact cells of *Salmonella typhimurium* by phospholipid vesicle fusion, *J. Bacteriol.* **137:**309–312.

Chou, P. Y., and Fasman, G. D., 1977, β-Turns in proteins, *J. Mol. Biol.* **115:**135–175.

Date, T., Zwizinski, C., Ludmerer, S., and Wickner, W., 1980, Mechanisms of membrane assembly. Effects of energy poisons on the conversion of soluble M13-coliphage procoat to membrane bound coat protein, *Proc. Natl. Acad. Sci. USA* **77:**827–831.

DiRenzo, J., and Inouye, M., 1979, Lipid fluidity-dependent biosynthesis and assembly of the outer membrane proteins of *E. coli.*, *Cell* **17:**155–161.

Fung, J. C., MacAlister, T. J., Weigand, R. A., and Rothfield, L. I., 1980, Morphogenesis of the bacterial division septum: Identification of potential sites of division in *lkyD* mutants of *Salmonella typhimurium*, *J. Bacteriol.* **143:**1019–1024.

Ghrayeb, J., and Inouye, M., 1984, Nine amino acid residues at the amino terminal of lipoprotein are sufficient for its modification, processing and localization in the outermembrane of *Escherichia coli*, *J. Biol. Chem.*, **259:**463–467.

Halegoua, S., and Inouye, M., 1979, Translocation and assembly of outer-membrane proteins in *Escherichia coli*. Selective accumulation of precursors and novel assembly intermediates caused by phenylethyl alcohol, *J. Mol. Biol.* **130:**39–61.

Hall, M. N., and Schwartz, M., 1982, Reconsidering the early steps of protein secretion, *Ann. Microbiol. (Inst. Pasteur)* **133A:**123–127.

Hall, M. N., Gabay, J., and Schwartz, M., 1983, Evidence for a coupling of synthesis and export of an outer membrane protein in *Escherichia coli*, *EMBO J.* **2:**15–19.

Hantke, K., and Braun, V., 1973, Covalent binding of lipid to protein. Diglyceride and amide linked fatty acid at the N-terminal end of the murein lipoprotein of the *Escherichia coli* outer membrane, *Eur. J. Biochem.* **34:**284–296.

Hirota, Y., Suzuki, H., Nishimura, Y., and Yasuda, S., 1977, On the process of cellular division in *Escherichia coli:* A mutant of *E. coli* lacking a murein-lipoprotein, *Proc. Natl. Acad. Sci. USA* **74:**1417–1420.

Hussain, M., Ichihara, S., and Mizushima, S., 1980, Accumulation of glyceride-containing precursor of the outer membrane lipoprotein in the cytoplasmic membrane of *Escherichia coli* treated with globomycin, *J. Biol. Chem.* **255:**3707–3712.

Hussain, M., Ichihira, S., and Mizushima, S., 1982, Mechanism of signal peptide cleavage in the biosynthesis of the major lipoprotein of the *Escherichia coli* outer membrane, *J. Biol. Chem.* **257:**5177–5182.

Ichihara, S., Hussain, M., and Mizushima, S., 1981, Characterization of new membrane lipoproteins and their precursors of *Escherichia coli*, *J. Biol. Chem.* **256:**3125–3129.

Ichihara, S., Hussain, M., and Mizushima, S., 1982, Mechanism of export of outer membrane lipoproteins through the cytoplasmic membrane in *Escherichia coli*, *J. Biol. Chem.* **257:**495–500.

Inouye, M. (ed.), 1979a, *Bacterial Outer Membranes: Biogenesis and Functions*, Wiley-Interscience, J. Wiley and Sons, New York.

Inouye, M., 1979b, Lipoprotein of the outer membrane of *Escherichia coli*, in: *Biomembranes*, Vol. 10 (L. A. Manson, ed.), Plenum Press, New York.

Inouye, M., 1979c, What is the outer membrane, in: *Bacterial Outer Membranes: Biogenesis and Function*, Wiley-Interscience, J. Wiley and Sons, New York.

Inouye, M., 1982, Lipoproteins from the bacterial outer membranes. Their gene structures and assembly mechanism, in: *Membranes and Transport*, Vol. 1 (A. N. Martonosi, ed.), Plenum Press, New York, pp. 289–297.

Inouye, M., and Halegoua, S., 1980, Secretion and membrane localization of proteins in *Escherichia coli*, *CRC Crit. Rev. Biochem.* **7:**339–371.

Inouye, M., Shaw, J., and Shen, C., 1972, The assembly of a structural lipoprotein in the envelope of *Escherichia coli*, *J. Biol. Chem.* **247:**8154–8159.

Inouye, S., Lee, N., Inouye, M., Wu, H. C., Suzuki, H., Nishimura, Y., Iketani, H., and Hirota, Y., 1977a, Amino acid replacement in a mutant lipoprotein of the *Escherichia coli* outer membrane, *J. Bacteriol.* **132:**308–313.

Inouye, S., Wang, S., Sekizawa, J., Halegoua, S., and Inouye, M., 1977b, Amino acid sequence for the peptide extension on the prolipoprotein of the *Escherichia coli* outer membrane, *Proc. Natl. Acad. Sci. USA* **74:**1004–1008.

Inouye, S., Soberon, X., Franceschini, T., Nakamura, K., Itakura, K., and Inouye, M., 1982, Role of positive charge on the amino-terminal region of the signal peptide in protein secretion across the membrane, *Proc. Natl. Acad. Sci. USA* **79:**3438–3441.

Inouye, S., Franceschini, T., Sato, M., Itakura, K., and Inouye, M., 1983a, Prolipoprotein signal peptidase of *Escherichia coli* requires a cysteine residue at the cleavage site, *EMBO J* **2:**87–91.

Inouye, S., Hsu, C-P. S., Itakura, K., and Inouye, M., 1983b, Requirement for signal peptide cleavage of *Escherichia coli* prolipoprotein, *Science,* **221:**59–61.

Inouye, S., Vlasuk, G. P., Hsuing, H., and Inouye, M., 1984, Effects of replacing and deleting glycine residues in the hydrophobic region of the *Escherichia coli* prolipoprotein signal peptide on its secretion and processing. *J. Biol. Chem.*, in press.

Inukai, M., and Inouye, M., 1983, Association of the prolipoprotein accumulated in the presence of globomycin with the *Escherichia coli* outer membrane, *Eur. J. Biochem.* **130:**27–32.

Inukai, M., Nakajima, M., Osawa, M., Haneishi, T., and Arai, M., 1978a, Globomycin, a new peptide antibiotic with spheroplast-forming activity. II. Isolation and physio-chemical and biological characterization, *J. Antibiotics* **31:**421–425.

Inukai, M., Takeuchi, M., Shimuzu, K., and Arai, M., 1978b, Mechanism of action of globomycin, *J. Antibiotics* **31:**1203–1205.

Inukai, M., Takeuchi, M., Shimuzu, K., and Arai, M., 1979, Existence of the bound form of prolipoprotein in *Escherichia coli* B cells treated with globomycin, *J. Bacteriol* **140:**1098–1101.

Inukai, M., and Inouye, M., 1983a, Association of the prolipoprotein accumulated in the presence of globomycin with the *Escherichia coli* outer membrane, *Eur. J. Biochem.* **130:**27–32.

Inukai, M., Masui, Y., Vlasuk, G. P., and Inouye, M., 1983b, Effects of inserting eight amino acid residues into the major lipoprotein on its assembly in the outer membrane of *Escherichia coli, J. Bacteriol.,* **155:**275–280.

Inukai, M., Nakamura, K., Ghrayeb, J., and Inouye, M., 1984, Apolipoprotein, an intermediate in the processing of the major lipoprotein of the *Escherichia coli* outer membrane, *J. Biol. Chem.,* **259:**757–760.

Ito, K., 1982, Purification of the precursor form of maltose-binding protein, a periplasmic protein of *Escherichia coli, J. Biol. Chem.* **257:**9895–9897.

Lai, J. S., and Wu, H. C., 1980, Incorporation of acyl moieties of phospholipid into murein lipoprotein in intact cells of *Escherichia coli* by phospholipid vesicle fusion, *J. Bacteriol.* **144:**451–453.

Lai, J. S., Philbrick, W. M., and Wu, H. C., 1980, Acyl moieties in phospholipids are the precursors for the fatty acids in the murein lipoprotein of *Escherichia coli, J. Biol. Chem.* **255:**5384–5387.

Lai, J. S., Philbrick, W. M., Hayashi, S., Inukai, M., Arai, M., Hirota, Y., and Wu, H. C., 1981, Globomycin sensitivity of *Escherichia coli* and *Salmonella typhimurium:* Effects of mutations affecting structures of murein lipoprotein, *J. Bacteriol.* **145:**657–660.

Lee, N., Yamagata, H., and Inouye, M., 1984, Inhibition of secretion of a mutant lipoprotein across the cytoplasmic membrane by the wild type lipoprotein of the *Escherichia coli* outer membrane, *J. Bacteriol.,* in press.

Lin, J. J. C., Kanawaza, H., Ozols, J., and Wu, H. C., 1978, An *Escherichia coli* mutant with an amino acid alteration within the signal sequence of outer membrane prolipoprotein, *Proc. Natl. Acad. Sci. USA* **75:**4891–4895.

Maeda, T., Glass, J., and Inouye, M., 1981, Accumulation of the prolipoprotein of the *Escherichia coli* outer membrane caused by benzyloxycarbonylalanine chloromethyl ketone, *J. Biol. Chem.* **256:**4712–4714.

Michaelis, S., and Beckwith, J., 1982, Mechanism of incorporation of cell envelope proteins in *Escherichia coli, Annu. Rev. Microbiol.* **36:**435–465.

Nakamura, K., and Inouye, M., 1979, DNA sequence of the gene for the outer membrane lipoprotein of *E. coli.* An extremely AT rich promoter, *Cell* **18:**1109–1117.

Nakamura, K., and Inouye, M., 1982, Construction of versatile expression cloning vehicles using the lipoprotein gene of *Escherichia coli, EMBO J.* **1:**771–775.

Nakamura, K., Masui, Y., and Inouye, M., 1982, Use of *lac* promoter-operator fragments as a transcriptional control switch for the expression of the constitutive *lpp* gene in *Escherichia coli, J. Mol. Appl. Genet.* **1:**289–299.

Rick, P. D., Neumeyer, B. A., and Young, D. A., 1983, Effect of altered lipid A synthesis on the synthesis of the OmpA protein in *Salmonella typhimurium, J. Biol. Chem.* **258:**629–635.

Rotering, H., and Braun, V., 1977, Lipid deficiency in a lipoprotein mutant of *Escherichia coli, FEBS Lett.* **83:**41–44.

Schwartz, M., Roa, M., and Debarboville, M., 1981, Mutations that effect *lamB* gene expression at a posttranscriptional level, *Proc. Natl. Acad. Sci. USA* **78:**2937–2941.

Smith, W., Tai, P. C., and Davis, B. D., 1978, Nascent peptide as sole attachment of polysomes to membranes in bacteria, *Proc. Natl. Acad. Sci. USA* **75:**814–817.

Tokunaga, M., Tokunaga, H., and Wu, H. C., 1982a, Post-translational modification and processing of *Escherichia coli* prolipoprotein *in vitro, Proc. Natl. Acad. Sci. USA* **79:**2255–2259.

Tokunaga, M., Loranger, J. M., Wolfe, P. B., and Wu, H. C., 1982b, Prolipoprotein signal peptidase in *Escherichia coli* is distinct from the M13 procoat protein signal peptidase, *J. Biol. Chem.* **257:**9922–9925.

Tokunaga, M., Loranger, J. M., and Wu, H. C., 1983, Isolation and characterization of an *Escherichia coli* clone overproducing prolipoprotein signal peptidase, *J. Biol. Chem.* **258:**12102–12105.

Vlasuk, G. P., and Inouye, S., 1983, Site-specific mutagenesis using synthetic oligodeoxyribonucleotides as mutagens, in: *Experimental Manipulation of Gene Expression* (M. Inouye, ed.), Academic Press, New York, pp. 291–303.

Vlasuk, G. P., Inouye, S., Ito, H., Itakura, K., and Inouye, M., 1983, Effects of complete removal of basic amino acid residues from the signal peptide on secretion of lipoprotein in *Escherichia coli, J. Biol. Chem.,* **258:**7141–7148.

Vlasuk, G. P., Inouye, S., and Inouye, M., 1984, Effects of replacing serine and threonine residues within the signal peptide on the secretion of the major outer membrane lipoprotein of *Escherichia coli, J. Biol. Chem.,* in press.

Walter, P., and Blobel, G., 1981, Translocation of proteins across the endoplasmic reticulum. III. Signal recognition protein (SRP) causes signal sequence-dependent and site-specific arrest of chain elongation that is released by microsomal membranes, *J. Cell Biol.* **91:**557–561.

Weigand, R. A., Vinci, K. D., and Rothfield, L. I., 1976, Morphogenesis of the bacterial division septum: A new class of septation-defective mutants, *Proc. Natl. Acad. Sci. USA* **73:**1882–1886.

Weinsink, J., and Witholt, B., 1981, Conversion of free lipoprotein to the murein-bound form, *Eur. J. Biochem.* **117:**207–212.

Wickner, W., 1979, The assembly of proteins into biological membranes: The membrane trigger hypothesis, *Annu. Rev. Biochem.* **48:**23–45.

Wickner, W., 1983, M-13 coat protein as a model of membrane assembly, *Trends Biochem. Sci.* March, pp. 90–94.

Wolfe, P. B., Silver, P., and Wickner, W., 1982, The isolation of homogeneous leader peptidase from a strain of *Escherichia coli* which overproduces the enzyme, *J. Biol. Chem.* **25:**7898–7902.

Wu, H. C., Hou, C., Lin, J. J. C., and Yem, D. W., 1977, Biochemical characterization of a mutant lipoprotein of *Escherichia coli, Proc. Natl. Acad. Sci. USA* **74:**1388–1392.

Wu, H. C., Lai, J. S., Hayashi, S., and Giam, C. Z., 1982, Biogenesis of membrane lipoproteins in *Escherichia coli, Biophys. J.* **37:**307–315.

Yamagata, H., Ippolito, C., Inukai, M., and Inouye, M., 1982, Temperature-sensitive processing of outer membrane lipoprotein in an *Escherichia coli* mutant, *J. Bacteriol.* **152:**1163–1168.

Zimmermann, R., and Wickner, W., 1983, Energetics and intermediates of the assembly of protein OmpA into the outer membrane of *Escherichia coli, J. Biol. Chem.* **258:**3920–3925.

23

Anchoring and Biosynthesis of a Major Intrinsic Plasma Membrane Protein: The Sucrase–Isomaltase Complex of the Small-Intestinal Brush Border

Giorgio Semenza

I. INTRODUCTION

Aspects of the biochemistry and physiopathology of oligo- and disaccharidases of the small-intestinal brush border membrane have been the object of previous reviews (Semenza, 1968, 1976, 1978, 1981a). I will concentrate here on the membrane positioning and anchoring, and on the biosynthesis and probable phylogeny of one of the major disaccharidases, i.e., the sucrase–isomaltase complex* (See also Semenza, 1979,a,b; 1981b; Semenza *et al.*, 1983; Brunner *et al.*, 1983; Hauser and Semenza, 1983).

The sucrase–isomaltase complex of the small-intestinal brush border is a major intrinsic protein of this membrane, accounting for approximately 10% of the total protein. It is composed of two subunits, each a *single* glycosylated polypeptide chain of approximately 120 kDa and 140 kDa, respectively. [Their glycosidic moieties are responsible for the ABO antigenic reactivity of these enzymes in man (Kelly and Alpers, 1973).] Sucrase–isomaltase plays a major role in the digestion of carbohydrates

*Abbreviations: S, sucrase; I, isomaltase; SI, sucrase–isomaltase; TID, 3-trifluoromethyl-3-(*m*[^{125}I]iodophenyl)diazirine; DAP, 3-(dimethyl-2-(acetimidoxyethyl)ammonio)propanesulfonate; DOC: deoxycholate.

Giorgio Semenza • Laboratorium für Biochemie der ETH, ETH-Zentrum, CH-8092 Zürich, Switzerland.

in the gut, accounting for approximately 80% of the maltase activity, nearly all of the isomaltase, and all of the sucrase activity of the brush border membrane. Together, with α-limit dextrinase (Taravel *et al.*, 1983), it splits α-limit dextrins to free glucose. The catalytic mechanism (Cogoli and Semenza, 1975) involves protonation of the glycosidic *O*, temporary formation of a glucopyranose–oxocarbonium ion (stabilized by an Asp^-) and, finally, liberation of α-glucopyranose. For detailed discussions on this mechanism, the reader is referred to another review (Semenza, 1976).

The two subunits of the complex (a maltase with sucrase activity and a maltase with isomaltase activity) are subjected to the same or to related biological control mechanism(s) as shown by the following: (1) their simultaneous appearance during development, both in man, in whose intestine both activities appear during intrauterine life (Dahlqvist and Lindberg, 1966), and in species in which both activities appear after birth, e.g., in the rat (Rubino *et al.*, 1964) and in the mouse (Arthur, 1968). (2) their simultaneous response to stimulation by sugars and/or corticosteroids *in vivo* (Lebenthal *et al.*, 1972; for review see Koldovský, 1981), (3) their simultaneous absence, lack of activities or other alterations, in sucrose–isomaltose malabsorption (Auricchio *et al.*, 1965; Semenza *et al.*, 1965; Preiser *et al.*, 1974; Dubs *et al.*, 1973; Gray *et al.*, 1976; Freiburghaus *et al.*, 1978; Cooper *et al.*, 1979; Skovbjerg and Krasilnikoff, 1981; Prader *et al.*, 1961; Kerry and Townley, 1965; Rey and Frézal, 1967) which is a *monofactorial* genetic disease (for reviews see Semenza, 1981a,b; Hadorn *et al.*, 1981), and (4) the constant ratios between sucrase and isomaltase activities in random samples of human peroral biopsies (Auricchio *et al.*, 1963).

These observations, although not allowing the identification of the common step(s) in the control of biosynthesis and/or membrane insertion of sucrase and isomaltase, do nevertheless indicate that these processes in the subunits of the sucrase–isomaltase complex are mutually related. As to the structures of the two subunits, their large size has discouraged sequencing. However, there is a strong indication that they must have at least some degree of homology; i.e. (1) the (limited) sequence of a part of the active sites (Quaroni and Semenza, 1976) is identical in S and I; and (2) fingerprints of the tryptic digests of the SI complex yield only a few more peptides than there are Lys + Arg residues (Cogoli *et al.*, 1973).

The two subunits are also similar functionally: (1) Sucrase and isomaltase have extensive overlaps in their substrate specificities; maltose, maltitol (Semenza, 1968 and papers quoted therein), and a number of aryl-α-glucopyranosides (Cogoli and Semenza, 1975) are split by both subunits. The major differences in substrate specificities are that sucrose is not split by isomaltase, and that the 1,6-α-glucopyranoside bonds in isomaltose, isomaltulose, and in α-limit dextrins are not split by sucrase. (2) Sucrase and isomaltase have extensive overlaps in their fully competitive inhibitors; common to both are D-glucose (Janett, 1974; Semenza and Balthazar, 1974), *tris*-hydroxylmethylaminomethane, lanthanides (Vanni *et al.*, unpublished results), D-1:5 gluconolactone (Cogoli and Semenza, 1975), nojirimycin, and deoxynojirimycin (Hanozet *et al.*, 1981). Competitive inhibitors that discriminate between sucrase and isomaltase are acarbose, which does not inhibit isomaltase (Müller *et al.*, 1980), and dextran, which does not inhibit sucrase (Kolínská and Semenza, 1967). (3) Conduritol-B-epoxide acts as an active-site-directed irreversible inhibitor of both sucrase and isomaltase (Quaroni *et al.*, 1974). (4) The secondary deuterium effect is approximately the same in sucrase and isomaltase (Cogoli and Semenza, 1975). (5) The values of the coefficients in the Hensch–Hammett equation are similar for both subunits (Cogoli and Semenza, 1975). (6) The pK_a' values of the groups in the active sites involved in catalysis and/or substrate binding, as deduced from double-logarithmic plots, are approximately the same in sucrase and isomaltase (Flückiger, 1973; Tellier *et al.*, 1979; Hanozet *et al.*, 1981; Vasseur *et al.*, 1982). (7) Both enzymes are activated by Na^+ (Auricchio *et al.*, 1965), (8) both enzymes have the same kinetic mechanism (Janett, 1974; Semenza and Balthazar, 1974), and (9) both enzymes have the same minimum catalytic mechanism (Cogoli and Semenza, 1975).

By and large it seems, therefore, that S and I are similar proteins, the differences being quantitative rather than qualitative. In addition to the differences mentioned above (molecular weight, overlapping but not identical substrate and inhibitor specificity), the following should be also mentioned: different stability towards heat (Dahlqvist, 1962; Kolinská and Semenza, 1967) alkali (Kolínská and Semenza, 1967; Cogoli *et al.*, 1973), trypsin digestion (Quaroni *et al.*, 1974), and somewhat different antigenic specificity (Nishi and Takesue, 1976; Sjöström, 1983).

Do the similarities between the subunits of the sucrase–isomaltase complex indicate common structural ancestry? Can one explain in a simple manner their relationship in biosynthetic control? Prior to formulating any hypothesis on the biosynthesis of this intrinsic membrane protein, it is necessary to discuss its positioning and anchoring in the brush border membrane.

II. GROSS POSITIONING

Most of the protein mass of the sucrase–isomaltase complex is located on the extracellular, luminal side of the membrane, as indicated by the following facts: (1) detergent treatment (Kessler *et al.*, 1978) of brush border membrane vesicles (which are better than 95% closed and right-side-out; Klip *et al.*, 1979) does not produce any apparent increase in sucrase activity, (2) papain treatment of these vesicles solubilizes sucrase activity totally, without affecting the active or passive permeability properties of the membrane (Tannenbaum *et al.*, 1977) or changing its apparent "unit membrane" structure (Nishi and Takesue, 1978). Functional criteria (discussed by Semenza, 1968) and morphological data (Nishi *et al.*, 1968) also show that sucrase–isomaltase is identical with some of the "lollipops" which one can see in negative-stained brush borders.

III. DETAILED POSITIONING

Whereas sucrase–isomaltase is easily solubilized by appropriate treatment with detergents or by controlled proteolytic digestion with papain or elastase, it is not brought into solution by aqueous buffers of low- or high-ionic strengths. By definition it is, therefore, an intrinsic membrane protein. How is it anchored to the brush border membrane? Much information was first obtained by comparing the chemical properties of Triton X-100-solubilized sucrase–isomaltase with those of the papain-solubilized protein, under the reasonable assumption that the former treatment did not produce any covalent modification and that the latter treatment cleaved off most of the protein mass from segment(s) responsible for the anchoring of sucrase–isomaltase to the membrane.

It was found (Brunner *et al.*, 1979) that the C-terminal regions of the subunits were unchanged by papain solubilization. Since papain acted exclusively from the outside of the membrane (as mentioned above), the C-termini of the subunits could not be located on the cytosolic side of the membrane. In fact, if any segment of either polypeptide chain were to span the membrane in its C-terminal region, solubilization by proteolytic treatment should have produced new C-termini. However, papain solubilization did involve covalent changes in the sucrase–isomaltase complex; it split a segment of approximately 20 kDa from the N-terminal region of the *isomaltase* subunit, leaving the N- and C-termini and the apparent size of the *sucrase* subunit intact. This showed that (1) the sucrase–isomaltase complex is anchored to the membrane via a segment of the isomaltase subunit, localized not far from its N-terminal region, (2) the sucrase subunit does not interact with the membrane directly, but only via the isomaltase subunit, and (3) the C-termini of both subunits and the N-terminus of the sucrase subunit are not involved in the interaction with the membrane (Figure 1a). The peripheral positioning of S was demonstrated further by its preferential solubilization after citraconylation of brush border membrane vesicles (Brunner *et al.*, 1979) by the fact that it was not labeled by TID, a photolabel which is confined to, and thus, reacting solely from within, the hydrophobic core of the membrane (Spiess *et al.*, 1982), and by the electron microscopic observation that antisucrase antibody conjugates are 4–5 nm more remote from the apparent "unit membrane" of the brush border than those formed by anti-isomaltase antibodies (Nishi and Takesue, 1976).

Contrary to sucrase, the isomaltase subunit *is* labeled by the hydrophobic photolabel TID, and the label is confined to that portion of isomaltase which is *not* solubilized from the membrane by papain treatment (Spiess *et al.*, 1982). This TID-labeled portion could be isolated; it is highly hydrophobic (see more below) and has Ala as the N-terminus. Its highly hydrophobic character was expected from the sequence of the N-terminal region of the isomaltase subunit (Figure 2; Frank *et al.*, 1978; Sjöström *et al.*, 1982).

The N-terminal sequence of the isomaltase subunit is worth discussing in some detail (Figure 2). The residue at positon 11 is glycosylated and at least the following neutral sugars are associated with this residue in 1 : 1 : 1 approximate molar ratios: galactose, glucose, and fucose (H. Wacker, unpublished). Since enzymatic glycosylation of membrane proteins is confined to the luminal compartment of the endoplasmic reticulum and of the Golgi membranes (for reviews see Hanover and Lennarz, 1981; Sabatini and Kreibich, 1976; Sabatini *et al.*, 1982), these data strongly suggest that residue 11 is located on the extracellular, luminal side of the membrane. The 1–10 segment, even if stretched, could barely be made to cross the membrane (and many of its residues are charged). Thus, it is unlikely that Ala-1 is located at a side of the membrane opposite to residue 11.

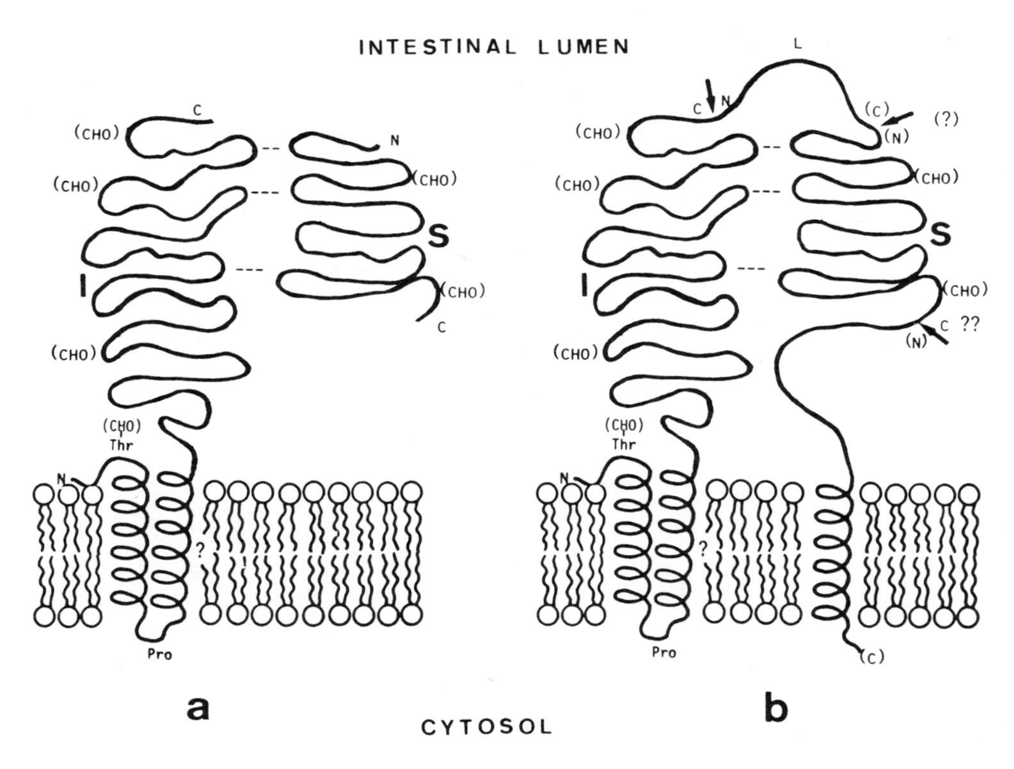

Figure 1. Suggested positioning of the (a) sucrase–isomaltase complex and of (b) pro-sucrase–isomaltase in the brush border membrane. It is not known whether the extracellular C-terminus of the sucrase subunit (S) is the product of posttranslational proteolysis, or not. CHO, carbohydrate chains, →, site of a proteolytic attack. From Semenza (1981b).

PARTIAL N-TERMINAL SEQUENCES OF ProSI AND OF ISOMALTASE AND SUCRASE POLYPEPTIDES

```
HOG ProSI     Ala-Arg-Lys-Ser/Thr(?)-Phe-Ser-Gly-Leu-Glu-Ile—X(CHO)—Leu-Ile-Val-Leu-Phe-Ala-Ile-Val-

HOG I         Ala-Arg-Lys-Lys-Phe-Ser-Gly-Leu-Glu-Ile—X(CHO)—Leu-Ile-Val-Leu-Phe-Ala-Ile-Val-Leu-Ser-Ile-Ala-Ile- Ala-Leu-Val-Val-Val—X—Ala-Ser(?)-Lys(?)—X—Pro-Ala-Val

RAT ProSI[a]  Ala-Lys-Lys-Lys(?)-Phe-Arg(?)-Ala-Leu-Glu-Ile(?)—X(CHO)—Leu-Ile-Val-Leu(?)-Phe-Ile-Ile-

RAT I[a]      Ala-Lys-Lys-Lys-Phe-Ser-Ala-Leu-Glu-Ile—X(CHO)—Leu-Ile-Val-Leu-Phe-Ile-Ile-Val-Leu-Ala-Ile-Ala-Ile- Ala-Leu(?)-Val-Leu-Val(?)-

RABBIT I[b]   Ala-Lys-Arg-Lys-Phe-Ser-Gly-Leu-Glu-Ile-Thr(CHO)-Leu-Ile-Val-Leu-Phe-Val-Ile-Val-Phe-Ile-Ile-Ala-Ile- Ala-Leu-Ile-Ala-Val-Leu-Ala—X—X—X—Pro-Ala-Val

HOG S         Ile-Lys-Leu-Pro-Ser-Asp-Pro-Ile-Pro-Thr-Leu-Arg-Val/Met-Glu-Val/Met-Lys/Thr-Tyr-His-Lys/Thr-Asp-Tyr-Val/Met-Leu-Glu- Phe—X—Arg-Tyr-Asp-Pro-Glu-Arg-

RAT S[a]      Ile-Lys-Leu-Pro-Ser(?)-Asn-Pro-Ile-Arg-Thr(?)-Leu-Arg(?)-Val-Glu-Val-Thr(?)-Tyr—X—Thr(?)-Asn-Arg-Val-Leu-Glu/Gln- Phe-Arg-Ile-Tyr-Arg-Ala-Glu—X—X—Gly-

RABBIT S      Ile-Asn-Leu-Pro-Ser-Glu/Thr-Pro-Glu/Thr-

              1       5         10        15        20        25        30        35
```

```
HUMAN GLYCOPHORIN[c]  -His-His-Phe-Ser-Glu-Pro-Glu-Ile-Thr-Leu-Ile-Ile-Phe-Gly-Val-Met-Ala-Gly-Val-Ile-
                        66              70              75              80              85
HOG GLYCOPHORIN[d]    -Gln-Asp-Phe-Ser-His-Ala-Glu-Ile-Thr-Gly-Ile-Ile-Phe-Ala-Val-Met-Ala-Gly-Leu-Leu-
                        56              80              85              90              95
HORSE GLYCOPHORIN[e]  -His-Asp-Phe-Ser-Gln-Pro-Val-Ile-Thr-Val-Ile-Ile-Leu-Gly-Val-Met-Ala-Gly-Ile-Ile
                        44              50              55              60
```

Figure 2. Partial N-terminal sequences of pro-SIs, I and S subunits from various species. ----, hydrophobic segment beginning at residue 12 in the pro-SIs and Is, ——, residues in the glycophorins which are identical with residues in rabbit I. From Sjöström *et al.* (1982). [a]From Hauri *et al.*, (1982). [b]From Frank *et al.* (1978); Sjöström *et al.* (1982). [c]From Furthmayr *et al.* (1978). [d]From Honma *et al.* (1980). [e]From Murayama *et al.* (1982).

Table 1. The Extent of Amidination by DAP of the NH_2-Terminal Amino Groups of Sucrase–Isomaltase Complex in Membranes[a]

Type of membrane subjected to amidination	Reaction conditions[b]	Extent of amidination[e] (%)	
	Time[c]	Isoleucine[f]	Alanine[f]
Sucrase–isomaltase-egg-lecithin proteoliposomes[d]	20	69	73
Right-side-out, sealed brush border membrane vesicles	10	58 ± 8	33 ± 4
Deoxycholate-treated brush border membrane fragments ("leaky")	10	46 ± 8	35 ± 6

[a] From Bürgi *et al.* (1983).
[b] Amidination was performed at 23°C in 100 mM Na borate buffer, pH 8.5 at initial concentrations of DAP of 40 mM.
[c] Reaction was terminated by the addition of ice-cold ethanolamine (70 mM).
[d] Approximate concentrations of sucrase–isomaltase and egg lecithin were 0.2 mg/ml and 2 mg/ml, respectively.
[e] (± . . . *SD*): Standard deviation as obtained from four experiments.
[f] Isoleucine is the N-terminus of the sucrase subunit, alanine is that of the isomaltase subunit (see Fig. 2).

Direct evidence for Ala-1 to occur indeed at the extracellular, luminal side of the brush border membrane was produced by the use of the slightly permeant, water soluble imidate DAP (Table 1). This reagent labels to the same extent this terminus, no matter whether it has access to both sides of the membrane (in DOC-extracted fragments) or to the outer side only (in sealed, right-side-out vesicles; Bürgi *et al.*, 1983).

Summing up the data reported in the previous paragraphs, it is clear that the S subunit does not interact with the membrane directly, whereas the I subunit does. The "anchoring segment" of the latter is located at the N-terminal region. The first 11 amino-acid residues of isomaltase sequence are located at the extracellular, luminal side. Since the "body" of the protein, including the C-terminus, is also located at the luminal side, the I polypeptide chain either does not cross the membrane, or does so an *even* number of times.

As our attempts to label the "anchor" from the cytosolic side failed, we tackled the problem of its positioning in the membrane indirectly, i.e., via the most likely secondary structure and mode of folding of the anchoring segment. As isolated after TID labeling and papain solubilization of SI, the "anchor" has an apparent size of approximately 6000–6500 Da (Spiess *et al.*, 1982). We thus know approximately one-half of its sequence (Figure 2). Its amino acid composition is, as expected, strongly hydrophobic (Spiess *et al.*, 1984).

CD-spectra of the "anchor" in detergents (cholate, SDS), in 2-chloroethanol, or in liposomes clearly showed it to have a very high α- or 3_{10}-helix content (75–85% to 100%, depending on the conditions; Spiess *et al.*, 1982). The occurrence of a proline residue approximately in the middle of the anchor (position 35, Fig. 2), and the fact that the (unidentified) residues 32–34 are hydrophilic in nature indicate that the anchor has a first stretch in a helix configuration (from residue 12 to approximately 31), then a β-turn around Pro-35, and then a helical domain again. Note that the approximately 20 hydrophobic residues which follow Thr-11 are sufficient to form an α-helix of about 3 nm or a 3_{10}-helix of about 4 nm in length, enough to span a hydrophobic layer 3.5 nm thick. Of the 20–25 amino-acid residues following Pro-35 approximately 20–25 must also be in a prevailing helical configuration if the total helical content of the anchor is to be 80% or more. It is thus most likely that the isomaltase polypeptide chain, i.e., its portion functioning as the "anchor," crosses the hydrophobic bilayer of the membrane twice, Pro-35 and neighboring residues being located at the cytosolic side of the membrane. Indeed, it is very difficult, if not impossible, to otherwise accommodate in the bilayer two helices of this length connected by a β-turn. Figure 1a summarizes schematically the state of our present knowledge on the structure and membrane anchoring of SI.

The information on the "stalk" segment is still fragmentary. It is a polypeptide segment of approximately 100 fairly hydrophilic amino-acid residues, which in a completely extended configuration would have a length of approximately 300 Å (Spiess *et al.*, 1984). Electron micrographs, after negative staining, show a gap of approximately 20–30 Å between the membrane surface and the globular part of SI (Nishi and Takesue, 1978; Brunner *et al.*, 1979). A gap of this size could accommodate papain and allow it to split the "stalk," thereby solubilizing the part of SI which protrudes from the membrane. If the negative stain does not grossly affect the secondary structure of a polypeptide chain (an assumption which may or may not hold true), the difference in gap size in the electron micrographs and the maximum possible length of the "stalk" indicates that the stalk has some kind of secondary structure.

IV. PRO-SUCRASE–ISOMALTASE, A FULLY ENZYMATICALLY ACTIVE PRECURSOR

Clearly, any proposed biosynthetic mechanism of sucrase–isomaltase must explain (1) the peripheral position of sucrase, (2) the occurrence of the hydrophobic "anchor" at the N-terminal region of a polypeptide chain, (3) the facing of the two N-terminal and of the two C-terminal amino acids, outside the cell, (4) the absence of any significant role of the C-terminal regions in the anchoring to the membrane, (5) the coordination in biosynthesis and assembly of the two subunits, and (6) their similar, if not homologous structure. In 1978, I presented (Semenza, 1978, 1979a,b; Frank *et al.*, 1978; Brunner *et al.*, 1979) the "one-chain precursor hypothesis" which can be formulated as follows: (1) an ancestral gene produced a one-polypeptide-chain, one-active-site enzyme splitting both maltose and isomaltose (a "simple isomaltase"), (2) (partial) gene duplication led to a gene coding for a long single polypeptide chain with two identical active sites, each endowed with enzymatic activity (a "double isomaltase"), (3) point mutation(s) and/or deletion(s) then changed one of these active sites from an isomaltase–maltase into a sucrase–maltase; a single chain carrying two similar, but not identical active sites was thus formed, an (active?) "precursor," the "pro-sucrase–isomaltase," (4) this single-chain pro-sucrase–isomaltase is synthesized, glycosylated, and inserted in the membrane of the endoplasmic reticulum, and finally transferred to the brush border membrane, and (5) posttranslational modification of this single, long polypeptide chain [by intracellular or extracellular, e.g., pancreatic, protease(s)] leads to the two subunits which make up the "final" SI complex; they remain associated with one another via interactions formed during the folding of the original single-chain pro-sucrase–isomaltase (Figure 1b).

On purpose, several details were left unspecified in the original hypothesis. To name a few: (1) Is single-chain pro-sucrase–isomaltase enzymatically active or not? (2) Is single-chain pro-sucrase–isomaltase synthesized *without* a hydrophobic segment near the carboxyl terminal and thus remains anchored to the membrane from the beginning of its biosynthesis solely via the hydrophobic segment that eventually belongs to the N-terminal region of the isomaltase subunit, or is single-chain pro-sucrase–isomaltase synthesized *with* an additional hydrophobic segment at the carboxyl terminal, with the C-terminal located either intra- or extracellularly, which would imply that *both* "final" extracellular C-terminals are the result of posttranslational proteolysis? (3) does single-chain pro-sucrase–isomaltase begin with the portion of the eventual I subunit, or with that of the S subunit? and (4) does the sucrase portion begin with a hydrophobic segment (an "anchor") of its own? As shown below, many of these questions have now been answered.

The existence of a single-polypeptide chain (pro-sucrase–isomaltase) of >260 kDa during biosynthesis, insertion, and transfer of "final" SI is now generally accepted. First, Hauri *et al.* (1979) found that *in vivo* injection of [^{3}H]-fucose lead to the appearance of a fast-labeled, high-molecular weight polypeptide in the Golgi membranes, which, after Triton solubilization, precipitated with antibodies directed against the "final" SI complex. This large polypeptide was split by pancreatic elastase into two bands having apparent molecular weights close to those of the subunits of "final" SI. The kinetics of the transfer of the label from the Golgi membranes into brush border membranes, with simultaneous splitting into two bands, was also compatible with a single-chain pro-sucrase–isomaltase appearing in the Golgi membranes, being transferred to, and then split in the brush border membrane.

Direct evidence for the single-chain pro-sucrase–isomaltase was then provided by observations made on hogs whose pancreas had been totally disconnected from the duodenum 3–4 days prior to sacrifice. The sucrase and isomaltase activities in the small-intestinal brush border membranes of these animals were associated with a single large (⩾260 kDa) polypeptide chain (which we call "pro-sucrase–isomaltase," although it is enzymatically fully active). Treatment with minute amounts of pancreatic proteases splits this chain into two bands of apparent molecular sizes similar to those of the subunits of the "final" sucrase–isomaltase complex; the cleavage occurred with little or no change in sucrase or isomaltase acitivity (Sjöström *et al.*, 1980).

Hauri *et al.* (1980) and later Montgomery *et al.* (1981) obtained from transplants of small intestine of fetal rats into adult animals, an enzymatically active large polypeptide. This pro-SI from rat transplants was purified recently (Hauri *et al.*, 1982). Small amounts of unconverted pro-SI are present in the small intestinal cells of adult, intact animals (Danielsen *et al.*, 1981a,b).

Finally, *in vitro* cell-free translation of pro-sucrase–isomaltase has been achieved (Figure 3; Wacker *et al.*, 1981). Pro-sucrase–isomaltase is the largest identified membrane polypeptide chain successfully translated thus far in a cell-free system from total RNA.

It thus seems beyond doubt that SI is indeed synthesized as pro-SI. The cell-free *in vitro* translation will hopefully allow one to tackle and solve problems related to the detailed mechanism of biosynthesis and membrane insertion. Prior to this, however, five questions on the positioning and structure of pro-sucrase–isomaltase have been investigated.

1. Does pro-sucrase–isomaltase begin with the isomaltase or the sucrase portion? Figure 2 shows the N-terminal sequences of rat (Hauri *et al.*, 1982) and hog (Sjöström

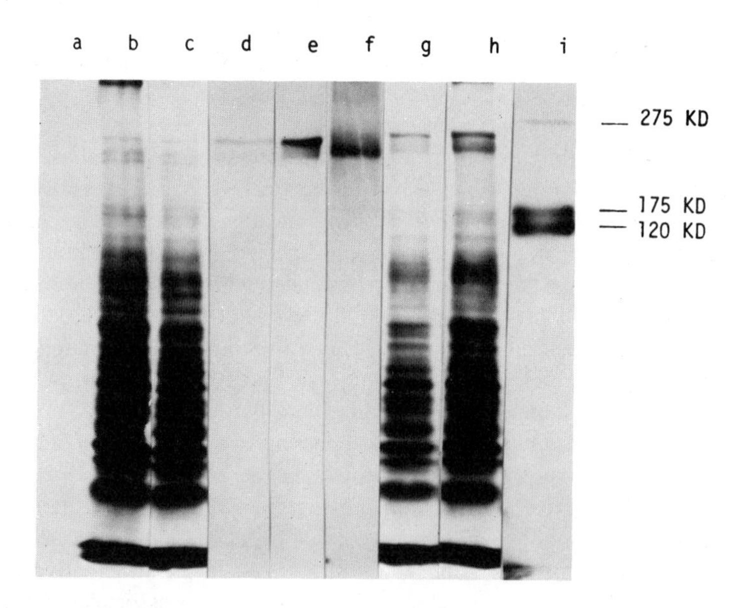

Figure 3. *In vitro* synthesis of pro-SI (or a precursor thereof). Fluorograph of ^{35}S-labeled polypeptides synthetized in a reticulocyte lysate pretreated with nuclease in the presence or absence of dog pancreas microsomes, in response to total RNA extracted from rabbit intestinal mucosa. After translation, the synthetized polypeptides were precipitated with anti-SI-antiserum (for details, see text). The SDS–polyacrylamide gels (6%) were exposed for fluorography for 4 days (a–d,g,h) or 2 days (e,f). (a) Control (no RNA, no microsomes), (b) translation mixture without microsomes, (c) as in (b) centrifugation after incubation (100,000*g* supernatant), (d) immunoprecipitate from (c), (e) mixture of the following three immunoprecipitates: (d) + immunoprecipitate from the 100,000*g* pellet (see (c)) + immunoprecipitate from (g) (see below), (f) translation mixture with microsomes, spun at 100,000*g* after the incubation (immunoprecipitate of the pellet), (g) translation mixture with microsomes, spun at 100,000 × *g* after the incubation (supernatant, no immunoprecipitation), (h) translation mixture with microsomes, (i) [^{125}I]-SI (120 kDa, sucrase subunit; 140 kDa, isomaltase subunit). From Wacker *et al.*, (1981).

et al., 1982) pro-sucrase–isomaltases. Clearly, they are *identical* (not merely homologous!) to those of the corresponding isomaltase subunits. This identity strongly indicates that the isomaltase position corresponds to the N-terminal part of pro-sucrase–isomaltase.

2. The *in vivo* (and some of the *in vitro*) proteolytic conversions of pro-SI to SI do not entail changes in either the enzymatic or immunological properties of the enzyme. This shows that the regions responsible for the catalytic functions or for the antigenic properties of SI are preformed in pro-SI. Presumably, the interactions, which eventually keep the two subunits of SI together, are also present in pro-SI.

3. The "loop" connecting the I and S portions in pro-SI probably does not contain detectable hydrophobic segment(s), in fact, it is possible to obtain, from Triton-solubilized pro-SI by controlled proteolytic treatment, a totally hydrophilic pro-SI, which does *not* show any shift in charge shift electrophoresis. This "hydrophilic" pro-SI is composed of a single polypeptide chain, of a size only slightly smaller than the original, amphipathic Triton-solubilized pro-SI. Clearly no region (I, loop, or S) in this water-soluble pro-SI has detectable hydrophobic properties, thus, the S portion does not have any "hydrophobic segment" or "anchor" at or near its N-terminal region (Sjöström and Norén, unpublished; Figure 1b).

4. Does pro-SI carry a (hydrophobic) segment beyond the eventual C-terminus of S? (See Figure 1b). At the time of writing it seems that this question should be answered positively. In fact, pro-SI binds covalently approximately 60–70% more of the hydrophobic photolabel TID than SI under comparable conditions (Sjöström *et al.*, 1984). It seems likely, therefore, that pro-SI possesses a segment beyond the C-terminal of the S region. This segment may act as an additional "anchor," since it is labeled from the hydrophobic core of the membrane by TID.

5. How is pro-SI anchored to the brush border membrane? Although most of the work has been carried out on "final" SI, the identity between the N-terminal sequences of pro-SI and those of the respective I subunits (Figure 2) strongly indicates that pro-SI also is anchored to the brush border membrane via a "double-helical" segment starting at residue 12. In addition (see above), pro-SI is likely to have one more hydrophobic anchor at the C-terminal region.

Summing up, it seems clear that the one-polypeptide-chain pro-SI hypothesis, which is now universally accepted, satisfactorily explains the positioning of "final" SI, the homology between the I and S, and the tightly related biological control mechanism of I and S. It is interesting to note that the two other glycosidase complexes of the brush border membrane, i.e., the glucoamylase and the β-glucosidase complexes, are also synthesized as a single gigantic polypeptide chain which is then split either by extracellular or by intracellular proteases (Sjöström *et al.*, 1983). Other enzymes, such as aminopeptidase N, which is also composed of two subunits, are synthesized and inserted as separate polypeptide chains, each carrying its own hydrophobic anchor (Sjöström *et al.*, 1983).

The advantage of synthesizing pro-SI as a single polypeptide chain and splitting it into the S and I of "final" SI, without any detectable change in enzymatic or other property is not immediately obvious. Perhaps the advantage is having to bring about one single contact, rather than two, between the ribosome-bound nascent polypeptide chain and the membrane of the rough endoplasmic reticulum.

V. BIOSYNTHESIS OF PRO-SI

Clearly, a polypeptide of the size of pro-SI is not the appropriate model to investigate general mechanisms in membrane-protein biosynthesis and insertion. If the "loop" or "hairpin" (Austen, 1979; DiRienzo *et al.*, 1978; Inouye and Halegoua, 1980; Engelman and Steitz, 1981; see also Sabatini *et al.*, 1982) or the "membrane trigger" (Wickner, 1979, 1980) mechanism is operating (see also Von Heijne and Blomberg, 1979; Von Heijne, 1980), a mechanism similar to that depicted in Figure 4 is most likely. It is, of course, not possible to make any firm statement as to the possible presence of a leader sequence in nascent pro-SI on the basis of expected differences in apparent molecular sizes amounting to approximately 1%, especially since pro-SI is a glycoprotein. However, preliminary observations from our laboratory indicate that pro-SI may indeed be synthesized as pre-pro-SI.

A computer-assisted search for homologies of the sequences in Figure 2 with other published sequences was carried out. The only significant homology found was between segment 68–80 of human glycophorin and segment 5–17 of the isomaltases which had >61% identity. If one restricted the comparison to segment 68–76 of glycophorin and segment 5–13 of rabbit isomaltase, the identity would become as high as 78%. The *analogy* between the isomaltase and glycophorin sequences extends much further, encompassing His-66 and His-67 in glycophorin at positions corresponding to Arg-3 and Lys-4 in rabbit isomaltase, and the long hydrophobic sequences starting at position 75 in glycophorin and position 12 in isomaltase. Note that the homology and analogy refer to parts of glycophorin and of isomaltase which are highly conserved in the species considered.

The significance of such long analogous segments containing highly homologous subsegments is not immediately obvious. The localization of the segment in the polypeptide chain in relation to the membrane is depicted in Figure 5. It can be noted that these segments are located at the transition between the extracellular medium and the hydrophobic membrane layer of $N_{out} \rightarrow C_{in}$ polypeptide chains. The positioning of the rest of the protein mass is different. In the case of glycophorin, the hydrophobic segment crosses the membrane only once and most of the extracellular protein mass is located at the N-terminal end, i.e., "before" this analogous segment. In the case of isomaltase, the hydrophobic anchor crosses the membrane twice and most of the extracellular mass is located at the C-terminal end, i.e., "after" this analogous segment. The significance of this segment is not known. It has been suggested (Sjöström *et al.*, 1983) that this sequence interacts with a collecting and/or transporting protein of the endoplasmic reticulum, and thus may be a part of the address for transport to the plasma membrane.

By pulse chase experiments of explants in organ cultures, Danielsen (1982) followed the sites of biosynthesis and glycosylation of pro-sucrase–isomaltase. Pro-SI was always associated with membranes, and did not occur in the cytosol, thus con-

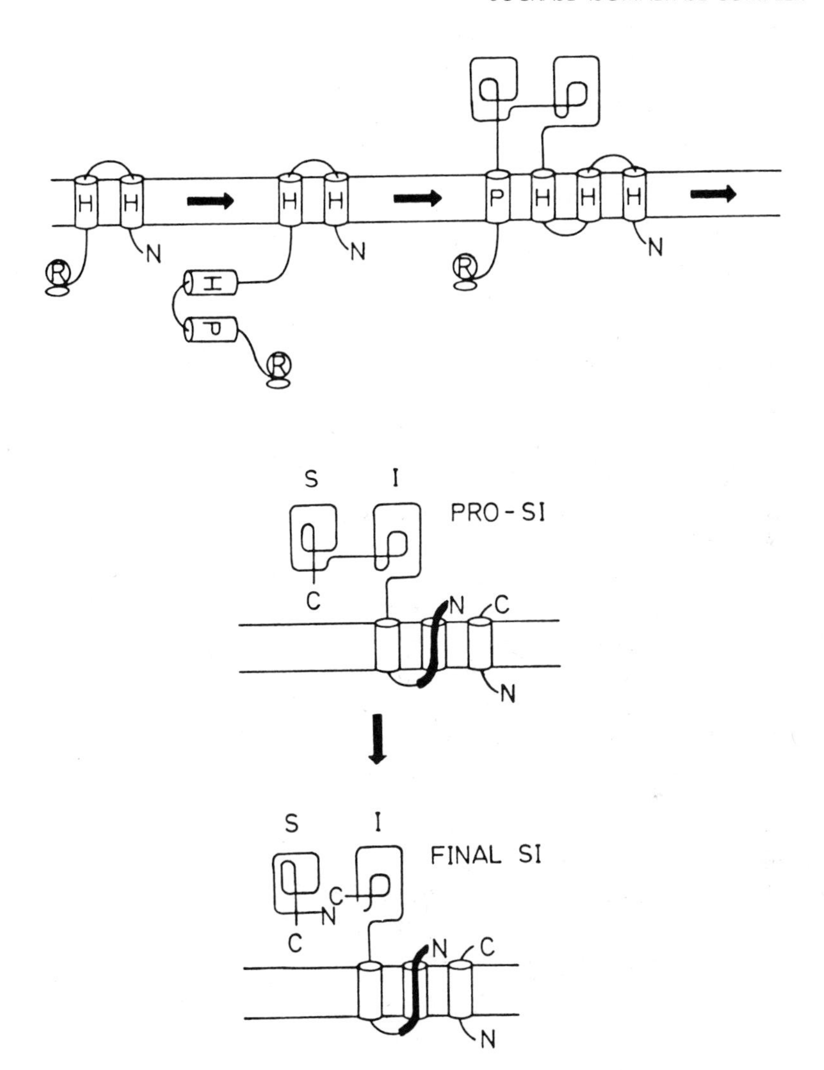

Figure 4. Suggested minimum mechanism of folding, membrane insertion and processing (by signalase and pancreatic protease(s)) of an as yet hypothetical pre-pro-sucrase–isomaltase. Note: the proposed scheme puts emphasis on the folding and mode of insertion only, without indicating any possible interaction with recognition systems. It is possible that (which is not indicated in the figure) the initial C-terminal segment of pro-SI interacts with the membrane. Also, we do not imply that either cleavage of the leader sequence or glycosylation occur after completion of the synthesis of the whole pro-SI. R, ribosome; H, hydrophobic; P, polar. Segments indicated with a bold line are those the sequence of which is known (Figure 2). From Sjöström *et al.* (1982).

firming the previous observations of Hauri *et al.* (1979). In all likelihood, therefore, pro-SI is synthesized on membrane-bound polyribosomes. Glycosylation must take place during or immediately after completion of the polypeptide chain, as shown by the early labeled form of pro-SI being sensitive to endo-β-*N*-acetylglucosaminidase H. The next labeled form, which has a somewhat larger apparent molecular size, is

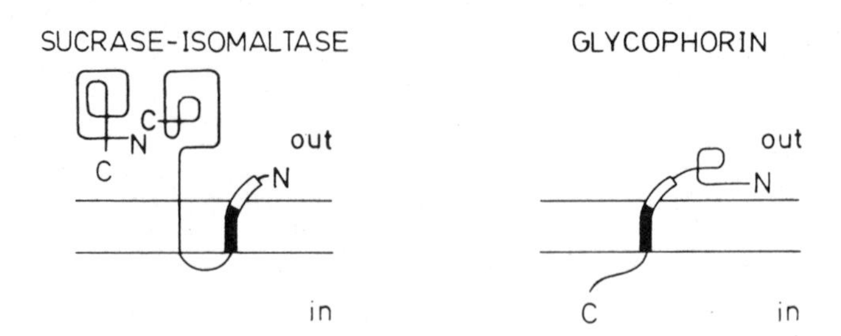

Figure 5. Models of SI and of glycophorin depicting the analogous segment (■□) in the membrane. The white part indicates the highly homologous segment. From Sjöström *et al.* (1982).

not susceptible to this enzyme; this indicates reglycosylation in agreement with the current views on protein glycosylation (Hanover and Lennarz, 1981). Shortly thereafter, i.e., 60–90 min after the beginning of biosynthesis, pro-SI reaches the microvillus membrane.

"Early" sucrase-isomaltases (i.e. fetal, in humans; Auricchio, 1983; Auricchio *et al.*, 1984 and the early-weaning forms, in rats; Kraml *et al.*, 1983) have somewhat different sugar compositions than the adult forms.

The route of pro-SI and of other brush border proteins from the Golgi membranes to the brush border membrane is not yet clear. In essence, the two routes, via basolateral membrane and/or via the terminal web, are still possible (Quaroni *et al.*, 1979a,b; Hauri, 1983). Work in Hauri's laboratory is likely to resolve this question soon.

ACKNOWLEDGMENTS

The reviewer's work was partially supported by the SNSF, Berne and by Nestlé Alimentana, Vevey. Sincere thanks are due to the reviewer's co-workers, who made this project possible, and particularly to biochemists associated with laboratories other than ours, who agreed to collaborate with us.

Note Added in Proof

Recently the enzyme responsible for the isomaltase activity and for much of the maltase activity in the small intestine of the Californian sea lion (*Zalophus californianus*) was isolated. The enzyme does not have sucrase activity and consists of a single, large polypeptide chain with two identical noninteracting active sites. This "double isomaltase" could thus mimic the enzyme presumed to occur in the phylogenetic tree of pro-sucrase-isomaltase (Wacker *et al.*, 1984).

REFERENCES

Arthur, A. B., 1968, Development of disaccharidase activity in the small intestine of suckling mouse, *N. Z. Med. J.* **67:**614–616.

Auricchio, S., 1983, "Fetal" forms of brush border enzymes in the intestine and meconium *J. Pediat. Gastroenterol. Nutr.* **2**(Suppl.1):164–171.

Auricchio, S., Rubino, A., Tosi, R., Semenza, G., Landolt, M., Kistler, H., and Prader, A., 1963, Disaccharidase activities in human intestinal mucosa, *Enzymol. Biol. Clin.* **3:**193–208.

Auricchio, S., Rubino, A., Prader, A., Rey, J., Jos, J., Frézal, J., and Davidson, M., 1965, Intestinal glucosidase activities in congenital malabsorption of disaccharidases, *J. Pediat.* **66:**555–564.

Auricchio, S., Caporale, C., Santamaria, F. and Skovbjerg, H., 1984, Fetal forms of oligoaminopeptidase, dipeptidylaminopeptidase IV, and sucrase in human intestine and meconium. *J. Pediat. Gastroenterol. Nutr.* **3:**28–36.

Austen, B. M., 1979, Predicted secondary structures of amino-terminal extension sequences of secreted proteins, *FEBS Lett.* **103:**308–318.

Brunner, J., Hauser, H., Braun, H., Wilson, K. J., Wacker, H., O'Neill, B., and Semenza, G., 1979, The mode of association of the enzyme complex sucrase–isomaltase with the intestinal brush border membrane, *J. Biol. Chem.* **254:**1821–1828.

Brunner, J., Wacker, H., and Semenza, G., 1983, Sucrase–isomaltase of the small-intestinal brush border membrane: Assembly and biosynthesis, in: *Methods in Enzymology* (S. Fleischer and B. Fleischer, eds.), **vol. 96J.,** Academic Press, New York, pp. 386–406.

Bürgi, R., Brunner, J., and Semenza, G., 1983, A chemical procedure for determining the sidedness of the NH_2-terminus in membrane proteins. The sucrase-isomaltase of the small intestinal brush border, *J. Biol. Chem.*, **258:**15114–15119.

Cogoli, A., and Semenza, G., 1975, A probable oxocarbonium ion in the reaction mechanism of small intestinal sucrase and isomaltase, *J. Biol. Chem.* **250:**7802–7809.

Cogoli, A., Eberle, A., Sigrist, H., Joss, Ch., Robinson, E., Mosimann, H., and Semenza, G., 1973, Subunits of the small-intestinal sucrase–isomaltase complex and separation of its enzymatically active isomaltase moiety, *Eur. J. Biochem.* **33:**40–48.

Cooper, B. T., Candy, D. C. A., Harries, J. T., and Peters, T. J., 1979, Subcellular fractionation studies of the intestinal mucosa in congenital sucrase–isomaltase deficiency, *Clin. Sci.*, **57:**181–185.

Dahlqvist, A., 1962, Specificity of the human intestinal disaccharidases and implications for hereditary disaccharidase intolerance, *J. Clin. Invest.* **41:**453–470.

Dahlqvist, A., and Lindberg, T., 1966, Development of the intestinal disaccharidase and alkaline phosphatase activities in the human foetus, *Clin. Sci.* **30:**517–528.

Danielsen, E. M., 1982, Biosynthesis of intestinal microvillar proteins. Pulse-chase labelling studies on aminopeptidase *N* and sucrase–isomaltase, *Biochem. J.* **204:**639–645.

Danielsen, E. M., Sjöström, H., and Norén, O., 1981a, Biosynthesis of intestinal microvillar proteins. Putative precursor forms of microvillus aminopeptidase and sucrase–isomaltase isolated from Ca^{2+}-precipitated enterocyte membranes, *FEBS Lett.* **127:**129–132.

Danielsen, E. M., Skovbjerg, H., Norén, O., and Sjöström, H., 1981b, Biosynthesis of intestinal microvillar proteins. Nature of precursor forms of microvillar enzymes from Ca^{2+}-precipitated enterocyte membranes, *FEBS Lett.* **132:**197–200.

Danielsen, E. M., Norén, O., and Sjöström, H., 1982, Biosynthesis of intestinal microvillar protein, *Biochem. J.* **204:**323–327.

DiRienzo, J. M., Nakamura, K., and Inouye, M., 1978, The outer membrane proteins of gramnegative bacteria: Biosynthesis, assembly and functions, *Annu. Rev. Biochem.* **47:**481–532.

Dubs, R., Steinmann, B., and Gitzelmann, R., 1973, Demonstration of an inactive enzyme antigen in sucrase–isomaltase deficiency, *Helv. Paediat. Acta* **28:**187–198.

Engelman, D. M., and Steitz, J., 1981, The spontaneous insertion of proteins into and across membranes: The helical hairpin hypothesis, *Cell* **23:**411–422.

Flückiger, R., 1973, *Untersuchungen über den Reaktionsmechanisms der Isomaltase,* Diplomarbeit, ETH Zürich.

Frank, G., Brunner, J., Hauser, H., Wacker, G., Semenza, G., and Zuber, H., 1978, The hydrophobic anchor of small-intestinal sucrase–isomaltase. N-terminal sequence of the isomaltase subunit, *FEBS Lett.* **96:**183–188.

Freiburghaus, A. U., Dubs, R., Hadorn, B., Gase, H., Hauri, H. P., and Gitzelmann, R., 1978, The brush border membrane in hereditary sucrase–isomaltase deficiency: Abnormal protein pattern and presence of immunoreactive enzyme, *Eur. J. Clin. Invest.* **7:**455–459.

Furthmayr, H., Galardy, R. E., Tomita, M., and Marchesi, V. T., 1978, The intramembranous segment of human erythrocyte glycophorin A, *Arch. Biochem. Biophys.* **185:**21–29.

Gray, G. M., Conklin, K. A., and Townley, R. R. W., 1976, Sucrase–isomaltase deficiency. Absence of an inactive enzyme variant, *N. Engl. J. Med.* **294:**750–753.

Hadorn, B., Green, J. R., Sterchi, E. E., and Hauri, H. P., 1981, Biochemical mechanisms in congenital enzyme deficiencies of the small intestine, *Clin. Gastroenterol.* **10:**671–690.

Hanover, J. A., and Lennarz, W. J., 1981, Transmembrane assembly of membrane and secretory glycoproteins, *Arch. Biochem. Biophys.* **211:**1–19.

Hanozet, G., Pircher, H.-P., Vanni, P., Oesch, B., and Semenza, G., 1981, An example of enzyme hysteresis. The slow and tight interaction of some fully competitive inhibitors with small intestinal sucrase, *J. Biol. Chem.* **256:**3703–3711.

Hauri, H. P., 1983, Biosynthesis and transport of plasma membrane glycoproteins in the rat intestinal epithelial cell: studies with sucrase-isomaltase in: Brush border membranes (R. Porter, and G. M. Collins, eds.) *Ciba-Symposium 95*, pp. 132–147.

Hauri, H. P., Quaroni, A., and Isselbacher, K., 1979, Biogenesis of intestinal plasma membrane: Posttranslational route and cleavage of sucrase–isomaltase, *Proc. Nat. Acad. Sci. USA* **76:** 5183–5186.

Hauri, H. P., Quaroni, A., and Isselbacher, K., 1980, Monoclonal antibodies to sucrase–isomaltase: Probes for the study of post-natal development and biogenesis of the intestinal microvillus membrane, *Proc. Natl. Acad. Sci. USA* **77:**6529–6633.

Hauri, H. P., Wacker, H., Rickli, E. E., Bigler-Meier, B., Quaroni, A., and Semenza, G., 1982, Biosynthesis of sucrase–isomaltase. Purification and NH_2-terminal amino acid sequence of the rat sucrase–isomaltase precursor (pro-sucrase–isomaltase) from fetal intestinal transplants, *J. Biol. Chem.* **257:**4522–4528.

Hauser, H., and Semenza, G., 1983, Sucrase-isomaltase: a stalked intrinsic protein of the brush border membrane, *CRC Critical Rev. Biochem.* **14:**319–345.

Honma, K., Tomita, M., and Hamada, A., 1980, Amino acid sequence and attachment sites of oligosaccharide units of porcine erythrocyte glycophorin, *J. Biochem.* **88:**1679–1691.

Inouye, M., and Halegoua, S., 1980, Secretion and membrane localization of proteins in Escherichia coli, *CRC Crit. Rev. Biochem.* **10:**339–371.

Janett, M., 1974, *Identifikation der durch Saccharase und Isomaltase Gespaltenen Bindung in Substrat. Steady State Kinetik der Isomaltase,* Diplomarbeit, ETH Zürich.

Kelly, J. J. and Alpers, D. H., 1973, Blood group antigenicity of purified human intestinal disaccharidases, *J. Biol. Chem.* **248:**8216–8221.

Kerry, K. R., and Townley, R. R. W., 1965, Genetic aspects of intestinal sucrase–isomaltase deficiency, *Australian J. Pediat.* **1:**223–235.

Kessler, M., Acuto, O., Storelli, C., Murer, H., Müller, M., and Semenza, G., 1978, A modified procedure for the rapid preparation of efficiently transporting vesicles from small intestinal brush border membranes, *Biochim. Biophys. Acta* **506:**136–154.

Klip, A., Grinstein, S., and Semenza, G., 1979, Transmembrane disposition of the phlorizin binding protein of intestinal brush border, *FEBS Lett.* **99:**91–96.

Koldovský, O., 1981, Developmental, dietary and hormonal control of intestinal disaccharidases in mammals (including man), in: Carbohydrate Metabolism and Its Disorders, Vol. 3, (J. P. Randle, D. F. Steiner, and W. J. Whelan, eds.), Academic Press, London, pp. 481–522.

Kolínská, J., and Semenza, G., 1967, Studies on intestinal sucrase and on intestinal sugar transport. V. Isolation and properties of sucrase–isomaltase from rabbit small intestine, *Biochim. Biophys. Acta* **146:**181–195.

Kraml, J., Kolínska, J., Kadleková, L., Zákostelecka, M. and Lojda, Z., 1983, Analytical isoelectric focusing of rat intestinal brush border enzymes: postnatal changes and effect of neuraminidase *in vitro*. *FEBS Lett.* **151:**193–196.

Lebenthal, E., Sunshine, P., and Kretchmer, N., 1972, Effect of carbohydrates and corticosteroids on activity of α-glucosidases in the intestine of the infant rat, *J. Clin. Invest.* **51:**1244–1250.

Montgomery, R. K., Sybicki, A. A., Forcier, A. G., and Grand, R. J., 1981, Rat intestinal microvillus membrane sucrase–isomaltase is a single high molecular weight protein and fully active enzyme in the absence of luminal factors, *Biochim. Biophys. Acta* **661**:346–349.

Müller, L., Junge, B., Frommer, W., Schmidt, D. T., and Truscheit, E., 1980, Acarbose (BAY g 5421) and homologous α-glucosidase inhibitors from Actinoplanaceae, in: *Enzyme Inhibitors* (U. Brodbeck, ed.), Verlag Chemie, Weinheim, FR Germany, pp. 109–122.

Murayama, J.-I., Tomita, M., and Hamada, A., 1982, Primary structure of horse erythrocyte glycophorin HA. Its amino acid sequence has a unique homology with those of human and porcine erythrocyte glycophorins, *J. Membr. Biol.* **64**:205–215.

Nishi, Y., and Takesue, Y., 1976, Electron microscope studies on the subunits of the small-intestinal sucrase–isomaltase complex, *J. Electr. Microsc.* **25(Abst.)**:197.

Nishi, Y., Takesue, Y., 1978, Localization of intestinal sucrase–isomaltase complex on the microvillus membrane by electron microscopy using nonlabeled antibodies, *J. Cell Biol.* **79**:516–525.

Nishi, Y., Yoshida, T. O., and Takesue, Y., 1968, Electron microscope studies on the structure of rabbit intestinal sucrase, *J. Mol. Biol.* **37**:441–444.

Prader, A., Auricchio, S., and Mürset, G., 1961, Durchfall infolge hereditären Mangels an intestinaler Saccharaseaktivität (Saccharose-Intoleranz), *Schweiz. Med. Wochztschr.* **91**:465–468.

Preiser, H., Menard, D., Crane, R. K., and Cerda, J. J., 1974, Deletion of enzyme protein from the brush border membrane in sucrase–isomaltase deficiency, *Biochim. Biophys. Acta* **363**:279–282.

Quaroni, A., Gershon-Quaroni, E., and Semenza, G., 1975, Tryptic digestion of native small-intestinal sucrase-isomaltase complex: isolation of the sucrase subunit, *Eur. J. Biochem.*, **52**:481–486.

Quaroni, A., and Semenza, G., 1976, Partial amino acid sequences around the essential carboxylate in the active sites of the intestinal sucrase–isomaltase complex, *J. Biol. Chem.* **251**:3250–3253.

Quaroni, A., Gershon, E., and Semenza, G., 1974, Affinity labeling of the active sites in the sucrase–isomaltase complex from small intestine, *J. Biol. Chem.* **249**:6424–6433.

Quaroni, A., Kirsch, K., and Weiser, M. M., 1979a, Synthesis of membrane glycoproteins in rat small-intestinal villus cells. Redistribution of L-(1,5,6-^{3}H)fucose-labelled membrane glycoproteins among Golgi, lateral basal and microvillus membranes *in vivo, Biochem. J.* **182**:203–212.

Quaroni, A., Kirsch, K., and Weiser, M. M., 1979b, Synthesis of membrane glycoproteins in rat small-intestinal villus cells. Effect of colchine on the redistribution of L-(1,5,6-^{3}H)fucose-labelled membrane glycoproteins among Golgi, lateral basal and microvillus membranes, *Biochem. J.* **182:** 213–221.

Rey, J., and Frézal, J., 1967, Les anomalies des disaccharidases, *Arch. Françaises Pédiat.* **24**:65–101.

Rubino, A., Zimbalatti, F., and Auricchio, S., 1964, Intestinal disaccharidase activities in adult and suckling rats, *Biochim. Biophys. Acta* **92**:305–311.

Sabatini, D. D., and Kreibich, G., 1976, Functional specialization of membrane-bound ribosomes in eukaryotic cells in: (A. Martonosi, ed.), *The Enzymes of Biological Membranes,* Vol. 2 John Wiley, New York, pp. 531–579.

Sabatini, D. D., Kreibich, G., Takashi, M., and Adesnik, M., 1982, Mechanisms for the incorporation of proteins in membranes and organelles, *J. Cell Biol.* **92**:1–22.

Semenza, G., 1968, Intestinal oligo- and disaccharidases, in: *Handbook of Physiology, Section 6, Alimentary Canal* (C. F. Code, ed.), American Physiological Society, Washington, D.C., pp. 2543–2566.

Semenza, G., 1976, Small-intestinal disaccharidases: Their properties and role as sugar translocators across natural and artificial membranes, in: *The Enzymes of Biological Membranes,* Vol. 3 (A. Martonosi, ed.), Plenum Publishing Corporation, New York, pp. 349–382.

Semenza, G., 1978, The sucrase–isomaltase complex, a large dimeric amphipathic protein from the small intestinal brush border membrane: Emerging structure–function relationships, in: *Structure and Dynamics of Chemistry* (P. Ahlberg and L.-O. Sundelöf, eds.), Symposia Universitatis Upsaliensis Annum Quingentesimum Celebrantis 12, Almqvist and Wiksell International, Sweden, 1978, Symposium in Uppsala, Sept. 22–27, pp. 226–240.

Semenza, G., 1979a, The mode of anchoring of sucrase–isomaltase to the small-intestinal brush border membrane and its biosynthetic implications, in: *Proceedings of the 12th FEBS Meeting, Dresden, 1978,* Vol. 53 (S. Rapoport, and T. Schewe, eds.), Pergamon Press, Oxford and New York, pp. 21–28.

Semenza, G., 1979b, Mode of insertion of the sucrase–isomaltase complex in the intestinal brush border membrane: Implications for the biosynthesis of this stalked intrinsic membrane protein, in: *Development of Mammalian Absorptive Processes,* Ciba-Symposium 70 (K. Elliot and J. Whelan, eds.) Elsevier-North-Holland, Amsterdam, pp. 133–146.

Semenza, G., 1981a, Intestinal oligo- and disaccharidases, from: *Carbohydrate metabolism and its disorders* (J. P. Randle, D. F. Steiner, and W. J. Whelan, eds.) vol. 3, Academic Press, London, pp. 425–479.

Semenza, G., 1981b, Molecular pathophysiology of small-intestinal sucrase–isomaltase, in: *Clinics in Gastroenterology* (W. B. Saunders, Co., Lts.) **10:**691–706.

Semenza, G., Auricchio, S., Rubino, A., Prader, A., and Welsh, J. D., 1965, Lack of some intestinal maltases in a human disease transmitted by a single genetic factor, *Biochim. Biophys. Acta* **105:**386–389.

Semenza, G., and Balthazar, A.-K., 1974, Steady state kinetics of rabbit intestinal sucrase: Kinetic mechanism, Na^+-activation, inhibition by Tris (hydroxymethyl)-amino-methane at the glucose subsite, with an appendix on interactions between enzyme inhibitors: A kinetic test for some simple cases, *Eur. J. Biochem.* **41:**149–162.

Semenza, G., Brunner, J., and Wacker, H., 1983, Biosynthesis and assembly of the largest and major intrinsic polypeptide of the small intestinal brush borders, in: *Brush Border Membranes,* (R. Porter and G. M. Collins, eds.) Ciba Symposium, **95:**92–107.

Sjöström, H. 1983, Discussion in: *Brush Border Membranes,* Ciba Symposium 95 (R. Porter, and G. M. Collins, eds.) pp. 130–131.

Sjöström, H., Norén, O., Christiansen, L., Wacker, H., and Semenza, G., 1980, A fully active, two-active site, single-chain-sucrase–isomaltase from pig small intestine, *J. Biol. Chem.* **255:**11332–11338.

Sjöström, H., Norén, O., Christiansen, L. A., Wacker, H., Spiess, M., Bigler-Meier, B., Rickli, E. E., and Semenza, G., 1982, Membrane anchoring and partial structure of small-intestinal pro-sucrase–isomaltase. A possible biosynthetic mechanism, *FEBS Lett.* **148:**321–325.

Sjöström, H., Norén, O., Danielsen, E. M., and Skovbjerg, H., 1983, Structure of microvillar enzymes in different phases of their life cycle, in: *Brush Border Membranes,* Ciba Symposium, (R. Porter and G. M. Collins, eds.), **95:**50–69.

Skovbjerg, H., and Krasilnikoff, P. A., 1981, Immunoelectrophoretic studies on human small intestinal brush border proteins. The residual isomaltase in sucrose intolerant patients, *Pediat. Res.* **15:**214–218.

Spiess, M., Brunner, J., and Semenza, G., 1982, Hydrophobic labeling, isolation and partial characterization of the NH_2-terminal membranous segment of sucrase–isomaltase complex, *J. Biol. Chem.* **257:**2370–2377.

Spiess, M., Brunner, J., and Semenza, G., 1984, A stalked intrinsic membrane protein. The stalk and anchor of the small-intestinal sucrase-isomaltase, submitted.

Tannenbaum, C., Toggenburger, G., Kessler, M., Rothstein, A., and Semenza, G., 1977, High-affinity phlorizin binding to brush border membranes from small intestine: Identity with (a part of) the glucose transport system, dependence on the Na^+-gradient, partial purification, *J. Supramol. Struct.* **6:**519–533.

Taravel, F. R., Datema, R., Woloszczuk, W., Marshall, J. J., and Whelan, W. J., 1983, Purification and characterization of a pig intestinal α-limit dextrinase, *Eur. J. Biochem.* **130:**147–153.

Tellier, C., Bertrand-Triadou, N., and Alvarado, F., 1979, Determination of the catalytic groups of intestinal brush-border sucrase by pH-variation studies, *Trans. Biochem. Soc. (London)* **7:**1071–1072.

Vasseur, M., Tellier, Ch., and Alvarado, F., 1982, Sodium-dependent activation of intestinal brush border sucrase: Correlation with activation by deprotonation from pH 5 to 7, *Arch. Biochem. Biophys.* **218:**263–274.

Von Heijne, G., 1980, Trans-membrane translocation of proteins. A detailed physico-chemical analysis, *Eur. J. Biochem.* **103:**431–438.

Von Heijne, G., and Blomberg, G., 1979, Transmembrane translocation of proteins, *Eur. J. Biochem.* **97:**175–181.

Wacker, H., Jaussi, R., Sonderegger, P., Dokow, M., Ghersa, P., Hauri, H.-P., Christen, Ph., and Semenza, G., 1981, Cell-free synthesis of the one-chain precursor of a major intrinsic protein complex of the small-intestinal brush border membrane (pro-sucrase–isomaltase), *FEBS Lett.* **136:**329–332.

Wacker, H., Aggeler, R., Kretchmer, N., O'Neill, B., Takesue, Y., and Semenza, G., 1984, A two-active site one-polypeptide enzyme: the isomaltase from sea lion small intestinal brush-border membrane. Its possible phylogenetic relationship with sucrase-isomaltase, *J. Biol. Chem.* **259,** in press.

Wickner, W., 1979, The assembly of proteins into biological membranes: The membrane trigger hypothesis, *Annu. Rev. Biochem.* **48:**23–45.

Wickner, W., 1980, Assembly of proteins into membranes, *Science* **210:**861–868.

24

Multifunctional Glucose-6-Phosphatase: A Critical Review

Robert C. Nordlie and Katherine A. Sukalski

I. INTRODUCTION

Glucose-6-phosphatase (D-glucose-6-phosphate phosphohydrolase; EC 3.1.3.9) was reviewed in some detail in the first edition of this work (Nordlie and Jorgenson, 1976). Our intention here is to focus on recent developments since the literature search for that earlier chapter was completed. Accordingly, we attempt to consider here the literature concerning glucose-6-phosphatase which has appeared since 1975. The authors were pleased to note a renewed interest in this complex enzyme, worldwide, more than 150 papers on the subject having appeared in the past seven years. A detailed consideration of the contents of all these papers is impossible in this limited space; we have therefore chosen to allude to many of them through the use of tables. Most of these are straightforward, descriptive studies, the essence of which is included in tables along with the literature references.

That glucose-6-phosphatase is a multifunctional enzyme capable not only of the hydrolysis of glucose-6-*P** but of its biosynthesis via potent phosphotransferase activities is now well established (see Nordlie, 1971, 1974, 1976). A very recent paper by Hefferan and Howell (1977) gives further, genetic evidence for this plurality of functions.

* Abbreviations used: glucose-6-*P*, D-glucose-6-phosphate; carbamyl-*P*, carbamyl phosphate; PP_i, inorganic pyrophosphate; P_i, inorganic orthophosphate; mannose-6-*P*, D-mannose-6-phosphate; cAMP, 3′,5′-cyclic AMP; Hepes, *N*-2-hydroxyethylpiperazine-*N*′-2-ethanesulfonic acid; DIDS, 4,4′-diisothiocyanostilbene-2,2′-disulfonic acid; SITS, 4-acetamido-4′-isothiocyanostilbene-2,2′-disulfonic acid.

Robert C. Nordlie and Katherine A. Sukalski • Department of Biochemistry, University of North Dakota School of Medicine, Grand Forks, North Dakota 58202.

For purpose of reference in the remainder of this chapter, some of the most extensively studied hydrolytic and synthetic functions of the enzyme are listed below:

$$\text{Glucose-6-}P + H_2O \rightarrow \text{glucose} + P_i \tag{1}$$

$$PP_i + \text{glucose} \rightarrow \text{glucose-6-}P + P_i \tag{2}$$

$$\text{Carbamyl-}P + \text{glucose} \rightarrow \text{glucose-6-}P + NH_3 + CO_2 \tag{3}$$

$$PP_i + H_2O \rightarrow 2\ P_i \tag{4}$$

$$\text{Carbamyl-}P + H_2O \rightarrow NH_3 + CO_2 + P_i \tag{5}$$

$$\text{Glucose-6-}P + {}^*\text{glucose} \rightleftarrows \text{glucose} + {}^*\text{glucose-6-}P \tag{6}$$

$$\text{Mannose-6-}P + \text{glucose} \rightleftarrows \text{glucose-6-}P + \text{mannose} \tag{7}$$

$$\text{Mannose-6-}P + H_2O \rightarrow \text{mannose} + P_i \tag{8}$$

Nucleoside tri- and diphosphates, phosphoenolpyruvate, 1,3-bisphosphoglycerate, and certain other compounds also may function as phosphoryl donors, while mannose, 3-*O*-methyl-D-glucose, 2-deoxyglucose, and a number of other sugars and polyols, may serve as phosphoryl acceptors (Nordlie, 1971, 1976).

As in our chapter in the earlier edition (Nordlie and Jorgenson, 1976), we will emphasize here interrelationships which exist between the catalytic characteristics of the glucose-6-phosphatase system and the biomembranes of which the enzyme is a part. In our earlier review (Nordlie and Jorgenson, 1976), we indicated that work with this complex enzyme seemed stymied at the phenomenological stage. And while there continues to be a lot of data-gathering, a significant body of recent work has served to focus many of these earlier observations mechanistically. Some of these underlying mechanistic hypotheses regarding glucose-6-phosphatase action and control will be discussed here in Section X.

Recent work, largely from our own laboratory, has brought new insights regarding physiological functions and regulation of both hydrolytic and synthetic activities of glucose-6-phosphatase. This chapter will conclude with a detailed discussion of these subjects.

II. REVIEWS

Reviews prior to 1976 are listed by Nordlie and Jorgenson (1976). Perhaps the most informative earlier reviews of catalytic characteristics of the multifunctional enzyme are in *The Enzymes* (Nordlie, 1971; Byrne, 1961) and *Vitamins and Hormones* (Ashmore and Weber, 1959). Our chapter in *Methods in Enzymology* (Nordlie and Arion, 1966) details methods for assay of various hydrolytic and synthetic activities of the enzyme. Physiological roles and regulation are considered in detail in *Current Topics in Cellular Regulation* (Nordlie, 1974).

More recently, studies of this enzyme have been reviewed in Japanese (Mizushima, 1977) and Russian (Ogorodnikova, 1981a). The senior author has discussed characteristics and functions of both hydrolytic and synthetic activities of the enzyme as well as other aspects of its cellular biology in three recent reviews (Nordlie, 1976, 1979, 1981), and has reviewed briefly its characteristics in relationship to its membranous nature (Nordlie, 1982a). A very comprehensive treatment of the kinetics of the glucose-6-phosphatase system has just been published (Nordlie, 1982b).

III. NEWER METHODS OF ASSAY

Teutsch (1978a) has described a histochemical method for demonstration of glucose-6-phosphatase which represents an improvement over the classical methods of Chiquoine (1953) and Wachstein and Meisel (1956). Improvement was achieved by optimizing reagent and substrate concentrations and conditions. The optimal medium contains 3.6 mM lead nitrate, 40 mM Tris-maleate buffer (pH 6.5), 10 mM glucose-6-*P*, and 300 mM sucrose. With the use of this technique, Teutsch (1978a,b) confirmed the differential zonal distribution of glucose-6-phosphatase in the rat hepatocyte (see Section VII). Benner *et al.* (1979) described the use of the electron microscope as a tool in the histochemical study of glucose-6-phosphatase, as did Shin *et al.* (1978).

Gierow and Jergil (1980) have utilized a modified method for measuring glucose-6-phosphatase in liver microsomes involving an improved assay of glucose released through coupled action of glucose oxidase and peroxidase. A newer method for assay of glucose-6-*P* hydrolysis based on measurement of P_i released by a heteropoly blue modification has been described by Zak *et al.* (1977). This method is considered superior to the usual Fiske–SubbaRow P_i determination in being much more sensitive at low P_i concentrations and relatively free from interference by normal metabolites.

An innovative approach to the measurement of glucose-6-*P* hydrolysis has been described by Kitcher *et al.* (1978). Basically, the technique uses [U-^{14}C]D-glucose-6-*P* as substrate, and involves the use of $ZnSO_4$ and $Ba(OH)_2$ at the termination of the reaction to precipitate P_i and unreacted [^{14}C] glucose-6-*P*. [^{14}C]Glucose then may be measured by scintillation spectrometry.

In the past, glucose-6-*P*:glucose phosphotransferase activity of glucose-6-phosphatase has usually been monitored by measuring the phosphoryation of [U-^{14}C]glucose (Segal, 1959; Hass and Byrne, 1960; Jorgenson and Nordlie, 1980). This approach has the disadvantage that relatively small amounts of [^{14}C]glucose-6-*P* must be separated quantitatively from large amounts of [^{14}C]glucose. Arion and Wallin (1973) have suggested an innovative and effective alternative. [^{14}C]Glucose-6-*P* is incubated with [^{12}C]glucose in the presence of the enzyme, and release of [^{14}C]glucose and P_i is measured. The rate of the glucose-6-*P*:glucose exchange reaction is then obtained as the difference between [^{14}C]glucose and P_i released.

In recent years, Arion and collaborators have emphasized the need to adjust for activity due to "disrupted" microsomes when studying membrane-related phenomena with untreated microsomal preparations. The importance of microsomal membrane integrity is especially emphasized in a very recent paper from their laboratory (Arion and Walls, 1982). Activity of "low-K_m mannose-6-P phosphohydrolase," i.e., that determined with 1 mM mannose-6-P, is used as a criterion of microsomal integrity (Arion *et al.*, 1976a). Mathematical equations for calculation of glucose-6-phosphatase activity due exclusively to "intact" microsomes in which the selective permeability barrier remains intact are presented by Arion and Walls (1982).

Jorgenson and Nordlie (1980) have studied glucose-6-phosphatase within its natural environment in the intact endoplasmic reticulum of isolated hepatocytes. Phosphate substrates gain access to the enzyme in the endoplasmic reticulum when plasma membranes of the isolated hepatocytes are made permeable by treatment with the polyene antibiotic filipin. The use of such preparations along with assays for the various activities of the enzyme based on P_i release, spectrophotometric assays, formation of ^{14}C-labeled sugar-phosphates from [U-^{14}C]D-glucose or [U-^{14}C]-3-*O*-methyl-D-glucose, and the release of ^{3}H from [2-^{3}H]D-glucose by modification of the technique of Katz *et al.* (1978) is described in detail by Jorgenson and Nordlie (1980). The use of 3-*O*-methyl-D-glucose, a specific substrate for phosphotransferase activity of glucose-6-phosphatase, makes possible the study of this activity in the presence of glucokinase and hexokinase, and has been used recently to study glucose-6-phosphatase phosphotransferase catalytic characteristics in isolated rat hepatocytes (Jorgenson and Nordlie, 1980). Gankema *et al.* (1981) have adapted this approach to the study of the regulation of metabolism of such compounds as fructose-1,6-P_2, *in situ*.

The mechanistic and metabolic implications of the studies of Arion and co-workers with isolated, intact microsomes and by Nordlie and colleagues with permeable hepatocytes are considered in detail in Sections X and XI.

IV. PURIFICATION AND PHYSICAL CHARACTERISTICS

The purification of glucose-6-phosphatase to apparent homogeneity from rat brain (Anchors and Karnovsky, 1975), and to a significant degree from rabbit liver (Bickerstaff and Burchell, 1980; Burchell and Burchell, 1982) and human placenta and liver (Reczek and Villee, 1982) has been reported.

In 1975, Anchors and Karnovsky (1975) described the first complete purification of glucose-6-phosphatase from any tissue. They obtained an apparently homogeneous, enzymatically active preparation from rat brain by a combination of sucrose gradient centrifugation, gel filtration on Sephadex G-100, and adsorption chromatography on DEAE-Sephadex in the presence of sodium deoxycholate. Their preparation represents a recovery of 0.015% of total brain protein with an enrichment of 9800-fold in glucose-6-phosphatase activity. A molecular weight of 28,000 was observed with sodium dodecylsulfate polyacrylamide gel; this value was found to exceed 70,000 when determined in the presence of 0.1% deoxycholate, however. The enzyme displayed catalytic characteristics similar to those of the liver and kidney enzyme. An involvement of the enzyme in slow-wave sleep (Karnovsky *et al.*, 1980) is suggested by the observations that (1) this enzyme protein acquired radioactivity from $^{32}P_i$ or glucose-6^{32}P more rapidly during sleep than during wakefulness (Anchors and Karnovsky, 1975), and (2) there is enhanced metabolic cycling between brain glucose and glucose-6-*P* during slow-wave sleep (Karnovsky *et al.*, 1980). The implication of the hydrolytic, but not the synthetic, function of glucose-6-phosphatase in this process has been suggested by the work of Anchors *et al.* (1977).

Reczek and Villee (1982) very recently have described the purification of microsomal glucose-6-phosphatase from human liver and placenta by the use of detergents and Sepharose affinity columns. Purification of the placental enzyme of as much as 600-fold was obtained with a yield of 2%. The liver enzyme was purified only five-fold with a recovery of but 0.4% of the activity present in autopsy specimens. Their preparations from both tissues displayed three bands on sodium dodecylsulfate-polyacylamide gel, indicating molecular weight in the range of 58,000–64,000. While the liver enzyme preparation exhibited a K_m for glucose-6-*P* of about 2.8 mM which is in good correspondence with the value for the microsomal enzyme (see Nordlie, 1971), that for the placental enzyme was nearly an order of magnitude smaller (Reczek and Villee, 1982), raising a question as to whether this preparation constitutes "glucose-6-phosphatase" in the classical sense. These preparations currently are being used in attempts to develop monoclonal antibodies to human glucose-6-phosphatase.

Interesting progress in the purification of glucose-6-phosphatase from rabbit liver microsomes presently is being made in the laboratory of the Burchells (Bickerstaff and Burchell, 1980; Burchell and Burchell, 1982) at Dundee. Sodium fluoride was employed to stabilize the enzyme in the early stages of solubilization and purification (see Burchell and Burchell, 1980). By a combination of polyethylene glycol and sodium cholate treatments, elution from phenyl-Sepharose columns with 1 M KCl, precipitation by centrifugation at 105,000*g* for 60 min, resuspension in buffer containing 2 M KCl, and recentrifugation, a 65-fold purification with a recovery of 31% was achieved. Their best preparations had a specific activity of 17 units/mg protein at 30°C, which represents a three-fold improvement over the best previous preparations of the liver enzyme. This preparation was not homogeneous, however, showing five major bands plus numerous minor contaminants on SDS–polyacrylamide disk gel electrophoresis. Further purification by a variety of techniques could not be achieved due to insolubility in the wide range of detergents tested.

In separate experiments, the Burchells (Burchell and Burchell, 1982) showed they could purify the smallest of these five major polypeptides (a doublet) in inactive form from heat-treated rabbit liver microsomes. They then produced antisera to this protein in chickens, and showed that this antisera would specifically immmunoprecipitate active glucose-6-phosphatase from cholate-solubilized microsomes. They concluded that one or both of these low molecular weight (approximately 18,500), heat-stable microsomal polypeptide(s) is/are the long-sought liver glucose-6-phosphatase. Separation of this (these) polypeptide(s) in an active form from their 65-fold purified preparation remains to be accomplished, but now appears eminently possible.

Other studies suggest that liver microsomal glucose-6-phosphatase may consist of both a catalytic unit and a substrate-specific translocase system (see Section X). The first steps in the direction of the isolation and characterization of this putative translocase have been made by Zoccoli and Karnovsky (1980) and Zoccoli *et al.* (1982). They have presented evidence that 4,4′-diisothiocyanostilbene 2,2′-disulfonic acid inhibits specifically microsomal glucose-6-*P* transport through binding to a microsomal protein of molecular weight 54,000.

Two recent studies from Russia, unavailable except in abstract form to the reviewer, suggest that isozymes of glucose-6-phosphatase may exist in microsomes from normal tissues and liver-derived tumor cells (Tretiakov, 1979). In these studies, glucose-6-*P* phosphohydrolase and PP_i-glucose phosphotransferase from liver differed from kidney, and it was suggested that different proteins are responsible for catalysis of these two reactions (Chernysheva and Fal, 1977). These conclusions are diametrically opposed to all other evidence regarding the duality of function of glucose-6-phosphatase to date, and would seem in need of further experimental examination.

V. RECENT MECHANISTIC STUDIES

Lowe and Potter (1982) have studied the stereochemical course of glucose-6-phosphatase-catalyzed phosphoryl transfer from D-glucose-[(R)-^{16}O, ^{17}O, ^{18}O] to D-glucose by NMR spectroscopy. They observed a retention of configuration at the phosphorus atom. They interpreted this observation in terms of double-displacement mechanism with a phosphoryl-enzyme intermediate. These observations and conclusions are consistent with earlier mechanistic conclusions based on kinetic (Segal, 1959; Hass and Byrne, 1960; Arion and Nordlie, 1964; Lueck *et al.*, 1972) and direct analytical (Feldman and Butler, 1969, 1972) considerations.

VI. NEWER REPORTS ON THE DISTRIBUTION OF GLUCOSE-6-PHOSPHATASE

Reports continue to accumulate concerning the presence of glucose-6-phosphatase in various organisms, organs, and subcellular structures of the cell. As we have pointed out in the past, a question often remains as to whether the discrete enzyme D-glucose-6-phosphate phosphohydrolase (EC 3.1.3.9) is involved or not. The great majority of these studies simply demonstrate the ability of some tissue or organelle preparation to hydrolyze glucose-6-*P* under mildly acidic conditions. We have, nonetheless, tabulated all of these reports in earlier reviews (see, especially, Nordlie, 1971; Nordlie and Jorgenson, 1976) and do so here as well (see Table 1).

References to studies of the rat liver enzyme and that of kidney are limited to a very few in Table 1, since the enzyme from these sources, especially that from liver microsomal preparations, is the subject of the great bulk of studies considered throughout this review.

The sources of the enzyme described in these reports, in the great majority of cases, duplicate those reported earlier for glucose-6-phosphatase (see Nordlie, 1971, 1974; Ashmore and Weber, 1959). A few are of special interest, at least to the reviewer, in regard to current concepts of physiological function and regulation of this enzyme and are discussed in some detail below.

Table 1. Recent Reports of the Presence of "Glucose-6-Phosphatase" in Various Tissues

Tissue	Organism	References
Liver	Human	Reczek and Villee (1982)
	Human fetus	Meisel *et al.* (1982)
	Human fetus, explants	Schwartz *et al.* (1974)
	Monkey	Benedetto *et al.* (1979) and Benedetto and Got (1980)
	Cat	Vergnes *et al.* (1981)
	Various birds, reptiles, amphibia, fishes, and molluscs	Vorhaben and Campbell (1979)
	Guinea pig (including fetal)	Band and Jones (1980a,b)
	Rabbit liver	Bickerstaff and Burchell (1980) and Burchell and Burchell (1982)
	Rainbow trout	Castilla *et al.* (1978) and Vernier and Sire (1978)
	Chicken, chick embryo	O'Neill and Langslow (1978) and Campbell and Langslow (1978)
	Alpaca	Arnao *et al.* (1981)
	Rat hepatocytes, isolated and in culture	Jorgenson and Nordlie (1980), Garfield and Cardell (1979), see also Section VIII-C.2.c
	Rat, hepatomas	Moller *et al.* (1977), Chernysheva and Fal (1977), see also Section VIII-C
Kidney	River lamprey	Murat *et al.* (1979)
	Rat	See Section VIII-C
Small intestine (mucosa)	Mouse	Chabot *et al.* (1978), Calvert *et al.* (1979), Menard (1980), and Menard and Malo (1982)
Pancreatic islets	Amphibia (salamander and frog)	Trandaburu (1977)
Brain and central nervous system	Rat	Anchors and Karnovsky (1975), Anchors *et al.* (1977), Karnovsky *et al.* (1980) Stephens and Sandborn (1976), Al-Ali and Robinson (1981), and Goldstein *et al.* (1975)
	Guinea pig	Rossowska and Dabrowiecki (1978)
	Lamprey, carp, frog, trout, chicken, rabbit, and cat	Ogorodnikova and Fomina (1977) and Ogorodnikova (1981b)
Placenta	Human	Reczek and Villee (1982), Bulienko *et al.* (1980), and Matalon *et al.* (1977)
	Human amniotic fluid	Boral *et al.* (1981)
Mononuclear phagocytes	Rabbit	Nichols and Setzer (1981)
Thyroid gland	Bovine	Hilderson *et al.* (1979a,b)
Epididymis	Mouse	Kanai *et al.* (1981)
	Rat	Llamas (1977)
Testis	Chinese hamster	Yokoyama and Chang (1977)
Vaginal epithelium	Rat	Kang and West (1982)
Lung (respiratory epithelium)	Rat	Antal-Ikov (1979)
Muscle	Various invertebrates and vertebrates	Surholt and Newsholme (1981) and Ogorodnikova and Legedinskaia (1980)
Slime mold (*Dictyostelium discoideum*)		Kelleher *et al.* (1978)

The slime mold *Dictyostelium discoidem* provides a relatively simple system for metabolic studies of the roles of glucose-6-phosphatase and glucokinase in cellular carbohydrate metabolism (see Kelleher *et al.*, 1978). It remains to be established whether the enzyme from this source exhibits the phosphotransferase capability characteristic of the enzyme from mammalian sources (see Nordlie, 1974).

The purification of the enzyme from rat brain to apparent homogeneity (Anchors and Karnovsky, 1975) represents a major accomplishment, and the multifunctional enzyme from this source would seem to be particularly attractive as a subject for further study as to structure and catalytic mechanism. The enzyme appears to be present in all the major areas of the brain which have been examined: cortex, hippocampus, striatum, hypothalamus, cerebellum, bulb, and spinal cord (Anchors and Karnovsky, 1975). Huang and Veech (1982) very recently have described an isotopic method for measuring quantitatively glucose-6-*P* dephosphorylation *in vivo,* in rat brain.

The alpaca, a ruminant indigenous to regions of high altitude in the Andes mountains of South America, constitutes an especially interesting model for comparative studies of roles and regulation of glucose-6-phosphatase (and other gluconeogenic and glycogenic enzymes) in mammalian carbohydrate homeostasis. Although a grazing ruminant like the sheep, the alpaca has a normal blood glucose level two to three times that of the sheep (100 mg% compared with 40 mg%). M. Villavicencio and his associates at San Marcos University in Lima, Peru, have characterized the enzyme from alpaca liver and kidney (Arnao *et al.*, 1981). They have found it to be multifunctional and are presently studying its bioregulation.

In recent studies of great interest, Surholt and Newsholme (1981) have reported the presence of glucose-6-phosphatase in 59 muscles from 34 species of annelids, molluscs, crustacea, insects (including diptera, hymenoptera, lepidoptera, orthoptera, coleoptera, and trichoptera), fishes, amphibia, birds, and mammals (including mouse, rat, and rabbit). Levels varied widely, from <0.1–8.0 μmol/min per g wet weight at 30°C. When checked, e.g., with flight muscle preparation from the moth, the pH optimum was 6.8 and K_m was 0.6 mM, similar to characteristics of classical mammalian microsomal glucose-6-phosphatase. Totally dissimilar to the enzyme of mammalian liver, kidney, and small intestinal mucosa (see Nordlie, 1974), however, the muscle enzyme appears confined exclusively to the cytosolic fraction of the cell (Surholt and Newsholme, 1981). A similar observation also recently was made with the enzyme of mollusc gastropod hepatopancreas (Vorhaben and Campbell, 1979). Whether the activity from these sources reflects the characteristics of classical "glucose-6-phosphatase," i.e., enzyme EC 3.1.3.9, or whether it constitutes a previously uncharacterized enzyme, is not clear at this time. The matter is under further investigation in Newsholme's laboratory. Surholt and Newsholme (1981) have suggested that this enzyme of muscle may function either to produce glucose from glucose-6-*P* derived from glycogen in muscle or to make possible "futile" substrate cycling at the glucose–glucose-6-*P* level in muscle, thus providing an increased sensitivity to bioregulation at that site.

The recent extensive purification of glucose-6-phosphatase from human placenta (Reczek and Villee, 1982) should provide a basis for the ultimate resolution of the relationship of the placental enzyme to that from liver in regard to multifunctionality and other commonality or diversity of catalytic characteristics.

VII. ZONAL DISTRIBUTION OF THE HEPATIC ENZYME

Once believed to be exclusively confined to the hepatic endoplasmic reticulum, glucose-6-phosphatase has now come to be recognized as present in nuclear, plasma, and mitochondrial membranes of the liver as well (see Nordlie and Jorgenson, 1976). Some interesting new interrelationships between these cellular structural elements have emerged in the past few years. And recent histochemical and biochemical studies indicate the heterogeneous distribution of this enzyme within various regions of the liver cell, as well.

Meher *et al.* (1981) made the very significant observation that when livers are preperfused with 0.25 M sucrose to which NaCl to 0.9% (wt./vol.) was added, or when livers are homogenized in this medium rather than 0.25 M sucrose without added electrolyte, a large fraction of glucose-6-phosphatase sediments as a mitochondrial-rough endoplasmic reticulum complex. They interpret their data as indicative of a physiological association between these two structures in the intact liver cell. The authors have confirmed these observations with rat, quail, and alpaca liver (Nordlie and Traxinger, unpublished observations; Villavicencio and Nordlie, unpublished observations), as has Katz with rat liver (Katz, unpublished observations).

This close physical association of glucose-6-phosphatase-rich rough endoplasmic reticulum with mitochondria would favor the metabolic channeling of carbamyl-*P* formed by mitochondrial carbamyl-*P* synthetase (see Lueck and Nordlie, 1970) and of PP_i produced via the mitochondrial fatty acid activating system (see Veech *et al.*, 1980; King *et al.*, 1982) to glucose phosphorylation via phosphotransferase activities of glucose-6-phosphatase of the endoplasmic reticulum. The possible significance of these observations is expanded upon in Section XI.

Katz, Jungermann, Teutsch, and their colleagues at Freiburg University have examined by histochemical and biochemical techniques the possibility that in rat parenchyma, as in rat kidney nephrons, there might be a zonal separation of pathways for glycolysis and for gluconeogenesis (Katz *et al.*, 1977; Teutsch, 1978a,b; Jungermann *et al.*, 1982; Teutsch *et al.*, 1978). Glucose-6-phosphatase was used as a marker for gluconeogenesis and glucokinase as an index of glycolysis. The periportal region, i.e., that around the terminal portal vessels ("Zone 1") of the liver acinus, was 2- to 3-fold richer in glucose-6-*P* phosphohydrolase than was the perivenous region, i.e., around the central vein ("Zone 3") of the liver acinus. More recent studies on metabolic cycling between glucose and glucose-6-*P* showed that, during a normal feeding agenda, the periportal zone should indeed catalyze net glucose output while the perivenous zone should mediate net glucose uptake. This they term their "Metabolic Zonation" model. Further support for this theory is found in the work of Schwarz (1978). All of these observations raise an intriguing question to the present authors regarding possible differential functions for hydrolytic and synthetic functions of multifunctional glucose-6-phosphatase in these two hepatocyte zones (see Nordlie, 1974; also see Section XI).

A heterogeneous distribution of hepatic glucose-6-phosphatase also is suggested by observations of Tongianai *et al.* (1977). They noted that subpopulations of hepatocytes fractionated by velocity sedimentation differed in glucose-6-phosphatase as well as triglyceride content.

VIII. REGULATION OF GLUCOSE-6-PHOSPHATASE ACTIVITIES

A. Inhibitors and Activators

1. Vanadate

Vanadate is the most potent inhibitor of glucose-6-phosphatase yet discovered (Singh *et al.*, 1981). Inhibition of both hydrolytic and synthetic activities of the enzyme of intact and detergent-dispersed microsomes and permeable hepatocytes was observed. Inhibition was competitive with phosphate substrates. K_i values, noted at pH 7.4 with activities of permeable hepatocytes, were 1.5 μM for glucose-6-*P* phosphohydrolase, 0.48 μM for PP_i-glucose phosphotransferase, and 1.0 μM for carbamyl-*P*:glucose phosphotransferase. This finding is supportive of the catalysis of all three activities by a single enzyme in the cell. Inhibition by ATP previously noted (Nordlie *et al.*, 1968) appears in large part to have been due to contamination by vanadate.

Vanadate concentrations up to 65 μM had no effect on rat hepatic glucokinase or hexokinase. In contrast, μM levels of Cr^{3+} inhibited the activities potently but exerted no effect on activities of glucose-6-phosphatase (Nordlie *et al.*, 1981). The possible physiological consequences of these observations, as well as the potential usefulness of vanadate and Cr^{3+} as probes in the study of the regulation of mammalian hepatic carbohydrate metabolism, have been pointed out (Singh *et al.*, 1981).

2. Cupric Ion

Johnson and Nordlie (1977) have observed concentration-dependent, activity-discriminant effects of Cu^{2+} on carbamyl-*P*:glucose phosphotransferase and glucose-6-*P* phosphohydrolase activities of glucose-6-phosphatase of intact and detergent-disrupted rat liver microsomes, intact nuclei, nuclear membranes with and without detergent-supplementation, and partially purified enzyme. Increasing levels of Cu^{2+} progressively increased the ratio of phosphotransferase to phosphohydrolase activity. This observation is consistent with the known hypoglycemic effects of excess Cu^{2+} administration (Ussolzew, 1935). Kinetic analysis indicated that Cu^{2+} may promote the displacement of the product of the phosphotransferase reaction, i.e., glucose-6-*P*, from the enzyme and thus account for its stimulating action on synthetic activity. The physiological implications of these observations (see Johnson and Nordlie, 1977) are presently undergoing further experimental examination with Cu^{2+}-deficient rats in this laboratory.

3. *Alkyl- and Phenylphosphate Esters*

The inhibitory effects of various alkyl- and phenylphosphate esters have been studied systematically at pH 6.0 and 7.5 with untreated and deoxycholate-disrupted microsomes (Walls and Lygre, 1980). Inhibition by P_i and monoethyl-*P* was competitive, while that by mono- and diphenyl-*P* was mixed. Monoalkyl phosphates were more effective than analogous di- and trialkyl phosphates, and deoxycholate potentiated inhibition. Mono- and diphenyl phosphates were more effective than triphenyl-*P*, and deoxycholate decreased the inhibitory effects. Membrane–enzyme interrelationships are discussed from the viewpoint of these observations by Walls and Lygre (1980).

4. *Polyamines and Histone 2A*

Got and her group have studied in detail glucose-6-phosphatase from monkey liver (Benedetto and Got, 1980; Benedetto *et al.*, 1979). Unlike the rat, intact microsomal preparations from this species show no latency with respect to activities of glucose-6-phosphatase (Benedetto *et al.*, 1979). That is, deoxycholate or Triton X-100 has no effects on activity. Nonetheless, both histone 2A and various polyamines activate both hydrolytic and carbamyl-*P*:glucose phosphotransferase activities of the enzyme (Benedetto and Got, 1979).

Polylysine also was seen by Vergnes *et al.* (1981) to increase glucose-6-phosphatase activity of cat liver microsomes without altering the K_m. Nordlie *et al.* (1979a) found that the polyamines spermine, spermidine, putrescine, and polylysine (mol. wt. 3400) activated extensively carbamyl-*P*:glucose phosphotransferase and glucose-6-*P* phosphohydrolase activities of both intact and deoxycholate-disrupted microsomes. The effect persisted in the presence of catalase, i.e., it was independent of H_2O_2 formation, and was eliminated by Ca^{2+}. K_m values for phosphate substrates, but not for glucose, were lowered by polyamines.

Both the Nordlie (Nordlie *et al.*, 1979a) and Got (Benedetto *et al.*, 1979; Vergnes *et al.*, 1981) groups attributed the effects of these polycationic compounds to their electrostatic interactions with phospholipids of the membrane of the endoplasmic reticulum, thus modifying membrane charge and probably membrane conformation. Thermotropic studies of glucose-6-phosphatase of microsomes, with and without polyamines, by Johnson and Nordlie (1980) supported this conclusion.

5. *N-Alkylmaleimides*

Wallin and Arion (1972) observed inhibitions of glucose-6-*P* phosphohydrolase by several sulfhydryl reagents (*N*-ethylmaleimide, 5-5′-dithio*bis*-2-nitrobenzoic acid, *para*-mercuriphenyl sulfonate, and Hg^{2+}) with intact rat liver microsomes. When microsomes were disrupted, however, no inhibition was seen. These observations constituted the original basis from which their phosphate–substrate transporter–catalytic unit concept of microsomal glucose-6-phosphatase action evolved (see Section X-C). In contrast, Colilla and Nordlie (1973) observed inhibitions of both phosphohydrolase and phosphotransferase activities by *N*-ethylmaleimide, iodosobenzoate, *para*-chloromercuriphenyl sulfonate, iodoacetate, and Hg^{2+} with both intact and detergent-disrupted microsomes and with partially purified microsomal glucose-6-phosphatase. Inhibition with partially purified enzyme was kinetically competitive with regard to glucose-6-*P*, carbamyl-*P*, and PP_i.

Recent work by Vakili and Banner (1979, 1981) appeared to resolve the differences between the observations of Wallin and Arion (1972) and Colilla and Nordlie (1973). A series of *N*-alkylmaleimides of varying alkyl chain length were employed. With intact microsomes, a biphasic effect was noted in time-dependency studies. A rapid early phase of inhibition was followed by a slower secondary phase. The former, but not the latter, could be eliminated by disruption of the microsomal membrane with detergent. With disrupted microsomes as the initial subject of study, inactivation by the sulfhydryl inhibitor was slower but ultimately complete. With both intact and disrupted preparations, inactivation of glucose-6-phosphatase was dependent on the length of the alkyl side chain. The authors interpreted these observations to suggest that there are at least two thiol groups essential for glucose-6-phosphatase activity, and that they are located in separate nonpolar regions of the membrane–enzyme system (Vakili and Banner, 1981). The authors further suggested that both are present in relatively hydrophobic environments, that of the membrane-dependent group much more so than the other. Interpreted in terms of the transporter–catalytic unit concept (see Section X-C), the former would be involved in substrate transport, the latter in overall catalysis. These same workers reached generally similar conclusions in other work with *para*-chloromercuribenzene sulfonate (Vakili and Banner, 1981).

6. *Pyridoxal Phosphate*

Gold and Widnell (1976) observed that pyridoxal-*P* was a linear noncompetitive inhibitor of glucose-6-phosphatase activity of intact rat liver microsomes. The effect is markedly reduced by nitrogen cavitation, treatment with deoxycholate, or glutaraldehyde fixation of microsomes, and is abolished by partial purification of the enzyme. The significance of these observations is discussed further in Section X. Carvo *et al.* (1974) had earlier noted inhibition of glucose-6-phosphatase by pyridoxal-*P*, and had described the effects of microsomal disruption thereon.

7. *Fructose-2,6-bisphosphate and Calmodulin*

In our continuing search for factors which may potentially regulate hydrolytic and synthetic activities of this enzyme discriminately, we recently examined the effects of fructose-2,6-P_2 and calmodulin (Singh and Nordlie, unpublished observations). The former was a gift of Dr. Simon Pilkis and the latter was provided by Dr. Ramji Khandelwal.

Studies with fructose-2,6-P_2 were carried out at 30°C and pH 7.4 with 40 mM Hepes buffer, 5 mM glucose-6-*P* or carbamyl-*P*, 180 mM D-glucose (phosphotransferase only), and 10 mg bovine serum albumin/1.5 ml reaction mixture. Liver microsomes from fed rats, supplemented to 0.2% with sodium deoxycholate, served as enzyme source. Incubations were for 10 min. Fructose-2,6-P_2 levels included were 0,

2, 4, 6, 8, 10, and 20 μM in assay mixtures. In no instance was activity altered from that of controls.

Studies with calmodulin proved equally negative. In these studies, both intact and deoxycholate-supplemented (to 0.2%, wt./vol.) microsomes were employed. Glucose-6-*P* phosphohydrolase, carbamyl-*P*:glucose phosphotransferase, and PP_i-glucose phosphotransferase were studied. The effects of concentrations of calmodulin, Ca^{2+}, and presence or absence of bovine serum albumin were studied systematically. Calmodulin levels tested ranged from 1 μg to 12 μg/1.5 ml assay mixture, and that of Ca^{2+} was varied from 0.01 to 10.0 mM. Microsomes from both fed and 48-hr fasted rats were studied. In no instance was there either a change in activity level from the normal or an alteration in degree of latency.

B. Control by Protein Phosphorylation–Dephosphorylation

Data to date regarding enzyme protein phosphorylation–dephosphorylation as a mechanism for control of glucose-6 phosphatase are equivocal. Prior to 1970, Greengard (Greengard and Dewey, 1967; Greengard, 1969) showed that glucagon or cAMP may prematurely trigger an increase in glucose-6-phosphatase in rat fetuses *in utero*. Later, Schwartz *et al.* (1974) demonstrated with human fetal liver explants in culture that dibutyryl cAMP produced a 4- to 8-fold increase in glucose-6-phosphatase levels within 24 hr. Theophylline markedly potentiated the effect, cycloheximide or actinomycin abolished the increase, and insulin diminished the response. Boxer *et al.* (1974) likewise observed stimulation by dibutyryl cAMP of glucose-6-phosphatase in cultured cells from human and rat liver.

In 1980, Band and Jones (1980a,b) reported the functional activation of glucose-6-phosphatase and elevation of overall gluconeogenic flux by glucagon in perfused fetal guinea pig liver.

Mechanisms for protein phosphorylation via microsomal protein kinase action have been demonstrated by Zetterqvist and Engström (1966). In a recent, interesting paper, Lam and Kasper (1980) described a PP_i-protein phosphotransferase tightly bound to the membrane of rat liver microsomal preparations.

Burchell and Burchell (1980) have noted a time- and temperature-dependent decline in rabbit liver microsomal glucose-6-phosphatase activity following supplementation with the detergent sodium cholate. This loss of activity was not seen in the absence of detergent treatment, and was completely prevented by the inclusion of sodium molybdate or NaF (but not EDTA) in the preparation. The Burchells (1980) interpreted these observations as indicative of the involvement of enzyme protein phosphorylation–dephosphorylation in the regulation of glucose-6-phosphatase action. They hypothesized that NaF precludes loss of activity through inhibition of phosphoprotein phosphatase activity endogenous to microsomes.

Begley and Craft (1981) published data which they believe are directly supportive of the activation of glucose-6-phosphatase of rat liver microsomes by phosphorylation via endogenous (or exogenous) protein kinase action. These observations, while potentially exciting, have been refuted on methodological and procedural grounds independently by three groups at Ithaca and Dundee (Burchell *et al.*, 1982) and Grand Forks (Singh *et al.*, 1983). Further, Singh *et al.* (1983) have raised the possibility that the effects seen by Burchell and Burchell (1980) may simply be a reflection of protection of the detergent-treated microsomal enzyme against progressive thermal inactivation through the binding of competitive inhibitors (fluoride is one such competitive inhibitor of glucose-6-phosphatase) to the active site. In addition to repeating the original observations with F^-, Singh *et al.* (1983) observed protection against loss of activity by the competitive inhibitors vanadate, molybdate, oxalate, bicarbonate, and P_i, and by the substrates glucose-6-*P*, PP_i, and carbamyl-*P*.

The intriguing question of whether glucose-6-phosphatase may be controlled through protein phosphorylation–dephosphorylation mechanisms remains just that for the time-being—an intriguing question. The fact that inhibitors of protein synthesis block responses to dibutyryl cAMP in cultured fetal hepatic explants (see above) suggests that the mechanism, if operative, may be quite complex.

C. Control by in Vivo Perturbations

More than 60 papers have appeared in the past seven years dealing with changes in levels of glucose-6-phosphatase in response to one type or another of *in vivo* perturbation. The effects of various drugs and toxic compounds, carcinogens, fasting and other dietary manipulations, pre- and post-partum development and aging, liver regeneration, weightlessness of space flight, irradiation, certain inborn errors, and deprivation or administration of various hormones all have been considered.

A major emphasis has been placed in recent years upon the phenomenon of "latency" of activities of glucose-6-phosphatase and on underlying mechanisms. The partial latency of the enzyme of isolated liver microsomes, as manifest by the enzyme's detergent sensitivity, was established shortly after the enzyme was discovered. Pioneering work of Nordlie and his associates demonstrated that the latency of the microsomal enzyme was both activity discriminant and hormone-sensitive. Latency increased along with an elevation in absolute level of enzymic activity in insulin deprivation (experimental diabetes or fasting) and decreased following glucocorticoid therapy. These initial observations have been reviewed in detail (Nordlie, 1971, 1974; Nordlie and Jorgenson, 1976). Mechanisms underlying this latency, which is substrate and activity disciminant, have been the subject of some controversy and ensuing vigorous experimental examination recently. The subject, along with a detailed description of proposed mechanisms underlying latency, is considered extensively in Section X. Some relevant new observations on the effects of fasting, diabetes, glucocorticoid administration, and thyroid hormone therapy on latency are described below.

Glucose-6-phosphatase, with its multiplicity of functions, is an inducible enzyme which is responsive to many factors. In general, those perturbations which tend to raise blood glucose levels effect an increase in functional glucose-6-phosphatase activity and those which serve to lower blood glucose concentration also depress glucose-6-phosphatase.

Brief synopses of experimental observations on responses of glucose-6-phosphatase to various *in vivo* perturbations follow, along with pertinent literature references to which the reader with interest in greater specific detail is directed. Many of the observations tabulated are repetitions or elaborations of results reported previously and considered in earlier reviews (see Nordlie, 1971, 1974; Ashmore and Weber, 1959; Nordlie and Jorgenson, 1976).

1. Effects of Drugs

a. Carbon Tetrachloride. Studies recently have been carried out with the enzyme of rat and mouse liver microsomes (Glende *et al.*, 1976; Hruszkewycz *et al.*, 1978; Sein and Chu, 1979; Masuda, 1981; Kanamura *et al.*, 1981; Paradist *et al.*, 1979; Benedetti *et al.*, 1977a,b, 1979; Bertone and Dianzani, 1977; Santiago *et al.*, 1977). Ingestion of CCl_4 led to a depression of hepatic microsomal glucose-6-phosphatase. This loss of activity correlated with the extent of peroxidation of microsomal membrane lipids. Anti-oxidants prevented this effect. The sequence of events appeared to involve the initial metabolism of CCl_4 in the endoplasmic reticulum to form free radicals which in turn initiated a sequence of events marked by the disruption of membranous components of the liver cell (Glende *et al.*, 1976; Hruszkewycz *et al.*, 1978). Recent work indicated that lipids or lipid-derived materials extracted from peroxidized liver microsomes inhibited extensively glucose-6-phosphatase of untreated microsomes (Hruszkewycz *et al.*, 1978; Benedetti *et al.*, 1977a,b, 1979; Bertone and Dianzani, 1977). Some work suggested that those inhibitory components may include fatty acid-derived aldehydes (Benedetti *et al.*, 1977a,b, 1979; Bertone and Dianzani, 1977).

Other studies indicated that the oxidation, specifically, of microsomal membrane phosphatidylcholine containing one saturated and one unsaturated fatty acid specifically correlated with the decline in glucose-6-phosphatase activity. This was interpreted to indicate that glucose-6-phosphatase and molecules of phosphatidylcholine having one saturated and one unsaturated fatty acid are in close apposition within the microsomal membrane (Santiago *et al.*, 1977). Earlier work of Garland and Cori (1972) likewise implicated phosphatidylcholine in glucose-6-phosphatase action.

b. Carcinogenic Materials. Feeding thioacetamide to rats produced massive hyperplasia, disorganization, and cirrhosis; glucose-6-phosphatase appeared in thioacetamide-enlarged nucleoli of hepatic parenchymal cells (Dasgupta *et al.*, 1979). Glucose-6-phosphatase markedly decreased as preneoplastic nodules appeared and hepatocarcinomas developed following ingestion of diethylnitrosamine (Elitze *et al.*, 1977; Wanson *et al.*, 1980; Bernaert *et al.*, 1979; Kitagawa and Pitot, 1975) or *N*-2-fluorenylacetamide (Kitagawa and Pitot, 1975).

c. p-Aminophenol. Ingestion by rats caused acute kidney tubular necrosis and markedly depressed glucose-6-phosphatase activity. No effect was seen on the enzyme *in vitro* (Crowe *et al.*, 1977).

d. Phenol, Dinitrophenol, Pentachlorophenol. Low concentrations stimulated kidney glucose-6-phosphatase; higher concentrations inhibited the enzyme in liver, brain, kidney, and gills of *Notopterus notopterus* (Verma *et al.*, 1981).

e. Phenobarbital. Administration caused a marked loss of glucose-6-phosphatase in livers of young, mature, and adult rats (Schmucker and Wang, 1980).

f. Sodium Fluoride. Large doses, i.p. (35 mg/kg body weight), of NaF produced a 2.4–3.4-fold increase in rat liver and kidney glucose-6-phosphatase levels. The effects were suppressed by pretreatment with actinomycin D or cycloheximide. The response was markedly suppressed by adrenalectomy and partially supressed by parathyroidectomy, and appeared to involve stimulation by NaF of adrenal function (Suketa *et al.*, 1980; Suketa and Sato, 1980).

g. Sodium Dodecylbenzene Sulfonate. Inclusion of this compound in the diet had no effect on rat liver glucose-6-phosphatase (Selmeci-Antal and Blaskovits, 1977).

h. Miscellaneous Compounds. The possible effects of phenocular DPG, 1,4-benzodiazepine, Endrin, hexachlorohexane, and Valekson on glucose-6-phosphatase of mouse, rat, or *Ophiocephalus punctatus* organs is not specified in information to the reviewer (Narbonne *et al.*, 1979; Goldvenko and Orliuk, 1979; Sastry and Sharma, 1978; Karnik *et al.*, 1981; Kuzminskaia *et al.*, 1977).

2. Effects of Dietary Manipulations

a. Fasting. Based on assays with untreated microsomes, mouse liver activity increased after 48 and 72 hr of fasting and mouse kidney enzyme increased after 48 hr, only. Inclusion of deoxycholate potentiated the fasting-induced responses of the liver enzyme but abolished that of kidney enzyme (Sein and Maw, 1978).

Responses to fasting were studied with isolated rat hepatocytes rendered permeable to the substrate by treatment with filipin. Activity increased following a 48-hr fast, correlative with an increase in latency, an increase in K_m, and a small increase in V_{max}. Following disruption of cells, the fasting-induced differences in K_m disappeared, but the disparity between V_{max} values was accentuated. These observations parallel responses noted earlier with isolated microsomes (Nordlie and Jorgenson, 1981).

b. High-Protein Diet Followed by Fasting. In other studies by Nath (1980), rats were fed a high-protein diet for five days and then shifted to a zero-protein diet. Liver and kidney glucose-6-phosphatase levels decreased one day after the dietary shift, but then increaesd and remained higher than day zero for the next two days.

3. Effects of Hormonal Manipulations

a. Experimental Diabetes and Insulin Administration. Glucose-6-phosphatase of the fetal rat liver was responsive to insulin 21.5 days following conception; insulin decreased glucose-6-phosphatase activity, showing the enzyme was sensitive to insulin beginning just prior to or just subsequent to birth (Porterfield, 1979). Injection of insulin into the chorioallantoic membrane of the incubating hen's egg somewhat depressed levels of glucose-6-phosphatase and caused an increase in K_m value (Campbell and Langslow, 1978). Injection of insulin lowered hepatic hexokinase but not glucose-6-phosphatase of the domestic chicken (O'Neill and Langslow, 1978). Injection of insulin did not affect levels of glucose-6-phosphatase in small intestine of suckling mouse (Menard and Malo, 1982).

Glucose-6-phosphatase of rat liver increased in diabetes and was depressed by insulin (Seni and Gandhi, 1981; Garfield and Cardell, 1979; Maddaiah *et al.*, 1981; Speth and Schulze, 1981; Zoccoli *et al.*, 1982; Arion *et al.*, 1976b; Jacquot and Felix, 1977; Nordlie *et al.*, 1979b). Diabetes was characterized by a marked increase in smooth endoplasmic reticulum of liver with an increased content of glucose-6-phosphatase. Increased glucose-6-phosphatase in diabetes was characterized by a marked increase in latency. Alterations in the lipid environment of the enzyme in the endoplasmic reticulum was indicated by thermotropic studies and interactions with 1-anilino-δ-naphthalene. An increase in the ratio of catalytic unit relative to translocase was suggested as an underlying mechanism explaining elevation in latency (see Section X). Activities of both glucose-6-*P* phosphohydrolase and carbamyl-*P*:glucose phosphotransferase of isolated rat liver nuclei increased 3 to 4-fold in alloxan diabetes, but latency was not affected, in contrast with the activities of isolated microsomes from these same rat livers (Nordlie *et al.*, 1979b).

b. Glucocorticoids. Renal glucose-6-phosphatase of fetal rat kidney was increased by administration of triamcinolone *in utero* (DeLaval *et al.*, 1981). Glucocorticoids by themselves increased glucose-6-phosphatase, but suppressed the response to dibutyryl cAMP under certain conditions in fetal rat liver explants in culture (Mizushima and Ishikawa, 1979).

Responses of rat liver glucose-6-phosphatase have been studied extensively (Speth and Schulze, 1981; Zoccoli *et al.*, 1982; Arion *et al.*, 1976b; Nordlie *et al.*, 1979b; Dobrosielski-Vergona and Widnell, 1982; Nordlie, 1981. Preincubation of rat liver microsomes with glucocorticoids, or injection of rats with glucocorticoids, increased glucose-6-phosphatase, possibly by interaction with the putative transport moiety of the enzyme. Apparent latency decreased. This appeared to reflect an increased substrate transport capability, which in turn was ascribed largely to an increased functional activation of putative translocase protein rather than an increase in its number (see also Section X). Directly contrasting responses were noted in comparative studies of responses of glucose-6-phosphatase activities of isolated hepatic nuclei compared with microsomes. Hydrocortisone administered to normal rats produced a 3–4-fold increase in nuclear glucose-6-phosphatase and accompanying carbamyl-*P*:glucose phosphotransferase without altering latency. In contrast, the glucocorticoid administered to alloxan-diabetic rats effected a reduction in activity with an increase in degree of latency.

c. Estrogens. An increase was seen in rat liver glucose-6-phosphatase after 5–10 injections of prednisolone acetate (Shin *et al.*, 1977).

d. Glucagon. First reported by Greengard (1969) and Greengard and Dewey (1967), the stimulatory effects of cAMP or dibutyryl cAMP on hepatic glucose-6-phosphatase, especially the fetal liver enzyme, are supported additionally by data obtained in recent years. That glucagon, triiodothyronine, and even glucocorticoids may function through this intermediate is supported by recent studies.

Significant increases in glucose-6-phosphatase in fetal guinea pig liver and in cultured human and rat liver explants possibly mediated through cAMP recently have been reported (Band and Jones, 1980a,b; Boxer *et al.*, 1974). In contrast, the chicken liver enzyme was unresponsive to glucagon (O'Neill and Langslow, 1978).

e. Thyroxine and Triiodothyronine. Both hormones increased glucose-6-phosphatase activity in fetal rat liver and adult rat and mouse liver (Dobrosielski-Vergona and Widnell, 1982; Llamas, 1977; Gold and Widnell, 1974; Paul and Dhar, 1980; Vernier and Sire, 1978). The increase was blocked by puromycin. The response appeared age-related; responses in old rats were larger than those in young rats. The T_3 effects appeared to be mainly due to an increase in catalytic unit rather than that of elevated translocase capability (Dobrosielski-Vergona and Widnell, 1982; see also Section X).

f. Somatotropin. Administration of this hormone increased both glucose-6-*P* phosphohydrolase and PP_i phosphohydrolase activities of mouse liver (Llamas, 1977).

g. Growth Hormone. At 21.5 days of gestation, growth hormone increased fetal rat liver glucose-6-phosphatase by about 50% (Porterfield, 1979).

h. Epinephrine. No effect was noted on liver glucose-6-phosphatase of rainbow trout (Vernier and Sire, 1978).

4. *Effects of Miscellaneous in Vivo Treatments*

a. Pre- and Postnatal Development and Aging. A two-fold increase in the surface area of the hepatic endoplasmic reticulum, attributable exclusively to smooth-surfaced endoplasmic reticulum, was seen as rats matured (Schmucker *et al.*, 1977; Siegel *et al.*, 1979). Hepatic glucose-6-phosphatase increased approximately two-fold in 2-day postmature fetus, compared with a 6.5-fold increase in the 2-day neonate (Portha *et al.*, 1978). These observations correlated with mobilization of fetal glycogen. Both glucose-6-*P* phosphohydrolase and PP_i-glucose phosphotransferase activities of rat liver glucose-6-phosphatase increased rapidly from low prenatal levels to maxima 2–5 days postpartum and then declined slowly to adult levels. Latencies of these two activities showed strikingly different age-related changes. That of the transferase peaked soon after birth and thereafter remained high. That of the hydrolase decreased slowly with aging. The neonatal overshoot in activity appeared due exclusively to the enzyme of the rough endoplasmic reticulum. PP_i-glucose phosphotransferase was much more latent in smooth than in rough endoplasmic reticulum (Goldsmith and Stetten, 1979). Responses of the kidney enzyme differed in some details from that of the liver (Goldsmith and Stetten, 1979).

b. Liver Regeneration. A remarkable decrease in rat hepatic glucose-6-phosphatase, compared with laparotomized controls, was seen 4 and 15 hr after partial hepatectomy, correlative with a dramatic decrease in blood glucose level (Bruscalupi *et al.*, 1979).

c. Flight in Space. Glucose-6-phosphatase was increased after an 18.5-day flight in space, and in simulated flight groups as well. These observations are interpreted as manifest through glucocorticoid level changes and chronic stress due to weightlessness (Nemet and Tigranian, 1982).

d. Irradiation. The complex effects of irradiation on rat liver glucose-6-phosphatase are described by Bernard *et al.* (1980).

e. Inborn Errors: Glycogenoses Types I, Ib, and Ic. Confirming evidence for the common identity of liver microsomal glucose-6-*P* phosphohydrolase, PP_i-glucose phosphotransferase, inorganic pyrophosphatase, and carbamyl-*P*:glucose phosphotransferase was provided by the observed deficiency of all four activities in human patients with Type I glycogenosis (Hefferan and Howell, 1977). The inappropriateness of measuring glucose-6-phosphatase in blood platelets for diagnosis of Type I glycogenosis is indicated by some recent work of Oka and collaborators (Oka *et al.*, 1978). The possible use of assay of human placental glucose-6-phosphatase in screening for Type I glycogenosis was suggested (Matalon *et al.*, 1977).

A new variant of Type I glycogenosis, Type 1b glycogenosis, recently was independently demonstrated by several groups (Bialek *et al.*, 1977; Nordlie, 1981; Narisawa *et al.*, 1978; Igarashi *et al.*, 1980; Lange *et al.*, 1980; Sann *et al.*, 1980, Kamoun, 1980). Clinical symptomatology was that characteristic of Type I. Glucose-6-phosphatase of liver appeared very low when assayed without added detergent, but normal when assayed with detergent-activated preparations. Phosphotransferase activity, in contrast, appeared unaffected. These observations suggested functional changes in liver microsomal glucose-6-phosphatase relating either to altered membrane conformation and/or a defect in the glucose-6-*P* transporter. A third variant, Type Ic, in which the primary lession appears to be a defect in the P_i/PP_i transport function, recently has been identified in our laboratory (Nordlie *et al.*, 1983b; see also Section X).

D. Newer Techniques for Study of the Bioregulation of Glucose-6-Phosphatase

1. Isolated, Permeable Hepatocytes

An innovative experimental approach to the study of latency of glucose-6-phosphatase has been developed in the authors' laboratory (Jorgenson and Nordlie, 1980) as an alternative to the use of isolated hepatic microsomes. This approach, utilizing isolated hepatocytes rendered permeable to phosphate substrates by treatment with filipin, has been applied to a consideration of responses of the enzyme to acute fasting (Nordlie and Jorgenson, 1981). While characteristics of the enzyme with this preparation varied somewhat from those noted with microsomal preparations, the fundamental mode and pattern of response with permeable cells paralleled those seen in earlier studies with microsomal suspensions.

2. Studies with Intact Microsomes

Arion and associates (Arion and Walls, 1982; Arion *et al.*, 1976a) have recently described in detail methods for adjusting data from studies with rat liver microsomal preparations so that activities due only to "intact" microsomes in which the permeability barrier to phosphate compounds has not been disrupted may be analyzed.

3. Studies with Cells in Culture

The use of cells in culture for study of the bioregulation of activities of glucose-6-phosphatase is developing rapidly. This approach has the distinct advantage of confining responses to those biochemical phenomena unique to the organ and indeed the specific cell type under study.

Studies dealing with the effects of cAMP and dibutyryl cAMP, glucagon, and glucocorticoids on glucose-6-phosphatase cited above, e.g., utilized human and rat fetal liver and kidney cells in culture (Boxer *et al.*, 1974; Schwartz *et al.*, 1974; Mizushima and Ishikawa, 1979; DeLaval *et al.*, 1981).

Recently, Chabot *et al.* (1978) studied hormonal responses of the enzyme in cultured adult mouse intestinal mucosal cells. Hepatocytes from normal and preneoplastic liver have been used to study glucose-6-phosphatase by Wanson *et al.* (1981) and Bernaert *et al.* (1979), while Moller *et al.* (1977) studied the enzyme in cells in culture from Chang rat hepatomas.

Dobrosielski-Vergona and Widnell (1982) point out the advantage of hepatocytes in culture for continuation of their interesting studies of hormonal regulation of hepatic glucose-6-phosphatase in rats of various ages. Spagnoli *et al.* (1981) described some preliminary results with this new model.

It would appear to the present authors that cells in culture will become the model of choice for future studies of the modes and mechanisms of regulation of multifunctional glucose-6-phosphatase-phosphotransferase.

IX. MEMBRANE PHOSPHOLIPIDS AND GLUCOSE-6-PHOSPHATASE ACTION

A. Synopsis of Relevant Earlier Work

The effects of phospholipases and phospholipid supplementation on activities of glucose-6-phosphatase have been studied in considerable detail in the past (see reviews by Nordlie, 1971, 1974; Nordlie and Jorgenson, 1976; Duck-Chong, 1976). These earlier reports have produced conflicting results, phospholipids having been reported as essential for glucose-6-phosphatase activity, necessary or unnecessary for stabilization of the enzyme, constituting a mechanism for the normal constraint of activities, inhibitory, a determinant of membrane morphology, and as a modulator of the enzyme's kinetic constants (see above reviews for primary references).

At least a partial reason for these apparently contradictory earlier observations would now appear to be variations in the morphological state of the preparation selected for study by each group. For example, Garland and Cori (1972) studied the enzyme partially purified from liver microsomal acetone powder while Snoke and Nordlie (1972) investigated the enzyme of isolated rat microsomes assayed in the absence and presence of membrane disruption by the detergent deoxycholate. The former group removed phospholipids by gel filtration in the presence of detergent and found monounsaturated phosphatidylcholine most effective as reactivating supplemental phospholipid. Snoke and Nordlie (1972), in contrast, found phosphatidylethanolamine much more effective than phosphatidylcholine in reactivating the enzyme of microsomes in the absence of deoxycholate. The effect was much less when detergent-disrupted microsomes were assayed. It would seem from these and like observations, and some recent findings regarding the mechanism of action of the microsomal glucose-6-phosphatase system (see Section X), that substrate transport and catalytic functions of the enzyme may differ in their phospholipid requirements.

B. Recent Studies of Phospholipid Requirements

Cater *et al.* (1975) have shown a partial loss of rat liver glucose-6-phosphatase following treatment with phospholipase C. The presence of albumin partially prevented this effect. These authors conclude, as others have, that microsomal glucose-6-phosphatase is at least partially phospholipid dependent.

Thermotropic studies of liver microsomal glucose-6-phosphatase corroberate the interrelationships between enzyme activity and membrane phospholipids (Johnson and Nordlie, 1980; Maddaiah *et al.*, 1981, Grinna, 1975, 1977). Aging (Grinna, 1977), polyamines (Johnson and Nordlie, 1980), and alloxan-induced diabetes (Maddaiah *et al.*, 1981), all of which altered glucose-6-phosphatase activity, also modulated thermotropic responses, presumably through modifications of the lipid environment of the membrane of the endoplasmic reticulum.

The work of Santiago *et al.* (1977), relating to the effects of membrane lipid peroxidation on microsomal glucose-6-phosphatase activity, implicated rather specifically phosphatidylcholine with one unsaturated fatty acid moiety as essential for glucose-6-phosphatase action.

Burlakova *et al.* (1979) studied the effects of alterations in rat liver microsomal membrane phospholipids on glucose-6-*P* phosphohydrolase and inorganic pyrophosphatase activities of the enzyme. Injection of the antioxidant Ionol, i.p., or delipidation of microsomal membranes *in vitro* by extraction with acetone–water solution were employed. A change in glucose-6-*P* phosphohydrolase activity was seen which correlated directly with the alteration in phosphatidylethanolamine content, while inorganic pyrophosphatase activity varied directly with membrane phosphatidylserine content. Hill coefficients indicated that these phospholipids serve as allosteric activators of glucose-6-phosphatase. These results were confirmed by later studies of Burlakova *et al.* (1980) in which membrane viscosity was measured along with enzymic activities following Ionol administration.

In studies with rat liver microsomes involving the use of purified phospholipid transfer proteins and a phosphatidylinositol-specific phospholipase C, Crain and Zilversmith (1981) found that membrane phosphatidylinositol and cholesterol had no apparent effect on glucose-6-phosphatase. They did, however, observe a direct correlation between microsomal membrane phosphatidylethanolamine and glucose-6-phosphatase, consistent with earlier observations of Snoke and Nordlie (1972). Schulze and Speth (1980) used phospholipase C and toluene-2,4-diisocyanite to probe rat liver microsomal glucose-6-phosphatase and membrane phospholipid interactions. They, likewise, found a correlation between activity and microsomal membrane phosphatidylethanolamine content.

Dyatlovitskaya *et al.* (1979a,b) also have used protein-mediated lipid exchange to study membrane phospholipid-glucose-6-phosphatase interrelationships. Microsomes from both normal rat livers and hepatomas were subjects of study. Studies were carried out without and with detergent-supplementation of microsomes. With intact microsomes (no detergent added), exogenous phosphatidylcholine had no effect, but phosphatidylethanolamine or phosphatidylserine prevented loss of glucose-6-phosphatase. In contrast, with deoxycholate-dispersed microsomes, phosphatidylcholine activated glucose-6-phosphatase but phosphatidylethanolamine did not.

In studies by Rossowska and Dabrowiecki (1978), hypoxia was found not to affect glucose-6-phosphatase of guinea pig brain while ischemia lowered the activity in the microsomal fraction with a concomitant increase in the cytosol. Alterations in microsomal phosphatidylcholine or phosphatidylethanolamine were shown not to be involved, and it was concluded that, unlike the liver microsomal enzyme, brain microsomal glucose-6-phosphatase is not phospholipid dependent.

C. Some Tentative Conclusions

The work of Dyatlovitskaya *et al.* (1979a,b), Crain and Zilversmith (1981), Schulze and Speth (1980), and Burlakova *et al.* (1979, 1980) is all consistent with earlier observations of Snoke and Nordlie (1972), Garland and Cori (1972), and others. Phosphatidylethanolamine appears essential for glucose-6-phosphatase activity with undisrupted microsomes, while phosphatidylcholine (with one unsaturated fatty acid moiety) seems necessary for activity with disrupted microsomal preparations.

These observations recently have been interpreted in terms of the transporter–catalytic unit hypothesis (see Section X) as indicative of the involvement of membrane phosphatidylethanolamine (or phosphatidylserine) in glucose-6-*P* transport and of the specific phosphatidylcholine of the microsomal membrane in catalytic unit function (Dyatlovitskaya *et al.*, 1979b; Schulze and Speth, 1980; Crain and Zilversmith, 1981).

The potential for discriminant control of activities of this multifunctional catalyst with various substrates is suggested by the varying requirements for activity with PP_i compared with glucose-6-*P* (Burlakova *et al.*, 1979). The significance of membrane phospholipids as enzymic activity determinants also is indicated by the suggestion of diminished glucose-6-phosphatase activity in hepatomas being due to low levels of essential phosphatidylcholine in the endoplasmic reticulum of such cells (Dyatlovitskaya *et al.*, 1979b). Further work in this interesting area is essential.

X. THE MOLECULAR BASIS OF LATENCY

A. Historical Considerations

Beaufay and de Duve (1954) were the first to demonstrate the detergent-sensitivity of liver microsomal glucose-6-phosphatase. Since that time, this phenomenon of detergent-sensitivity or "latency," as it is now generally termed, of the enzyme has been studied in great detail. "Latent activity" refers to that activity which is not expressed during assay unless microsomal or other membranous preparations are treated with detergent or otherwise disrupted. Equation (9) is generally used to calculate percent latency.

$$\text{Percent latency} = 100 \times \frac{\text{activity with detergent–activity without detergent}}{\text{activity with detergent}} \quad (9)$$

Latency is a variable parameter, fluctuating with the nature of the membranous preparation studied, the species, the activity of glucose-6-phosphatase-phosphotransferase considered, and the hormonal and nutritional state of the experimental subject. This matter of differential latency has been discussed at length in earlier reviews (see Nordlie, 1971, pp. 562–564, 1974, pp. 46–47, 71–80, 1981, pp. 297–303, 1982a, pp. 264–267; Nordlie and Jorgensen, 1976, pp. 479–485). Space limitations do not permit a reiteration of earlier work here, and the interested reader is directed to the indicated reviews as a preface to the discussion to follow. Some additional, relevant observations regarding alterations in latency of the enzyme are included in Section VIII-C.

Latency or "crypticity" of the enzyme relates to the fact that this catalyst is sequestered in a special, phospholipid-rich environment of cellular biomembranes (endoplasmic reticulum, nucleus, plasmalemma, and probably mitochondria; see Nordlie and Jorgenson, 1976) which through one mechanism or another exerts discriminant constraints on the various activities of the enzyme. Some key questions regarding this phenomenon of latency of glucose-6-phosphatase follow: (1) Is latency a normal, physiological phenomenon or does it arise in part or *in toto* as an artifact of preparation and isolation of cellular membranous components?, (2) Does it relate to accessibility of substrates to the enzyme through the biomembrane?, (3) Is latency related to the effectiveness of enzyme–substrate binding, once the substrate is available to the enzyme?, and (4) Is reactivity of the enzyme–substrate complex, once formed, sensitive to membrane-associated effects?

Work since 1954 in a number of laboratories indicates that *all* of the above apply with the various activities of glucose-6-phosphatase (see earlier reviews).

De Duve (1962) first suggested that orientation of glucose-6-phosphatase in the membrane of the endoplasmic reticulum may relate to function. Stetten and Burnett (1967), Segal and Washko (1959), Nordlie and co-workers (see reviews, Section II), and others have elaborated upon this concept, and have invoked it to explain differential patterns of latency accompanying variations in hormonal and dietary state of the experimental subject. For example, experimental diabetes or acute fasting leads to an increase in the rat liver microsomal enzyme activity also characterized by an elevation in latency and an increase in apparent K_m value, while glucocorticoid administration lowers latency of the microsomal enzyme from normal rats and returns the diabetes-induced increase in latency to normal. These observations with microsomes recently have been confirmed by Nordlie *et al.* (1979b) who, however, found quite different patterns of response of the enzyme of isolated hepatic *nuclei* from these same rats (relevant observations are summarized in Section VIII-C.3). Gunderson and Nordlie (1975) earlier delineated fundamental differences between latency of glucose-6-phosphatase of nuclei and isolated microsomes.

B. The "Membrane-Sidedness" and "Constrained–Unconstrained Conformer" Hypotheses

These and like observations constituted the basis from which Nordlie formalized two mechanistic theories to explain this variable, discriminant, and alterable latency of glucose-6-phosphatase. These hypotheses, the "membrane-sidedness theory" and a concept involving membrane-regulated interconversions of constrained and unconstrained conformers of the enzyme, are presented in detail elsewhere (see especially Nordlie, 1971, 1974, pp. 77–80, 1981; Nordlie and Jorgenson, 1976), and will not be repeated here in the interest of brevity.

C. The Translocase–Catalytic Unit Hypothesis

During the past seven years, a major focus has been placed upon the question of latency by Arion and his colleagues at Cornell, who have proposed an alternative mechanistic hypothesis, the "translocase–catalytic unit" concept of glucose-6-phosphatase action. This concept was initially suggested in 1972 by Wallin and Arion (1972) to explain observed differences in sensitivity to inhibition by sulfhydryl reagents of glucose-6-phosphatase in untreated compared with disrupted microsomal preparations, and more recently has been supported by a number of other observations. It has not previously been reviewed in detail and accordingly it will be considered at some length here.

1. The Hypothesis

Wallin and Arion (1972) proposed that glucose-6-phosphatase of isolated rat liver microsomes might consist of more than a single membrane component. A sulfhydryl-reagent-sensitive component, a glucose-6-*P*-specific permease, was postulated to translocate the hexose phosphate from its point of formation in the cytosol to the phosphohydrolase within the lumen of the endoplasmic reticulum. The concept, in more expanded narrative and diagrammatic form, along with a summary of supportive evidence, was published in 1975 (Arion *et al.*, 1975). Following five further years of study, the concept in its present state of development was published in 1980 (Arion *et al.*, 1980b). As presently constituted, the hypothesis takes into account not only glucose-6-*P* transport but that of the substrates PP_i, carbamyl-*P*, and glucose, and the products P_i and glucose as well. A schematic representation of the system as presently envisiaged is given in Figure 1.

The scheme in Figure 1 is based upon the assumption that the glucose-6-phosphatase system behaves in intact endoplasmic reticulum as it has been observed to behave with intact isolated rat liver microsomal preparations. The system is proposed to include a fundamental catalytic unit of broad specificity [see Eq. (1–8)] located on the lumenal side of the endoplasmic reticulum. Further, it is proposed that the membrane of the endoplasmic reticulum is of itself impermeable to phosphate compounds, and that permeation of selected phosphate substrates is effected by transporters or "translocases" of high specificity (see T_1 and T_2 in Figure 1). Translocase T_1 displays nearly absolute specificity for glucose-6-*P* at physiological levels. It serves to transport

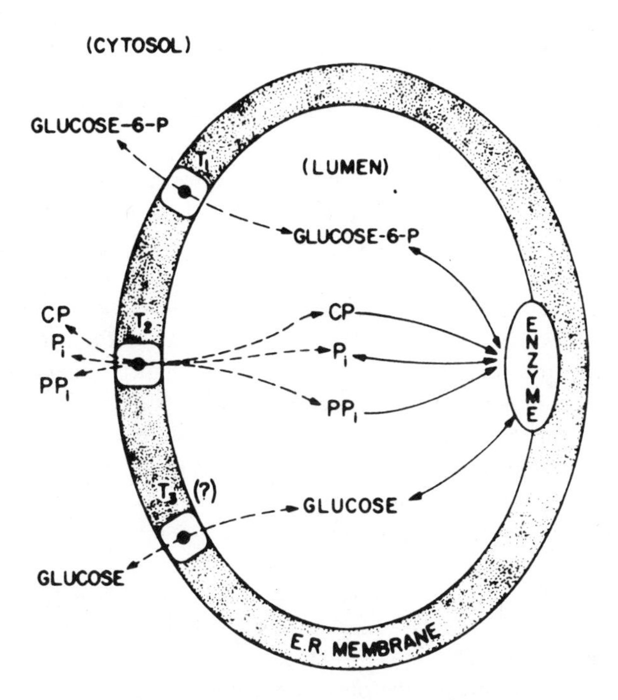

Figure 1. Schematization of the glucose-6-phosphatase system of the hepatic endoplasmic reticulum. The system is proposed to consist of a catalytic unit of broad specificity ("enzyme") located on the lumenal side of the endoplasmic reticulum membrane ("E. R. membrane") operating in conjunction with phosphate–substrate-specific translocase units T_1, which is specific for glucose-6-*P*, T_2, which is functional with PP_i and carbamyl-*P* ("CP") (and also with product P_i), and possibly T_3, which is functional with D-glucose. (From Arion *et al.*, 1980b, by permission of the American Society of Biological Chemists, Inc.).

glucose-6-*P*, formed in the cytosol via gluconeogenesis, glycogenolysis, or glucokinase action, to the lumen of the endoplasmic reticulum where it then may be hydrolyzed to glucose and P_i by catalytic unit action.

A critical feature of the mechanism is the establishment of a pool of substrate (glucose-6-*P*) within the lumen of the microsomes (endoplasmic reticulum) at a steady-state concentration lower than that of the medium (cytosol). The larger apparent K_m for glucose-6-*P* with intact rather than with disrupted microsomes, as well as the phenomenon of latency itself, relate to this circumstance. The rate of glucose-6-*P* efflux from microsomes is believed to limit glucose-6-*P* synthesis via phosphotransferase activities.

To account for limited reactivity with PP_i or carbamyl-*P* seen *in vitro* with intact microsomes, a second transporter, T_2, is hypothesized. Removal of products of hydrolysis of glucose-6-*P*, i.e., P_i and glucose, is essential if the system is to function continuously. In this regard, the existence of a third translocase, T_3, for glucose transport has been tentatively suggested. That translocase T_2 serves physiologically to remove P_i from the lumen of the endoplasmic reticulum to the cytosol is supported by results of inhibition studies which show a mutual competition among P_i, PP_i, and carbamyl-*P* for a common site with intact microsomal preparations. In contrast with such preparations, inhibition between glucose-6-*P* and P_i has been found to be of the linear noncompetitive type with intact microsomes, indicating a noncommonality of transport mechanism for P_i and glucose-6-*P* (Arion *et al.*, 1980b; see however Vianna and Nordlie, 1969; Nordlie and Jorgenson, 1981). Glucose-6-*P*, P_i, PP_i, and carbamyl-*P* all compete for the catalytic unit of disrupted microsomes.

Under presumed physiological conditions, the system was suggested by Arion *et al.* (1980b) to function primarily as a glucose-6-*P*-specific hydrolase rather than as a glucose-6-*P*-synthesizing system via PP_i-glucose or carbamyl-*P*:glucose phosphotransferase, a conclusion with which the present authors do not concur (see Sections X-E and XI-C).

2. *Supportive Evidence from the Arion Group*

Evidence in support of this interesting concept has been presented by Arion and his associates in a series of complex, sophisticated papers beginning in 1972 (Wallin and Arion, 1972, 1973; Arion *et al.*, 1972a,b, 1976a,b, 1980a,b; Arion and Wallin, 1973; Ballas and Arion, 1977; Nilsson *et al.*, 1978; Lange *et al.*, 1980; Arion and Walls, 1982). Because of the complex and voluminous nature of this supportive evidence, we must limit ourselves here to some of the salient features and direct the interested reader to the primary references, especially Arion *et al.* (1975, 1980b).

Much of the supportive evidence has accrued from comparative kinetic analyses carried out with isolated "intact" and disrupted microsomal preparations. These studies have shown the existence of a highly selective microsomal permeability barrier. Glucose-6-*P*, but not mannose-6-*P* (except at high concentration), thus may gain access to the catalytic unit only through the intervention of the specific glucose-6-*P* carrier T_1 when microsomes are intact. With disrupted microsomal preparations, glucose-6-*P*, mannose-6-*P*, PP_i, carbamyl-*P*, P_i, etc. all are competitive for the enzyme's active site, while with microsomes with the permeability barrier intact, P_i, carbamyl-P, and PP_i compete with one-another for T_2, but not with glucose-6-*P*. Interactions of the system with mannose-6-*P* and mannose, and with phlorizin also may be rationalized in terms of the translocase–catalytic unit concept. Phlorizin selectively inhibits glucose-6-*P* phosphohydrolase and PP_i-glucose phosphotransferase reactions of microsomes with intact permeability barriers. K_m for glucose-6-*P*, but not PP_i, is altered by detergent treatment. Catalysis may be selectively inactivated by mild heating in acid medium without the destruction of the microsomal permeability barrier.

Perhaps some of the most convincing evidence for the transporter concept comes from the work of Arion *et al.* (1976a) who demonstrated a direct correlation between EDTA uptake and "low K_m mannose-6-*P* phosphohydrolase" in rat liver microsomes treated with progressively increasing levels of detergent. In other studies, a selective effect of limited proteolysis or diazobenzene was seen with intact microsomes, but not disrupted microsomes, suggesting the existence of a membrane protein acting as a glucose-6-*P*-specific carrier (Nilsson *et al.*, 1978). These workers indicate, however, that definitive evidence for the translocase concept must come from (1) reconstitution studies, and (2) direct measurement of the glucose-6-*P* pool within the microsomal vesicle. Ballas and Arion (1977) have had limited success with the latter approach.

Further, circumstantial support for the translocase–catalytic unit concept evolves from its ability to rationalize differential latencies seen with various phosphate substrates and with microsomes from animals in various hormonal, dietary, and genetic states. For example (see Figure 2), the enhancement in latency seen in conjunction with the diabetes- or fasting-induced increase in hepatic microsomal glucose-6-phosphatase is suggested to be due to an increase in the number of functional catalytic units with little or no concomitant increase in transporters. Thus, transport becomes increasingly rate limiting in insulin-deprivation relative to the normal condition, the intramicrosomal glucose-6-*P* pool is lowered, and latency is accordingly increased. In contrast, the diminution in latency, with accompanying apparent increase in glucose-6-*P* phosphohydrolase noted in untreated microsomal preparations from glucorticoid-treated compared with untreated rats, would involve an increase in transporters but not catalytic units (see Figure 2).

Type Ib glycogenosis (see Section VIII-C.4.e) may be rationalized in terms of an inborn metabolic error in which the transporter specific for glucose-6-*P* is absent or markedly diminished from normal. In such a condition, the latency of glucose-6-*P* phosphohydrolase would be selectively increased rather markedly in a manner somewhat analogous to insulin-deprivation, since glucose-6-*P* transport to the catalytic unit would be minimal in the absence of membrane disruption (Lange *et al.*, 1980).

3. *Recent Supportive Evidence from Other Groups*

Recent experimental results from a number of laboratories around the world either provide further supportive evidence for, or are consistent with, the translocase–catalytic unit hypothesis. Gratzl (1975) interprets results of his studies of inhibition by galactose-6-*P* of glucose-6-phosphatase of microsomes in the absence and presence of membrane disruption as supportive of the translocase concept. Studies involving the use of phos-

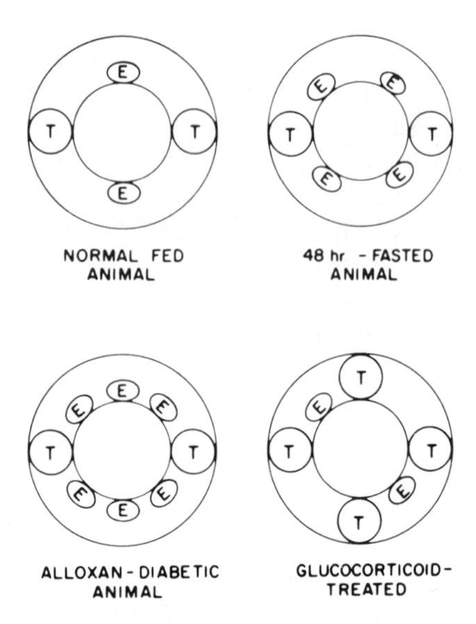

Figure 2. Schematic representation of the postulated quantitative relationships between the glucose-6-*P* phosphohydrolase ("*E*") and translocase ("*T*") components of liver microsomes from livers of rats under various experimental conditions. *T* and *E* correspond with T_1, and "enzyme," respectively, in Figure 1. An individual *E* is equivalent to approximately 0.15 units of phosphohydrolase activity per mg microsomal protein, and *T* is equal to approximately 0.35 units of translocase activity per mg microsomal protein. (From Arion *et al.*, 1976b, by permission of the American Society of Biological Chemists, Inc.)

pholipid transfer proteins and phospholipases also support this hypothesis and suggest that different phospholipids may be required for transport and catalytic functions (Dyatlovitskaya *et al.*, 1979a,b; Crain and Zilversmith, 1981; Schulze and Speth, 1980). Type Ib glycogenosis, characterized by an elevated latency of human hepatic glucose-6-phosphatase, has been observed independently by four groups in three countries (Bialek *et al.*, 1977; Narisawa *et al.*, 1978; Sann *et al.*, 1980; Lange *et al.*, 1980; Nordlie, 1981; Igarashi *et al.*, 1980). Kamoun (1980) has suggested that an alteration in anomeric specificity in regard to glucose-6-*P* uptake may be involved in this condition.

Gold and Widnell (1976) found that destruction of microsomal integrity desensitized microsomal glucose-6-phosphatase to inhibition by pyridoxal-*P*. Very recently, studies from that same laboratory (Dobrosielski-Vergona and Widnell, 1982) involving the responses of glucose-6-phosphatase to triiodothyronine and triamcinolone in rats of various ages have been interpreted in terms of the translocase–catalytic unit concept.

Zoccoli and Karnovsky (1980) studied the effects of the anion transport inhibitors 4-acetamido-4′-isothiocyanostilbene-2,2′-disulfonic acid (SITS) and 4,4′-diisothiocyanostilbene-2,2′-disulfonic acid (DIDS) on activities of glucose-6-phosphatase in untreated and detergent-disrupted microsomes. They concluded that SITS or DIDS inhibits glucose-6-*P* hydrolysis in intact microsomes by blocking specifically glucose-6-*P* transport across the microsomal surface. In more recent studies from that laboratory (Zoccoli *et al.*, 1982), these studies have been extended. Inhibition by [^{14}C]-DIDS has been characterized by its binding to a microsomal membrane protein of molecular weight 54,000, which is believed to be involved in microsomal glucose-6-*P* transport. They also studied the effects of fasting, glucocorticoid administration, and experimental diabetes, and found effects altering latency to be manifest through an increase in functional translocase activity rather than an increase in number of such transporter molecules (see also Sections X-C.4 and X-F).

Several studies by Speth and Schulze at Giessen (Speth and Schulze, 1981, 1982; Schulze and Speth, 1980), involving the use of antibodies to glucose-6-phosphatase, controlled proteolysis, thermotropic technique, phospholipase C, toluene-2,4-diisocyanate, and hormonally-induced manipulations *in vivo,* give limited support to the translocase concept (but see also Section X-C.4). Meissner and Allen (1981) have presented evidence recently for the existence of two types of rat liver microsomes differing in their permeability to glucose and other small molecules. This important work supports permeation through one or more channels (which they also term "permeases"), one of which, they believe, may be involved with, among other compounds, glucose-6-*P* (see also Section X-C.4).

4. *Some Reservations*

As pointed out above, Nilsson *et al.* (1978) described the relative effects of proteolytic digestion and diazobenzene sulfonate-inhibition on microsomal glucose-6-phosphatase in the presence and absence of membrane disruption. They concluded that "the simplest explanation (for their observations) is that there is a protein carrier in the microsomal membrane which transports glucose-6-*P* from the medium to the lumen, where it is hydrolyzed, and that diazobenzene sulfonate and proteases attack this carrier." They further point out, however, that "it should be noted that even though the results presented, as well as a large number of other studies, are consistent with the multi-component model of microsomal glucose-6-phosphatase, the definitive evidence must come either from isolation and reconstitution of the glucose-6-P carrier or actual measurement of the pool of glucose-6-*P* within the microsomal vesicle." Neither of these has been totally accomplished to date, although progress in these directions has been reported (Zoccoli *et al.*, 1982; Ballas and Arion, 1977).

The work of Meissner and Allen (1981) supports the concept that selective permeation of small molecules through the microsomal membrane is important. Their work supports the existence of two types of rat liver microsomes with varying permeability. Their studies, however, indicate that permeation of Type A microsomal vesicles is via a *single* permease or "channel" which selects against solutes on the basis of size and charge, and that entrance of a variety of uncharged and charged molecules including glucose-6-*P* is via this mechanism. In their hands, DIDS inhibited transport of several solutes, including D-glucose and glucose-6-*P*.

Benedetto *et al.* (1979) observed no latency of glucose-6-phosphatase with microsomes prepared from monkey liver, and concluded that their observations were not consistent with the translocase–catalytic unit hypothesis.

Goldsmith and Stetten (1979) studied both PP_i-glucose phosphotransferase and glucose-6-*P* phosphohydrolase of rat liver (and kidney) microsomes from rats both prenatally and postnatally. Maximal activities, with a latency of glucose-6-*P* phosphohydrolase much above that of the adult level, were seen at days 2–5 following birth (liver enzyme). They rationalized these and related observations in terms of altered membrane conformation as well as in terms of the translocase concept.

In a recent paper, Zakim and Edmondson (1982) described the analysis of the guinea pig liver microsomal glucose-6-phosphatase system by both pre-steady-state and steady-state kinetic approaches. They interpreted their results as contraindicative of the existence of a translocase. Production of glucose from glucose-6-*P* via the reaction, enzyme + glucose-6-*P* $\rightleftarrows$ enzyme-*P* + glucose, was not significantly affected by the presence or absence of detergent, indicating that accessibility of glucose-6-*P* to the catalytic site was not altered by membrane-disruption. In contrast, the reaction of enzyme-*P* + H_2O to give enzyme + P_i was detergent sensitive, consistent with the earlier conclusion of Lueck *et al.* (1972). Arion and Walls (1982) have questioned these results, however, on the basis that freezing microsomes in 0.25 M mannitol, as employed by Zakim and Edmonson, destroys the microsomal permeability barrier.

The use of pH 6.1 or 6.25 for most of the earlier kinetic work supportive of the translocase–catalytic unit concept concerns the present authors, who currently are reinvestigating the system at pH 7.4 or 7.0 (see Nordlie and Jorgenson, 1981; Jorgenson and Nordlie, 1980). Significant differences in patterns of inhibition, studied in the absence and presence of detergents, e.g., at pH 7.0 compared with 6.1, already have become apparent (Nordlie and Traxinger, unpublished observations).

Studies of Pollak *et al.* (1971) suggested that membrane-phospholipid-related constraints on the catalytic behavior of glucose-6-phosphatase-phosphotransferase are a distinct possibility. They found that adding phospholipid extracts back to delipidated microsomes ("reticulosomes") increased the latency of activities of glucose-6-phosphatase, and that the effect was greater with PP_i- than with glucose-6-*P*-involving reactions. More recent studies of Gunderson and Nordlie (1973, 1975) with nuclei and isolated nuclear membranes, demonstrated that alterations in membrane morphology lead to alterations in latency of activities of glucose-6-phosphatase. Studies by Johnson and Nordlie (1980) and by Nordlie *et al.* (1979a) indicate that polyamines may selectively affect activities of glucose-6-phosphatase through membrane-associated perturbations. For example, these effects may be manifest *even after* disruption of the microsomal membrane permeability barrier.

An alternative to the translocase concept to rationalize observed latency of microsomal glucose-6-phosphatase has been suggested by Drabowiecki (1979).

Gratzl (1975) observed competitive inhibition by galactose-6-*P* of glucose-6-*P* hydrolysis by rat liver microsomes both in the absence and presence of disruption of the vesicles by exposure to high pH, with a decrease in apparent K_i value from 23 mM to 10 mM following disruption. He interpreted this and other observations as supportive of the translocase concept (see Section X-C.3). In contrast, we would interpret these findings as inconsistent with the involvement of a transporter specific for glucose-6-*P*. Linear noncompetitive inhibition would be anticipated with intact vesicular preparations were glucose-6-*P* to enter the microsomal lumen via a carrier while galactose-6-*P*, inhibitory at the catalytic site within the lumen, were to enter otherwise than by this carrier (see Arion *et al.*, 1980b).

Hori and Takahashi (1977) and Takahashi and Hori (1978, 1981) have reported evidence indicating that latency of glucose-6-*P* dehydrogenase of rat liver microsomes relates basically to the limited penetrability of the substrate NADP rather than glucose-6-*P*.

Very recent studies in the authors' laboratory (Nordlie *et al.*, 1983b) have revealed the existence of a third variant of glycogenosis Type I—Type Ic. A major problem here appears to be a congenital defect in translocase T_2. Exogenously added P_i, PP_i, or carbamyl-*P* was non-interactive with glucose-6-phosphatase of intact liver microsomes from this patient. Nonetheless, P_i generated within the microsomal lumen *via* glucose-6-*P* hydrolysis appeared to find a pathway of egress independent of T_2, since progressive inhibition of glucose-6-*P* hydrolysis by buildup of P_i produced in the hydrolytic reaction was not seen. Several alternative explanations for these observations, involving modifications in the translocase/catalytic unit model of Arion *et al.* (1980b), were suggested.

Schulze and Speth (1980) and Speth and Schulze (1981, 1982) conclude, from their studies referred to in Section X-C.3, that the functional glucose-6-phosphatase system may consist of two components, but that ". . . these components represent at the time of reaction a topographical unit which traverses the microsomal membrane in a precise spatial arrangement." Their recent, immunological work (Speth and Schulze, 1982) supports the concept that ". . . the phosphohydrolase cannot be exposed on the lumenal surface of the microsomal membrane but is buried within the hydrophobic part of the bilayer, as suggested in and expected for a phospholipid-dependent integral membrane protein." They suggest that aspects of both the translocase concept of Arion and the membrane conformational concept championed by Nordlie, Stetten, and others may be included in bioregulation of glucose-6-phosphatase.

D. A Proposed Synthesis of the Membrane Conformational and Translocase–Catalytic Unit Concepts

Despite the concerns outlined briefly above, the credibility of the translocase concept is supported by its consistency with an ever-increasing body of observations. The concept, we believe, must be accepted at this time as a highly-credible working hypothesis, subject to further experimental examination. However, that membrane conformational alterations also may be involved in regulation of the enzyme's activity, both at the transporter and at the catalytic unit level, also seems firmly established. The senior author (Nordlie, 1981) first suggested in 1981 that aspects of *both* the translocase hypothesis and the membrane conformational model may be operative in the living cell. This conclusion was based on some of the considerations outlined immediately above, and especially, it derives from the observations of Arion and

Nordlie (1967) that the diminution in latency of microsomal glucose-6-phosphatase in response to glucocorticoids could be obtained in the presence as well as absence of inhibition of protein biosynthesis by actinomycin D. More recently, Zoccoli *et al.* (1982) similarly have concluded that ". . . . the increase in translocase activity observed in microsomes from fasted, triamcinolone-treated, and diabetic rats *cannot* be ascribed to increased numbers of translocase molecules, but rather to increased functional activity of the translocase protein." These, and other observations from the authors' laboratory with permeable isolated hepatocytes described immediately below, support the idea that *both* transport and membrane conformation serve to control glucose-6-phosphatase *in situ*. Conformational effects on both transport and catalysis, *per se*, are supported by data at hand (see below).

E. Behavior of the Enzyme in Intact Endoplasmic Reticulum of Isolated Permeable Hepatocytes

Jorgenson and Nordlie (1980) recently developed a technique for studying activities of glucose-6-phosphatase *in situ*, utilizing isolated hepatocytes the plasmalemma of which was rendered permeable to phosphate substrates by treatment with the polyene compound filipin. This approach was prompted by earlier work of Gunderson and Nordlie (1973, 1975) in which marked differences were seen in latency and associated catalytic properties of glucose-6-phosphatase-phosphotransferase of isolated nuclear membrane fragments compared with intact nuclei. Special techniques employed have

Table 2. Latencies of Phosphohydrolase and Phosphotransferase Activities of Filipin-Treated Rat Hepatocytes[a]

Activity	Number of experiments	Latency (%)
Phosphohydrolase		
Glucose-6-*P* phosphohydrolase	7	27 ± 3
Mannose-6-*P* phosphohydrolase	1	54
Phosphotransferase		
Carbamyl-*P*-dependent		
Spectrophotometric assay	3	58 ± 5
3-*O*-methyl[U-^{14}C]glucose assay	3	55 ± 2
[2-^{3}H]glucose assay	3	55 ± 6
Glucose-6-*P*:3-*O*-methyl[U-^{14}C]glucose	1	53
Mannose-6-*P* dependent		
Spectrophotometric assay	1	79
3-*O*-methyl[U-^{14}C]glucose assay	1	82
[2-^{3}H]glucose assay	1	90
PP_i-dependent		
Spectrophotometric assay	2	16
3-*O*-methyl[U-^{14}C]glucose assay	2	20
[2-^{3}H]glucose assay	2	14

[a] From Jorgenson and Nordlie (1980). See that reference for further details.

Table 3. Latencies of Phosphohydrolase and Phosphotransferase Activities of Isolated Microsomes[a]

Activity	Latency (%)
Glucose-6-*P* phosphohydrolase	51
Mannose-6-*P* phosphohydrolase	95
Carbamyl-*P*:glucose phosphotransferase	83
Carbamyl-*P*:3-*O*-methylglucose phosphotransferase	87
PP_i-glucose phosphotransferase	73

[a] From Jorgenson and Nordlie (1980). Latency values are those determined with microsomes isolated from liver homogenates. Latency values with microsomes derived from isolated hepatocytes were 40% for glucose-6-*P* phosphohydrolase, 74% for carbamyl-*P*:glucose phosphotransferase, and 80% for carbamyl-*P*:3-*O*-methylglucose phosphotransferase. See Jorgenson and Nordlie (1980) for further details.

been referred to in Section III, and are described in detail by Jorgenson and Nordlie (1980). With such cell preparations with the endoplasmic reticulum and nuclei intact, it is believed that observed activities of glucose-6-phosphatase may more closely approximate the *in vivo* condition than with microsomes—isolated fragments of endoplasmic reticulum.

With permeable hepatocyte preparations, latencies of various phosphohydrolase and phosphotransferase activities of glucose-6-phosphatase were consistently lower than with microsomes isolated from either liver homogenates or from hepatocytes. Percent latency values are presented in Table 2 (permeable hepatocytes) and Table 3 (microsomes). Latencies of only 27%, 18%, 54%, and 56%, respectively, were seen with glucose-6-*P* phosphohydrolase, PP_i-glucose phosphotransferase, mannose-6-*P* phosphohydrolase, and carbamyl-*P*:glucose phosphotransferase with permeable hepatocytes. Corresponding values with microsomes isolated from liver homogenates were 51%, 73%, 95%, and 83%. Filipin had no effect on activities of isolated microsomes. K_m values determined with these cell preparations agreed reasonably well with corresponding values from studies with detergent-activated microsomes. K_i values for D-glucose, P_i, bicarbonate, Cl^-, mannose-6-*P*, phosphopyruvate, and carbamyl-*P* all agreed with values determined with disrupted microsomes, or were intermediate between such values and higher values observed with intact microsomes (Nordlie and Jorgenson, 1981). Inhibition of glucose-6-*P* hydrolysis by P_i was purely competitive with permeable hepatocyte preparations, as was inhibition by carbamyl-*P*, Cl^-, PP_i, and a number of other compounds.

Latency of glucose-6-*P* phosphohydrolase increased from 27–38% following a 48-hr fast and was accompanied by an increase in K_m from 2.6–3.4 mM and a small elevation in V_{max}, observations reflective of responses seen earlier with isolated microsomes (Nordlie and Jorgenson, 1981).

These and related observations were interpreted by Jorgenson and Nordlie (1980) as consistent with *either* the transporter hypothesis or the membrane conformation concept. In regard to the latter, they pointed out that ". . . it need only be postulated that alterations take place in membrane conformation accompanying the fragmentation of endoplasmic reticulum and isolation of microsomes. Activity-discriminant constraints on the various activities, normally present to a degree within the intact cell, may thus be altered with resultant changes in patterns and degree of latency observed with microsomes." In contrast, it is suggested that if ". . . one accepts the concept of a transporter–catalytic unit system, then enhanced latency of activities of microsomes would suggest simply the rather extensive (physical or functional) loss during isolation of microsomes of transport (capability) for glucose-6-*P* and for carbamyl-*P*, PP_i, and mannose-6-*P*, the latter three more than the first."

Latencies of only 17% and 56%, respectively, were seen with PP_i and carbamyl-*P* phosphotransferase activities with permeable cells. These and related observations lead to the conclusion that, ". . . when provided with adequate levels of substrates, a significant portion of phosphotransferase as well as glucose-6-*P* phosphohydrolase activities of glucose-6-phosphatase may be manifest in the cell" (Jorgenson and Nordlie, 1980).

F. Some Tentative Conclusions

As additional experimental data have become available, the original transporter concept has been modified (compare Figure 8 of Arion *et al.*, 1980b, with Figure 1 of Arion *et al.*, 1975). Our feeling is that, at this time, some further refinements in this model may be in order. These modifications must take into account (1) apparent *activation* of translocase activity independent of protein biosynthesis indicated by the earlier work of Arion and Nordlie (1967; see Nordlie, 1981, 1982a) and supported strongly by the recent work of Zoccoli *et al.* (1982), Goldsmith and Stetten (1979), Schulze and Speth (1980), and others (see Section X-C.4), (2) quantitative and qualitative differences in latencies and related characteristics of the enzyme seen in isolated permeable hepatocytes compared with microsomes (see Section X-E), and (3) modifications by such membrane-perturbants as polyamines and phospholipids seen even with disrupted microsomes (see Sections IX and X-C.4).

In these regards, two crucial questions in need of further, critical experimental examination arise: (1) Is the [^{14}C]-DIDS-labeled microsomal protein isolated and identified by Zoccoli *et al.* (1982) the translocase *per se,* as envisiaged by Arion *et al.* (1975, 1980b)? Or is it a membrane protein which is involved in some way in controlling selectively the passage of small molecules (including glucose-6-*P*) through some microsomal membrane "channel" as conceived by Meissner and Allen (1981), for example?, and (2) Is indeed the adjustment of glucose-6-phosphatase data obtained with microsomes, to take into account activity due to disrupted microsomes normally contaminating the best of preparations, based on low-K_m mannose-6-*P* phosphohydrolase activity as advocated by Arion (Arion and Walls, 1982; see also Section III) the correct way to proceed? Or does this mannose-6-phosphatase activity really reflect a small, vestigial residual of that mannose-6-*P* phosphohydrolase *normally present* in intact endoplasmic reticulum *in situ,* as the recent work of Jorgenson and Nordlie (1980; Nordlie and Jorgenson, 1981) would suggest? (Perhaps activities with those microsomes normally isolated in the "disrupted" condition more nearly reflect the *in vivo* behavior of the enzyme within intact cellular biomembranes than does the behavior of the enzyme in "intact" microsomes)?

Hopefully, the present review will serve to stimulate enhanced experimental activity in this intriguing area of membrane biochemistry.

The glucose-6-phosphatase story has come a long way, mechanistically, since our last review in *The Enzymes of Biological Membranes* in 1976 (Nordlie and Jorgenson, 1976). Rational bases for interpretation of previously confusing data are at hand. A synthesis of features of various earlier concepts now seems reasonable. Further experimental testing of the various features of the glucose-6-phosphatase system, as outlined above, is sure to intensify in coming years, and finally, a complete understanding of the complex behavior of this complex system seems eminently obtainable.

XI. METABOLIC CONSIDERATIONS

The major objective of this chapter is a consideration of characteristics of the glucose-6-phosphatase system *per se*. However, it seems inappropriate to conclude without a brief mention of the potential metabolic consequences of many of the newly defined, membrane-related features of this complex system described in detail above.

A. Muscle

As discussed briefly in Section VI, Surholt and Newsholme (1981) have described the presence of glucose-6-phosphatase in muscle from a number of vertebrates and invertebrates. While the precise nature of this cytosolic "glucose-6-phosphatase" is not yet clear, it appears not to be due to action of nonspecific acid or alkaline phosphatase. The authors suggest the possibility that in some species (but not the honey bee or wasp) the enzyme might function to make available to other tissues glucose from muscle glycogen possibly formed via gluconeogenesis. More attractive is their suggestion that the presence of the phosphatase makes possible fine metabolic regulation in insect flight muscle via substrate cycling between glucose and glucose-6-*P*.

B. Brain

Several recent studies show unequivocally the presence of glucose-6-phosphatase in brain (Colilla *et al.*, 1975; Anchors and Karnovsky, 1975; Huang and Veech, 1982; see also Section VI). Brain may account for as much as 20% of total body glucose utilization. These observations indicate that earlier estimations of brain metabolic rates based on unopposed hexokinase activity may be erroneously high (Huang and Veech, 1982). The findings further suggest that brain glucose metabolism may be regulated by controlled cycling between glucose and glucose-6-*P*. Huang and Veech (1982) hypothesize that PP_i-glucose phosphotransferase also may be involved in glucose phosphorylation in brain, thus saving on ATP. In contrast, Anchors *et al.* (1977) have questioned a synthetic role for the brain enzyme. Anchors and Karnovsky (1975) have seen differences in labeling of a brain phosphoprotein which they identified as glucose-6-phosphatase, in sleep compared with wakefulness, and have suggested that the differences may relate to an increased glycogen content of the brain in slow wave sleep (Karnovsky *et al.*, 1980).

Clearly, the roles and regulation of brain glucose-6-phosphatase require further study.

C. *Liver*

Katz and Rognstad (1976), Stalmans (1977), Hue (1982), and Hers (1976, 1982) recently have published instructive reviews of the regulation of mammalian carbohydrate metabolism including the consideration of metabolic cycling between liver glucose-6-*P* and glucose. Cycling at this site, as well as mechanisms for the glucose concentration-dependent stimulation of the flux from hepatic glucose to glycogen and diversion of gluconeogenesis from glucose production to glycogen synthesis, have been studied in detail by Hue and Hers (1974), Clark *et al.* (1973), Rognstad *et al.* (1973), Katz *et al.* (1978, 1979), Issekutz (1977), Bontemps *et al.* (1978), Shikama and Ui (1978), El-Refai and Bergman (1976), and Bergman and El-Refai (1975). Interesting metabolic models incorporating cycling between glucose and glucose-6-*P* as well as glycogenesis and glycogenolysis have been described recently for mammalian liver (El-Refai and Bergman, 1976; Bergman and El-Refai, 1975; Anderson and Wright, 1980; and Anderson *et al.*, 1980) and for the slime mold *Dictyostelium discoidem* (Kelleher *et al.*, 1978).

Hue and Hers (1974) have suggested a model (their "pull" concept) to explain the glucose concentration-dependent flux from hepatic glucose to glycogen, controlled by substrate concentration alone. Inactivation of glycogen phosphorylase and resulting activation of hepatic glycogen synthase are essential features of the concept, and it follows that an elevation in hepatic glucose level should result in lowered hepatic glucose-6-*P* concentration. In contrast, El-Refai and Bergman (1976), with the aid of computer-assisted metabolic modeling, have suggested a "push" mechanism wherein a metabolic push through the hepatic glucose-6-*P*–glucose couple is fundamental, and an increase in the glucose-6-*P* pool in response to a glucose load is inherent. Recent studies by Nordlie *et al.* (1980) strongly support this latter concept, as significant increases in glucose-6-*P* levels in perfused livers from fed or fasted rats consistently were seen in response to added glucose loads. These studies would seem to establish hepatic glucose–glucose-6-*P* as a prime site for discriminant metabolic control.

Nordlie and Sukalski (1982) have further considered the mechanism whereby glucose-enhanced flux through the glucose–glucose-6-*P* site may be effected in the presence of elevated glucose-6-*P*. Since glucose-6-*P* hydrolysis at physiological concentrations exceeds glucokinase plus hexokinase by several fold, the need for constraints on the phosphatase activity is essential. Constraint by a partial latency, inhibition by the alternate substrate glucose, and competitive inhibition by P_i, HCO_3^-, and Cl^- all seem indicated. Further, the need for hepatic glucose phosphorylative capacity above and beyond that attributable to glucokinase plus hexokinase also was demonstrated by stoichiometric and kinetic analysis in these same studies (Nordlie and Sukalski, 1982).

The need for this auxiliary mechanism or mechanisms for hepatic glucose phosphorylation supplementally to glucokinase or hexokinase also has been demonstrated in studies of concentration-dependent glucose uptake by perfused livers from fed or 48-hr fasted rats (Alvares and Nordlie, 1977) and diabetic rats (Nordlie *et al.*, 1983a); involving [2-^{3}H]-D-glucose phosphorylation in hepatocytes from rats fasted for various durations of time (Singh and Nordlie, 1983), seen *in vivo* in normal and methylprednisolone-treated dogs (Issekutz, 1977), and from considerations based on kinetic analysis (Nordlie and Sukalski, 1982; McCraw, 1968; McCraw *et al.*, 1967; Singh and Nordlie, unpublished observations). That such a system may function in isolated perfused rat liver with 3-*O*-methyl-D-glucose as well as glucose as substrate, and that it may procede uninhibited by the potent glucokinase- and hexokinase-inhibitor *N*-acetylglucosamine following a 48-hr fast has been shown very recently in this laboratory (Nordlie *et al.*, 1982a,b).

That this system involves a K_m for glucose (ca. 40–60 mM) several-fold larger than that of glucokinase has been shown by kinetic analysis (see Nordlie and Sukalski, 1982; Singh and Nordlie, 1983; Nordlie, 1974).

The possibility that PP_i-glucose phosphotransferase activity of glucose-6-phosphatase may play this role has been suggested earlier by Lawson *et al.* (1976), Lawson and Veech (1979), McCraw (1968), McCraw *et al.* (1967), Issekutz (1977), El-Refai and Bergman (1976), Bergman and El-Refai (1975), Anderson and Wright (1980), Nordlie (1971, 1974, 1976, 1981), and Nordlie and Jorgenson (1976). A body of evidence has accumulated recently which further strongly supports the involvement of phosphotransferase activities of glucose-6-phosphatase [see Eq. (2 and 3)]. This supportive evidence includes the following considerations:

(1) Consistency of the apparent K_m for hepatic glucose phosphorylation with that of the phosphotransferase activity of glucose-6-phosphatase (see Nordlie and Sukalski, 1982; Singh and Nordlie, 1983; Nordlie, 1974).

(2) Ubiquitous distribution of these activities compared with that of glucokinase which appears absent in a number of species (Herrman and Nordlie, 1972; Nordlie, 1974). In this regard, interesting studies from the laboratory of Katz (Golden *et al.*, 1982; Riesenfeld *et al.*, 1981; Wals and Katz, 1981) indicate glucose phosphorylation by a system unsaturated even at the 60 mM glucose level by hepatocytes from liver of Japanese quail where the cytosolic, high-K_m glucokinase is absent.

(3) The very high V_{max} for these activities compared with glucokinase.

(4) Positive response of glucose-6-phosphatase-associated phosphotransferase to diabetes.

(5) Function with 3-*O*-methyl-D-glucose (which is inactive with glucokinase or hexokinase) as substrate and insensitivity to inhibition by *N*-acetylglucosamine, a potent inhibitor of glucokinase and hexokinase.

(6) Observed behavior of these activities in isolated, permeable rat liver hepatocytes where latency is much less than that associated with isolated microsomes (see Section X-E).

(7) Observations of Lawson *et al.* (1976) and Lawson and Veech (1979) that hepatic glucose, PP_i, glucose-6-*P*, and P_i constitute a metabolic control point based on "far from equilibrium" considerations.

The hypothesized role of these phosphotransferases in hepatic glucose phosphorylation has been discussed in earlier reviews (see Nordlie, 1971, 1974, 1981; and Nordlie and Jorgenson, 1976) and in a more recent paper in *Trends in Biochemical Science* (Nordlie, 1976).

One major problem with this hypothesis, the apparent extensive latency of phosphotransferase activities with PP_i or carbamyl-*P* seen earlier with isolated microsomal preparations, seems to have been resolved by our recent work with permeable hepatocytes (Jorgenson and Nordlie, 1980; Nordlie and Jorgenson, 1981). A second problem remains, that of availability of phosphate donor(s) at K_m levels. However, several recent studies indicate that this concern may soon be resolved. Veech and his colleagues (Veech *et al.*, 1980; King *et al.*, 1982) have shown that hepatic PP_i levels as high as ca. 5 mM may be realized in liver following the administration, i.p., of fatty acids to rats. Cohen *et al.* (1980) have demonstrated the availability of mitochondrially-generated carbamyl-*P* for extramitochondrial reaction. Lueck and Nordlie (1970) earlier had demonstrated glucose phosphorylation from carbamyl-*P* of mitochondrial origin via microsomal carbamyl-*P*:glucose phosphotransferase. The recent findings of Meher *et al.* (1981; see also Section VII), showing an apparent physical continuity between rat hepatic mitochondria and rough endoplasmic reticulum, would seem to strengthen the possibility of metabolic channeling of either carbamyl-*P* formed by mitochondrial carbamyl-*P* synthetase or PP_i formed in mitochondrial fatty acid activation to glucose phosphorylation via phosphotransferase activity of glucose-6-phosphatase of the endoplasmic reticulum. All of these considerations, we believe, should counter the arguments against such function for the phosphotransferase raised recently by Arion *et al.* (1980b). With sufficient levels of PP_i or carbamyl-*P* and glucose, the glucose-6-phosphatase system as depicted in Figure 1 should function as effectively in the direction of glucose-6-*P* formation as it would under other conditions in the direction of glucose-6-*P* hydrolysis. That inhibition by P_i, *per se,* does not constitute a prohibitive barrier has been shown by earlier calculations (Nordlie, 1974). Very recent observations by Katz, Teutsch, and Jungerman regarding zonal separation within the liver of gluconeogenesis from glycolysis (see Section VII) suggest the interesting possibility that glucose-6-*P* phosphohydrolase and phosphotransferase activities of this enzyme might, respectively, be the predominant function of multifunctional glucose-6-phosphatase in Zone 1 and Zone 3 of the liver.

Recent findings suggest interesting possibilities regarding the potential differential regulation of various activities of glucose-6-phosphatase. Earlier studies had indicated the potential for differential regulation of glucose-6-*P* phosphohydrolase and PP_i-glucose or carbamyl-*P*: glucose phosphotransferase of liver microsomes or nuclei by such factors, e.g., as phlorizin (Lygre and Nordlie, 1969), palmityl CoA (Nordlie *et al.*, 1967), Cu^{2+} (Johnson and Nordlie, 1977), polyamines (Nordlie *et al.*, 1979a), and by various hormonal manipulations (see Section VIII-C). The translocase concept described in detail in Section X provides an ideal focus for discriminant control of these activities. For example (see Figure 1), a modifier affecting translocase T_1 could modulate activities involving glucose-6-*P* discriminantly while causing no direct effect of activities with PP_i and carbamyl-*P*. In fact, just such an effect has been seen with the inhibitor DIDS by Zoccoli *et al.* (1982). In contrast, modulation of putative translocase T_2 could activate or inhibit PP_i- or carbamyl-*P*:glucose phosphotransferase without affecting either glucose-6-*P* hydrolysis (except indirectly through alteration in lumenal glucose-6-*P* concentration) or rate of efflux of glucose-6-*P*. Metabolic consequences of such actions would be very complex. Modulations through direct effects of inhibitors, covalent modifications of membrane transporter proteins, or membrane conformational alterations all could be involved as underlying mechanisms for control at this level, as at the catalytic unit, *per se*.

ACKNOWLEDGMENTS

Experimental work from our laboratory cited above was supported in part by Research Grant AM 07141 from the National Institutes of Arthritis, Metabolic, and Digestive Diseases, U.S. Public Health Service. The diligence of our graduate student and postdoctoral collaborators and Melissa Nordlie's invaluable assistance with the literature search are gratefully acknowledged.

REFERENCES

Al-Ali, S. Y., and Robinson, N., 1981, Ultrastructural demonstration of glucose-6-phosphatase in cerebral cortex, *Histochemistry* **72:**107–111.

Alvares, F. L., and Nordlie, R. C., 1977, Quantitative correlation of glucose uptake and phosphorylation with the activities of glucose-phosphorylating enzymes in perfused livers of fasted and fed rats, *J. Biol. Chem.* **252:**8404–8414.

Anchors, J. M., and Karnovsky, M. L., 1975, Purification of cerebral glucose-6-phosphatase—an enzyme involved in sleep, *J. Biol. Chem.* **250:**6408–6416.

Anchors, J. M., Haggerty, D. F., and Karnovsky, M. L., 1977, Cerebral glucose-6-phosphatase and the movement of 2-deoxy-D-glucose across cell membranes, *J. Biol. Chem.* **252:**7035–7041.

Anderson, J. P., and Wright, B. E., 1980, Kinetic models of glycogen metabolism in normal rat liver, Morris hepatoma 7787 and host liver, *Int. J. Biochem.* **12:**361–369.

Anderson, J. P., Rotenberg, S. A., Morris, H. P., and Wright, B. E., 1980, Glycogen metabolism in normal liver and Morris hepatoma 7787 in mealfed rats, *Int. J. Biochem.* **12:**371–378.

Antal-Ikov, A. L., 1979, Incidence of glucose-6-phosphatase in the cells of rat respiratory epithelium, *Folia Morphol.* **27:**67–71.

Arion, W. J., and Nordlie, R. C., 1964, Liver microsomal glucose-6-phosphatase, inorganic pyrophosphatase, and pyrophosphate-glucose phosphotransferase. II. Kinetic studies, *J. Biol. Chem.* **239:**2752–2757.

Arion, W. J., and Nordlie, R. C., 1967, Biological regulation of inorganic pyrophosphate-glucose phosphotransferase and glucose-6-phosphatase. Activation by triamcinolone, *in vivo,* in the presence of actinomycin D, *J. Biol. Chem.* **242:**2207–2211.

Arion, W. J., and Wallin, B. K., 1973, Kinetics of the glucose-6-phosphate-glucose exchange activity and glucose inhibition of glucose-6-phosphatase of intact and disrupted rat liver microsomes, *J. Biol. Chem.* **248:**2372–2379.

Arion, W. J., and Walls, H. E., 1982, The importance of membrane integrity in kinetic characteristics of the microsomal glucose-6-phosphatase system, *J. Biol. Chem.* **257:**11217–11220.

Arion, W. J., Carlson, P. W., Wallin, B. K., and Lange, A. J., 1972a, Modifications of hydrolytic and synthetic activities of liver microsomal glucose-6-phosphatase, *J. Biol. Chem.* **247:**2551–2557.

Arion, W. J., Wallin, B. K., Carlson, P. W., and Lange, A. J., 1972b, The specificity of glucose-6-phosphatase of intact liver microsomes, *J. Biol. Chem.* **247:**2558–2565.

Arion, W. J., Wallin, B. K., Lange, A. J., and Ballas, L. M., 1975, On the involvement of a glucose-6-phosphate transport system in the function of microsomal glucose-6-phosphatase, *Mol. Cell. Biochem.* **6:**75–83.

Arion, W. J., Ballas, L. M., Lange, A. J., and Wallin, B. K., 1976a, Microsomal membrane permeability and the hepatic glucose-6-phosphatase system—interactions of the system with D-mannose 6-phosphate and D-mannose, *J. Biol. Chem.* **251:**4901–4907.

Arion, W. J., Lange, A. J., and Ballas, L. M., 1976b, Quantitative aspects of relationship between glucose 6-phosphate transport and hydrolysis for liver microsomal glucose-6-phosphatase system—selective thermal inactivation of catalytic component *in situ* at acid pH, *J. Biol. Chem.* **251:**6784–6790.

Arion, W. J., Lange, A. J., and Walls, H. E., 1980a, Microsomal membrane integrity and the interactions of phlorizin with the glucose-6-phosphatase system, *J. Biol. Chem.* **255:**6–13.

Arion, W. J., Lange, A. J., Walls, H. E., and Ballas, L. M., 1980b, Evidence of the participation of independent translocation for phosphate and glucose-6-phosphate in the microsomal glucose-6-phosphatase system. Interactions of the system with orthophosphate, inorganic pyrophosphate, and carbamyl phosphate, *J. Biol. Chem.* **255:**10396–10406.

Arnao, I., Ore, R., and Villavicencio, M., 1981, Some properties of Alpaca-liver microsomal glucose-6-phosphatase, *Fed. Proc.* **40:**1776.

Ashmore, J., and Weber, G., 1959, The role of hepatic glucose-6-phosphatase in the regulation of carbohydrate metabolism, in: *Vitamins and Hormones,* Vol. 17 (R. S. Harris, G. F. Marrian, and K. V. Thimann, eds.), Academic Press, New York, pp. 91–132.

Ballas, L. M., and Arion, W. J., 1977, Measurement of glucose-6-phosphate penetration into liver microsomes. Confirmation of substrate transport in the glucose-6-phosphatase system, *J. Biol. Chem.* **252:**8512–8518.

Band, G. C., and Jones, C. T., 1980a, Activation by glucagon of glucose-6-phosphatase activity in the liver of the fetal guinea pig, *Biochem. Soc. Trans.* **8:**586–587.

Band, G. C., and Jones, C. T., 1980b, Functional activation by glucagon of glucose-6-phosphatase and gluconeogenesis of the perfused liver of the fetal guinea pig, *FEBS Lett.* **22:**190–194.

Beaufay, H., and de Duve, C., 1954, Le Systeme Hexose-phosphatasique. VI. Essais de demembrement des microsomes porteurs de glucose-6-phosphatase, *Bull. Ste. Chim. Biol.* **36:**1551–1568.

Begley, P. J., and Craft, J. A., 1981, Evidence for protein phosphorylation as a regulatory mechanism for hepatic microsomal glucose-6-phosphatase, *Biochem. Biophys. Res. Commun.* **103:**1029–1034.

Benedetti, A., Casini, A. F., Ferrali, M., and Comporti, M., 1977a, Inactivation of microsomal glucose-6-phosphatase coupled to the peroxidation of liver microsome lipids in compartmentalized systems, *Boll. Soc. Ital. Biol. Sper.* **53:**1385–1390.

Benedetti, A., Casini, A. F., Ferrali, M., and Comporti, M., 1977b, Studies on the relationships between carbon tetrachloride induced alterations of liver microsomal lipids and impairment of glucose-6-phosphatase activity, *Exp. Mol. Pathol.* **27:**309–323.

Benedetti, A., Casini, A. F., Ferrali, M., and Comporti, M., 1979, Extractions and partial characterization of dialysable products originating from the peroxidation of liver microsomal lipids and inhibiting microsomal glucose-6-phosphatase activity, *Biochem. Pharmacol.* **28:**2909–2918.

Benedetto, J. P., and Got, R., 1980, Effects of basic proteins of low molecular weight on the phosphohydrolase and phosphotransferase activities of microsomal glucose-6-phosphatase in adult monkey hepatocytes, *Biochim. Biophys. Acta* **614:**400–406.

Benedetto, J. P., Martel, M. B., and Got, R., 1979, Glucose-6-phosphatase microsomique d'hepatocytes de singe, *Biochimie* **61:**1125–1132.

Benner, U., Hacker, H. J., and Bannasch, P., 1979, Electron microscopical demonstration of glucose-6-phosphatase in native cryostat sections fixed with glutaraldehyde through semipermeable membranes, *Histochemistry* **65:**41–47.

Bergman, R. N., and El-Refai, M., 1975, Dynamic control of hepatic glucose metabolism: Studies by experiment and computer simulation, *Ann. Biomed. Eng.* **3:**411–432.

Bernaert, D., Penasse, W., and Wanson, J. C., 1979, Glucose-6-phosphatase activity in isolated and cultured hepatocytes from control and diethylnitrosamine-treated rats. A biochemical and cytochemical approach, *Arch. Int. Physiol.* **87:**777–778.

Bernard, P., Neveux, Y., Rocquet, G., and Drouet, J., 1980, Studies of microsomal glucose-6-phosphatase on liver of irradiated rats, *Enzyme* **25**:250–257.

Bertone, G., and Dianzani, M. U., 1977, Inhibition by aldehydes as a possible further mechanism for glucose-6-phosphatase inactivation during CCl_4-poisoning, *Chem. Biol. Interact.* **19**:91–100.

Bialek, D. S., Sharp, H. L., Kane, W. J., Elders, J., and Nordlie, R. C., 1977, Latency of glucose-6-phosphatase in type IB glycogen storage disease, *J. Pediat.* **91**:838.

Bickerstaff, G. F., and Burchell, B., 1980, Studies on the purification of glucose-6-phosphatase from rabbit liver microsomal fraction, *Biochem. Soc. Trans.* **8**:389–390.

Bontemps, F., Hue, L., and Hers, H.-G., 1978, Phosphorylation of glucose in isolated rat hepatocytes. Sigmoidal kinetics explained by the activity of glucokinase alone, *Biochem. J.* **174**:603–611.

Boral, R. C., Das, H., and Bose, S. K., 1981, Glucose-6-phosphatase in human amniotic fluid, *Indian J. Exp. Biol.* **19**:1181–1182.

Boxer, J., Kirby, L. T., and Hahn, P., 1974, The response of glucose-6-phosphatase in human and rat fetal liver cultures to dibutyryl cyclic AMP, *Proc. Soc. Exp. Med.* **145**:901–903.

Bruscalupi, G., Leoni, S., Spagnuolo, S., and Trentalance, A., 1979, Activity of glucose-6-phosphatase during liver regeneration, *Boll. Soc. Ital. Biol. Sper.* **55**:445–449.

Bulienko, S. D., Skvarko, S. I., and Fogel, P. I., 1980, RNA content and glucose-6-phosphatase activity in subcellular placental fractions in women with an uncomplicated pregnancy, *Akush. Ginekol. (Mosk.)* **12**:9–11.

Burchell, A., and Burchell, B., 1980, Stabilization of partially-purified glucose-6-phosphatase by fluoride: Is enzyme inactivation caused by dephosphorylation? *FEBS Lett.* **118**:180–184.

Burchell, A., and Burchell, B., 1982, Identification and purification of a liver microsomal glucose-6-phosphatase, *Biochem. J.* **205**:567–573.

Burchell, A., Burchell, B., Arion, W. J., and Walls, H. E., 1982, A critical evaluation of the possible modulation of hepatic microsomal glucose-6-phosphatase activity by protein phosphorylation, *Biochem. Biophys. Res. Commun.* **107**:1046–1052.

Burlakova, E. B., Gvakhariia, V. C., Glushchenko, N. N., and Dupin, A. M., 1979, Role of lipids in regulation of microsomal glucose-6-phosphatase activity after modification of microsomes *in vivo* and *in vitro*. Effect of phospholipid effectors on the activity of glucose-6-phosphatase, *Biokhimia* **44**:1111–1116.

Burlakova, E. B., Gvakhariia, V. C., Glushchenko, N. N., and Dupin, A., 1980, Role of microviscosity of the lipid phase of microsomal membranes in regulation of enzymatic activity of glucose-6-phosphatase under microsomal modifications *in vivo* and *in vitro*, *Biokhimia* **45**:387–391.

Byrne, W. L., 1961, Glucose-6-phosphatase and phosphoserine phosphatase, in: *The Enzymes*, 2nd Ed., Vol. 5 (P. D. Boyer, H. Lardy and K. Myrback, eds.), Academic Press, New York, pp. 73–78.

Calvert, R., Malka, D., and Menard, D., 1979, Developmental pattern of glucose-6-phosphatase activity in the small intestine of the mouse fetus, *Histochemistry* **63**:209–220.

Campbell, R. S., and Langslow, D. R., 1978, The effect of insulin administration *in vivo* on liver glucose-6-phosphatase in chick embryos, *Biochem. Soc. Trans.* **6**:149–152.

Carvo, M., Maddaiah, V. T., Collipp, P. J., and Chen, S. Y., 1974, Hepatic microsomal membrane: Activation of glucose-6-phosphatase, *FEBS Lett.* **39**:102–104.

Castilla, C., Paris, H., and Murat, J. C., 1978, Similar distribution of the activities of neutral alpha-glucosidase (gamma-amylase) and glucose-6-phosphatase in subcellular fractions from rat and trout livers, *C. R. Soc. Biol. (Paris)* **172**:968–971.

Cater, B. R., Trivedi, P., and Hallinan, T., 1975, Inhibition of glucose-6-phosphatase by pure and impure C-type phospholipases. Reactivation by phospholipid dispersions and protection by serum albumin, *Biochem. J.* **148**:279–294.

Chabot, J. G., Menard, D., and Hugon, J. S., 1978, Organ culture of adult mouse intestine. IV. Stimulation of glucose-6-phosphatase *in vitro*, *Histochemistry* **57**:33–45.

Chernysheva, M. D., and Fal, V.Ia., 1977, Change in the structure of glucose-6-phosphatase in Zajdela ascitic hepatoma cells, *Vestn. Akad. Med. Nauk. SSSR* **3**:30–32.

Chiquoine, A. D., 1953, The distribution of glucose-6-phosphatase in the liver and kidney of the mouse, *J. Histochem. Cytochem.* **1**:429–439.

Clark, D. G., Rognstad, R., and Katz, J., 1973, Isotopic evidence for futile cycles in liver cells, *Biochem. Biophys. Res. Commun.* **54:**1141–1148.

Cohen, N. S., Cheung, C.-W., and Raijman, L., 1980, The effects of ornithine on mitochondrial carbamyl phosphate synthesis, *J. Biol. Chem.* **255:**10248–10255.

Colilla, W., and Nordlie, R. C., 1973, Effects of sulfhydryl reagents on synthetic and hydrolytic activities of multifunctional glucose-6-phosphatase, *Biochim. Biophys. Acta* **309:**328–338.

Colilla, W., Jorgenson, R. A., and Nordlie, R. C., 1975, Mammalian carbamyl phosphate: Glucose phosphotransferase and glucose-6-phosphate phosphohydrolase: Extended tissue distribution, *Biochim. Biophys. Acta* **377:**117–125.

Crain, R. C., and Zilversmith, D. B., 1981, Lipid dependence of glucose-6-phosphate phosphohydrolase: A study with purified phospholipid transfer proteins and phosphatidylinositol-specific phospholipase C, *Biochemistry* **20:**5320–5326.

Crowe, C. A., Madsen, N. P., Tange, J. D., and Calder, I. C., 1977, Loss of kidney microsomal glucose-6-phosphatase activity following acute administration of P-aminophenol, *Biochem. Pharmacol.* **26:**2069–2071.

Dabrowiecki, Z., 1979, Kinetics and topology of rat liver microsomal enzymes synthesizing glycerophospholipids, *Fed. Proc.* **38:**471.

Dasgupta, A., Chatterjee, R., and Chowdhury, J. R., 1979, Nucleolar succinate dehydrogenase and glucose-6-phosphatase activities during hepatocellular carcinogenesis in thioacetamide fed rat, *Indian J. Exp. Biol.* **17:**1380–1381.

De Duve, C., 1962, Endoplasmic reticulum, in: *CIBA Foundation Symposia on Enzymes and Drug Action* (J. L. Mongar and A. V. S. de Reuck, eds.), Little, Brown and Company, Boston, pp. 505–506.

DeLaval, E., Moreau, E., and Geloso, J. P., 1981, Hormonal control of glucose-6-phosphatase and phosphoenolpyruvate carboxykinase activities in the fetal rat kidney, *Pediat. Res.* **15:**138–142.

Dobrosielski-Vergona, K., and Widnell, C. C., 1982, Age-related differences in the response of hepatic microsomal glucose-6-phosphatase to triiodothyronine and triamcinolone in the rat, *Endocrinology* **111:**953–958.

Duck-Chong, C. G., 1976, A reassessment of the phospholipid dependence of membrane-bound enzymes with special reference to glucose-6-phosphatase and (Na^+-K^+)-dependent adenosine triphosphatase, *Enzyme* **121:**174–192.

Dyatlovitskaya, E. V., Lemenovskaya, A. F., and Bergelson, L. D., 1979a, Study of lipid dependence of glucose-6-phosphatase from rat liver and hepatoma using lipid exchange proteins, *Biokhimia* **44:**498–503.

Dyatlovitskaya, E. V., Lemenovskaya, A. F., and Bergelson, L. D., 1979b, Use of protein-mediated lipid exchange in the study of membrane-bound enzymes. The lipid dependence of glucose-6-phosphatase, *Eur. J. Biochem.* **99:**605–612.

Elitze, M., Jung, A., and Jackisch, R., 1977, Cytoplasmic changes in level and distribution of glucose-6-phosphatase activities from rat liver during diethylnitrosamine-induced carcinogenesis, *Chem. Biol. Interact.* **18:**295–308.

El-Refai, M., and Bergman, R. N., 1976, Simulation study of control of hepatic glycogen synthesis by glucose and insulin, *Am. J. Physiol.* **231:**1608–1619.

Feldman, F., and Butler, L. G., 1969, Detection and characterization of the phosphorylated form of microsomal glucose-6-phosphatase, *Biochem. Biophys. Res. Commun.* **36:**119–125.

Feldman, F., and Butler, L. G., 1972, Protein-bound phosphoryl histidine: A probable intermediate in the microsomal glucose-6-phosphate/inorganic pyrophosphatase reaction, *Biochim. Biophys. Acta* **268:**698–710.

Gankema, H. S., Laanen, E., Groen, A. K., and Tager, J. M., 1981, Characterization of isolated rat-liver cells made permeable with filipin, *Eur. J. Biochem.* **119:**409–414.

Garfield, S. A., and Cardell, R. R., Jr., 1979, Hepatic glucose-6-phosphatase activities and correlated ultrastructural alterations in hepatocytes of diabetic rats, *Diabetes* **28:**664–679.

Garland, R. C., and Cori, C. F., 1972, Separation of phospholipids from glucose-6-phosphatase by gel chromatography. Specificity of phospholipid reactivation, *Biochemistry* **11:**4712–4718.

Gierow, P., and Jergil, B., 1980, A spectrophotometric method for the determination of glucose-6-phosphatase activity, *Anal. Biochem.* **101:**305–309.

Glende, E. A., Jr., Hruszkewycz, A. M., and Recknagel, R. U., 1976, Critical role of lipid peroxidation in carbon tetrachloride-induced loss of aminopyrine demethylase, cytochrome P-450, and glucose-6-phosphatase, *Biochem. Pharmacol.* **25:**2163–2170.

Gold, G., and Widnell, C. C., 1974, Reversal of age-related changes in microsomal enzyme activities following the administration of triamcinolone, triiodothyronine and phenobarbital, *Biochim. Biophys. Acta* **334:**75–85.

Gold, G., and Widnell, C. C., 1976, Relationship between microsomal membrane permeability and the inhibition of hepatic glucose-6-phosphatase by pyridoxal phosphate, *J. Biol. Chem.* **251:**1035–1041.

Golden, S., Riesenfeld, G., and Katz, J., 1982, Carbohydrate metabolism of hepatocytes from starved Japanese Quail, *Arch. Biochem. Biophys.* **213:**118–126.

Goldsmith, P. K., and Stetten, M. R., 1979, Different developmental changes in latency for two functions of a single membrane bound enzyme—glucose-6-phosphatase activities as a function of age, *Biochim. Biophys. Acta* **583:**133–147.

Goldstein, G. W., Wolinsky, J. S., Csejtey, J., and Diamond, I., 1975, Isolation of metabolically active capillaries from rat brain, *J. Neurochem.* **25:**715–717.

Goldvenko, N.Ia., and Orliuk, E. I., 1979, Effect of 1,4-benzodiazepine tranquilizers on the activity of the hepatocyte hydroxylating complex and glucose-6-phosphatase in white rats, *Vopr. Med. Khim.* **25:**28–31.

Gratzl, M., 1975, Membrane effects on hepatic microsomal glucose-6-phosphatase, *Hoppe-Seyler's Z. Physiol. Chem.* **356:**861–865.

Greengard, O., 1969, The hormonal regulation of enzymes in prenatal and postnatal rat liver. Effects of adenosine-3′,5′-(cyclic)-monophosphate, *Biochem. J.* **115:**19–24.

Greengard, O., and Dewey, H. K., 1967, Initiation by glucagon of the premature development of tyrosine aminotransferase, serine dehydratase, and glucose-6-phosphatase in fetal rat liver, *J. Biol. Chem.* **242:**2986–2991.

Grinna, L. S., 1975, Multiple thermal discontinuities in glucose-6-phosphatase activity, *Biochim. Biophys. Acta* **403:**388–392.

Grinna, L. S., 1977, Age-related alterations in membrane lipid and protein interactions: Arrhenius studies of microsomal glucose-6-phosphatase, *Gerontology* **23:**342–349.

Gunderson, H. M., and Nordlie, R. C., 1973, The fully-active nature of synthetic and hydrolytic activities of glucose-6-phosphatase of intact nuclear membrane, *Biochem. Biophys. Res. Commun.* **52:**601–607.

Gunderson, H. M., and Nordlie, R. C., 1975, Carbamyl-phosphate: Glucose phosphotransferase and glucose-6-phosphate phosphohydrolase of nuclear membrane. Interrelationships between membrane integrity, enzymic latency, and catalytic behavior, *J. Biol. Chem.* **250:**3552–3559.

Hass, L. F., and Byrne, W. L., 1960, The Mechanism of glucose-6-phosphatase, *J. Am. Chem. Soc.* **82:**947–954.

Hefferan, P. M., and Howell, R. R., 1977, Genetic evidence for the common identity of glucose-6-phosphatase, pyrophosphate-glucose phosphotransferase, carbamyl phosphate-glucose phosphotransferase and inorganic pyrophosphatase, *Biochim. Biophys. Acta* **496:**431–435.

Herrman, J. L., and Nordlie, R. C., 1972, The catalytic potential of carbamyl-phosphate: Glucose phosphotransferase: A species distribution study, *Arch. Biochem. Biophys.* **152:**180–186.

Hers, H.-G., 1976, The control of glycogen metabolism in the liver, *Annu. Rev. Biochem.* **45:**167–190.

Hers, H.-G., 1982, Non-hormonal control of glycogen synthesis, in: *Short-Term Regulation of Liver Metabolism* (L. Hue and G. Van de Werve, eds.), Elsevier/North-Holland, Amsterdam, pp. 105–117.

Hilderson, H. J., Lagrou, A., and Dierick, W., 1979a, Hydrolysis of glucose-6-phosphate in bovine thyroid tissue, *Arch. Int. Physiol. Biochim.* **87:**185–186.

Hilderson, H. J., Voets, R., De Wolf, J. J., Lagrous, A., Van Dessel, G. A., and Dierick, W. S., 1979b, Occurrence and subcellular localization of glucose-6-phosphatase in bovine thyroid, *Arch. Int. Physiol. Biochem.* **87:**729–740.

Hori, S. H., and Takahashi, T., 1977, Latency of microsomal hexose-6-phosphate dehydrogenase activity, *Biochim. Biophys. Acta* **496:**1–11.

Hruszkewycz, A. M., Glende, E. A., Jr., and Recknagel, R. O., 1978, Destruction of microsomal cytochrome P-450 and glucose-6-phosphatase by lipids extracted from peroxidized microsomes, *Toxicol. Appl. Pharmacol.* **46:**695–702.

Huang, M.-T., and Veech, R. L., 1982, The quantitative determination of the *in vivo* dephosphorylation of glucose-6-phosphate in rat brain, *J. Biol. Chem.* **257:**11358–11363.

Hue, L., 1982, Futile cycles and regulation of metabolism, in: *Metabolic Compartmentation* (H. Sies, ed.), Academic Press, New York, pp. 71–97.

Hue, L., and Hers, H.-G., 1974, Utile and futile cycles in the liver, *Biochem. Biophys. Res. Commun.* **58:**540–548.

Igarashi, Y., Otomo, H., Narisawa, K., and Tada, K., 1980, A new variant of glycogen storage disease type 1: Probably due to a defect in the glucose-6-phosphate transport system, *J. Inherit. Metabol. Dis.* **2:**45–49.

Issekutz, B., Jr., 1977, Studies on hepatic glucose cycles in normal and methylprednisolone-treated dogs, *Metabolism* **26:**157–170.

Jacquot, R. L., and Felix, J. M., 1977, Multihormonal control of glucose-6-phosphatase activity in rat fetal liver, in: *Hormonal Receptors in Digestive Tract Physiology* (S. Bon-fils, ed.), North-Holland, Amsterdam, p. 198.

Johnson, W. T., and Nordlie, R. C., 1977, Differential effects of Cu^{2+} on carbamoyl phosphate: Glucose phosphotransferase and glucose-6-phosphate phosphohydrolase activities of multifunctional glucose-6-phosphatase, *Biochemistry* **16:**2458–2466.

Johnson, W. T., and Nordlie, R. C., 1980, Stimulation of glucose-6-phosphatase by polyamines is a membrane-mediated event, *Life Sci.* **26:**297–302.

Jorgenson, R. A., and Nordlie, R. C., 1980, Multifunctional glucose-6-phosphatase studied in permeable isolated hepatocytes, *J. Biol. Chem.* **255:**5907–5915.

Jungermann, K., Heilbronn, R., Katz, N., and Sasse, D., 1982, The glucose/glucose-6-phosphate cycle in the periportal and perivenous zone of rat liver, *Eur. J. Biochem.* **123:**429–436.

Kamoun, P. P., 1980, Is type 1B glycogenosis related to an anomeric preference for glucose-6-phosphate uptake by hepatic microsomes? *Med. Hypoth.* **6:**1135–1139.

Kanai, K., Asada-Kubota, M., and Kanamura, S., 1981, Ultrastructural localization of glucose-6-phosphatase activity in the cells of the epididymis of the mouse, *Experientia* **37:**509–511.

Kanamura, S., Kanai, K., and Asada-Kubota, M., 1981, Demonstration of injured hepatocytes after carbon tetrachloride administration by loss of histochemical glucose-6-phosphatase reaction, *Exp. Pathol.* **20:**68–70.

Kang, Y. H., and West, W. L., 1982, Ultrastructural localization of glucose-6-phosphatase and alkaline phosphatase in the vaginal epithelium of rat, *J. Morphol.* **171:**1–10.

Karnik, A. B., Thakore, K. N., Nigam, S. K., Babu, K. A., Lakkad, B. C., Bhatt, D. K., Kashyap, S. K., and Chatterjee, S. K., 1981, Studies on glucose-6-phosphatase, fructose-1,6-diphosphatase activity, glycogen distribution and endoplasmic reticulum changes during hexachlorocyclohexane induced hepatocarcinogenesis in pure inbred Swiss mice, *Neoplasma* **28:**575–584.

Karnovsky, M. L., Burrows, B. L., and Zoccoli, M. A., 1980, Cerebral glucose-6-phosphatase and the movement of 2-deoxyglucose during slow wave sleep, in: *Cerebral Metabolism and Neural Function* (J. V. Passonneau, R. A. Hawkins, W. D. Lust, and F. A. Welsh, eds.), Williams and Wilkins, Baltimore, pp. 359–366.

Katz, J., and Rognstad, R., 1976, Futile cycles in the metabolism of glucose, in: *Current Topics in Cellular Regulation,* Vol. 10 (B. L. Horecker and E. R. Stadtman, eds.), Academic Press, New York, pp. 237–289.

Katz, N., Teutsch, H. F., Sasse, D., and Jungermann, K., 1977, Heterogeneous distribution of glucose-6-phosphatase in microdissected periportal and perivenous rat liver tissue, *FEBS Lett.* **76:**226–230.

Katz, J., Wals, P. A., and Rognstad, R., 1978, Glucose phosphorylation, glucose-6-phosphatase, and recycling in rat hepatocytes, *J. Biol. Chem.* **253:**4530–4536.

Katz, J., Golden, S., and Wals, P. A., 1979, Glycogen synthesis by rat hepatocytes, *Biochem. J.* **180:**389–402.

Kelleher, J. K., Kelly, P. J., and Wright B. E., 1978, A kinetic analysis of glucokinase and glucose-6-phosphatase in *dictyostelium, Mol. Cell. Biochem.* **19:**67–73.

King, M. T., Uyeda, K., and Veech, R. L., 1982, Effects of short chain fatty acids on liver content of pyrophosphate, calcium and intermediary metabolites, *Fed. Proc.* **41:**1439.

Kitagawa, T., and Pitot, H. C., 1975, The regulation of serine dehydratase and glucose-6-phosphatase in hyperplastic nodules of rat liver during diethylnitrosamine and *N*-2-fluorenylacetamide feeding, *Cancer Res.* **35:**1075–1084.

Kitcher, S. A., Siddle, K., and Luzio, J. P., 1978, A method for the determination of glucose-6-phosphatase activity in rat liver with (U-^{14}C) glucose-6-phosphate as substrate, *Anal. Biochem.* **88:**29–36.

Kuzminskaia, U. A., Bersan, L. V., and Pismennaia, M. V., 1977, Activity of glucose-6-phosphate metabolism enzymes in the livers of rats with experimental valekson poisoning, *Bull. Eksp. Biol. Med.* **84:**677–679.

Lam, K. S., and Kasper, C. B., 1980, Pyrophosphate: Protein phosphotransferase: A membrane-bound enzyme of endoplasmic reticulum, *Proc. Natl. Acad. Sci. USA* **77:**1927–1931.

Lange, A. J., Arion, W. J., and Beaudet, A. L., 1980, Type 1B glycogen storage disease is caused by a defect in the glucose-6-phosphate translocase of the microsomal glucose-6-phosphatase system, *J. Biol. Chem.* **255:**8381–8384.

Lawson, J. W. R., and Veech, R. L., 1979, Effects of pH and free Mg^{2+} on the K_{eq} of the creatine kinase reaction and other phosphate hydrolyses and phosphate transfer reactions, *J. Biol. Chem.* **254:**6528–6537.

Lawson, J. W. R., Guynn, R. W., Cornell, N., and Veech, R. W., 1976, A possible role for pyrophosphate in the control of hepatic glucose metabolism, in: *Gluconeogenesis: Its Regulation in Mammalian Species* (R. W. Hanson and M. A. Mehlman, eds.), John Wiley and Sons, New York, pp. 481–512.

Llamas, R., 1977, Comparative effects of L-3,3′5-triiodothyronine on the activity of glucose-6-phosphatase in the liver and the adipose tissue of the epididymus of the rat. Changes in the body and the epididymal fat originated by the hormone, *Gac. Med. Mex.* **113:**533–537.

Lowe, G., and Potter, B. V., 1982, The stereochemical course of phosphoryl transfer catalyzed by glucose-6-phosphatase, *Biochem. J.* **201:**665–668.

Lueck, J. D., and Nordlie, R. C., 1970, Carbamyl phosphate: Glucose phosphotransferase activity of hepatic microsomal glucose-6-phosphatase at physiological pH, *Biochem. Biophys. Res. Commun.* **39:**190–196.

Lueck, J. D., Herrman, J. L., and Nordlie, R. C., 1972, The general kinetic mechanism of microsomal carbamyl phosphate: Glucose phosphotransferase, glucose-6-phosphatase, and other associated activities, *Biochemistry* **11:**2792–2799.

Lygre, D. G., and Nordlie, R. C., 1969, Rabbit intestinal glucose-6-phosphate phosphohydrolase and inorganic pyrophosphate-glucose phosphotransferase: Inhibition by phlorizin, *Biochim. Biophys. Acta* **185:**360–366.

Maddaiah, V. T., Stemmer, C. L., Clejan, S., and Collipp, P. J., 1981, Alloxan-diabetic rats. Thermotropic effects on kinetics and interactions with deoxycholate and 1-anilino-8-naphthalene sulfonate, *Biochim. Biophys. Acta* **657:**106–121.

Masuda, Y., 1981, Carbon tetrachloride-induced loss of microsomal glucose-6-phosphatase and cytochrome P-450 *in vitro, Jpn. J. Pharmacol.* **31:**107–116.

Matalon, R., Michals, K., Justice, P., and Deanching, M. N., 1977, Glucose-6-phosphatase activity in human placenta: Possible detection of heterozygote for glycogen-storage disease Type I, *Lancet* **1:**1360–1361.

McCraw, E. F., 1968, The effect of fasting on glucose utilization in the isolated perfused rat liver, *Metabolism* **17:**833–837.

McCraw, E. F., Peterson, M. J., and Ashmore, J., 1967, Autoregulation of glucose metabolism in the isolated perfused rat liver, *Proc. Soc. Exp. Biol. Med.* **126:**232–235.

Meher, P. J., Spycher, M. A., and Meyer, U. A., 1981, Isolation and characterization of rough endoplasmic reticulum associated with mitochondria from normal rat liver, *Biochim. Biophys. Acta* **646:**283–297.

Meisel, M., Amon, I., Amon, K., and Huller, H., 1982, Investigation of glucose-6-phosphatase in the liver of the human fetus, *Int. J. Biol. Res. Pregnancy* **3:**73–76.

Meissner, G., and Allen, R., 1981, Evidence for two types of rat liver microsomes with differing permeability to glucose and other small molecules, *J. Biol. Chem.* **256:**6413–6422.

Menard, D., 1980, Ultrastructural localization of intestinal glucose-6-phosphatase activity during the postnatal development of the mouse, *Histochemistry* **67:**53–64.

Menard, D., and Malo, C., 1982, Lack of effect of cortisone, thyroxine and insulin on the developmental pattern of mouse intestinal glucose-6-phosphatase activity, *Experientia* **38:**111–112.

Mizushima, Y., 1977, Glucose-6-phosphatase, *Tanpakushitsu Kakusan Koso* **22:**1524–1529.

Mizushima, Y., and Ishikawa, M., 1979, The role of glucocorticoids in the regulation of glucose-6-phosphatase activity in fetal rat liver, *Biochem. Pharmacol.* **28:**2279–2283.

Moller, P. C., Yokoyama, M., and Chang, J. P., 1977, Ultracytochemical localization of glucose-6-phosphatase in Chang rat hepatoma *in vivo* and *in vitro, J. Natl. Cancer Inst.* **58:**1401–1405.

Murat, J. C., Plisetskaya, E. M., and Soltitskaya, L. P., 1979, Glucose-6-phosphatase activity in kidney of the river lamprey *(lampetra fluviatilis, L.), Gen Comp. Endocrinol.* **39:**115–117.

Narbonne, J. F., Bourdichon, M., and Daubeze, M., 1979, Effect of chronic phenoclor DP6 ingestion of some liver enzymatic activities in rats, *Toxicol. Eur. Res.* **2:**103–110.

Narisawa, K., Igarashi, Y., Otomo, H., and Tada, K., 1978, A new variant of glycogen storage disease type I probably due to system defect in the glucose-6-phosphate transporter, *Biochem. Biophys. Res. Commun.* **83:**1360–1364.

Nath, N., 1980, Time-course of changes of liver tryptophan pyrrolase (tryptophan oxygenase) and liver and kidney glucose-6-phosphatase in rats shifted from high to zero-protein diets, *J. Nutr. Sci. Vitaminol.* **26:**261–269.

Nemet, S. H., and Tigranian, R. A., 1982, Activity of various liver enzymes in rats following a flight aboard the Cosmos-936 biosatellite, *Kosm. Biol. Aviakosm. Med.* **16:**77–80.

Nichols, B. A., and Setzer, P. Y., 1981, Cytochemical localization of glucose-6-phosphatase in rabbit mononuclear phagocytes during differentiation, *J. Histochem. Cytochem.* **29:**317–320.

Nilsson, O. S., Arion, W. J., DePierre, J. W., Dallner, G., and Ernster, L., 1978, Evidence for the involvement of a glucose-6-phosphate carrier in microsomal glucose-6-phosphatase activity, *Eur. J. Biochem.* **82:**627–634.

Nordlie, R. C., 1971, Glucose-6-phosphatase, hydrolytic and synthetic activities, in: *The Enzymes,* 2nd Ed., Vol. 4 (P. D. Boyer, ed.), Academic Press, New York, pp. 543–609.

Nordlie, R. C., 1974, Metabolic Regulation of multifunctional glucose-6-phosphatase, in: *Current Topics in Cellular Regulation,* Vol. 8 (B. L. Horecker and E. R. Stadtman, eds.), Academic Press, New York, pp. 33–117.

Nordlie, R. C., 1976, Multifunctional hepatic glucose-6-phosphatase and the "tuning" of blood glucose levels, *Trends Biochem. Sci.* **1:**199–202.

Nordlie, R. C., 1979, Multifunctional glucose-6-phosphatase: Cellular biology, *Life Sci.* **24:**2397–2404.

Nordlie, R. C., 1981, Multifunctional glucose-6-phosphatase characteristics and functions, in: *Regulation of Carbohydrate Formation and Utilization in Mammals* (C. M. Veneziale, ed.), University Park Press, New York, pp. 291–314.

Nordlie, R. C., 1982a, Multifunctional glucose-6-phosphatase of endoplasmic reticulum and nuclear membrane, in: *Membranes and Transport,* Vol. 1 (A. V. Martonosi, ed.), Plenum Press, New York, pp. 263–286.

Nordlie, R. C., 1982b, Kinetic examination of enzyme mechanisms involving branched reaction pathways. Including a detailed consideration of multifunctional glucose-6-phosphatase, in: *Methods in Enzymology,* Vol. 87 (D. L. Purich, ed.), Academic Press, New York, pp. 319–353.

Nordlie, R. C., and Arion, W. J., 1966, Glucose-6-phosphatase, in: *Methods in Enzymology,* Vol. 9 (W. A. Woods, ed.), Academic Press, New York, pp. 619–625.

Nordlie, R. C., and Jorgenson, R. A., 1976, Glucose-6-phosphatase, in: *The Enzymes of Biological Membranes,* Vol. 2 (A. Martonosi, ed.), Plenum Press, New York, pp. 465–491.

Nordlie, R. C., and Jorgenson, R. A., 1981, Latency and inhibitability by metabolites of glucose-6-phosphatase of permeable hepatocytes from fasted and fed rats, *J. Biol. Chem.* **256:**4768–4771.

Nordlie, R. C., and Sukalski, K. A., 1982, The regulation of metabolic flux through the glucose ⇄ glucose-6-phosphate cycle in liver, in: *Developments in Biochemistry—The Biochemistry of Metabolic Processes* (F. W. Stratman, D. L. F. Lennon, and R. N. Zahlten, eds.), Elsevier Biomedical, New York, pp. 125–138.

Nordlie, R. C., Hanson, T. L., and Johns, P. T., 1967, Differential effects of palmityl coenzyme A on liver microsomal inorganic pyrophosphate-glucose phosphotransferase and glucose-6-phosphate phosphohydrolase, *J. Biol. Chem.* **242:**4144–4148.

Nordlie, R. C., Hanson, T. L., Johns, P. T., and Lygre, D. G., 1968, Inhibition by nucleotides of liver microsomal glucose-6-phosphatase, *Proc. Natl. Acad. Sci. USA* **60:**590–597.

Nordlie, R. C., Johnson, W. T., Cornatzer, W. E., Jr., and Twedell, G. W., 1979a, Stimulation by polyamines of carbamyl-*P*:glucose phosphotransferase and glucose-6-phosphate phosphohydrolase activities of multifunctional glucose-6-phosphatase, *Biochim. Biophys. Acta* **585:**12–23.

Nordlie, R. C., Meeks, F. A., and Stepanik, P. L., 1979b, Responses of nuclear glucose-6-phosphatase to diabetes and to hydrocortisone administered to normal and diabetic rats differ from those of the microsomal enzyme, *Biochim. Biophys. Acta* **586:**433–441.

Nordlie, R. C., Sukalski, K. A., and Alvares, F. L., 1980, Responses of glucose-6-phosphate levels to varied glucose loads in the isolated perfused rat liver, *J. Biol. Chem.* **255:**1834–1838.

Nordlie, R. C., Stepanik, P. L., Johnson, W. T., and Jorgenson, R. A., 1981, Differential effects of Cr^{3+}, Cu^{2+} and VO_4^{3-} on glucose-6-phosphatase and glucokinase, *Fed. Proc.* **40:**1674.

Nordlie, R. C., Sukalski, K. A., and Singh, J., 1982a, Identity of enzymes involved in glucose $\rightleftarrows$ glucose-6-phosphate cycling in liver, *Fed. Proc.* **41:**509.

Nordlie, R. C., Sukalski, K. A., Singh, J., and Alvares, F. L., 1982b, Mechanisms of hepatic glucose phosphorylation, Abst. 12th Int. Congr. Biochem., p. 121.

Nordlie, R. C., Alvares, F. L., and Sukalski, K. A., 1982, Stimulation by 3-mercaptopicolinate of net glucose uptake by perfused livers from diabetic rats, *Biochim. Biophys. Acta* **719:**244–250.

Nordlie, R. C., Sukalski, K. A., Muñoz, J. M., and Baldwin, J. J., 1983b, Type 1c, a novel glycogenosis. Underlying mechanism, *J. Biol. Chem.* **258:**9739–9744.

Ogorodnikova, L. G., 1981a, Glucose-6-phosphatase activity in tissues and its role in representatives of different classes of vertebrates, *Zh. Evol. Biokhim. Fiziol.* **17:**199–204.

Ogorodnikova, L. G., 1981b, Glycogen content and glucose-6-phosphatase activity in the ligated frog ischiatic nerve. *Zh. Evol. Biokhim. Fiziol.* **17:**531–534.

Ogorodnikova, L. G., and Fomina, E. B., 1977, Glucose-6-phosphatase activity of the pia mater of several representative vertebrates, *Zh. Evol. Biokhim. Fiziol.* **13:**340–343.

Ogorodnikova, L. G., and Legedinskaia, I. I., 1980, Glycogen content and activities of phosphorylase and glucose-6-phosphatase in the fast and slow muscles of representatives of different classes of vertebrates, *Zh. Evol. Biokhim. Fiziol.* **16:**125–132.

Oka, Y., Mitsuyama, T., Nagai, B., Arashima, S., Ohkubo, I., and Matsuda, I., 1978, Glucose-6-phosphatase activity in liver and blood platelets of two patients with glycogen storage disease type I, *Clin. Chim. Acta* **87:**319–326.

O'Neill, I. E., and Langslow, D. R., 1978, The action of hydrocortisone, insulin, and glucagon on chicken liver hexokinase and glucose-6-phosphatase and on the plasma glucose and free fatty acid concentrations, *Gen. Comp. Endocrinol.* **34:**428–437.

Paradist, L., Negro, F., Panagini, C., and Torrielli, M. V., 1979, Interference of antioxidants E/O of some free radical "scavengers": With the activity of glucose-6-phosphatase after administration of carbon tetrachloride, *Boll. Soc. Ital. Biol. Sper.* **55:**1877–1883.

Paul, A. K., and Dhar, A., 1980, Effect or thyroxine on the hepatic glycogen and glucose-6-phosphatase activity of developing rats, *Horm. Metabol. Res.* **12:**261–264.

Pollak, J. K., Malor, R., Morton, M., and Ward, K. A., 1971, The reconstitution of microsomal membranes, in: *Autonomy and Biogenesis of Mitochondria and Chloroplasts* (N. K. Boardman, A. W. Linnane, and R. M. Smillie, eds.), North-Holland, Amsterdam, pp. 27–41.

Porterfield, S. P., 1979, The effects of growth hormone, thyroxine and insulin on dinucleotide phosphate dehydrogenase, glucose-6-phosphatase and glycogen phosphorylase in fetal rat liver, *Horm. Metabol. Res.* **11:**444–448.

Portha, B., Le Provost, E., Picon, L., and Rosselin, G., 1978, Postmaturity in the rat: Phosphorylase, glucose-6-phosphatase and phosphoenolpyruvate carboxykinase activities in the fetal liver, *Horm. Metabol. Res.* **10:**141–144.

Reczek, P. R., and Villee, C. A., Jr., 1982, A purification of microsomal glucose-6-phosphatase from human tissue, *Biochem. Biophys. Res. Commun.* **107:**1158–1165.

Riesenfeld, G., Wals, P. A., Golden, S., and Katz, J., 1981, Glucose, amino acids, and lipogenesis in hepatocytes of Japanese Quail, *J. Biol. Chem.* **256:**9973–9980.

Rognstad, R., Clark, D. G., and Katz, J., 1973, *Biochem. Biophys. Res. Commun.* **54:**1149–1156.

Rossowska, M., and Dabrowiecki, Z., 1978, Effect of hypoxia and ischemia on the activity of glucose-6-phosphatase in the guinea-pig brain, *J. Neurochem.* **30:**1203–1204.

Sann, L., Mathieu, M., Bourgeois, J., Bienvenu, J., and Bethenod, M., 1980, *In vivo* evidence for defective activity of glucose-6-phosphatase in type 1B glycogenosis, *J. Pediat.* **96:**691–694.

Santiago, E., Lopez-Moratalla, N., and Lopez-Zabalza, M. J., 1977, Glucose-6-phosphatase inactivation and peroxidation of phosphatidylcholine in microsomal membranes, *Rev. Exp. Fisiol.* **33:**191–196.

Sastry, K. V., and Sharma, S. K., 1978, The effect of *in vivo* exposure of endrin on the activities of acid, alkaline and glucose-6-phosphatase in liver and kidney of *Ophiocephalus (Channa) punctatus, Bull. Environ. Contam. Toxicol.* **20:**456–460.

Schmucker, D. L., and Wang, R. K., 1980, Effects of animal age and phenobarbital on rat liver glucose-6-phosphatase activity, *Exp. Gerontol.* **15:**7–13.

Schmucker, D. L., Mooney, J. S., and Jones, A. L., 1977, Age-related changes in the hepatic endoplasmic reticulum: A quantitative analysis, *Science* **197:**1005–1007.

Schulze, J. U., and Speth, M., 1980, Investigations on the possible involvement of phospholipid in the glucose-6-phosphate transport system of rat-liver microsomal glucose-6-phosphatase, *Eur. J. Biochem.* **106:**505–514.

Schwartz, A. L., Raiha, N. C., and Rall, T. W., 1974, Effect of dibutyryl cyclic AMP on glucose-6-phosphatase activity in human fetal liver explants, *Biochim. Biophys. Acta* **343:**500–509.

Schwarz, G., 1978, Quantitative investigations of the zonal distribution of SDH, G6Pase and malic enzyme activity in liver parenchyma, *Acta Histochem.* **62:**133–141.

Segal, H. L., 1959, Some consequences of the non-competitive inhibition by glucose of rat liver glucose-6-phosphatase, *J. Am. Chem. Soc.* **81:**4047–4050.

Segal, H. L., and Washko, M. E., 1959, Studies of liver glucose-6-phosphatase. III. Solubilization and properties of the enzyme from normal and diabetic rats, *J. Biol. Chem.* **234:**1937–1941.

Sein, K. T., and Chu, N., 1979, Liver and kidney glucose-6-phosphatase levels in carbon tetrachloride and DDT-administered mice, *Enzyme* **24:**72–74.

Sein, K. T., and Maw, T. T., 1978, The effects of fasting on glucose-6-phosphatase of mouse liver and kidney, *Enzyme* **23:**70–72.

Selmeci-Antal, M., and Blaskovits, A., 1977, The effects of sodium dodecylbenzene sulphonate and various diets on rat liver glucose-6-phosphatase and glucose-6-phosphate dehydrogenase activities, *Nutr. Metabol.* **21:**244–246.

Seni, R., and Gandhi, B. S., 1981, Hormonal effects of glucose-6-phosphatase in the liver and kidney of rats of different ages, *Indian J. Biochem. Biophys.* **18:**291–294.

Shikama, H., and Ui, M., 1978, Glucose load diverts hepatic gluconeogenic product from glucose to glycogen *in vivo, Am. J. Physiol.* **235:**E354–E360.

Shin, T. S., Chung, I. H., and Kim, S. S., 1977, Prednisolone and glucose-6-phosphatase activity in liver cells, *Yonsei. Med. J.* **18:**9–18.

Shin, T. S., Chung, I. H., and Kim, S. S., 1978, Electron microscopy on activity and localization of glucose-6-phosphatase in liver cells, *Yonsei, Med. J.* **19:**1–10.

Siegel, S. R., Oh, W., and Fisher, D. A., 1979, Fructose 1,6-diphosphatase and glucose-6-phosphatase in newborn rats with intrauterine growth retardation, *Early Hum. Dev.* **3:**43–49.

Singh, J., and Nordlie, R. C., 1983, The progressive effects of fasting on glucose phosphorylation by isolated rat hepatocytes: The involvement of a high-$K_{0.5}$ enzyme, *FEBS Lett.* **150:**325–328.

Singh, J., Nordlie, R. C., and Jorgenson, R. A., 1981, Vanadate: A potent inhibitor of multifunctional glucose-6-phosphatase, *Biochim. Biophys. Acta* **678:**477–482.

Singh, J., Martin, R. E., and Nordlie, R. C., 1983, Glucose-6-phosphatase: Is activity regulated by phosphorylation/dephosphorylation? *Can. J. Biochem. Cell Biol.* **61:**1085–1089.

Snoke, R. E., and Nordlie, R. C., 1972, Comparative studies of the responses of rat liver microsomal glucose-6-phosphatase and inorganic pyrophosphate-glucose phosphotransferase to phospholipase C treatment and phospholipid supplementation, *Biochim. Biophys. Acta* **258:**188–205.

Spagnoli, D., Dobrosielski-Vergona, K., and Widnell, C. C., 1981, The maintenance of glucose-6-phosphatase activity in primary cultures of rat hepatocytes, *J. Cell Biol.* **91:**272a.

Speth, M., and Schulze, H. U., 1981, Hormone-induced effects on the rat liver microsomal glucose-6-phosphatase systems *in vitro, Biochem. Biophys. Res. Commun.* **99:**134–141.

Speth, M., and Schulze, H. U., 1982, Accessibility of glucose-6-phosphate: Phosphohydrolase to antibody attack in modified microsomal vesicles, *FEBS Lett.* **144:**140–144.

Stalmans, W., 1977, The role of the liver in the homeostasis of blood glucose, in: *Current Topics in Cellular Regulation,* Vol. 11 (B. L. Horecker and E. R. Stadtman, eds.), Academic Press, New York, pp. 51–97.

Stephens, H. R., and Sandborn, E. B., 1976, Cytochemical localization of glucose-6-phosphatase activity in the central nervous system of the rat, *Brain Res.* **113:**127–146.

Stetten, M. R., and Burnett, F. F., 1967, Some properties of variously activated microsomal glucose-6-phosphatase, inorganic pyrophosphate and inorganic pyrophosphate-glucose phosphotransferase. Shift in pH optimum, *Biochim. Biophys. Acta* **139:**138–147.

Suketa, Y., and Sato, M., 1980, Changes in glucose-6-phosphatase activity in liver and kidney of rats treated with a single large dose of fluoride, *Toxicol. Appl. Pharmacol.* **52:**386–390.

Suketa, Y., Sato, M., and Kura, M., 1980, Effect of fluoride administration on renal glucose-6-phosphatase activity in rats, *Experientia* **36:**438–439.

Surholt, B., and Newsholme, E. A., 1981, Maximum activities and properties of glucose-6-phosphatase in muscles from vertebrates and invertebrates, *Biochem. J.* **198:**621–629.

Takahashi, T., and Hori, S. H., 1978, Intramembranous location of rat liver microsomal hexose-6-phosphate dehydrogenase and membrane permeability to its substrates, *Biochim. Biophys. Acta* **524:**262–276.

Takahashi, T., and Hori, S. H., 1981, Alterations in the latency of hepatic microsomal hexose-6-phosphate dehydrogenase under various *in vivo* and *in vitro* conditions, *Biochim. Biophys. Acta* **672:**165–175.

Teutsch, H. F., 1978a, Improved method for the histochemical demonstration of glucose-6-phosphatase activity, *Histochemistry* **57:**107–117.

Teutsch, H. F., 1978b, Quantitative determination of G6Pase activity in histochemically defined zones of the liver acinus, *Histochemistry* **58:**281–288.

Teutsch, H. F., 1981, Chemomorphology of liver parenchyma. Qualitative histochemical distribution patterns and quantitative sinusoidal profiles of G6Pase, G6PDH and malic enzyme activity and of glycogen content, *Prog. Histochem. Cytochem.* **14:**1–92.

Teutsch, H. F., Sasse, D., Katz, N., and Jungermann, K., 1978, Qualitative and quantitative chemomorphology of the liver glucose-6-phosphatase, *Verh. Anat. Ges.* **72:**665–667.

Tongianai, R., Bibbiani, C., and Malvaldi, G., 1977, Dry weight, triglyceride content and glucose-6-phosphatase activity of hepatocytes separated by sedimentation, *Acta Histochem.* **59:**232–238.

Trandaburu, T., 1977, Fine structural localization of glucose-6-phosphatase activity in the pancreatic islets of two amphibian species *(Salamandra salamandra L.* and *Rana esculenta L.), Acta Histochem.* **59:**246–253.

Tretiakov, A. V., 1979, Properties and isoenzyme contents of glucose-6-phosphatase of normal and tumor cell microsomes, *Vopr. Onkol.* **25:**33–36.

Ussolzew, S., 1935, Über den Einfluss von Kupferdarreichungen auf den Blutzuckerspiegel, *Biochem. Z.* **276:**431–433.

Vakili, B., and Banner, M., 1979, Inhibition of rat liver glucose-6-phosphatase by *N*-alkylmaleimides, *Biochem. Soc. Trans.* **7:**168–170.

Vakili, B., and Banner, M., 1981, The effects of *N*-alkylmaleimides on the activity of rat liver glucose-6-phosphatase, *Biochem. J.* **194:**319–325.

Veech, R. L., Cook, G. A., and King, M. T., 1980, Relationship of free cytoplasmic pyrophosphate to liver glucose content and total pyrophosphate to cytoplasmic phosphorylation potential, *FEBS Lett.* **117:**K65–K72.

Vergnes, O., Martel, M. B., and Got, R., 1981, Comparison between the effect of polycations and detergent on the specificity of cat liver microsomal glucose-6-phosphatase, *Int. J. Biochem.* **13:**1265–1268.

Verma, S. R., Rani, S., and Dalela, R. C., 1981, Effect of phenolic compounds on *in vivo* activity of glucose-6-phosphatase in certain tissues of *Notopterus notopterus, Toxicol. Lett.* **9:**27–33.

Vernier, J. M., and Sire, M. F., 1978, *In vivo* effects of adrenaline, glucocorticoids and fasting on glycogen phosphorylase and glucose-6-phosphatase activity and glycogen content in rainbow trout liver, *Gen. Comp. Endocrinol.* **34:**370–376.

Vianna, A. L., and Nordlie, R. C., 1969, The inhibition by physiological orthophosphate concentrations of hydrolytic and synthetic activities of liver glucose-6-phosphatase, *J. Biol. Chem.* **244:**4027–4034.

Vorhaben, J. E., and Campbell, J. W., 1979, Subcellular localization of glucose-6-phosphatase in animal tissues, *Comp. Biochem. Physiol.* **62B:**85–87.

Wachstein, M., and Meisel, E., 1956, On the histochemical demonstration of glucose-6-phosphatase, *J. Histochem. Cytochem.* **4:**592.

Wallin, B. K., and Arion, W. J., 1972, The requirement for membrane integrity in the inhibition of hepatic glucose-6-phosphatase by sulfhydryl reagents and taurocholate, *Biochem. Biophys. Res. Commun.* **48:**694–699.

Wallin, B. K., and Arion, W. J., 1973, Evaluation of the rate-determining steps and the relative magnitude of the individual rate constants for the hydrolytic and synthetic activities of the catalytic component of liver microsomal glucose-6-phosphatase, *J. Biol. Chem.* **248:**2380–2386.

Walls, M. A., and Lygre, D. G., 1980, Inhibition by orthophosphate esters of glucose-6-phosphatase, *Can. J. Biochem.* **58:**673–676.

Wals, P. A., and Katz, J., 1981, Glucokinase in bird liver. A membrane bound enzyme, *Biochem. Biophys. Res. Commun.* **100:**1543–1548.

Wanson, J. C., Bernaert, D., May, C., Deschuyteneer, M., and Prieels, J. P., 1980, Isolation and culture of adult rat hepatocytes and preneoplastic nodules from diethylnitrosamine treated livers: Glucose-6-phosphatase distribution, albumin synthesis and hepatic binding protein activity, *Ann. N.Y. Acad. Sci.* **349:**413–415.

Wanson, J. C., Penasse, W., Bernaert, D., and Popowski, A., 1981, Glucose-6-phosphatase distribution in isolated and cultured adult rat hepatocytes, *Eur. J. Cell. Biol.* **24:**88–96.

Yokoyama, M., and Chang, J. P., 1977, Cytochemical study of glucose-6-phosphatase in Chinese hamster testis, *Biol. Reprod.* **17:**265–268.

Zak, B., Epstein, E., and Baginski, E. S., 1977, Determination of liver microsomal glucose-6-phosphatase, *Ann. Clin. Lab. Sci.* **7:**169–177.

Zakim, D., and Edmondson, D. E., 1982, The role of the membrane in the regulation of activity of microsomal glucose-6-phosphatase, *J. Biol. Chem.* **257:**1145–1148.

Zetterqvist, O., and Engström, L., 1966, Isolation of ^{32}P-phosphohistidine from different rat-liver cell fractions after incubation with ^{32}P-adenosine triphosphate, *Biochim. Biophys. Acta* **113:**520–530.

Zoccoli, M. A., and Karnovsky, M. L., 1980, Effect of two inhibitors of anion transport on the hydrolysis of glucose-6-phosphatate by rat liver microsomes. Covalent modification of the glucose-6-*P* transport component, *J. Biol. Chem.* **255:**1113–1119.

Zoccoli, M. A., Hoopes, R. R., and Karnovsky, M. L., 1982, Rat liver microsomal glucose-6-P translocase. Effect of physiological status on inhibition and labelling by stilbene disulfonic acid derivatives, *J. Biol. Chem.* **257:**11296–11300.

25

The β-Adrenergic Receptor: Elucidation of Its Molecular Structure

Robert G. L. Shorr, Robert J. Lefkowitz, and Marc G. Caron

I. HISTORICAL PERSPECTIVE

Neurotransmitters and drugs interact with specific receptors in order to produce a cellular response. This recognition of specific agonists by receptors is the first step in an amplification process resulting in physiological modulation of homeostasis. The concept of a receptor-mediated mechanism for drug action was first proposed by Langley (1878, 1905). In 1906, examination by Dale of the action of various ergot alkaloid compounds in physiological preparations led to the observation of a number of compounds which would inhibit the excitatory effects of epinephrine without affecting its inhibitory actions. These types of observations provided support for the earliest notions of specific sites for catecholamine receptor interactions. In 1948, Ahlquist categorically demonstrated using a series of several sympathomimetic amines one order of potency in stimulating vasoconstriction and excitation of the uterus and ureters, and a different order of potency for these compounds for stimulation of vasodilation, for inhibition of uterine tone, and for stimulation of the heart. Ahlquist (1948) proposed that these actions of catecholamines were in fact mediated by two distinct populations of receptors which he termed α and β, and that in many cases, both receptors are present in the same organ or tissue. This proposal of distinct

Robert G. L. Shorr, Robert J. Lefkowitz, and Marc G. Caron • Howard Hughes Medical Institute Research Laboratories, Departments of Medicine (Cardiology), Biochemistry, and Physiology, Duke University Medical Center, Durham, North Carolina 27710.

populations of adrenergic receptors has been strongly supported by the synthesis of a large number of potent agonists and antagonists that are clearly specific for α- or β-receptor-mediated functions. More recent pharmacological studies have resulted in the further subclassification of these α- and β-receptors. Lands *et al.* (1967) subdivided the β-adrenergic responses into two subtypes termed β_1 and β_2. This division is based on the relative potency of epinephrine and norepinephrine in eliciting the appropriate physiological response in a variety of tissues. In β_1 systems, epinephrine and norepinephrine are nearly equipotent, while in β_2 systems, epinephrine is significantly more potent than norepinephrine. A number of compounds have been found to show varying degrees of selectivity between the β_1- and β_2-receptor subtypes (De Lean *et al.*, 1981), supporting this concept.

The effects of α-receptors have also been shown to be mediated by two populations of receptor termed α_1 and α_2. These receptors have also been classified as presynaptic or postsynaptic (Langer, 1979), although they are probably best distinguished by their relative affinities for α-adrenergic antagonists rather than by their anatomic location. Thus, α_1-receptors prefer the antagonist prazosin to yohimbine, while the converse is true for α_2-receptors (Hoffman *et al.*, 1981).

II. ADRENERGIC RECEPTORS AND ADENYLATE CYCLASE

Adenylate cyclase has been identified in a wide variety of tissues as a plasma-membrane-bound enzyme which catalyzes the conversion of ATP to cyclic AMP (Sutherland *et al.*, 1962). In early studies, it was demonstrated that stimulation of this enzyme was subject to hormonal regulation (Sutherland and Rall, 1960; Sutherland *et al.*, 1968) and specific drug effects. For the case of receptors whose pharmacological properties were known in detail, such as the β-adrenergic receptors, it was invariably found that the specificity of adenylate cyclase stimulation followed the pharmacological specificity for a receptor-regulated physiological effect. This naturally led to the hypothesis of a hormonally responsive adenylate cyclase system composed of at least two distinct units believed to be receptor and catalytic moieties (Robinson *et al.*, 1967). During the early 1970's, Rodbell *et al.* (1971) discovered the requirement of hormonally regulated adenylate cyclase systems for guanine nucleotides and predicted the existence of a nucleotide-binding regulatory site, probably acting as an intermediate coupling moiety between receptor and cyclase catalytic units (Birnbaumer *et al.*, 1970). α_2-Adrenergic receptors, as well as several other hormone receptors, were later found to inhibit the enzyme (Jakobs, 1979) through a similar but distinct class of guanine-nucleotide-binding proteins.

III. MOLECULAR COMPONENTS OF ADRENERGIC-RESPONSIVE ADENYLATE CYCLASE SYSTEMS

In this section, we will attempt to briefly outline the current understanding of receptor-mediated adenylate cyclase modulation in order to establish a background for the review of data concerning the molecular structure of adrenergic receptors, particularly those from β_1 and β_2 systems.

Beginning in the 1970's, direct binding studies of receptors linked to adenylate cyclase became possible utilizing radiolabeled hormones and drugs (Lefkowitz *et al.*, 1970). Shortly thereafter, binding of labeled adrenergic antagonist ligands to plasma-membrane preparations demonstrated characteristics expected for the receptors based on pharmacological and physiological studies (Levitzki *et al.*, 1974; Atlas *et al.*, 1974; Lefkowitz *et al.*, 1974; Aurbach *et al.*, 1974). A number of receptors were identified using this technique (Lefkowitz *et al.*, 1976; Cuatrecasas, 1974). The availability of these ligands for the characterization of α- and β-receptors greatly contributes to the study of these proteins in cell-free plasma-membrane preparations and permits detailed binding studies using a wide variety of receptor-directed compounds (Williams and Lefkowitz, 1976; Mukherjee *et al.*, 1976; Chenieux-Guicheney *et al.*, 1978). Examination of the data obtained in several systems has led to the elaboration of a number of essentially convergent models of receptor-mediated adenylate cyclase regulation. These models have been well reviewed elsewhere (Stadel *et al.*, 1980, 1982; Levitzki and Helmreich, 1979; Jakobs, 1979; Rodbell, 1980; Ross and Gilman, 1980; Weiland and Molinoff, 1981; Swillens and Dumont, 1981; Birnbaumer *et al.*, 1980; De Lean *et al.*, 1980; Abramowitz *et al.*, 1979) and will only be briefly outlined here.

In general, at least three distinct protein entities have been proposed to interact in the hormonal regulation of adenylate cyclase (Citri and Schramm, 1980). These include receptor, GTP-binding protein, and the catalytic unit of adenylate cyclase. As summarized in Figure 1, hormone interaction with receptor is believed to result in a probable conformational change in receptor, allowing fruitful interaction with the GTP-binding protein, the release of GDP, and the binding of GTP to this protein. This interaction results in the GTP-binding protein now capable of interacting with the catalytic unit of adenylate cyclase to enhance the formation of cAMP from ATP with

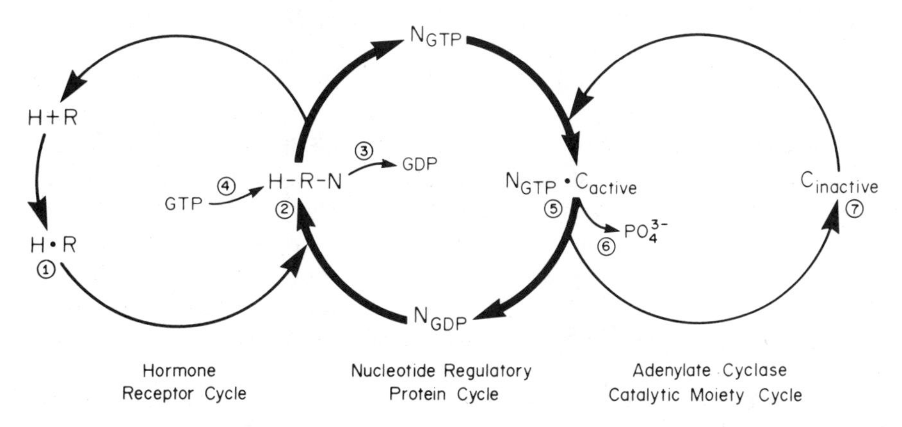

Figure 1. Schematic model for activation of adenylate cyclase by β-adrenergic agonists. H, hormone or agonist drug; R, β-adrenergic receptor; N, nucleotide regulatory protein; C, catalytic moiety of adenylate cyclase.

subsequent hydrolysis of the triphosphate nucleotide (Cassel and Sellinger, 1978). In these situations where hormone binding is believed to modulate enzyme activity (Jakobs, 1979), the GTP-binding protein is believed to be the transducer across the membrane of the hormone-regulated event. A similar GTP-binding event in enzyme modulation is suggested for both stimulatory and inhibitory receptors although distinct GTP-binding proteins appear to be involved (Rodbell, 1980). The hydrolysis of GTP to GDP is believed to signal the turnoff of the stimulatory cycle, although the location of this activity is not yet known and may require a separate protein or interaction of the GTP-binding protein and adenylate cyclase.

Of the several components of the adenylate cyclase system, much effort has focused on the biochemical characterization of the receptors and GTP-binding proteins. Some progress has also been made in the examination of adenylate cyclase itself (Neer and Salter, 1980; Guillon *et al.*, 1981; Neer, 1977). Recently, a guanine nucleotide-binding regulatory protein from rabbit liver membranes has been obtained in highly purified form (Sternweis *et al.*, 1981). The protein, termed G/F or N, is assayed by a reconstitution procedure utilizing plasma membranes prepared from a mutant lymphoma cell line which lacks the GTP-binding regulatory functionality of the hormone-responsive adenylate cyclase system (Ross *et al.*, 1978; Sternweis and Gilman, 1979; Northup *et al.*, 1980). This protein, on examination by SDS-PAGE, is found to be comprised of three polypeptides of 35,000, 45,000, and 52,000 M_r (Northup *et al.*, 1980). These proteins can be further separated by a number of techniques and the stimulatory activity appears to be associated with the 45,000-M_r polypeptide which can be selectively labeled with [^{32}P]-NAD and cholera toxin which has been demonstrated to stimulate adenylate cyclase. The current concept of the nature of these proteins and their relationship to adenylate cyclase has been extensively reviewed elsewhere (Pfeuffer, 1979; Spiegel *et al.*, 1979; Schleifer *et al.*, 1980; Ross, 1981; Levitzki, 1978; Limbird, 1981; Stadel *et al.*, 1981).

IV. PURIFICATION OF β-ADRENERGIC RECEPTORS

We have already described the existence of different receptor types (α and β) as well as their further subdivision into α_1, α_2, β_1, and β_2 subclasses, based on detailed pharmacological characterization. An important question in understanding the pharmacological distinction of receptor subtypes and subclasses is the structural basis of such differences. Towards this end, several laboratories have developed potent, active site-directed photoaffinity (Lavin *et al.*, 1981, 1982; Burgermeister *et al.*, 1982; Gozlan *et al.*, 1982; Rashidbaigi and Ruoho, 1981, 1982) or affinity (Atlas *et al.*, 1976; Pitha *et al.*, 1980) probes for the β-adrenergic receptor. A number of these compounds in their radiolabeled forms have been utilized in conjunction with SDS-PAGE to examine the subunit size of β-receptors in a variety of tissue-membrane preparations. We have additionally sought to compare β_1- and β_2-receptor subtypes by achieving their complete purification from convenient model systems such as avian (β_1) and amphibian (β_2) erythrocyte plasma membranes (Shorr *et al.*, 1982a,b). We believe such purification is necessary before rigorous structural comparisons between receptor subtypes and subclasses can be made.

As alluded to in the introduction to this chapter and reviewed elsewhere (Ross and Gilman, 1980; Weiland and Molinoff, 1981; Swillens and Dumont, 1981), radiolabeled agonists and antagonists have served to delineate many of the properties of receptors in cell-free particulate (plasma-membrane) preparations. An important aspect of receptor identification is the establishment of a correlation between ligand binding and receptor function. Such correlations have been clearly established for the β-adrenergic receptor and adenylate cyclase. Thus, a principal criterion for identification of a putative purified protein as a receptor must be demonstration of characteristic binding. Ultimate proof, of course, of receptor identification lies in the reconstitution of receptor function (cyclase stimulation) in an artificial environment.

A. Solubilization of β-Adrenergic Receptors

Once identified in a particular tissue by radioligand binding or functional assay, the first step in the purification of membrane proteins is the dissolution of the membrane with detergent with the subsequent solubilization of biological activity. Biological activity for the solubilized β-receptor is defined here as the ability to interact specifically and characteristically with β-adrenergic ligands. This is usually demonstrated by competition-binding assays using radiolabeled antagonists such as $[^3H]$dihydroalprenolol (Mukherjee *et al.*, 1975) or ^{125}I-labeled cyanopindolol (Engel *et al.*, 1981). Nonspecific binding is defined by including in the assays an excess concentration of an unlabeled competitor (agonist or antagonist). To date, the only detergent demonstrated to be useful for the solubilization of β-adrenergic receptors in a biologically active form is the plant glycoside digitonin (Caron and Lefkowitz, 1976). Using digitonin, we have solubilized β_1 and β_2 receptors, respectively, from plasma-membrane preparations of turkey and frog erythrocytes (Caron and Lefkowitz, 1976; Vauquelin *et al.*, 1977; Pike and Lefkowitz, 1978). Typically, purified plasma-membrane preparations contain 400–600 fmol receptor/mg protein for frog erythrocytes with a similar concentration in turkey erythrocyte preparations. For the frog erythrocyte preparation, as much as 80% of the total receptor activity can be solubilized with a 1–2% digitonin solution (Caron and Lefkowitz, 1976). For the turkey erythrocyte preparation, solubilization of receptor with 2% digitonin routinely results in the solubilization of 25–30% of the total binding activity (Pike and Lefkowitz, 1978; Shorr *et al.*, 1982a). The reasons for this disparity are unclear. Additionally, for the β_1-receptor of turkey erythrocytes, solubilization with digitonin results in an increased affinity for agonists (2 to 50-fold over particulate preparations) without any changes in antagonist affinities (Pike and Lefkowitz, 1978). This effect is not observed in frog erythrocyte β_2-receptor preparations.

Strauss *et al.* (1979) have further examined the efficacy of various batches of digitonin from several suppliers in solubilizing biologically active β-adrenergic receptors. It is unclear why different preparations of digitonin should possess different efficacies in solubilization of receptors, but this may be due to the presence of deleterious contaminants in various batches of detergent. Other sources from which β-adrenergic receptors have been solubilized with digitonin include rat lung (Benovic *et al.*, 1982, 1983), canine lung (Homcy *et al.*, 1983), and human placental membranes (Shocken *et al.*, 1980). It should be pointed out, however, that these plasma-membrane preparations are heterogeneous in their receptor populations; i.e., they contain mixtures of both β_1- and β_2-receptor subtypes (Shocken *et al.*, 1980; Dickinson and Nahorski, 1981). It is of interest that, as observed for the β_1- and β_2-receptors of frog and turkey erythrocytes, β_2-receptors are solubilized to a much greater extent in these tissues and in a more stable form than β_1-receptors. Reasons for the differential ability of digitonin to solubilize these receptors are uncertain, but this may be due to differences in membrane association, stability, and interactions with the detergent.

Recently, Homcy and Sylvester (1982) have reported the solubilization of β-receptors from canine lung plasma-membrane preparations with 8 M urea. Such receptors, however, reveal a three- to four-fold decreased affinity for β-adrenergic ligands when compared to particulate membrane or digitonin-solubilized preparations. Although primarily relating to methodology, a major obstacle in purification of receptors is the development of suitable large-scale rapid membrane preparations. This has been a particularly acute problem for the turkey erythrocyte (Durieu-Trautmann *et al.*, 1980). We have approached this problem by the utilization of nitrogen-cavitation technology, whereby the cells are saturated with nitrogen under 800 psi and lysed by release to atmospheric pressure. These procedures are performed in the presence of protease inhibitors (bacitracin, EDTA, PMSF, STI, benzamidine) and preparations are two- to four-fold enriched in receptor over previously described methods (Shorr *et al.*, 1982b). Interestingly, this increase in recovery of particulate receptor-binding activity is not dependent on the presence of protease inhibitors and probably reflects an overall increase in the yield of membranes. However, unlike membranes prepared in a more conventional way and described elsewhere (Stadel *et al.*, 1980), these membranes do not demonstrate high-affinity nucleotide-sensitive agonist binding, probably due to gross nucleotide contamination. Nonetheless, these procedures are adequate for purification purposes since they are very rapid, and we have processed as much as 8 liters of whole turkey blood in a day.

B. Affinity Chromatography of β-Adrenergic Receptors

We have approached the purification of the β_2-adrenergic receptor from frog erythrocytes in two ways. First, as shown in Table 1, two affinity chromatography cycles followed by ion-exchange chromatography on DEAE-Sepharose 6B-CL and a final affinity chromatography procedure can be used. For such affinity chromatography, we have utilized the ligand alprenolol, a potent antagonist ($K_d \cong$ 1–2 nM) which is nonselective between β_1- and β_2-receptor subtypes. Techniques for the immobilization of this ligand have been developed by Vauquelin *et al.* (1977, 1978, 1979) and Caron *et al.* (1979). Both gels have been demonstrated to serve as suitable supports for affinity chromatography (Vauquelin *et al.*, 1977, 1978, 1979; Caron *et al.*, 1979) and behave in a biospecific fashion. Solubilized α-receptors derived from liver membranes or adenylate cyclase activity solubilized from turkey erythrocytes (Vauquelin *et al.*, 1978) are not retained on the columns (Wood *et al.*, 1979).

Receptor activity is nearly completely retained by the affinity column and is selectively eluted with β-receptor agonists or antagonists in high yield (approximately 50%). As shown in Table 1, an approximate 100-fold purification is obtained from initial detergent extracts for the frog erythrocyte β_2-adrenergic receptor.

As shown in Table 1, repetitive affinity chromatography and DEAE-Sepharose chromatography results in an approximate 5500-fold purification of receptor from frog erythrocyte membranes with a final specific activity of 8000–10,000 pmol binding sites/mg protein. Typically, 2–5 μg of protein are obtained from 3 liters of digitonin extract prepared from 700–900 ml of packed erythrocytes.

While the techniques described above result in the preparation of purified protein, they fail to provide sufficient material for complete characterization. For this reason, we have developed a second approach to receptor purification. This combines affinity chromatography and size-exclusion high-performance liquid chromatography (HPLC). In this approach, after detergent solubilization, receptor is chromatographed once on Sepharose-alprenolol and subsequently twice on size-exclusion HPLC. These columns are typically one TSK 3000 SW (60 cm) and two TSK 4000 (60 cm) columns tandem linked. As shown in Table 2, combination of these two techniques results in the rapid preparation of approximately 12 μg of protein from 200–300 ml of packed frog erythrocytes with a final specific activity of 11,800 pmol binding sites mg/protein.

When competition-binding isotherms are generated for these preparations using [^{3}H]-DHA and various concentrations of adrenergic agents (Figure 2), typical β_2-adrenergic receptor-binding properties are observed. This is exemplified in the figure by the order of potency (affinity) of isoproterenol, epinephrine, norepinephrine, and the maintenance of stereoselectivity characteristic of these adrenergic receptor-directed compounds. A comparison of ligand affinities between membrane-bound, solubilized, and purified preparations shows no significant differences in K_d. This demonstration

Table 1. Summary of Purification of the β-Adrenergic Receptor of Frog Erythrocytes[a]

Step	Yield at each step (%)	Overall yield (%)	Specific activity (pmoles/mg)	Purification (fold)
Crude frog erythrocyte membranes	100	100	0.15	1
Digitonin extract	70	70	1.0	6.6
First alprenolol-gel pass	50–70	35–50	106	706
Second alprenolol-gel pass	50	18–30	1470	9800
DEAE-Sepharose	60–80	10–15	n.d.[b]	n.d.[b]
Final alprenolol-gel pass	30–50	4–8	8000–10,000	55,000

[a] All steps were performed as described previously (Shorr *et al.*, 1981). Throughout the procedures, the receptor activity was followed by [^{3}H]-DHA binding. [^{3}H]-DHA binding in alprenolol-gel eluates was measured after the sample had been desalted on Sephadex G50. This purification procedure has been found to yield 2–5 μg of measurable protein at the end of the four steps starting with ~700 ml of packed frog erythrocytes. Data taken from Shorr *et al.* (1981).
[b] n.d. = not determined.

Table 2. Summary of Purification of β-Adrenergic Receptor of Frog Erythrocytes by Affinity and High-Performance Liquid Chromatography[a]

Step	Activity (pmoles)	Yield at each step (%)	Overall yield (%)	Specific activity, (−)[^{3}H]dihydro-alprenolol bound (pmoles/mg)	Purification (fold)	
					Each step	Overall
Detergent extract of frog erythrocyte membranes	396	100	100	1.9	1	1
Eluate of alprenolol affinity gel	245	62	62	136	72	72
First HPLC	206	84	52	3416	25	1800
Second HPLC	122	59	31	11,800	3.5	6300

[a] Typically, membranes from 200–300 ml of frog erythrocytes were solubilized with digitonin as described previously (Caron and Lefkowitz, 1976). The affinity chromatography and HPLC steps were performed as described in Shorr *et al.* (1982a). Protease inhibitors at concentrations of 5 μg/ml soybean trypsin inhibitor, 100 μg/ml bacitracin, 1 mM benzamidine, 0.1 mM EDTA, and 0.01 mM phenylmethylsulfonyl fluoride were included up to the first HPLC step. The experiment shown is representative of five such experiments. The overall purification from crude frog erythrocyte membranes (0.15 pmoles/mg) is ~80,000-fold. Data taken from Shorr *et al.* (1982a).

of ligand-binding properties consistent with known pharmacology is one of the strongest criteria for the identification of a putative purified receptor.

For the turkey erythrocyte β_1-adrenergic receptor, nearly 1000-fold purification is obtained with about 60% yield of receptor activity (see Table 3) on elution of the Sepharose-alprenolol affinity support with a gradient of adrenergic agent (Shorr *et al.*, 1982b). The purified peptides obtained after HPLC fractionation also demonstrate ligand-binding specificity which is virtually identical to that of soluble receptor preparations (Shorr *et al.*, 1982b). Vauquelin *et al.* (1977, 1978, 1979) and Strosberg *et al.* (1980) have also reported the application of affinity chromatography to the β-adrenergic receptor solubilized with digitonin from turkey erythrocyte membranes utilizing a different alprenolol agarose adsorbent. The differences between the two types of alprenolol agarose are in the chemistry of side-arm spacing and ligand attachment. They have reported a 2000-fold purification of receptor in a single step. To elute the adsorbed receptor activity, however, high concentrations of salt (1 M NaCl) in the presence of radioactive ligand were required, and it was necessary to stabilize activity by addition of 0.1% γ-globulin to the eluting buffers. Under these conditions, 25–30% of adsorbed receptor activity was eluted and apparent dissociation constants (K_d) for adrenergic agents and stereoselectivity were maintained (Vauquelin *et al.*, 1977). Recently, Homcy *et al.* (1983) have reported the use of a different affinity chromatography support (acebutolol-Sepharose) for the purification of the receptor from canine lung. Successive passes of the solubilized receptor activity over this affinity support yielded a 150 to 300-fold purification at each step with purified receptor material approaching theoretical specific activity (10,800 pmoles/mg).

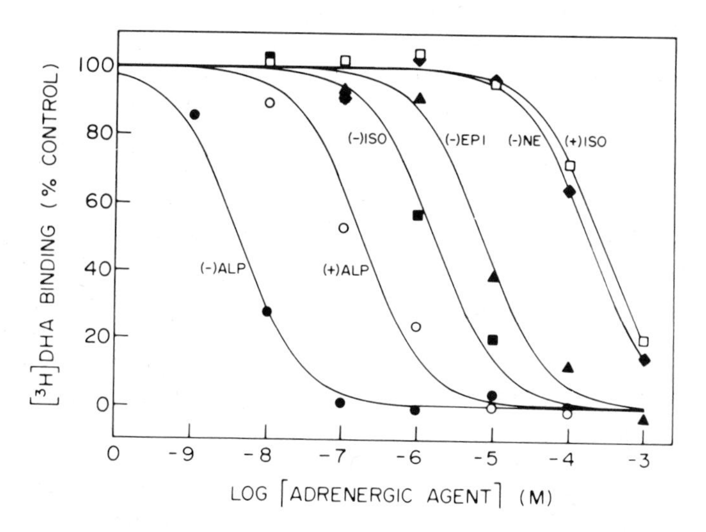

Figure 2. Specificity of [^{3}H]dihydroalprenolol-binding activity to a purified β-adrenergic receptor preparation. Competition by various β-adrenergic agonists and antagonists for [^{3}H]dihydroalprenolol binding to a purified receptor preparation. The purified receptor preparation used in these studies was obtained as described previously (Shorr *et al.*, 1981) by sequential affinity chromatography with a DEAE-Sepharose step between the second and third affinity chromatography steps. The material obtained show only a single broad band at 58,000 M_r when radioiodinated and electrophoresed on SDS gels. The receptor preparation was incubated in 0.2% digitonin, 100 mM NaCl, and 10 mM Tris-HCl pH 7.4 at 4° in the presence of 4 nM [^{3}H]dihydroalprenolol and various concentrations of adrenergic agents for 1 hr at 4°C. Bound ligand was determined by the Sephadex G50 chromatography method (Caron and Lefkowitz, 1976). The data represent the average of triplicate determinations. Taken from Shorr *et al.* (1981).

Table 3. Purification of Turkey Erythrocyte β-Adrenergic Receptor by Affinity Chromatography and High-Performance Liquid Chromatography[a]

Sample	Activity (pmoles)[b]	Yield (% overall)	Yield (% step)	Specific activity[b] (pmoles/mg protein)	Purification (fold)	
					Overall	Step
Digitonin extract	199	100	100	1.15	1	1
Alprenolol Sepharose eluate	71.6	60	60	1108	963	963
First HPLC pass	62.1	52	87	4968	4320	4.5
Second HPLC pass	35.5	30	58	18,733	16,289	3.7

[a] Typically, 100–200 pmoles of soluble β-adrenergic receptor was prepared from 200–300 ml of purified turkey erythrocyte membranes solubilized with 300 ml of 2% digitonin, 100 μg/ml bacitracin, 10 μl/ml soybean trypsin inhibitor, 3 × 10^{-5} M phenylmethylsulfonyl fluoride, 3 × 10^{-4} M benzamidine, 100 mM NaCl/10 mM Tris-HCl pH 7.2, and applied to a column (200 ml) of alprenolol-Sepharose and eluted with a linear gradient (300 ml) of 0–40 μM (±)alprenolol in 100 mM NaCl and 10 mM Tris-HCl pH 7.2, 0.1% digitonin. Steric exclusion HPLC on the pooled concentrated material was performed as described in Shorr *et al.* (1982b). Protease inhibitors were included until the first HPLC pass at concentrations stated above. Data shown here represent the average of two such preparations. Final receptor specific activities represent ~33,000-fold purification of receptor from membrane preparations. Data taken from Shorr *et al.* (1982b).

[b] As measured by [^{3}H]-DHA binding.

An additional approach to the purification of receptors is based upon the utilization of antibodies directed toward receptor, as affinity supports (Venter and Fraser, 1983). Putative monoclonal antibodies directed towards β_1- and β_2-receptors (Venter *et al.*, 1981; Fraser and Venter, 1980) or polyclonal antibodies obtained from patients believed to possess circulating antibodies directed towards β_2-adrenergic receptors (Venter *et al.*, 1980) have been used. In this approach, purified antibodies prepared by ammonium sulfate precipitation or preparative isoelectric focusing were coupled to cyanogen bromide-activated Sepharose 4B. Turkey erythrocyte ghosts were surface labeled with $[^{131}I]$iodine/lactoperoxidase and β_1-receptors solubilized with 0.5% digitonin. After preparative isoelectric focusing, samples containing putative ^{131}I-labeled receptors were then applied to the antibody Sepharose column and absorbed protein eluted with SDS sample buffer. For preparations using the putative circulating autoantibodies, affinity adsorbents were prepared utilizing protein A Sepharose 4B-CL (Fraser and Venter, 1980), and the receptors labeled with $[^3H]$-NHNP-NBE, a β-adrenergic affinity label, were solubilized with Triton X-100 and chromatographed as described above. A primary difficulty with this approach, however, is the total inability to conclusively identify, by direct radioligand-binding studies, the nature of the putative adrenergic receptor proteins eluted from the immunoadsorbent columns and the inability so far to generate useable amounts of receptor protein for further characterization.

V. MOLECULAR IDENTIFICATION OF β-ADRENERGIC RECEPTORS

A. Visualization of the Purified Receptor Peptide(s)

In order to investigate the molecular identity of β-adrenergic receptors, intensive effort has focused on the purification of these receptors to apparent homogeneity and on the development of suitable photoaffinity and affinity probes. As described above, we and others have been successful in the development of an affinity support utilizing the compound alprenolol. This potent receptor antagonist is nonselective between β_1- and β_2-receptor subtypes and can therefore be utilized for affinity purification of both proteins. As demonstrated in Table 1, use of such an affinity support in a repetitive fashion, coupled with ion-exchange chromatography on DEAE-Sepharose 6B-CL can result in the preparation of purified β_2-adrenergic receptor from frog erythrocyte membranes. This material, on iodination with $[^{125}I]$-Na, chloramine T, and SDS-PAGE autoradiography, reveals a major band of protein centered at about 58,000 M_r in addition to material migrating at lower molecular weight. However, two-dimensional SDS-PAGE of such preparations reveals that only the 58,000 M_r band coincides with receptor-binding activity. Additionally, as will be discussed below, this band can be specifically labeled with the photoaffinity probe $[^{125}I]$paraazidobenzylcarazolol (Lavin *et al.*, 1982) in particulate and purified preparations. These data then strongly support the identification of this peptide at 58,000 M_r as containing the β-adrenergic binding site of the frog erythrocytes.

As these techniques were found successful for the purification of the β_2-adrenergic receptor, we applied them directly to the purification of the β_1-adrenergic receptor from the turkey erythrocyte. It is important to note again, however, that β_1-receptors are solubilized to a lesser extent than β_2-receptors accounting, in part, for the overall lower recovery of receptor protein, as shown in Table 2. When purified to apparent homogeneity, such turkey erythrocyte β_1-adrenergic receptor preparations display specific activities of ~18,000 *p*mol/mg protein.

When these purified preparations are radiolabeled with [^{125}I]-Na and chloramine T and subjected to SDS-PAGE, autoradiography demonstrates two major bands of iodinated protein at apparent molecular sizes of 40,000 M_r and 45,000 M_r (Figure 3A). These bands, on quantitation by slicing and counting or densitometry, are consistently present in a stoichiometry of 3–4 : 1 for the 40,000- and 45,000-M_r protein, respectively. Additionally, shown in the figure, is a trace band of about 30,000 M_r. The levels of this protein are variable from preparation to preparation and increase upon aging of the preparations but can be reduced by the use of protease inhibitors, particularly EDTA. No difference is observed in subunit pattern on SDS-PAGE in the absence of reducing conditions.

In order to establish whether one or both of these peptides contain the ligand-binding site of the β_1-adrenergic receptor from turkey erythrocytes, we utilize the photoaffinity compound ^{125}I-labeled paraazidobenzylcarazolol (*p*ABC). In these experiments, particulate or purified receptor preparations of turkey erythrocyte β_1-receptor are incubated with [^{125}I]-*p*ABC in the presence and absence of competing unlabeled adrenergic agents, photolyzed and subjected to SDS-PAGE. As shown in Figure 3B, for particulate preparations, two polypeptides are labeled with apparent molecular sizes of 40,000 and 45,000 M_r and in a stoichiometry of 3–4 : 1, respectively. As shown in Figure 3C, identical results are obtained in purified preparations and labeling can be clearly demonstrated to be β_1-receptor specific by inhibition of labeling of the peptide by adrenergic agents. These data suggest that both peptides contain a binding site for β-adrenergic ligands.

In order to further assess this possibility, attempts were made to separate these polypeptides to apparent homogeneity by repeated HPLC. Those fractions containing primarily one or the other polypeptide are pooled and rechromatographed leading to their complete separation. Once obtained in homogeneous form, the detailed patterns of ligand binding to these two proteins are examined using the reversible antagonist [^{125}I]-CYP and characteristic adrenergic agents. In both cases, a typical β_1-adrenergic order of potency: (−)isoproterenol > norepinephrine ≅ (−)epinephrine is observed. In addition, in both cases, stereoselectivity of (−) isomers over (+) isomers is maintained for both agonists and antagonists such as (−)alprenolol and (±)metoprolol competed for binding with expected potencies. These results strongly suggest that both isolated peptides contain pharmacologically similar β_1-adrenergic binding sites which presumably represent two distinct and separate forms of the same receptor.

We have sought to preliminarily examine the relationship between these two proteins primarily to assess the possibility that the 40,000-M_r protein is a degradation product of the 45,000-M_r polypeptide by treatment of each purified preparation with protesases to assess similarities or differences in protein structure. Such protease digestion of native folded proteins reflects differences in glycosylation and configuration as well as protein sequence, which would determine accessibility of various

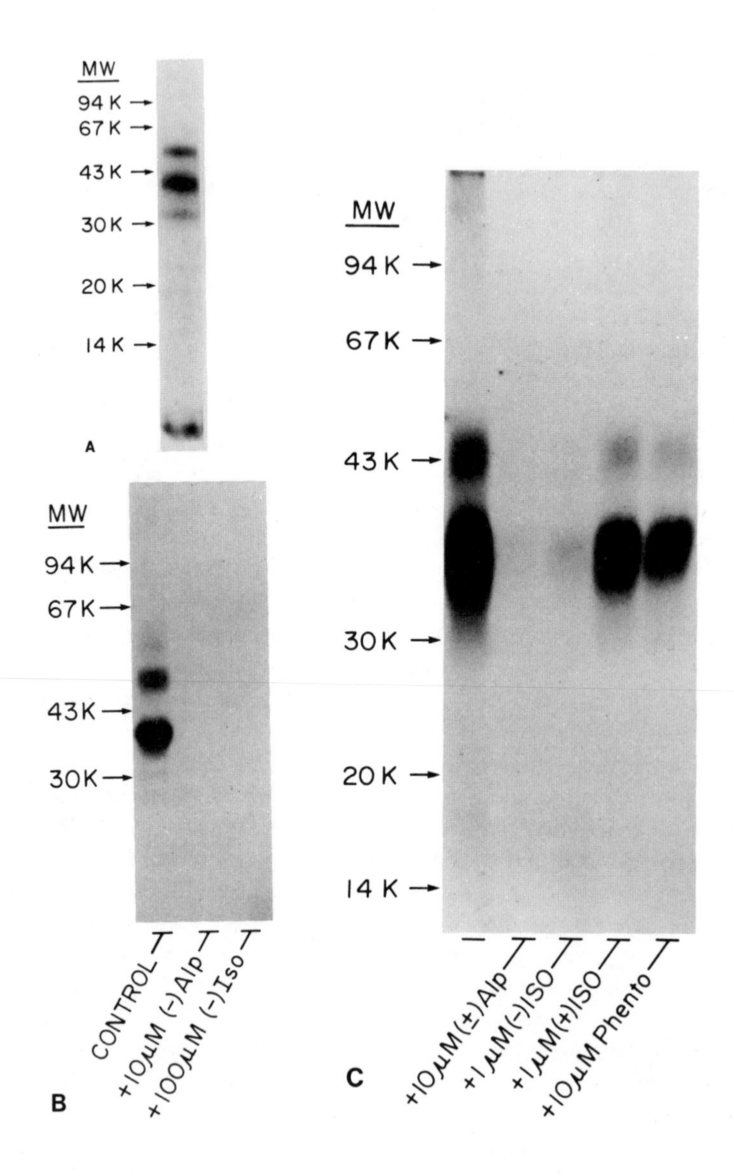
MW
94 K
67 K
43 K
30 K
20 K
14 K
A
MW
94K
67K
43K
30K
B
CONTROL
+10μM (-) Alp
+100μM (-) Iso
MW
94 K
67 K
43 K
30 K
20 K
14 K
C
+10μM(±)Alp
+1μM(-)ISO
+1μM(+)ISO
+10μM Phento

sites for proteolytic cleavage. In order to visualize these digestion products, aliquots of [^{125}I]-*p*ABC or ^{125}I-labeled receptor preparations were digested and subjected to SDS-PAGE autoradiography. The results demonstrate differences as well as similarities between these two receptor polypeptides and suggest that the 40,000-M_r polypeptide is probably not simply derived from the 45,000-M_r protein by degradation during receptor preparations (Shorr *et al.*, 1982b). The precise relationship between these two receptor forms remains unknown.

Durieu-Trautmann *et al.* (1980) have recently reported on the visualization of turkey erythrocyte β-adrenergic receptors. In their study, as described above, digitonin extracts were chromatographed on Sepharose alprenolol prepared as in Vauquelin *et al.* (1977) with a subsequent 2000-fold purification. On ^{125}I-labeling with chloramine T, [^{125}I]-Na, and subsequent SDS-PAGE autoradiography, four major bands were visualized at 170,000, 33,000, 30,000, and 52,000 M_r. The 52,000-M_r protein was identified as a contaminant and no difference was observed in subunit profiles on SDS-PAGE in the presence or absence of reducing agents. As reported earlier by these workers (Vauquelin *et al.*, 1978, 1979), purified receptor preparations presumably retain characteristic binding specificity for adrenergic ligands. Specific activities, however, do not exceed 4900 pmoles binding sites per mg protein, and protease inhibitors are not present during receptor purification. Viewed in the light of the work described

Figure 3. SDS-PAGE pattern of purified iodinated and photoaffinity-labeled turkey erythrocyte β-adrenergic receptor. (A) SDS-PAGE of affinity and HPLC-purified radiolabeled receptor. Turkey erythrocyte solubilized receptor preparation was purified as described in Shorr *et al.* (1982b) by affinity chromatography and one high-pressure steric exclusion chromatography step. Those fractions containing receptor activity were pooled and an aliquot (6 pmoles) radiolabeled with [^{125}I]-Na and chloramine T as described in (Shorr *et al.*, 1982b). After an additional HPLC run (to separate labeled protein contaminants and radiolabeled detergent), those fractions corresponding to receptor activity were pooled and an aliquot (25,000 cpm) was subjected to SDS-PAGE and autoradiography (three days of exposure). Arrows indicate the relative mobility of the various iodinated proteins used as molecular weight (MW) standards. (B) Photoaffinity labeling of intact erythrocyte membranes with [^{125}I]-pABC. Turkey erythrocyte membranes (25–30 pM receptor concentration) were incubated with [^{125}I]-*p*ABC (25–30 pM) for 90 min at 25°C in the absence (control) and presence of 10^{-5} M (−)alprenolol or 10^{-4} M (−)isoproterenol. At the end of the incubation, the samples were washed three times with 0.5% BSA in 75 mM Tris-HCl, 25 mM $MgCl_2$ pH 7.4, and once with the same Tris buffer alone. Photolysis was carried out for 90 sec as described previously (Lavin *et al.*, 1982). The pattern is that of an 8% polyacrylamide slab gel, exposed to Kodak XAR-5 film for three days with Cronex-plus intensifying screens. (C) Pharmacological specificity of [^{125}I]-*p*ABC covalent incorporation into partially purified receptor preparations. A partially purified receptor preparation (affinity chromatography eluate, 1.5 pmoles) freed of alprenolol, was aliquoted and incubated with [^{125}I]-*p*ABC (1.2 nM) in a total volume of 1 ml for each sample and containing 5 μg/ml leupeptin, 10^{-4} M EDTA for 24 hr at 4°C in the presence and absence of 10 μM (±)alprenolol, 1 μM (−)isoproterenol (Iso), 1 μM (+)isoproterenol, or 10 μM phentolamine (Phento). Each sample was then desalted on Sephadex G50 columns equilibrated in 2.0 mM Tris-HCl and 0.05% digitonin pH 6.8 to separate unbound [^{125}I]-*p*ABC. Desalted material (void volume fractions, 2 ml) was exposed to ultraviolet light for 2.5 min at 12 cm from the source as described previously (Lavin *et al.*, 1982). Samples were then lyophilized and taken up in 100 μl of 10% SDS, 5% β-mercaptoethanol, and 10% glycerol to give a final concentration of 40 mM Tris-HCl pH 6.8, 1% digitonin. Samples were loaded onto a 7.5–15% gradient pore SDS-PAGE and electrophoresed at 20 mA overnight. Incorporation was revealed by exposing the dried gel on Kodak XAR-5 X-ray film for 96 hr at −90°C. Results are taken from (Shorr *et al.*, 1982b).

here and the observation of turkey erythrocyte receptor susceptibility to degradation to a peptide of approximately 30,000–34,000 M_r, the results obtained in the study of Durieu-Trautmann *et al.* (1980) probably reflect the proteolytic degradation of partially purified receptor.

As mentioned previously, Fraser and Venter (1980) have developed an approach for receptor purification utilizing autoantibodies and monoclonal antibodies prepared to partially purified receptor preparations. In these studies, subunit compositions were examined by radiolabeling with $[^{125}I]$-Na and SDS-PAGE as well as by affinity labeling with $[^{3}H]$-NHBE-NBE. For turkey erythrocyte preparations, three components of 70,000, 31,000, and 22,000 M_r were visualized by slicing and counting of SDS gels prepared utilizing receptor eluted from the monoclonal antibody columns. When receptor was surface labeled in the membrane using $[^{125}I]$-Na and lactoperoxidase, and bound material was eluted from the immunoaffinity column with propranolol and subjected to SDS-PAGE, gel slicing revealed only the 70,000-M_r component. In our studies with turkey erythrocyte β_1-adrenergic receptors, we have consistently observed a nonreceptor contaminant band at 67,000–70,000 M_r after the initial Sepharose alprenolol chromatography. The relationship of any of the peptides purified by Venter and Fraser (1983) to receptor is unclear since no binding studies or specific activity values utilizing these preparations have been presented. In similar studies, the same group of investigators (Venter and Fraser, 1983; Venter *et al.*, 1981) have also examined the subunit profile of preparations from bovine and canine lung-plasma membranes. These tissues contain approximately 80% of their total receptor population as the β_2-subtype. Iodination and SDS-PAGE gel slicing reveal a major peptide at 59,000 M_r (Venter and Fraser, 1983; Venter *et al.*, 1981). Recently, another group (Homcy *et al.*, 1983) has purified a receptor peptide from canine lung to apparent homogeneity. The molecular weight of the purified peptide was estimated to be 52,000 by SDS-PAGE and protein staining.

B. Photoaffinity Labeling of β-Adrenergic Receptors

Our group has utilized $[^{125}I]$-*p*ABC to photoaffinity label the β_1- and β_2-adrenergic receptors of a wide variety of mammalian and nonmammalian tissues. Some of these results are summarized in Table 4. The salient conclusions to emerge from these studies are the following. In mammalian systems, both β_1- and β_2-adrenergic receptor-binding sites reside on peptides of 60,000–65,000 M_r. These peptides are virtually identical in overall size. Thus, the determinants of β_1- and β_2-subtype specificity must reside in some subtle alterations in the peptide structure (Benovic *et al.*, 1983; Stiles *et al.*, 1983).

The β-adrenergic receptor appears to be exquisitely sensitive to the action of endogenous metaloproteases that cleave it to peptides of ~55,000, 40,000–45,000, and 30,000 M_r. Each of these peptides appear to contain an intact ligand binding site to which the photoaffinity reagent can bind. This susceptibility to proteolysis likely accounts for the peptide heterogeneity observed in photoaffinity labeling studies unless great care is taken to include appropriate inhibitors of metaloproteases such as EDTA (Benovic *et al.*, 1983; Stiles *et al.*, 1983).

Another point to be emphasized is the uniqueness of the apparent structure of avian erythrocyte β-adrenergic receptors. The peptides identified generally have 40,000–50,000 M_r. Although this may also simply be a reflection of proteolysis from a larger precursor of ~60,000 M_r, we have been unable to demonstrate it. Thus, the peptide size of this β-adrenergic receptor may indeed differ substantially from that of mammalian β-adrenergic receptors. Whether this represents the structural correlate of the often emphasized point that the detailed pharmacology of avian β_1-adrenergic receptors differs from that of mammalian β_1-adrenergic receptors remains to be determined.

Table 4. β-Adrenergic Receptor Polypeptides Identified by Photoaffinity-Labeling Techniques

Species	Tissue	Receptor subtype	Peptides identified
Frog	Erythrocytes	β_2	58,000 (Shorr *et al.*, 1982a; Lavin *et al.*, 1981)
			63,000 (Rashidbaigi and Ruoho, 1982)
	Heart	β_2	62,000 (Stiles *et al.*, 1983)
Turkey	Erythrocytes	β_1	45,000[a], 40,000 (Lavin *et al.*, 1982; Shorr *et al.*, 1982b)
			43,500 (Rashidbaigi and Ruoho, 1981)
			50,000[b], 40,000 (Burgermeister *et al.*, 1982)
Duck	Erythrocytes	β_1	48,500[a], 45,000 (Rashidbaigi and Ruoho, 1981)
Pigeon	Erythrocytes	β_1	52,000, 45,000[a] (Rashidbaigi and Ruoho, 1982)
Rat	Erythrocytes	β_2	65,000 (Stiles *et al.*, 1983)
			62,500 (Rashidbaigi and Ruoho, 1982)
	Reticulocytes	β_2	65,000 (Stiles *et al.*, 1983)
	Lung	β_1 (20%)	62,000 (Benovic *et al.*, 1983)
		β_2 (80%)	
	Heart	β_1	62,000 (Stiles *et al.*, 1983)
Rabbit	Lung	β_1 (80%)	65,000, 45,000, 38,000 (Lavin *et al.*, 1982)
		β_2 (20%)	
	Heart	β_1	62,000 (Stiles *et al.*, 1983)
Hamster	Heart	β_1	62,000 (Stiles *et al.*, 1983)
Dog	Lung	β_1 (20%)	52,000, 39,000[a] (J. L. Benovic *et al.*, unpublished observations)
		β_2 (80%)	50,000–55,000 (Homcy *et al.*, 1983)
	Heart	β_1	62,000 (Stiles *et al.*, 1983)
Guinea pig	Lung	β_1 (20%)	62,000 (J. L. Benovic *et al.*, unpublished observations)
		β_2 (80%)	
Cultured cells (mouse)	S49 lymphoma	β_2	65,000, 55,000 (Rashidbaigi *et al.*, 1982, 1983)
Human	Heart	β_1	62,000 (Stiles *et al.*, 1983)

[a] Denotes minor components labeled with the various photoaffinity probes. In several of the tissues alluded to in the table, initial studies with β-adrenergic photoaffinity probes suggested heterogeneity in the labeling pattern obtained (Lefkowitz *et al.*, 1983). More recently, it has become clear that inclusion of proteinase inhibitors, in particulate EDTA in membrane preparations, reduces greatly the occurrence of lower M_r bands (Benovic *et al.*, 1983; Stiles *et al.*, 1983). In the few instances where more than one value is given for the photoaffinity-labeled receptor peptides, i.e., avian erythrocytes, rabbit and dog lung, it indicates that the labeling pattern in most of these tissues was not affected in the presence of proteinase inhibitors as described in Benovic *et al.* (1983) and Stiles *et al.* (1983). In most tissues, at least mammalian tissues, the binding subunits of both β_1- and β_2-adrenergic receptors appear to reside on peptides of similar overall 62,000–65,000 M_r.

Rashidbaigi and Ruoho (1981) have examined the size of the β-adrenergic receptor ligand-binding subunit utilizing the photoaffinity probe [^{125}I]*p*-azidobenzylpindolol, a receptor antagonist. This ligand has also been shown to label two bands of 45,000 and 48,500 M_r in duck erythrocyte plasma membranes which contain receptor of the β_1-subtype (Rashidbaigi and Ruoho, 1982). The same ligand has been utilized in frog erythrocyte plasma-membrane preparations where a single broad band of ~60,000 M_r was shown to be specifically labeled. This receptor is of the β_2-subtype. Burgermeister *et al.* (1982) have also examined the turkey erythrocyte β-adrenergic receptor utilizing the compound [^{125}I]azidocyanopindolol. This compound is shown to incorporate specifically into two proteins of 45,000 and 50,000 M_r. These results are in good agreement with those obtained with the photoactive [^{125}I]*p*-azidobenzylcarazolol cited above.

^{3}H-Labeled, receptor-directed affinity probes have also been utilized in β-adrenergic receptor characterization studies. Atlas and Levitzki (1978), using *N*-[2-hydroxy-3-(1-naphthyloxy)-propyl]-*N'*-bromoacetylethylenediamine, have demonstrated incorporation of this label into two proteins of 37,000 and 41,000 M_r for turkey erythrocyte plasma-membrane preparations. Pitha *et al.* (1980) have demonstrated the incorporation of a [^{3}H]bromoacetyl analog of alprenolol into a 58,000-M_r region on SDS gel of frog erythrocyte plasma-membrane preparations. Wrenn and Homcy (1980) have developed a tritiated β-adrenergic antagonist possessing a photosensitive moiety, [^{3}H]acebutolol-azide. Recently, this ligand has been shown to incorporate covalently into the 50,000–55,000 M_r-receptor peptide of dog lung (Homcy *et al.*, 1983). A number of additional photoaffinity and affinity probes have been developed for β-receptors (Lucas *et al.*, 1979; Terasaki *et al.*, 1979; Gozlan *et al.*, 1982), most of which have not yet been labeled or utilized for structural studies; these will not be reviewed here.

C. *Radiation Inactivation Studies of β-Adrenergic Receptors*

An additional approach to the examination of β-adrenergic receptors is the characterization of native receptor size in membrane preparations by radiation inactivation. This is a useful technique for the determination of molecular size and the examination of interactions between proteins of a given receptor or enzyme complex in purified and particulate preparations (Kempner and Schlegel, 1979; Kempner *et al.*, 1980). The method of analysis is based upon random probable hits and destruction of a given protein. The rationale of inactivation is that the deposition of ionization energy within a protein molecule results in loss of biological activity and, since the ionization occurs randomly in the mass of the protein, the larger the protein the greater the change of a hit and inactivation. Samples are routinely exposed to a beam of electrons (Kempner and Schlegel, 1979) and after irradiation, surviving biological activity are assayed and data examined using the equation

$$A = A_o e^{-\mu D}$$

where A is the measured activity, A_o the activity prior to irradiation, D the radiation dose (in rads), and μ a factor proportional to the mass of the particular functional unit. For an activity due to a single-sized component, a plot of ln (A/A_o) vs. D, results in a straight line with slope = μ. Kepner and Macey (1963) have reduced this equation to

$$M_r = \frac{6.4 \times 10^{11}}{D_{37}}$$

where D_{37} represents the dose required to inactivate the measured activity to 37% of control values. However, for irradiations at −135°C, Kempner and Schlegel (1979) have determined that the target size can be calculated from the relation: molecular weight-17.9 × $10^{11}u$. Linearity of radiation-induced decay is usually demonstrated by application of doses to inactivation of 80–95% of the measured activity. In the case of receptor studies where reversible ligand-binding assays are used, it is additionally necessary to establish that irradiation truly results in loss of active units and not in a change of the dissociation constant for a given ligand.

Previously, Nielsen *et al.* (1981) have examined various aspects of the adenylate cyclase system of turkey erythrocytes using this technique. Plasma membranes were irradiated at various doses ranging from 1–36 Mrads and surviving receptor activity was assayed for each dose using [^{125}I]hydroxybenzylpindolol. This compound is a potent nonsubtype selective β-receptor antagonist. Scatchard plots were constructed for each irradiation dose, and the dissociation constant (K_d), and number of receptor sites calculated for each sample. Estimates of receptor antagonist-binding site size by this technique results in an apparent molecular weight for the β_1-adrenergic receptor of this system of 90,000 daltons.

We have recently compared the molecular sizes of particulate and purified preparations of β_1- (turkey erythrocyte) and β_2- (frog erythrocyte) receptors utilizing radiation inactivation (Shorr *et al.*, 1983). In these studies, plasma membranes or purified preparations were irradiated from 12–96 Mrads, receptor activity assayed with [^{3}H]-DHA or [^{125}I]-CYP, and estimates of molecular size calculated as in Kempner and Schlegel (1979). When frog erythrocyte membrane preparations were irradiated and saturation-binding curves generated for each dose using [^{125}I]-CYP, levels of receptor activity were found to decrease with increasing amounts of radiation. Computer analysis of the data and calculation of dissociation constants (K_d) reveals no significant change in K_d with radiation dose. Calculation of the total number of receptors (B_{max}) for each saturation curve, and plotting of data as described above results in a monophasic inactivation line with a predicted molecular size of 54,000. This is in direct agreement with results obtained by purification and photoaffinity-labeling studies of the frog erythrocyte plasma-membrane β_2-adrenergic receptor (described above). Moreover, identical slopes and molecular size estimates were obtained for purified β_2-adrenergic receptor preparations. As a further precaution in the utilization of this technique, a protein of known molecular size, alkaline phosphatase (*E. coli*), was included along with receptor preparations, subjected to the irradiation procedures, and assayed (Stadman, 1961). This enzyme has been purified to homogeneity (Stadman, 1961) and shown to be a dimer of 84,000 M_r composed of two 43,000-M_r subunits (Lazdunshi and Lazdunshi, 1969). Our experiments routinely yield an estimate of 45,000–49,000 M_r in close agreement with published values observed by other techniques.

As in the case of the frog erythrocyte β_2-receptor, irradiation of turkey erythrocyte membrane preparations and construction of [^{125}I]-CYP saturation curves reveal a radiation-dependent loss of binding activity. Calculation of apparent size for the β_1-receptor of turkey erythrocyte membrane results in an estimate of 45,000–49,000 M_r for particulate and slightly higher, 50,000 M_r, for purified preparations. These results, while in close agreement with results obtained by photoaffinity labeling (Lavin *et al.*, 1981) and purification studies (Shorr *et al.*, 1982b), differ from those reported by Nielsen *et al.* (1981) for β_1-receptors from the same source. The reasons for this lack of agreement are not clear but may reflect differences in membrane preparations, i.e., level of protease contamination and lack of inhibitors, and conditions of receptor assay. This is particularly important since radiation inactivation measurements indicate only the size of the functional unit being measured (Kempner *et al.*, 1980). Thus, assay conditions such as temperature and the ligands utilized which can affect the function being measured can alter the estimates of molecular size.

Recently, Fraser and Venter (1982) have examined the β_2-adrenergic receptor of mammalian lung (canine and bovine) by radiation inactivation. Purification and affinity-labeling experiments with these receptors has suggested a molecular size of 59,000 daltons, as determined by SDS-PAGE (Venter and Fraser, 1982). In radiation-inactivation experiments, an estimated receptor size of 109,000 ± 5300 daltons was obtained. This data was interpreted as indicating a dimeric structure of two identical subunits for this mammalian β_2-adrenergic receptor. In this study, however, sample temperatures were maintained at approximately −50°C rather than −135°C, and no temperature correction factor as proposed by Kempner and Schlegel (1979) was used. In addition, radiation doses did not exceed 5 Mrads and receptor activity was not inactivated below 45% of control values. This is of interest since the authors claim to have calculated receptor sizes from the equation

$$M_r = \frac{6.4 \times 10^{11}}{D_{37}}$$

where, as described above, D_{37} is the radiation dose required to reduce the measured activity to 37% of the control value. A further complication in this study is the pharmacological heterogeneity of mammalian lung tissue. In particular, bovine lung preparations have been demonstrated to contain β-adrenergic receptors in an apparent ratio of 80% β_2 and 20% β_1 (K. E. J. Dickinson and S. Nahorski, personal communication).

As described above, in our homogeneous particulate and purified β-adrenergic receptor preparations, molecular sizes are obtained by target analysis methods which are in close agreement to those obtained by purification and affinity-labeling experiments. In addition, the values we have reported for receptor-specific binding activities, after complete purification, approach theoretical values predicted for a single-binding site per receptor subunit (Shorr *et al.*, 1982a,b). These data support our identification of the purified proteins as representing the receptors in the membrane and also suggest a single (antagonist) ligand-binding site per receptor subunit.

VI. SUMMARY AND PERSPECTIVES

Thus, from the results presented in this brief review, it would appear that the binding subunit of both β_1- and β_2-adrenergic receptors (at least in mammalian tissues) reside on similar peptides of 62,000–65,000 M_r. This contention is supported by several lines of investigation. As reviewed here, photoaffinity-labeling studies in many tissues in the presence of proteinase inhibitors result in the covalent labeling of a peptide of 62,000–65,000 M_r. In several systems, purification studies, using as the main step affinity chromatography procedures, yield peptides which appear to have identical M_rs to those labeled by photoaffinity labeling (Shorr *et al.*, 1982a; Benovic *et al.*, 1982; Homcy *et al.*, 1983). Moreover, estimates of functional receptor size by radiation inactivation of both particulate and purified preparations (Shorr *et al.*, 1984) have yielded the same molecular size as that determined by both purification and photoaffinity labeling.

The obvious exception to the contention that the binding subunit of both β_1- and β_2-receptors reside on similar M_r peptides is the receptor from the avian erythrocyte. As mentioned previously, it is not known whether the 40,000 and 45,000–50,000 peptides represent different forms of the same receptor, different receptors, or degradation products of a larger M_r peptide.

The data reviewed here strongly suggest that the different peptides isolated or identified by photoaffinity labeling do indeed represent the ligand-binding site of these receptors. These peptides bind reversible ligands with appropriate affinity and specificity and moreover, are covalently labeled by the varous photoaffinity probes with appropriate β-adrenergic specificity. However, it has not previously been possible to determine whether these peptides, in addition to carrying the "binding function" of the receptor, also carry its "activating function," i.e., the ability to interact with the effector system. Recently, we have developed procedures for reconstituting receptors of graded purity into lipid vesicles and for fusing these receptor-containing vesicles with a cell (*Xenopus laevis* erythrocyte) which lacks β-adrenergic receptors and β-adrenergic responsiveness of its adenylate cyclase (Cerione *et al.*, 1983). These experiments have indicated that the β-adrenergic receptor peptides purified from different sources are able to confer β-adrenergic responsiveness to the hybrid system. The extent of the newly established response is directly related to the amount of purified receptor reconstituted regardless of its extent of purification. This suggests that the peptides obtained by purification do indeed carry both functions of the receptor, binding, and activation, and that a necessary accessory component is not lost during purification. With these techniques at hand, it will now be possible to explore the structure-function relationships of these polypeptides.

ACKNOWLEDGMENT

We would like to express our sincere thanks to Lynn Tilley for her superb assistance in the preparation of this manuscript.

REFERENCES

Abramowitz, J., Iyengar, R., and Birnbaumer, L., 1979, Guanyl nucleotide regulation of hormonally-responsive adenylyl cyclases, *Mol. Cel. Endocrinol.* **16:**129–146.

Ahlquist, R. P., 1948, A study of the adrenotropic receptor, *J. Physiol. (London)* **153:**586–599.

Atlas, D., and Levitzki, A., 1978, Tentative identification of beta-adrenoreceptor subunits, *Nature* **272:**370–371.

Atlas, D., Steer, M. L., and Levitzki, A., 1974, Stereospecific binding of propranolol and catecholamines to the beta-adrenergic receptor, *Proc. Natl. Acad. Sci. USA* **71:**4246–4248.

Atlas, D. M., Steer, M. L., and Levitzki, A., 1976, Affinity label for beta-adrenergic receptor in turkey erythrocytes, *Proc. Natl. Acad. Sci. USA* **73:**1921–1925.

Aurbach, G. D., Fedak, S. A., Woodward, C. J., Palmer, J. S., Hauser, D., and Troxler, F., 1974, The beta-adrenergic receptor: Stereospecific interaction of an iodinated beta blocking agent with a high affinity site, *Science* **186:**1223–1224.

Benovic, J. L., Shorr, R. G. L., Heald, S. L., Lavin, T. N., Caron, M. G., and Lefkowitz, R. J., 1982, Purification and photoaffinity labeling of a mammalian $beta_2$ adrenergic receptor, *Fed. Proc.* **41:**1161.

Benovic, J. L., Stiles, G. L., Lefkowitz, R. J., and Caron, M. G., 1983, Photoaffinity labeling of mammalian beta-adrenergic receptors: Metal-dependent proteolysis explains apparent heterogeneity, *Biochem. Biophys. Res. Commun.* **110:**504–511.

Birnbaumer, L., Pohl, S. L., Krans, M. L., and Rodbell, M., 1970, The actions of hormones on the adenyl cyclase system, *Adv. Biochem. Psychoparmacol.* **3:**185–208.

Birnbaumer, L., Bearer, C. F., and Iyengar, R., 1980, A two-state model of an enzyme with an allosteric regulatory site capable of metabolizing the regulatory ligand: Simplified mathematical treatment of transient and steady state kinetics of an activator and its competitive inhibition as applied to adenylyl cyclases, *J. Biol. Chem.* **255:**3552–3557.

Burgermeister, W., Hekman, M., and Helmreich, J. M., 1982, Photoaffinity labeling of the beta-adrenergic receptor with azide derivatives of iodocyanopindolol, *J. Biol. Chem.* **257:**5306–5311.

Caron, M. G., and Lefkowitz, R. J., 1976, Solubilization and characterization of the beta-adrenergic receptor of frog erythrocytes, *J. Biol. Chem.* **251:**2374–2384.

Caron, M. G., Srinivasan, Y., Pitha, J., Kiolek, K., and Lefkowitz, R. J., 1979, Affinity chromatography of the beta-adrenergic receptors, *J. Biol. Chem.* **254:**2923–2927.

Cassel, D., and Sellinger, Z., 1978, Mechanism of adenylate cyclase activation through the beta-adrenergic receptor: Catecholamine-induced displacement of bound GDP by GTP, *Proc. Natl. Acad. Sci. USA* **75:**4155–4159.

Cerione, R. A., Strulovici, B. Benovic, J. L., Strader, C. D., Caron, M. G., and Lefkowitz, R. J., 1983, Reconstitution of beta-adrenergic receptors into lipid vesicles: Affinity chromatography purified receptors convey catecholamine responsiveness to a heterologous adenylate cyclase system, *Proc. Natl. Acad. Sci. USA,* **80:**4899–4903.

Chenieux-Guicheney, P., Dausse, J. P., Meyer, P., and Schnitt, H., 1978, Inhibition of [^{3}H]dihydroalprenolol binding to rat cardiac membranes by various beta blocking agents, *Br. J. Pharmacol.* **63:**177–182.

Citri, Y., and Schramm, D., 1980, Resolution, reconstitution and kinetics of the primary action of a hormone receptor, *Nature* **287:**297–300.

Cuatrecasas, P., 1974, Membrane receptors, *Annu. Rev. Biochem.* **43:**169–214.

Dale, H. H., 1906, On some physiological actions of ergot, *J. Physiol. (London)* **34:**163–206.

De Lean, A., Stadel, J. M., and Lefkowitz, R. J., 1980, A ternary complex model explains the agonist-specific binding properties of the adenylate cyclase coupled beta adrenergic receptor, *J. Biol. Chem.* **255:**7108–7117.

De Lean, A., Hancock, A. A., and Lefkowitz, R. J., 1981, Validation and statistical analysis of a computer modeling method for quantitative analysis of radioligand binding data for mixtures of pharmacological receptor subtypes, *Mol. Pharmacol.* **21:**5–16.

Dickinson, K. E. J., and Nahorski, S. R., 1981, Identification of solubilized $beta_1$ and $beta_2$ adrenoceptors in mammalian lung, *Life Sci.* **29:**2527–2533.

Durieu-Trautmann, O., Delavier-Klutchko, C., Vauquelin, G., and Strosberg, A. D., 1980, Visualization of the turkey erythrocyte beta-adrenergic receptor, *J. Supramol. Struct.* **13:**411–419.

Engel, G., Hoyer, D., Berthold, R., and Wagner, H., 1981, (±)[^{125}I]Cyanopindolol, a new ligand for beta-adrenoceptors: Identification and quantitation of subclasses of beta-adrenoceptors in guinea pig in *Naunyn-Schmiedebergs Arch. Pharmacol.* **317:**277–285.

Fraser, C. M., and Venter, J. C., 1980, Monoclonal antibodies to beta-adrenergic receptors: Use in purification and molecular characterization of beta receptors, *Proc. Natl. Acad. Sci. USA* **77:**7034–7038.

Fraser, C. M., and Venter, J. C., 1982, The size of the mammalian lung beta$_2$-adrenergic receptor as determined by target size analysis and immunoaffinity chromatography, *Biochem. Biophys. Res. Commun.* **109:**21–29.

Gozlan, H., Homburger, V., Lucas, M., and Bockaert, J., 1982, Photoaffinity labeling of beta-adrenergic receptors in C6 glioma cells: Presence of a nucleophilic group in the receptor, *Biochem. Pharmacol.* **31:**2879–2886.

Guillon, G., Cantau, B., and Jard, S., 1981, Effects of thiol-protecting reagents on the size of solubilized adenylate cyclase and on its ability to be stimulated by guanyl nucleotides and fluoride, *Eur. J. Biochem.* **117:**401–406.

Hoffman, B. B., Dukes, D. F., and Lefkowitz, R. J., 1981, Alpha-adrenergic receptor subtypes in liver membranes: Delineation with subtype selective radioligands, *Life Sci.* **28:**265–272.

Homcy, C. J., and Sylvester, D., 1982, Characterization of the mammalian beta$_2$ receptor in 8 M urea, *Biochem. Biophys. Res. Commun.* **108:**504–509.

Homcy, C. J., Rockson, S. G., Countaway, J., and Egan, D. A., 1983, Purification and characterization of the mammalian beta$_2$ adrenergic receptor, *Biochemistry* **22:**660–668.

Jakobs, K. H., 1979, Inhibition of adenylate cyclase by hormones and neurotransmitters, *Mol. Cell. Endocrinol.* **16:**147–156.

Kempner, E. S., and Schlegel, W., 1979, Size determination of enzymes by radiation inactivation, *Anal. Biochem.* **92:**2–10.

Kempner, E. S., Miller, J. H., Schlegel, W., and Hearon, J. Z., 1980, The functional unit of polyenzymes: Determination by radiation inactivation, *J. Biol. Chem.* **255:**6826–6831.

Kepner, G. R., and Macey, R. I., 1963, Membrane enzyme systems: Molecular size determinations by radiation inactivation, *Biochim. Biophys. Acta* **163:**188–203.

Lands, A. M., Arnold, A., McAuliff, J. P., Cuduena, F. P., and Brown, T. G., 1967, Differentiation of receptor systems by sympthomimetic amines, *Nature (London)* **214:**597–598.

Langer, S. Z., 1979, Presynaptic adrenoceptors and regulation of release, in: *The Release of Catecholamines from Adrenergic Neurons* (D. M. Paton, ed.), Pergamon Press, New York, pp. 59–85.

Langley, J. N., 1878, On the physiology of salivary secretion: Part 2. The mutual antagonism of atropin and pilocarpin having especial reference to their relations in the sub-maxillary gland of the cat, *J. Physiol. (London)* **1:**339–342.

Langley, J. N., 1905, On the reaction of cells and nerve endings to certain poisons, chiefly as regards the reaction of striated muscle to nicotine and to curari, *J. Physiol. (London)* **33:**374–376.

Lavin, T. N., Heald, S. L., Jeffs, P. W., Shorr, R. G. L., Lefkowitz, R. J., and Caron, M. G., 1981, Photoaffinity labeling of the beta-adrenergic receptor, *J. Biol. Chem.* **256:**11944–11950.

Lavin, T. N., Nambi, P., Heald, S. L., Jeffs, P. W., Lefkowitz, R. J., and Caron, M. G., 1982, [^{125}I]Para-azidobenzylcarazolol, a photoaffinity label for the beta-adrenergic receptor: Characterization of the ligand and photoaffinity labeling of beta$_1$ and beta$_2$ adrenergic receptors, *J. Biol. Chem.* **257:**12332–12340.

Lazdunshi, C., and Lazdunshi, M., 1969, Zn^{2+}- and Co^{2+}-alkaline phosphatases of *E. coli:* A comparative kinetic study, *Eur. J. Biochem.* **7:**294–300.

Lefkowitz, R. J., Roth, J., Pricer, W., and Pastan, I., 1970, ACTH receptors in the adrenal: Specific binding of ACTH-^{125}I and relation and to adenyl cyclase, *Proc. Natl. Acad. Sci. USA* **65:**745–752.

Lefkowitz, R. J., Mukherjee, C., Coverstone, M., and Caron, M. G., 1974, Stereospecific [^{3}H](−)alprenolol binding sites, beta-adrenergic receptors and adenyl cyclase, *Biochem. Biophys. Res. Commun.* **69:**703–710.

Lefkowitz, R. J., Caron, M. G., Limbird, L., Mukherjee, C., and Williams, L. T., 1976, Membrane-bound hormone receptors, in: *The Enzymes of Biological Membranes* (A. Martonosi, ed.), Plenum Press, New York, pp. 283–310.

Lefkowitz, R. J., Stadel, J. M., and Caron, M. G., 1983, Adenylate cyclase-coupled beta-adrenergic receptors: Structure and mechanisms of activation and densensitization, *Annu. Rev. Biochem.*, **52:**159–186.

Levitzki, A., 1978, The mode of coupling of adenylate cyclase to hormone receptors and its modulation by GTP, *Biochem. Pharmacol.* **27**:2083–2088.

Levitzki, A. and Helmreich, E. J. M., 1979, Hormone-receptor adenylate cyclase interactions, *FEBS Lett.* **101**:213–219.

Levitzki, A., Atlas, D., and Steer, M. L., 1974, The binding characteristics and number of beta-adrenergic receptors on the turkey erythrocyte in *Proc. Natl. Acad. Sci. USA* **71**:2773–2776.

Limbird, L., 1981, Activation and attenuation of adenylate cyclase: The role of GTP-binding proteins as macromolecular messengers in receptor-cyclase coupling, *Biochem. J.* **195**:1–13.

Lucas, M., Homburger, V., Dolphin, A., and Bockaert, J., 1979, *In vitro* and *in vivo* kinetic analysis of the interaction of a norbornyl derivative of propranolol with beta-adrenergic receptors of brain and C6 glioma cells: An irreversible or slowly reversible ligand, *Mol. Pharmacol.* **15**:588–597.

Mukherjee, C., Caron, M. G., Coverstone, M., and Lefkowitz, R. J., 1975, Identification of beta-adrenergic receptors in frog erythrocyte membranes with [^{3}H](−)alprenolol, *J. Biol. Chem.* **250**:4869–4876.

Mukherjee, C., Caron, M. G., Mullikin, D., and Lefkowitz, R. J., 1976, Structure activity relationship of beta-adrenergic receptors: Determination by direct binding studies, *Mol. Pharmacol.* **12**:16–31.

Neer, E. J., 1977, Solubilization and characterization of adenylyl cyclase: Approaches and problems, in: *Receptors and Hormone Action,* Vol. 1 (B. W. O'Malley and L. Birnbaumer, eds.), Academic Press, New York, pp. 463–484.

Neer, E. J., and Salter, R. S., 1980, Modification of adenylate cyclase structure and function by ammonium sulfate, *J. Biol. Chem.* **256**:5497–5503.

Nielsen, T. B., Lad, P. M., Preston, M. S., Kempner, E., Schlegel, W., and Rodbell, M., 1981, Structure of the turkey erythrocyte adenylate cyclase system, *Proc. Natl. Acad. Sci. USA* **78**:722–726.

Northup, J. K., Sternweis, P. C., Smigel, M. D., Schleifer, L. S., Ross, E. M., and Gilman, A. G., 1980, Purification of the regulatory components of adenylate cyclase, *Proc. Natl. Acad. Sci. USA* **77**: 6516–6520.

Pfeuffer, T., 1979, Guanine nucleotide-controlled interactions between components of adenylate cyclase, *FEBS Lett.* **101**:85–89.

Pike, L. J., and Lefkowitz, R. J., 1978, Agonist-specific alterations in receptor binding affinity associated with solubilization of turkey erythrocyte membrane beta-adrenergic receptors, *Mol. Pharmacol.* **14**:370–375.

Pitha, J., Zjawiony, J., Nasrin, N., Lefkowitz, R. J., and Caron, M. G., 1980, Potent beta-adrenergic antagonist possessing chemically reactive group, *Life Sci.* **27**:1791–1798.

Rashidbaigi, A., and Ruoho, A. E., 1981, Iodoazidobenzylpindolol, a photoaffinity probe for the beta-adrenergic receptor, *Proc. Natl. Acad. Sci. USA* **78**:1609–1613.

Rashidbaigi, A., and Ruoho, A. E., 1982, Photoaffinity labeling of beta-adrenergic receptors: Identification of the beta receptor binding site(s) from turkey, pigeon and frog erythrocytes, *Biochem. Biophys. Res. Commun.* **106**:139–148.

Rashidbaigi, A., Ruoho, A. E., Green, D., and Clark, R. B., 1982, Photoaffinity labeling of the beta-adrenergic receptor of lymphoma S-49, *Fed. Proc.* **41**:1327.

Rashidbaigi, A., Ruoho, A. E., Green, D., and Clark, R. B., 1983, Photoaffinity labeling of the β-adrenergic receptor from cultured lymphoma cells with [^{125}I]iodoazidobenzylpindolol: Loss of the label with desensitization, *Proc. Natl. Acad. Sci. USA* **80**:2849–2853.

Robison, G. A., Butcher, R. W., and Sutherland, E. W., 1967, Adenyl cyclase as an adrenergic receptor, *Ann. N.Y. Acad. Sci.* **139**:703–723.

Rodbell, M., 1980, The role of hormone receptors and GTP-regulatory proteins in membrane transduction, *Nature (London)* **284**:17–22.

Rodbell, M., Birnbaumer, L., Pohl, S. L., and Krans, M. J., 1971, The glucagon-sensitive adenyl cyclase system in plasma membranes of rat liver: V. An obligatory role of guanyl nucleotides in glucagon action, *J. Biol. Chem.* **246**:1877–1882.

Ross, E., 1981, Physical separation of the catalytic and regulatory proteins of hepatic adenylate cyclase, *J. Biol. Chem.* **256**:1949–1953.

Ross, E. M., and Gilman, A. G., 1980, Biochemical properties of hormone-sensitive adenylate cyclase, *Annu. Rev. Biochem.* **49**:533–564.

Ross, E. M., Howlett, A. C., Ferguson, K. N., and Gilman, A. G., 1978, Reconstitution of hormone-sensitive adenylate cyclase activity with resolved components of the enzyme, *J. Biol. Chem.* **253:**6401–6412.

Schleifer, L. S., Garrison, J. C., Sternweis, P. C., Northup, J. K., and Gilman, A. G., 1980, The regulatory component of adenylate cyclase from uncoupled S49 lymphoma cells differs in charge from the wild type protein, *J. Biol. Chem.* **255:**2641–2643.

Shocken, D., Caron, M. G., and Lefkowitz, R. J., 1980, The human placenta—a rich source of beta-adrenergic receptors: Characterization of the receptors in particulate and solubilized preparations, *J. Clin. Endocrinol. Metabol.* **50:**1082–1088.

Shorr, R. G. L., Lefkowitz, R. J., and Caron, M. G., 1981, Purification of the beta-adrenergic receptor: Identification of the hormone binding subunit, *J. Biol. Chem.* **256:**5820–5826.

Shorr, R. G. L., Heald, S. L., Jeffs, P. W., Lavin, T. N., Strohsacker, M. W., Lefkowitz, R. J., and Caron, M. G., 1982a, The beta-adrenergic receptor: Rapid purification and covalent labeling by photoaffinity crosslinking, *Proc. Natl. Acad. Sci. USA* **79:**2778–2782.

Shorr, R. G. L., Strohsacker, M. W., Lavin, T. N., Lefkowitz, R. J., and Caron, M. G., 1982b, The $beta_1$-adrenergic receptor of the turkey erythrocyte: Molecular heterogeneity revealed by purification and photoaffinity labeling, *J. Biol. Chem.* **257:**12341–12350.

Shorr, R. G. L., Kempner, E. S., Strohsacker, M. W., Nambi, P., Lefkowitz, R. J., and Caron, M. G., 1984, Determination of the molecular size of frog and turkey beta adrenergic receptors by radiation inactivation, *Biochemistry* **23:**747–752.

Spiegel, A. M., Downs, R. W., and Aurbach, G. D., 1979, Separation of a guanine nucleotide regulatory unit from the adenylate cyclase complex with GTP affinity chromatography, *J. Cyclic Nucleotide Res.* **5:**3–17.

Stadel, J. M., De Lean, A., and Lefkowitz, R. J., 1980, A high affinity agonist beta-adrenergic receptor complex is an intermediate for catecholamine stimulation of adenylate cyclase in turkey and frog erythrocyte membranes, *J. Biol. Chem.* **255:**1436–1441.

Stadel, J. M., Shorr, R. G. L., Limbird, L., and Lefkowitz, R. J., 1981, Evidence that a beta-adrenergic receptor associated guanine nucleotide regulatory protein conveys GTP-γS dependent adenylate cyclase activity, *J. Biol. Chem.* **256:**8718–8723.

Stadel J. M., De Lean, A., and Lefkowitz, R. J., 1982, Molecular mechanisms of coupling in hormone-receptor-adenylate cyclase systems, *Adv. Enzymol.* **53:**1–43.

Stadman, T. C., 1961, Alkaline phosphatases, in: *The Enzymes,* Vol. 5, 2nd Ed. (P. D. Boyer, H. Lardy, and K. Murback, eds.), Academic Press, New York, pp. 55–71.

Sternweis, P. C., and Gilman, A. G., 1979, Reconstitution of catecholamine-sensitive adenylate cyclase: Reconstitution of the uncoupled variant of the S49 lymphoma cell, *J. Biol. Chem.* **254:**3333–3340.

Sternweis, P. C., Northup, J. K., Smigel, M. D., and Gilman, A., 1981, The regulatory component of adenylate cyclase: Purification and properties, *J. Biol. Chem.* **256:**11517–11526.

Stiles, G. L., Strasser, R. H., Lavin, T. N., Jones, L. R., Caron, M. G., and Lefkowitz, R. J., 1983, The cardiac beta-adrenergic receptor: Structural similarities of $beta_1$ and $beta_2$ receptor subtypes demonstrated by photoaffinity labeling, *J. Biol. Chem.,* **258:**8443–8449.

Strauss, W. L., Geetha, G., Fraser, C. M., and Venter, J. C., 1979, Detergent solubilization of mammalian cardiac and hepatic beta-adrenergic receptors, *Arch. Biochem. Biophys.* **196:**566–573.

Strosberg, A. D., Vauquelin, G., Durieu-Trautmann, O., Delavier-Klutchko, C., Bottari, S., and Andre C., 1980, Towards the chemical and functional charaterization of the beta-adrenergic receptor, *Trends Biochem. Sci.* **5:**11–14.

Sutherland, E. W., and Rall, T. W., 1960, The relation of adenosine-3′,5′ phosphate and phosphorylase to the actions of catecholamines and other hormones, *Pharmacol. Rev.* **12:**265–299.

Sutherland, E. W., Rall, T. W., and Mennon, T., 1962, Adenyl cyclase: I. Distribution, preparation, and properties, *J. Biol. Chem.* **237:**1220–1227.

Sutherland, E. W., Robison, G. A., and Butcher, R. W., 1968, Some aspects of the biological role of adenosine 3′,5′-monophosphate (Cyclic AMP), *Circulation* **37:**279–306.

Swillens, S., and Dumont, J. E., 1981, A unifying model of current concepts and data on adenylate cyclase activation by beta-adrenergic agonists, *Life Sci.* **27:**1013–1028.

Terasaki, W. L., Linden, J., and Brooker, G., 1979, Quantitative relationship between beta-adrenergic receptor number and physiologic responses as studied with a long-lasting beta-adrenergic antagonist, *Proc. Natl. Acad. Sci. USA* **76:**6401–6405.

Vauquelin, G., Geynet, P., Hanoune, J., and Strosberg, A. D., 1977, Isolation of adenylate cyclase-free beta-adrenergic receptor from turkey erythrocyte membranes by affinity chromatography, *Proc. Natl. Acad. Sci. USA* **74:**3710–3714.

Vauquelin, G., Geynet, P., Hanoune, J., and Strosberg, A. D., 1978, Purification of beta-adrenergic receptors: The search for an adequate affinity adsorbent, *Life Sci.* **23:**1791–1796.

Vauquelin, G., Geynet, P., Hanoune, J., and Strosberg, A. D., 1979, Affinity chromatography of the beta-adrenergic receptor from turkey erythrocytes, *Eur. J. Biochem.* **98:**543–556.

Venter, J. C., and Fraser, C. M., 1983, Beta-adrenergic receptor isolation and characterization with immobilized drugs and monoclonal antibodies, *Fed. Proc.* **42:**273–278.

Venter, J. C., Fraser, C. M., and Harrison, L. C., 1980, Autoantibodies to $beta_2$ adrenergic receptors: A possible cause of adrenergic hyporesponsiveness in allergic rhinitis and asthma, *Science* **207:**1361–1363.

Venter, J. C., Fraser, C. M., Soiefer, A. I., Jeffrey, D. R., Strauss, W. L., Charlton, R. R., and Greguski, R., 1981, Autoantibodies and monoclonal antibodies to beta-adrenergic receptors: Their use in receptor purification and characterization, *Adv. Cyclic Nucleotide Res.* **14:**135–143.

Weiland, G., and Molinoff, P. B., 1981, Quantitative analysis of drug-receptor interactions: I. Determination of kinetic and equilibrium properties, *Life Sci.* **29:**313–330.

Williams, L. T., and Lefkowitz, R. J., 1976, Alpha-adrenergic receptor identification by [^{3}H]dihydroergocryptine, *Science* **192:**791–793.

Wood, C. L., Caron, M. G., and Lefkowitz, R. J., 1979, Separation of alpha- and beta-adrenergic receptors by affinity chromatography, *Biochem. Biophys. Res. Commun.* **88:**1–8.

Wrenn, S. M., and Homcy, C. J., 1980, Photoaffinity label for the beta-adrenergic receptor: Synthesis and effects on isoproterenol-stimulated adenylated cyclase, *Proc. Natl. Acad. Sci. USA* **77:**4449–4452.

26

Ionic Channels and Their Metabolic Control

P. G. Kostyuk

I. INTRODUCTION

Extensive investigations performed during recent decades on many excitable cells firmly establish the main principles concerning the mechanisms of active cellular responses to external stimuli. These principles can be stated as follows.

1. The ionic content of the cytoplasm largely differs from that of the extracellular medium. Such a deviation from thermodynamic equilibrium is related partly to specific binding of ions by intracellular structures, but mainly to the activity of the surface membrane of the cell through which the influx and efflux of ions take place. The main component of this activity is produced by enzymatic transport systems localized in the surface membrane pumping out certain ions and sucking in the others. In order to maintain the uneven distribution of ions, a considerable amount of energy liberated in the course of cell metabolism should be utilized.

2. The principal meaning of uneven distribution of ions on both sides of the surface membrane is that in conditions of relatively low passive membrane permeability, it creates for the cell a considerable reserve of potential energy in the form of transmembrane electrochemical gradients. In order to transform it into an active response, it is only necessary that a way should be open by external stimulus for the movement of ions through the membrane. Existing transmembrane electrochemical gradient will immediately produce a flow of these ions which, one way or another, bring into action the subsequent chain of physicochemical and biochemical cell reactions.

P. G. Kostyuk • A. A. Bogomoletz Institute of Physiology, Academy of Sciences of the Ukrainian SSR, Kiev, USSR.

Studies carried out on squid giant axons are of great importance in the statement of these principles as they open the way for direct measurement of electric ionic currents through the excitable membrane. The ideas that arise from these studies about the presence in the membrane of special molecular structures, ionic pores or channels forming specific pathways for the transmembrane flow of ions, prove to be extremely fruitful and lay the foundation for an extensive series of further investigations which not only confirm the existence of such channels but give quantitative functional characteristics for many of them. Discovery of specific toxins selectively blocking certain ionic currents gives independent evidence for the correctness of these ideas. It was established that ionic channels, together with a functional component determining its ability to select one or another penetrating ions ("selectivity filter"), have also another component which opens or blocks the way of ions through the channel in response to external influences ("gating mechanism"). In some cases, the gating mechanism responds to a change in the transmembrane electric field (electrically-operated ionic channels), in others, to the presence of some physiologically-active substances (chemically-operated ionic channels).

Resulting from all these works, a general scheme of functioning of excitable membranes has been developed, according to which, ionic channels and cell metabolism are operationally independent in the mechanism of maintaining cell excitability. Metabolism only creates prerequisites for the generation of transmembrane ionic currents by forming transmembrane electrochemical gradients. This scheme looked universal for a long time. However, recently, experimental data have been obtained which convincingly show that, in certain cases, the functioning of ionic channels in the membrane may be under direct control of intracellular metabolic processes. Before presenting these data, the functional peculiarities of different types of ionic channels and, in the first place, of electrically-operated calcium channels should be elucidated.

II. Peculiarities of Different Types of Membrane Ionic Channels

A. Selectivity

Sodium and potassium currents through the corresponding electrically-operated ionic channels were a subject of thorough examination and characteristics of their functioning are now widely known (see reviews by Khodorov, 1975; Hille, 1978; Hille and Schwarz, 1979; etc.). Apart from sodium and potassium gradients, a high transmembrane gradient of calcium ions is effectively maintained in excitable cells. The content of Ca^{2+} ions in the extracellular medium is relatively small. However, the free level of these ions inside the cell is extremely low owing to the presence of binding systems having a considerable preference for divalent cations as compared to monovalent ones. Therefore, transmembrane flux of Ca^{2+} ions can be a no less effective carrier of electric current than the flux of sodium or potassium ions. Moreover, Ca^{2+} current may be of particular importance for cell activity taking into account specific physicochemical properties of these ions, due to the fact that they can effectively bind to large organic molecules and regulate those biochemical reactions, in which the latter are involved. The search for specific calcium channels began later than that for sodium and potassium channels, however, it turned out to be quite successful. The systems of specific calcium ionic channels have been discovered first in the muscle fiber membrane (see reviews by Reuter, 1973; Hagiwara, 1973), and then in the neuronal membrane. The somatic membrane of nerve cell proved to be an especially convenient object for the study of calcium channels. In contrast to the axonal membrane, it can generate full-scale action potentials in Na-free solution which depend on the extracellular concentration of Ca^{2+} ions (as well as of Sr^{2+} or Ba^{2+} ions). Besides the somatic membrane, such channels are also well presented in the membrane of presynaptic endings where an inward calcium current created by these channels is a coupling factor between depolarization of the membrane and transmitter release (Katz and Miledi, 1967a–c). However, because of small dimensions of the terminal, the introduction of a microelectrode and direct measurement of the calcium current becomes possible only in exceptional cases (Llinas *et al.*, 1972, 1976). Meanwhile, the development of a perfusion technique for isolated nerve cells (Kostyuk *et al.*, 1975) allows reliable separation of calcium currents from other types of ionic currents and precise measurement of their characteristics under any desirable influences upon the external and internal side of the cell membrane. Initially applied to giant molluscan neurons (Kostyuk *et al.*, 1977; Kostyuk and Krishtal, 1977a,b), it proved to be quite universal and became widely used for the perfusion of amphibian (Veselovsky *et al.*, 1977a) and mammalian neurons (Veselovsky *et al.*, 1979) and neuroblastoma cells (Veselovsky *et al.*, 1977b; Kostyuk *et al.*, 1978). Modifications of the intracellular perfusion technique have also been developed by Lee *et al.* (1978) and Takahashi and Yoshii (1978). The method of the latter authors was successfully used for the perfusion of oocytes and the separation of calcium current in the egg-cell membrane.

The findings obtained in these studies give detailed characteristics of the differences in the functional organization of calcium and sodium ionic channels which allow the cell to govern separately the inward transmembrane fluxes of these cation species. First, these differences concern the arrangement of the "selectivity filter," the information about which has been obtained from the studies of the passage of "unusual" cations through the channels that differ in their physicochemical properties from normal current-carrying ions. The investigations of the relative permeability of electrically-operated sodium channels to such ions performed on a variety of excitable membranes show great similarity between the channels. Table 1 presents data about relative permeabilities of the sodium channels in various excitable membranes to several monovalent

Table 1. Relative Selectivity of Sodium Channel to Different Monovalent Cations

Object	Na^+	Li^+	$N_2H_5^+$	NH_4^+	K^+	Reference
Squid axon	1.0	1.10			0.08	Chandler and Meves (1965)
Frog axon	1.0	0.93	0.59	0.16	0.09	Hille (1971, 1972)
Frog muscle fiber	1.0	0.96	0.31	0.11	0.05	Campbell (1976)
Rat neuron soma	1.0	0.79	0.43	0.33	0.18	Kostyuk *et al.* (1981a)
Snail neuron soma	1.0	1.04	0.44		0.1	Kostyuk and Shuba (1982)

cations obtained from measurements of equilibrium potential shifts of the transmembrane current at corresponding changes in the ionic composition of the extracellular medium. The sequences obtained are quite similar and have only small quantitative differences. They can be well explained proceeding from the idea about the selectivity filter as a pore of a certain cross section (3.1 × 5.1 Å) passing ions of a given group together with some amount of hydrating water molecules with the help of steric selection (Hille, 1972). The formation of hydrogen bonds between ions and the oxygen atoms in the channel wall can contribute additionally to their passage through the channel.

Table 2 demonstrates data about the permeability of calcium channels in the somatic membrane to divalent cations. Because these channels pass only inward-going currents (see below), the values presented were obtained by comparing the maximum currents under equimolar replacements of corresponding ions in the extracellular medium.

The sequence of preferable permeation of divalent cations through the calcium channels is similar for both objects but the values differ quantitatively. In both cases, calcium channels do not show noticeable permeability towards monovalent ions. The disagreement can be stipulated by the fact that not only differences in channel permeability but also some other factors, e.g., changes in near-membrane ion concentration due to the presence of surface charges on the membrane, may influence the maximum current values. For detailed assessment of calcium channel properties, measurements of concentration dependences of currents created by different carrier ions have been performed. They show that such dependences have a clear-cut saturation and fit well to the Langmuir isotherm. From these measurements, the dissociation constants K_d and the values $pK = -\log K_d$ for the complexes formed by divalent cations with a binding site in the channel were calculated. The corresponding pK values after correction for the change in the near-membrane concentration of ions caused by fixed charges on the outer membrane surface are presented in Table 3. The table also illustrates the effect of other divalent cations on the calcium channels which, unlike Ba^{2+}, Sr^{2+}, and Ca^{2+} ions, block the channels. Such blocking effect can also be described by Langmuir's isotherm and is due to a competitive binding to the same binding site in the channel. The binding characteristics of divalent cations to carboxylic group (glycine) in aqueous solution are given in this table for comparison.

Obviously, divalent cations in respect to their interaction with calcium channels form a continuous series close to that for their binding to a single carboxylic group. All divalent cations that weakly bind to the latter (weaker than Ca^{2+} ions), pass through the channel better the weaker they are bound; the cations that bind stronger, block the channel more actively the stronger they are bound.

Table 2. Relative Values of Currents Carried through Calcium Channels by Different Divalent Cations

Site	Ba^{2+}	Sr^{2+}	Ca^{2+}	Mg^{2+}	Reference
Snail neuron soma	2.8	2.6	1.0	0.2	Doroshenko *et al.* (1978)
Rat neuron soma	1.8	1.3	1.0		Kostyuk *et al.* (1981b)

Table 3. pK of Binding of Different Divalent Cations in Calcium Channels

Site	Ba^{2+}	Ca^{2+}	Co^{2+}	Ni^{2+}	Cd^{2+}	Reference
Snail neuron	1.0	2.0	3.2	3.5	4.3	Kostyuk *et al.* (1982)
Rat neuron	0.6	1.3	2.5	2.6	4.3	Kostyuk and Mironov (1982)
Carboxylic group	0.8	1.4	3.2	3.2	4.8	Martell and Smith (1977)

These results allow us to conclude that the organization of the "selectivity filter" in the calcium channel differs, in principle, from that in the sodium channel. The ability of calcium channels to select certain divalent cations is based on the characteristics of their binding in the channel (apparently by a carboxylic group); such a binding is a necessary step in the passage of ions through the channel.

The question arises why calcium channels do not pass monovalent cations (for example, Na^+) that bind weakly to a carboxylic group but are close in size to that of permeant Ca^{2+} cations. A clue for the understanding of this phenomenon may be given by the fact that calcium channels reversibly transform into sodium channels under complete removal of divalent cations from the extracellular medium by introduction of calcium-chelating agents (Kostyuk and Krishtal, 1977b). Following such a modification, the relative permeability (P) of channels to monovalent cations becomes $P_{Na} : P_{Li} : P_{N_2H_5} : P_{NH_4} : P_K = 1.0 : 0.98 : 0.47 : 0.42 : 0.26$, i.e., practically similar to that for natural sodium channels (see Table 1). At the same time, these modified channels not only retain the capability to be blocked by pharmacological blockers of the calcium channels which do not affect normal sodium channels (derivatives of verapamil and 1,4-dihydropyridine), but become even more sensitive to their action (Kostyuk and Shuba, 1982).

Thus, there are reasons to believe that the presence of divalent cations, *per se*, in the external medium is a factor that predetermines the elimination of monovalent cations from the calcium channels. One could think that high probability of the presence of bound divalent cation in the "selectivity filter" of these channels is the reason why monovalent cations cannot pass across them. However, a study of the concentration dependence of the modifying action of the lowering of divalent cations concentration in the medium upon selectivity of calcium channels (performed by introduction into the extracellular medium of different calcium-chelating or calcium-binding substances varying in free calcium level from 10^{-6} to 10^{-9} M) has shown that the pK values for the modifying action of divalent cations differ from the pK for the binding of these cations in the "selectivity filter" by more than four units (Kostyuk *et al.*, 1983a). Therefore, one has to suggest the presence of an additional structure in the external mouth of a calcium channel which binds divalent cations in a highly selective manner. This structure in a complex state prevents monovalent cations (by a still unknown mechanism) from entering the channel. Schematic presentation of this suggestion is shown in Figure 1.

B. *Sensitivity to Changes in Intracellular Ionic Composition*

A change in the intracellular content of carrier ions can be easily made in the course of intracellular perfusion. It will alter the ionic current flowing through the

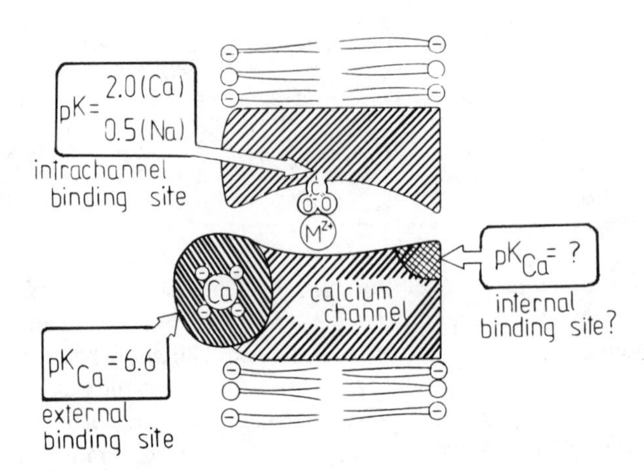

Figure 1. Schematic representation of presumed organization of a calcium channel providing its ionic selectivity. It contains a highly effective outer binding site which regulates channel selectivity towards monovalent cations and intrachannel binding site determining selectivity to different divalent cations. The presence of a third internal binding site is speculative.

corresponding channels due to the change in its electrochemical gradient. Numerous investigations of this kind made on sodium or potassium channels confirmed this prediction. The changes of the currents corresponded well to those predicted by the Nernst equation. Quite different behavior is characteristic of calcium channels. The use of intracellular perfusion techniques for varying the intracellular free calcium content (by introduction of Ca-EGTA or Ca-EDTA buffers of corresponding composition) has shown that an increase of internal calcium to 10^{-7} M results in a reversible block of the inward calcium current (Kostyuk and Krishtal, 1977b; Doroshenko and Tsyndrenko, 1978). Therefore, it is impossible to produce a reversal of the calcium current by elevation of intracellular Ca^{2+} content; in all cases, this current may have only an inward-going direction. This peculiarity of calcium channels seems to be closely associated with their behavior at prolonged depolarization of the surface membrane. Sodium and potassium channels become inactivated at sustained depolarization; the steady-state level and speed of their inactivation are potential dependent. Calcium currents also decline at sustained depolarization of the membrane. However, in this case, the decline is proportional to the inward calcium current itself but not to the membrane potential. For the first time, this was shown in two-pulse experiments on the somatic membrane of snail neurons (Tillotson, 1979; Eckert and Tillotson, 1981). Starting from the data about the effect of artificial increase of intracellular free calcium, the authors suggest that, under these conditions, the inactivation of calcium channels is also determined by the increase of internal free calcium. The study of the kinetics of calcium-current decline at sustained depolarization has shown the presence of at least two exponents with time constants of several dozens and several hundreds of msec (Shakhovalov, 1977). This circumstance was assessed by some investigators as an indication of the possibility of both current-dependent and potential-dependent inactivation of calcium channels (Fox, 1981; Plant and Standen, 1981; Brown *et al.*,

1981). However, more detailed consideration of this question on perfused snail neurons indicate that the fast component of the calcium-current decline represents an artifact caused by parallel activation of a nonspecific outward current (carried, probably, through Ca-activated potassium channels by Tris ions or protons). When such a nonspecific current was completely removed (by increasing intracellular pH to 8.3), the fast component of the current decline completely disappeared; it was also practically absent in the case when Ba^{2+} ions but not Ca^{2+} were ion carriers through the channel (Doroshenko *et al.*, 1984). Therefore, it seems that true inactivation of the calcium channels in the somatic membrane is a slow process caused by entry of Ca^{2+} ions.

The described feature of the calcium channels can be explained by the presence of one more specific binding site located at the inner mouth of the channel. However, this explanation entails substantial difficulties because a very effective binding (K_d is about 10^{-8} M) must be assumed for such a site. If such a deep potential well would exist in the channel, its conductance would be negligible (currents not exceeding 10^{-15} A). Meanwhile, experimental measurement of unitary calcium-channel conductance in snail somatic membrane gives a value of 0.5 p*S* using fluctuation analysis (Krishtal *et al.*, 1981), and even more in patch clamp experiments (Lux and Nagy, 1981; Reuter *et al.*, 1982). These data suggest a more complex mechanism of the action of internal Ca^{2+} ions on the calcium channel.

C. $[Ca^{2+}]_i$-Activated Channels

There is one more type of ionic channel whose functioning largely depends on binding of divalent cations. Activation of potassium conductance upon increasing $[Ca^{2+}]_i$ has been found by Meech and Strumwasser (1970) in snail neurons. These channels essentially differ in their functional properties from "fast" and "slow" potential-dependent potassium channels. It was supposed that they are, in fact, chemically-operated channels activated by binding of Ca^{2+} ions to their inner mouth. As regards external influences, these channels strikingly resemble calcium channels. It was even suggested that they are a result of modification of calcium channels while calcium ions are moving through them (Hofmeier and Lux, 1978). However, more thorough examination of the activity of these channels in conditions of intracellular perfusion show that they are electrically operated channels which require Ca^{2+} ions at the inner side of the membrane as a cofactor for retaining the ability to be activated (Doroshenko *et al.*, 1979; Kostyuk *et al.*, 1980; Lux *et al.*, 1981).

III. Metabolic Regulation of Ionic Channels

A. Activity of Calcium Channels Dependent on Metabolism

A suggestion that the activity of electrically-operated calcium channels can be under direct metabolic control was first stated based on the studies of inward calcium currents in heart muscle fibers. It is known that such currents can be potentiated under catecholamine action; surplus current is potential and time dependent, and, consequently, is also created by electrically operated calcium channels. As was shown by Reuter (1974), in parallel to such potentiation, an increase of 3′,5′-cAMP inside the

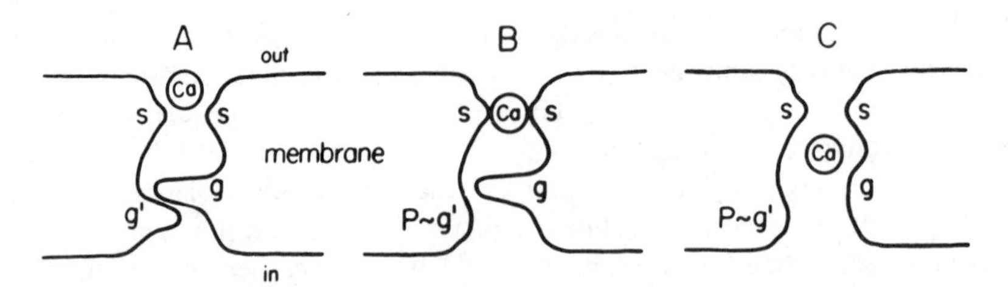

Figure 2. Hypothetical scheme of regulation of Ca channels in cardiac muscle. The channel contains a selectivity filter (s) and two gates; one is the voltage-dependent and provides the activation of conductance (g), the other one is voltage-independent gate but is dependent on phosphorylation by cAMP-dependent protein kinase (g′). A—without phosphorylation and depolarization the channel is closed. B—channel is phosphorylated but closed by voltage-dependent gates. C—phosphorylated channel is open by membrane depolarization (Reuter, 1979).

fibers takes place. Taking into consideration modern ideas about the role of cyclic nucleotides and cAMP-catalyzed protein phosphorylation (see Greengard, 1978; and others), it was assumed that activation of additional calcium channels is mediated through cAMP synthesis by membrane adenylate cyclase, which is, in turn, activated by catecholamines. The cAMP-mediated phosphorylation of calcium-channel proteins is a factor maintaining their ability for activation, as if calcium channels have an additional gating mechanism directly operated by cellular biochemical processes. Figure 2 depicts this suggestion schematically (Reuter, 1979). It was substantiated by the finding that introduction of cAMP into cardiac muscle fiber is also accompanied by a similar effect (Tsien, 1973).

Possible participation of cAMP in the modulation of calcium conductance in the nerve-cell membrane was proposed in the work by Klein and Kandel (1978). The authors in experiments on *Aplysia* neurons blocked the delayed outward potassium current by TEA and recorded the remaining inward calcium current. Serotonin prolonged this current; similar effect on the inward calcium current was exerted by intracellular cAMP injection or by incubation of the ganglia in solutions containing phosphodiesterase blockers. A serotonin-evoked intensification of the calcium current was also observed by Deterre *et al.* (1981) in snail neurons. This current was dependent on the external concentration of Ca^{2+} ions and blocked by Co^{2+} ions. Intracellular EGTA injection blocking the Ca-dependent potassium conductance (see below) did not affect this current; it could be reproduced either by intracellular cAMP injection or by extracellular application of phosphodiesterase inhibitors. A study of adenylate cyclase activity in the homogenate of ganglion neurons has shown that it increases by 50–100% under serotonin action.

The calcium channels in perfused nerve cells are especially suitable for the study of their functioning dependent on metabolism. During intracellular perfusion by saline solution, the calcium currents rapidly decrease for several minutes in large snail neurons and for several dozens of seconds in smaller mammalian neurons (Kostyuk and Krishtal, 1977a; Fedulova *et al.*, 1981; Byerly and Hagiwara, 1982), whereas sodium and potassium currents remain practically unaffected. Kinetic parameters of the calcium currents do not change in this case; the speed of the calcium-current decline slows down considerably when the temperature is lowered. Obviously, some cytoplasmic factor necessary for normal functioning of just calcium channels is washed out from the cell in the course of its perfusion, and in the absence of this factor, the channels become inactivated.

The intracellular perfusion technique extremely facilitates the search for such a factor. Introduction of cAMP together with ATP and Mg^{2+} ions to the perfusing solution not only prevents the further decline of the calcium conductance but restores it to a considerable extent. Introduction of either ingredient into the cell exerts some, but weak, stabilizing effects. The maximum effect was observed on a perfused *Helix* neuron at the cAMP concentration of about 10^{-4} M, though restoration was also seen with its micromolar concentrations. Optimal concentration was 2 mM ATP and 3 mM Mg^{2+} (Doroshenko *et al.*, 1982). After reaching the maximum, the calcium conductance begins to decline again, but much more slow than during perfusion with saline solution only. Intracellular administration of cGMP does not influence the decline of calcium conductance.

Evidence that membrane-bound enzymes are not destroyed during intracellular perfusion was obtained in experiments where they could be affected by introduction of corresponding enzyme activators or inhibitors. For instance, the introduction of 10 mM fluoride (adenylate cyclase activator) + ATP and Mg^{2+} in the intracellular solution restores the calcium conductance. On the contrary, addition of small quantities of Cu^{2+} ions (blockers of adenylate cyclase) interrupts the restorative effect. These experiments may be interpreted in such a way that adenylate cyclase preserving its function and activated by fluoride can still enhance the intracellular cAMP concentration in order to keep the calcium channels activated on the condition that ATP is present in the cell in required quantities. Large concentrations of internal fluoride, on the contrary, evokes fast and irreversible inactivation of the calcium conductance (Kostyuk *et al.*, 1975). This effect seems to be also metabolically mediated; it slows down abruptly at lowering the temperature, however, its mechanism is not yet clear.

It should be noted that in experiments with cAMP injection into intact snail nerve cells (by pressure or iontophoresis from a microelectrode), just the opposite effect has been observed, i.e., a reversible suppression of the calcium conductance (Kononenko and Kostyuk, 1982). In this case, the mechanism of cAMP action could be possibly more complex, for example, by liberation of calcium ions from the intracellular stocks. In perfused neurons, the level of intracellular free Ca^{2+} is stabilized. Therefore, direct cAMP effects may be displayed more clearly. In large neurons of *Aplysia,* similar injection of cAMP from a microelectrode potentiated the calcium-dependent inward current (Pellmar, 1981).

The evidence presented about the participation of intracellular cAMP in maintaining the activity of calcium channels in the somatic membrane provided a way to explain the described blocking effect of the increase in the intracellular level of free calcium on the calcium conductance. Introduction of exogenous cAMP, together with ATP into the neuron perfused with saline containing elevated free Ca^{2+} level, completely abolishes its blocking effect; the cell can endure a very strong elevation of $[Ca^{2+}]_i$ without loss of calcium conductance. On the other hand, reduction in intracellular free Ca^{2+} level by addition of calcium-chelating substances to the perfusate increases the restorative effect of cAMP on the calcium conductance (Doroshenko *et al.*, 1982). It is known that Ca^{2+} ions activate cell phosphodiesterase (which destroys cAMP) in micromolar concentrations by means of calmodulin; their removal, on the contrary, inhibits its activity (Rasmussen *et al.*, 1979). The activation of phosphodiesterase in the cell and the corresponding reduction of cAMP level during the enrichment of the cell by Ca^{2+} ions seem to be an essential (if not principal) factor underlying the suppression of membrane-calcium conductance. Intracellular administration of 2 mM theophylline also slowed down the development of calcium block, though less effectively than introduction of cAMP. Theophylline at such concentrations probably does not depress completely the phosphodiesterase activity.

The participation of calmodulin in the process triggered by the elevation of intracellular free calcium is in agreement with the data about the effect of calmodulin blockers (triphtazine and R24571) on the speed of calcium-conductance decline. Intracellular administration of these substances considerably reduces its velocity (Doroshenko and Martynyuk, 1983).

These experimental findings do not support directly the idea that intracellular cAMP exerts its influence on the calcium conductance through the change in the activity of membrane phosphoprotein kinases phosphorylating membrane proteins essential for the function of calcium channels, though such a possibility is most likely.

The cAMP-dependent protein kinase is known to be a nonactive complex consisting of two regulatory and two catalytic subunits. After joining cAMP, the complex dissociates into individual subunits, but only free catalytic subunit reveals phosphotransferase activity (Glass and Krebs, 1980). Active catalytic subunit of the cAMP-dependent protein kinase can be obtained now as a homogenous protein (Kuo *et al.*, 1971); the properties of the enzyme isolated from various tissues and animal species are practically identical. Based on these data, the studies on the effects of intracellularly administered catalytic subunit (CS) on calcium current have been undertaken on *Helix* neurons in conditions of intracellular perfusion. CS of the cAMP-dependent protein kinase with molecular weight 35,000 was obtained from rabbit myocardium; its specific activity was about 20 nmole ^{32}P/mg protein/min. Addition of 0.7 μmole CS + ATP to the perfusing solution stopped the decline of calcium conductance and restored it to the initial level. The restored calcium current remained practically unchanged in amplitude for prolonged (several hours) intracellular perfusion of the cell with CS-containing solution. Removal of ATP from the intracellular solution resulted in fast reduction of calcium conductance; subsequent introduction of ATP partly restored it (Kostyuk *et al.*, 1983b).

Slow velocity of the changes described could be accounted for either by slow entry of CS molecules (due to their large molecular weight compared with inorganic cations) into the intracellular medium through the pore in the surface membrane or by their slow reconstruction in the corresponding site of activity.

Based on the described experiments, there are reasons to assume that the level of calcium conductance in the somatic membrane does depend on continuous phosphotransferase activity of the cAMP-dependent protein kinase. This conclusion is in good agreement with the data that injection of CS of the cAMP-dependent protein kinase isolated from bovine heart muscle into *Aplysia* neurons facilitates the generation of calcium-action potentials (Kaczmarek *et al.*, 1980). In addition, in this work, an increase of the membrane input resistance was observed in several cases that could be associated with changes in the function of potassium channels (see below).

So far, there is no information for the somatic membrane which membrane proteins are phosphorylated under these conditions. However, the cAMP-dependent phosphorylation of a 23,000-dalton protein (so-called calciductin) was shown on vesicles formed from cardiac sarcolemma which took place in parallel with an increase of potential-dependent Ca^{2+} uptake (Rinaldi *et al.*, 1981).

Thus, calcium channels appear to be highly dynamic structures whose functioning requires continuous metabolic support. If the latter is failing, dephosphorylation of the corresponding proteins by cell phosphatases makes the channels inactive ("sleeping"). The question of whether this statement can be extended to all types of calcium channels in various excitable membranes is open for the present and needs a special experimental study. Data have been recently obtained that two populations of electrically-operated calcium channels can be distinguished in the somatic membrane of rat spinal ganglia neurons. They differ in their kinetics and potential dependence as well as in inactivation properties. A more slowly developing component has the same properties as the above described calcium currents in the somatic membrane of mollusc neurons; it rapidly decreases in the course of cAMP and ATP washout. However, a weaker fast component proves to be more stable to cell perfusion, and intracellular cAMP injections do not considerably influence its amplitude. Inactivation of this component is potential dependent (as inactivation of sodium or potassium channels), whereas the inactivation of the slow component depends on the amplitude of inward calcium current (Veselovsky and Fedulova, 1983). It is possible that such "fast" calcium channels are similar to other types of ionic channels and do not require a continuous metabolic support for their activity. According to Stefani (1981), inactivation of calcium current in frog skeletal muscle fibers also shows a fast development and potential dependence, and no Ca^{2+} entry into the fiber is needed for its inactivation.

B. Data on Metabolic Dependence of the Activity of Ca-Dependent Potassium Channels

Studies of changes in Ca-dependent potassium current made on perfused snail neurons have shown that intracellular introduction of CS of the cAMP-activated protein kinase elicits its increase. Both CS inactivation and lowering of Ca^{2+} content in the extracellular solution from 10 to 1 mM abolish this effect. As a matter of fact, the CS introduction reproduces the effect of an increase in intracellular Ca^{2+} content; therefore, the authors suggest that CS enhances the sensitivity of a binding group in the potassium channel to the action of these ions (de Peyer *et al.*, 1982). Activation of Ca-dependent potassium conductance was also observed in intact snail neurons during cAMP injection from a microelectrode under pressure or iontophoretically (Kononenko *et al.*, 1983). However, Aldenhoff *et al.* (1979) have reported the suppression of Ca-activated potassium conductance during intracellular injection of cAMP through a microelectrode.

C. Possible Metabolic Dependence of the Activity of Other Types of Ionic Channels

A detailed investigation of serotonin-induced increase of Ca^{2+} entry into identified *Aplysia* neurons was conducted by Klein and Kandel (1980) who showed that metabolic regulation of the activity of certain *potential-dependent potassium channels* may underlie this effect. Using cells with blocked calcium current, the authors found that serotonin causes in them a suppression of outward currents. This effect does not appear to involve the known types of potassium channels (channels of fast and delayed potassium current and Ca-activated potassium channels), but is associated with the activity of special channels analogous to M-channels described by Brown and Adams (1980). These channels are insensitive to the increase of intracellular Ca^{2+}, do not inactivate and cannot be blocked by Ba^{2+}, but are potentiated by serotonin. Therefore, the authors designated the outward current as S-current (Klein *et al.*, 1982). Recordings of single potassium channel activity in *Aplysia* neurons by patch clamp confirmed that extracellular application of serotonin or intracellular cAMP injection results in cessation of their activity (Siegelbaum *et al.*, 1982). The authors came to the conclusion that cAMP-mediated suppresssion of the activity of these channels by serotonin prolongs the phase of depolarization of the membrane during action potential and, thus, prolongs Ca^{2+} entry into the cell. They have also suggested that cAMP-dependent protein kinases may phosphorylate some proteins in the described potassium channels, thus, unlike their effect on the calcium channels, making these channels *inactive*. Castellucci *et al.* (1980) injected CS of the cAMP-dependent protein kinase from bovine myocardium into the corresponding *Aplysia* neurons and found both prolongation of Ca^{2+} entry into the cell and facilitation of the outward potassium current. In addition, transmitter release from the presynaptic endings was also increased. However, the conclusion about possible cAMP-dependent inactivation of potassium channels in certain nerve cells should not be extended to all varieties of neurons, since R15 neuron of the *Aplysia* displayed serotonin-dependent *elevation* of potassium conductance in the somatic membrane. The additional current had a reversal potential of -75–80 mV and altered like a potassium current with a change in the extracellular K^+ content. External application of 8-benzilthio-cAMP (a derivative of cAMP penetrating through the cell membrane and resistant to phosphodiesterase action) had the same effect as serotonin (Drummond *et al.*, 1980; Benson *et al.*, 1980). An increase in adenylate cyclase activity and accumulation of internal cAMP occurred in parallel with the current increase. Analogous elevation of potassium conductance was observed by Treistman (1981) during injection of 5-guanyl-imidodiphosphate into R15 neuron, the substance which activates adenylate cyclase and stimulates accumulation of cAMP in the cell. Intracellular injection of protein kinase inhibitor isolated from the homogenate of rabbit skeletal muscle blocked the serotonin-induced increase of potential-dependent potassium conductance (Adams and Levitan, 1982).

As it was mentioned above, in experiments with the washout of cAMP or its artificial introduction into the cell, no changes in potential-dependent functions of sodium channels were found commensurate with those of calcium channels. Nevertheless, such procedures may exert definite effects on the sodium channels that could be mediated through some intermediate metabolic processes. It is known that during sustained perfusion of giant axon, the sodium conductance of its membrane and correspondingly its electrical excitability start to decrease. This process can be stopped and even reversed by introduction of tyrosine + (ATP + cAMP + Mg^{2+}) complex into the axon. Addition of small amounts of intracellular proteins or the axoplasm contributed to the restoration of disturbed membrane function (Matsumoto and Sakai, 1979). The authors suggest that microtubular proteins contain tyrosine–tubulin ligase which restores the disturbed relation between cytoskeletal structures and the surface membrane; phosphorylation of some membrane proteins is essential for such restoration. In cultured mammalian neurons, it has been shown that breakdown of microfilaments by cytochalasin B selectively disturbs the membrane-sodium conductance without changing its other characteristics (Fukuda *et al.*, 1981). In the Ranvier node of frog nerve fiber, the external application of dibutyryl-cAMP or a highly effective blocker of phosphodiesterase produces a simultaneous slow (for dozens of minutes) increase of both inward and outward components of the total transmembrane current evoked by membrane depolarization. However, when outward current is blocked by TEA, the changes of a pure inward current are insignificant indicating a complicated character of this phenomenon (Seelig and Kendig, 1982).

Finally, experiments with intracellular cAMP injection (or extracellular application of dibutyryl-cAMP) show that, in many nerve cells of molluscs, this action is accompanied by depolarization of the membrane not associated with the activation of potential-dependent ionic conductance (Liberman *et al.*, 1975). Besides, in many cells, inward transmembrane currents are also induced (Akopyan *et al.*, 1979). A detailed investigation of this effect was performed on identified snail neurons by Kononenko and Mironov (1980), Kononenko (1980a), and Kononenko *et al.* (1983). In these neurons, cAMP injection (as well as external application of theophylline; Kononenko, 1981) brought about the generation of a relatively slowly developing inward current with reversal potential close to zero (at normal ionic composition of the extracellular medium). Tolbutamide (protein kinase inhibitor) and lowering of temperature considerably decreases the speed of current development while theophylline, on the contrary, increases its amplitude. cGMP injection does not produce any effects. Changes of holding potential within wide limits do not affect the generation of the "cAMP-induced current." A study of the shifts of current-voltage curves of this current has shown that it is associated with the increase of membrane conductance, predominantly, for sodium ions as well as for potassium and calcium ions. By these characteristics, the "cAMP-induced current" closely resembles that produced by *chemically operated ionic channels,* which under natural cellular activity might be activated by some transmitters or hormonal substances exerting their influence through the activation of membrane adenylate cyclase and elevation of intracellular cAMP level. It should be noted that, in snail neurons, similar changes in the membrane conductance can be induced by external application to the neuronal surface of a peptide factor isolated from the snail brain homogenate with molecular weight close (but not identical) to oxytocin which stimulates the generation of cellular periodic activity (Kononenko, 1980b). It is possible that the observed effects of intracellular cAMP injection may be the components of this natural reaction.

In connection with the above stated facts it is interesting to note that Akopyan *et al.* (1980), in experiments on dialyzed *Planorbis* neurons, observed the cAMP-induced changes in cellular responses to acetylcholine. However, these changes are of an inhibiting and not of an activating nature.

IV. CONCLUSION

Experimental data discussed in the present chapter undoubtedly indicate the existence of complex functional relations between cellular metabolic processes and ionic channels of the surface membrane. Cellular enzyme systems cannot be regarded only as the source of energy for the transmembrane transfer of ions and the creation of corresponding electrochemical gradients. In a number of cases, the activity of ionic channels itself is under direct metabolic control whose disturbance may somehow influence this activity.

The fact deserves attention that the presence of direct metabolic control can be convincingly shown for ionic channels whose activation or transfer of ions across them depend on Ca^{2+} ions (calcium and Ca-dependent potassium channels). This property seems to reflect the leading role of calcium ions in the coupling between membrane and cytoplasmatic processes and their effective participation in the regulation of enzymatic systems. The recurrent influence of these systems on the activity of corresponding ionic channels forms a feedback system which is extremely essential for successful activity of the cell as an integrated system. A possible role of enzymatic processes in the control of other systems of ionic conductance in the surface membrane is also not excluded. However, it should be kept in mind that changes in the activity of ionic channels mediated through other cellular changes may take place, and the corresponding suggestions should be thoroughly proved by experiment.

REFERENCES

Adams, W. B., and Levitan, I. B., 1982, Intracellular injection of protein kinase inhibitor blocks the serotonin-induced increase in K^+ conductance in Aplysia neuron R15, *Proc. Natl. Acad. Sci. USA* **79**:3877–3880.

Akopyan, A. R., Piskunova, G. M., and Chemeris, N. K., 1979, Dialyzed neurone as a model for study of cyclic nucleotides effect on membrane electrical parameters, *Dokl. AN SSSR (Moscow)* **246**:997–999.

Akopyan, A. R., Chemeris, N. K., Iljin, V. I., and Veprintsev, B. N., 1980, Serotonin, dopamine and intracellular cyclic AMP inhibit the responses of nicotinic cholinergic membrane in snail neurons, *Brain Res.* **201**:480–484.

Aldenhoff, J. B., Hofmeier, G., and Lux, H. D., 1979, Depolarizing action of injected cAMP in Helix neurones, *Pflügers Arch.* **382**(Suppl.):92.

Benson, J., Drummond, A., and Levitan, I., 1980, Cyclic AMP mediates a serotonin-induced increment of K^+ conductance in Aplysia neuron R15, *Neurosci. Lett.* **19**(Suppl. 5):67.

Brown, A. M., Morimoto, K., Tsuda, Y., and Wilson, D. L., 1981, Calcium current-dependent and voltage-dependent inactivation of calcium channels in Helix aspersa, *J. Physiol. (London)* **320**:193–218.

Brown, D. A., and Adams, P. R., 1980, Muscarinic suppression of a novel voltage-sensitive K^+ current in a vertebrate neurone, *Nature* **283**:673–676.

Byerly, L., and Hagiwara, S., 1982, Calcium currents in internally perfused nerve cell bodies of Limnea stagnalis, *J. Physiol. (London)* **322**:503–528.

Campbell, D. T., 1976, Ionic selectivity of the sodium channel of frog skeletal muscle, *J. Gen. Physiol.* **67**:295–307.

Castellucci, V. F., Kandel, E. R., Schwartz, J. H., Wilson, F. D., Nairn, A. L., and Greengard, P., 1980, Intracellular injection of the catalytic subunit of cyclic AMP-dependent protein kinase simulates facilitation of transmitter release underlying behavioral sensitization in Aplysia, *Proc. Natl. Acad. Sci. USA* **77**:7492–7496.

Chandler, W. K., and Meves, H., 1965, Voltage clamp experiments on internally perfused giant axons, *J. Physiol. (London)* **180**:788–820.

Deterre, P., Paupardin-Tritsch, D., Bockaert, J., and Gerschenfeld, H. M., 1981, Role of cyclic AMP in a serotonin-evoked slow inward current in snail neurones, *Nature* **290**:783–785.

Doroshenko, P. A., and Martynyuk, A. E., 1983, Effect of calmodulin blockers on inhibition of potential-dependent calcium conductance by intracellular calcium ions in a nerve cell, *Dokl. AN SSSR (Moscow)* **273**:N4.

Doroshenko, P. A., and Tsyndrenko, A. Y., 1978, Effect of intracellular calcium on the calcium inward current, *Neurophysiology (Kiev)* **10**:203–205.

Doroshenko, P. A., Kostyuk, P. G., and Tsyndrenko, A. Y., 1978, Separation of potassium and calcium channels in the somatic membrane of a nerve cell, *Neurophysiology (Kiev)* **10**:645–653.

Doroshenko, P. A., Kostyuk, P. G., and Tsyndrenko, A. Y., 1979, A study of the TEA-resistant outward current in the somatic membrane of perfused nerve cells, *Neurophysiology (Kiev)* **11**:460–468.

Doroshenko, P. A., Kostyuk, P. G., and Martynyuk, A. E., 1982, Extracellular metabolism of adenosine 3′,5′-cyclic monophosphate and calcium inward current in perfused neurones of Helix pomatia, *Neuroscience* **7**:2125–2134.

Doroshenko, P. A., Kostyuk, P. G., and Martynyuk, A. E., 1984, Inactivation of calcium current in the somatic membrane of snail neurons, *Gen. Physiol. Biophys.* **3**:1–17.

Drummond, A. H., Benson, J. A., and Levitan, I. B., 1980, Serotonin-induced hyperpolarization of an identified Aplysia neuron is mediated by cyclic AMP, *Proc. Natl. Acad. Sci. USA* **77**:5013–5017.

Eckert, R., and Tillotson, D., 1981, Calcium-mediated inactivation of the calcium conductance in caesium-loaded giant neurones of Aplysia californica, *J. Physiol. (London)* **314**:265–280.

Fedulova, S. A., Kostyuk, P. G., and Veselovsky, N. S., 1981, Calcium channels in the somatic membrane of the rat dorsal root ganglion neurons, effect of cAMP, *Brain Res.* **214**:210–214.

Fox, A., 1981, Voltage-dependent inactivation of a calcium channel, *Proc. Natl. Acad. Sci. USA* **78**:953–956.

Fukuda, J., Kameyama, M., and Yamaguchi, K., 1981, Breakdown of cytoskeletal filaments selectively reduces Na and Ca spikes in cultured mammal neurones, *Nature* **294**:82–85.

Glass, D. B., and Krebs, E. G., 1980, Protein phosphorylation catalyzed by cyclic AMP-dependent and cyclic GMP-dependent protein kinases, *Annu. Rev. Pharmacol. Toxicol.* **20**:363–388.

Greengard, P., 1978, *Cyclic Nucleotides, Phosphorylated Proteins, and Neuronal Function,* Raven Press, New York.

Hagiwara, S., 1973, Ca spike, *Adv. Biophys.* **4**:71–102.

Hille, B., 1971, The permeability of the sodium channel to organic cations in myelinated nerve, *J. Gen. Physiol.* **58**:599–619.

Hille, B., 1972, The permeability of the sodium channel to metal cations in myelinated nerve, *J. Gen. Physiol.* **59**:637–658.

Hille, B., 1978, Ionic channels in excitable membranes. Current problems and biophysical approaches, *Biophys. J.* **22**:283–294.

Hille, B., and Schwarz, W., 1979, K channels in excitable cells as multi-ion pores, *Brain Res. Bull.* **4**:159–162.

Hofmeier, G., and Lux, H. D., 1978, Inversely related behavior of potassium and calcium permeability during activation of calcium-dependent outward currents in voltage-clamped snail neurones, *J. Physiol. (London)* **287**:28P–29P.

Kaczmarek, L. K., Jennings, K. R., Strumwasser, F., Nairn, A. C., Walter, U., Wilson, F. D., and Greengard, P., 1980, Microinjection of catalytic subunit of cyclic AMP-dependent protein kinase enhances calcium action potentials of bag cell neurons in cell culture, *Proc. Natl. Acad. Sci. USA* **77**:7487–7491.

Katz, B., and Miledi, R., 1967a, The timing of calcium action during neuromuscular transmission, *J. Physiol. (London)* **189**:535–544.

Katz, B., and Miledi, R., 1967b, A study of synaptic transmission in the absence of nerve impulses, *J. Physiol. (London)* **192:**407–436.

Katz, B., and Miledi, R., 1967c, Tetrodotoxin and neuromuscular transmission, *Proc. Roy. Soc. B* **167:**8–22.

Khodorov, B. I., 1975, *General Physiology of Excitable Membranes* (in Russian), Nauka Publishing House, Moscow.

Klein, M., and Kandel, E. R., 1978, Presynaptic modulation of voltage-dependent Ca^{2+} current: Mechanisms for behavioral sensitization in Aplysia californica, *Proc. Natl. Acad. Sci. USA* **75:**3512–3516.

Klein, M., and Kandel, E. R., 1980, Mechanism of calcium modulation underlying presynaptic facilitation and behavioral sensitization in Aplysia, *Proc. Natl. Acad. Sci. USA* **77:**6912–6916.

Klein, M., Camardo, J., and Kandel, E. R., 1982, Serotonin modulates a specific potassium current in the sensory neurons that show presynaptic facilitation in Aplysia, *Proc. Natl. Acad. Sci. USA* **79:**5713–5717.

Kononenko, N. I., 1980a, Ionic mechanism of the transmembrane current evoked by injection of cyclic AMP into identified Helix pomatia neurons, *Neurophysiology (Kiev)* **12:**339–343.

Kononenko, N. I., 1980b, Application of a factor modulating the electrical activity in bursting neuronal soma of the snail, Helix pomatia, *Dokl. AN SSSR (Moscow)* **250:**1490–1493.

Kononenko, N. I., 1981, Theophylline effect on electrical activity of the RPa2 neurone in Helix pomatia, *Neurophysiology (Kiev)* **13:**655–657.

Kononenko, N. I., and Kostyuk, P. G., 1982, Effect of intracellular cyclic adenosine monophosphate injection on calcium current in identified Helix neurons, *Neurophysiology (Kiev)* **14:**290–297.

Kononenko, N. I., and Mironov, S. L., 1980, Effect of intracellular injection of cyclic AMP on electrical characteristics of identified neurons of Helix pomatia, *Neurophysiology (Kiev)* **12:**332–338.

Kononenko, N. I., Kostyuk, P. G., and Shcherbatko, A. D., 1983, The effect of intracellular cAMP injections on stationary membrane conductance and voltage- and time-dependent ionic currents in identified snail neurons, *Brain Res.*, **268:**321–338.

Kostyuk, P. G., and Krishtal, O. A., 1977a, Separation of sodium and calcium currents in the somatic membrane of mollusc neurones, *J. Physiol. (London)* **270:**545–568.

Kostyuk, P. G., and Krishtal, O. A., 1977b, Effects of calcium and calcium-chelating agents on the inward and outward currents in the membrane of mollusc neurones, *J. Physiol. (London)* **270:**569–580.

Kostyuk, P. G., and Mironov, S. L., 1982, Theoretical description of calcium channels in the neuronal membrane, *Gen. Physiol. Biophys.* **1:**289–305.

Kostyuk, P. G., and Shuba, Ya. M., 1982, A study of monovalent cation selectivity of calcium EDTA-modified channels, *Neurophysiology (Kiev)* **14:**491–498.

Kostyuk, P. G., Krishtal, O. A., and Pidoplichko, V. I., 1975, Effect of internal fluoride and phosphate on membrane currents during intracellular dialysis of nerve cells, *Nature* **257:**691–693.

Kostyuk, P. G., Krishtal, O. A., and Pidoplichko, V. I., 1977, Asymmetrical displacement currents in nerve cell membrane and effect of internal fluoride, *Nature* **267:**70–72.

Kostyuk, P. G., Krishtal, O. A., and Pidoplichko, V. I., 1978, Ionic currents in the neuroblastoma cell membrane, *Neuroscience* **3:**327–332.

Kostyuk, P. G., Doroshenko, P. A., and Tsyndrenko, A. Y., 1980, Calcium-dependent potassium conductance studied on internally dialysed nerve cells, *Neuroscience* **5:**2187–2192.

Kostyuk, P. G., Veselovsky, N. S., and Tsyndrenko, A. Y., 1981a, Ionic currents in the somatic membrane of rat dorsal root ganglion neurons. I. Sodium currents, *Neuroscience* **6:**2423–2430.

Kostyuk, P. G., Veselovsky, N. S., and Fedulova, S. A., 1981b, Ionic currents in the somatic membrane of rat dorsal root ganglion neurons. II. Calcium currents, *Neuroscience* **6:**2431–2437.

Kostyuk, P. G., Mironov, S. L., and Doroshenko, P. A., 1982, Energy profile of the calcium channels in the membrane of mollusc neurons, *J. Membr. Biol.* **70:**181–189.

Kostyuk, P. G., Mironov, S. L., and Shuba, Ya. M., 1983a, Two ion-selecting filters in the calcium channels of the somatic membrane of mollusc neurons, *J. Membr. Biol.*, **76:**83–93.

Kostyuk, P. G., Doroshenko, P. A., Martynyuk, A. E., Kursky, M. D., and Vorobets, Z. D., 1983b, Intracellular injection of catalytic subunit of cAMP-dependent protein kinase recovers the membrane calcium current in the snail nerve cell, *Dokl. AN SSSR (Moscow)* **271:**756–758.

Krishtal, O. A., Pidoplichko, V. I., and Shakhovalov, Yu. A., 1981, Conductance of the calcium channel in the membrane of snail neurones, *J. Physiol. (London)* **310:**423–434.

Kuo, J. F., Wyatt, G. R., and Greengard, P., 1971, Cyclic nucleotide-dependent protein kinases. IX. Partial purification and some properties of guanosine 3′,5′-monophosphate-dependent protein kinases from various tissues and species of Arthropoda, *J. Biol. Chem.* **246:**7159–7167.

Lee, K. S., Akaike, N., and Brown, A. M., 1978, Properties of internally perfused, voltage-clamped, isolated nerve cell bodies, *J. Gen. Physiol.* **71:**489–507.

Liberman, E. A., Minina, S. V, and Golubtsov, K. V., 1975, Study of the metabolic synapse. I. Effect of intracellular microinjection of 3′,5′-AMP (in Russian), *Biofizika (Moscow)* **20:**451–456.

Llinas, R., Blinks, J. R., and Nicholson, C., 1972, Calcium transient in presynaptic terminal of squid giant synapse: Detection with aequorin, *Science* **176:**1127–1129.

Llinas, R., Steinberg, I. Z., and Walton, K., 1976, Presynaptic calcium currents and their relation to synaptic transmission: Voltage clamp study in squid giant synapse and theoretical model for the calcium gate, *Proc. Natl. Acad. Sci. USA* **73:**2918–2922.

Lux, H. D., and Nagy, K., 1981, Single channel Ca^{2+} currents in Helix pomatia neurons, *Pflügers Arch.* **391:**252–254.

Lux, H. D., Neher, E., and Marty, A., 1981, Single channel activity associated with the calcium dependent outward current in Helix pomatia, *Pflügers Arch.* **389:**293–295.

Martell, A. E., and Smith, R. M., 1977, *Critical Stability Constants,* Plenum Press, New York.

Matsumoto, G., and Sakai, H., 1979, Restoration of membrane excitability of squid giant axons by reagents activating tyrosine-tubulin ligase, *J. Membr. Biol.* **50:**15–22.

Meech, R. W., and Strumwasser, F., 1970, Intracellular calcium injection activates potassium conductance in Aplysia nerve cells, *Fed. Proc.* **29:**834.

Pellmar, T. C., 1981, Ionic mechanism of a voltage-dependent current elicited by cyclic AMP, *Cell. Mol. Neurobiol.* **1:**87–97.

de Peyer, J. E., Cachelin, A. B., Levitan, I. B., and Reuter, H., 1982, Ca^{2+}-activated K^+ conductance in internally perfused snail neurons is enhanced by protein phosphorylation, *Proc. Natl. Acad. Sci. USA* **79:**4207–4211.

Plant, T. D., and Standen, N. B., 1981, Calcium current inactivation in identified neurones of Helix aspersa, *J. Physiol (London)* **321:**273–285.

Rasmussen, H., Clayberger, C., and Gustin, M. C., 1979, The messenger function of calcium in cell activation, in: *Secretory Mechanisms,* in Symp. Soc. Exp. Biol. No. XXXIII, Cambridge, pp. 161–197.

Reuter, H., 1973, Divalent cations as charge carrier in excitable membranes, in: *Progress in Biophysics and Molecular Biology,* Vol. 36 (J. A. V. Butler and D. Noble, eds.), Pergamon Press, Oxford and New York, pp. 1–43.

Reuter, H., 1974, Localization of beta adrenergic receptors and effects of noradrenaline and cyclic nucleotides on action potentials, ionic currents and tension in mammalian cardiac muscle, *J. Physiol. (London)* **242:**429–451.

Reuter, H., 1979, Properties of two inward membrane currents in the heart, *Annu. Rev. Physiol.* **41:**413–424.

Reuter, H., Stevens, C. F., Tsien, R. W., and Yellen, G., 1982, Properties of single calcium channels in cardiac cell culture, *Nature* **297:**501–504.

Rinaldi, M. L., Le Peuch, C. J., and Demaille, J. G., 1981, The epinephrine-induced activation of the cardiac slow Ca^{2+} channel is mediated by the cAMP-dependent phosphorylation of calciductin, a 23,000 M_r sarcolemmal protein, *FEBS Lett.* **129:**277–281.

Seelig, T. L., and Kendig, J. J., 1982, Cyclic nucleotide modulation of Na^+ and K^+ currents in the isolated node of Ranvier, *Brain Res.* **245:**144–147.

Shakhovalov, Yu. A., 1977, Quantitative description of the Na and Ca inward currents in molluscan neurones by Hodgkin–Huxley equations, *J. Physiol. (London)* **270:**562–567.

Siegelbaum, S. A., Camardo, J. S., and Kandel, E. R., 1982, Serotonin and cyclic AMP close single K^+ channels in Aplysia sensory neurones, *Nature* **299:**413–417.

Stefani, E., 1981, Kinetics of calcium currents in frog skeletal muscle fibres, Abstracts VII International Biophysical Congress, Mexico City, August 1981, p. 202.

Takahashi, K., and Yoshii, M., 1978, Effects of internal free calcium upon the sodium and calcium channels in the tunicate egg analysed by the internal perfusion technique, *J. Physiol. (London)* **279:**519–549.

Tillotson, D., 1979, Inactivation of Ca conductance dependent on entry of Ca ions in molluscan neurones, *Proc. Natl. Acad. Sci. USA* **76:**1497–1500.

Treistman, S. N., 1981, Effect of adenosine 3′,5′-monophosphate on neuronal pacemaker activity: A voltage clamp analysis, *Science* **211:**59–61.

Tsien, R. W., 1973, Adrenaline-like effects of intracellular iontophoresis of cyclic AMP in cardiac Purkinje fibres, *Nature New Biol.* **245:**120.

Veselovsky, N. S., and Fedulova, S. A., 1983, Two types of calcium channels in the neuronal somatic membrane of spinal ganglia in rats (in Russian), *Dokl. AN SSSR (Moscow)* **268:**747–750.

Veselovsky, N. S., Kostyuk, P. G., Krishtal, O. A., and Pidoplichko, V. I., 1977a, Separation of ionic currents responsible for a generation of the action potential in isolated neurones of frog spinal ganglia, *Neurophysiology (Kiev)* **9:**638–640.

Veselovsky, N. S., Kostyuk, P. G., Krishtal, O. A., Naumov, A. P., and Pidoplichko, V. I., 1977b, Transmembrane ionic currents in neuroblastoma cells, *Neurophysiology (Kiev)* **9:**641–643.

Veselovsky, N. S., Kostyuk, P. G., and Tsyndrenko, A. Y., 1979, Separation of ionic currents responsible for a generation of action potential in a somatic membrane of spinal ganglion neurones in newborn rats, *Dokl. AN SSSR (Moscow)* **249:**1466–1469.

Index